Studies in Classification, Data Analysis, and Knowledge Organization

AF862523

Springer-Verlag Berlin Heidelberg GmbH

Titles in the Series

H.-H. Bock and P. Ihm (Eds.)
Classification, Data Analysis, and
Knowledge Organization
(out of print)

M. Schader (Ed.)
Analyzing and Modeling Data
and Knowledge

O. Opitz, B. Lausen, and R. Klar (Eds.)
Information and Classification
(out of print)

H.-H. Bock, W. Lenski, and M.M. Richter (Eds.)
Information Systems and Data Analysis
(out of print)

E. Diday, Y. Lechevallier, M. Schader, P. Bertrand,
and B. Burtschy (Eds.)
New Approaches in Classification and Data Analysis
(out of print)

W. Gaul and D. Pfeifer (Eds.)
From Data to Knowledge

H.-H. Bock and W. Polasek (Eds.)
Data Analysis and Information Systems

E. Diday, Y. Lechevallier and O. Opitz (Eds.)
Ordinal and Symbolic Data Analysis

Rüdiger Klar · Otto Opitz (Eds.)

Classification and Knowledge Organization

Proceedings of the 20th Annual Conference
of the Gesellschaft für Klassifikation e. V.,
University of Freiburg, March 6–8, 1996

With 167 Figures
and 80 Tables

Springer

Prof. Dr. Rüdiger Klar
Institut für Medizinische Biometrie
und Medizinische Informatik
University of Freiburg
Stefan-Meier-Straße 26
D-79104 Freiburg
Germany
e-mail: klar@ami1.ukl.uni-freiburg.de

Prof. Dr. Otto Opitz
Lehrstuhl für Mathematische Methoden
der Wirtschaftswissenschaften
University of Augsburg
D-86135 Augsburg
Germany
e-mail: otto.opitz@wiso.uni-augsburg.de

Cataloging-in-Publication Data applied for

Die Deutsche Bibliothek - CIP-Einheitsaufnahme

Classification and knowledge organization : with 80 tables / Rüdiger Klar ; Otto Opitz (ed.). - Berlin ; Heidelberg ; New York ; Barcelona ; Budapest ; Hong Kong ; London ; Milan ; Paris ; Santa Clara ; Singapore ; Tokyo : Springer, 1997
(Studies in classification, data analysis, and knowledge organization)
ISBN 978-3-540-62981-8

ISBN 978-3-540-62981-8 ISBN 978-3-642-59051-1 (eBook)
DOI 10.1007/ 978-3-642-59051-1

Originally published by Springer-Verlag Berlin Heidelberg New York in 1997

Hardcover design: Erich Kirchner, Heidelberg

SPIN 10548610 43/2202-5 4 3 2 1 0 – Printed on acid-free paper

Preface

This volume presents 71 articles dealing with models and methods of data analysis and classification, statistics and stochastics, information systems and text analysis as well as manifold applications. These articles are selected from about 160 papers presented at the 20th Annual Conference of the Gesellschaft für Klassifikation. This conference was organized by R. Klar, at the Abteilung für Medizinische Informatik of the University of Freiburg from March 6 to 8, 1996. Based on the submitted and revised papers eight sections have been arranged, where the number of papers in the sections is given in parentheses:

1. Data Analysis and Classification (10)
2. Neural Networks and Pattern Recognition (4)
3. Statistical Models and Methods (11)
4. Information Systems: Design and Implementation (7)
5. Text Analysis and Information Retrieval (10)
6. Applications in Medicine (15)
7. Applications in Economics and Social Sciences (7)
8. Applications in Archaeology, Biology, Linguistics and Dialectometry (7)

This grouping doesn't separate strictly, but it shows how theoretical aspects, applications and interdisciplinarities are interrelated in many respects. For convenience of the reader the contents of this volumne is briefly summarized.

1. Data Analysis and Classification

This section presents ten articles dealing with different problems of data analysis and classification, especially with several aspects of *multidimensional scaling* (MDS), *three-way data analysis*, *stochastic aspects in clustering* and *regression of ordinal data.*

The survey article of *M. W. Trosset* considers an approach to define and classify MDS problems as optimization problems. Based on certain fundamental ideas that often occur in the design of MDS algorithms numerical methods for MDS are described as algorithms for some special numerical optimization problems. *R. Mathar* tackles a special problem of multidimensional scaling: the most algorithms often get stuck at local minima of a well-defined STRESS function and not at a global one. He proposes a hybrid algorithm consisting of both a local optimization phase and a global optimization step and demonstrates the high success probability of the suggested method.

H. Haan, J. Timmer, R. Mathar and *J. Krüger* use the MDS as an exploratory tool to discover structures in vast data sets. They consider different MDS models and analyze a neuro-data set of recordings from visual cortex of monkeys.

In the focus of attention of the survey paper of *P. M. Kroonenberg* about three-way data is the question to which areas the three-way data analysis has expanded. After a short and historical introduction he presents several techniques used in the three-way data analysis and demonstrates how to use these techniques with three-way data by examples.

H. Hofmann considers the reliability of computer orientated graphical methods for data analysis. She presents several problems occuring when analyzing hard data and introduces several solutions of the considered problems.

Graphical techniques such as parallel coordinates and the grand tour are presented by *E. J. Wegman* and *Q. Luo* for cluster analysis of high dimensional data. Their use for inverse regression and clustering is illustrated.

The following two articles deal with stochastic aspects in clustering. *H. H. Bock* presents probability-based clustering methods assuming class-specific distributions for the random data vectors. Some clustering models are considered where each of the underlying classes is characterized by an uniform distribution in a convex and finite domain. Corresponding clustering criteria and the maximum likelihood ratio test statistic for testing the existence of such clustering structures are derived. Fixed point clustering is a method for the analysis of non-distance based data where outliers could be present or the number of clusters is unknown, e.g. in clusterwise regression. In the paper of *C. Hennig* the relation of fixed point clustering to parameter estimation in stochastic models is discussed in a very general way.

The next two papers are concerned with ordinal regression. *A. Hilbert* derives a metric approach for the regression of ordinal data if the independent variables have an ordinal scale and the response variable has a metric scale. Further he describes some statistical properties of the estimated parameters. Also the article of *H. Paul* considers the problem of ordinal data in linear regression. But he analyzes the situation that cardinal data are measured on ordinal scale, and therefore he proposes a rescaling of the data. The aim of the presented method is to approximate these "ordinal-cardinal" data to their original values.

2. Neural Networks and Pattern Recognition

The aim of the survey article of *K. Hornik* is to introduce neural networks and to describe the terminology of these neural networks with algebraic and statistical terms. Further he discusses some of the most important porperties of the nets. *S. Simon, F. Schwenker* and *G. Palm* propose a hierarchical neural network for classification tasks. Their network consists of two different types of neural networks: a learning vector quantization network and a radial basis function network. With these two components they are able to solve a high dimensional classification problem more faster than traditional networks are able to do this.

G. Nakhaeizadeh, C. C. Taylor and *G. Kunisch* consider neural networks and the problem that real-world data sets change over the times. They propose a network which is able to capture these dynamic aspects of data sets and demonstrate the success of their net on simulated and real data from the credit industry. Whenever a great amount of data is recorded over a certain time and complex pattern have to be found traditional methods get problems. *G. Guimarães* and *A. Ultsch* propose to use extended context-free grammars, so-called Definitive Clause Grammars, to solve the described problem.

3. Statistical Models and Methods

This section contains eleven papers which are concerned with some different problems of statistical estimation and testing as well as their relations to special applications in data analysis, classification and prediction.

A. Satorra analyzes multivariate regression models with error in variables assuming normal distributions of the variables. He formulates conditions for the asymptotic efficiency of estimates, the asymptotic validity of standard errors and chi-square goodness of fit tests. Results of Dham and Fuller are extented to multiple-group data and general type of multivariate regression models.

The problem of outliers in Bayesian autoregressive and conditional heteroscedasticity models with an AR structure in the mean is discussed in the paper of *W. Polasek* and *S. Jin.* For estimation of the outliers a "location shift" regression model is used. *R. Vonthein* considers classical and Bayesian regression models. For observations badly explained by the model appropriate diagnostics are presented which use an influence measure depending on a definition of influence. The author illustrates how to derive and to apply the diagnostics in special cases.

Bayesian non-linear latent variable models with squares and interactions of latent regressor variables, observed regressors and with missing data are specified in the paper of *J. Wittenberg* and *G. Arminger.* The posterior distributions of the parameters and the latent variables are estimated using Markov chain Monte Carlo methods such as Gibbs sampler and Metropolis-Hastings algorithm.

J. S. Pai suggests an one-step optimization procedure for parameter estimation in autoregressive fractionally integrated moving average processes using the simulated annealing method. Comparisons are performed to the two-step procedure of Haslett and Raftery.

B. J. A. Mertens and *D. J. Hand* deal with some reflections on the assessment of classification rules for high-dimensional data. Methods of discrimination in high dimensions are explained, and it is shown how these methods have been implemented to allow for a comparison on unbiased estimates of error rate.

Methods of identification analysis have been developed for directional statistics. To show the actual implementation of these methods the contribution of *F. Streit* uses a well-known data set working out the details of computations.

The utility of a test for a diagnostic or therapeutic decision can be improved by chronologically repeated realizations. Based on several conditions and scores *G. Tusch* pursues the aim to find cut-off values for an average minimum number of tests.

Minimum volume sets are sets with high mass concentration. *W. Polonik* uses these sets for estimating level sets of densities, for testing multimodality and for other problems of data analysis and prediction.

Finally, *J. Dreesman* deals with Markov random fields to describe the distribution of spatially arranged random variables. Assuming nonstationarity of the mean and the covariance spatially varying coefficients are introduced. So the author obtains more appropriate models for applications and further insight into data structures.

4. Information Systems: Design and Implementation

Many branches of computer science – such as software, knowledge and data engineering – are dealing with the *design and implementation of information systems.* Therefore this subject has many facets. Some of these are considered in the following seven papers. The covered spectrum extends from case-based modelling to knowledge representation and data integration.

D. Krampe presents some aspects of CBModeler, a case-based assistant for the conceptual modeling of information systems. An introduction to case-based design is also given. Another approach to case-based modelling is described by *H. Czap.* He introduces CBSE, a case-based software engineering tool which supports the development of customer specific applications, as far as they show same basic functionality.

The paper by *W. Lenski* and *E. Wette-Roch* demonstrates in the concrete setting of an information system on mathematical logic that philosophical aspects must be taken into account for the design. The authors show that besides syntactical relationships also semantical and pragmatical aspects have to be observed and clearly distinguished from each other.

R. Kiel and *M. Schader* show how a method can be implemented which allows the user of a knowledge-based system to alter and re-evaluate previously handled queries. Such a dialog with the user can not be managed by simple if/then rules which are typically used in knowledge-based systems.

In the contribution of *J. Hunstock*, *O. Rauh* and *E. Stickel* a methodology for data integration that is driven by cost/benefit considerations is summarized. The authors introduce a tool that supports all stages of the data integration process.

Probabilistic logic (PL) is a well-suited framework for representing und processing partial information (PI) in decision support systems. *F. Dellmann* discusses the problems in PL arising from conditional probabilities and the applicability of signomial programming algorithms to the optimization problems resulting from nonlinear PI.

A. Geyer-Schulz investigates the possibility of learning strategies for managing new and innovative products with genetic programming methods. For this purpose he applies a simple genetic algorithm to a simulation of a market based on a diffusion modell, a firm, and its competition.

5. Text Analysis and Information Retrieval

This section is devoted to various papers presenting methods and software tools for the *analysis of unstructured textual material* as well as to contributions concerned with the abilities and limits of *information retrieval tools*, especially with regard to the faster and faster growing Internet.

H. Klein gives an overview on the field of text analysis as well as definitions of terms. Furthermore, he introduces a classification of text analysis software and discusses the implications of the qualitative and quantitative approach in the research process.

The paper of *R. Popping* and *C. W. Roberts* is focused on network approaches to quantitative and qualitative text analysis. They provide introductions to four network text analysis methodologies as well as to computer programs developed for each method.

In his survey paper *U. Kelle* deals with the variety of computer-based techniques to aid the qualitative text analysis in interpretive sociology and ethnography, focusing especially on the building of typologies. He also investigates, how computer-aided methods can stimulate progress in the field of qualitative methodology. Presenting different tools of the text analysis package WINMAX, *U. Kuckartz* shows the advantages of bringing together procedures of qualitative and quantitative content analysis. His focus is on the process of classification and typification of qualitative data.

In the paper of *J. Strübing* the basic statements of grounded theory are sketched and translated into requirements for qualitative data analysis (QDA) software. The Windows version of ATLAS/ti is introduced as one approach to technically support grounded theory based analytical work. *T. Richards* and *L. Richards* distinguish the textual and conceptual levels of work in QDA and discuss the relation between them. It is argued that the present QDA software hardly bridges the gap between the two levels; also requirements for making the bridge are discussed.

In the field of information retrieval, *C. Carranza* and *W. Lenski* sretrieval systems (IRS) are inflexible in the sense that the solving process for satisfying an information need is fixed. The authors' approach is to improve the retrieval effectiveness by applying methods from artificial intelligence. An introduction to the CoBRA system is given. IRS are also considered in the paper of *A. Kaiser*. He shows that there is a great need for intelligent information retrieval tools (IIRT), especially for finding relevant information in the World Wide Web. Two prototypes of IIRT as well as an approach for a case-based retrieval tool are presented.

Another consequence of the increasing amounts of data in advanced network environments, such as the Internet, is that the design of the IRS' user interface becomes an important factor. *R. Dässler* and *A. Otto* introduce a virtual reality interface which allows an interactive exploration of abstract information spaces.

Finally, *B. A. Nardi*, *V. O'Day* and *E. J. Valauskas* analyze the work of librarians and their interactions with information, technology and patrons in order to find ideas for software agents on the Internet. The authors state that services of experienced librarians cannot be replaced by automated search engines; the future rather points to a new kind of information ecology, where librarians and software tools work together.

6. Applications in Medicine

This section deals with numerical as well as semantical classification methods applied to medicine. *A. T. McCray* gives an overview of the large international UMSL project (Unified Medical Language System). She explained the huge and complex metathesaurus containing nearly all available medical classifications and terminologies which are combined by a semantical network in order to structure the conceptual complexity in biomedical terminologies. *M. Schoop* and *D. Schoop* emphasize a more computer linguistical approach in the European GALEN project for analysing differences in representing medical classifications by a compositional concept representation language. The formal modelling of medical concepts within this type of representation is described in more detail by *J. Bernauer, M. Schoop* and *D. Schoop*, who consider especially the medical highly important part-whole relation. *M. Romacker, K. Schnattinger, U. Hahn* and *R. Klar* present a knowledge based model for understanding medical texts (pathology reports) by representing terminological structures with LOOM in contrast to most other grammar oriented approaches. *J. Stausberg, M. Laghmouch* and *C. Moraga* explain a medical data dictionary containing all relevant medical terminologies in order to be included in hospital information systems and accessed by SQL. Their ER data model is implemented on an ORACLE database and successfully applied.

The numerical classification methods in medicine are mainly applied in diagnostic or epidemiological field. *C. Begg* and *E. S. Venkatraman* compare two ROC curves with a new distribution free procedure in order to test the hypothesis that two curves (and not two areas under the curves) are identical. *W. Sauerbrei* and *T. L. Diepgen* use k-fold cross validation to a more realistic assessment of diagnostic classifications achieved by CART (Classification and Regression Trees) or logistic regression with back elimination. They applied this improvements to develope a new diagnostic score of atopic dermatitis.

A. Brodbeck, *J. R. Iglesias-Rozas* and *A. Zell* applied neural networks (NN) to the classification of a group of brain tumors and compared this method with discriminant analysis, which turned out to work significantly worse than the best NN. An other NN application in medicine is presented by *C. P. Janson* and *D. D. Komitowski* who classified histological images of tumors for a better standardized pathological grading.

O. Gefeller reviews different statistical methods to adjust the attributable risk for confounding and effect modification. He describes several simulation studies for the adjustments and illustrates the different methods by applying to epidemiological heart infarction data. In addition to this adjusting methods *M. Land* and *O. Gefeller* present partitioning methods for improving the application of attributable risks to more epidemiological situations. *A. Pfahlberg*, *O. Gefeller*, *U. Cleffmann* and *K. F. Klmel* present the methods and results of two cross-sectional studies concerning the change of sun protection knowledge and behaviour before and after public health campaigns. They used a log-linear model to improve the pre-post comparison because some relevant variables were not identical in the two data sets.

M. Schumacher describes the interdisciplinary Center of Data Analysis and Model Building of the University Freiburg which at the time develops solutions for medical, financial and some other questions mainly based on methods for time series and statistics of stochastic process.

W. Sauerbrei shows that the two common methods for classification schemes in survival analysis (prognostic index by Cox regression and CART) suffer from over fitting if the final models are based on many factors. He recommends a validation dataset (if available) or resampling methods to solve this problem and he illustrates these effects using data from a large breast cancer study.

B. Lausen extends the CART methods via additive combination of trees and generalized linear models. He applies this new approach to longitudinal nutritional survey data and discusses the results.

7. Applications in Economics and Social Sciences

Economics and social sciences present manifold applications of data analysis and classification. This section consists of seven papers dealing with *two-mode classification*, *fuzzy clustering*, *neural networks* or *correspondence analysis* to solve problems in application fields like *marketing research*, *advertising research*, *strategic planning* and *social mobility*.

D. Baier, *W. Gaul* and *M. Schader* consider a new two-mode overlapping clustering method which can be used for simultaneous benefit segmentation and market structuring. Various applications of this procedure to conjoint data are used for comparisions. The paper of *M. Schwaiger* presents a new approach of constructing the grand matrix to use one-dimensional clustering methods for two-mode classification. This method is applied to obtain more detailed information about cognitive communication effects in advertising research.

U. Bankhofer and *A. Hilbert* investigate the market of statistical software packages for Microsoft Windows. Beside the analysis of the numerical accuracy of the software packages, a segmentation of the market using a two-mode classification method is performed. Another application of two-mode clustering is presented by *R. Lasch*. When using the management method benchmarking, appropriate benchmarking-partners must be identified. Two-mode classification methods can help to find benchmarking-partners even outside the own industry.

The problem of scenario analysis is to select some few consistent, but quite different pictures of the future. *M. Missler-Behr* uses the fuzzy-c-means algorithm to find an appropriate number of final scenarios and to determine and interpret the final scenarios.

K.-P. Wiedmann and *H.-H. Jung* demonstrate the efficiency of neural networks to realize a market segmentation in the automobile industry. In addition, they present a methodological comparison between the conventional and the neural discriminant analysis.

The paper of *M. Greenacre* and *J. Blasius* considers the application of correspondence analysis to square tables using a modification of the centering and weighting of the correspondence matrix. The methodology is illustrated by an example of social mobility.

8. Applications in Archaeology, Biology, Linguistics and Dialectometry

This section contains some papers with applications of data analysis, clustering, stochastic and statistical methods in several different areas.

R. R. Laxton introduces the standard model of seriation for chronologically ordered archaeological provenances from their profiles. Ideally several seriated orders should be produced and submitted as possible candidates for the chronological order of the provenances together with a measure of their relative merits. So a Bayesian approach which is connected with certain difficulties of implementation seems to be indicated.

Sequencing hybridization is a fast tool to determine a DNA sequence from the unordered list of certain tuples. Using the Chen-Stein method for Poisson process approximation *G. Reinert* approximates the probability that a given DNA sequence is uniquely recoverable from its list.

O. Eulenstein and *M. Vingron* introduce the problem of comparing phylogenetic trees under the assumption of gene duplication events. To resolve the contradiction between the trees two known biologically important measures are discussed, which are equivalent according to a conjecture of Mirkin, Muchnik and Smith.

S. Naumann and *K. Gieseking* propose a method for generating grammars for natural language which applies an inductive learning algorithm to large corpora. For structures not describable by this grammar, rule hypotheses are generated, evaluated and integrated into the grammar. On the basis of texts *P. Filzmoser* investigates the grammatical distribution of words distinguished by codes. The relative frequencies of codes are analyzed by fuzzy-clustering and the graphical presentation of results gives an insight into the data.

Similarity relations between geographically fixed dialects outline fundamental problems in handling synonym features. The paper of *G. Schiltz* describes several similarity indexes actually used in dialectometry. Using data from the bavarian-suebian language atlas *A. Wilhelm* and *M. Sander* show the application of interactive graphical methods to these data using the software packages REGARD and MANET.

The organizers of the conference are very indepted to several institutions that supported the conference: Especially the invitation of excellent plenary and survey speakers would not have been possible without the financial support provided by the *Deutsche Forschungsgemeinschaft* and other private enterprises and public institutions such as Ciba-Geigy, Gesellschaft Sozialwissenschaftlicher Infrastruktureinrichtungen (GESIS) e. V. / Informationszentrum Sozialwissenschaften, Bonn; DIGITAL, Dorner KG, DSE, GRAS (Graphische Systeme GmbH), Philips Medizin-Systeme, Prompt!, SAG, Siemens, Softcon and SUN.

We ary very grateful to all colleagues who chaired a session during the conference or reviewed some papers for this proceedings volume. We gratefully acknowledge the help and support given by the members of the program committee as well as the active cooperation of all participants and authors. We like to emphasize the excellent work of all involved assistants and secretaries of the Unversities of Freiburg and Augsburg who organized the conference and prepared the final program, abstracts and this proceedings volume, especially from Freiburg Dipl. Biol. A. Glockmann, Mrs. D. Kocher, Mrs. F. Kömm, Mrs. M. Voigt, Dipl. Ing. S. Büchele, Dipl. Phys. F. Nold, Dr. U. Schrader, Dr. W. Vach, Dr. A. Zaiß and from Augsburg Dr. U. Bankhofer, Dipl.-Math.oec. A. Hilbert, I. Betz.

We hope that this volume will find interested readers and may encourage further research.

Augsburg,
Freiburg, December 1996

R. Klar
O. Opitz

Contents

Section 3: Statistical Models and Methods

Section 4: Information Systems: Design and Implementation

Section 5: Text Analysis and Information Retrieval

Section 6: Applications in Medicine

Section 7: Applications in Economics and Social Sciences

Section 8: Applications in Archaeology, Biology, Linguistics and Dialectometry

Section 1

Data Analysis and Classification

Probability Models for Convex Clusters

Hans H. Bock

Institut für Statistik, Technical University of Aachen
D-52056 Aachen, Germany

Summary: We consider partitional clustering methods for n objects with a fixed number m of classes. It is well-known that suitable clustering criteria can be obtained in the framework of probabilistic models that assume class-specific distributions for n random vectors $X_1, ..., X_n \in R^p$. This paper proposes three special clustering models where each of the underlying classes $C_1, ..., C_m$ is characterized by a uniform distribution in a convex and finite domain of $\mathbf{R}^p$. We derive the corresponding clustering criteria and the maximum likelihood ratio test statistics for testing the existence of such clustering structures in the observed data. All criteria are based on the determination of convex hulls of data points in $\mathbf{R}^p$.

1. Introduction

Consider a set $\mathcal{O} = \{1, ..., n\}$ of n objects whose properties are characterized by n observed p-dimensional vectors $x_1, ..., x_n \in \mathbf{R}^p$. A clustering problem arises if there is a conjecture that the set $\mathcal{O}$ is heterogeneous and hides a classification structure with, say, m non-overlapping classes. We are then faced with the problem to reconstruct these classes from the data, i.e. to find a partition $\mathcal{C} = (C_1, ..., C_m)$ of $\mathcal{O}$ such that the classes $C_1, ..., C_m \subset \mathcal{O}$ are as 'homogeneous' as possible (with respect to the observed data).

In this paper we adopt a probabilistic approach and consider the data as realizations of n random vectors $X_1, ..., X_n$ in $\mathbf{R}^p$ and the m unknown classes are characterized by class-specific distributions for the vectors X_k. In contrast to well-documented approaches where these distributions are taken from a parametric family such as the multinormal or loglinear one (e.g., Bock 1974, 1987, 1996a,b,c), we will consider here uniform distributions defined over finite *convex* sets D_i of $\mathbf{R}^p$. In Section 2 we define four probability models describing either homogeneity or a clustering structure and derive corresponding clustering criteria and optimum class estimators which reside on the convex hulls of data points in $\mathbf{R}^p$. Section 3 presents statistical tests for checking the existence of such a clustering structure. Section 4 comments briefly on properties of the test statistics, on computational aspects and on probability distributions.

Throughout the paper, we denote by $\mathcal{K}$ the system of all compact (thus finite) convex subsets $D \subset \mathbf{R}^p$ with a non-empty interior and by $\lambda_p(D)$ the p-dimensional Lebesgue measure (volume) of a set $D \subset \mathbf{R}^p$.

2. Probabilistic clustering models involving convex sets

2.1 The homogeneity model H_0

In a first step we consider the case of a homogeneous sample and suppose that the vectors $X_1, ..., X_n$ are independent, all with the same uniform distribution $U(K)$ in an unknown convex set $K \in \mathcal{K}$ with the density function:

$$f(x) = \frac{1}{\lambda_p(K)} \mathbb{1}_K(x) := \begin{cases} \frac{1}{\lambda_p(K)} & x \in K \\ 0 & x \notin K. \end{cases} \tag{2.1}$$

For estimating the unknown convex support K on the basis of a sample $x_1, ..., x_n$ we use the maximum likelihood method, thus maximize the joint density (likelihood function) of $X_1, ..., X_n$:

$$L(K; x_1, ..., x_n) = \prod_{k=1}^{n} \frac{\mathbb{1}_K(x_k)}{\lambda_p(K)^n} = \begin{cases} \frac{1}{\lambda_p(K)^n} & \text{if } x_k \in K \text{ for all } k \\ 0 & \text{else} \end{cases} \tag{2.2}$$

Let $H_n := H(x_1, ..., x_n)$ denote the convex hull of all n data points, a polytope in $\mathbf{R}^p$ with a certain number V_n of vertices. By the convexity of K, the likelihood (2.2) is positive if and only if $H_n \subseteq K$. Thus maximizing (2.2) amounts to minimizing the volume $\lambda_p(K)$ over all convex sets $K \in \mathcal{K}$ with $H_n \subseteq K$ such that the optimum is attained by the convex hull $\hat{K}_n = H_n$. In particular, we obtain:

$$L_1^* := \max_{K \in \mathcal{K}} L(K; x_1, ..., x_n) = \frac{1}{\lambda_p(H_n)^n}. \tag{2.3}$$

Remark 2.1: The convex hull $\hat{K}_n = H_n$ always underestimates the true support K. Therefore Ripley and Rasson (1977) and Rasson (1979) have proposed to estimate K by a dilatation $\tilde{K}_n := g(H_n) + c \cdot (H_n - g(H_n))$ of H_n where $g(H_n) \in \mathbf{R}^p$ is the centroid of H_n and $c = \{(n+1)/[n+1-E[V_{n+1}]]\}^{1/2}$ a constant factor chosen such that $E[\lambda_p(\tilde{K}_n)] = \lambda_p(K)$ (in fact, for computational reasons they propose to use the constant $c = \sqrt{n/(n-v_n)}$ where v_n is the observed number of vertices of H_n). Moore (1984) has investigated the estimation of K in a decision theoretic (Bayesian) framework and proved that $\tilde{K}_n$ is asymptotically equivalent to some Bayes estimator.

Remark 2.2: Efficient algorithms for calculating the convex hull of n points in $\mathbf{R}^p$ can be found, e.g., in Eddy (1977), Preparata and Shamos (1986), Miller and Stout (1988), Rasson, Hardy and Weverbergh (1988), Sedgewick (1988) or Borgwardt (1993). For a general class of partition optimization problems yielding disjoint convex hulls see Barnes et al. (1989).

2.2 The fixed-partition clustering model H_m

Probabilistic clustering models formalize the intuitive idea that data points x_k belonging to the same class should either (a) be close to the other members of this class, (b) close to some class representative such as a centroid, a class-specific hyperplane etc. or (c) be located in a distinct region of $\mathbf{R}^p$ with a high density of sample points. In this paper we will consider clustering structures where each class C_i is characterized by a uniform distribution

$U(D_i)$ in a finite convex set $D_i \subset \mathbf{R}^p$ such that all data points X_k with $k \in C_i$ are concentrated on D_i.

There are basically three different ways of incorporating m class-specific distributions into an overall clustering model (Bock 1996a): *fixed-partition models* (to be considered in this section), *mixture models* (see the next sections) and *random-partition models* (which lead, in our convex clusters case, to the same criteria as mixture models). For a fixed known number $m \geq 2$ of classes we define a *fixed-partition clustering model* H_m as follows: There exists a fixed partition $\mathcal{C} = (C_1, ..., C_m)$ of the set of objects $\mathcal{O}$ with m non-empty disjoint classes $C_1, ..., C_m$ and a system $\mathcal{D} = (D_1, ..., D_m)$ with m pairwise disjoint convex sets $D_1, ..., D_m \in \mathcal{K}$ (with positive volumes) such that for all $i = 1, ..., m$:

$$X_k \quad \sim \quad U(D_i) \qquad \text{for all } k \in C_i. \tag{2.4}$$

The model 'parameters' $\mathcal{C}, \mathcal{D}$ are unknown and will be estimated by the maximum likelihood method. The likelihood function for (2.4) is given by

$$L_m(\mathcal{C}, \mathcal{D}) \quad := \quad \prod_{i=1}^{m} \prod_{k \in C_i} \frac{\mathbb{1}_{D_i}(x_k)}{\lambda_p(D_i)} \quad = \quad \prod_{i=1}^{m} \frac{1}{\lambda_p(D_i)^{|C_i|}} \cdot \prod_{k \in C_i} \mathbb{1}_{D_i}(x_k)$$

where $|C_i|$ denotes the size of the class C_i. $L_m(\mathcal{C}, \mathcal{D})$ is positive if and only if $H(C_i) \subseteq D_i$ for all i where $H(C_i) := H(\{x_k | k \in C_i\})$ is the convex hull of the data points that belong to the class C_i. Since $D_1, ..., D_m$ are non-overlapping, the convex hulls $H(C_1), ..., H(C_m)$ are necessarily non-overlapping too. Moreover, any decrease in D_i increases L_m such that maximizing $L_m(\mathcal{C}, \mathcal{D})$ is equivalent to:

$$\prod_{i=1}^{m} \lambda_p(D_i)^{|C_i|} \quad \to \quad \min_{(\mathcal{C}, \mathcal{D})} \tag{2.5}$$

where minimization is over all pairs of systems $(\mathcal{C}, \mathcal{D})$ such that $H(C_i) \subseteq D_i$ for all i and $(\star)$ the convex hulls $H(C_1), ..., H(C_m)$ are pairwise disjoint, each with a positive volume. Minimizing (2.5) for a *fixed* m-partition $\mathcal{C}$ with respect to all $\mathcal{D}$ fulfilling the constraint $(\star)$ yields the optimum (i.e., minimal) sets $\hat{D}_i := H(C_i)$. This shows, after taking logarithms, that the maximization of $L_m(\mathcal{C}, \mathcal{D})$ reduces to the combinatorial optimization problem:

$$g_m(\mathcal{C}) \quad := \quad \sum_{i=1}^{m} |C_i| \cdot \log \lambda_p(H(C_i)) \to \min_{\mathcal{C} \text{ with } (*)} \tag{2.6}$$

and we may formulate:

Theorem 1: *If an* m*-partition* $\mathcal{C}^* = (C_1^*, ..., C_m^*)$ *of* $\mathcal{O}$ *minimizes (2.6) then the system* $\hat{\mathcal{D}} := (\hat{D}_1, ..., \hat{D}_m) = (H(C_1^*), ..., H(C_m^*))$ *of convex hulls* *maximizes the likelihood $L_m(\mathcal{C},\mathcal{D})$ with maximum value:*

$$L_m^* := \max_{(\mathcal{C},\mathcal{D})} L_m(\mathcal{C},\mathcal{D}) = e^{-g_m(\mathcal{C}^*)} = \prod_{i=1}^{m} \frac{1}{\lambda_p(H(C_i^*))^{|C_i^*|}} \tag{2.7}$$

Remark 2.3: In order to make sure that $\hat{D}_i \in \mathcal{K}$, i.e. $\lambda_p(H(C_i)) > 0$ for all i, each class C_i must necessarily contain at least $p+1$ elements of $\mathcal{O}$. This additional constraint must be incorporated into the optimization problem (2.6). As a consequence, the method makes sense only if the total number n of objects is (considerably) larger than $m(p+1)$. A similar remark applies to the clustering criteria in the subsequent sections.

2.3 The mixture model H_m^{mix}

As an alternative to the clustering model H_m the *mixture model* H_m^{mix} assumes that the random vectors $X_1,...,X_n$ are independent and identically distributed, each with the mixture density:

$$f(x;\mathbf{p},\mathcal{D}) := \sum_{i=1}^{m} p_i \cdot \frac{\mathbb{1}_{D_i}(x)}{\lambda_p(D_i)} \qquad \text{for } x \in \mathbf{R}^p. \tag{2.8}$$

As before, $D_1,...,D_m \in \mathcal{K}$ are supposed to be m unknown, pairwise disjoint, convex subsets of $\mathbf{R}^p$ and $p_1,...,p_m \geq 0$ are m relative frequencies compiled in the probability vector $\mathbf{p} = (p_1,...,p_m)$ with $\sum_{i=1}^m p_i = 1$.

It is well-known that the previous model is motivated by the idea that n objects are randomly and independently sampled from a super-population Π which is subdivided into m mutually exclusive sub-populations $\Pi_1,...,\Pi_m$ with relative frequencies $p_1,...,p_m$ and each Π_i is characterized by a uniform distribution $U(D_i)$. With this interpretation, the subsets $C_i \subset \mathcal{O}$ formed by those objects which belong to Π_i, generate a random (non-observable or non-observed) partition $\mathcal{C} = (C_1,...,C_m)$ of $\mathcal{O}$ with *at most* m non-empty classes. Insofar the model H_m^{mix} can be termed a *random-partition* model as well. The statistical analysis of the mixture model H_m^{mix} proceeds by estimating the unknown parameters $\mathbf{p} = (p_1,...,p_m)$ and $\mathcal{D}$ (in fact, the latent partition $\mathcal{C}$ is not much emphasized in classical mixture models; cf. Bock 1996a,b). Adopting again the maximum likelihood approach, we find that the likelihood function $L_m^{mix}(\mathbf{p},\mathcal{D})$ is positive if and only if the condition $(\star\star)$ $\{x_1,...,x_n\} \subset D := D_1 + \cdots + D_m$ holds, so we have to solve:

$$L_m^{mix}(\mathbf{p},\mathcal{D}) = \prod_{k=1}^{n} f(x_k;\mathbf{p},\mathcal{D}) = \prod_{i=1}^{m} \left[\frac{p_i}{\lambda_p(D_i)}\right]^{|C(D_i)|} \to \max_{\mathbf{p},\mathcal{D} \text{ with } (\star\star)} \tag{2.9}$$

where $C(D_i) := \{k \in \mathcal{O} | x_k \in D_i\}$ collects the objects whose data points belong to D_i, with $\sum_{i=1}^m |C(D_i)| = n$.

First we minimize (2.9) with respect to the vector $\mathbf{p}$ for a fixed support system $\mathcal{D}$. Writing $\alpha_i := |C(D_i)|/n$, this is equivalent to maximizing $\prod_{i=1}^m p_i^{\alpha_i}$, a problem with the well-known solution $\hat{p}_i = \alpha_i = |C(D_i)|/n$, $i = 1, ..., m$. Substitution into (2.9) and taking the negative logarithm yield the reduced problem:

$$\begin{aligned} \tilde{g}_m(\mathcal{D}) &:= -\log\left[n^n \cdot L_m^{mix}(\hat{\mathbf{p}}, \mathcal{D})\right] \\ &= \sum_{i=1}^{m} |C(D_i)| \cdot \log \frac{\lambda_p(D_i)}{|C(D_i)|} \quad \to \min_{\mathcal{D} \text{ with } (**)} . \end{aligned} \tag{2.10}$$

Since any shrinkage of $D_i \in \mathcal{K}$ which maintains the class $C_i = C(D_i)$ decreases the criterion $\tilde{g}_m(\mathcal{D})$, any solution $\mathcal{D}^* = (D_1^*, ..., D_m^*)$ of (2.10) must necessarily fulfil the stationarity conditions:

$$D_i^* = H(C(D_i^*)) \qquad \text{for } i = 1, ..., m \tag{2.11}$$

that can be written as $C_i^* = C(D_i^*)$, $D_i^* = H(C_i^*)$ as well. Combining these conditions with the previous results, we are led to formulate a *combinatorial* maximization problem:

$$g_m^{mix}(\mathcal{C}) := \sum_{i=1}^{m} |C_i| \cdot \log \frac{\lambda_p(H(C_i))}{|C_i|} \quad \to \min_{\mathcal{C} \text{ with } (***)} . \tag{2.12}$$

where minimization is over all partitions $\mathcal{C} = (C_1, ..., C_m)$ with m classes that $(\star\star\star)$ have disjoint convex hulls $H(C_1), ..., H(C_m)$, all with a positive volume $\lambda_p(H(C_i))$ (see also Remark 2.3).

The following theorem shows that the m.l. estimates may be found by solving (2.12):

Theorem 2: *Let $\mathcal{C}^* = (C_1^*, ..., C_m^*)$ be an $m-$partition minimizing the clustering criterion $g_m^{mix}(\mathcal{C})$, (2.12). Then a maximum likelihood estimate for $\mathcal{D}$ and $\mathbf{p}$ is given by the support system $\mathcal{D}^* := (D_1^*, ..., D_m^*) = (H(C_1^*), ..., H(C_m^*))$ comprizing the convex hulls of the optimum classes C_i^* and by $\mathbf{p}^* = (p_1^*, ..., p_m^*)$ with the relative class sizes $p_i^* := |C_i^*|/n$, $i = 1, ..., m$. The maximum likelihood value is given by:*

$$\begin{aligned} L_m^{mix,*} &:= \max_{\mathbf{p}, \mathcal{D} \text{ with } (**)} L_m^{mix}(\mathbf{p}, \mathcal{D}) = L_m^{mix}(\mathbf{p}^*, \mathcal{D}^*) \\ &= n^{-n} \cdot \prod_{i=1}^{m} \left[\frac{|C_i^*|}{\lambda_p(H(C_i^*))}\right]^{|C_i^*|} . \end{aligned} \tag{2.13}$$

Proof: The formal proof combines the reasoning after (2.9) with the following chain of (in-)equalities:

$$\begin{aligned} \tilde{g}_m(\mathcal{D}) &= \sum_{i=1}^{m} |C(D_i)| \cdot \log \frac{\lambda_p(D_i)}{|C(D_i)|} \\ &\overset{(i)}{\geq} \sum_{i=1}^{m} |C(D_i)| \cdot \log \frac{\lambda_p(H(C(D_i)))}{|C(D_i)|} \end{aligned}$$

$$= g_m^{mix}((C(D_1), ..., C(D_m))) \geq \min_{\mathcal{C} \text{ with } (***)} g_m^{mix}(\mathcal{C})$$

$$= g_m^{mix}(\mathcal{C}^*) = \sum_{i=1}^{m} |C_i^*| \cdot \log \frac{\lambda_p(H(C_i^*))}{|C_i^*|} \overset{(ii)}{=} \tilde{g}_m(\mathcal{D}^*)$$

where *(i)* follows from the inclusion $H(C(D_i)) \subseteq D_i$ and *(ii)* derives from the definition $D_i^* := H(C_i^*)$ which implies $C_i^* = C(D_i^*)$ by convexity.

Remark 2.4: It is plausible that the optimum m-partition $\mathcal{C}^*$ provides, as a by-product, an approximation to the true, but random and unobserved partition $\mathcal{C}$ of the objects mentioned in the beginning of this section. In fact, this ad-hoc clustering corresponds to the application of a Bayesian assignment rule with estimated 'parameters' D_i^* (as it is typical for parametric mixture models as well). However, the random classification $\mathcal{C}$ does not show up in the underlying marginal mixture model (2.8).

2.4 The pseudo-mixture model H_m^{uni}

In the framework of discrimination problems, Hardy and Rasson (1982), Baufays and Rasson (1984) and Rasson et al. (1988) considered the following *pseudo-mixture model* that will be denoted by H_m^{uni} here: The random vectors $X_1, ..., X_n$ are independent and identically distributed, all with the same *uniform* distribution over the *union* $D := D_1 + \cdots + D_m$ of m unknown convex disjoint subsets $D_1, ..., D_m \in \mathcal{K}$ of $\mathbf{R}^p$, thus with the density:

$$f_D(x) := \frac{1}{\lambda_p(D)} \cdot \mathbb{I}_D(x) = \sum_{i=1}^{m} \frac{\lambda_p(D_i)}{\lambda_p(D)} \cdot f_{D_i}(x). \tag{2.14}$$

Obviously, this density can be interpreted as a mixture (2.8) of m uniform densities $f_{D_i}(\cdot)$ with relative frequencies by $p_i := \lambda_p(D_i)/\lambda_p(D)$. However, in contrast to H_m^{mix}, the model H_m^{uni} has only one set $\mathcal{D} = (D_1, ..., D_m)$ of free 'parameters'. It has found useful applications in spatial pattern recognition and remote sensing, e.g., for LANDSAT data (see Rasson et al. 1993a,b).

Considering (2.14) as a clustering model, we have to reconstruct the cluster regions $D_1, ..., D_m \in \mathcal{K}$ from the data $x_1, ..., x_n$, e.g. by maximizing the likelihood:

$$L_m^{uni}(\mathcal{D}) = \prod_{k=1}^{n} f_D(x_k) = \frac{1}{\lambda_p(D)^n} \cdot \prod_{k=1}^{n} \mathbb{I}_D(x_k) \to \max_{\mathcal{D}}. \tag{2.15}$$

The function $L_m^{uni}(\mathcal{D})$ is positive if and only if all x_k belong to the union D, and obviously $(\star\star\star\star)$ the classes $C_i(\mathcal{D}) := C(D_i) := \{k \in \mathcal{O} | x_k \in D_i\}$ that are non-empty have disjoint convex hulls. Denoting by Δ the family of all systems $\mathcal{D} = (D_1, ..., D_m)$ of m non-overlapping subsets $D_i \in \mathcal{K}$ such that their union D contains all x_k's, (2.15) is equivalent to maximizing the total volume:

$$\lambda_p(D) = \sum_{i=1}^{m} \lambda_p(D_i) \to \max_{\mathcal{D} \in \Delta}. \tag{2.16}$$

The solution is given by:

Theorem 3: *Let $\mathcal{C}^{**} = (C_1^{**}, ..., C_m^{**})$ be a partition of $\mathcal{O}$ with m classes that minimizes the 'sum-of-volumes criterion':*

$$t_m(\mathcal{C}) := \sum_{i=1}^{m} \lambda_p(H(C_i)) \rightarrow \min_{\mathcal{C} \text{ with } (****)} \tag{2.17}$$

under the constraint $(\star\star\star\star)$ that $H(C_1), ..., H(C_m)$ are non-overlapping. Then the convex hulls $D_i^{**} := H(C_i^{**})$ of all data points belonging to C_i^{**} $(i = 1, ..., m)$ provide the solution $\mathcal{D}^{**} = (D_1^{**}, ..., D_m^{**})$ of (2.15), i.e., a maximum likelihood estimate for $\mathcal{D}$. The maximum likelihood value is:

$$L_m^{uni,*} \quad := \quad \max_{\mathcal{D} \in \Delta} L_m^{uni}(\mathcal{D}) \quad = \quad \frac{1}{\lambda_p(D^{**})^n}. \tag{2.18}$$

Proof: For a given system $\mathcal{D} \in \Delta$ of convex sets we denote by $\mathcal{C}(\mathcal{D}) := (C_1(\mathcal{D}), ..., C_m(\mathcal{D}))$ the partition of $\mathcal{O}$ with classes $C_i(\mathcal{D}) := \{k \in \mathcal{O} | x_k \in D_i\}$. By the convexity of D_i and the definition of $C_i(\mathcal{D})$ we have $H(C_i(\mathcal{D})) \subseteq D_i$ for all i. Therefore the following (in-)equalities hold for all $\mathcal{D} \in \Delta$:

$$\begin{aligned} \lambda_p(D) \quad &= \quad \sum_{i=1}^{m} \lambda_p(D_i) \geq \sum_{i=1}^{m} \lambda_p(H(C_i(\mathcal{D}))) \;=\; t_m(\mathcal{C}(\mathcal{D})) \\ &\geq \quad t_m(\mathcal{C}^{**}) = \sum_{i=1}^{m} \lambda_p(H(C_i^{**})) \;=\; \sum_{i=1}^{m} \lambda_p(D_i^{**}) \;=\; \lambda_p(D^{**}) \end{aligned}$$

where $D^{**} = D_1^{**} + \cdots + D_m^{**}$.

Remark 2.5: A generalization of H_m^{uni} to the case of non-uniform distributions over the convex supports $D_1, ..., D_m$ is described by Rasson et al. (1993a).

3. Validation and testing problems

Every clustering algorithm produces a classification of n data points whether or not these data show a 'relevant' grouping structure. Therefore we need methods for checking if the configuration $x_1, ..., x_n$ points to a 'significant' grouping structure of the objects and/or if the classification $\mathcal{C}^*$ obtained from the clustering algorithm is sufficiently well distinguished from a partition that can be obtained from a 'random', nonstructured or 'homogeneous' data. Problems of this kind can be tackled, on the one hand, by descriptive methods combined with a substance-related interpretation of the calculated clusters, and, on the other hand, by formal statistical hypothesis tests in the framework of a probabilistic clustering model. Disregarding the first approach, we emphasize here the second one and derive the *maximum likelihood ratio test statistics* for testing the homogeneity hypothesis H_0 against each of the three convex clustering models H_m, H_m^{mix} and H_m^{uni} described in the previous section.

3.1 Testing for the fixed-partition model H_m

First we consider the problem of testing the hypothesis H_0 that our data $x_1, ..., x_n$ were obtained from a 'homogeneous' population as described by the uniform distribution (2.1) with an (unknown) convex set $K \in \mathcal{K}$ in $\mathbf{R}^p$ against the alternative that there is a clustering structure described by the fixed-classification model H_m, (2.4), with m non-overlapping convex stets $D_1, ..., D_m \in \mathcal{K}$. Since (2.3) and (2.7) provide the maximum likelihood under these hypotheses, we obtain the following maximum likelihood ratio:

$$Q_m := \frac{\max_{(\mathcal{C},\mathcal{D})} L_m(\mathcal{C},\mathcal{D})}{\max_{K\in\mathcal{K}} L(K; x_1, ..., x_n)} = \frac{L_m^*}{L_1^*} = \frac{\lambda_p(H_n)^n}{\prod_{i=1}^m \lambda_p(H(C_i^*))^{|C_i^*|}} \quad (3.1)$$

where $\mathcal{C}^* = (C_1^*, ..., C_m^*)$ is an optimum m-partition of the objects (i.e. maximizing the criterion $g_m(\mathcal{C})$, (2.6)) and $\mathcal{D}^* = (D_1^*, ..., D_m^*)$ the corresponding system of convex hulls $D_i^* = H(C_i^*)$. Taking logarithms leads to the *maximum likelihood ratio test (MLRT)*:

$$T_m := \log Q_m^{1/n} = \quad (3.2)$$

$$= -\sum_{i=1}^m \frac{|C_i^*|}{n} \cdot \log \frac{\lambda_p(H(C_i^*))}{\lambda_p(H_n)} \begin{cases} > c & \text{decide for clustering } H_m \\ \le c & \text{accept uniformity } H_0. \end{cases}$$

where the threshold $c = c(\alpha) > 0$ must be determined from the desired type I error probability α (e.g., $\alpha = 0.05$) and the distribution of T_m under H_0.

Remark 3.1: The criterion T_m (as well as the following criteria (3.4), (3.6) and (3.7)) compares the convex hull $H_n = H(x_1, ..., x_n)$ to the m class-specific convex hulls (polytopes) $D_i^* = H(C_i^*)$ contained in it (where each of the V_n vertices of H_n is a boundary point of some D_i^*): Informally, the MLRT (3.2) accepts the clustering hypothesis H_m (and recommends the optimum classification $\mathcal{C}^*$) if these polytopes D_i^* have, *on the average*, a sufficiently small volume in comparison to the total volume of H_n.

3.2 Testing for the mixture model H_m^{mix}

In a similar way we obtain the MLRT for testing the uniformity hypothesis H_0 versus the mixture model H_m^{mix} with m disjoint convex sets $D_1, ..., D_m$. If $\mathcal{C}^* = (C_1^*, ..., C_m^*)$ denotes an m-partition of the objects that minimizes the clustering criterion g_m^{mix}, (2.12), and $\mathcal{D}^* = (D_1^*, ..., D_m^*)$ is the corresponding system of convex hulls $D_i^* = H(C_i^*)$, the MLR is obtained from (2.3) and (2.13):

$$Q_m^{mix} := \frac{\max_{\mathbf{p},\mathcal{D} \text{ with } (**)} L_m^{mix}(\mathbf{p},\mathcal{D})}{\max_{K\in\mathcal{K}} L(K; x_1, ..., x_n)}$$

$$= \frac{L_m^{mix,*}}{L_1^*} = \frac{\lambda_p(H_n)^n}{n^n} \cdot \prod_{i=1}^m \left[\frac{|C_i^*|}{\lambda_p(H(C_i^*))}\right]^{|C_i^*|} \quad (3.3)$$

The corresponding *mixture MLRT* can therefore be formulated as follows:

$$\begin{aligned} T_m^{mix} &:= \log(Q_m^{mix})^{1/n} = \qquad (3.4)\\ &= \sum_{i=1}^{m} \frac{|C_i^*|}{n} \cdot \log\left[\frac{|C_i^*|}{n} / \frac{\lambda_p(H(C_i^*))}{\lambda_p(H_n)}\right] \quad \begin{cases} > c & \text{decide for } H_m^{mix} \\ \leq c & \text{accept uniformity } H_0. \end{cases} \end{aligned}$$

with an $(1-\alpha)$-percentage point $c = c(\alpha) > 0$ obtained from the H_0 distribution of T_m^{mix}.

Remark 3.2: There is an interesting interpretation of T_m^{mix} in terms of a *Kullback-Leibler divergence*: If we denote by $q_i := \lambda_p(H(C_i^*))/\lambda_p(H_n)$ the relative size of $D_i^* = H(C_i^*)$ inside H_n, the percentage of H_n covered by all D_i^*'s is given by $\gamma_n := \sum_{i=1}^m q_i = \lambda_p(D^*)/\lambda_p(H_n) \in [0,1]$ and the numbers $\tilde{q}_i := q_i/\gamma_n$, $i = 1, ..., m$, define a probability distribution $\tilde{\mathbf{q}}$ over $\{1, ..., m\}$, in analogy to the distribution $\hat{\mathbf{p}}$ of the relative class sizes $\hat{p}_i = |C_i^*|/n$. It appears that the statistic T_m^{mix} is related to the Kullback-Leibler divergence $I(\mathbf{p}|\tilde{\mathbf{q}}) := \sum_{i=1}^m p_i \log(p_i/\tilde{q}_i) \geq 0$ of those distributions as follows:

$$T_m^{mix} \quad = \quad I(\mathbf{p}|\tilde{\mathbf{q}}) - \log \gamma_n.$$

This decomposition shows that, for a given coverage γ_n, the criterion T_m^{mix} favours the decision for H_m^{mix} the more the more dissimilar the distributions $\hat{\mathbf{p}}$ and $\tilde{\mathbf{q}}$ will be with respect to this divergence measure.

3.3 Testing for the uniformity model H_m^{uni}

Finally, we test the homogeneity model H_0 against the pseudo-mixture clustering model H_m^{uni} that assumes a uniform distribution on the union $D = D_1 + \cdots + D_m$ of m disjoint convex sets from $\mathcal{K}$: The MLR is provided by (2.3) and (2.18):

$$\begin{aligned} Q_m^{uni} &:= \frac{\max_{\mathcal{D} \in \Delta} L_m^{uni}(\mathcal{D})}{\max_{K \in \mathcal{K}} L(K; x_1, ..., x_n)} \\ &= \frac{L_m^{uni,*}}{L_1^*} = \left[\frac{\lambda_p(H_n)}{\lambda_p(D^{**})}\right]^n = \left[\frac{\lambda_p(H_n)}{\sum_{i=1}^m \lambda_p(H(C_i^{**}))}\right]^n \qquad (3.5) \end{aligned}$$

This expression leads to an intuitive MLR clustering test:

$$\begin{aligned} T_m^{uni} &:= -\log(Q_m^{uni})^{1/n} \qquad (3.6)\\ &= \sum_{i=1}^{m} \frac{\lambda_p(H(C_i^{**}))}{\lambda_p(H_n)} \quad \begin{cases} < c & \text{decide for clustering } H_m^{uni} \\ \geq c & \text{accept uniformity } H_0 \end{cases} \end{aligned}$$

with a critical threshold $0 < c < 1$. In fact, denote by $G_n := H_n - D^{**} = H_n - (D_1^{**} + \cdots + D_m^{**}) \subset \mathbf{R}^p$ that part of H_n that is *not* covered by the convex hulls $D_i^* = H(C_i^*)$. Then (3.6) is equivalent to the *'gap' or 'empty* *space test'*:

$$\epsilon_m \quad := \quad 1 - T_m^{uni} = \frac{\lambda_p(G_n)}{\lambda_p(H_n)} \quad \begin{cases} > \tilde{c} & \text{decide for clustering } H_m^{uni} \\ \leq \tilde{c} & \text{accept uniformity } H_0. \end{cases} \tag{3.7}$$

where $\tilde{c} = 1 - c$. Thus the test decides for the clustering alternative H_m^{uni} if in H_n the empty space (gap) G_n *between* the convex hulls D_i^{**} is too large. This latter test has been proposed by Rasson and Kubushishi (1994) for the case of $m = 2$ classes and is well-known in the one-dimensional case.

4. Remarks and conclusions

The clustering criteria and test statistics derived in the previous sections share various useful and desirable properties that are related to the invariance of convexity under regular linear transformations (LT) of $\mathbf{R}^p$: (1) They provide linearly equivariant estimates $\hat{D}_i$ for the underlying cluster regions D_i, i.e., when transforming all data points x_k with the same LT, the estimates $\hat{D}_i$ are transformed in the same way. (2) The resulting optimum classes $C_i^* \subset \mathcal{O}$, say, are invariant with respect to LTs of the data. (3) The H_0 distribution of all test criteria is invariant with respect to all LTs of the data (such that, e.g., the same H_0-distribution results for all ellipsoidal supports $K \in \mathcal{K}$ as for the special choice of the unit ball $K^* = \{x \in \mathbf{R}^p | \; ||x|| \leq 1\}$).

The exact solution of combinatorial optimization problems involving convex hull clusters such as (2.6), (2.12) and (2.17) is far beyond any computer capacity for $p \geq 3$, say, such that approximate solutions must be found. A general approach is provided by pairwise exchange algorithms which start from an initial classification $\mathcal{C}^{(0)}$ comprizing convex classes which can be obtained, e.g., by the well-known k-means algorithm for the variance criterion (cf. Bock 1996a,c). Other methods proceed by 'peeling' successively a given class C_i and assigning the peeled objects successively to the neighbouring classes. Hardy (1996) describes a hierarchical split and merge strategy. In the case of two-dimensional data the geometrical simplicity of $\mathbf{R}^2$ allows a fast determination of convex hulls. In this case and for $m = 2$ classes, it is even possible to determine exactly the optimum partition for moderate sample sizes since any two-partition of $\{x_1, ..., x_n\}$ with convex classes is largely determined by two pairs of neigbouring data points on the boundary of the convex hull H_n that are located at the transition between both classes.

The probabilistic problems related to the (finite-sample or asymptotic) H_0 distribution of the clustering test statistics derived in Section 3 are far from being resolved. Whilst there exist numerous articles dealing with the convex hull of random points in $\mathbf{R}^p$, there are no exact or asymptotic probability results that relate simultaneously to two or several convex hulls and could be applied for our situation. Analytical results might be obtained for the special case of the 'empty space test' described in section 3.3, at least for $m = 2$ classes, for other cases we must recur to simulations. This problem is under investigation (see also Rasson et al. 1994).

References:

BARNES, E. R.; HOFFMAN, A. J. and ROTHBLUM, U. G. (1989): Optimal partitions having disjoint convex and conic hulls. Ructor Research Report no. 30-89, September 1989, Rutgers Center for Operations Research, Rutgers University, New Brunswick, NJ. 31 pp.

BAUFAYS, P. and RASSON, J.-P. (1984): Une nouvelle règle de classement utilisant l'enveloppe convexe et la mesure de Lebesgue. *Statistique et Analyse des Données 9 (2), 31-47.*

BOCK, H. H. (1974): *Automatische Klassifikation.* Vandenhoeck & Ruprecht, Göttingen, 1974.

BOCK, H. H. (1987): On the interface between cluster analysis, principal component analysis, and multidimensional scaling. In: H. Bozdogan and A.K. Gupta (eds.): *Multivariate statistical modeling and data analysis.* Reidel, Dordrecht, 1987, 17-34.

BOCK, H. H. (1996a): Probability models and hypotheses testing in partitioning cluster analysis. In: P. Arabie, L. Hubert and G. De Soete (eds.): *Clustering and Classification.* World Science Publishers, River Edge/NJ, 1996, 377-453.

BOCK, H. H. (1996b): Probabilistic models in cluster analysis. *Computational Statistics and Data Analysis 22* (in press).

BOCK, H. H. (1996c): Probability models in partitional cluster analysis. In: A. Ferligoj, and A. Kramberger (Eds.): Developments in data analysis. FDV, Metodoloski zvezki, 12, Ljubljana, Slovenia, 1994, 3-25.

BORGWARDT, K. H. (1993): Average complexity of a gift-wrapping algorithm for determining the convex hull of randomly given points. Report No. 478, Institut für Mathematik, Universität Augsburg, 51 pp.

EDDY, W. F. (1980): A new convex hull algorithm for planar sets. *ACM Trans. Math. Software 3 (4), 398-403*; the algorithm: *411-412.*

HARDY, A. (1996): A split and merge algorithm for cluster analysis. Paper presented at the 5th Conference of the IFCS, March 1996, Kobe/Japan.

HARDY, A. and RASSON, J.-P. (1982): Une nouvelle approche des problèmes de classification automatique. *Statistique et Analyse des Donnés 7 (2), 41-56.*

MILLER, R. and STOUT, Q. F. (1988): Efficient parallel convex hull algorithms. *IEEE Trans. Computers 37, 1605-1618.*

MOORE, M. (1984): On the estimation of a convex set. *Ann. Stat. 12, 1090-1099.*

PREPARATA, F. P. and SHAMOS, M. I.(1986): *Computational geometry. An introduction.* Springer-Verlag, New York.

RASSON, J. P. (1979): Estimation de formes convexes du plan. *Statistique et Analyse des Données 1, 31-46.*

RASSON, J.-P.; HARDY, A. and WEVERBERGH, D. (1988): Point process, classification and data analysis. In: H.H. Bock (ed.): *Classification and related methods of data analysis.* Proc. 1st Conf. IFCS, Aachen, 1987. North Holland/Elsevier, Amsterdam, 1988, 245-256.

RASSON, J.-P.; GRANVILLE, V. and ORBAN-FERAUGE, F. (1993a): How to match visual classification using non-parametric intensities? - A behavioral answer. Proc. IGARSS, vol. 3. Tokyo, Japan, 1993, 1521-1523.

RASSON, J.-P.; ORBAN-FERAUGE, F. and GRANVILLE, V. (1993b): ¿From a natural to a behavioral classification rule. Report 93/25. Publications du Département de Mathématique, Facultés Universitaires de Namur, Belgique. 21 pp.

RASSON, J.-P. and KUBUSHISHI, T. (1994): The gap test: an optimal method for determining the number of natural classes in cluster analysis. In: E. Diday, Y. Lechevallier, M. Schader, P. Bertrand, and B. Burtschy (eds.): *New approaches in classification and data analysis.* Springer-Verlag, Heidelberg, 1994, 186-193.

RIPLEY, B. D. and RASSON, J. P. (1977): Finding the edge of a Poisson forest. *J. Appl. Prob. 14, 483-491.*

SEDGEWICK, R. (1988): *Algorithms.* Addison-Wesley Publ. Comp., Reading, Mass.

Multidimensional Scaling: Analyzing Multineuronal Spike Patterns

H. Haan[1+2], J. Timmer[2+3], R. Mathar[4], J. Krüger[1]

[1] Institut für Biophysik, Arbeitsgruppe Hirnforschung, D-79104 Freiburg

[2] Fakultät für Physik, D-79104 Freiburg

[3] Zentrum für Datenanalyse und Modellbildung, D-79104 Freiburg

[4] Institut f. Statistik und Wirtschaftsmathematik, RWTH Aachen, D-52056 Aachen

Summary: Multidimensional scaling (MDS) is used as an exploratory tool to discover structure in vast neuro-data sets. Examples are given for recordings from visual cortex of monkeys, using an array of 30 microelectrodes. Different models of MDS are applied and compared.

1. Introduction

To understand higher brain functions, one has first to find structures in neuro-data without considering circumstances of the experiment and thereafter to try to relate these structures to the circumstances. Multidimensional scaling may be capable to reveal relevant structures. A brief abstract of part of these results have been given (Haan et al., 1995). For an overview of MDS see Hartung (1992).

2. Multidimensional scaling

The aim of MDS is to embed $\frac{n*(n-1)}{2}$ given pairwise dissimilarities δ_{ij} as a configuration of n points $X = x_1, \ldots, x_n$, thus, that the pairwise distances of the points d_{ij} match the dissimilarities δ_{ij} best. If the embedding is performed to $R^k, k \leq 3$, the configuration can be visualized and the structure of the points be investigated. In fact, only the dissimilarities enter the computation, which, in most applications, are euclidian distances of high dimensional vectors, called objects $\mathcal{O}_1, \ldots, \mathcal{O}_n$, or other experimental measurements thuch as similarities, confusion probabilities, interaction rates, correlation coefficients or other measures of proximity or dissociation of the most diverse kind. The embedding is performed by different models of MDS. We discuss

- a weighted least-squares-model STRESS, where the configuration X is achieved by optimization,
- for a special case, the model "classical scaling", an analytical computation of X,
- and the means how to assess the quality of the embedding.

2.1 MDS by optimization of STRESS

The demand, that each pairwise dissimilarity matches each distance best, is expressed in the weighted least-squares-model STRESS:

$$\rho(X) = \frac{1}{\sum_{i<j} \delta_{ij}} \sum_{i<j} \frac{(\delta_{ij} - d_{ij})^2}{\delta_{ij}^l}, \tag{1}$$

where the exponent l of the weighting factor $w_{ij} = \delta_{ij}^{-l}$ allows to favour large dissimilarities to variable extents.
For the computation of distances d_{ij} a Minkowski-norm $\mathcal{L}_p$ is applied:

$$d_{ij} = (\sum_{l=1}^{k} |x_i(l) - x_j(l)|^p)^{\frac{1}{p}}, \tag{2}$$

most common is $\mathcal{L}_2$.
The configuration X is achieved by optimization of STRESS. We found best convergence using gradient-based methods (e.g. quasi-Newton), actually when the gradient was approximated.

2.2 MDS by classical scaling

For special conditions the configuraion X can be computed analytically. As shown in (3) and (4), these conditions are:

- euclidian distances in (2)
- performing a suitable centering
- squaring each dissimilarity and each distance:

Let Δ be the matrix of all dissimilarities δ_{ij} and $D(X)$ be the matrix of all distances d_{ij}, then, with a variety of matrix norms $\|\cdot\|$, classical scaling is written as "centered squared STRESS":

$$\text{minimize } \| E_n(\Delta^{(2)} - D^{(2)}(X))E_n \| . \tag{3}$$

E_n is the centering-operator:

$$E_n = \begin{pmatrix} 1-\frac{1}{n} & -\frac{1}{n} & \cdots & -\frac{1}{n} \\ -\frac{1}{n} & 1-\frac{1}{n} & \cdots & -\frac{1}{n} \\ \vdots & \vdots & & \vdots \\ -\frac{1}{n} & -\frac{1}{n} & \cdots & 1-\frac{1}{n} \end{pmatrix}. \tag{4}$$

The exponent of the matrices in (3) denoting the squared distances and dissimilarities.

Mathar (1985) has shown, that if rank$(A = -\frac{1}{2}E_n\Delta^{(2)}E_n) \leq k$, and A is nonnegative definite, then the best configuration is computed by spectral decomposition of $A = T\Lambda T'$:

$$X = \left((\sqrt{\lambda_1^+}t_1, \ldots, \sqrt{\lambda_k^+}t_k) \right), \tag{5}$$

where the λ_i^+ denote the sorted eigenvalues > 0 in descending order.

There is a relation between classical scaling and principal component decomposition, but beside the different interpretation, the outstanding difference is that principal component decomposition urgently requires the knowledge of the objects $\mathcal{O}_i$ themselves, whereas the embedding by MDS is achieved only with the aid of the dissimilarities δ_{ij}.

2.3 Assesment of the quality of the embedding

The assesment of the quality of the embedding by different models cannot be done by comparison of STRESS of the final configuration X, because the different models incorporate different normalizations and weighting factors. We propose to plot the distances d_{ij} versus the dissimilarities δ_{ij} in a scatter plot and use the correlation coefficient of the resulting distribution of points to asses the quality of the embedding.

3. Application of MDS to neuro-data

3.1 Exploratory analysis of neuronal spike patterns

Recordings from striate cortex of anaesthetized monkeys were taken with 30 microelectodes during visual stimulation by light bars moving in 16 different orientations through the receptive fields. The multineuronal patterns of responses to each of the 16 stimuli – multivariate time series – consist of 10 time bins of 30 ms duration at each of 30 loci, each containing the number of spike counts recorded. The patterns of responses are high-dimensional objects $\mathcal{O}_i \in R^{300}$. Responses to equal stimuli (20 repetitions each) are expected to cluster in R^{300} and those to different stimuli are likely to form different clusters. To investigate the structure of the objects $\mathcal{O}_i$, their pairwise dissimilarities are embedded as a configuration of points $X = x_1, \ldots, x_n \in R^3$ and inspected as projections to R^2.

Figure 1 shows the embedding by classical scaling. The orientations of the light bars are visualized by arrow-symbols.

The structure in Fig. 1 shows a clear circular topology. Note that this result has been obtained without relying on the nature of the stimuli. If the stimuli are known, the interpretation of this structure is in agreement with classical neurophysiology.

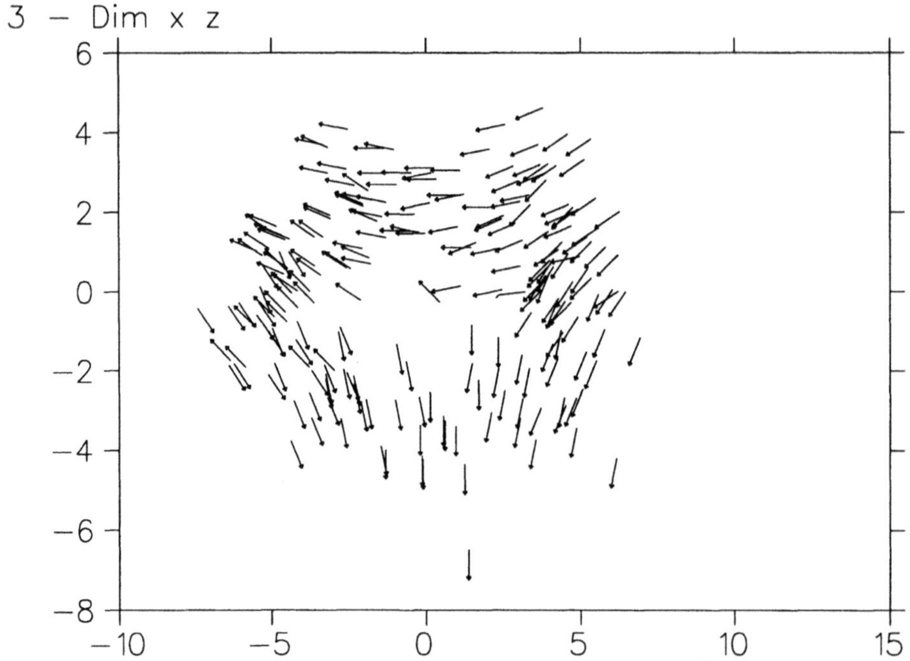

Figure 1: Embedding of dissimilarities of 320 objects to R^3 (16 stimuli, each repeated 20 times). One projection to the plane is shown. The angles of the movement directions of the stimuli have been devided by 2, so that equal orientation but opposite moving directions are symbolized by arrows at right angles to each other.

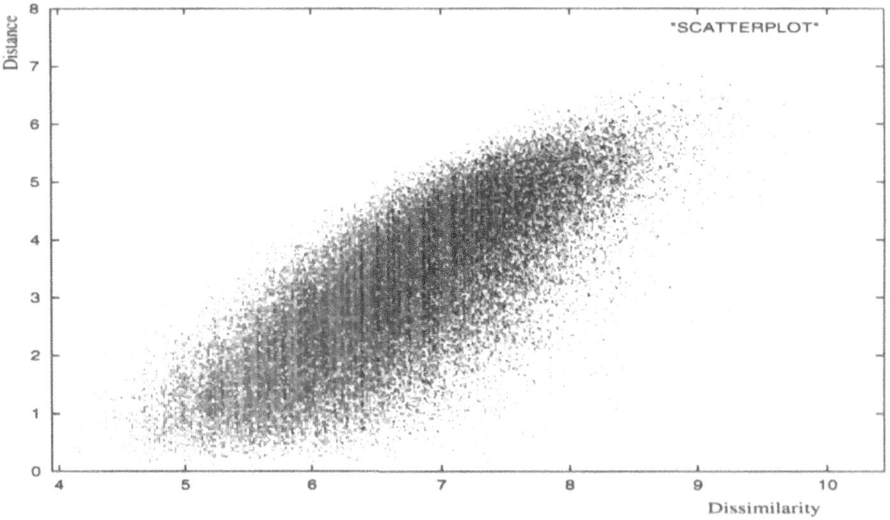

Figure 2: Scatterplot of each pairwise distance d_{ij} over each pairwise dissimilarity δ_{ij}. The correlation coefficient is 0.75.

Figure 2 shows the scatter plot of distance d_{ij} over dissimilarity δ_{ij} . The correlation coefficient of 0.75 and the clear annular structure seen in Fig. 1 suggest that a low number of dimensions is sufficient to express the structure of the objects. The model STRESS was less successful with this data set, since it renders smaller dissimilarities with higher fidelity, so that the variability of responses to equal stimuli is overemphasized.

3.2 Exploratory analysis of local field potentials

We recorded spikes and local field potentials (LFPs) from the visual cortex of awake monkeys with 30 permanently implanted microelectrodes.

The multivariate LFPs are highly correlated. MDS was applied to find brief deviations of the LFP of some electrodes.

Usually classical analysis, for instance cross correlation is performed, gaining for n LFP-electrodes $\frac{n*(n-1)}{2}$ cross correlograms for a glidin window across the data set.

By defining a suitable measure of dissimilarity between pairs of LPFs, MDS allowed to detect instances of differences between LFPs at different electrodes. For each time bin, the judgement of $\frac{30*(30-1)}{2}$ dissimilarities was replaced by the inspection of one plot.

4. Discussion

The choice of the applied model strongly depends on the nature of the data. Classical scaling emphasizes larger dissimilarities and tends to suppress structural details within clusters of objects belonging to equal stimuli.

So showing structure even in data where a variation of objects belonging to equal stimuli is present. STRESS would show better results when a large amount of very accurate data are to be visualized. For the analysis described in chap. 2.2, classical scaling is optimal, because the computing time is much less then that of the optimization of STRESS.

References:

HAAN, H. et al. (1995): Multineuronal Spike Patterns: Visualization of relations by Multidimensional Scaling. In: N. Elsner and R. Menzel (eds.): *Göttingen Neurobiology Report 1995.* G. Thieme Verlag Stuttgart, 475.

HARTUNG, J. et al. (1992): Multivariate Statistik. R. Oldenbourg Verlag München.

MATHAR, R. (1985): The best euclidian fit to a given distance matrix in prescribed dimensions. *Linear Algebra and Its Applications, 67, 1–6.*

Fixed Point Clusters and Their Relation to Stochastic Models

C. Hennig[1]

[1] Institut für Mathematische Stochastik,
Universität Hamburg, Bundesstr. 55, D-20146 Hamburg, Germany

Summary: The idea of Fixed Point Clusters (FPC) will be introduced in this paper. An FPC is a set of data points which is a fixed point of a procedure consisting of a statistical parameter estimation and a classification of data in "near" and "outlying" with respect to the estimator. The relation of Fixed Point Clustering to parameter estimation in stochastic models will be discussed. FPCs use stochastic models as orientation rather than as basis. They are mainly a tool for explorative data analysis and model building. FPCs are a useful alternative to other clustering approaches at least in clusterwise Linear Regression.

1. Fixed Point Clusters - The General Case

The idea of Fixed Point Clusters is as follows: Given a dataset from $(\mathbb{R}^d)^n$, we call a subset of the data points a "cluster" if one can define an area, where these data points appear highly concentrated compared to the points which lie in the neighborhood outside this area. How can we find such areas? We take a subset of the data points and classify every point in the whole dataset with respect to the criterion, if it lies "near" or "far" to the points of our subset. In other words, we carry out an outlier classification. Here this means to compute an area which represents the points of the subset as a whole and to look if each data point of the dataset lies inside or outside this area. We call the subset of data points an FPC if it reproduces itself (i.e. it is a fixed point) with respect to this procedure. Thus, all points of an FPC are classified to lie "near" to themselves, all other points are classified to lie outside. "Themselves" here means a structure which represents the "general tendency" of the points in the subset.

The crucial point is how to assign such a structure to a set of data points. If we imagine a stochastic model as an adequate description of the points belonging to a common cluster, it is natural to estimate the parameters of the model and compute an area where points generated by the model would lie with high probability. In many stochastic models this can be done by computing those data points which have small deviation from the predicted value by the model. This will be formalized using a distance between a data point and a parameter estimator.

Definition: Let $(z_1, \ldots, z_n)$, $z_i \in \mathbb{R}^d, i = 1, \ldots, n$ denote the dataset, $g \in \{0,1\}^n$ be an indicator vector where $g_i = 1$ means that z_i belongs to the set indicated by g.

Let $\hat{\theta}$ be some parameter estimator, i.e. $\hat{\theta}$ maps from the space of datasets into some parameter space Θ. $\hat{\theta}(g)$ denotes $\hat{\theta}$ applied to the points indicated by g. $D : I\!R^d \times \Theta \mapsto [0, \infty)$ denotes a distance between a data point and a parameter estimator. Let $c > 0$ be a tuning constant.

Then g is a **Fixed Point Cluster Vector** (FPCV), iff g is fixed point of

$$f : \{0,1\}^n \mapsto \{0,1\}^n, \tag{1}$$

$$f_i(g) := 1\{D(z_i, \hat{\theta}(g)) < c\}, \quad i = 1, \dots, n. \tag{2}$$

c will be chosen such that

$$P_{\hat{\theta}(g)}\{D(z, \hat{\theta}(g)) < c\} \text{ is large.} \tag{3}$$

Example: In the one-dimensional location setup, one could choose the sample mean or median as $\hat{\theta}$ and $D(x,t) := |x - t|$. The choice of c then depends on the expected variance of the clusters. If one wants to find clusters with different and unknown variances, $\hat{\theta} = (\hat{t}, \hat{s}^2)$ is a more reasonable choice, where $\hat{t}$ is some location estimator and $\hat{s}^2$ is some scale estimator. Then one could choose $D(x, (t, s^2)) := \frac{(x-t)^2}{s^2}$.

Computation: The natural approach to compute a fixed point is the usual fixed point algorithm:

Step 1: Choose $g^0 \in \{0,1\}^n$.

Step 2: $g^{j+1} := f(g^j)$.

Step 3: $g^j = f(g^j)$ or $j := j + 1$, step 2 again.

In general, convergence of this algorithm is not clear. It can be forced by replacing step 2 by

Step 2b: $g_i^{j+1} := f_i(g^j)$ for $g_i^j = k, \quad g_i^{j+1} := g_i^j$ else,

$k = 0$ or $k = 1$ fixed. Then, on the other hand, the result is not necessarily an FPCV. If step 2b is used alternating with $k = 0$ and $k = 1$, then the algorithm has been proven to converge towards an FPCV at least in the Regression setup (Hennig (1995)).

In order to find all FPCVs contained in a dataset, one theoretically has to check the fixed point property for every subset of the data. This is impossible in practice. Thus, one has to start the algorithm many times with randomly chosen starting sets. Depending on the concrete clustering problem, it has to be investigated if there is a high probability to get every "relevant" cluster by this method.

However, it is always possible to check if a particular set of data points forms an FPC. One can also find a "familiar" cluster by using it as starting point for the algorithm.

2. Application to Clusterwise Linear Regression

In clusterwise Linear Regression, data points should be treated as "similar", if they follow the same linear relation between an independent variable

$x \in I\!R^p$ and a dependent variable $y \in I\!R$. There are different possibilities to model such a situation. In the *Finite Mixture Model* the independent variables and the cluster membership of each point are assumed to be randomly generated:

Finite Mixture Model:

$$H(x,y) := \sum_{j=1}^{s} \pi_j H_j(x,y), \quad \pi_j > 0 \forall j = 1,\ldots,s, \quad \sum_{j=1}^{s} \pi_j = 1, \tag{4}$$

where the distribution functions H_j are defined by $x \sim G_j$ with some c.d.f. G_j on $I\!R^p$,

$$y = x'\beta_j + e, \qquad e \sim F_j, \quad j = 1,\ldots,s, \tag{5}$$

where $\beta_j \in I\!R^p$, $E_{F_j}(e) = 0$, e stochastically independent of x. To specify the estimators and the constant for the definition of Fixed Point Clusters, let $F_j = \Phi(0, \sigma_j^2)$ the c.d.f. of the Normal distribution with mean 0 and variance σ_j^2. An intercept parameter can be included by setting the $p+1$th component of each x_i equal to 1.

Remark: Although the theoretical results on Regression FPCs use the Finite Mixture Model, it would be also possible to work with fixed independent variables or fixed but unknown cluster membership of each points ("*Fixed Partition Model*").

Cluster analysis in such a situation is difficult. There are no adequate distances between two points, because two points can always be fitted perfectly by a suitable Regression parameter β. The components of the model can be estimated by a Maximum Likelihood approach (DeSarbo and Cron (1988)), but this approach is not robust against model deviations (e.g. the presence of outliers), the number of clusters has to be assumed as known (DeSarbo and Cron suggest the use of an information criterion) and the model is not generally identifiable which implies that some parameters cannot be estimated consistently (Hennig (1996)).

Fixed Point Clusters are useful in this case, because they do not force a partition of the dataset. Thus, they can exclude outliers and can make various suggestions if the parameters are not uniquely defined because of identifiability problems.

Definition: To define Fixed Point Clusters in the Regression setup, let $z_i := (x_i, y_i)$, $i = 1,\ldots,n$ a single observation, $X := (x_1,\ldots,x_n)'$, $y := (y_1,\ldots,y_n)'$,

$$\hat{\theta} := (\hat{\beta}, \hat{\sigma}^2), \text{ where} \tag{6}$$

$$\hat{\beta}(X,y) := (X'X)^{-1}X'y, \quad \hat{\sigma}^2(X,y) := \tfrac{1}{n-p-1}\sum_{i=1}^{n}(y_i - x_i'\hat{\beta}(X,y))^2, \tag{7}$$

$$D(x,y,\beta,\sigma^2) := \tfrac{(y-x'\beta)^2}{\sigma^2}. \tag{8}$$

To choose c, the distribution of D would be useful under the hypothesis, that β and σ^2 would be the parameter estimators for the mixture component containing (x, y). This distribution is unknown since denominator and nominator are dependent, because (x, y) was used to compute the estimators. As an approximation, a high quantile of the χ^2-distribution is a reasonable choice for c.

An FPC is "relevant" in this situation (concerning the discussion on the algorithm in the previous section), if it corresponds to a component of the mixture distribution. Simulations show, that if the proportions π_i of the components are not too small and the dimension p of the independent variables is not too large, one can get all relevant FPCVs with a moderate number of algorithm runs. The result of such a Fixed Point Cluster analysis will be illustrated as follows.

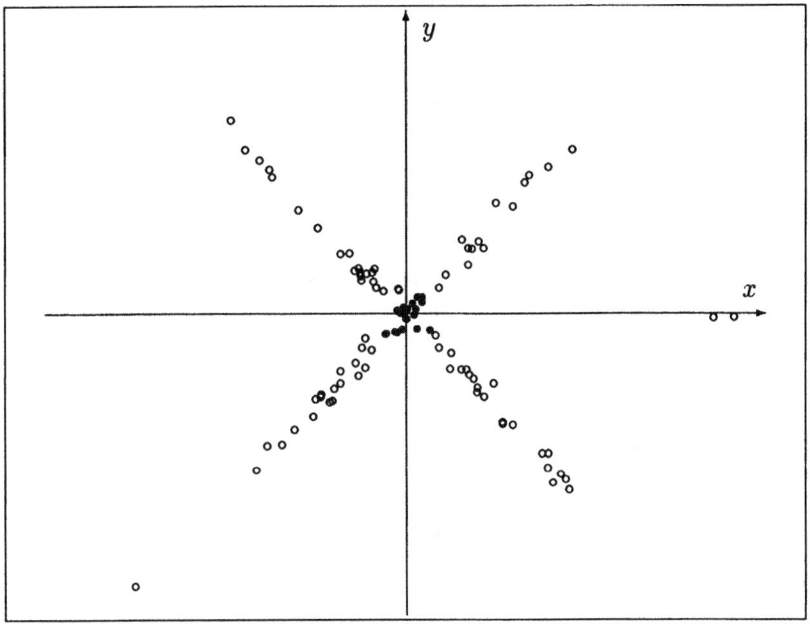

Figure 1: Artificial dataset

Example: Figure 1 shows an artificial generated dataset, where the points 1-50 were generated according to

$$x \sim \Phi(0,1), \quad y = x + e, \quad e \sim \Phi(0, 0.01), \tag{9}$$

the points 51-98 were generated according to

$$x \sim \Phi(0,1), \quad y = -x + e, \quad e \sim \Phi(0, 0.01), \tag{10}$$

and the points 99 and 100 are outliers. $c = 10$ and 150 runs of the algorithm yielded 8 clusters of size larger than 6. A minimum cluster size has to be specified, since one often gets many very small and useless FPCs. Especially all sets of points which can be fitted perfectly are trivially FPCs.

One cluster contains all points, a second contains all points except the point with lowest y-value. These FPCs can be left out from the interpretation of the results, since FPCs, which contain almost all points, usually exist. But they are only useful for outlier analysis if almost the whole dataset is homogenous. The existence of additional clusters, which were found more than once by the algorithm, indicates that this is not the case.

The remaining six clusters contain either the points 1-50 or the points 51-98. None of them contain the outliers. The two FPCs found most often from this six contain 53 and 64 points, respectively. Their parameter estimators (including intercept component) are

$$\hat{\beta}_1 = (0.998, -0.001), \quad \hat{\sigma}_1^2 = 0.013, \tag{11}$$

$$\hat{\beta}_2 = (-1.003, -0.007), \quad \hat{\sigma}_2^2 = 0.028. \tag{12}$$

The points represented by bullets lie in the intersection of these FPCs. They bias $\hat{\sigma}_2^2$, since they were almost all generated by the first component. The other clusters only differ from these two in what they contain from the "intersection" in the center. This can be used in the analysis of how clear the single points can be assigned to the components.

3. The Cluster Concept of Fixed Point Clusters

So far, FPCs have been defined and illustrated. The following section summarizes some ideas of the meaning of the concept "cluster", defined as above. Particular attention will be paid to the relation between FPCs and stochastic modelling. The notion of FPCs is inspired by stochastic models, but it is not a usual model based procedure. Here models offer a possibility to construct structure in the perception of a dataset, but they are not assumed to be true, not even approximately true.

3.1 Fixed Point Clusters are Probabilistic

The main role of stochastic models for FPCs is that the points belonging to a single FPC are treated as if coming from a common model. In the fixed point iteration, all of them and no other are used to compute the parameter estimators, none of them but all others are classified as outliers with respect to their parameter estimators. The common model for a single FPC has to be "homogenous" in some sense, i.e. it must be considered as "cluster-generating". For example, a Normal distribution can generate a "cluster" in contrast to a bimodal distribution with well seperated areas of high density.

For the general idea it does not matter where the remaining points come from, if some points are classified to belong to an FPC (though it does matter for the quality of the parameter estimators). Thus, the models H for which FPCs are supposed to be adequate only need to have a single component H_0 generated by a distribution from the class which defines the estimators:

$$H := \epsilon H_0 + (1-\epsilon)H^*, \quad 0 < \epsilon \leq 1. \tag{13}$$

This formalization is familiar in Robust Statistics, if the data is treated as a combination of "good data" and outliers. ("*ϵ-Contamination Neighborhood*" of H_0, usually $\epsilon < \frac{1}{2}$. The latter is not necessary for FPCs.) The points modelled by the arbitrary H^* need not even be considered as coming from any stochastic model. The Finite Mixture Regression Model is only the special case, in which FPCs would be adequate for every data point.

Remark: Unfortunately this cannot be used to investigate the adequacy of a certain model for the points of a particular FPC, since the algorithm treats all points as generated by the model until they are classified as outliers. In the given example the whole dataset forms an FPC and only in comparison with the other FPCs in the dataset it becomes clear that a homogenous model is not reasonable for it. If a Linear Regression model is not adequate for any part of the data, there nevertheless will be FPCs.

According to that, stochastic models should be used to assess the quality of the method. A first approach to do this for the Linear Regression case is to formulate the method as a functional on the space of probability distributions of the domain of the data points. The values of the functional are sets of indicator functions (or their corresponding parameters), each corresponding to an FPC. Then the key question is: What is the result for the distributions of (13) ?

It is proven (Hennig (1995)) that for a homogenous population ($\epsilon = 1$) the functional is Fisher consistent, i.e. there exists an unique FPC with the same parameters as the model H_0 (σ^2 standardized with a proper constant).

If the population is not homogenous, H^* usually will generate some points which are not recognized as outliers with respect to H_0. This causes a bias on the parameter estimates of the FPC corresponding to H_0. The existence of such an FPC and bounds for the bias can be proven if H_0 and H^* are well separated.

Consequently Fixed Point Clustering is at least a "probabilistic" concept.

3.2 Fixed Point Clusters are not Model Based

A procedure is considered as "model based" in this context, if it assumes the whole dataset to be generated by a member of some class of stochastic distributions and, additionally, a decision problem concerning some distribution parameters is solved under this assumption.

Some differences between FPCs and model based methods:

- The data is not assumed to be generated by a common stochastic model.
  - For a single FPC only a part of the data is considered: The data modelled by H_0 in the contamination model (13).
  - Stochastic models (including the contamination model) are not

treated as "true" but as suggestions to give adequate images for at least parts of the data.

- FPCs are not a solution of a certain decision problem. They do not optimize a target criterion formalized in the context of a certain model.
- Overlap and inclusion of FPCs as well as points, which do not belong to any FPC, are possible. The result of a Fixed Point Cluster analysis gives neither a model nor a class of models for the whole dataset nor a probabilistic classification for single data points.

Consequently, FPCs offer a different point of view on the data and their relation to stochastic models than model based procedures. The question of adequecy of a stochastic model is treated as open. FPCs give various possibilities for interpretations of parts of the data but they do not force decisions between them.

For example, classification methods based on a Fixed Partition Model classify each point to a particular cluster. This is not necessarily adequate. In the data of figure 1 one would loose the information, that the points in the intersection are fitted good from both mixture components.

Another example: If there were an FPC in the data example which contained the outliers and some small part of the other points, the decision whether the outliers are "outlying" to the "good" data or whether they are a part of an own mixture component could not be made on the basis of the data alone. A model based procedure would base a decision on an uncertain assumption.

Contrarily to that, FPCs are more an explorative tool. They make suggestions for considering points as belonging together, i.e. generated by a common model. This information can be used only to get an idea from the present data, but it can also be used to build a model for a later survey, to decide on significant extern factors defining a cluster membership or to validate the results of a model based clustering.

3.3 Modeling the Self Organization of Perception

There is another way of motivating FPCs. It is related to the idea of cognition in self organization theory. Heinz von Förster (1976) formalizes human cognitive actions as recursive operators on a domain of "observables", i.e. a cognitive action operates with its own output:

$$\text{obs}_{i+1} = \text{COORDINATION}(\text{obs}_i). \tag{14}$$

Cognitive processes are treated as the recursive computation of a reality depending on the way the subject operates. This is illustrated by von Förster (1973)[1] as follows:

[1]The German translation of both papers is included in von Förster (1993)

This idea can also be explained by biological arguments. Von Förster refers to Piaget (1975) and Maturana (1970).

Von Förster states that for an external observer the "objects" of a person cannot be distinguished from the fixed points[2] of the operation (14). This means: The observer constitutes an object as "recognized" by the person observed if the coordinations of actions of the person concerning the object are stable.

Fixed Point Clustering models such a process. The clusters (or patterns) of a dataset can be viewed as the "objects" to be recognized. The function f defines the action the algorithm carries out with the data points which play the role of the initial observables. An FPC is a stable point of this process.

References:

DESARBO, W. S. and CRON, W. L. (1988): A Maximum Likelihood Methodology for Clusterwise Linear Regression. *Journal of Classification 5, 249-282.*

HENNIG, C. (1995): Fixed Point Cluster Analysis for Regression Data: Computation and Weak Consistency, Preprint No. 95-2, Institut für Mathematsche Stochastik, Universität Hamburg.

HENNIG, C. (1996): Identifiability of Finite Linear Regression Mixtures, Preprint No. 96-6, Institut für Mathematsche Stochastik, Universität Hamburg.

MATURANA, H. R. (1970): Neurophysiology of Cognition. In: P.L. Garvon (ed.): *Cognition: A Multiple View, Spartan Books, New York, 3-23.*

PIAGET, J. (1975): L'Equilibration des structures cognitives, P.U.F., Paris.

VON FÖRSTER, H. (1973): On Constructing a Reality. In: W.F.E. Preiser (ed.): *Environmental Design Research, vol. 2, Stroudberg, 35-46.*

VON FÖRSTER, H. (1976): Objects: Tokens for Eigen-Behaviors. *ASC Cybernetic Forum 8, 91-96.*

VON FÖRSTER, H. (1993): Wissen und Gewissen (ed. by S.J. Schmidt), Suhrkamp, Frankfurt/Main.

[2]Von Förster uses the older term "Eigenvalue"

A Metric Approach for Ordinal Regression

Andreas Hilbert

Institut für Statistik und Mathematische Wirtschaftstheorie,
Universität Augsburg, D-86135 Augsburg

Summary: This paper presents a metric approach for the regression of ordinal variables. In contrast to most other studies, the problem of independent, ordinal variables with a dependent variable that is a metric scale is analyzed. For this situation, some properties of the estimated parameters of the model are described.

1. Introduction

Different models were proposed in the past for the analysis of metric and/or nominal variables in linear regression. The analysis of ordinal variables in linear regression has received little attention. One approach to analyze ordinal variables was proposed by Anderson (1984). His model is derived from the classical linear regression model and studies the situation where the response variable of the model is an ordinal scale. Another possibility to analyze ordinal variables in the sense of regression was proposed by Young et al. (1976) and Gifi (1990): the so-called optimal scaling methodology. This method is a nonlinear approach based on canonical regression to consider ordinal variables (either as regressor or as response variable) and rescales each variable so that an optimum of a least-squares loss function (equivalent to the canonical correlation between the variables) is reached. However, no statements about closed form solution or statistical properties of the solution are made.

The following study will present an alternative formulation of how ordinal variables can be embedded into a linear regression model. A closed form solution is given as well as some statistical properties of the solution.

2. The Dummy Variable Regression

The basic model considered here is the **classical metric linear regression model**, described in most statistical books (e.g. Fahrmeir and Hamerle (1984)). It describes the population of a random exogenous variable Y, where it is assumed that the observations $y_1, \ldots, y_N$ on Y constitute a random sample with a linear conditional mean made up of P other endogenous variables $x_1, \ldots, x_P$. Both, the exogenous and the endogenous variables are metric scales. Defining $\mathbf{y} = (y_1, \ldots, y_N)^T$, $\epsilon = (\epsilon_1, \ldots, \epsilon_N)^T$ and $\boldsymbol{\beta} = (\beta_0, \ldots, \beta_P)^T$, the classical regression model is described by

$$\mathbf{y} = \mathbf{X} \cdot \boldsymbol{\beta} + \epsilon \tag{1}$$

where $\mathbf{X} = (X_{np}) \in \mathbb{R}^{N \times (P+1)}$ is the so-called design matrix. X_{np} describes the value of the n-th observation on the p-th variable. Moreover, the first column of $\mathbf{X}$ contains only 1's to embed a constant effect (the intercept) by the parameter β_0 into the model. Furthermore, the following assumptions must be satisfied:

- The matrix $\mathbf{X}$ is a nonstochastic matrix with full rank and - as the sample size N becomes infinitely large - $\lim_{N\to\infty}(\mathbf{X}^T\mathbf{X}/N) = \mathbf{V}$, where V is a finite and nonsingular matrix.
- The vector $\boldsymbol{\epsilon}$ consists of unobservable random errors which satisfy the property

$$\boldsymbol{\epsilon} \sim N(\mathbf{0}, \sigma^2 \cdot \mathbf{I}) \tag{2}$$

The model (1) can only consider metric scales. If categorical variables are analyzed, so-called **dummy-coding** can be used: X_{n0} remains unchanged and the other X_{np} $(p = 1, \ldots, P)$ of the design matrix $\mathbf{X}$ must be replaced by a row-vector $\mathbf{X}_{np}$ which describes the chosen category of the p-th variable at the n-th observation. The vector $\mathbf{X}_{np} = (X_{np2}, \ldots, X_{npM_p})$ is defined by

$$X_{npm} = \begin{cases} 1 & \text{if variable } p \text{ has the category } m \text{ at observation } n \\ 0 & \text{otherwise} \end{cases} \tag{3}$$

for $n = 1, \ldots, N$, $p = 1, \ldots, P$ and $m = 2, \ldots, M_p$, where M_p is the number of categories of the variable p. By defining the vector $\mathbf{X}_{np}$ this way, the first category of each variable is chosen as the reference category with no effects on the response variable y. Other coding procedures such as effect-coding are also possible but not considered here.

Because of the redefinition of the matrix $\mathbf{X}$ in (3), the parameter vector $\boldsymbol{\beta}$ must also be redefined. A single parameter β_p, describing the effect of the p-th variable to the response variable y, must be replaced by a vector $\boldsymbol{\beta}_p = (\beta_p^2, \ldots, \beta_p^{M_p})^T$, where β_p^m is the effect of category m of variable p to the response variable y.

Using this redefinition of the design matrix $\mathbf{X}$ and the parameter vector $\boldsymbol{\beta}$, the **dummy variable regression model** is described - equivalent to the classical regression model - by equation (1) .

3. The Ordinal Regression

Now, it is demonstrated how it is possible to embed ordinal variables into the **dummy variable regression model** of the previous section.

Whereas the coefficients β_p^m of nominal variables are not ordered, the corresponding coefficients of ordinal variables should be put in order. The following postulation ensures the monotonicity of the coefficients:

$$\beta_p^m \leq \beta_p^{m+1} \qquad \begin{aligned} \forall m &= 1, \ldots, M_p - 1 \\ \forall p &= 1, \ldots, P \end{aligned} \tag{4}$$

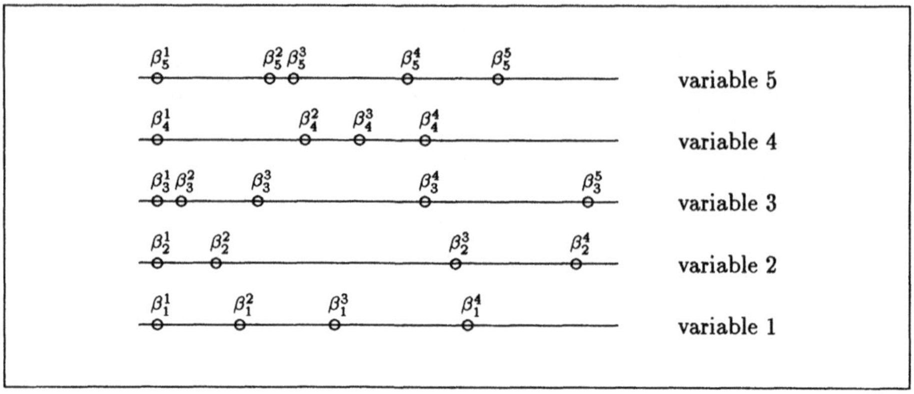

Figure 1: Example of the coefficients β_p^m for five ordinal variables

Figure 1 gives an example of coefficients β_p^m for five ordinal variables.

Because inequality (4) does not imply any other conditions, the coefficients of the different variables have their own structure.

The monotonicity of the coefficients can alternatively be expressed by

$$\begin{aligned} \beta_p^1 &= \Delta_p^1 \\ \beta_p^2 &= \Delta_p^1 + \Delta_p^2 \\ &\vdots \\ \beta_p^{M_p} &= \Delta_p^1 + \Delta_p^2 + \ldots + \Delta_p^{M_p} \end{aligned} \tag{5}$$

$$\text{with} \qquad \Delta_p^m \geq 0 \qquad \forall m = 1, \ldots, M_p \tag{6}$$

and the normalization condition

$$\beta_p^1 \equiv 0 \tag{7}$$

for all variables $p = 1, \ldots, P$.

Under consideration of the normalization condition (7), the reparametrization (5) can be formulated by matrices for one variable p as follows:

$$\begin{pmatrix} \beta_p^2 \\ \vdots \\ \beta_p^{M_p} \end{pmatrix} = \begin{pmatrix} 1 & 0 & 0 \\ \vdots & \ddots & 0 \\ 1 & \cdots & 1 \end{pmatrix} \cdot \begin{pmatrix} \Delta_p^2 \\ \vdots \\ \Delta_p^{M_p} \end{pmatrix}, \tag{8}$$

or, equivalently,

$$\boldsymbol{\beta}_p = \mathbf{C}_p \cdot \boldsymbol{\Delta}_p \qquad \forall p = 1, \ldots, P, \tag{9}$$

where the matrix $\mathbf{C}_p$ has the dimension $(M_p - 1) \times (M_p - 1)$ and the column vector $\boldsymbol{\Delta}_p$ the dimension $(M_p - 1)$.

Summarizing the equations of the form (9) for all variables $p = 1, \ldots, P$ and taking into account that a coefficient β_0 respectively Δ_0 for the constant effect has to be introduced, the following equation results:

$$(\beta_0, \boldsymbol{\beta}_1, \ldots, \boldsymbol{\beta}_P)^T = \begin{pmatrix} 1 & 0 & \cdots & 0 \\ 0 & \mathbf{C}_1 & & \\ \vdots & & \ddots & \\ 0 & & & \mathbf{C}_P \end{pmatrix} \cdot \begin{pmatrix} \Delta_0 \\ \boldsymbol{\Delta}_1 \\ \vdots \\ \boldsymbol{\Delta}_P \end{pmatrix} \tag{10}$$

respectively $\boldsymbol{\beta} = \mathbf{C} \cdot \boldsymbol{\Delta}$.

Using equation (10) into the original dummy variable regression model (see section 2), a **metric regression model for ordinal variables** can be expressed by:

$$\begin{aligned} \mathbf{y} &= \mathbf{X} \cdot \boldsymbol{\beta} + \epsilon \\ &= \mathbf{X} \cdot \mathbf{C} \cdot \boldsymbol{\Delta} + \epsilon \\ &= \mathbf{Z} \cdot \boldsymbol{\Delta} + \epsilon \end{aligned} \tag{11}$$

subject to

$$\mathbf{A} \cdot \boldsymbol{\Delta} \geq \mathbf{0}, \tag{12}$$

where $\mathbf{A}$ is a suitable chosen matrix to describe the condition (6). The full rank of the matrix $\mathbf{Z}$, a necessary condition for this regression model, is ensured by equation (7).

In contrast to the original regression model for nominal variables, this model takes ordinal variables and the monotonicity of their coefficients into account. Unfortunately, the simple estimation of the regression model is made more difficult by the condition (12).

4. Ordinal Regression with Common Scales

Moreover, a model with a smaller number of parameters arises when different variables has the same structure in their coefficients. Figure 2 illustrates this situation. Variables 1 and 4 have the same structure in their coefficients, and so do variables 3 and 5: the distances between the different categories within a variable are the same. Only variable 2 does not share its structure with any other variable.

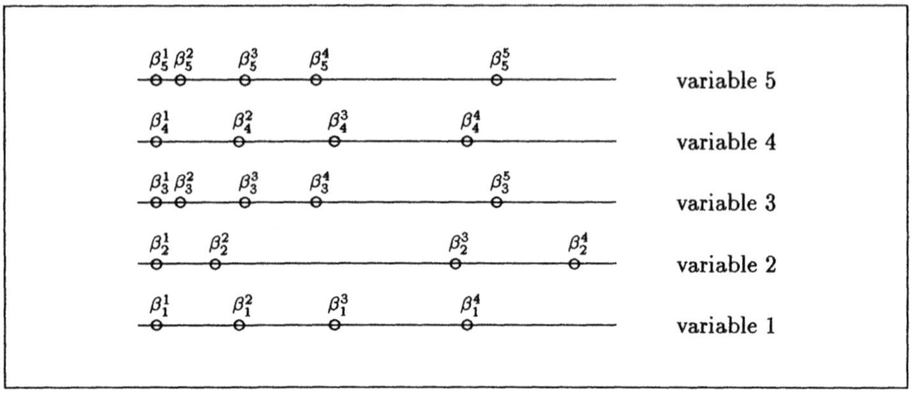

Figure 2: Example of ordinal variables where variables 1 and 4, and variables 3 and 5 have a common scale.

Equivalent to equation (4), the situation above can be described as follows:

$$\begin{array}{lllll} \beta_p^m & \leq & \beta_p^{m+1} & \forall m & = 1,\ldots,M_p-1 \\ & & & \forall p & = 1,\ldots,P \\ & & \text{and} & & \\ \beta_p^m & = & \beta_{p'}^m & \forall m & = 1,\ldots,M_p \\ & & & \forall p,p' & \in Q_k \end{array} \tag{13}$$

with

$$\bigcup_{k=1}^{K} Q_k = \{1,\ldots,P\}$$

The sets Q_k divide the variables in K groups allowing for only one scale per group. In Figure 2 there are $K = 3$ groups: two groups with two variables ($Q_1 = \{1,4\}, Q_2 = \{3,5\}$) and one group with only one variable ($Q_3 = \{2\}$). Using the known reparametrization and the matrix formulation above, the following equation results for all variables p of group k ($k = 1,\ldots,K$):

$$\begin{pmatrix} \beta_p^2 \\ \vdots \\ \beta_p^{M_k} \end{pmatrix} = \begin{pmatrix} 1 & 0 & 0 \\ \vdots & \ddots & 0 \\ 1 & \cdots & 1 \end{pmatrix} \cdot \begin{pmatrix} \Delta_k^2 \\ \vdots \\ \Delta_k^{M_k} \end{pmatrix} \tag{14}$$

respectively

$$\boldsymbol{\beta}_p = \mathbf{C}_k \cdot \boldsymbol{\Delta}_k \qquad \forall p \in Q_k, \tag{15}$$

where the matrix $\mathbf{C}_k$ has the dimension $(M_k - 1) \times (M_k - 1)$, the column vector $\boldsymbol{\Delta}_k$ the dimension $(M_k - 1)$, and M_k is the number of categories of all variables p which belong to group Q_k.

If, for example, variables 1 and 2 have an equal structure, both variables can be described by the same matrix $\mathbf{C}_1$, so that, equivalent to (10), the following equation results when summarizing equation (15) for all variables:

$$(\beta_0, \boldsymbol{\beta}_1, \ldots, \boldsymbol{\beta}_P)^T = \begin{pmatrix} 1 & 0 & \cdots & 0 \\ 0 & \mathbf{C}_1 & & \\ \vdots & \mathbf{C}_1 & & \\ \vdots & & \ddots & \\ 0 & & & \mathbf{C}_K \end{pmatrix} \cdot \begin{pmatrix} \Delta_0 \\ \boldsymbol{\Delta}_1 \\ \vdots \\ \boldsymbol{\Delta}_K \end{pmatrix} \tag{16}$$

respectively

$$\boldsymbol{\beta} = \mathbf{C} \cdot \boldsymbol{\Delta}.$$

[5pt] For this situation, an regression model results equivalent to the one described in equation (12). The only difference is that the dimension of the matrices and the dimension of the parameter vector has to be modified.

5. Parameter Estimation

The model for metric regression for ordinal variables and the common scales model can be summarized in the same way:

$$\begin{aligned} \mathbf{y} &= \mathbf{X} \cdot \boldsymbol{\beta} + \boldsymbol{\epsilon} \\ &= \mathbf{X} \cdot \mathbf{C} \cdot \boldsymbol{\Delta} + \boldsymbol{\epsilon} \\ &= \mathbf{Z} \cdot \boldsymbol{\Delta} + \boldsymbol{\epsilon} \end{aligned} \tag{17}$$

subject to

$$\mathbf{A} \cdot \boldsymbol{\Delta} \geq \mathbf{0} \tag{18}$$

where $\mathbf{C}$ and $\mathbf{A}$ are suitable chosen matrices to describe the different models. Generally, this leads to the following constrained quadratic programming problem

$$\begin{aligned} &\text{Minimize} \quad (\mathbf{y} - \mathbf{Z} \cdot \boldsymbol{\Delta})^T (\mathbf{y} - \mathbf{Z} \cdot \boldsymbol{\Delta}) \\ &\text{subject to} \qquad \mathbf{A} \cdot \boldsymbol{\Delta} \geq \mathbf{0}. \end{aligned} \tag{19}$$

and its so-called **inequality constrained least squares estimator (ICLS)** solution.

The numerical solution of (19) can be obtained by several algorithms, for example by the Cottle-Dantzig algorithm (see Cottle and Dantzig (1974)). For more information about the feasible algorithms see, e.g., Liew (1976).

To describe statistical properties of the ICLS, the closed form of the solution is needed, which can be derived from the well-known Kuhn-Tucker conditions (cp. Kuhn and Tucker (1951)) of this convex minimization problem (19). Fomby et al. (1984, p.102–105), develop the following closed form solution of the ICLS:

$$\mathbf{\Delta}^{+} = \begin{cases} \hat{\mathbf{\Delta}} & \text{if } \mathbf{A}\cdot\hat{\mathbf{\Delta}} \geq \mathbf{0} \\ \hat{\mathbf{\Delta}} + (\mathbf{Z}^T\mathbf{Z})^{-1}\mathbf{A}_2^T\left[\mathbf{A}_2(\mathbf{Z}^T\mathbf{Z})^{-1}\mathbf{A}_2^T\right]^{-1}(\mathbf{0} - \mathbf{A}_2\hat{\mathbf{\Delta}}) & \text{if } \mathbf{A}_2\cdot\hat{\mathbf{\Delta}} < \mathbf{0} \\ \hat{\mathbf{\Delta}} + (\mathbf{Z}^T\mathbf{Z})^{-1}\mathbf{A}^T\left[\mathbf{A}(\mathbf{Z}^T\mathbf{Z})^{-1}\mathbf{A}^T\right]^{-1}(\mathbf{0} - \mathbf{A}\hat{\mathbf{\Delta}}) & \text{if } \mathbf{A}\cdot\hat{\mathbf{\Delta}} < \mathbf{0}, \end{cases} \quad (20)$$

where $\hat{\mathbf{\Delta}}$ is the **ordinary least squares estimator** with $\hat{\mathbf{\Delta}} = (\mathbf{Z}^T\mathbf{Z})^{-1}\mathbf{Z}^T\mathbf{y}$. Furthermore, the matrix $\mathbf{A}_2$ is defined by

$$\mathbf{A} = \begin{pmatrix} \mathbf{A}_1 \\ \mathbf{A}_2 \end{pmatrix} \quad \text{with} \quad \begin{matrix} \mathbf{A}_1\cdot\hat{\mathbf{\Delta}} \geq \mathbf{0} \\ \mathbf{A}_2\cdot\hat{\mathbf{\Delta}} < \mathbf{0}, \end{matrix} \quad (21)$$

where, without loss of generality, the first J^* constraints are satisfied and the remaining $J - J^*$ constraints are violated, with J as the number of constraints described by the matrix $\mathbf{A}$.

Unfortunately, this closed form is not *always* defined. But if the following sufficient condition is satisfied, then (20) is a feasible solution for the ICLS (cp. Fomby et al. (1984, p. 104–105)):

$$\begin{matrix} \left[\mathbf{A}_1(\mathbf{Z}^T\mathbf{Z})^{-1}\mathbf{A}_2^T\right]\cdot\left[\mathbf{A}_2(\mathbf{Z}^T\mathbf{Z})^{-1}\mathbf{A}_2^T\right]^{-1}(\mathbf{0} - \mathbf{A}_2\hat{\mathbf{\Delta}}) \geq \mathbf{0} & \text{if } \mathbf{A}_2\cdot\hat{\mathbf{\Delta}} < \mathbf{0} \\ \left[\mathbf{A}(\mathbf{Z}^T\mathbf{Z})^{-1}\mathbf{A}^T\right]^{-1}(\mathbf{0} - \mathbf{A}\hat{\mathbf{\Delta}}) \geq \mathbf{0} & \text{if } \mathbf{A}\cdot\hat{\mathbf{\Delta}} < \mathbf{0} \end{matrix} \quad (22)$$

If $\mathbf{A}\cdot\hat{\mathbf{\Delta}} \geq \mathbf{0}$ holds, $\hat{\mathbf{\Delta}}$ is always a feasible solution.

Regarding the statistical properties of the ICLS estimator, Wirnhier (1985, p. 97–108), derives the following statement:

If some regular conditions are hold, the ICLS estimator $\mathbf{\Delta}^+$ (20) converges in probability to the ordinary least squares estimator $\hat{\mathbf{\Delta}}$ (OLS) providing the sample size N becomes infinitely large, i.e., $N \to \infty$. Furthermore, the ICLS is consistent and asymptotically distributed like

$$N\left(\mathbf{\Delta}, \sigma^2(Z^TZ)^{-1}\right).$$

Important to these statistical properties is the fact that the real - but unknown - parameter vector $\mathbf{\Delta}$ satisfies the following condition:

$$\mathbf{A}\cdot\mathbf{\Delta} > \mathbf{0} \quad (23)$$

However, inequality (23) makes sense because the different categories should be separated. $\Delta_p^m = 0$ means the corresponding categories respectively their parameters are – a priori – equal. Consequently, they could be combined before the analysis and one of them could be removed.

If in (23) the equality is permitted, then a more complicated distribution for the parameter $\boldsymbol{\Delta}^+$ follows (see, e.g., Wirnhier (1985)).

6. Conclusions

The main result of this work is to demonstrate how it is possible to embed ordinal variables into the ordinary **dummy variable regression model**, originally derived for nominal scales. Furthermore, the considered ordinal variables are – in contrast to other studies – **independent** regressors and not dependent response variables of the model. Additionally, some properties of this model are derived. The corresponding so-called **inequality constraint least squares estimator** is formulated and its closed form as well as its statistical properties are discussed.

References:

ANDERSON, J. A. (1984): Regression and Ordered Categorical Variables. *Journal of Royal Statistical Society, B 46, 1–30.*

COTTLE, R. W. and DANTZIG, G. B. (1974): Complementary Pivot of Mathematical Programming. In: G.B. Dantzig and B.C. Eaves (eds.): *Studies in Optimization*, Vol. 10, Mathematical Association of America, Washington.

ESCOBAR, L. A. and SKARPNESS, B. (1984): A Closed Form Solution for the Least Squares Regression Problem with Linear Inequality Constraints. *Commun. Stat. Theor. Methods, 13 (9), 1127–1134.*

FAHRMEIR, L. and HAMERLE, A. (1984): Multivariate Statistische Verfahren, DeGruyter, Berlin.

FOMBY, T. B.; HILL, R. C. and JOHNSON, S. R. (1984): Advanced Econometric Methods, Springer, New York.

GIFI, A. (1990): Nonlinear Multivariate Analysis, Wiley, Chichester.

KUHN, H. W., and TUCKER, A. W. (1951): Nonlinear Programming. In: J. Neyman (ed.): *Proceedings of the Second Berkeley Symposium on Mathematical Statistics and Probability*, University of California Press, Berkeley, 481–492.

LIEW, C. K. (1976): Inequality Constrained Least Squares Estimation. *Journal of the American Statistical Association, 71, 746–751.*

WIRNHIER, F. X. (1985): Lineare Regression mit linearen Parameterbedingungen und unbekannter Fehlerkovarianzmatrix, Dissertation, München.

YOUNG, F. W.; DE LEEUW, J. and TAKANE, Y. (1976): Regression with Qualitative and Quantitative Variables: An Alternating Least Squares Method with Optimal Scaling Features. *Psychometrika, 41, 505–529.*

Graphical Stability of Data Analysing Software

Heike Hofmann

Institut für Mathematik,
Universität Augsburg, D-86135 Augsburg, Germany

Summary: Using computer orientated methods for data analysis it gets necessary to think of the question whether the used software is reliable. This is especially interesting for graphics, where it is sometimes very difficult to check the results arithmetically. Therefore it must be guaranteed, that the user may rely on the representation. According to this the expression of "graphical stability" can be explained as a possibility to summarize some of the fundamental qualities of reliable graphics.

1. Arithmetical basis of graphical representation

The reliability of the used software should be important to every user. Recent research relating to this point, however, mainly concentrates on alphanumeric input and output.

1.1 The situation

The situation is based on the following dataset, published by Leland Wilkinson in the "Statistics Quiz" in 1985. The variables serve two important criterions. On the one hand they turn the software to the limits of its processing abilities due to a concentrated collection of well-known numerical difficulties but on the other hand a precision of eight digits does not demand too much of a programme, either. A special problem is the variable "LITTLE" with its comparatively high mean value combined with a small variation.

LABEL	X	ZERO	MISS	BIG	LITTLE	HUGE	TINY	ROUND
ONE	1	0	•	99999991	0.99999991	1E+12	1E-12	0.5
TWO	2	0	•	99999992	0.99999992	2E+12	2E-12	1.5
THREE	3	0	•	99999993	0.99999993	3E+12	3E-12	2.5
FOUR	4	0	•	99999994	0.99999994	4E+12	4E-12	3.5
FIVE	5	0	•	99999995	0.99999995	5E+12	5E-12	4.5
SIX	6	0	•	99999996	0.99999996	6E+12	6E-12	5.5
SEVEN	7	0	•	99999997	0.99999997	7E+12	7E-12	6.5
EIGHT	8	0	•	99999998	0.99999998	8E+12	8E-12	7.5
NINE	9	0	•	99999999	0.99999999	9E+12	9E-12	8.5

Figure 1: "nasty" dataset of L. Wilkinson

1.2 A problem

One problem already appears with many common statistics programmes when you look at histograms. In the following one can see the results of three packages:

- **JMP** scale:

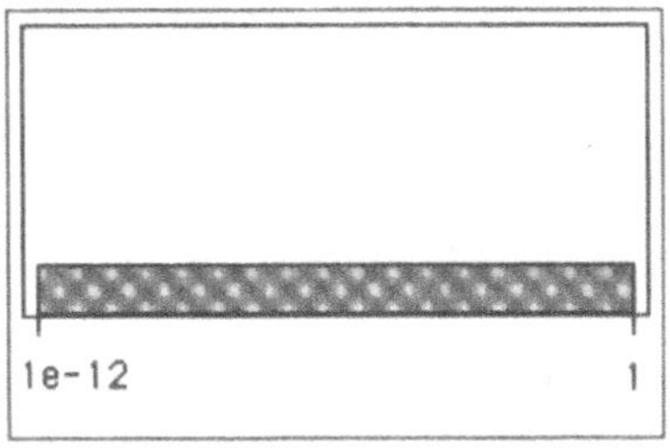

Figure 2: Histogram referring to variable "TINY". The real range comes to $8 \cdot 10^{-12}$, but in the graphic the range almost comes to 1

- **Data Desk** division of intervals:

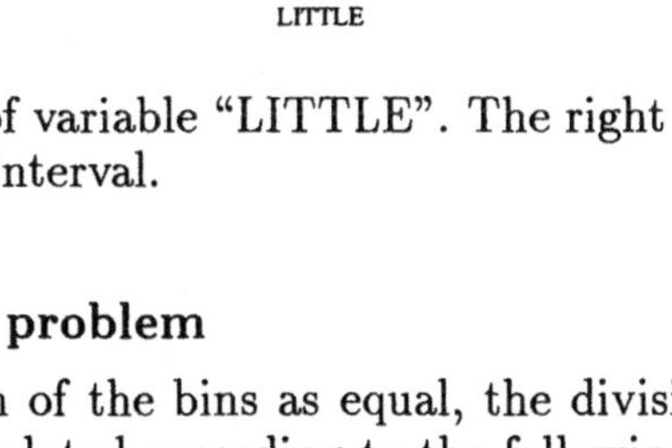

Figure 3: Histogram of variable "LITTLE". The right most bin corresponds to a closed interval.

1.3 Analysis of the problem

Considering the width of the bins as equal, the division of the bins in the histogram can be calculated according to the following function:

$$f : \mathbb{R} \rightarrow J, J \subset \mathbb{N} \quad \textit{index set}$$
$$f(x) = \lfloor \frac{x - L}{R - L} \cdot bins \rfloor + 1 \tag{1}$$

In this connection *bins* is defined as the number of bins in a histogram, R as the right limit and L as the left one.
Please notice that the bins in this case result in semiclosed intervals.

- **MANET** division of intervals:

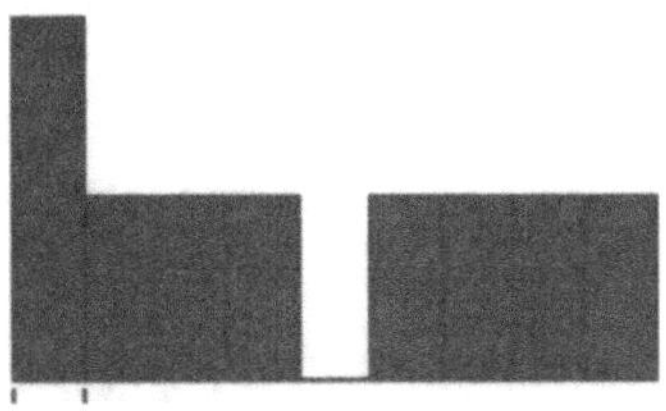

Figure 4: Histogram of variable "LITTLE". The first bin contains too many values, while in the fifth one some values are missing.

The conditioning number κ, which can be assumed as a measure for the behaviour of the function results presupposing small changes of the input data, of the inner term

$$\phi(x) = \frac{x - L}{R - L} \cdot bins \tag{2}$$

is gained by the following calculation:

$$\begin{aligned} \kappa(x) &= \frac{x}{\phi(x)} \cdot \phi'(x) = \\ &= \frac{x}{\frac{x-L}{R-L} \cdot bins} \cdot \frac{1}{R - L} \cdot bins = \\ &= \frac{x}{x - L} = 1 + \frac{L}{x - L} \end{aligned} \tag{3}$$

For the variable "LITTLE" this value comes to $\approx 10^8$.

Because of that the problem is ill-conditioned and therefore there does not exist a theoretical solution at all. One has to deal with the errors in a different way.

Doing this the additonal calculation of controlling values provides a solution in assuming that different function results for the value x and its closest neighbours $x \cdot (1 + \varepsilon)$ and $x \cdot (1 - \varepsilon)$ at least indicate the possiblity of a problem. $\varepsilon \approx 1 \cdot 10^{-19}$ (precision of machine using floating point arithmetic according to IEEE-standard)

1.4 Attempt of a solution

Marking the spot in the display window where this calculated value exceeds a certain limit (in the best case fixed by the user himself) one's attention is drawn to the problem and therefore one is free to react properly.

Session with MANET:

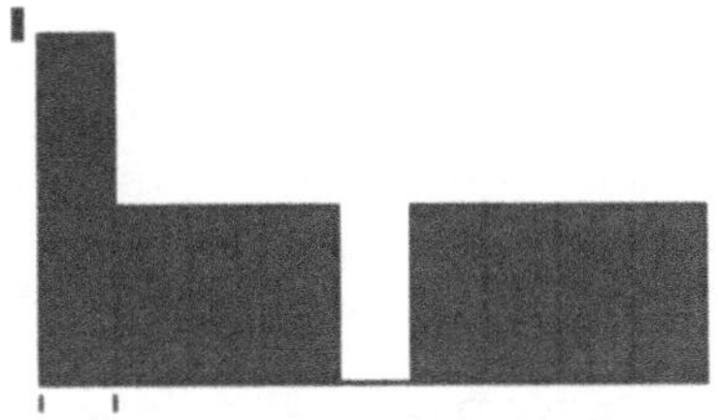

Figure 5: Histogram of variable "LITTLE". In the upper left corner a mark is shown as warning for possible numerical problems.

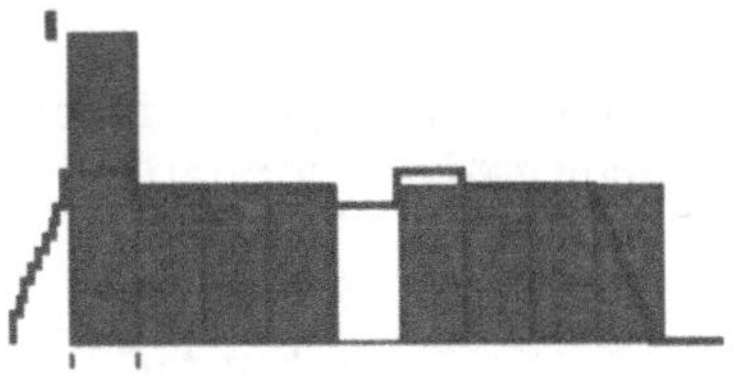

Figure 6: An average-shifted histogram is layed over the former graphic indicating the existence of the problem.

Figure 7: Offset-displacement of the histogram solving the former problem.

2. Discrete screen visualisation

The display resolution gives another problem when visualising data. While dealing with arithmetical data the user can expect at least eight digits of precision when handling the input and output of data and even double precision at internal calculations. On the screen, however, the number of different figures is reduced to only a few hundred.

Example:
Every bin has the same height, though representing different values. To achieve this result one had not had to go as far as 99999991 but values of 991 had been enough.

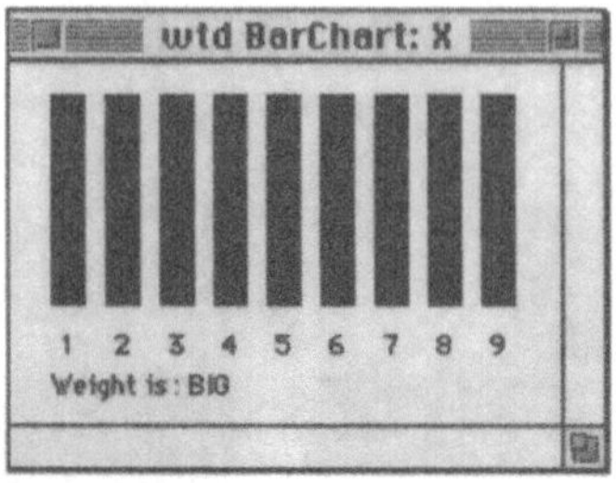

Figure 8: Barchart of variable X

Goal:
The deviation of the underlying values should not be too large.

2.1 Calculation of the error

As a measure of this deviation Tufte suggested following quotient in his book "The Visual Display of Quantitative Information"

$$liefactor = \frac{shown\ effect}{effect\ in\ the\ data} \tag{4}$$

If this factor differs less than 5% from 1 (i.e. between 0.95 and 1.05), one can assume graphical stability in a representation of data. In this connection the "effect" of a value must be handled more detailed. When representing real values in a graphic the calculation of scaling factors in horizontal and vertical direction gets necessary. Doing this the data is transformed into a representable state. In this first step there is no loss of information (apart from some slight rounding errors). The second step, rounding to integer values gives a source of error, therefore a suitable measure for the occurring error are following terms:

$$\delta_i := \left| \frac{\hat{x}_i}{x_i} - 1 \right|$$

Defining x_i as the real value, $\hat{x}_i$ as the effect ($\hat{x}_i = round(x_i)$) and δ_i as the relative error in component i.
The quotient of two effects therefore is:

$$\frac{\frac{\hat{x}_1}{x_1}}{\frac{\hat{x}_2}{x_2}} = \frac{\pm\delta_1 + 1}{\pm\delta_2 + 1} \tag{5}$$

On condition that the total sum of errors is limited to $\sum_i \delta_i < \varepsilon$ one gets dependent on the permissible variation of the total error ε and the direction of the deviation, intervals as an estimation of the real error.

2.2 Total error of deviation

In case of a d dimensional mosaic plot with variables $X_1, ..., X_d$ the terms of the relative error are of the following form:

$$\left| \frac{h_{i_1,...,i_d} \cdot b_{i_1,...,i_d} - h^x_{i_1,...,i_d} \cdot b^x_{i_1,...,i_d}}{h^x_{i_1,...,i_d} \cdot b^x_{i_1,...,i_d}} \right|$$

with $h_{i_1,...,i_d}$ and $b_{i_1,...,i_d}$ defined as height and width of the bin, which includes the combination $(i_1, ..., i_d)$ of each variable $X_1, ..., X_d$.

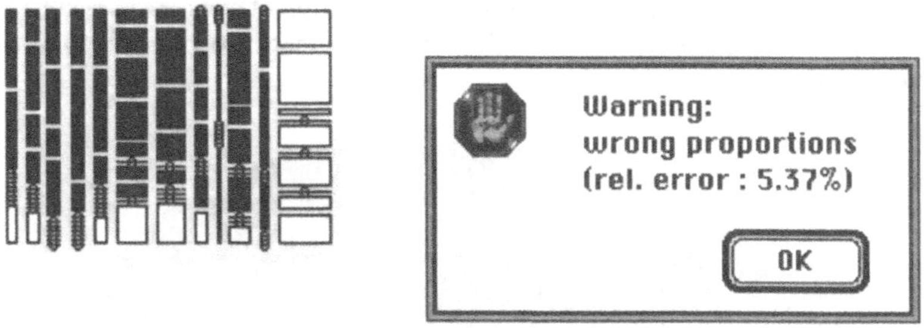

Figure 9: Two dimensional mosaic plot in MANET and additional warning window

2.3 Structural exceptions

A further aspect of discrete representation gets especially clear with global linking, where small groups of individuals in graphics do not become apparent. In the following example (Fig. 10) of the Titanic dataset all surviving children are marked. When having no further information one could assume there weren't either any children in the first class or none of them had survived (in fact there were six of them and all survived). To indicate such

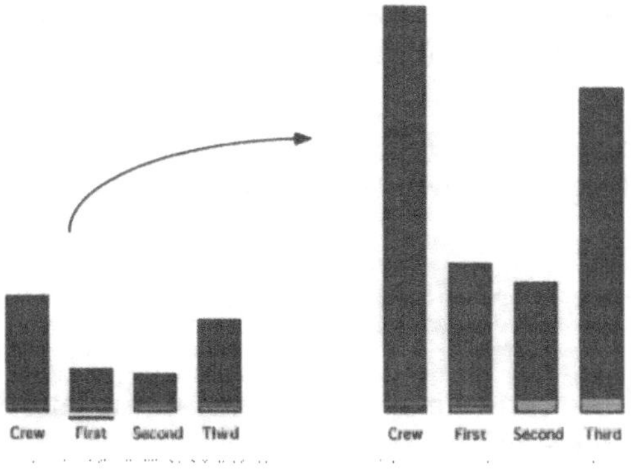

Figure 10: Barchart of Titanic's passengers, separated according to class

structural problems to the user the affected bin is underlined. The other way round, when a whole bin is marked though part of the data does not belong to the marked group, is equally treated.

3. Distortion

Additional dimensions to lower dimensional data or vanishing point perspective representation cause further posssiblities of graphical inaccuracies and therefore demonstrate facts which differ from the values they are based on. The following example from the SPIEGEL magazine (Fig. 11) shows a projection of values onto a hemisphere. So the values become distorted both in vertical and horizontal direction. Low temperatures in the center of the graphic get oversized while the values at the margin especially positive ones are scaled down.

Figure 11: Spiegel 7/96,"kalt erwischt": averages of daily temperatures in january in Berlin during the years from 1920 to 1996

Making the difference more obvious Fig. 12 shows two bins from the upper graphic representing values with the same absolute value.

Figure 12: Two bins from the upper graphic, in the left -3°C in 1950, on the right side +3°C in 1994

On condition that the left bin has a positive deviation, the right bin a negative one it is possible to estimate the lie factor with following calculation:

$$\begin{aligned} x_1 = 140(pixel) &\hat{=} 1994 \\ x_2 = 240(pixel) &\hat{=} 1950 \\ \frac{\frac{\hat{x}_1}{x_1}}{\frac{\hat{x}_2}{x_2}} &= \frac{x_1}{x_2} == \frac{140}{240} \approx 0.58 \end{aligned}$$

Therefore the liefactor ε is greater than 40%.

4. Conclusion

To be able to guarantee graphical stability several aspects have to be taken into account beforehand. One has to check that the displayed values are the result of numerical stable calculations and that the errors due to the discrete representation or projections do not escalate. More than ever the demands on software has been growing in coping with the increasing stream of information while at the same time the accompanying errors should be kept low. While such problems have not been solved satisfyingly software developers should think of mechanisms providing the users with the opportunity to check the correctness of the received results.

References:

TUFTE, E. (1983): The Visual Display of Quantitative Information. Graphics Press, Chesire, Connecticut

SAWITZKI, G. (1994): Testing Numerical Reliability of Data Analysis Systems. *Computational Statistics & Data Analysis*

WILKINSON, L. (1985): Statistics Quiz. *Systat, Evanston*

UNWIN, A. R.; HAWKINS, G.; HOFMANN, H. and SIEGL, B. (1996): Interactive graphics for data sets with missing values MANET. To be published in *Journal of Computational and Graphical Statistics.*

THEUS, M. (1996): MANET — Extensions to Interactive Statistical Graphics for Missing Values. To be published in: Proceedings of the NTTS 95

Recent Developments in Three-Way Data Analysis: A Showcase of Methods and Examples

P. M. Kroonenberg

Department of Education, Leiden University
Wassenaarseweg 52, 2333 AK Leiden, The Netherlands

Summary: In this paper an compact idiosyncratic overview will be provided of the areas into which three-way data analysis has expanded. The historical introduction will be followed by a scheme presenting an indication of the techniques involved. Then four condensed examples will give a feel of the scope of applications, while the final section is devoted to publicly available programs to perform the analyses.

1. Historical overview

Three-way analysis of continuous data originated in psychology during the *sixties.* Its founder Tucker (e.g. 1966) conceived the basic ideas for three-mode component and factor analysis models, developed algorithms to estimate the parameters, and published several, primarily psychological, applications. During the *seventies* three-way analysis expanded into multidimensional scaling primarily due to Carroll (e.g. Carroll and Chang (1970)), while Harshman (e.g. Harshman and Lundy (1984a,b)), working with similar models, extended the scope of component models. At the same time Bentler and co-workers (e.g. Bentler and Lee (1979)) developed structural-equation formulations for the three-mode common factor model. In the *eighties*, Kroonenberg and De Leeuw (1980) developed new algorithms for Tucker's models, and the former extended interpretational aspects of the technique and published several applications emphasizing interpretation, primarily within the social and behavioural sciences (e.g. Kroonenberg (1983a)). Flury's (1988) book summarises the development of his work on multi-sample common principal component methods. In France a new approach called STATIS (Structuration des Tableaux à Trois Indices de la Statistique) was developed by Escoufier and co-workers (especially, Lavit (1988)). Carroll, Arabie, DeSarbo and co-workers (e.g. Carroll and Arabie (1983), DeSarbo and Carroll (1985)) took three-way analysis to the clustering and unfolding domain, and Basford and McLachlan (e.g. 1985) introduced three-mode clustering. Carroll et al. (1980) also introduced constraints into three-way analysis. In the mean time the methods gradually filtered into other disciplines, like agriculture (e.g. Basford et al., 1991), ecology (e.g. Beffy (1992), spectroscopy (initially independently from the developments in psychology; e.g. Leurgans and Ross, 1992), and chemistry (e.g. Smilde, 1992). In the first part of this decade, much effort has gone into developing constrained three-way analyses especially from Kiers, Ten Berge and co-workers (e.g. Kiers, 1991, 1992; Krijnen, 1993). Franc (1992) extended French linear vector space thinking to the three-way area. Another strand is the expansion of the techniques into the analysis of three-way contingency tables (e.g. Carlier and Kroonenberg, 1996) and three-way analysis-of-variance designs (e.g. Van Eeuwijk and Kroonenberg, submitted).

The names mentioned above are an unfair selection of some of the protagonists, and many other persons have contributed to the developments in this area. An annotated bibliography upto 1983 is Kroonenberg (1983b) and the sequel to this ever-expanded bibliography is available from the author which covers virtually the whole field, both theory and applications, and everyone is invited to contribute papers, programs, and applications.

2. Three-way data and the method chart

2.1 Three-way data

In their basic form, most data seem to come in one of three broad classes, i.e. *profile data* (scores of subjects on variables), *similarity data* (judgements of similarity between two stimuli), and *preference data* (rankings of subjects on variables); here, the terms subjects, variables, and stimuli are used as generic terms. From these basic forms many other forms can be derived, such as means, covariances, frequency counts, etc., while data of one class can be transformed into another class, e.g., indirect similarities can be derived from profile data. Profile data are the bread-and-butter data in many substantive areas, similarity ratings are very prominent in several areas in psychology and marketing, while preference data enjoy limited popularity, probably because they are far more difficult to analyse and carry relatively less information. A further distinction is that between data with a *dependence structure* and those with an *interdependence structure*. With the former kind, the (mostly profile) data are split up in two (or more) groups, one of which is designated as the set of variables to be explained or predicted from the other set, while in the latter kind there is no such distinction, and the interrelationships between the varables is the object of study. A final distinction is that between 'true' three-mode data (or *repeated measures data*), which are fully crossed, barring missing data, and multiple-sets data. The former type of data are the 'true' or 'ordinary' three-mode data, because three different kinds of entities or modes are generally involved, one of which might be a time mode. When a *time mode* is involved, many techniques discussed here do not explicitly include this design feature in their analysis, but rather use time as an interpretational feature. True multivariate time series models within the Box-Jenkins approach (e.g. Chatfield, 1989), while falling essentially within the present framework, are not dealt with in this paper. *Multiple-sets data* are generally two-mode three-way data derived from raw data, rather than consisting of the raw data themselves. Typically one may think of cross-product matrices, covariance matrices, and often (dis)similarity matrices. Such data cannot always be analysed in their raw form with three-way methods, because they originate, for instance, from different samples, but after transformation to two-mode three-way data they can be analysed with such methods.

2.2 Three-way methods

With respect to methods, one may distinguish between those that explicitly use a *stochastic framework*, rely heavily on distributional assumptions, and therefore more or less by default treat subjects as replications. The other group consists of techniques which are primarily *data-analytic* in orientation, can be considered to address populations, and pay attention to *individual differences*. A further distinction that is often useful is that between *direct modelling* and *indirect modelling* techniques (e.g. Kruskal, 1984). The former techniques attempt to model the three-way data directly, while the latter techniques try to fit models to derived three-way matrices, such as covariance (correlation) matrices or cross-product matrices, and thus are often used for multiple-sets data.

2.3 Three-way method chart

In this section a three-part chart of methods is presented. Obviously a thorough discussion of this chart with proper references would require a full book. The chart is here presented as an appetiser of which topics one might come across in the three-way field. A more extended discussion of component models is contained in Kroonenberg (1992), while the other parts are still in my portfolio. In a way the Sage booklet by Arabie et al. (1987) can be seen as a treatment of Part III of the Method Chart.

3. Examples

In this section we will present four condensed examples to illustrate some more recent or less well-known techniques. The examples come from such diverse disciplines as agriculture (interactions from three-way ANOVA), demographics (three-way correspondence analysis), psychology (three-mode covariance structure analysis), and sensory perception (individual differences in orientation scaling).

3.1 Interactions from three-way ANOVA: Dutch maize data

In agriculture, three-way data summaries often take the form of three-way tables with one way consisting of different varieties of a crop, while the other two ways contain two environmental factors such as years and locations. This kind of summary is a direct consequence of the interpretation of the phenotype as the joint product of genotype and environment, where the environment encompasses everything that is non-genetic. Such three-way tables may be modelled with three-way analysis of variance, but due to the generally moderate to large number of entities in the ways, facilities to model two-way and three-way interactions multiplicatively should be considered.

METHOD CHART

Part I: Profile data

Dependence techniques: General linear model methods

two-block multiple regression, three-mode redundancy analysis;

Interdependence techniques: Component methods

three-mode component analysis, parallel factor analysis, three-mode correspondence analysis, latent class analysis, spatial evolution analysis;

Dependence and interdependence techniques

multi-set canonical correlation analysis, procrustes analysis, multi-set discriminant analysis;

Clustering methods I

three-way mixture method clustering.

Part II: Covariance models for profile data

Stochastic covariance (or structural equations) models

Repeated-measures methods

invariant factor analysis,three-mode common factor analysis, additive and multiplicative modelling of multivariable-multioccasion matrices;

Cross-sectional methods

simultaneous factor analysis;

Exploratory covariance model methods

Repeated-measures methods

(quasi-) three-mode component analysis;

Cross-sectional methods

simultaneous component analysis, indirect fitting with component analysis methods;

Part III: Similarity and preference data

Multidimensional scaling models

individual differences scaling, general euclidean models, three-way multidimensional scaling;

Clustering methods II

individual differences clustering, three-way ultrametric trees, synthesized clustering;

Unfolding models

three-way unfolding

3.1.1 Theory

In this section we will concentrate on modelling the three-way interaction with a three-way decomposition. There are various three-way generalizations of the two-way singular value decomposition, but the one most satisfactory for our purposes was that due to Tucker (1966). First we arrange the three-way ANOVA interaction parameters, $\alpha\beta\gamma_{ijk}$, in a three-way array, and decompose that array according to the Tucker3 model to give

$$\alpha\beta\gamma_{ijk} = \Sigma_p \Sigma_q \Sigma_r a_{ip} b_{jq} c_{kr} g_{pqr} + e_{ijk}$$

Only the most important multiplicative terms for each of the ways will generally be retained in the model. For testing multiplicative three-way interaction not very much theory has been developed. Boik (1990) presents a likelihood ratio test for the first term, including a table of critical values for comparatively small three-way tables. In cases covered by Boik, his test can be used otherwise we advise to use a procedure similar to one of the procedures for two-way tables, attributing *df* equal to the number of independent parameters to terms that stand out in the amount of three-way interaction described. This usually concerns only the first three-way component for each way. Occassionally second terms are involved.

For an $I \times J \times K$ three-way table of raw data, a three-way decomposition with P components for the first way, Q components for the second way, and R components for the third way, implies the estimation of $I \times P + J \times Q + K \times R$ scores plus $P \times Q \times R$ singular values. Owing to rotational invariance the number of constraints is equal to $P^2 + Q^2 + R^2$. The difference between parameters estimated and constraints imposed then results in $(I \times P + J \times Q + K \times R + P \times Q \times R) - (P^2 + Q^2 + R^2)$ df for the fitted model. When considering the three-way interaction instead of the raw data, I, J, and K should be reduced by one.

3.1.2 Example: Dutch maize variety trials

In the Netherlands there is an ongoing programme of testing maize varieties which aims to select the best varieties for cultivation and use under Dutch conditions. The present data set consists of 6 varieties (Brutus, Splenda, Markant, Vivia, Dorina, Irla) planted in four regions of the Netherlands characterised by the soil and location (Southern Sand, Central Sand, Northern Sand, and River Clay). Mean values on the attribute Percentage Dry Matter Content – DMC – for these varieties were available for the years 1980 through 1987 with the exclusion 1983 which was deleted due to missing data in one location. In other words, the data can be arranged as a 6 (varieties) by 4 (locations) by 7 (years) three-way table.

Table 1 shows the results of the straightforward three-way ANOVA for percentage dry matter content of the maize. The variance explained of the variety-by-year and the three-way interaction are roughly of the size of the explained variance by the varieties themselves, but they have, of course, much smaller mean squares due their larger degrees of freedom. One con-

Source	Degrees of Freedom	Sum of Squares	Mean Square
Variety	5	80	16.04
Site	3	962	320.73
Year	6	1195	199.11
Site × Year	18	851	47.30
Variety × Site	15	17	2.76
Variety × Year	30	65	2.16
Var. × Site × Year	90	80	0.89
1 × 1 × 1-solution	12	31	2.61
2 × 2 × 2-solution (given 1 × 1 × 1)	12	18	1.49
Deviations	66	31	0.47
Total	167	3520	

Table 1: Analysis of Variance for the Maize Data

clusion is that the first two terms of the decomposition of the three-way interaction are significant (for details see Van Eeuwijk and Kroonenberg, submitted).

With a three-way singular value decomposition of the three-way interaction effects table using two components for each of the ways, $P = Q = R = 2$, it was possible to identify the contrasts as responsible for the whole of the three-way interaction. The interaction patterns are summarised in Table 2.

The three-way interaction could be partially interpreted as a kind of correction on the Location by Year interaction. The three-way interactive pattern shows that the IJK-interaction is of a contrast type, i.e. it is caused by specific varieties reacting in a specific way at specific locations in contrast with the behaviour of some other varieties.

3.2 Three-way correspondence analysis: French canton data

Most three-way methods require data to be at least interval scaled. In this section we will analyse the situation where we have three categorical variables with more than a few categories. Carlier and Kroonenberg (1996) described in detail both the theory and the interpretation of three-way correspondence analysis (three-way CA). With the technique it is possible to produce measures and graphical displays of the dependence in a three-way

	Splenda			Brutus/Vivia			Dorina		
	South Sands	North Sands	Central Sands	South Sands	North Sands	Central Sands	South Sands	North Sands	Central Sands
1980	-	+		+	-				
1981				+	-	-	-		+
1985	+	-		-	+				
1986				-	+	+	+		-

Table 2: Interactive Patterns from Two-Dimensional Multiplicative Solution of Three-Way Interaction

tables, and as such it shares and extends many properties of ordinary (two-way) correspondence analysis.

3.2.1 Theory

The basic starting point is a three-way contingency table with orders I, J and K relative frequencies, p_{ijk}. To measure the deviations from the three-way independence model in such a table (thus taking into account all interactions), Pearson's *mean-square contingency coefficient*, Φ^2, or *Inertia* is an appropriate measure.

$$\Phi^2 = \frac{X^2}{n} = \sum_{i,j,k} \frac{(p_{ijk} - p_{i..}p_{.j.}p_{..k})^2}{p_{i..}p_{.j.}p_{..k}}. \tag{1}$$

It can be shown that the total inertia may be partitioned as follows

$$\begin{aligned} \Phi^2 &= \sum_{ij} p_{i..}p_{.j.}\left(\frac{p_{ij.} - p_{i..}p_{.j.}}{p_{i..}p_{.j.}}\right)^2 + \sum_{ik} p_{i..}p_{..k}\left(\frac{p_{i.k} - p_{i..}p_{..k}}{p_{i..}p_{..k}}\right)^2 \\ &+ \sum_{jk} p_{.j.}p_{..k}\left(\frac{p_{.jk} - p_{.j.}p_{..k}}{p_{.j.}p_{..k}}\right)^2 + \sum_{ijk} p_{i..}p_{.j.}p_{..k}\left(\frac{p_{ijk} - {}_{\alpha}p_{ijk}}{p_{i..}p_{.j.}p_{..k}}\right)^2 \\ &= \Phi^2_{IJ} + \Phi^2_{JK} + \Phi^2_{IK} + \Phi^2_{IJK}, \end{aligned} \tag{2}$$

where ${}_{\alpha}p_{ijk}$ is implicitly defined. This is clearly an additive definition of the interaction in a three-way array. Equation (2) shows that the global measure of dependence, Φ^2, can be split into separate measures of dependence: there are three measures for the dependence due to each two-way margin which are identical to those used in two-way correspondence analysis, and one measure for the three-way interaction. Such a partitioning is the first step in the analysis of a three-way table.

The three-way analogue of the singular value decomposition, especially the Tucker3 model, will be used to model the dependence. There is a subtle difference in the present usage of the Tucker3 model in that, analogous to two-way correspondence analysis, orthonormality is defined with respect to *weighted* metrics defined by $\{p_{i..}\}$, $\{p_{.j.}\}$, and $\{p_{..k}\}$, respectively, moreover, a weighted least-squares criterion is used.

One of the attractive features of using the additive approach over a multiplicative (or loglinear modeling) one, is that one single decomposition of the global dependence is made, and that the marginal dependence can directly be modeled and assessed from the global decomposition. The contributions of the marginal dependences to the global dependence can be evaluated without having to construct special decompositions for lower-order interactions as was necessary in the previous example. Moreover, such interactions can be portrayed in the same plot as the global dependence. To portray the dependence we will again use *interactive biplots*, in which the markers of two of the three ways (here: (i, j))are combined and plotted in the same figure with the markers of the remaining way. Assuming j is an ordered mode, trajectories can be drawn in the biplot by connecting, for each i, the points (i, j) in their proper order. This will greatly facilitate interpretation, especially if j is a time mode.

3.2.2 Example: Changes over time in the Languedoc work force

During the census of 1954, 1962, 1968 and 1975, the people of 42 cantons in Languedoc-Roussillon (Southern France) were asked to state their profession. Their occupations could be grouped into seven major occupational classes: Farmers (*AF*), Agricultural labourers (*AL*), Owners of small and medium-sized businesses (*SB*), Professionals and senior managers (*PS*), Middle managers (*MM*), Employees (white-collar workers - *WC*), Labourers (blue-collar workers - *BC*), Employees in the service sector (*SE*), Other occupations (*OO*). Full details as well as the data themselves can be found in a special issue of *Statistique et Analyse des Données* (1985, *10* (1), especially p. 11-15).

To evaluate the results of the analysis we will first examine the table with the partitioning of the χ^2-variance to evaluate the sizes of the different interactions. Of the total χ^2-variance, the absolute largest amount is explained by canton-by-occupation interaction (57%) followed by the occupation-by-time interaction (22%). If the degrees-of-freedom are taken into account as well, the occupation-by-time interaction has by far the largest contribution per *df*, which indicates that the occupational distributions have undergone considerable changes over time. Also the canton-by-occupation has a sizeable contribution per *df* showing that there is considerable diversity among the cantons. The smaller cantons-by-time interaction contains the differential increase and decrease of the cantons, primarily a trek from the rural areas to the towns. The three-way interaction is not large, and it has by far the smallest contribution per *df*, and it will not be discussed further.

To describe the patterns in the data set, we have fitted a Tucker3 model with 5 components for the cantons, 5 components for the occupations and 2 components for the time mode. This model fits very well leaving only 8% unexplained, and Table 3 shows that the canton-by-occupation and the occupation-by-time interactions are very well explained (96% and 95%), re-

					Results Fitting 3-Way Model		
Source	*df*	X^2_{Tot}	% of X^2_{Tot}	X^2_{Tot}/ *df*	X^2_{Err}	% of X^2_{Err}	X^2_{Fit}/ X^2_{Tot}
Main effects					3	≈0%	-
Cant. x Occup.	328	114307	57%	348	4571	29%	96%
Canton x Time	123	16278	8%	132	1793	11%	89%
Occup. x Time	24	43469	22%	1811	2191	14%	95%
C x O x T	984	25249	13%	26	7441	47%	60%
Total	1459	199303	100%	137	16001	100%	92%

Table 3: Cantons: Global and Marginal Quality Indexes: Chi-square Variances

spectively. The relatively unimportant three-way interaction has the smallest fit (60%). One of the strong points of the method proposed here is that it takes into account these two-way interactions within the framework of a single model which is fitted to the deviations from the three-way independence model.

To be able to display the results in interactive biplots we need to select of a reference mode. As we intend to study the changes in the distributions of the cantons over time, the occupations have to be chosen as reference mode. In the interactive biplot, the four canton-occasions points of each canton i (or interactive markers, $(i,k), k = 1,..,4$) are connected by a line ending in an arrow head for 1975; such lines are called *trajectories*. Trajectories can be interpreted in terms of the distributions of occupations in the cantons at each occasion. Here we will only look at the first two axes of the interactive biplot. The complete example is contained in Carlier and Kroonenberg (1996). The interactive plot of dimensions one and two (Figure 1) explains 71% of the inertia. In principle the biplot displays the complete global dependence, but it can also be used to study the three two-way interactions, because these interactions can be derived from the global dependence by (weighted) averaging of the coordinates. In other words, by computing weighted means of coordinates over one way, and displaying these means in the same figure as the global dependence figure, we can at the same time interpret the two-way interactions and assess what the global dependence contains over and above the two-way interactions. Here we will only look at the canton-by-occupation interaction.

In Figure 1 the centroids of the canton trajectories together with the occupation points indicate the canton-by-occupation interaction and they are displayed in the same graph as the global dependence itself by marking them with the abbreviations of the cantons on the trajectories. Moreover, the biplot with the canton centroids and the occupations can be interpreted exactly as the comparable two-way biplot from two-way correspondence analysis. Some of the more extreme features of the interaction as evident from

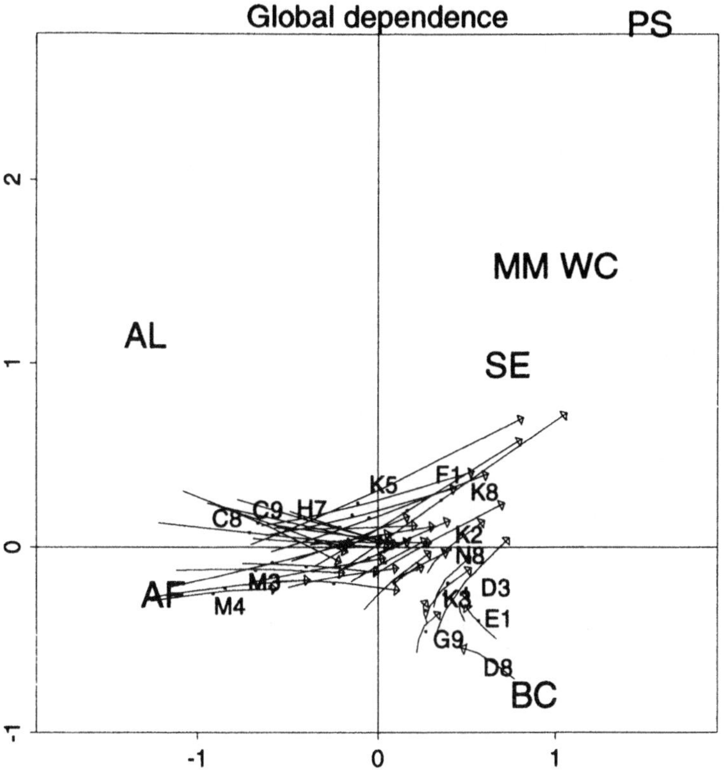

Figure 1: Interactive Biplot for Global Dependence of Cantons Data

Figure 1 are the following.

The cantons Fournels (M4) and Chateauneuf de Randon (M3) are very rural with a marked presence of independent farmers (AF). Similarly Lézignan-Corbières (C8), Narbonne (C9), and Capestang (H7) are very rural but with a predominance of agricultural labourers (AL) – possibly due to the viticulture in those areas. La Grand'Combe (D8), St Ambroix (E1), Ganges (K3), and Sumène (G9) are primarily industrial cantons with around 70 – 50%, of the work force occupied as blue collar workers, especially in the coal mines. Montpellier (K8), Nîmes (F1) and Les Matelles (K5) (a suburb of Montpellier) have a strong tertiary flavour with around 30% of the work force employed in the tertiary sector.

3.3 Exploratory three-mode covariance structure analysis: Personality-judgement data

In the literature, information in three-way data sets is often only reported in terms of covariances or correlations between the variables. A prevalent case is that of multitrait-multimethod (MTMM) matrices. There exists an

Methods	Ratings	Extraversion P	T	S	Impulsivity P	T	S	L.Anxiety P	T	S	L.Motivation P	T	S
	Peers	1.											
Extraversion	Teacher	.6	1.										
	Self	.5	.5	1.									
	Peers	.4	.3	.1	1.								
Impulsivity	Teacher	.4	.6	.4	.6	1.							
	Self	.3	.4	.4	.4	.5	1.						
	Peers	.4	.2	.1	.2	.2	.0	1.					
Lack of	Teacher	.3	.3	.1	.1	.1	-.1	.7	1.				
Test Anxiety	Self	-.0	-.1	.0	.1	.1	-.2	.4	.3	1.			
	Peers	.3	.1	-.1	.5	.2	.2	.5	.4	.1	1.		
Lack Academic	Teacher	.2	-.1	-.2	.3	.1	.1	.5	.6	.2	.7	1.	
Ach.Motivation	Self	.2	.0	.0	.4	.1	.4	.4	.4	.0	.6	.6	1.

Table 4: Correlations of Four Personality Variables Measured by Peers, Teachers, and Self Ratings

extensive literature on analyzing MTMM matrices following the stochastic line of three-way analysis, especially in the field of the analysis of covariance structures (e.g. Bentler and Lee, 1979; see Browne (1984) for a review). The procedures proposed are primarily confirmatory and statistical, and generally based upon specific distributional assumptions.

Kiers et al. (1992) developed a least-squares algorithm for Tucker's (1966) Method III, which allows an exploratory analysis of MTMM matrices by three-mode component analysis models. Just as in Bentler and Lee's analysis, subject component scores can no longer be found, and their role is taken over by what they call a (structured) loading matrix: the supermatrix containing the correlations between the individual differences components and the traits for each method.

As a (simplistic) illustration we will re-analyze an MTMM correlation matrix, analysed several times by Bentler et al. (e.g. 1979). As formulated in Bentler and Lee (1979, p. 93), correlations were available among four personality variables, each measured by Peers, Teachers and Self-ratings in a sample of 68 fifth-graders from two classrooms of a middle-class public elementary school (A sample size which is really too small for the stochastic approach). The traits measured were Extraversion (Extr), test anxiety, Impulsivity (Imp), and academic achievement motivation. For ease of interpretation we will reverse the scores on test anxiety and academic achievement motivation and label these variables Lack of test Anxiety (L.Anx) and Lack of Motivation (L.Mot), respectively. As a result, all sizeable correlations are positive, see Table 4 which contains a rearranged and rounded version of the original correlation matrix. The rearrangement was based on the results to be presented, and most of the large scale patterns can be seen from this table.

In an attempt to stay as close as possible to Bentler and Lee's original analyses, at first a solution was determined with 2 method components, 3 trait components, and 5 individual differences components (a $2 \times 3 \times 5$-solution). The solution accounted for 77% of the variability, but the fifth individual differences component accounted for only 5 percent, with an unclear pattern. Therefore, a $2 \times 3 \times 4$-solution was also determined. This solution explained 72% and will be presented here.

Table 5 contains the orthonormal components for the traits and the methods, the core matrix, and the loading matrices. The two method components indicate to what extent Teachers, Peers, and Self-ratings concur, and to what extent Teachers and Peers differ from the Self-ratings, respectively. The first method component also shows that the correlations tend to be somewhat smaller for the Self-ratings. The first two trait components show the classic pattern for generally correlated traits but divided into two blocks (here: Extraversion/Impulsivity and Lack of Test Anxiety/Lack of Academic Achievement Motivation) with lower correlations between blocks than within blocks (see also Table 4). The third trait component indicates that the situation is somewhat more complex, in particular that there is also some correlation between Lack of Test Anxiety and Extraversion, and between Impulsivity and Lack of Academic Achievement Motivation (again see Table 4).

From the size of the core elements, we see that all large core elements (1.93; 1.50; 1.90) refer to what the methods measure in the same way, as they pertain to the first method component. In other words, the major variability is due to traits rather than methods: Peer judgements, Teachers judgements, and subjects' Self-ratings concur on the general pattern among the four traits. Finally, the only other largish element (0.72) in the core matrix pertains to the Self-ratings versus Peer-and-Teacher contrast. It is rather difficult to see what this contrast precisely entails from just looking at the components, probably because the effect itself is rather small. Further information about this can be derived from the loading matrices. There we see in the fourth subject component the contrast between especially Peers and the Self-ratings with the Teachers siding largely with the Peers. However, an interpretation of the observed difference is not very clear, and the difference is also difficult to trace in the original correlation matrix.

Certain aspects of the interpretation would benefit from rotations, especially the trait components. The inverse transformations, however, immediately make the core matrix less interpretable because before rotation there are only four large elements in the core matrix while after rotation there will be hardly any large elements, but many medium-sized ones.

Clearly, the above description is a long way from a real substantive interpretation and carries little theoretical content. However, it may serve to illustrate that an exploratory analysis can be performed with the Tucker models. Furthermore, it should be noted that thc outcomes of the analyses have a distinctly different flavour from Bentler and Lee's results. In particular, the question may be raised whether their consistent split between

A. Components

Trait	T1	T2	T3	Method	M1	M2
	Trait Components				Method Components	
Impulsivity	.61	.37	.34	Teacher	.61	-.24
Extraversion	.52	.47	-.49	Peers	.59	-.51
L. Ac.Ach. Motivation	.50	-.53	.51	Self	.53	.83
L. Test Anxiety	.33	-.61	-.62			
Proportion Variability	.37	.22	.13		.63	.09

B. Core Matrix

Method Components	Trait Components	S1	S2	S3	S4
		Subject Components			
	T1 (Common)	1.93	.31	.01	-.02
M1	T2 (I&E vs. L.A&L.M)	-.39	1.50	.19	-.15
(Shared)	T3 (I&L.M vs. E&L.A)	-.04	-.26	1.09	-.01
	T1 (Common)	-.24	.23	-.06	.72
M2	T2 (I&E vs. L.A&L.M)	.10	.15	-.01	.37
(P&T vs. S)	T3 (I&L.M vs. E&L.A)	.32	.02	.27	.33
Proportion Variability Accounted for		.33	.21	.11	.06

C. Variable-Component Correlations

Method		S1	S2	S3	S4			S1	S2	S3	S4
		Subject Components						Subject Components			
	E	.7	.4	-.2	-.2		E	.5	.6	-.3	-.1
S1	I	.7	.1	.4	-.4	S2	I	.6	.5	.2	-.3
(Peers)	L.M	.7	-.4	.2	-.1	(Teacher)	L.M	.7	-.6	.2	.0
	L.A	.7	-.4	-.4	.1		L.A	.7	-.4	-.4	.1
	E	.3	.7	-.3	.3						
S3	I	.5	.5	.3	.5						
(Self)	L.M	.6	-.3	.4	.3						
	L.A	.2	-.4	-.6	-.1						

Table 5: Three-Mode Principal Component Solution with four Subject Components, three Trait Components, and two Method Components

Peers and Teacher judgements is adviseable, and our analysis suggests that a Peers/Teacher factor and a Self-ratings factor might be worth contemplating.

3.4 Individual differences in orientation scaling: Grigg's pain data

In the behavioural sciences it is not uncommon that data are collected as similarities. People are requested to express either directly or indirectly how similar two stimuli are. In the example, people were requested to indicate how similar certain sensations related to pain were. The question is whether the subjects perceived pain in a similar manner, and how the pain sensations grouped together. The unpublished data were kindly supplied by Dr. L. Grigg (University of Queensland, Australia).

3.4.1 Theory

In the present case these similarities are converted to dissimilarities. This makes the values comparable to distances, and in fact we will treat the dissimilarities as if they were *squared* distances. As is shown in the MDS-literature, double centring squared distances gives scalar products which can in turn analysed by scalar-product models, such as INDSCAL and IDIOSCAL (see Carroll and Chang (1970)). The INDSCAL model assumes that there exists a common stimulus configuration, which is shared by all judges (subjects), and that this configuration has the same structure for all judges, except that they may attach different importance (salience) to each of the (fixed) axes of the configuration. This results in some judges having configurations which are stretched out more along one of the axes. The IDIOSCAL model is similar except that each judge may put the axes of the common configuration under a different angle, and thus they may orient the common space in a different way.

Recently, Ten Berge et al. (1994) showed that the TUCKALS2 algorithm (see section 4.1.4) is an efficient way to estimate the IDIOSCAL model (see section 4.1.2). On the other hand, Arabie et al. (1987) indicate that this model "[..] has empirically yielded disappointing results in general" (p.45). In this section we will present an application of the IDIOSCAL model, but at the same time show that also in this data set interpreting the IDIOSCAL model is not worth the effort compared to interpreting the INDSCAL model.

3.4.2 Example: Grigg's pain data

From the original 41 subjects, sixteen were chosen for this example on the basis of the fit of the IDIOSCAL model to their data during a preliminary analysis. The analysis reported here is a two-component solution with a fit of 39.2%, indicating that the data are still very noisy. Figure 2 shows the stimulus space. Violent pains, such as shooting, burning, cramping, intense pain are in the same region of the space, less dramatic ones, such are mild, moderate, and annoying are also located near each other, as are tiring, miserable, and distressing. One identify directions in the space, for instance, by specifying an axis going from mild to intense and one from more short

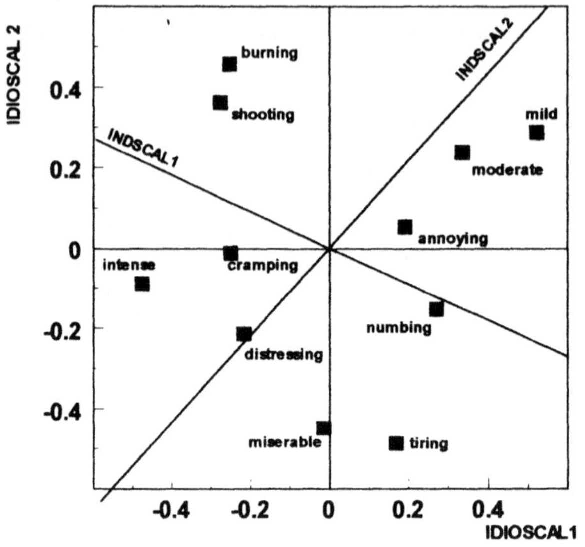

Figure 2: Grigg's Pain Data. IDIOSCAL Stimulus Space

sensations to longer lasting sensations.

The subject weights are shown in the left-hand panel of Table 6. The table provides the weights allocated to the first dimension and to the second dimension as well as the "individual orientation" expressed as a cosine between the two dimensions as well as the angle this represents. Clearly the subjects fall in two groups, those that put the dimensions under an acute angle and those that put them at an obtuse angle. Proof enough for an individual differences of orientation scaling it seems. However, one problem is that for identifiability of the model, we had to assume that the stimulus space was orthogonal. To check whether this was problematic we performed an INDSCAL analysis (with our TRILIN program (see section 4.1.4)). This analysis provided a fit of 38.3%, hardly worse than the previous analysis, and given that its interpretation is more straightforward it is clearly to be preferred. The additional complexity of the IDIOSCAL model was only apparent in this case and the results support Arabie et al.'s conclusion.

In Figure 2 we have drawn the orientation of the two INDSCAL axes, which have an inner product of -.31 and thus make an angle of 108 degrees. In the right hand panel of Table 6 we see the subject weights. Of course, they reflect the two groups found earlier. Staying with the basic INDSCAL interpretation we see that one group of subjects (1,2,3,10,11,12,13,14) tends to emphasize the axis of burning, shooting, intense, cramping pain in contrast with mild, numbing, and tiring. The other group of subjects (5,6,7,8,9,15,16) contrast mild and moderate pain with intense, tiring, distressing and miserable, and

Type of Subject	IDIOSCAL Subject Weights (1,1)	(2,2)	(1,2)	IDIOSCAL Cosines (1,2)	INDSCAL Subject Weights (1,1)	(2,2)
Control	.71	.15	-.27	-.82	.45	.03
Chronic Pain	.59	.20	-.22	-.64	.38	.05
Chronic Pain	.66	.25	-.18	-.46	.37	.13
Control	.53	.22	-.15	-.44	.31	.09
Control	.66	.09	-.03	-.13	.29	.19
Chronic Pain	.41	.23	-.24	-.79	.28	.02
Control	.50	.14	-.06	-.22	.24	.14
RSI Pain	.42	.32	.27	.75	.07	.42
RSI Pain	.53	.32	.19	.45	.16	.38
Chronic Pain	.34	.45	.17	.44	.09	.36
RSI Pain	.52	.24	.15	.44	.16	.33
RSI Pain	.45	.29	.12	.34	.15	.31
Chronic Pain	.35	.47	.08	.19	.14	.30
Control	.51	.30	.08	.22	.19	.30
Control	.25	.32	.08	.32	.08	.23
RSI Pain	.52	.09	.04	.16	.22	.19

Table 6: Grigg's Pain Data. IDIOSCAL Subject Weights and Cosines and INDSCAL Subject Weights (sorted with respect to INDSCAL weights)

place burning and shooting somewhere in the middle.

In the original design the subjects consisted of three groups: chronic pain sufferers, repetitive-strain-injury sufferers, and a control group. If the information available is correct then the division into two groups runs right through the design groups. Unfortunately, we have been unable to contact the original researcher to confirm the placement of the subjects in the groups or to request additional information on the subjects, which might shed light on the relationship between perceived and semantic pain expressions.

4. Software: 3WAYPACK (version 2)

The analyses presented in this paper have been carried out with the author's program package 3WAYPACK (Kroonenberg, 1994, 1996). This collection of programs has specifically designed for the analysis of three-way data. It consists of an menu-based interface, called INTERFACE3 (IF3), a preprocessing program, PREPROC3, three main analysis programs, TUCK-

ALS2 (Tucker2 model), TUCKALS3 (Tucker3 model), and TRILIN (Parafac model), and three output processing or postprocessing programs, ROTATE (rotating components), RESIDUAL (analysing residuals), and JOINTPLT (constructing joint (bi)plots).

References:

ARABIE, P.; CARROLL, J. D. and DESARBO, W. S. (1987): *Three-way scaling and clustering.* Sage, Beverly Hills.

BASFORD, K. E.; KROONENBERG, P. M. and DELACY, I. H. (1991): Three-way methods for multiattribute genotype by environment data: An illustrated partial survey. *Field Crops Research, 27, 131-157.*

BASFORD, K. E. and MCLACHLAN, G. J. (1985): The mixture method of clustering applied to three-way data. *Journal of Classification, 2, 109-125.*

BEFFY, J. L. (1992): Application de l'analyse en composantes principales à trois modes pour l'étude physico-chimique d'un écosystème lacustre d'altitude: Perspective en écologie. *Revue Statistique Appliquée, 40(1), 37-56.*

BENTLER, P. M. and LEE, S. Y. (1979): A statistical development of three-mode factor analysis. *British Journal of Mathematical and Statistical Psychology, 32, 87-104.*

BOIK, R. J. (1990): A likelihood ratio test for three-mode singular values: Upper percentiles and an application to three-way anova. *Computational Statistics and Data Analysis, 10, 1-9.*

BROWNE, M. W. (1984): The decomposition of multitrait-multimethod matrices. *British Journal of Mathematical and Statistical Psychology, 37, 1-21.*

CARLIER, A. and KROONENBERG, P. M. (1996): Decompositions and biplots in three-way correspondence analysis. *Psychometrika, 61, 355-373.*

CARROLL, J. D. and ARABIE, P. (1983): INDCLUS: an individual differences generalization of the ADCLUS model and the MAPCLUS algorithm. *Psychometrika, 48, 157-169.*

CARROLL, J. D. and CHANG, J. J. (1970): Analysis of individual differences in multidimensional scaling via an N-way generalization of "Eckart-Young" decomposition. *Psychometrika, 35, 283-319.*

CARROLL, J. D.; PRUZANSKY, S. and KRUSKAL. J. B. (1980): CANDELINC: A general approach to multidimensional analysis of many-way arrays with linear constraints on parameters. *Psychometrika, 45, 3-24.*

CHATFIELD, C. (1989): The analysis of time series: An introduction (4th edition). Chapman and Hall, London.

DESARBO, W. S. and CARROLL, J. D. (1985): Three-way metric unfolding via alternating weighted least squares. *Psychometrika, 50, 275-300.*

FLURY, B. (1988): Common principal components and related multivariate models. Wiley, New York.

FRANC, A. (1992): Étude algebrique des multitableaux: Apports de l'algèbre tensorielle [An algebraic study of multi-way tables: Contributions of tensor algebra.] Unpublished doctoral thesis, Université de Montpellier II, France.

HARSHMAN, R. A. and LUNDY, M. E. (1984a): The PARAFAC model for three-way factor analysis and multidimensional scaling. In: H. G. Law, C. W. Snyder Jr., J. A. Hattie, and R. P. McDonald (eds): *Research methods for multimode data analysis. Praeger, New York, 122-215.*

HARSHMAN, R. A. and LUNDY, M. E. (1984b): Data preprocessing and the extended PARAFAC model. In: H. G. Law, C. W. Snyder Jr., J. A. Hattie, and R.P. McDonald (eds): *Research methods for multimode data analysis. Praeger, New York, 216-284.*

KIERS, H. A. L. (1991): Hierarchical relations among three-way methods. *Psychometrika, 56, 449-470.*

KIERS, H. A. L. (1992): TUCKALS core rotations and constrained TUCKALS modelling. *Statistica Applicata, 4, 659-667.*

KIERS, H. A. L.; KROONENBERG, P. M. and TEN BERGE, J. M. F. (1992): An efficient algorithm for TUCKALS3 on data with large numbers of observation units. *Psychometrika, 57, 415-422.*

KRIJNEN, W. P. (1993): The analysis of three-way arrays by constrained parafac methods. DSWO Press, Leiden.

KROONENBERG, P. M. (1983a): Three-mode principal component analysis: Theory and applications. DSWO Press, Leiden.

KROONENBERG, P. M. (1983b): Annotated bibliography of three-mode factor analysis. *Britsh Journal of Mathematical and Statistical Psychology, 36, 81-113.*

KROONENBERG, P. M. (1992): Three-mode component models. *Statistica Applicata, 4, 619-634.*

KROONENBERG, P. M. (1994): The TUCKALS line: A suite of programs for three-way data analysis. *Computational Statistics and Data Analysis, 18, 73-96.*

KROONENBERG, P. M. (1996): 3WAYPACK User's manual (Version 2). Department of Education, Leiden University, Leiden.

KROONENBERG, P. M. and DE LEEUW, J. (1980): Principal component analysis of three-mode data by means of alternating least-squares algorithms. *Psychometrika, 45, 69-97.*

KRUSKAL, J. B. (1984): Multilinear methods. In: H. G. Law, C. W. Snyder Jr., J. A. Hattie, and R. P. McDonald (eds): *Research methods for multimode data analysis. Praeger, New York, 36-62.*

LAVIT, C. (1988): *Analyse conjointe de tableaux quantitatifs* [Simultaneous analysis of several quantitative matrices]. Masson, Paris.

LEURGANS, S. E. and ROSS, R. T. (1992): Multilinear models: Application in spectroscopy (with discussion): *Statistical Science, 7*, 289-319.

Statistique et Analyse des Données, 1985, *10* (1).

VAN EEUWIJK, F. A. and KROONENBERG, P. M. (Submitted): Multiplicative decompositions of interactions in three-way ANOVA, with applications to plant breeding. (*Biometrics*)

TEN BERGE, J. M. F.; BEKKER, P. A. and KIERS, H. A. L. (1994): Some clarifications of the TUCKALS2 algorithm applied to the IDIOSCAL problem. *Psychometrika, 59*, 193-201.

TUCKER, L. R. (1966): Some mathematical notes on three-mode factor analysis. *Psychometrika, 31, 279-311.*

A Hybrid Global Optimization Algorithm for Multidimensional Scaling

Rudolf Mathar

Institut für Statistik,
RWTH Aachen, D-52056 Aachen, Germany

Summary: Local search algorithms in Multidimensional Scaling (MDS), based on gradients or subgradients, often get stuck at local minima of STRESS, particularly if the underlying dissimilarity matrix is far from being Euclidean. However, in order to remove ambiguity from the model building process, it is of paramount interest to fit a suggested model best to a given data set. Hence, finding the global minimum of STRESS is very important for applications of MDS.
In this paper a hybrid iteration scheme is suggested consisting of a local optimization phase and a genetic type global optimization step. Local search is based on the simple and fast majorization approach. Extensive numerical testing shows that the presented method has a high success probability and clearly outperforms simple random multistart.

1. Why Global Optimization is Important

The purpose of MDS is to fit Euclidean interpoint distances to given dissimilarities. Depending on the underlying model corresponding benefit criteria can become quite complicated, for an overview see, e.g., de Leeuw and Heiser (1980) and Cox and Cox (1994). In its simplest and most intuitive form, however, the aim of MDS is to

$$\text{minimize} \left\{ \sigma^2(\mathbf{X}) = \sum_{i<j} w_{ij}(\delta_{ij} - d_{ij}(\mathbf{X}))^2 \right\} \quad \text{over} \quad \mathbf{X} \in \mathbb{R}^{n \times k}, \tag{1}$$

where δ_{ij} are given symmetric dissimilarities between objects $\mathcal{O}_1, \dots, \mathcal{O}_n$. $\mathbf{X} = (\mathbf{x}_1, \dots, \mathbf{x}_n)'$ denotes so called *configurations*, i.e., n points $\mathbf{x}_1, \dots, \mathbf{x}_n \in \mathbb{R}^k$, and $d_{ij}(\mathbf{X}) = \|\mathbf{x}_i - \mathbf{x}_j\|$ is defined as the Euclidean distance between $\mathbf{x}_i$ and $\mathbf{x}_j$. $w_{ij} \geq 0$ are given weights, which by setting $w_{ij} = 0$ allows for including the case of missing values δ_{ij}. The benefit function of the weighted least squares scaling problem (1) is usually called STRESS. It is known to have a lot of stationary points which are not global minimizers. Fitting a low-dimensional ($k = 2$ or 3) metric coordinate model $\mathbf{X}$ to the dissimilarities allows for a graphical representation of the objects in a Euclidean space, and hence for visual inspection of conspicuous structures. Moreover, even positions of objects can be estimated if rotational and translational indeterminancy can be removed (see example 1 below).

Once having the model and the benefit function fixed, it is of paramount importance to fit the model to the data as close as possible. For MDS this means to determine the global minimum of STRESS. A lot of attempts has

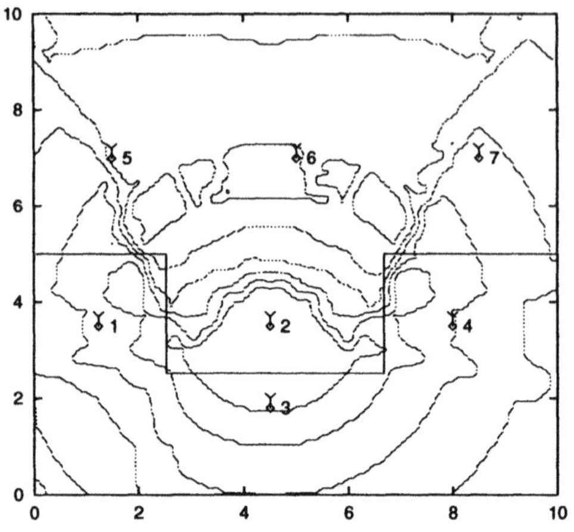

Figure 1: Isoclines of equal signal strength of base station 2.

been made to improve local search procedures by including smoothing, improved step sizes, or second order information, see, e.g., de Leeuw and Stoop (1984), Glunt et al. (1993), Kearsley et al. (1994), Mathar and Groenen (1991). In contrast, relatively few papers deal with the global optimization problem, e.g., Groenen (1993), Mathar and Zilinskas (1993). The purpose of the following examples is to show that striving for a global minimum of STRESS is most important for certain applications of MDS.

Example 1. This example deals with an interesting application of MDS to mobile communication in cellular radio networks, for a more detailed description see Hellebrandt and Mathar (1996). In a cellular network communication is established by radio channels. A certain area is covered by the transmission ranges of a number of base stations using noninterfering channels. Fig. 1 shows an area of 10×10 km with $b = 7$ base stations located at the indicated positions. Furthermore, the solid line depicts the track of a mobile moving from left to right with two right and two left quarter turns at a speed of 100 km/h. Radio wave propagation is usually quite irregular because of shadowing and reflections. The average signal strength isoclines of base station 2 for instance are shown as dotted lines in Fig. 1.

According to the GSM-standard a mobile measures the signal power of the 6 best received surrounding base stations on a scale from 0 to 63 each 0.48 second. These data are used to estimate the mobile's position over time.

Let $\mathbf{z}_1, \ldots, \mathbf{z}_b \in \mathbb{R}^2$ denote the locations of b base stations in an aerea of interest. For a fixed point of time let $\delta_1, \ldots, \delta_b$ be the signal strength measurements at a certain unknown location $\mathbf{x}$. $d(\mathbf{x}, \mathbf{z}_i) = \|\mathbf{x} - \mathbf{z}_i\|$, $i = 1, \ldots, b$, denotes the Euclidean distance from the base stations. In a somewhat simplifying approach we assume that the expected signal strength of base station i at position $\mathbf{x}$ can be described by an invertible function F as $F(d(\mathbf{x}, \mathbf{z}_i))$. Free space propagation, e.g., is described by $F(d) = cd^{-\alpha}$ with α ranging

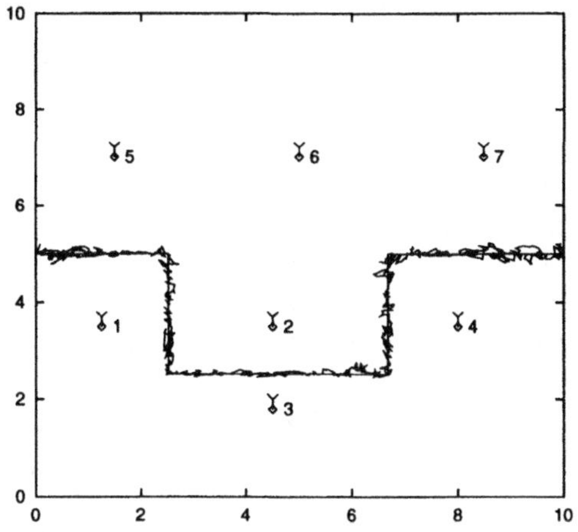

Figure 2: Estimated track of the mobile.

between 2 and 5, depending on the environment and the used frequency. Estimating the position $\mathbf{x}$ of a mobile means to

$$\text{minimize} \sum_{i=1}^{b} \left(d(\mathbf{x}, \mathbf{z}_i) - F^{-1}(\delta_i)\right)^2 \quad \text{over} \quad \mathbf{x} \in \mathbb{R}^2. \tag{2}$$

Numerical experiments showed that there exist local minimizers of (2) at completely irrelevant positions, in Fig. 2 even north of the upper row of base stations. This shows that it is very important to obtain the global minimizer as a reliable estimate of the mobile's position. Using appropriate initial values and a linear regression based smoothing procedure the track of the mobile was estimated as shown in Fig. 2.

Example 2. Ten different Cola brands were tasted by 38 students in an experiment reported by Green et al. (1988). Each pair was judged on its dissimilarity on a 9 point scale (1: very similar, ..., 9: completely different). The accumulated dissimilarities are represented in Table 1 (see also Groenen et al. (1995)).

Simple multistart from 9 random initials with a gradient-based local optimization procedure ended up with 9 different minimizing configurations as shown in Fig. 3. Stopping at an arbitrary local minimum and interpreting the data on the basis of the achieved configuration leads to different results, as this example shows. Hence, it is necessary to find the global minimum, and a corresponding configuration which explains the data best.

The existence of so many local minima is motivated Fig. 4. $\sigma^2(\mathbf{X})$ is plotted with variables $x_{1,1}$ and $x_{10,2}$ and all other variables fixed. Actually, 152 different points of attraction, probably local minima, were found numerically in a series of 1000 trials.

	Pepsi	Coke	Classic Coke	Diet Pepsi	Diet Slice	Diet 7-Up	Dr. Pepper	Slice	7-Up	Tab
Pepsi	0									
Coke	127	0								
Classic Coke	169	143	0							
Diet Pepsi	204	235	243	0						
Diet Slice	309	318	326	285	0					
Diet 7-Up	320	322	327	288	155	0				
Dr. Pepper	286	256	258	259	312	306	0			
Slice	317	318	318	312	131	164	300	0		
7-Up	321	318	318	317	170	136	295	132	0	
Tab	238	231	242	194	285	281	256	291	297	0

Table 1: Dissimilarities between 10 cola brands accumulated over 38 judges.

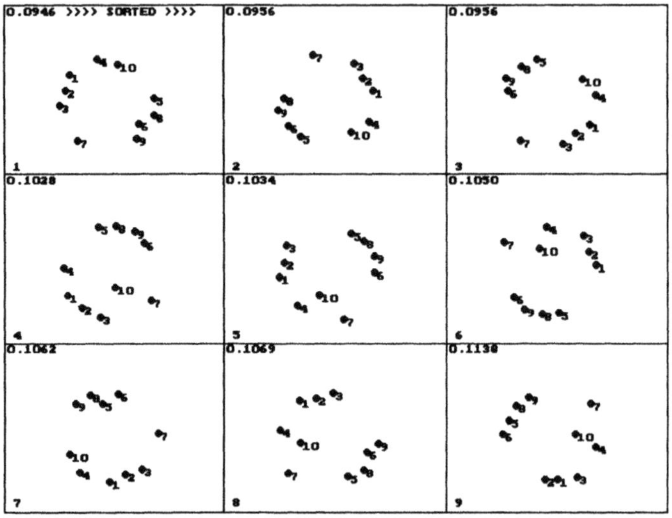

Figure 3: 9 local minimizers from random multistart.

2. Local Minimization by Majorization

Any local search procedure can be used as a part of the following hybrid iteration scheme. Because of the random disturbances added in the genetic type steps, however, it is advisable to keep the local updates as simple and fast as possible. The Guttman updates used in SMACOF serve this purpose perfectly. For ease of reference a short description of the underlying majorization principle is given here, for a comprehensive review see de Leeuw (1988).

Let $\varphi : \mathcal{M} \times \mathcal{M} \to \mathbb{R}$ be a real valued function. A function $\hat{\varphi} : \mathcal{M} \times \mathcal{M} \to \mathbb{R}$

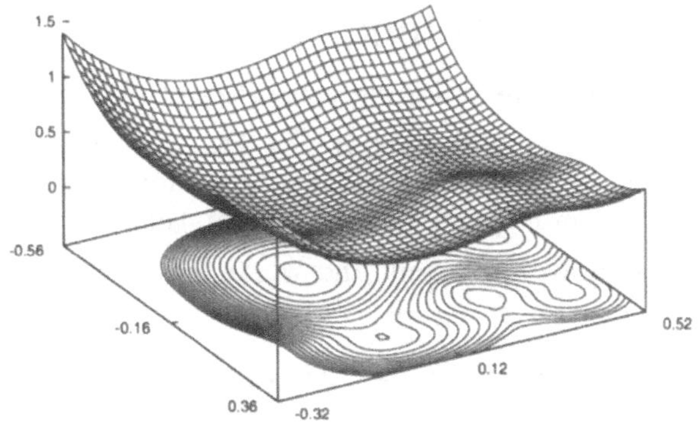

Figure 4: Surface plot of two variables $x_{1,1}$ and $x_{10,2}$.

is called a *majorizing function*, if for all $x, y \in \mathcal{M}$

$$\varphi(x) \leq \hat{\varphi}(x, y) \quad \text{and} \quad \varphi(x) = \hat{\varphi}(x, x).$$

Let $\bar{x} \in \mathcal{M}$ be a solution of $\min_{x \in \mathcal{M}} \hat{\varphi}(x, y)$ when y is fixed. Then

$$\varphi(\bar{x}) \leq \hat{\varphi}(\bar{x}, y) \leq \hat{\varphi}(y, y) = \varphi(y),$$

such that by iterating this step

$$x_{m+1} = \min_{x \in \mathcal{M}} \hat{\varphi}(x, x_m) \tag{3}$$

yields a sequence $\{x_m\}_{m \in \mathbb{N}_0}$ of updates with decreasing values $\varphi(x_m)$. With $\mathcal{M} = \mathbb{R}^{n \times k}$, and $\varphi = \sigma^2$ iteration (3) describes the basic SMACOF-steps. To obtain an explicit form of the updates in (3) we start with the following representation of $\sigma^2(\mathbf{X})$

$$\sigma^2(\mathbf{X}) = \text{const} + \text{tr}(\mathbf{X}'\mathbf{V}\mathbf{X}) - 2\text{tr}(\mathbf{X}'\mathbf{B}(\mathbf{X})\mathbf{X})$$

with $\mathbf{V} = (v_{ij})_{1 \leq i,j \leq n}$, $v_{ij} = -2w_{ij}$ for $i \neq j$, and $v_{ii} = -\sum_{j \neq i} v_{ij}$, furthermore, $\mathbf{B}(\mathbf{X}) = (b_{ij}(\mathbf{X}))_{1 \leq i,j \leq n}$, $b_{ij}(\mathbf{X}) = -2w_{ij}\delta_{ij}/d_{ij}(\mathbf{X})$ for $i \neq j$, and $b_{ii}(\mathbf{X}) = -\sum_{j \neq i} b_{ij}(\mathbf{X})$. It can be shown that $\mathbf{B}(\mathbf{X})\mathbf{X}$ is a subgradient of $r(\mathbf{X}) = \text{tr}(\mathbf{X}'\mathbf{B}(\mathbf{X})\mathbf{X})$ (see Mathar and Meyer (1994)). Hence, for all $\mathbf{Y} \in \mathbb{R}^{n \times k}$ it follows that

$$\begin{aligned} \sigma^2(\mathbf{X}) &\leq \text{const} + \text{tr}(\mathbf{X}'\mathbf{V}\mathbf{X}) - 2[\text{tr}(\mathbf{Y}'\mathbf{B}(\mathbf{Y})\mathbf{Y}) + \text{tr}(\mathbf{Y}'\mathbf{B}(\mathbf{Y})(\mathbf{X} - \mathbf{Y}))] \\ &= \text{const} + \text{tr}(\mathbf{X}'\mathbf{V}\mathbf{X}) - 2\text{tr}(\mathbf{Y}'\mathbf{B}(\mathbf{Y})\mathbf{X}) = \hat{\sigma}^2(\mathbf{X}, \mathbf{Y}). \end{aligned}$$

Minimizing $\hat{\sigma}^2(\mathbf{X},\mathbf{Y})$ over $\mathbf{X}$ is easy, since $\hat{\sigma}^2(\mathbf{X},\mathbf{Y})$ is quadratic in $\mathbf{X}$. Solving the stationary equations yields a solution $\mathbf{X} = \mathbf{V}^+\mathbf{B}(\mathbf{Y})\mathbf{Y}$ with the Moore-Penrose inverse $\mathbf{V}^+$ of $\mathbf{V}$. Iteration scheme (3) now reads as

$$\mathbf{X}_{m+1} = \mathbf{V}^+\mathbf{B}(\mathbf{X}_m)\mathbf{X}_m, \quad m \in \mathbb{N}_0. \tag{4}$$

De Leeuw (1988) proves the following properties of (4).

Theorem 1. a) $\sigma^2(\mathbf{X_m})$ is decreasing. b) $\|\mathbf{X}_{m+1} - \mathbf{X}_m\| \to 0$ $(m \to \infty)$ on the subspace of centered configurations, provided $\mathbf{V}$ is irreducible. c) If $\mathbf{X}_m \to \mathbf{X}^*$ and $\mathbf{X}^*$ is an isolated local minimum, then convergence is linear.

Obviously, (4) is easy to implement and works fast, even for large numbers n of objects.

3. A Hybrid, Genetic Type Optimization Scheme

Genetic and evolutionary algorithms are well established methods for finding global optima of highly complicated benefit functions (see, e.g., Goldberg (1989)). For the purpose of minimizing STRESS a population at time $t \in \mathbb{N}_0$ consists of N_t configurations, i.e.,

$$\mathcal{A}(t) = (\mathbf{X}_1(t), \ldots, \mathbf{X}_{N_t}(t)), \quad \mathbf{X}_i(t) \in \mathbb{R}^{n \times k}.$$

$\mathcal{A}(t+1)$ is developed from $\mathcal{A}(t)$ by selection, crossover, and mutation in the following way.

- Start with a random population $\mathcal{A}(t_0)$.
- Iterate the following steps over $t \in \mathbb{N}$:
  1. Apply a local optimization procedure (e.g. the above described majorization scheme) to each configuration $\mathbf{X}_i(t)$, $i = 1, \ldots, N_t$. Stop after a prescribed number of iterations, ending up with
  $$\mathcal{A}(t) = (\mathbf{X}_1^*(t), \ldots, \mathbf{X}_{N_t}^*(t)).$$
  2. Let $\sigma^2_{\max}(t)$ denote the largest STRESS-value occurring in $\mathcal{A}(t)$. Determine
  $$p_i(t) = \frac{v(\sigma^2_{\max}(t) - \sigma^2(\mathbf{X}_i^*(t)))}{\sum_{j=1}^{N_t} v(\sigma^2_{\max}(t) - \sigma^2(\mathbf{X}_j^*(t)))}, \quad i = 1, \ldots, N_t.$$
  with an increasing function v.
  3. Select independently at random N_{t+1} pairs from $\mathcal{A}^*(t) \times \mathcal{A}^*(t)$, according to the distribution $p_{ij}(t) = p_i(t)p_j(t)$, $i, j = 1, \ldots, N_{t+1}$.
  4. Crossover is carried out by forming convex combinations of each pair $(\mathbf{X}^*, \mathbf{Y}^*)$ selected in step 3, namely
  $$\mathbf{X}^+(t) = \alpha(t)\mathbf{X}^* + \beta(t)\mathbf{Y}^*,$$
  where $\alpha(t), \beta(t) \geq 0$ and $\alpha(t) + \beta(t) = 1$.

5. Finally, mutation is implemented by

$$\mathbf{X}_i(t+1) = \mathbf{X}_i^+(t) + \mathbf{R}_i(t), \quad i = 1, \ldots, N_{t+1},$$

with i.i.d. random matrices $\mathbf{R}_i(t) \in \mathbb{R}^{n \times k}$.

Extensive numerical testing showed that this hybrid algorithm works very well, both considering success probability and computing time. Only a small number of local iteration steps may be used, if the best configuration is always retained in the next generation. In this case the actual optimum is iterated to high accuracy as long as it is not outperformed by a member of the next generation, which rarely happens in later generations.

We have applied the above algorithm, including the "always keep the best" strategy to the Cola-data introduced in example 2. The parameter scheme was as simple as $N_t = N = 9$ (constant population size), $v(x) = x$ (selection probabilities proportional to $\sigma^2_{\max} - \sigma^2(\mathbf{X}_i^*(t))$, $\alpha(t) = \beta(t) = 1/2$ (crossover by taking 50% from each partner), and $\mathbf{R}_i(t)$ with independent uniformly on $(-0.05, 0.05)$ distributed components. The algorithm has been implemented with a graphical interface. Fig. 5 shows a population of 9 configurations in $\mathbb{R}^2$, after stopping the algorithm when step 4 was completed. An overlaying window in the middle visualizes the actual selection and crossover step.

By means of this program the global optimum of STRESS was determined in a long pre-phase. Then the above described hybrid algorithm was started 100 times with 60 and 100 local iterations for each generation, respectively. Table 2 shows the percentage of trials containing the global optimum in generation $i = 1, \ldots, 10$. It can be seen that the success probability of hitting

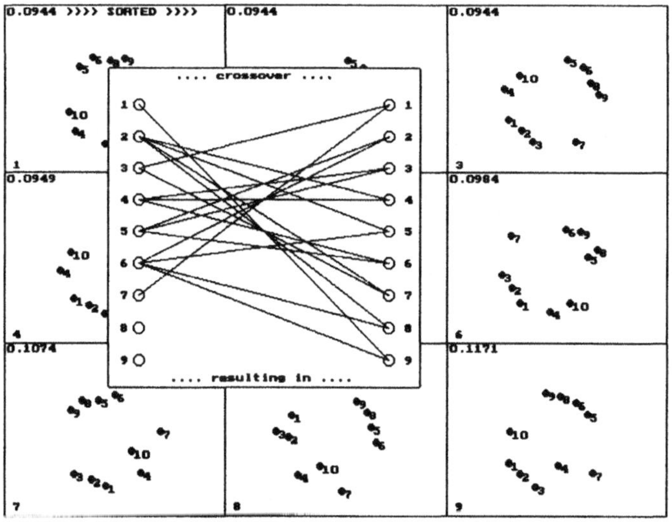

Figure 5: Screen showing the actual population and an overlaying window representing the actual crossover step.

Generation	1	2	3	4	5	6	7	8	9	10
60 local it.	3%	22%	47%	68%	76%	85%	90%	92%	96%	97%
100 local it.	8%	34%	61%	82%	92%	95%	98%	99%	99%	100%

Table 2: Percentage of runs containing the optimum configuration in generation i.

the global minimum is very high after 7 generations, even with only 100 local iterations.

Simple random multistart is considered as a concurrent method. It was implemented by the same program as above by setting $N_t = 1$, and starting with random initial configurations. Hence, the local iteration scheme is exactly the same in both approaches, and in either case runtime is nearly proportional to the number of local iterations. The success probability of multistart was estimated as 4% in a preseries of 1000 trials. Repeating random multistart 10 times with 900 local iterations each, yields an overall success probability of $1-0.96^{10} = 0.3352$ at the same computing effort as the above hybrid algorithm with population size 9 using 100 local iterations for each. If only 100 local iterations are applied for each initial, 90 random starts can be carried out. This yields a success probability of $1 - 0.96^{90} = 0.975$, however, at much less accuracy for the final configuration.

In summary, the suggested hybrid algorithm clearly outperforms random multistart, both considering success probability per unit runtime and accuracy of the final solution.

4. Conclusions

A hybrid algorithm is suggested, alternating between a local optimization phase and a genetic type step. Numerical testing shows that this method is superior to random multistart (100% success probability versus 34% in an example with 20 variables). In order to understand the algorithm's performance and optimize corresponding parameters, investigations on the convergence behavior are necessary. A theorem along the lines of the theory on Simulated Annealing, stating that

$$\lim_{t \to \infty} P(\mathbf{X}_{\text{opt}}(t) \in \mathcal{S}_{\text{opt}}) = 1,$$

where $\mathcal{S}_{\text{opt}}$ denotes the set of globally optimum configurations, will be aimed at in future work.

References:

COX, T. F. and COX M. A. A. (1994): Multidimensional Scaling. Chapman and Hall, London.

DE LEEUW, J. (1988): Convergence of the majorization method of multidimensional scaling. *Journal of Classification, 5, 163–180.*

DE LEEUW, J. and HEISER, W. (1980): Multidimensional scaling with restrictions on the configuration. In: P.R. Krishnaiah (ed.): *Multivariate Analysis, Vol. V.* North-Holland, Amsterdam, 501-522.

DE LEEUW, J. and STOOP, I. (1984): Upper bounds of Kruskal's stress. *Psychometrika 49, 391–402.*

GLUNT, W.; HAYDEN, T. L. and RAYDAN, M. (1993): Molecular conformations from distance matrices. *Journal of Computational Chemistry, 14, 114–120.*

GOLDBERG, D. E. (1989): Genetic Algorithms. Addison-Wesley, Reading, Mass.

GREEN, F. J.; CARMONE, F. J. and SMITH, S. M. (1989): Multidimensional Scaling: Concepts and Applications. Allyn and Bacon, Boston.

GROENEN, P. J. F. (1993): The Majorization Approach to Multidimensional Scaling, Some Problems and Extensions. DSWO-Press, Leiden.

GROENEN, P. J. F.; MATHAR, R. and HEISER, W. (1995): The majorization approach to multidimensional scaling for Minkowski distances. *Journal of Classification, 12, 3–19.*

HELLEBRANDT, M.; MATHAR, R. (1996): Estimating position and velocity of mobiles in celular radio networks. To appear: *IEEE Transactions on Vehicular Technology.*

KEARSLEY, A. J.; TAPIA, R. A. and TROSSET, M.W . (1994): The solution of the metric stress and sstress problems in multidimensional scaling using Newton's method. Technical Report, Dept. of Computational and Applied Mathematics, Rice University Houston, Texas, TR94-44.

MATHAR, R. and GROENEN, P. (1991): Algorithms in convex analysis applied to multidimensional scaling. In: E. Diday and Y. Lechevallier (eds.): *Symbolic-Numeric Data Analysis and Learning.* Nova Science Publishers, Commack, NY, 45–56.

MATHAR, R. and MEYER, R. (1994): Algorithms in convex analysis to fit l_p-distance matrices. *Journal of Multivariate Analysis, 51, 102–120.*

MATHAR, R. and ZILINSKAS, A. (1993): On global optimization in two-dimensional scaling. *Acta Applicandae Mathematicae, 33, 109–118.*

Ordinal Regression

Henning Paul

Lehrstuhl für Mathematische Methoden der Wirtschaftswissenschaften,
Universität Augsburg, D-86135 Augsburg, Germany

Summary: Many ordinal data come from data which are originally scaled cardinal. Nevertheless in some cases they are treated with methods, which require cardinal data. For such data a rescaling is allowed and useful. In this paper a method for rescaling is presented, which is based on information from highly correlated attributes and points out the places where the ordinal data are most probabely closer together than at other places. At these places adjacent categories are grouped together into one category. This metod is illustrated in a comparison of linear regression of the rescaled values and the original measured ordinal values.

1. Data model and basic assumptions

The present paper deals with data coming from a regression model of the following type:

$$\begin{aligned} y_i &= f(x_i) + u_i \quad \text{for} \quad 1 \leq i \leq n \\ &\text{with} \quad y_i, u_i \in \mathbb{R} \\ &\text{and} \quad x_i \in \mathbb{R}^k \quad k \in \mathbb{N} \\ &\text{where} \quad u_i \quad \text{is a random variable with} \quad E(U_i) = 0 \end{aligned} \tag{1}$$

The function f does not have to be linear, but the type of function has to be known. In general k may be larger than 1 but just for simplicity in the following parts k will be set to 1. Although all x-values are any real numbers, they are supposed to be ranked in ascending order, $x_i \leq x_{i+1}$, without any of generality.

We assume that all x_i and y_i are measured as ordinal scaled values p_i and q_i and not received as cardinal values. It is not necessary that the values differ in ranks. The observed values p_i and q_i may be considered to be based on latent trait variables x and y which can not be observed themselves.

In principle the measurement may contain errors too, which has the same effect as the effect caused by the random variable U_i in the data model. Therefore, all errors are supposed to be included in the random variable u_i given in (1), which is assumed to be symmetrically distributed, e.g., normal distributed with expectation $E(U_i) = 0$.

Thus we have

$$x_i < x_{i+1} \Rightarrow p_i \leq p_{i+1},$$

but because of the effect of the u_i for some $(i, i+1)$ we may observe

$$x_i < x_{i+1} \quad \text{and} \quad y_i > y_{i+1}$$

Now we can formulate two useful definitions:

Def. Rank disturbance
Pairs of values (p_i, q_i) and (p_{i+m}, q_{i+m}) with $p_i \leq p_{i+m}$ and $q_i > q_{i+m}$ form a so called *rank disturbance.*

Def. Rank jumping
Pairs of values (p_i, q_i) and (p_{i+m}, q_{i+m}) with $p_i \leq p_{i+m}$ and $q_i > q_{i+m}$ form a so called *rank jumping of height m.*

From the model given above result three conclusions, which are important for the procedure presented in this paper:

1. There exist no linear function which matches the differences $x_{i+m} - x_i$ or $y_{i+m} - y_i$ to the differences in ranks of the related pair $p_{i+m} - p_i$ respective $q_{i+m} - q_i$. Regrouping of objects can possibly improve this situation.

2. In general, an ascending sequence in x_i or p_i does not result in an ascending sequence in y_i or q_i respectively. Therefore, in general, we will have some rank disturbances.

3. If we suppose that U_i is normally distributed we can calculate the probability of a rank disturbance:

$$P(y_{i+m} < y_i) = \Phi((f(x_{i+m}) - f(x_i))/(\sqrt{2}\sigma)) \tag{2}$$

E.g. for $\lim_{x_i \to x_{i+m}} P(y_{i+m} < y_i) \to 0.5$

Reversely for a discrete, finite set of possible differences Δx, $\Delta x = x_{i+1} - x_i$ for any $i, 1 \leq i < n$, follows

$$P(x_{i+1} - x_i = \Delta x \mid y_{i+1} < y_i) = \frac{P(y_{i+1} < y_i \mid \Delta x)P(\Delta x)}{\sum_{\Delta x} P(y_{i+1} < y_i \mid \Delta x)P(\Delta x)} \tag{3}$$

and thus $P(\Delta x \mid \textit{rank disturbance})$ is maximal when $\Delta x \to 0$.

Hence it is obvious from conclusion 3., equ.(3), that for a pair (q_i, q_{i+1}) with a rank disturbance the probability of a small difference $x_{i+1} - x_i$ is higher than the probability of a big difference $x_{i+1} - x_i$. Therefore, corresponding ordinal values p_i and p_{i+1} or the related objects, o_i and o_{i+1}, respectively, should be grouped in one common class. The class or rank of object o_i when m classes exist will be called $p_i^{(m)}$.

When two objects o_i and o_{i+1}, are grouped in one common class, the ranks of both objects is equal: $p_i^{(m)} = p_{i+1}^{(m)}$. Furthermore, the number of classes has to be reduced by one, what leads to a similar reduction for all class or rank numbers greater than i: $p_j^{(m)} = p_j^{m+1} - 1$ for all $j > i$.

2. The procedure of grouping

The rescaling from ordinal to cardinal values can be done by different methods, Gifi(1990), e.g., optimizes a loss function which considers the goodness-of-fit of the rescaled values to special reference values, Anderson (1984) assumes an underlying transformation distribution. Both procedures give good results, when used appropriately. But Anderson needs additional information, that is not always available. Gifi uses the existing information only implicitly, and does not use all available informations in some cases.

In this paper an intermediate way is suggested. It is supposed that no possible type of transformation function is known but as much information as possible is used explicitly for the grouping. The values to be grouped in one common class are determined by rank disturbances.

As mentioned in the previous section we have the following variables and notions:

- n objects $o_i, 1 \leq i \leq n$.
- unobservable, latent cardinal data $(x_i, y_i), 1 \leq i \leq n$, with x in ascending order $x_i \leq x_{i+1}$, x and y highly correlated,
- the corresponding observed ordinally measured values $(p_i, q_i), 1 \leq i \leq n$, with p in ascending order, $p_i \leq p_{i+1}$,
- in all cases where $p_i = p_{i+1}$, the objects o_i and o_{i+1} have the same rank, e.g. $k : r(o_i) = r_i = k = r(o_{i+1}) = r_{i+1}, k \in \mathbb{N}, 1 \leq k \leq m$, m is the number of different ranks resp. classes.

Either one of the following functions or suitable other functions can be taken to decide which neighboured classes will be grouped together:

$$\max_i \{q_i - q_{i+1}\} \tag{4}$$

$$\max_i \{m \mid q_{i+m} < q_i\}$$

$$\max_i \left|\left\{j \mid j < i \wedge q_j^{(m)} > q_i^{(m)}\right\} \cup \left\{j \mid j > i \wedge q_j^{(m)} < q_i^{(m)}\right\}\right|$$

$$\max_i \{ \sum_{j<i \wedge q_j^{(m)} > q_i^{(m)}} (q_j^{(m)} - q_i^{(m)}) + \sum_{j>i \wedge q_j^{(m)} < q_i^{(m)}} (q_i^{(m)} - q_j^{(m)})\}$$

In most numerical experiments the last function gives the best indication for the pair of object (o_i, o_{i+1}) to be grouped in one common class. Therefore we use it as decision function D.

The grouping consists of three steps. First step is a mere organisation step, in the second step new rank numbers $p_j^{(m)}$ are assigned of the to all objects o_j with higher ranks than $p_i^{(m)}$, and in the third step the new values $q_i^{(m)}$ for all elements in the newly created class are calculated as the median value of all elements belonging to the new class. These steps can be summarized as:

$$
\begin{aligned}
i_l &= \min\{j \mid p_j^{(m)} = p_i^{(m)}\} \\
i_u &= \max\{j \mid p_j^{(m)} = p_{i+1}^{(m)}\} \\
i_{med_l} &= (i_l + i_u)/2 \\
i_{med_u} &= (i_l + i_u + 1)/2
\end{aligned} \tag{5}
$$

m has to be reduced by one : $m := m - 1$

$$
p_j^{(m)} = \begin{cases} p_j^{m+1} & \text{for} \quad j < i \\ p_j^{m+1} - 1 & \text{for} \quad j > i \end{cases} \tag{6}
$$

$$
q_j^{(m)} = \begin{cases} q_j^{m+1} & \text{for} \quad j < i_l \wedge j > i_u \\ (q_{i_{med_l}}^{m+1} + q_{i_{med_u}}^{m+1})/2 & \text{for} \quad i_l \leq j \leq i_u \end{cases} \tag{7}
$$

The last part of this procedure step is to decide when the grouping procedure has to be stopped, which is important for the quality of the results. A fairly good suggestion is to use an elbow criterion on the the decision function. When the slope of this function becomes relatively small, the grouping procedure should be stopped.

At the end of the procedure or after every grouping step a rescaling of the $p_i^{(m)}$ values is done. Depending on the information on the x_i values this can be done in two different ways. If $x_i < x_{i+1}$ is known and the assigned values p_i having maintained this order, also $p_i \leq p_{i+1}$, then all objects o_i from one class $k = p_i^{(m)}$ should be rescaled to $\hat{x}_i^{(m)}$ equally distributed on the interval $[g(k - 0.5), g(k + 0.5)]$. If these conditions do not hold, then all objects o_i of the class $k = p_i^{(m)}$ should simply be rescaled to one single value $\hat{x}_i = h(k)$. In both cases the values $\hat{x}_1$ and $\hat{x}_n$ may be chosen arbitrarily or set to $\hat{x}_1 = x_1$ and $\hat{x}_n = x_n$ if these values are known. The linear function g and h may be formulated depending on the number of classes m with the constants $\hat{a}^{(m)}$ and $\hat{b}^{(m)}$

$$
\begin{aligned}
\hat{x}_i &= h(p_i^{(m)}) \\
&= \hat{a}^{(m)} + \hat{b}^{(m)} \cdot p_i^{(m)} \\
\textit{with} & \\
\hat{b}^{(m)} &= \frac{\hat{x}_n - \hat{x}_1}{p_n^{(m)} - p_1^{(m)}} \\
\hat{a}^{(m)} &= \hat{x}_1 - \hat{b}^{(m)} \cdot p_1^{(m)}
\end{aligned} \tag{8}
$$

3. Results

To show the effects of the method presented in this paper, we constructed a set of test data (x_i, y_i) by simulation, see Figure 1.

$i = p_i$	x_i	$x_i - x_{i-1}$	y_i	$y_i - y_{i-1}$	q_i
1	1.000	1.000	2.236	2.236	3.0
2	1.671	0.671	1.605	-0.630	1.0
3	2.613	0.943	5.208	3.603	4.0
4	3.397	0.783	6.641	1.434	7.0
5	3.736	0.339	1.893	-4.748	2.0
6	3.755	0.019	7.486	5.592	8.0
7	3.871	0.117	6.122	-1.363	5.0
8	4.414	0.543	8.825	2.703	13.0
9	5.319	0.905	10.653	1.828	16.0
10	6.190	0.871	7.810	-2.843	9.0
11	6.662	0.472	13.533	5.722	18.0
12	6.736	0.074	8.268	-5.264	10.0
13	6.780	0.044	6.551	-1.718	6.0
14	7.189	0.409	8.638	2.088	12.0
15	8.022	0.834	11.475	2.837	17.0
16	8.949	0.927	9.324	-2.151	15.0
17	9.553	0.604	13.903	4.579	20.0
18	9.714	0.161	8.919	-4.984	14.0
19	9.720	0.005	13.557	4.638	19.0
20	10.000	0.280	8.451	-5.107	11.0

Figure 1: Simulated data

The x-values where simulated by:

$$x_i \quad = \quad x_{i-1} + d \cdot |sin(\frac{i}{2})|$$

with d chosen so that $x_1 = 1$ and $x_n = 10$. The related y-values where simulated by

$$y_i \quad = \quad a + b \cdot x_i + u_i$$

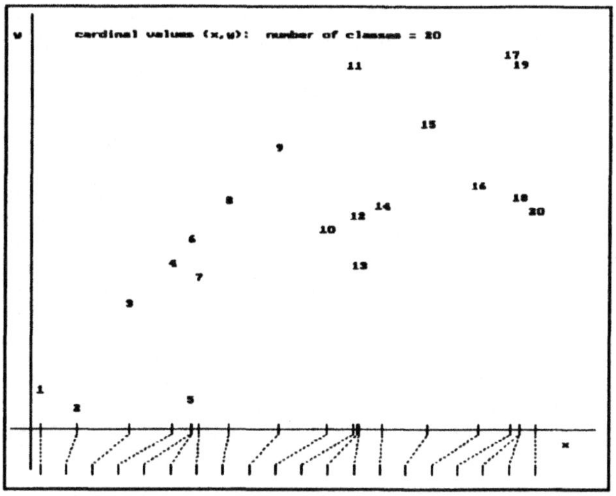

Figure 2: True cardinal values (x,y)

with u_i is a random $N(0, \sigma)$ distributed variable and $a = 2, \quad b = 1, \quad \sigma = 3.$

The following picture, Figure 2, shows the starting configuration using the true cardinal (x, y) - values. With these data the regression values $\hat{a} = 2.290$, $\hat{b} = 0.976$ and $R(x, y) = 0.759$ have been calculated.

The most interesting results of the grouping steps are presented in the Figure 3. Column D contains the decision value, that was the result of the calculation of the grouping criteria. The column L contains a measure for the goodness of the regrouping, what can be considered as loss function for the procedure, i.e:

$$L = \sum_{i=1}^{n} (x_i - \hat{x}_i)^2.$$

The $R(*, *)$ columns contain the Bravais Pearson type correlation coefficients for the variables included in the brackets. As an example of the use of the rescaled values $\hat{x}^{(m)}$ the regression coefficients $\hat{a}^{(m)}$ and $\hat{b}^{(m)}$, using $\hat{x}^{(m)}$ and y are calculated too.

Because of the difference in the D-value become relatively small, the procedure can be stopped at step $m = 12$, that means by having 12 groups. The ordinal values $(p^{(12)}, q^{(12)})$ show a fairly well rank order in both attributes, see Figure 4.

m	$R(p^{(m)}, q)$	$R(\hat{x}^{(m)}, q)$	$R(\hat{x}^{(m)}, y)$	L	$\hat{a}^{(m)}$	$\hat{b}^{(m)}$	D	i
20	0.7564	0.7564	0.7335	0.344	2.871	0.943	58.0	11
19	0.7684	0.7651	0.7438	0.318	2.657	0.977	49.0	19
18	0.7793	0.7725	0.7505	0.175	2.593	0.959	41.0	12
17	0.7900	0.7791	0.7614	0.160	2.339	0.992	30.7	9
16	0.7951	0.7796	0.7606	0.227	2.205	1.017	21.0	17
15	0.8098	0.7902	0.7720	0.168	1.985	1.013	15.0	4
14	0.8108	0.7867	0.7653	0.128	2.212	1.004	6.0	18
13	0.8203	0.7970	0.7754	0.125	2.057	0.985	6.0	6
12	0.8134	0.7853	0.7613	0.052	2.160	0.985	5.0	15
11	0.8222	0.7893	0.7722	0.188	1.787	1.003	4.5	10
10	0.8294	0.7973	0.7815	0.325	1.364	1.057	4.0	1
9	0.8350	0.7995	0.7771	0.150	1.980	1.002	3.4	8
8	0.8272	0.7846	0.7642	0.233	1.857	1.030	0.0	19

Figure 3: Values from each procedure step m

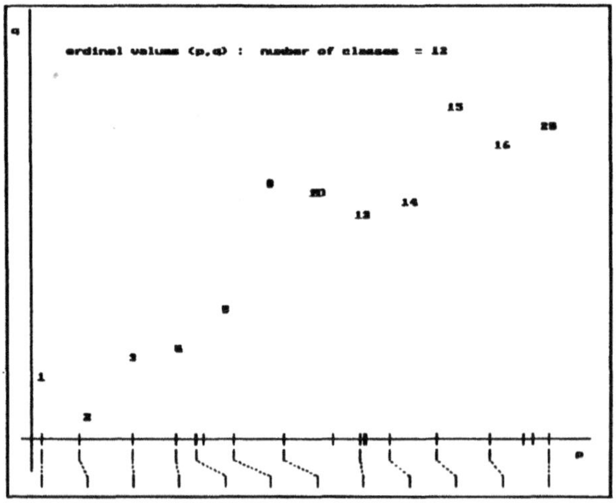

Figure 4: Ordinal values $(p^{(12)}, q^{(12)})$ resulting from 12 groups

Figure 5 shows the development of the rescaling. In the upper lines the rescaled values $\hat{x}^{(m)}$ are shown depending on the number of classes m, while the last line shows the true x - values.

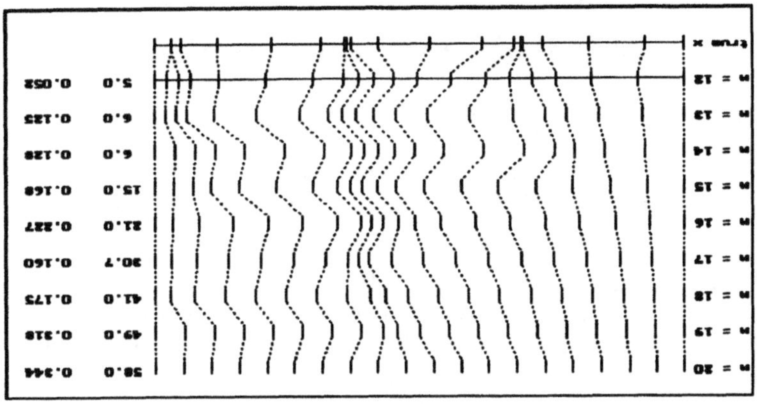

Figure 5: Rescaled values $\hat{x}^{(m)}$

4. Conclusions

The aim of the presented rescaling method is to approximate ordinaly measured cardinal data to their original cardinal values. A calculation, for example the linear regression, with these corrected values gives better results than the calculation with the ordinal data. The table in Figure 4 shows in the columns $\hat{a}^{(m)}$ and $\hat{b}^{(m)}$ the estimated values based on the different rescaling results depending on different classes. Using the ordinal data, the least square estimator for the parameters a and b are $\hat{a} = 2.871$ and $\hat{b} = 0.943$, while, e.g., $\hat{a}^{(12)} = 2.160$ and $\hat{b}^{(12)} = 0.985$. As the simulation used $a = 2$ and $b = 1$, the improvement by the rescaling is immediately visible.

References:

ANDERSON, J. A. (1984): Regression and ordered categorical variables. *J. R. Statist. Society B, 46, 1-30.*

GIFI, A. (1990): Nonlinear Multivariate Analysis. John Wiley & Sons. Chichester.

TUTZ, G. (1990): Modelle für kategoriale Daten mit ordinalem Skalenniveau; parametrische und nichtparametrische Ansätze. Vandenhoeck und Rupprecht , Göttingen.

WEISS, D. J.(editor) (1983): New Horizons in Testing; Latent Trait Test Theory and Computerized Adaptive Testing. Academic Press, New York, London.

Numerical Algorithms for Multidimensional Scaling

Michael W. Trosset

Department of Statistics,
University of Arizona, Tucson AZ 85721, United States

Summary: Multidimensional scaling (MDS) is a collection of techniques for constructing geometric configurations of points from information about interpoint distances. Important applications of MDS are routinely made in such diverse disciplines as psychophysics, in which the points typically correspond to sensory stimuli, and computational chemistry, in which the points typically correspond to atoms in a protein molecule. This paper surveys several fundamental issues in the computation of MDS configurations. Emphasis is placed on the formulation of the optimization problems that define different approaches to MDS and on the numerical algorithms that have been developed for solving these problems.

1. Introduction

The phrase *multidimensional scaling* (MDS) describes a vast collection of graphical techniques for visualizing the structure of certain types of data. Consider a finite collection of objects, e.g. cities, about which one has obtained information about the dissimilarities between pairs of objects, e.g. travel distances between cities. Then MDS attempts to construct a geometric configuration of points in such a way that points correspond to objects and interpoint distances correspond to dissimilarities.

Several excellent surveys have catalogued the considerable diversity of MDS methods. Carroll and Arabie (1980) proposed a very general taxonomy. De Leeuw and Heiser (1982) surveyed techniques for constructing Euclidean configurations; Arabie (1991) subsequently surveyed techniques for constructing non-Euclidean configurations. The purpose of the present paper is to survey some of the fundamental computational methods employed by MDS algorithms that construct Euclidean configurations. Since our survey must be brief, we restrict attention to a fairly narrow class of MDS problems. This restriction is made more palatable by the twin observations that the problems that we consider are central to any definition of MDS and that a careful exposition of the algorithms for even these familiar problems has not previously been collected in a single forum.

Our approach will be to define and classify MDS problems as optimization problems. Numerical algorithms for MDS can then be viewed as algorithms for numerical optimization problems with certain specific characteristics. It is not our intent to enumerate all of the myriad (many inferior) algorithms that have been proposed for the MDS problems of interest. Instead, we offer a thorough explication of certain fundamental ideas that often occur in the design of MDS algorithms. The reader will then be equipped to recognize and catalogue algorithms not mentioned in this survey, and even to juxtapose the ideas that we discuss to create new algorithms.

2. Preliminaries

For the purposes of this survey, a symmetric $n \times n$ matrix $\Delta = (\delta_{ij})$ is a *dissimilarity matrix* if $\delta_{ij} \geq 0$ and $\delta_{ii} = 0$. An $n \times n$ matrix $D = (d_{ij})$ is a p-dimensional Euclidean *distance matrix* if there exists a configuration of points $x_1, \ldots, x_n \in \Re^p$ such that d_{ij} is the Euclidean distance between x_i and x_j. We denote the set of all such matrices by $\mathcal{D}_n(p)$. We store the coordinates of x_i in row i of the $n \times p$ *configuration matrix* X and write $D(X)$ to denote the interpoint distances of X. Then, given a dissimilarity matrix Δ and a target dimension p, the purpose of (2-way) MDS is to construct a configuration matrix X for which the interpoint distances $D(X)$ somehow approximate the dissimilarities Δ.

2.1 Two Instructive Examples

To motivate certain considerations, it will help to refer to specific examples. The following examples not only illustrate some of the diverse problems to which MDS has been applied, but ultimately suggest that different applications may require different numerical algorithms.

Example 1: A famous psychophysical data set was collected by Ekman (1954), who presented all $m = 91$ pairs of $n = 14$ color stimuli that varied only in hue. Subjects rated each pair from 0 ("no similarity at all") to 4 ("identity"). Let s_{ij} denote the average similarity rating of pair ij and set $\delta_{ij} = 4 - s_{ij}$. Shepard (1962b) used MDS to recover "the familiar 'color circle' in two-dimensional space," which was viewed as a great triumph for the technique.

Example 2: Consider a molecule that contains n atoms. Nuclear magnetic resonance (NMR) spectroscopy can be used to measure distances between pairs of atoms. These measurements are the dissimilarities. Typically, roughly one third of the $m = n(n-1)/2$ interatomic distances cannot be measured, resulting in missing data. MDS is then used to construct a representation of the molecule in $\Re^3$.

We remark on several important differences between Examples 1 and 2. First, the number of objects in Example 1 is quite small ($n = 14$ color stimuli), as is typical of psychological applications of MDS. In contrast, the number of objects in Example 2 is rather large: a protein molecule will typically contain hundreds of atoms. It is apparent that the number of objects will directly affect the number of variables in the resulting optimization problems, so that computational efficiency will be of greater importance in applications like Example 2 than in applications like Example 1.

Second, the dissimilarities in Example 1 were obtained by averaging the perceptions of 31 human subjects. Respecting the imprecision of human observation, these perceptions were recorded on a crude, 5-point Likert scale. In contrast, the dissimilarities in Example 2 were obtained by NMR spectroscopy, which is considerably more precise than human observation. If we regard the data in Example 1 to be less precise than the data in Example 2, then it is natural to strive for more precise solutions in the latter case.

Finally, human perception is easily distorted. It is well known, for example, that human observers tend to underestimate large dissimilarities. In contrast, NMR spectroscopy is a relatively objective measurement procedure that is considerably less biased and less variable than human observation. Hence, we may be content to accept the dissimilarity data in Example 2 at face value, whereas we may be inclined to transform the dissimilarity data in Example 1 in an effort to mitigate the effects of distortion.

2.2 Taxonomy

We are now ready to examine some numerical algorithms for MDS. To explicate our taxonomy, it is necessary to distinguish between dissimilarities that represent observed data and dissimilarities that serve as free variables in optimization problems. We denote the former by Δ^0 and the latter, sometimes called *disparities*, by Δ. Our fundamental distinction is between problems in which we require $\Delta = \Delta^0$ and problems in which Δ is free to vary subject to constraints determined by Δ^0. We consider the former in Section 3, the latter in Section 4.

3. MDS with Fixed Dissimilarities

We first consider the case of fixed dissimilarities, sometimes called *metric* MDS. For $\Delta = \Delta^0$ fixed, we endeavor to solve optimization problems of the form

$$\begin{array}{ll} \text{minimize} & \rho(D, \Delta) \\ \text{subject to} & D \in \mathcal{D}_n(p), \end{array} \tag{1}$$

where ρ is some measure of discrepancy between distance matrices and dissimilarity matrices. A variety of measures have been studied in the MDS literature; we restrict attention to three of the most popular.

3.1 STRAIN

The so-called STRAIN criterion characterizes the first rigorous approach to MDS, proposed and studied by Torgerson (1952, 1958). This approach has been called *principal coordinate analysis* by Gower (1966) and is often called *classical* MDS. It is based on a well-known solution to the following problem from classical distance geometry.

Embedding Problem: Given an $n \times n$ dissimilarity matrix Δ and a target dimension p, determine if $\Delta \in \mathcal{D}_n(p)$. If it is, find an $n \times p$ configuration matrix X such that $\Delta = D(X)$.

The embedding problem was independently solved by Schoenberg (1935) and by Young and Householder (1938). Today, the standard solution is a slight variant introduced by Torgerson (1952).

Embedding Solution: Let $\tau : \Re^{n \times n} \to \Re^{n \times n}$ denote the linear *double centering* operator, defined by

$$b_{ij} = -\frac{1}{2}\left(a_{ij} - \bar{a}_{i\cdot} - \bar{a}_{\cdot j} + \bar{a}_{\cdot\cdot}\right).$$

Let $\Delta * \Delta = (\delta_{ij}^2)$ denote the Hadamard product of Δ with itself. Then $\Delta \in \mathcal{D}_n(p)$ if and only if $\tau(\Delta * \Delta) \in \Omega_n(p)$, the positive semidefinite matrices of rank $\leq p$. Furthermore, if X is any $n \times p$ configuration matrix such that $XX' = \tau(\Delta * \Delta)$, then $D(X) = \Delta$.

This solution suggests transforming D and Δ before measuring the discrepancy between them. The STRAIN criterion is therefore the objective function in the following optimization problem:

$$\begin{array}{ll} \text{minimize} & \|\tau(D * D) - \tau(\Delta * \Delta)\|_F^2 \\ \text{subject to} & D \in \mathcal{D}_n(p), \end{array} \tag{2}$$

where $\|\cdot\|_F$ denotes the Frobenius norm.

From a computational perspective, the great advantage of formulating MDS as Problem (2) is that one can then compute an explicit global solution. One simply computes the spectral decomposition

$$\tau(\Delta * \Delta) = Q\Lambda Q',$$

then sets

$$\bar{\lambda}_i = \left\{ \begin{array}{cc} \max(\lambda_i, 0) & i = 1, \ldots, p \\ 0 & i = p+1, \ldots, n \end{array} \right\}$$

and

$$X = Q\bar{\Lambda}^{1/2}.$$

Then $D(X)$ is a global minimizer of Problem (2), so that classical MDS is computationally trivial.

3.2 STRESS and SSTRESS

The popular STRESS and SSTRESS criteria were proposed in analogy to least squares regression. If one regresses the dissimilarities on the distances, then one obtains the STRESS criterion,

$$\rho(D, \Delta) = \|D - \Delta\|_F^2;$$

if one regresses the squared dissimilarities on the squared distances, then one obtains the smoother SSTRESS criterion,

$$\rho(D, \Delta) = \|D * D - \Delta * \Delta\|_F^2.$$

In analogy to weighted least squares, one typically allows a weighted norm in both cases.

In contrast to classical MDS, the minimization of STRESS or SSTRESS requires an iterative algorithm for numerical optimization. To manage the constraint that $D \in \mathcal{D}_n(p)$, it is standard to write $D = D(X)$. This eliminates the constraint (at the cost of complicating the objective function), resulting in an unconstrained optimization problem in the configuration coordinate variables. The new objective function is

$$\sigma_r(X) = 2\sum_{i<j} w_{ij} \left[\left(\sum_{k=1}^{p} (x_{ik} - x_{jk})^2 \right)^{r/2} - \left(\delta_{ij}^2\right)^{r/2} \right]^2,$$

where $r = 1$ corresponds to STRESS and $r = 2$ corresponds to SSTRESS.

Notice that the trick of writing $D = D(X)$ introduces both translational and rotational indeterminancy—if X is a solution, then so is any isometric transformation of X. Because connected sets of solutions degrade the performance of most numerical optimization algorithms, most algorithms for minimizing STRESS or SSTRESS employ some device for removing indeterminancy.

Numerous algorithms have been proposed for minimizing both STRESS and SSTRESS. A detailed critique of the most efficient algorithms for each criterion was offered by Kearsley, Tapia, and Trosset (1994). Here, we focus on two especially illustrative approaches.

Guttman (1968) noted that the stationary equation for the STRESS criterion can be written as $X = C(X)X$ and proposed the iterative substitution algorithm defined by $X^{k+1} = C(X^k)X^k$. A convergence analysis of this algorithm was supplied by de Leeuw (1977), de Leeuw and Heiser (1980), and de Leeuw (1988). If each $w_{ij}\delta_{ij} > 0$, then the Guttman sequence is globally convergent to a connected set of local stationary points.

An implementation of Guttman's algorithm is widely available in the popular SMACOF-I Fortran program documented by Heiser and de Leeuw (1986). In this implementation, each configuration coordinate is free to vary and indeterminancy is removed by centering the configuration and rotating to principal components. The algorithm terminates when an iterate fails to decrease σ_1 by at least a specified amount. This stopping criterion has a poor reputation in the numerical optimization community because it is prone to premature termination when the objective function is very flat near solutions.

It is widely known that Guttman's algorithm can also be written as a weighted gradient algorithm. Historically, gradient algorithms have been viewed as especially appropriate for MDS. Although second-order methods exhibit faster local convergence than first-order methods, Kruskal (1977) argued that this consideration was not compelling because "high accuracy solutions are almost never needed in this field due to the substantial random error which we typically find in the data." Of course, while this argument may apply to Example 1, it clearly does not apply to Example 2—one should note that applications of MDS to molecular conformation postdate Kruskal's remarks.

Noting that the problem of minimizing σ_r is a reasonably pleasant nonlinear least squares problem, Kearsley, Tapia, and Trosset (1994) undertook to develop an efficient second-order method for solving it. One of the advantages of this approach is that it simultaneously addresses both the STRESS and SSTRESS criteria. Their algorithm is a standard modification of Newton's method that uses analytic derivatives and incorporates a trust region globalization strategy. Indeterminancy is removed by fixing certain configuration coordinates and the algorithm terminates when either the gradient or the step length becomes sufficiently small in a certain relative sense.

Kearsley, Tapia, and Trosset (1994) conducted a set of numerical experiments designed to compare the performance of SMACOF-I and the Newton method for minimizing the STRESS criterion with weights $w_{ij} = 1$. The problems were large ($n = 394, 566, 122; p = 3$), designed to suggest Example 2. Predictably, Newton exhibited better local behavior than SMACOF-I, although at a cost of more cpu time. SMACOF-I exhibited better global behavior than Newton, although perhaps this is not of great practical importance due to the fact that an excellent initial configuration is usually available in the form of the classical solution.

Unlike SMACOF-I, Newton consistently found the same minimum STRESS value when started from different initial configurations—even when started from configurations at which SMACOF-I had terminated. This finding strongly suggests that SMACOF-I *does* have a tendency to premature termination and that second-order information and sophisticated stopping criteria are required to accurately minimize the STRESS criterion. Furthermore, this conclusion challenges the prevailing belief that the STRESS criterion often exhibits numerous local minimizers. Studies by Groenen (1993) and others have alleged empirical evidence of local minimizers using SMACOF-I as a local search procedure. However, it has not been analytically verified that any of these configurations actually are minimizers. How often SMACOF-I is trapped by a local minimizer—as opposed to how often it prematurely terminates—remains an open question of pressing importance.

4. MDS with Variable Dissimilarities

We now modify Problem (1) by allowing Δ to vary subject to constraints determined by the fixed dissimilarity matrix Δ^0. This results in optimization problems with a particular structure, which we explore in Section 4.1. In Section 4.2 we impose bound constraints on Δ, motivated by Example 2. In section 4.3 we impose order constraints (the case of nonmetric MDS), motivated by Example 1.

4.1 Optimization Strategies

The optimization problems that result from allowing Δ to vary are generally of the form

$$\begin{array}{ll} \text{minimize} & f(a,b) \\ \text{subject to} & a \in A,\ b \in B. \end{array} \tag{3}$$

These are problems in two sets of variables with no mixed constraints. If one fixes either a or b and optimizes the other, then the resulting subproblem is much easier to solve than Problem (3). (For example, if one fixes Δ and optimizes D, then one recovers the case of metric MDS.) In this section we describe two important strategies for exploiting this type of structure.

Variable Alternation: As detailed by de Leeuw (1993), the following simple strategy has been applied to a variety of statistical problems. In the context of MDS, it is usually described as the method of alternating least squares (ALS).

1. Fix b_0 and set $k = 1$.
2. Do until convergence:
   (a) $a_k = \text{argmin } f(a, b_{k-1})$
   (b) $b_k = \text{argmin } f(a_k, b)$
   (c) $k = k + 1$

It can usually be established that every accumulation point of a sequence produced by variable alternation is a fixed point, i.e. a pair (a_*, b_*) for which $a_* = \text{argmin } f(a, b_*)$ and $b_* = \text{argmin } f(a_*, b)$. However, there is no general guarantee that (a_*, b_*) solves Problem (3). Some appreciation for what can go wrong is obtained by examining the following simple example. Consider the problem of minimizing the Euclidean distance between the closed convex set

$$B = \{(x, y) : x \in [-1, 1], y = 1\}$$

and the closed nonconvex set

$$A = \left\{(x, y) : x \in [-1, 1], y = \left(1 - \sqrt{1 - x^2}\right)/2\right\}.$$

The projection of $a_* = (0, 0) \in A$ into B is the point $b_* = (0, 1)$. Furthermore, a_* is the point in A nearest b_*. Hence, (a_*, b_*) is a fixed point of variable alternation, despite the fact that a_* is the point in A that is *farthest* from B! Fortunately, in practice variable alternation often does converge to a local solution. A more practical limitation of the variable alternation strategy is that the speed of convergence is usually very slow.

Variable Reduction: Less common than variable alternation, the following strategy is usually employed when one of the subproblems can be solved explicitly.

1. For $\bar{a}$ fixed, define the *value function* by $b(\bar{a}) = \text{argmin } f(\bar{a}, b)$.
2. Minimize the *variable projection functional* $\bar{f}(a) = f(a, b(a))$.

Variable reduction eliminates one set of variables, albeit at the cost of complicating the objective function. The potential difficulty with this strategy is that the value function may not be differentiable, although when it is the derivatives of $\bar{f}$ usually have a very simple relation to the derivatives of f. Variable reduction also requires the user to use a possibly complicated numerical optimization algorithm to minimize the variable projection functional; however, a willingness to do so often leads to more efficient algorithms.

The distinction between variable alternation and variable reduction is potentially confusing because both strategies necessitate solving a sequence of subproblems of the form

$$\begin{aligned} &\text{minimize} \quad f(\bar{a}, b) \\ &\text{subject to} \quad b \in B. \end{aligned} \tag{4}$$

To discern the distinction, one should examine the sequence $\{a_k\}$. By definition, variable alternation always generates $b_k = \text{argmin } f(a_k, b)$ followed by $a_{k+1} = \text{argmin } f(a, b_k)$. In variable reduction, $\{a_k\}$ is obtained by employing some iterative numerical optimization algorithm to minimize $\bar{f}(a)$; the solution of Problem (4) is merely part of the computation required to evaluate $\bar{f}$.

A simple example is instructive. Consider the problem of minimizing the (squared) Euclidean distance between the closed convex sets

$$A = \{(x, y) : x \in (-\infty, \infty), y = 2\}$$

and

$$B = \left\{(x, y) : x^2 + y^2 \le 1\right\}.$$

Here, both subproblems have explicit solutions: the projection of $b = (x, y) \in B$ into A is given by $a(b) = (x, 2)$ and the projection of $a \in A$ into B is given by $b(a) = a/\|a\|$. The unique global solution is given by $a_* = (0, 2)$ and $b_* = (0, 1)$. In what follows, we evaluate performance by measuring the convergence of $\{a_k = (x_k, 2)\}$ to a_*, i.e. of $\{x_k\}$ to 0.

For this example, a theorem of Cheney and Goldstein (1959) guarantees that variable alternation will converge to the global solution. Let $b_0 = (1, 0)$. Then it is easily calculated that

$$x_k = \left[\frac{3}{4^k - 1}\right]^{1/2} \quad \text{and} \quad \lim_{k \to \infty} \frac{x_{k+1}}{x_k} = \frac{1}{2},$$

so that $\{x_k\}$ is linearly convergent.

To implement a variable reduction strategy, we write

$$\bar{f}(a) = \left\| a - \frac{a}{\|a\|} \right\|^2 = [\|a\| - 1]^2 = x^2 - 2(4 + x^2)^{1/2} + 5.$$

We can find a stationary point of this function by applying Newton's method, which has local quadratic convergence. The practical consequences of this acceleration are dramatic: from $x_1 = 1$, variable alternation produces $x_4 \doteq 1.085 \times 10^{-1}$ whereas variable reduction produces $x_4 \doteq 7.373 \times 10^{-11}$.

4.2 Bound Constraints

Consider Example 2. If δ_{ij}^0 is measured by NMR spectroscopy, then bounds on δ_{ij} can be inferred from the known accuracy of this much-studied measurement procedure. If δ_{ij}^0 is missing, then reasonable bounds on δ_{ij} can be inferred from the triangle inequality. We are thus motivated to allow the δ_{ij} to vary subject to constraints of the form

$$l_{ij} \leq \delta_{ij} \leq u_{ij},$$

resulting in the optimization problem

$$\begin{aligned} &\text{minimize} && \rho(D, \Delta) \\ &\text{subject to} && D \in \mathcal{D}_n(p),\ \Delta \in \square(L, U). \end{aligned} \tag{5}$$

The Data Box Algorithm proposed by Glunt, Hayden, and Raydan (1993) is now easily described: let ρ denote the STRESS criterion and apply variable alternation. The alternating subproblems are:

- The metric STRESS problem, which the authors solve by writing $D = D(X)$ and applying their own Spectral Gradient Method.
- Projection into the data box, for which an explicit formula is available.

It is also easy to suggest alternatives to the Data Box Algorithm. Obviously, one can substitute any metric STRESS algorithm for the Spectral Gradient Method. One can also use variable reduction instead of variable alternation: one simply substitutes the explicit projection formula into the objective function, resulting in an unconstrained optimization problem in X. Finally, one might contemplate modifying the problem itself by minimizing STRAIN instead of STRESS. Then *both* subproblems have explicit solutions, leading to a variety of possible algorithms.

4.3 Order Constraints

Consider Example 1. If the δ_{ij}^0 are measured by human perception, then the dissimilarities between pairs of objects may be distorted. It was to mitigate the effects of such distortion that Shepard (1962a,b) proposed allowing all monotone increasing transformations of the δ_{ij}^0. This is equivalent to using only the rank information in Δ^0; hence, ordinal dissimilarity data can be accommodated and the resulting techniques are termed *nonmetric* MDS.

The first rigorous formulation of nonmetric MDS was supplied by Kruskal (1964a,b), who recognized that Shepard's permitted transformations of Δ^0 define a closed convex set of Δ. Suppose that

$$\delta^0_{i_1 j_1} \leq \cdots \leq \delta^0_{i_m j_m}$$

and let $\mathcal{M}(\Delta^0)$ denote the set of dissimilarity matrices for which

$$\delta_{i_1 j_1} \leq \cdots \leq \delta_{i_m j_m}.$$

Then it seems natural to formulate nonmetric MDS as the following optimization problem:

$$\begin{aligned} &\text{minimize} \quad \rho(D, \Delta) \\ &\text{subject to} \quad D \in \mathcal{D}_n(p),\ \Delta \in \mathcal{M}(\Delta^0). \end{aligned} \tag{6}$$

Unfortunately, for the STRAIN, STRESS and SSTRESS criteria, Problem (6) has a degenerate global solution obtained by placing each $x_i \in \Re^p$ at the same location, resulting in $D(X) = 0 \in \mathcal{M}(\Delta^0)$. To preclude this solution, we modify Problem (6) by imposing a *nondegeneracy constraint*, e.g.

- $\|D\|_F^2 \geq c > 0$ • $\|D * D\|_F^2 \geq c > 0$
- $\|\Delta\|_F^2 \geq c > 0$ • $\|\Delta * \Delta\|_F^2 \geq c > 0$

Such constraints are usually managed by introducing a penalty term in the objective function, e.g. by setting $\rho(D, \Delta) = \|D - \Delta\|^2 / \|D\|^2$. This is usually justified by interpreting the penalty term as a scaling factor that normalizes the solution, as in Kruskal (1964a) and Kruskal and Carroll (1968). For our purposes, however, it is merely one of various possible approaches to constrained optimization.

We can now conclude our survey by providing succinct descriptions of two popular algorithms for 2-way nonmetric MDS. It should be noted that both of the programs that we mention are also capable of 3-way nonmetric MDS.

ALSCAL: This algorithm was proposed by Takane, Young, and de Leeuw (1977) and is widely available as a procedure in the Statistical Package for the Social Sciences (SPSS).

- Divide SSTRESS by $\|\Delta * \Delta\|_F^2$ to preclude degenerate solutions.
- Write $D = D(X)$ to manage the constraint $D \in \mathcal{D}_n(p)$.
- Apply variable alternation to minimize

$$f(X, \Delta) = \frac{\|D(X) * D(X) - \Delta * \Delta\|_F^2}{\|\Delta * \Delta\|_F^2}$$

subject to $\Delta \in \mathcal{M}(\Delta^0)$.

- One subproblem is metric MDS with the SSTRESS criterion; the other subproblem can be solved explicitly.

KYST: This program is distributed by AT&T Bell Laboratories and was documented by Kruskal, Young, and Seery (1973).

- Divide STRESS by $\|D\|_F^2$ to preclude degenerate solutions.
- Write $D = D(X)$ to manage the constraint $D \in \mathcal{D}_n(p)$.
- Let π denote projection into $\mathcal{M}(\Delta^0)$. This is isotonic regression (Barlow, Bartholomew, Bremner, and Brunk, 1972), for which various finite algorithms exist.
- Minimize the variable projection functional

$$\bar{f}(X) = \frac{\|D(X) - \pi D(X)\|_F^2}{\|D(X)\|_F^2}$$

by a gradient method.

Acknowledgements

This material was presented in a plenary lecture at the 20th Annual Conference of the Gesellschaft für Klassifikation e.V., held at the University of Freiburg, March 6–8, 1996. I thank the organizers for inviting me to participate and J. Douglas Carroll, Patrick Groenen, and Rudolf Mathar for many helpful discussions.

References:

ARABIE, P. (1991): Was Euclid an unnecessarily sophisticated psychologist? *Psychometrika, 56, 567–587.*

BARLOW, R. E.; BARTHOLOMEW, J. M.; BREMNER, J. M. and BRUNK, H. D. (1972): *Statistical Inference Under Order Restrictions.* John Wiley & Sons, New York.

CARROLL, J. D. and ARABIE, P. (1980): Multidimensional scaling. *Annual Review of Psychology, 31, 607–649.*

DE LEEUW, J. (1977): Applications of convex analysis to multidimensional scaling. In: J. R. Barra, F. Brodeau, G. Romier, and B. van Cutsem (eds.): *Recent Developments in Statistics*, North-Holland Publishing Co., Amsterdam, 133–145.

DE LEEUW, J. (1988): Convergence of the majorization method for multidimensional scaling. *Journal of Classification, 5, 163–180.*

DE LEEUW, J. and HEISER, W. (1980): Multidimensional scaling with restrictions on the configuration. In: P. R. Krishnaiah (ed.): *Multivariate Analysis*, North-Holland Publishing Co., Amsterdam, 501–522.

DE LEEUW, J. and HEISER, W. (1982): Theory of multidimensional scaling. In: P.R. Krishnaiah and I.N. Kanal (eds.): *Handbook of Statistics*, North-Holland Publishing Co., Amsterdam, 285–316.

EKMAN, G. (1954): Dimensions of color vision. *Journal of Psychology, 38, 467–474.*

GLUNT, W.; HAYDEN, T. L. and RAYDAN, M. (1993): Molecular conformations from distance matrices. *Journal of Computational Chemistry, 14, 114–120.*

GOWER, J. C. (1966): Some distance properties of latent root and vector methods in multivariate analysis. *Biometrika, 53, 315–328.*

GROENEN, P. J. F. (1993): *The Majorization Approach to Multidimensional Scaling.* DSWO Press, Leiden.

GUTTMAN, L. (1968): A general nonmetric technique for finding the smallest coordinate space for a configuration of points. *Psychometrika, 33, 469–506.*

HEISER, W. J. and DE LEEUW, J. (1986): *SMACOF-I.* Technical Report UG-86-02, Department of Data Theory, University of Leiden, Leiden.

KEARSLEY, A. J.; TAPIA, R. A. and TROSSET, M. W. (1994): *The Solution of the Metric STRESS and STRESS Problems in Multidimensional Scaling Using Newton's Method.* Technical Report TR94-44, Department of Computational & Applied Mathematics, Rice University, Houston, TX. To appear in *Computational Statistics.*

KRUSKAL, J. B. (1964a): Multidimensional scaling by optimizing goodness of fit to a nonmetric hypothesis. *Psychometrika, 29, 1–27.*

KRUSKAL, J. B. (1964b): Nonmetric multidimensional scaling: a numerical method. *Psychometrika, 29, 28–42.*

KRUSKAL, J. B. (1977): Multidimensional scaling and other methods for discovering structure. In: K. Enslein, A. Ralston, and H.S. Wilf (eds.): *Statistical Methods for Digital Computers*, John Wiley & Sons, New York, 296–339.

KRUSKAL, J. B. and CARROLL, J. D. (1968): Geometrical models and badness-of-fit functions. In: P.R. Krishnaiah (ed.): *Multivariate Analysis*, Academic Press, New York, 639–671.

KRUSKAL, J. B.; YOUNG, F. W. and SEERY, J. B. (1973): *How to use KYST, a very flexible program to do Multidimensional Scaling and Unfolding.* Technical Report, AT&T Bell Laboratories, Murray Hill, NJ.

SCHOENBERG, I. J. (1935): Remarks to Maurice Fréchet's article "Sur la définition axiomatique d'une classe d'espaces distanciés vectoriellement applicable sur l'espace de Hilbert". *Annals of Mathematics, 38, 724–732.*

SHEPARD, R. N. (1962a): The analysis of proximities: multidimensional scaling with an unknown distance function. I. *Psychometrika, 27, 125–140.*

SHEPARD, R. N. (1962b): The analysis of proximities: multidimensional scaling with an unknown distance function. II. *Psychometrika, 27, 219–246.*

TAKANE, Y.; YOUNG, F. W. and DE LEEUW, J. (1977): Nonmetric individual differences multidimensional scaling: an alternating least squares method with optimal scaling features. *Psychometrika, 42, 7–67.*

TORGERSON, W. S. (1952): Multidimensional scaling: I. Theory and method.

Psychometrika, 17, 401–419.

TORGERSON, W. S. (1958): *Theory and Methods of Scaling.* John Wiley & Sons, New York.

YOUNG, G. and HOUSEHOLDER, A. S. (1938): Discussion of a set of points in terms of their mutual distances. *Psychometrika, 3, 19–22.*

High Dimensional Clustering Using Parallel Coordinates and the Grand Tour

Edward J. Wegman, Qiang Luo

Center for Computational Statistics George Mason University Fairfax, VA 22030 USA

Summary: In this paper, we present some graphical techniques for cluster analysis of high-dimensional data. Parallel coordinate plots and parallel coordinate density plots are graphical techniques which map multivariate data into a two-dimensional display. The method has some elegant duality properties with ordinary Cartesian plots so that higher-dimensional mathematical structures can be analyzed. Our high interaction software allows for rapid editing of data to remove outliers and isolate clusters by brushing. Our brushing techniques allow not only for hue adjustment, but also for saturation adjustment. Saturation adjustment allows for the handling of comparatively massive data sets by using the α-channel of the Silicon Graphics workstation to compensate for heavy overplotting.

The grand tour is a generalized rotation of coordinate axes in a high-dimensional space. Coupled with the full-dimensional plots allowed by the parallel coordinate display, these techniques allow the data analyst to explore data which is both high-dimensional and massive in size. In this paper we give a description of both techniques and illustrate their use to do inverse regression and clustering. We have used these techniques to analyze data on the order of 250,000 observations in 8 dimensions. Because the analysis requires the use of color graphics, in the present paper we illustrate the methods with a more modest data set of 3848 observations. Other illustrations are available on our web page.

1. Introduction

Visualization of high-dimensional, multivariate data has enjoyed a considerable development with the introduction of the grand tour by Asimov (1985) and Buja and Asimov (1985) and the parallel coordinate display by Inselberg (1985) and Wegman (1990). The former technique is an animation methodology for viewing two-dimensional projections of general d-dimensional data where the animation is determined by a space-filling curve through all possible orientations of a two-dimensional coordinate system in d-space. Viewed as a function of time, the grand-tour animation reveals interesting projections of the data, projections that reveal underlying structure. This in turn allows for the construction of models of data's underlying structure.

The parallel coordinate display is in many senses a generalization of a two-dimensional Cartesian plot. The idea is to sacrifice orthogonal axes by drawing the axes parallel to each other in order to obtain a planar diagram in which each d-dimensional point has a unique representation. Because of elegant duality properties, parallel coordinate displays allow interpretations of statistical data in a manner quite analogous to two-dimensional Cartesian scatter plots. Wegman (1991) formulated a general d-dimensional form of the grand tour and suggested using the parallel coordinate plot as a visualization tool for the general d-dimensional animation. We have found this combination of multivariate visualization tools to be extraordinarily effective in the exploration of multivariate data. In Section 2, we briefly describe parallel coordinate displays including interpretation of parallel coordinate displays for detecting clusters. In Section 3, we describe the generalized d-dimensional grand tour and a partial sub-dimensional grand tour. In Section 4, we discuss brushing with hue and saturation including a discussion of perceptual considerations for visual presentation. Finally we close in Section 5 with a sequence of illustrations of the use of these techniques to remove noise and isolate clusters in a five-dimensional data set.

2. Parallel Coordinate and Parallel Coordinate Density Plots

The parallel coordinate plot is a geometric device for displaying points in high- dimensional spaces, in particular, for dimensions above three. As such, it is a graphical alternative to the conventional scatterplot. The parallel coordinate density plot is closely related and addresses the situation in which there would be heavy overplotting. In this circumstance, the parallel coordinate plot is replaced with its density and so is much more appropriate for very large, high-dimensional data sets. In place of the conventional scatter plot which tries to preserve orthogonality of the d-dimensional coordinate axes, draw the axes as parallel. A vector $(x_1, x_2, \ldots, x_d)$ is plotted by plotting x_1 on axis 1, x_2 on axis 2 and so on through x_d on axis d. The points plotted in this manner are joined by a broken line. The principal advantage of this plotting device is that each vector $(x_1, x_2, \ldots, x_d)$ is represented in a planar diagram in which each vector component has essentially the same representation.

The parallel coordinate representation enjoys some elegant duality properties with the usual Cartesian orthogonal coordinate representation. Consider a line $\mathcal{L}$ in the Cartesian coordinate plane given by $\mathcal{L}$: $y = mx + b$ and consider two points lying on that line, say $(a, ma + b)$ and $(c, mc + b)$. Superimpose Cartesian coordinate axes t, u on the xy parallel axes so that the y parallel axis has the equation $u = 1$. The point $(a, ma + b)$ in the xy Cartesian system maps into the line joining $(a, 0)$ to $(ma + b, 1)$ in the tu coordinate axes. Similarly, $(c, mc + b)$ maps into the line joining $(c, 0)$ to $(mc + b, 1)$. A straightforward computation shows that these two lines intersect at a point (in the tu plane) given by $\bar{\mathcal{L}} : (b(1 - m)^{-1}, (1 - m)^{-1})$. This point in the parallel coordinate plot depends only on m and b, the parameters of the original line in the Cartesian plot. Thus $\bar{\mathcal{L}}$ is the dual of $\mathcal{L}$ and one has the interesting duality result that points in Cartesian coordinates map into lines in parallel coordinates while lines in Cartesian coordinates map into points in parallel coordinates. This duality is discussed in further detail in Wegman (1990).

The point-line, line-point duality seen in the transformation from Cartesian to parallel coordinates extends to conic sections. The most significant of these dualities from a statistical point of view is that an ellipse in Cartesian coordinates maps into a hyperbola in parallel coordinates. A distribution which has ellipsoidal level sets would have hyperbolic level sets in the parallel coordinate presentation. It should be noted that the quadratic form does not describe a locus of points, but a locus of lines, a line conic. The notion of a line conic is, perhaps, a strange notion. By this is meant a locus of lines whose coordinates satisfy the equation for a conic. These may be more easily related to the usual notion of a conic when it is realized that the envelope of this line conic is a point conic. As mentioned there is a duality between points and lines and between conics and conics. It is worthwhile to point out two other nice dualities. Rotations in Cartesian coordinates become translations in parallel coordinates and vice versa. Perhaps more interesting from a statistical point of view is that points of inflection in Cartesian space become cusps in parallel coordinate space and vice versa. Thus the relatively hard-to-detect inflection point property of a function becomes the notably more easy to detect cusp in the parallel coordinate representation. Inselberg (1985) discusses these properties in detail.

Since ellipses map into hyperbolas, one has an easy template for diagnosing uncorrelated data pairs. With a completely uncorrelated data set, one would expect the 2-dimensional scatter diagram to fill substantially a circumscribing circle. The parallel coordinate plot would approximate a figure with a hyperbolic envelope. As the correlation approaches negative one, the hyperbolic envelope would deepen so that in the limit one would have a pencil of lines, what is called by Wegman (1990) the cross-over effect.

Most importantly for the present paper, it should be noted that clustering is easily diagnosed using the parallel coordinate representation. The individual parallel coordinate axes represent one-dimensional projections of the data. Thus, separation between or among sets of data on any one axis or between any pair of axes represents a view of the data which isolates clusters. An elementary view of this idea is seen in Figure 1, where we illustrate the appearance of three distinct clusters in a four dimensional space. Because of the connectedness of the multidimensional parallel coordinate diagram, it is usually easy to see whether or not this clustering propagates through other dimensions.

Some of the data analysis features of the parallel coordinate representation include the ability to diagnose one-dimensional features such as **marginal densities**, two-dimensional features such as **correlations** and **nonlinear structures**, and multi-dimensional features such as **clustering, hyperplanes**, and the **modes**. These interpretations are discussed in more detail in Wegman (1990) while parallel coordinate density plots are discussed in Miller and Wegman (1991).

3. The Grand Tour Algorithm in d-space

Let $\boldsymbol{e}_j = (0, 0, \ldots, 0, 1, 0, \ldots, 0)$ be the canonical basis vector of length d. The 1 is in the jth position. The $\boldsymbol{e}_j$ are the unit vectors for each coordinate axis in the initial position. We want to do a general rigid rotation of these axes to a new position with basis vectors $\boldsymbol{a}_j(t) = (a_1^j(t), a_2^j(t), \ldots, a_d^j(t))$, where, of course, t is a time index. The strategy then is to take the inner product of each data point, say $\boldsymbol{x}_i$, $i = 1, \ldots, n$ with the basis vectors, $\boldsymbol{a}_j(t)$. The j subscript on $\boldsymbol{a}_j(t)$ means that $\boldsymbol{a}_j(t)$ is the image under the generalized rotation of the canonical basis vector $\boldsymbol{e}_j$. Thus the data vector $\boldsymbol{x}_i$ is $(x_1^i, x_2^i, \ldots, x_d^i)$, so that the representation of $\boldsymbol{x}_i$ in the $\boldsymbol{a}_j$ coordinate system is

$$\boldsymbol{y}_i(t) = (y_1^i(t), y_2^i(t), \ldots, y_d^i(t)), i = 1, \ldots, n$$

where

$$y_j^i(t) = \sum_{k=1}^{d} x_k^i a_k^j(t),$$

$j = 1, \ldots, d$ and $i = 1, \ldots, n$. The vector $\boldsymbol{y}_i(t)$ is then the linear combination of basis vectors representing the ith data point in the rotated coordinate system at time t.

The goal thus is to find a generalized rotation, Q, such that $\boldsymbol{Q}(\boldsymbol{e}_j) = \boldsymbol{a}_j$. We can think of Q as either a function or as a matrix Q where $\boldsymbol{e}_j \times Q = \boldsymbol{a}_j$. We implement this by choosing Q as an element of the special orthogonal group denoted by SO(d) of orthogonal $d \times d$ matrices having determinant of +1. In order to find a continuous, space-filling path through the Grassmannian manifold of d-flats, we must find a continuous, space-filling path through the SO(d).

In general d-dimensional space, there are $d - 2$ axes orthogonal to each two-flat. Thus rather than rotating around an axis as we are used to in ordinary three-dimensional space, we must rotate in a plane in d-dimensional space. The generalized rotation matrix, Q, is built up from a series of rotations in individual two-flats. In d-space, there are d canonical basis vectors and, thus, $\binom{d}{2} = \frac{1}{2}(d^2 - d)$ distinct two-flats formed by the canonical basis vectors. We let $R_{ij}(\theta)$ be the element of SO(d) which rotates in the $e_i e_j$ plane through an angle of θ. We define Q by

$$\boldsymbol{Q}(\boldsymbol{\theta}_{1,2}, \boldsymbol{\theta}_{1,3}, \ldots, \boldsymbol{\theta}_{d-1,d}) = \boldsymbol{R}_{12}(\boldsymbol{\theta}_{1,2}) \times \cdots \times \boldsymbol{R}_{d-1,d}(\boldsymbol{\theta}_{d-1,d}).$$

There are $p = \frac{1}{2}(d^2 - d)$ factors. The restrictions on θ_{ij} are $0 \leq \theta_{ij} \leq 2\pi$, $1 \leq i < j \leq d$. The vector $(\theta_{1,2}, \theta_{1,3}, \ldots, \theta_{d-1,d})$ can thus be thought of as a point on a p-dimensional torus. This is the origin of the description of this method as the torus method. The individual factors $R_{ij}(\theta)$ are $d \times d$ matrices given by

$$\begin{bmatrix} 1 & \cdots & 0 & \cdots & 0 & \cdots & 0 \\ \vdots & & \vdots & & \vdots & & \vdots \\ 0 & \cdots & \cos(\theta) & \cdots & -\sin(\theta) & \cdots & 0 \\ \vdots & & \vdots & & \vdots & & \vdots \\ 0 & \cdots & \sin(\theta) & \cdots & \cos(\theta) & \cdots & 0 \\ \vdots & & \vdots & & \vdots & & \vdots \\ 0 & \cdots & 0 & \cdots & 0 & \cdots & 1 \end{bmatrix}$$

where the cosine and sine entries are in the ith and j^{th} columns and rows.

The final step in the algorithm is to describe a space filling path on the p-dimensional torus, T^p. This can be done by a mapping α: $R \to T^p$ given by

$$\alpha(t) = (\lambda_1 t, \lambda_2 t, \ldots, \lambda_p t)$$

where $\lambda_1, \ldots, \lambda_p$ is a sequence of mutually irrational real numbers and the $\lambda_i t$ are interpreted modulo 2π. The composition of α with Q will describe a space filling path in $\mathrm{SO}(d)$. Thus our final algorithm is given by

$$\boldsymbol{a}_j(t) = \boldsymbol{e}_j \times \boldsymbol{Q}(\lambda_1 t, \ldots, \lambda_p t)$$

The canonical unit vector for each coordinate axis at time t described by the grand tour algorithm is an orthogonal linear combination, $\boldsymbol{a}_j(t) = \boldsymbol{e}_j \times \boldsymbol{Q}(t)$, of the original unit vectors. This has several important implications for the utility of this methodology. First, it should be immediately clear that one can do a grand tour on any subset of the original coordinate axes simply by fixing the appropriate two-planes in the rotation matrix given above by Q. That is, if we wish the jth variable to not be included in the grand tour rotation, we simply put a 1 in the $\boldsymbol{Q}_{jj}(t)$ entry with 0 in the remaining positions in the jth row and the jth column. Thus it is straightforward to do a partial grand tour. The interest in doing a partial grand tour will be discussed in the next section.

The second important implication relates to the connection with the parallel coordinate display. An immediate concern with the parallel coordinate display is the preferential ordering of the axes. In our discussion above we indicated that the axis for variable one is adjacent to the axis for variable two, but not for variable three. In general the axis for variable j is adjacent to the axes for variables $j-1$ and $j+1$ but for no other axes. It is easy to see pair wise relationships for adjacent variables, but less easy for non-adjacent variables. Wegman (1990) has a substantial discussion on methods for considering all possible permutations. This concern is immaterial when one does the grand tour since eventually $\boldsymbol{a}_j(t) = \boldsymbol{e}_j \times \boldsymbol{Q}(t) = \boldsymbol{e}_i$, for every i. Thus eventually every possible permutation of the axes will appear in the grand tour.

4. Some Additional Visualization Devices

4.1 Brushing with Hue and Saturation

A powerful method in high interaction graphics is the brushing technique. The idea is to isolate clusters or other interesting subsets of a data set by, in effect, painting that subset with a color. This is usually done in two settings: 1) with co-plots and 2) with animations. The brushed color becomes an attribute of the data point and is maintained in all representations. The idea of co-plots is that a particular data set may be presented in more that one way, for example in a scatter plot matrix or say with a scatter plot, a histogram and a dot plot. Points colored the same way in all presentations allow the data analyst the ability to track coherent clusters or subsets of the data through different representations. Of course, with an animation, the coloring allows the data analyst to follow clusters or subsets of the data through the time evolution of the animation.

In general, colors may assume a range of saturations depending on the relative proportion of gray to chroma. We implement the following device. We de-saturate the hue with black so that the brushing color is nearly black when desaturated. This by itself would not be fully useful. However, when points are overplotted, we add the hue components. Say, for example, we use an approximate one and one half percent hue component of blue. This would mean (on an eight bit scale) approximately 2 bits of blue and 0 bits each of red and green. Thus if approximately 67 observations were overplotted at a given pixel, that pixel would be fully saturated with blue. Fewer observations mean a less saturated color. The level of saturation of the brushing color is controllable by the user. Larger data sets suggest lower saturation levels. The level of saturation thus reflects the degree of overplotting. This device is in essence a way of creating a parallel coordinate (or any other kind of) density plot. (See Miller and Wegman, 1991). The advantage of this technique for creating a density plot is that it does not depend on smoothing algorithms so that individual data points are still resolvable.

The addition of saturations is implementable in hardware on Silicon Graphics workstations by means of the α-channel. The α-channel is a hardware device for blending pixel intensities and has its primary use for transparency algorithms. However, by blending pixels intensities of the same color, we can in effect add the pixel intensities and achieve brushing with hue and saturation with no speed penalty whatsoever. This technique is incredibly powerful in resolving structure in large data sets with heavy overplotting as we hope to illustrate in the next section.

4.2 Some Perceptual Issues

Brushing with hue and saturation leads to an interesting question concerning perception of the resulting plots. When viewed against a black background, the low saturation observations, i.e. those that are not heavily overplotted, blend with that background. This is quite useful when trying to understand the internal structure of the high density regions of the plot. Our usual technique is to brush with a white (actually a very dark gray) color. Then the internal structure appears as white in the highest density regions as illustrated in Figure 4. The resulting plot looks rather like an x-ray of the internal structure of the data set.

When viewed against a white background, the low saturation level observations are nearly black and so are quite visible. This is extremely useful when looking for outliers, which would tend to be invisible against the black background. The white background is also extremely useful for data editing. Our implementation on the Silicon Graphics workstations supports a scissors feature so that we can prune away low density regions. This feature allows for rapid visual data editing which may be useful for eliminating outliers, transcription errors, and data with missing values that could impair the ability to reach sensible conclusions. Obviously, when using a white background, it is important to brush with some hue, since when brushed with a gray the result would be that highest density regions would be white again blending with the background. Because the apparent brightness of normal hues (red, blue, green, etc.) is lower than the apparent brightness of white, the internal structure of the data set is less apparent with a white background than with a black background. Thus it is clear that both backgrounds have their utility depending on the task at hand.

4.3 Visual Regression and Clustering Decision Rules

The combination of hue and saturation brushing and the partial grand tour creates a device for visual regression and clustering decision rules. Consider a response variable of interest, let us say, for example, profit in a financial setting. Let us suppose we wish to answer the following question, "What combination of customer demographics variables is likely to cause the corporation to lose money?" We brush the profit variable as follows: for negative profits, we brush the observations red, for positive profits we brush them green. Where the variables overlap, the combination of red and green sum to yellow. Where there are observations primarily leading to losses, the result will be generally red and where profits, primarily green. Since we are interested in the covariates leading to profit, we fix the profit variable so that it does not enter into the partial grand tour rotation. We may rotate on any combination of explanatory covariates we wish. For example, we may have data on customer's average account balance, sex, race, age and annual income. While all of these may affect the profitability of the corporation, the prohibitions against discrimination on the basis of race and sex would lead us to generate decision rules which do not consider these factors. Similarly, some data may be extremely difficult or expensive to collect. Thus while it may be an extremely helpful covariate, it may be missing so often that its value is substantially diminished in forming decision rules.

The partial grand tour is done on the explanatory covariates of interest while keeping the response variable and any other explanatory covariates that we wish to exclude fixed. Because the grand tour automatically forms orthogonal linear combinations of desired explanatory covariates, the color coding allows us in effect to see the response variable in terms of the orthogonal linear combinations of the explanatory variables. Thus when we see a linear combination of explanatory variables that is intensely red in our example, we know that this is a combination of variables which leads to a negative profit. We can thus isolate the range of the linear combination of covariates that is colored red and this will be a component of the decision tree in terms of demographic variables that causes the organization to lose money. We can then edit that particular cluster of observations from our data set and resume the partial grand tour. Repeating this process recursively allows us to determine a sequence of decision rules that isolate customers likely to cause financial loss to the organization. Because this methodology is so intensively dependent on color, it is not possible to easily illustrate these techniques in this paper. However, we have included an example based on this idea as well as other examples in our web server at URL

http://www.galaxy.gmu.edu/images/gallery/research_arcade.html.

This method thus leads to a high interaction technique for rapidly identifying a decision rule based on visual display. The rules are sophisticated in the sense that they need not be simple binary decision rules and they need not be based on simply the original covariates.

5. An Example

In this example we would like to consider a synthetic dataset about the geometric features of pollen grains. There are 3848 observations on 5 variables. This data is the 1986 ASA Data Exposition dataset, made up by David Coleman of RCA Labs. The data set is available from STALIB at URL=http://www.stat.cmu.edu/datasets/. Figure 2 is the scatterplot matrix for this five-dimensional data. Note that in all presentations the data appear to have elliptical contours. This is true even when all five variables are rotated through the grand tour. This is suggestive of the fact that the point cloud is sampled from a five-dimensional multivariate distribution with ellipsoidal level sets, perhaps a multivariate normal. Figure 3 is the corresponding parallel coordinate display with pure black on a white background. In this display each of the five variables have been rescaled so as to fill the parallel coordinate axes. Note that in the parallel coordinate display, the variables exhibit hyperbolic envelopes, the dual of elliptical contours in Cartesian plots. This confirms our observation of the ellipsoidal level sets.

Figure 4 represents a highly desaturated version of the same parallel coordinate plot, this time white on a black background. With this desaturated view, it is clear that there is an interesting internal structure buried in the noise. Figure 5 represents a partially pruned view with much of the noise removed. Figure 6 is the result of a second pruning edit in which the internal structure is fully revealed. In both Figures 5 and 6 the data have been rescaled to fill the axes. Figure 7 represents the results of grand tour in which it is clear from the gaps seen in axes two, three and four that this data forms six clusters separable in at least three dimensions of the five. Our software permits this edit to be accomplished in less than three minutes. Figure 8 displays the edited data in a scatter plot display. The 99 remaining points from the original 3848 points are perfectly isolated from the noise and spell the word EUREKA. The six letters are of course the six clusters isolated in Figure 7. The 99 points are only about 2.7% of the data set and yet we were able to isolate these points in six clusters using the techniques described in this paper.

Acknowledgment

This work was supported by the U.S. Army Research Office under Grant DAAH04-94-G-0267. The original SGI versions of ExplorN and Explor4 were developed in the Center for Computational Statistics under Dr. Daniel B. Carr's NSF Grant DMS-9107188. The objective included implementation of methods previously described by Dr. Carr and Dr. Edward Wegman in a powerful computing environment. The software also reflects the insights and contributions of students carrying out the implementation: Ji Shen, Barna Takacs and most notably Qiang Luo. The version of ExplorN shown in this paper emphasizes subsequent enhancements provided by Qiang Luo and stresses facets of ExplorN most intimately connected to Dr. Wegman's research. We thank Shan-chuan Li for converting our manuscript to TEX form.

References:

ASIMOV, D. (1985): The grand tour: a tool for viewing multidimensional data, *SIAM J. Sci. Statist. Comput., 6, 128-143.*

BUJA, A. and ASIMOV, D. (1985): Grand tour methods: an outline, *Computer Science and Statistics: Proceedings of the Seventeenth Symposium on the Interface,* 63-67, (D. Allen, ed.), New York: North Holland Publishing Company.

INSELBERG, A. (1985): The plane with parallel coordinates, *The Visual Computer, 1, 69-91.*

MILLER, J. J. and WEGMAN, E. J. (1991),: Construction of line densities for parallel coordinate plots, *Computing and Graphics in Statistics, (A. Buja and P. Tukey, eds.), 107-123, Springer-Verlag: New York.*

WEGMAN, E. J. (1990): Hyperdimensional data analysis using parallel coordinates, *J. American Statist. Assoc., 85, 664-675.*

WEGMAN, E. J. (1991): The grand tour in k-dimensions, *Computing Science and Statistics: Proceedings of the 22nd Symposium on the Interface, 127-136.*

Section 2

Neural Networks and Pattern Recognition

A Symbolic Representation for Patterns in Time Series Using Definitive Clause Grammars

G. Guimarães, A. Ultsch

Fachbereich Mathematik/ FG Informatik Philipps Universität Marburg,
D-35032 Marburg
e-mail:guimas, ultsch@informatik.uni-marburg.de
http://www.mathematik.uni-marburg.de/~ wina/

Summary: Recently in different application domains, like medicine, industrial processes, etc., a great amount of data is recorded over a certain time period. Often quite complex patterns are searched among a background of irrelevant data, such that diagnoses becomes feasible. In the theory of formal languages complex temporal patterns are described by syntactic pattern recognition methods, where a grammatical inference is deduced. We propose to use extended context-free grammars, called Definitive Clause Grammars, to describe context dependence in primitive pattern sequences. Then we propose to use temporal expressions to build up the temporal patterns. It seems to us, that the ontological primitives usually used in AI-systems are too complex and too exact to real applications.

1. Introduction

In several application domains, as medicine, industrial processes, meteorology, a great amount of data is recorded over a time period. Often the aim is to make a diagnosis of the observed system. For example, consider an EEG recording to diagnose different sleep stages, or a chemical plant that remains in different process states, or the development of hail cells that possibly origin severe hailstorms, etc.. In all these cases several types of processes are observed and problem specific diagnoses are searched for. Human beings after a training phase often develop the ability to recognise complex patterns in multivariate time series. The reason lies in their technical knowledge concerning the process and in their experience dealing with standard and non-standard being able to make a diagnosis of the system from time series differing in their time scales.

Such complex temporal patterns are very hard to handle using technical systems. Classical approaches in the field of pattern recognition (PR) are very useful for feature extraction, where no temporal context has to be considered (DUDA/HART 73, TOU/GONZALEZ 74). In order to interpret temporal patterns in time series temporal dependencies between the primitive patterns (features) have to be specified. Syntactic PR views complex patterns as sentences of primitive patterns. Thus, techniques for syntactic PR strongly lie on the theory of formal languages (FU 82, GONZALEZ/THOMASON 78).

New approaches on adaptive PR and neurocomputing have recently been developed (BISHOP 95, PAO 89, SCHLAKOFF 92) and enable a "connection" between both approaches. In this paper we will show a way to extend adaptive PR-methods with Artificial Intelligence (AI) techniques.

Our final aim is to use techniques from AI and Artificial Neural Networks (ANN) to automatically and exploratively generate symbolic descriptions of temporal patterns in multivariate time series. The main idea of this paper is to propose an intuitive and simple description formalism for the interpretation and analysis of the primitive patterns in their causal and temporal context. Complex temporal patterns in time series, as considered here, have to be seen in a temporal context. This requires context sensitive knowledge. This means that context-free grammars are not powerful enough to parse context dependency in languages. Therefore we choose a powerful extention of context-free grammars, called definitive clause grammars (DCGs). The advantage of DCGs, beside their context-dependency, lies in an easy implementation of DCG rules as logic statements. Such an implementation enables an efficient parsing using a theorem prover, like a Prolog-Interpreter.

2. Properties of Definitive Clause Grammars (DCGs)

The main applications of DCGs has been for many years in natural language parsing systems (BOLC 87, NAUMANN/LANGER 94). A good introduction to this formalism can be found in (PEREIRA/WARREN 80). Before starting a descripition of the main characteristics of DCGs we want to give a brief introduction to the main aspects of the theory of formal language and point out the relationship between different grammar types and DCGs. In the theory of formal language properties and complexity of languages are studied. All languages are related in a hierarchy known as Chomsky hierarchy. At the bottom of the hierarchy we find regular (rg-) languages, for example $L = \{a^n b,\ n > 0\}$, that can be recognized by simple deterministic finite automata. Next we have more powerful languages in the Chomsky hierarchy called context-free (cf-) languages. The most famous example of a cf-language is the language $L = \{a^n b^n,\ n > 0\}$ that generates words like "((()))()". Parsers eg. stackmachines are used to parse cf-languages. For the implementation of most computational languages cf-languages suffice. They have lower complexity than context- dependent (cd-) languages as no context in sentences can be handled. Cd- languages are very useful, since languages that require some context knowledge, as natural languages, can be described by cd-languages. For example consider $L = \{a^n b^n c^n,\ n > 0\}$, where the b's are seen in a context with a equal number of a's and c's. It is shown that

$$rg - L(type3) \subset cf - L(type2) \subset cd - L(type1). \tag{1}$$

DCGs are a powerful extension of cf-grammars. Basically, DCGs are built up by cf-rules. See Fig. 1 for the classification of DCGs in the Chomsky hierarchy.

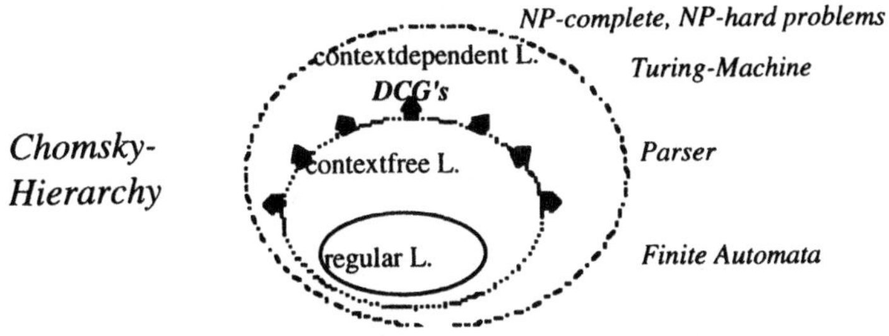

Figure 1: Chomsky hierarchy and DCGs

In order to provide context-dependency into the grammar a DCG extends a cf- grammar by augmenting non-terminals with arguments. DCGs extend cf- grammars in three important ways (PEREIRA/WARREN 80):

1. DCGs provide *context-dependency* in a grammar, such that a word in a text may depend on the context in which that word occurs in that text.
2. DCGs allow arbitrary *tree structures* that are built up in the course of parsing providing a representation of meaning of a text.
3. DCGs allow *extra conditions.*

The advantage of DCGs, beside their integration of context-dependency, lies in their efficient implementation of DCG-rules as logic statements by definitive clauses or "horn clauses". Now the problem of parsing a word of a language is transformed into the problem of proving a theorem in terms of a Prolog interpreter.

3. The Representation of Time in Artificial Intelligence

Pattern Recognition (PR) is an automatic process that includes extraction, identification, classification and description of patterns in data gathered from real and simulated environments. For time series this means that an automatic generation of a symbolic description of time series is searched for requiring an explicit knowledge representation. Classical approaches in the field of PR are very useful for the recognition of primitive patterns where no temporal context has to be considered (DUDA/HART 73, TOU/GONZALEZ 74). But also advanced techniques, like ANN, can be used for this task. Let us suppose that we have already classified the main primitive patterns for the time series using classical approaches or ANNs. The recognition of primitive patterns normally includes an additional step, the feature extraction process. All features are calculated from each window (sequence) generated from the time series using the well-known sliding windows technique. Let the number of sequences be $(N - M)/d + 1$, where N is the size of the time series, M is the size of the sequence and d is the delay for the next sequence.

$X_{(n-m)/d+1} = \{(x_{nm})|m \in \{1,...,M\}, n \in \{1,...,(N-M)/d+1\}, M < N\}$ describes the set of sequences. Then each sequence $(x_i) \in X_{(n-m)/d+1}$, $i \in \{1,...,(N-M)/d+1\}$ will be characterized by k features with

$$x_{ik} \in \Re := \{x_{ik}|\Re^M \mapsto \Re^K, x_{ik} \in [0,1], i \in \{1,...,(N-M)/d+1\}\}. \quad (2)$$

Each feature stands for a statistical property of the sequence. Primitive patterns can be interpreted as a mapping of the features into a set of clusters. This means, that our time series is partitioned into several similar time sequences. Let c be the number of detected primitive patterns. For each feature vector we get the following mapping

$$f : \Re^K \mapsto \aleph, f(m_n) = C_j, C_j \neq \emptyset, j \in \{1,...,c\}, n \in \{1,...,(M-N)/d+1\} \quad (3)$$

Then for each time series we get a sequence of primitive patterns, which lenght may vary in a certain range for each sequence in a pattern. An example for a primitive pattern is "heart frequency is ascending". In order to interpret temporal patterns in time series on a symbolic level temporal dependencies between the primitive patterns have to be specified. Therefore in AI a temporal reference made up by a set of temporal elements, called Ontological Primitives (OP), has to be build up. The main concepts for OPs are time points as proposed by (MCDERMOTT 82) or time intervals as proposed by (ALLEN 84) or a combination of both. For an overview of the main concepts on temporal reasoning, concerning logical formalisms in time in AI, ontological primitives and concepts related with reasoning about action, see (VILA 94). In our approach we use the more complex representation formalism within the concept of OPs, the Interval Calculus. The main disadvantage of the Interval Calculus proposed by Allen lies in a too complex and too specific representation to handle real-world applications. This means that for high frequency data like heart rates the interval limits are too hard. Here we need some kind of fuzzy expressions like "more or less simultaneous". We also believe that only a few OPs are needed to give a plenary description of multivariate time series. Besides this, we need to keep the description as simple as possible, since our final aim lies in the automatic extraction of rules for time series. Therefore we propose to use only some basic temporal expressions:

1. "and" for expressing sequences of alternative patterns
2. "or" for expressing alternative patterns
3. "immediate follow" enabling a description of successive states, whereas a state comprises a sequence of primitive patterns
4. "follows" describing successive states with a time interval
5. "more or less simultaneous" describing primitive patterns that occur nearly simultaneously

```
temppattern --> sequences, or, sequences.
temppattern --> sequences.
sequences --> sequence(M1,M2,T1,T2), and, sequences.
sequences --> sequence(M1,M2,T1,T2).
sequence(M1,M2,T1,T2) --> simpleseq(M1,M12,T11,T12),follow,
         simpleseq(M21,M2,T21,T22),distance,interval(T1,T2).

sequence(M1,M2,T1,T2) --> simpleseq(M1,M2,T1,T2).
simpleseq(M1,M2,T1,T2) --> primitive_pattern(M1),
         immediatefollow, primitive_pattern(M2).
simpleseq(M1,M2,T1,T2) --> sametime(M1,M2,T1,T2).
simpleseq(M1,infinite,0,0) --> primitive_pattern(M1).
sametime(M1,M2,T1,T2) --> primitive_pattern(M1), simultaneous,
         primitive_pattern(M2).
interval(tanf,tend) --> ['[u, o]'].
or --> [or].
and --> [and].
follow --> [follows].
distance --> [after].
immediatefollow --> [follows_immediately].
simultaneous --> [more_or_less_simultaneous].
primitive_pattern(m) --> [m].
```

Figure 2: A simple DCG for temporal patterns in time series

4. A Definitive Clause Grammar (DCG) for a Symbolic Representation of Temporal Patterns in Multivariate Time Series

We propose a new representation formalism for complex temporal patterns in time series using DCG rules. DCGs are a powerful extension of cf-grammars. This means, that basically DCGs are built up by cf-rules, but in order to provide context-dependency, an extension of the cf-rules by augmenting non-terminals with arguments is made. Generally a cf-grammar is defined by a quadruple G = (N,T,R,S), whereas N is a finite set of nonterminal symbols, T is a finite set of terminal symbols and $T \cap N = \emptyset$, S is a start symbol, and R is a finite set of rules, such that $R \subseteq N \times (T \cup N)^*$. In our case nonterminals describe temporal patterns or sequences or primitive patterns at different levels. The terminals are composed by the OPs, like "or", "follows", "more or less simultaneous" and by primitive patterns. See Fig. 2 for a simple DCG. In our formalism a pattern can be described through alternative temporal patterns. Each temporal pattern consists in a sequence of primitive patterns. Primitive patterns also can occur more or less simultaneously.

5. Conclusion

The recognition of complex temporal patterns in time series requires the integration of several methods, as statistical and signal processing PR, syntactic PR as well as new approaches like AI-methods and ANN. In order to handle time series with properties like instationarity, noise, nonlinearity, etc. and to realize an automatic temporal knowledge acquisition classical PR-methods have to be extended by new techniques, like AI and ANN. Specially in risk application domains, like medicine, chemical plants or meteorological prediction, a symbolic description and an explanation of the diagnoses are indispensable.

Our final aim is to use techniques from AI and ANN to automatically and exploratively generate symbolic descriptions of temporal patterns in multivariate time series. We propose an intuitive and simple description formalism for the interpretation and analysis of the features in their causal and temporal context. Therefore we use a powerful extension of context-free grammars, called Definitive Clause Grammars (DCGs). We suggest to use only some basic temporal expressions, like the OPs "or", "follows", "more or less simultaneous".

Further investigations lie in the field of ANN and rule extraction algorithms. First, we want to use unsupervised Neural Networks to extract the main primitive patterns out of the time series. This includes the development of special visualization tools for ANN, such that sequences of primitive patterns can be detected in multivariate time series. Second, special rule extraction algorithms for temporal knowledge conversion have to be developed.

Acknowledgements

We would like to thank D. Korus and S. Farsch that presently work at our department for their helpful discussions that in part contributed to the results presented in this paper.

References:

ALLEN, J. (1984): Towards a General Theory of Action and Time. *Artificial Intelligence, 23, 123-154.*

BISHOP, C. M. (1995): Neural Networks for Pattern Recognition, Clarendow Press, Oxford.

BOLC, L. (1987): Natural Language Parsing Systems. Springer Verlag.

BOW, T. (1984): Pattern Recognition, Marcel Dekker, New York.

DUDA, O. and HART, P. E. (1973): Pattern Classification and Scene Analysis, John Wiley and Sons, Inc. New York.

FU, S. (1982): Syntactic Pattern Recognition and Applications, Prentice-Hall, Englewood-Cliffs, N. J.

GONZALEZ, R. C. and THOMASON, M. G. (1978): Syntactic Pattern Recognition, Addison-Wesley.

KANAL, L. N. and DATTATREYA, G. R. (1992): Pattern Recognition. in S.C. Shapiro (ed): Encyclopedia of Artificial Intelligence, Vol. 2, John Wiley & Sons.

MCDERMOTT, D (1982): A Temporal Logic for Reasoning about Processes and Plans. *Cognitive Science, 6, 101-155.*

NAUMANN, S. and LANGER, H. (1994): Parsing: Eine Einführung in die maschinelle Analyse natürlicher Sprache, B.G. Teubner, Stuttgart.

PAO, Y.-H. (1994): Adaptive Pattern Recognition and Neural Networks. Addison-Wesley, New York.

PEREIRA, F. and WARREN, D. (1980): Definitive Clause Grammars for Language Analysis – A Survey of the Formalism and a Comparison with Augmented Transition Networks. *Artificial Intelligence 13, 231-278.*

TOU, J. T. and GONZALEZ, R. C. (1974): Pattern Recognition Principles, Addison-Wesley.

SCHALKOFF, R. (1992): Pattern Recognition: Statistical, Structural and Neural Approaches, John Wiley & Sons, New York.

STERLING, L. and SHAPIRO, E. (1986): The Art of Prolog. MIT Press.

ULTSCH, A. and SIEMON, H. P. (1990): Kohonens Self Organizing Feature Maps for Exploratory Data Analysis, Proc. Intern. Neural Networks, Kluwer Academic Press, Paris, 305 - 308.

ULTSCH, A. (1991): Konnektionistische Modelle und ihre Integration mit wissensbasierten Systemen, Forschungsbericht Nr. 396, Institut für Informatik, Universität Dortmund, Februar, Dortmund, Habilitationsschrift.

VILA, L. (1994): A Survey on Temporal Reasoning in Artificial Intelligence. *Ai Communications, Vol. 7, Nr.1, March, 4-28.*

Neural Networks: A Statistician's (Possible) View

K. Hornik

Institut für Statistik und Wahrscheinlichkeitstheorie
Technische Universität Wien, A-1040 Wien, Austria
Kurt.Hornik@tuwien.ac.at

Summary: Within the past few years, neural networks (NNs) have emerged as a popular, rather general-purpose means of data processing and analysis. As in most applications they are employed to perform rather standard statistical tasks like regression analysis and classification, one might wonder what is really new about them. We shed some light on this issue from a statistician's point of view by "translating" neural network terminology into more familiar terms and then discussing some of their most important properties. Particular attention is given to "supervised" classification, i.e., discriminant analysis.

1. Introduction

Within the past few years, neural networks (NNs) have emerged as a popular, rather general-purpose means of data processing and analysis. Success stories include e.g. using them for recognition of speech, speakers, fingerprints, and handwritten characters (zip codes), to classify medical images, as decision support for loan applications, to forecast electric load and financial time series (e.g. stock market data), and so on. But clearly, all the above tasks are standard statistical tasks of regression analysis and classification, and have been worked on with "conventional" methods for quite some time now. Hence the question arises whether there is anything special about neural networks that allows them to perform better in these problems.

In answering this question, one of the basic problems is that the NN community has created its own terminology which heavily relies on biological notions, which typically impedes communication with "outsiders" as e.g. conventional statisticians. Hence, in the past few years, it has become quite attractive and popular to look at NNs from an "outsider's" (e.g., a statistician's) point of view, to translate "neural" terminology into more familiar terms, and to compare their performance to conventional methods on benchmark problems. Among the many excellent examples are in particular Ripley (1993) and Ripley (1996).

In doing so, one has to pay attention as to which "neural networks" one refers to. Clearly, the human brain consists of neurons and hence is a (biological) NN, but not the subject of interest here where we will be concerned with *computational* NNs. Even without giving precise definitions, we can exhibit at least two of their basic characteristics.

- They are built from simple processing elements (PEs) called *nodes*, *units*, or *neurons*, and contain certain adjustable parameters ("weights").
- They use rules to modify these parameters upon the presentation of data, i.e., to "learn from the environment".

As far as we are concerned, another key feature is that these systems should (eventually) be *realizable in special-purpose hardware* rather than be simulated on conventional digital computers. I.e., in some sense NNs are the revival of analog computing with the new ingredient of computational intelligence by adaptivity (on-line learning).

However, typically "neural networks" is used synonymously for "multi-layer perceptrons (MLPs) trained by Error Back-Propagation", although many other NN paradigms exist, see e.g. Haykin (1994). In this case, the neurons compute maps of the form $x \mapsto \sigma(a'x + \theta)$, where x is a vector of input signals, a and θ are the "weights" and the "bias" of the neuron, respectively, $'$ denotes the transpose such that $a'x$ is the inner product of a and x, and σ is the "activation function", typically a sigmoidal function (monotonically nondecreasing with limits 0 and 1 at infinity) such as the logistic function $t \mapsto 1/(1 + e^{-t})$. If such units are grouped into "layers" and connected in a feed-forward manner (in a way that only units in adjacent layers are connected), one obtains an MLP which implements an input-output relation of the form $x \mapsto f(x, \omega)$, where ω contains all the adjustable parameters and f is determined by the network architecture. Given a sample of input-target training patterns (x_i, y_i), the network is then trained to (approximately) minimize the mean square error

$$E(w) = \langle |y - f(x, \omega)|^2 \rangle$$

(where $\langle \cdot \rangle$ could be the sample or population average) by gradient descent. The name "Error Back-Propagation" comes from the fact that due to the particular structure of the MLP, the gradient can be computed by first propagating the input forward through the network to give a network output, computing an error signal as the difference between output and target, and then propagating the errors backwards in order to adapt the weights Rumelhart et al. (1986).

But clearly, this amounts to "approximating" the conditional expectation $\mathbb{E}(y|x)$ by the family $f(\cdot, \omega)$ of model functions. Hence, we see that "using a neural network" can usually be translated into more familiar terminology as fitting a particular, typically nonlinear, parametric model by nonlinear least squares regression; more precisely, by an algorithm which attempts to do so. (In fact, Back-Propagation, as a gradient descent algorithm, compares rather unfavorably with more sophisticated numerical minimization algorithms, with respect to both stability and speed of convergence.)

In discriminant analysis, one typically uses the vector of class indicators as targets such that the posterior class probabilities are estimated. In this sense (under uniform loss), "neural networks approximate the Bayes classifier".

When evaluating the performance of NNs used as described above, one needs to investigate the following issues.

- How "rich" is the class of MLPs?
- How well can MLPs be trained?

The first question concerns the computational power of the MLP model and will be considered in Section 2. The second issue asks about the needs in terms of sample size and computational requirements to achieve certain levels of performance and will be dealt with in Section 3. Section 4 concludes this paper.

2. Computational Power

Let us look at the representational capabilities of MLPs first. The first general results were obtained independently and almost simultaneously by Cybenko (1989), Funahashi (1989), and Hornik et al. (1989), and have meanwhile substantially been generalized, cf. in particular Hornik (1991, 1993). It was shown that MLPs can approximate any reasonable function arbitrarily well, provided only that enough hidden units are available for internal computations and that their activation function be non-polynomial. To give a more precise statement, let d be the input dimension and, for $A \subseteq \mathbb{R}^d$ and $\Theta \subseteq \mathbb{R}$, let $\mathcal{N}(\sigma; A, \Theta)$ be the set of all functions on $\mathbb{R}^d$ of the form

$$x \mapsto y = \sum_{i=1}^{k} \beta_i \sigma(a_i' x + \theta_i),$$

i.e., the span of $\{\sigma(a'x + \theta), a \in A, \theta \in \Theta\}$ under finite linear combinations. This is the set of all mappings from $\mathbb{R}^d$ to $\mathbb{R}$ which can be implemented by a single hidden layer perceptron with input-to-hidden weights $a_i \in A$ and hidden unit biases $\theta_i \in \Theta$, respectively; for sake of simplicity, all hidden units have the same activation function σ and the output unit is linear. The following result is theorem 2 in Hornik (1993).

> Let σ be *essentially bounded and non-polynomial* on some non-degenerate compact interval Θ, and let A contain a neighborhood of the origin. Then for all compactly supported finite measures μ on $\mathbb{R}^d$ and $1 \le p < \infty$, $\mathcal{N}(\sigma; A, \Theta)$ contains a subset which is dense in $L^p(\mu)$.

Similar results can be given for uniform approximation of continuous functions on compacta. Hence, MLPs can basically implement any reasonable function (e.g., $g \in L^p(\mu)$), and for this reason are feasible competitors in non-parametric statistical inference.

Unfortunately, it is much harder to obtain results describing how well classes of functions can be approximated by MLPs of certain "complexity". Barron (1993) shows that if the function g to be approximated has $g(0) = 0$ and a Fourier transform $\hat{g}$ with finite first absolute moment and σ is sigmoidal, then there is always a function g_k implemented by a single hidden layer perceptron with k hidden units which have σ as their activation function such that

$$\|g - g_k\|_{L^2(B_r,\mu)} \le \frac{2\sqrt{\mu(B_r)}}{\sqrt{k}} \int |a\hat{g}(a)|\, da, \tag{1}$$

where B_r denotes the ball of radius r centered at the origin and μ an arbitrary measure; similar results can be obtained for uniform approximation Barron (1992). (1) means that the MSE scales like $1/k$. As this is independent from d, it has been argued that this result shows that MLPs can avoid the "curse of dimensionality"; however, it is clear that d enters via the integrability condition on $\hat{g}$. In fact, we should emphasize that as the classes of MLP model functions are nonlinear in some of their parameters, the number of free parameters is not necessarily the correct complexity measure, see also Section 3.

MLPs with Heaviside activation functions $H(u) = (1+\mathrm{sgn}(u))/2$ are of substantial interest to computer scientists and engineers from a non-statistical perspective. In this context, the units which compute the map $x \mapsto H(a'x + \theta)$ are usually referred to as linear threshold gates; the corresponding MLPs are called (linear) threshold circuits. These can be used to define interesting non-trivial complexity classes of boolean functions. Let C_n be a circuit (network) with n boolean inputs and a single boolean output. Consider the following classes of network sequences $(C_n, n \ge 1)$:

TC^0_d: C_n has depth $\le d$ and $\le \mathrm{poly}(n)$ linear threshold gates of arbitrary fan-in

AC^0: C_n has depth $O(1)$ and $\le \mathrm{poly}(n)$ AND, OR and NOT-gates.

Then one can show that e.g. PARITY, MAJORITY, MULTIPLICATION, and SORTING are in $\bigcup_d TC^0_d$, but not in AC^0. In fact, MULTIPLICATION is in $TC^0_3 \setminus TC^0_2$ and ADDITION is in $TC^0_2 \setminus TC^0_1$, cf. e.g. Bruck (1990) and Bruck and Smolensky (1992) and the references therein. These results do not only shed some light on the computational power of MLPs, but also show that standard problems can be realized in simple special-purpose NN hardware.

If feedback connections between units are allowed, one obtains the so-called recurrent MLP model. Such networks can perform sequential computations and hence can internally deal with *time*, which is important in many cognitive tasks encountered in applications, such as speech, vision, and control.

Language recognition is a typical task where the correct classification of temporal patterns of arbitrary (finite) length is required. The following model for recognizing formal languages L, i.e., families of 0-1 strings, is considered and rigorously analyzed in Siegelmann and Sontag (1994, 1995). The net evolves according to

$$u(t+1) = \sigma(Au(t) + Bx(t) + c), \qquad t = 0, 1, \ldots$$

where $u(t)$ and $x(t)$ are the network activations and inputs at time t, respectively, and σ performs coordinate-wise application of the saturated-linear function (ramp function)

$$\sigma_i(\alpha) = \begin{cases} 0 & \text{if } \alpha < 0, \\ \alpha & \text{if } 0 \leq \alpha \leq 1, \\ 1 & \text{if } \alpha > 1. \end{cases}$$

There are two binary inputs, one representing the data line over which the patterns are presented, and the other the so-called validation line which takes the value '1' when an input is applied, and '0' otherwise. Similarly, two units are taken as output units for data and validation. The initial state $u(0)$ of the net is always taken to be a zero and an equilibrium state, i.e., $\sigma(c) = 0$. The network classifies an input word $\omega = \omega_1 \cdots \omega_k$ in time τ if the output sequences are of the form

$$u_{\text{data}} = \underbrace{0 \cdots 0}_{\tau-1} \eta(\omega) 0 \cdots, \qquad u_{\text{validation}} = \underbrace{0 \cdots 0}_{\tau-1} 10 \cdots,$$

where $\eta(\omega)$, the network's "final response" to the input pattern ω, is either 0 or 1.

Results on the computational capabilities of this model, seen from the point of view of classical formal language theory, are given in Siegelmann and Sontag (1994, 1995). The following table shows which classes of languages can be recognized ("Recognition Power"), given constraints on the weights (A, B, and c) and the recogition time τ.

Weights	Recognition Power	
A, B, c	Arbitrary Time	Polynomial Time
integer	regular	regular
rational	recursive	(usual) P
real	arbitrary	analog P

This shows that with integer weights, all finite automata can be simulated; with rational weights, recurrent networks are computationally equivalent, up to polynomial time, to Turing machines. "Analog P" is a new complexity class which contains many languages not in (usual) P, but most likely does not contain NP.

3. Learning

In the neural networks context, "learning" is a process by which the free parameters of the system are adapted through a continuing process of stimulation by the environment in which the network is embedded. (In addition, units may be allocated or freed.) The three basic classes of learning paradigms are as follows.

Supervised Learning is performed under the supervision of an external "teacher" who presents target patterns,

Reinforcement Learning involves the use of a "critic" that evolves through a trial-and-error process,

Unsupervised Learning is performed in a self-organized way in that *no* external teacher or critic is required.

Back-Propagation as introduced in Section 1 is an example of supervised learning. Reinforcement learning is typically employed in control applications; unsupervised methods include e.g. on-line learning algorithms for Principal Component Analysis (Baldi and Hornik, 1995) or Kohonen's self-organizing feature maps (Kohonen, 1989, 1995). In statistical terminology, the data-based adjustment of parameters is typically referred to as estimation, and unsupervised algorithms for classification tasks perform cluster analysis.

How well does Back-Propagation work? To answer this, let us introduce the following rather general framework for supervised learning. Suppose that for each training pattern z and each element w of a (possibly infinite-dimensional) parameter space $\mathcal{W}$ we can compute a "loss" $\ell(z, w)$, and that the goal is to minimize the expected loss $E(w) = \mathbb{E}\,\ell(z, w)$, where the expectation is taken over an unknown distribution in the pattern space Z. In the case of Back-Propagation, the z are random input-target pairs (x, y), and $\ell(z, w) = |y - w(x)|^2$, where the w correspond to the functions $f(\cdot, \omega)$ implemented by MLPs with a certain architecture and certain weights ω.

A natural idea is to estimate the unknown $E(w)$ by its sample average

$$E_n(w) = \frac{1}{n} \sum_{i=1}^{n} \ell(z_i, w),$$

minimize E_n by a suitable algorithm, and hope that the w thus obtained approximately minimizes E. This approach is called *Empirical Risk Minimization* (ERM), and is exactly what Back-Propagation is trying to accomplish by on- or off-line gradient descent. Of course, the first question is how well this works, i.e., how easily MLP-type empirical risks can be minimized. Unfortunately, it turns out that BP does not perform too well. Gradient descent algorithms are generally known to exhibit rather poor convergence behavior, and in addition it turns out that the error surfaces associated with MLPs are typically flat and "full" of local minima. This results in rather high computational costs. One might argue that these costs could significantly be reduced by using special-purpose hardware ("neural chips"), but one can in fact show that the problem of finding an optimal MLP weight configuration is very hard (e.g. the Threshold Circuit Loading and Minimization problems are NP-complete, see Judd (1990), Lin and Vitter (1991).

Let us leave these computational issues aside and look at the question how "well" an MLP learns by minimizing E_n. Early NN applications basically consisted in generating look-up tables, e.g. implementing boolean functions such as XOR based on an *exhaustive* set of training patterns. In such a finite space scenario, it is obviously desirable to have E_n as small as possible (because $E_n = E$ for all sufficiently large n) which can be accomplished by choosing a large enough network architecture. However, in a statistical framework where the data is only a random sample from an underlying infinite population, an exact fit on the data set (zero bias) will result in rather poor performance on previously unseen patterns (large variance; in NN terminology, in bad *generalization* performance). I.e., in addition to minimizing the empirical risk as a function of the weights for a fixed architecture, one also needs to control the "complexity" of the employed architecture in order to maintain a certain bias-variance balance (to avoid over- and under-fitting). In statistical terminology, one needs to perform *model selection*.

A huge variety of such methods exist, ranging from cross-validation (Stone, 1974) over applications of the MDL principle (Rissanen, 1986) to "neural" generalizations of the classical (linear) AIC (Akaike, 1973); the so-called Network Information Criterion (NIC) presented in Murata et al. (1994) argues that the correct complexity measure is not the number of parameters of the network, but given by trace(SQ^{-1}), where S is the expected outer product of the gradient of ℓ (with respect to ω) and Q is the expected Hessian. Asymptotically, these methods are all equivalent. In addition, Bayesian methods have been proposed, as e.g. "Occam's Razor" (MacKay, 1992a, 1992b, 1992c). However, all these methods typically allow to control the "size" of the network, usually the number of hidden units in a single hidden layer perceptron, which is not necessarily the "right" complexity measure.

A general result for the consistency of Empirical Risk Minimization (in the sense that if w_n is a minimizer of E_n, then $E(w_n)$ and $E_n(w_n)$ both converge to the minimum of $E(w)$) was obtained by Vapnik and Chervonenkis (1991). If the loss functions are uniformly bounded, then consistency holds if and only if the empirical risks E_n converge to the actual risks E uniformly over $\mathcal{W}$, in the sense that

$$\lim_{n\to\infty} \mathbb{P}\left(\sup_{w\in\mathcal{W}}(E(w) - E_n(w)) > \epsilon\right) = 0, \qquad \forall\epsilon > 0.$$

To give necessary conditions for such a uniform one-sided convergence, let us restrict our attention to the case of binary classification with 0-1 loss, i.e., $\ell(z, w)$ is the indicator function of $y \neq w(x)$, where both y and $w(x)$ take values in $\{0, 1\}$ only. In this case, a necessary condition for the above (irrespective of the underlying pattern distribution) is that the *Vapnik-Chervonenkis* (VC) dimension of $\mathcal{W}$, symbolically VCdim($\mathcal{W}$), is finite.

This quantity is defined as follows. For each sequence of n input patterns, $(w(x_1), \ldots, w(x_n))$ is a binary vector of length n. The logarithm of the number of different such vectors, with arbitrary patterns $x_1, \ldots, x_n$ and $w \in \mathcal{W}$, is called the *growth* function $G_{\mathcal{W}}(n)$, and the maximal n such that $G_{\mathcal{W}}(n) = n \ln 2$ is the VC dimension of $\mathcal{W}$. (Or, more directly, VCdim($\mathcal{W}$) is the maximal size of a subset X_0 of the input space X which can completely be shattered by $\mathcal{W}$, i.e., for each subset S of X_0 there exists a function $w = w_S$ in $\mathcal{W}$ which is 1 on S and 0 on S^c.)

The VC dimension measures how "rich" a set of function is, in a very natural sense. However, it is not necessarily equal to the number of free parameters in (the Euclidean dimension of) the set of model functions $\mathcal{W}$. Its crucial role in modern learning theory is further substantiated by the characterization of PAC learnability. The PAC (Probably Approximately Correct) model Valiant (1984) quantifies the generalization performance of a binary classification system (such as MLPs with binary outputs) as follows. A learning algorithm λ is a rule which uses the training patterns $z^n = (z_1, \ldots, z_n)$ to produce an element $w_n = \lambda(z^n)$ in $\mathcal{W}$. Of course, such a rule should have the property that for most training patterns, the probability of misclassification $E(w_n)$ of test patterns is low, provided of course that training and test patterns are drawn (independently) from the same distribution. Without going into technical details, a learning algorithm is probably asymptotically correct if one can always find a finite $n = n(\delta, \epsilon)$, the *sample complexity* of the algorithm, such that $\mathbb{P}(E(w_n) > \epsilon) < \delta$ for all $n \geq n(\delta, \epsilon)$, irrespective of the underlying distribution of input patterns (and the target decision function t which is taken from some class $\mathcal{T}$ and assumed to generate the target patterns).

One can then show the following. If the targets are in fact generated as $y = w(x)$ for some w in $\mathcal{W}$, then the sample complexity of any learning algorithm λ which is consistent in the sense that $E_n(w_n) = 0$ satisfies

$$n_\lambda(\delta, \epsilon) \leq \frac{8}{\epsilon}\left(\ln\left(\frac{4}{\delta}\right) + \mathrm{VCdim}(\mathcal{W}) \ln\left(\frac{48}{\epsilon}\right)\right),$$

(Anthony, 1994). Conversely, if $\mathcal{T}$ is sufficiently rich, then one also has the lower bound

$$n_\lambda(\delta, \epsilon) \geq \max\left(\frac{\mathrm{VCdim}(\mathcal{W}) - 1}{32\epsilon}, \frac{\ln(1/\delta)}{\epsilon}\right).$$

If the target patterns are not generated by a target function $y = w(x)$, $w \in \mathcal{W}$, one can generalize the above to approximating what is best possible in $\mathcal{W}$.

Hence, the VC dimension of a class of "hypotheses" $\mathcal{W}$ plays a crucial role in its ability to learn (generalize) well. Ideally, one should thus strive for classes which can achieve certain rates of approximation with minimal VC dimension. Unfortunately, for MLPs both quantities are hard to compute. Note that there is no a priori reason for believing that the VC dimension of an MLP is e.g. proportional to the number of free parameters. E.g., Sontag (1992) has shown that for the real-analytic function

$$\psi(u) = \frac{1}{\pi}\arctan(u) + \frac{\cos(u)}{7(1+u^2)} + \frac{1}{2},$$

an MLP with 2 real valued inputs, 2 hidden units with activation function ψ and a linear threshold output unit has *infinite* VC dimension. For the standard (logistic) sigmoid $\sigma(t) = 1/(1+e^{-t})$ it has recently been shown (Karpinski and Macintyre, 1994) that the VC dimension of an MLP with activation function σ is bounded by $O(m^4)$, where m is the total number of weights in the network (in fact, the same bound holds for a large class of activation functions which satisfy a certain Pfaffian differential equation). For piecewise polynomial activation functions and boolean outputs, the VC dimension is at most $O(m^2)$ (Goldberg and Jerrum, 1993).

To sum up, the theoretical results which are currently available on the learning performance of MLPs trained by Back-Propagation (and suitable methods for controlling their complexity) are still rather dissatisfactory and can by no means substantiate possible superiority over more conventional approaches.

We have seen that from the perspective of learning theory, one should look for systems for which computationally simple learning algorithms exist and which have "small" VC dimension. The Support Vector (SV) Machines (Vapnik, 1995) implement such an idea: the input patterns are mapped into a high-dimensional feature space through some nonlinear mapping, and separated by a hyperplane in this space. More precisely, the decision functions used are of the form (assuming $-1/1$ rather than $0/1$ targets)

$$\operatorname{sign}\left(\sum_{\text{support vectors}} y_i \alpha_i K(x, x_i) - b \right),$$

where K is an "inner product", e.g. $K(x, x_i) = \sigma(\gamma x_i' x + \delta)$ for suitably chosen γ and δ, and the "support vectors" are those data vectors x_i which show up in the above expression (i.e., for which the α_i are nonzero) and are determined by quadratic programming. One can show that such classes of decision functions have rather small VC dimension. In the neural network case, one obtains an MLP with one hidden layer and a $-1/1$ output layer, and both architecture and weights can be obtained in a computationally efficient way. Of course, the computational power of such networks remains to be investigated, which should be a rather challenging task for future research.

4. Conclusion

We have seen that the "standard" use of neural networks, i.e., the supervised training of multi-layer perceptrons by Error Back-Propagation and its variants, amounts to (non-parametric) estimation of conditional expectations. Although the MLP is an attractive and powerful model for computation, little is known quantitatively about its approximation capabilities and its learning performance. Nevertheless, such systems work rather well in many applications, and certainly nowadays should be one of the standard tools in applied statistics.

References:

AKAIKE, H. (1973): Information theory and an extension of the maximum likelihood principle. In Petrov, B. N. and Csáki, F. (eds.), *Second International Symposium on Information Theory*, pp. 267–281. Budapest, Hungary: Akademiai Kiado.

ANTHONY, M. (1994): *Probabilistic analysis of learning in artificial neural networks: The PAC model and its variants.* Tech. Rep. NC-TR-94-3, NeuroColt Technical Report Series.

BALDI, P. and HORNIK, K. (1995): Learning in linear neural networks: a survey. *IEEE Transactions on Neural Networks*, **NN-6**(4), 837–858.

BARRON, A. R. (1992): Neural net approximation. In Narendra, K. (ed.), *Proceedings of the 6th Yale Workshop on Adaptive Learning Systems*, pp. 69–72. New Haven: Yale University.

BARRON, A. R. (1993): Universal approximation bounds for superpositions of a sigmoidal function. *IEEE Transactions on Information Theory*, **IT-39**, 930–945.

BRUCK, J. (1990): Harmonic analysis of polynomial threshold functions. *SIAM Journal on Discrete Mathematics*, **3**, 168–177.

BRUCK, J. and SMOLENSKY, P. (1992): Polynomial threshold functions, AC^0 functions and spectral norms. *SIAM Journal on Computing*, **21**, 33–42.

CYBENKO, G. (1989): Approximation by superpositions of a sigmoidal function. *Mathematics of Control, Signals, and Systems*, **2**, 303–314.

FUNAHASHI, K. (1989): On the approximate realization of continuous mappings by neural networks. *Neural Networks*, **2**, 183–192.

GOLDBERG, P. and JARRUM, M. (1993): Bounding the Vapnik-Chervonenkis dimension of concept classes parametrized by real numbers. In *Proceedings of the 6th Annual ACM Conference on Computational Learning Theory*, pp. 361–369. New York: ACM Press.

HAYKIN, S. (1994): *Neural Networks: A Comprehensive Foundation.* New York: Macmillan College Publishing.

HORNIK, K. (1991): Approximation capabilities of multilayer feedforward networks. *Neural Networks*, **4**, 251–257.

HORNIK, K. (1993): Some new results on neural network approximation. *Neural Networks*, **6**, 1069–1072.

HORNIK, K.; STINCHCOMBE, M. and WHITE, H. (1989): Multilayer feedforward networks are universal approximators. *Neural Networks*, **2**, 359–366.

JUDD, J. S. (1990): *Neural Network Design and the Complexity of Learning.* Cambridge, MA: MIT Press.

KARPINSKI, M. and MACINTYRE, A. (1994): *Polynomial bounds for the VC dimension of sigmoidal neural networks.* Tech. Rep. 85116-CS, University of Bonn.

KOHONEN, T. (1989): *Self-Organization and Associative Memory.* Berlin: Springer, 3rd edn.

KOHONEN, T. (1995): *Self-organizing maps.* Berlin: Springer.

LIN, J.-H. and VITTER, J. S. (1991): Complexity results on learning by neural nets. *Machine Learning*, **6**, 211–230.

MACKAY, D. J. C. (1992a): Bayesian interpolation. *Neural Computation*, **4**(3), 415–447.

MACKAY, D. J. C. (1992b): The evidence framework applied to classification networks. *Neural Computation*, **4**(5), 720–736.

MACKAY, D. J. C. (1992c): A practical bayesian framework for backpropagation networks. *Neural Computation*, **4**(3), 448–472.

MURATA, N.; YOSHIZAWA, S. and AMARI, S.-I. (1994): Network information criterion—determining the number of hidden units for an artificial neural network model. *Neural Networks*, **5**(6), 865–872.

RIPLEY, B. (1993): Statistical aspects of neural networks. In Barndorff-Nielsen, O. E., Jensen, J. L., and Kendall, W. S. (eds.), *Networks and Chaos—Statistical and Probabilistic Aspects*, vol. 50 of *Monographs on Statistics and Applied Probability*, pp. 40–123. London: Chapman and Hall.

RIPLEY, B. (1996): *Pattern Recognition and Neural Networks.* Cambridge University Press.

RISSANEN, J. (1986): Stochastic complexity and modeling. *The Annals of Statistics*, **14**(3), 1080–1100.

RUMELHART, D. E.; HINTON, G. E. and WILLIAMS, R. J. (1986): Learning representations by backpropagating errors. *Nature*, **323**, 533–536.

SIEGELMANN, H. T. and SONTAG, E. D. (1994): Analog computation, neural networks, and circuits. *Theor. Comp. Sci.*, **131**, 331–360.

SIEGELMANN, H. T. and SONTAG, E. D. (1995): On the computational power of neural nets. *J. Comp. Syst. Sci.*, **50**, 132–150.

SONTAG, E. D. (1992): Feedforward nets for interpolation and classification. *J. Comp. Syst. Sci.*, **45**, 20–48.

STONE, M. (1974): Cross-validatory choice and assessment of statistical predictions. *Journal of the Royal Statistical Society Series B*, **36**, 111–147.

VALIANT, L. G. (1984): A theory of the learnable. *Communications of the ACM*, **27**(11), 1134–1142.

VAPNIK, V. N. (1995): *The nature of statistical learning theory.* New York: Springer.

VAPNIK, V. N. and CHERVONENKIS, A. Y. (1991): The necessary and sufficient conditions for consistency of the method of empirical risk minimization. *Pattern Recognition and Image Analysis*, **1**(3), 284–305.

Dynamic Supervised Learning: Some Basic Issues and Application Aspects

G. Nakhaeizadeh[1], C. C. Taylor[2], G. Kunisch[3]

[1] Daimler-Benz, Forschung und Technik, D-89013 Ulm, Germany

[2] Department of Statistics, University of Leeds, Leeds LS2 9JT, UK

[3] Abteilung Stochastik, Universität Ulm, D-89069 Ulm, Germany

Summary: This paper discusses ideas for adaptive learning which can capture dynamic aspects of real-world datasets. Although some of these ideas have a general character and could be applied to any supervised algorithm, here we focus attention on the linear, logistic and quadratic discriminant. The classifiers use Statistical Process Control (SPC) to appropriately update the rule or modify it by modifying the "training data". These methods are tried out on simulated data and real data from the credit industry.

1. Introduction

In classical supervised learning, the available examples (the training data) are usually used to learn a classifier which can be applied later to determine the class of new, unseen examples. In many practical situations in which the environment changes, this procedure ceases to work. In the project StatLog this situation was encountered in the case of fault diagnosis of automatic transmissions (Michie *et al.*, 1994, chapter 9); the generated rules from a certain learning data set were no longer valid after six months due to changes in the parameters of the diagnosis device. Generally speaking, the application of a discrimination algorithm to classify new, unseen examples will be problematic if one of the following events occurs after the "learning phase":

1. The number of attributes changes. We experienced this in a credit-scoring application that certain customer information can no longer be stored due to legal requirements. Conversely one may be able to get more information than before about the customers, *i.e.* one can consider more attributes.
2. The number of attributes remains the same but the reliability and relevance of the variables changes over the time. In this situation the distribution of at least one class is gradually changing over time. As mentioned above, the changes of the payment behaviour of customers in the case of credit-scoring or the changes of the diagnosis device in the case of automatic transmissions are examples of this problem.

To summarize, we have learned from StatLog that the extracted rules or dependencies can be no longer valid over time due to different changes which can occur. The following solutions may be used to solve this problem.

Relearning. This approach generally does not use any of the information of the old learning dataset. If the old observations are discarded, it can take a long time to re-acquire enough new examples with known class values for relearning. For example, in the case of credit-scoring, it will take months before one can be sure whether a new customer is a good or a bad risk.

Incremental learning. There are some attempts in the literature to apply incremental learning to capture the dynamic aspects of the datasets; see, for example, Utgoff (1989, 1994) who has developed an algorithm that produces the same decision tree as would be found if all the examples had been available at once. An improved algorithm (ITI) can now handle continuous variables, multiple classes, noise *etc.*. These approaches, however, cannot cover the above mentioned aspects, since some of the old data should be downweighted or discarded as no longer being representative of the current population.

Capturing dynamic aspects of datasets is not only a problem specific to symbolic ML-algorithms. We think that further research is necessary as well to enable statistical and neural network algorithms to handle such dynamic aspects.

This paper considers the situation described in item 2 above, namely that the distributions of the classes may change over time. In the next sections we discuss ways in which the process can be monitored in a similar manner to applications in quality control. We also discuss ways of updating the classifier either by explicitly changing the rule, which has been learned, or by changing the learning data (substituting or adding more recent observations). These ideas are implemented using linear, quadratic and logistic discrimination on simulated and real data.

2. Monitoring Process

In order to monitor the performance of the allocation rule, for example to detect changes in the distribution of the data, it is necessary to observe values regularly. This is what we understand by the expression "process monitoring". We have two main reasons why we propose to take batches of new test samples to monitor the classification process. On the one hand noise is averaged and therefore reduced, on the other hand the presentation of the information is more transparent.

The following quantities can be monitored using a time series plot, with appropriate warning and action limits indicating that the process has changed:

1. the average (maximum) probability of class membership of the batch;
2. the class means and covariances of the most recent batch of observations;
3. current error rates e_{cur} (assuming that the class is known) and the average error rates $\bar{e}$ of previous batches (the error rate of a batch is the number misclassified divided by the number in the batch);
4. the estimated prior group probabilities (only if the observed data is acquired by mixture sampling).

If we consider the **probabilities of class membership** it is only important to compare the highest values. If there are two or more values close to the maximum, the allocation rule is nearly indifferent between these classes. As a monitoring value we could use the average difference between highest and second-highest group-probability of the last recent batch or the average probability of maximum class membership. If this latter value is drifting down towards $1/k$, where there are k groups then adaptation may be required.

Tests are available to determine if the **means and variances** have changed and the corresponding test statistics for instance could be monitored. Tests of the mean and variance of each class can use Hotelling's T^2 and Box's M-test respectively; see Mardia *et al.* (1979). If a deviant batch is detected we could carry out further diagnostics to investigate in which way the attributes have changed. Warning limits can be calculated if the distributions are known and only depend on the numbers of the attributes and the significance level. Although normal distribution assumptions are necessary for these tests, they can also be used for nonparametric classification methods just for monitoring purposes. Another possible omnibus test for a change in the distribution would be a multivariate version of the Kolmogorov-Smirnov test.

To monitor **error rates** we can measure the distance between previous averaged error rates and the current error rate. As a distance comparison we could take the standard deviation of the mean error rate similar to the Shewhart control chart in statistical process control. Standard errors for estimated error rates are available for normal-based discrimination procedures; see McLachlan (1992).

3. Modelling of Learning

In this section we discuss ways of updating the classifier either by explicitly changing the rule which has been learned, or by changing the learning data by substituting or adding more recent observations. Suppose that we examine the data in batches and that we detect a change which requires adaptation in the learned rule. Any classification algorithm can be totally relearned from recent observations but more interestingly, we can consider how best to re-use previously learned information.

3.1 Learning from Error Rates

One method is to introduce weights dependent of the performance of the classifier. Suppose that we have a current rule which is used on the next batch of observations and results in an error rate of e_{cur}. If the mean error rate, $\bar{e}$, and e_{cur} are very different (relative to the standard deviation s), this suggests that the current rule should be updated. One approach is to obtain a revised rule which is a weighted average of the current discriminant rule and the a rule which was obtained by using only the most recent data with some weights depending of $e_{cur}, \bar{e}$ and s or some function thereof (see figure 1).

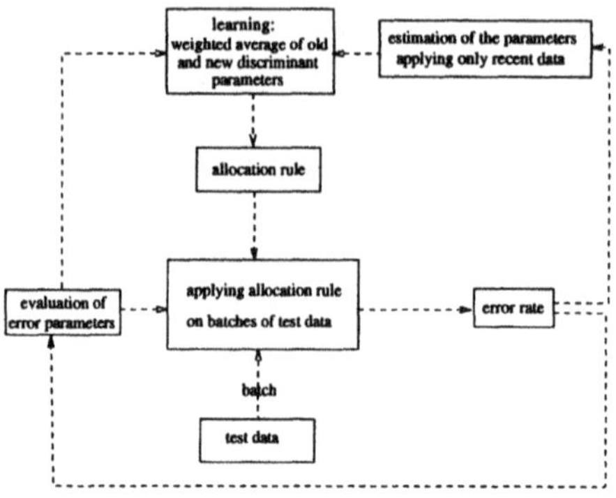

Figure 1: Learning by error reaction

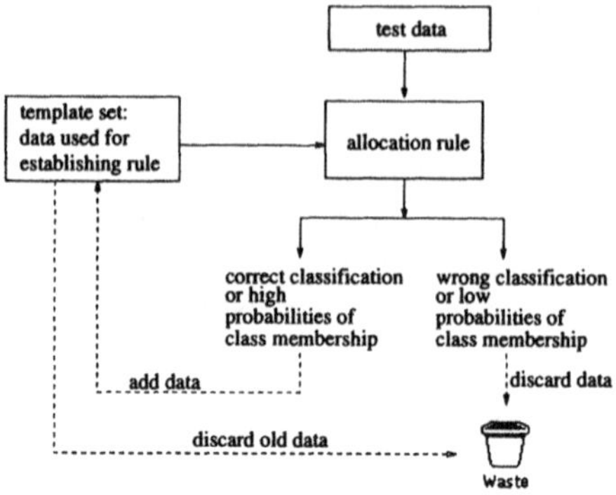

Figure 2: Moving data window

3.2 Learning by Data Selection

We can use a "similarity" rule to edit observations in the current training set which are "unusual" – *i.e.* they are close in feature space, but have a class label different from their recently observed neighbours. Alternatively, the older observations can be eliminated or a kind of moving window with a predefined number of observations, possibly representative and new ones, could be used. Figure 2 illustrates a possible implementation whereby old data are discarded and new data are included in the "template set" used for establishing a rule. This system will have some limitations. For example, if there are drifting populations then new data will be incorrectly classified, but should nevertheless be included in the template set. This point is taken up in an other approach with a 1-nearest neighbour algorithm, where a "record of usefulness" is kept as well as an age decay for each example and by changing that value the classification routine is given a dynamic aspect (see Nakhaeizadeh *at al.* (1996) for further details).

4. Algorithm Details

We used different implementations for linear and logistic discrimination on the one hand and quadratic discrimination on the other hand. One almost common part of both implementations is the initial cross-validation that we used for estimation of the mean error rate and its standard deviation. Another common aspect is that the data set was split into two parts, the training data set and the test data set. In real world applications the training data set would be a data set that has been constructed before applying the discriminant rule, maybe in the conventional way. Then the test data set was split in batches of almost the same size. This was done to simulate the time flow and mimic the real-life situation in which observations are initially classified using the current rule, and then after a period their true class becomes known and they become part of the accumulated training data.

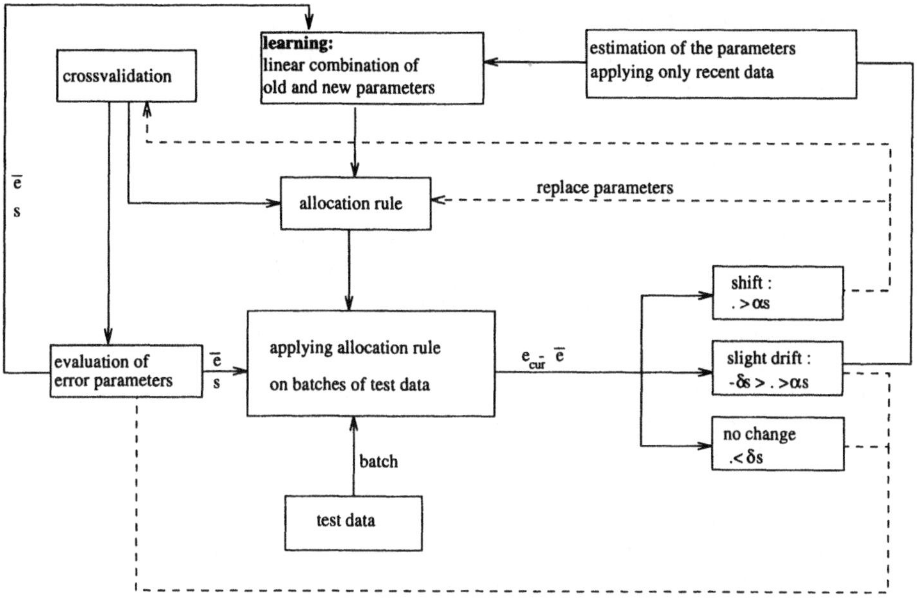

Figure 3: Flow of control in upating rule procedure

4.1 Experiments in Linear and Logistic Discrimination

Ten-fold cross-validation is used for the training data to get an unbiased estimate of the error rule (mean error $\bar{e}$ and standard deviation s). So it is assumed that the training data does not contain any kind of change. Then the discriminant function is established from the whole training data set. There is no built-in learning effect such as taking only those parts of data that have a less than average error rate. As soon as a new batch of test data is available the actual error rate e_{cur} is calculated using the current (most up-to-date) rule. Then the actual error rate e_{cur} is compared to some function of $\bar{e}$ and s:

- **action limit**: $\bar{e} + \alpha s \quad \alpha > 0$
- **operation limit**: $\bar{e} - \delta s \quad \delta \geq -\alpha$

Figure 3 shows the flow of control in this approach. From now on there are three cases available:

error rate is bigger than action limit: $e_{cur} \quad > \quad \bar{e} + \alpha s$
It is assumed that there was a major change in the population. That means that the data for building the discriminant function is no longer valid and should therefore no longer be used for establishing the rule. Instead the discriminant function should be completely relearned using only the current batch (see restrictions below), because there is no more up-to-date data than the last observations. In order to get a comparison for future observations and batches a new error rule has to be calculated. That is, because after a major shift it may happen that the allocation rule works with another

mean error rate and standard deviation, *e.g.* worse than before. To make sure that future shifts can be detected the error rule has to be adapted. In our implementation we executed a new 10-fold cross-validation to deal with this.

There are some problems that can happen at this stage, particularly if the size of the current batch is quite small :

- Robustness (stability) of the estimation of the new allocation rule: Especially in the case of small batch sizes and lots of parameters to be estimated (as in quadratic discrimination) caution is necessary to make sure that the parameters of the allocation rule are not levered by outliers but fit the data.
- Representativeness of the current batch is not given: Sometimes it may occur that the current batch is not truly representative for the current objects which we wish to classify, but contains a large proportion of "unusual" observations. This means that the batch is highly contaminated by outliers and/or atypical data points. In particular, if the distribution of one class is thick-tailed (Cauchy-distribution) this type of shortcoming may arise.
- Numerical instabilities : Within the algorithm some covariance matrices and their inverses have to be computed. In particular categorical data is highly proned to singularity problems that may arise, if in a small part of the data set a certain value is not observed. Another problem appearing at logistic discrimination is that the convergence of some numerical computation can be very slow or even nonexistent.

In our calculations we used batch sizes of 50 in case of the simulated data and 1000 in case of categorical real data. In addition we used the last two current batches (if available) to relearn the rule.

error rate smaller than operation limit: $e_{cur} \quad < \quad \bar{e} - \delta s$

In this case the classifier works better than expected, so that there is no need for changing the discriminant function. Only the error is added to the previous error rates and the error rule as well as $\bar{e}$ and s is updated. If there is no major change, it is to be expected that the mean error rate gives a good prediction of future error rates, when applying the classifier on new unseen examples. Also the long term trend can be visualized by this index.

error rate between operation and action limit: $\bar{e} - \delta s \leq e_{cur} \leq \bar{e} + \alpha s$

This is the normal stage, where the learning process is placed. The following actions are done within here:

- estimation of the discriminant rule applying only the recent batch: The parameters of the discriminant function β_{new} only based on the newest batch are calculated in order to get the trend of the most recent data.
- linear combination of old and new parameters of the discriminant function $(\beta_{old}, \beta_{new})$ with a weight depending on the error rate:

$$\beta_{rule} = \gamma(e_{cur}, \bar{e}, s)\beta_{new} + (1 - \gamma(e_{cur}, \bar{e}, s))\beta_{old} \quad \text{for} \quad 0 \leq \gamma(.) \leq 1$$

where

$$\gamma(e_{cur}, \bar{e}, s) = \frac{e_{cur} - (\bar{e} - \delta s)}{(\alpha + \delta)s} \quad \text{where} \quad \alpha \geq 0 \quad \text{and} \quad \alpha > -\delta.$$

There are many other ways weights could be introduced; we used the function γ, because it was the continuous and linear extension of the two previously described cases. By the parameters α and δ we could easily control action and operation limits and decide in which way we wanted to react to changes. In our experiments we tried values $\delta \in [-1, 1.5]$ and $\alpha \in [1, 4]$. However we think that these parameters are data specific, so that we suggest trying several values. By using a small value for α (e.g. 0.5, ,1, 1.5) the likelihood of re-estimating the whole discriminant function is increased, so that even small shifts and deviant batches suggest a completely new rule. The higher the value for α the smaller the chance to detect shifts, but tending to update the rule by the learning process. The parameter δ controls the operation limit, i.e. the limit, when the adaptation of the rule starts. With increasing values the following of trends is enhanced, because the starting value for the learning process is decreased and the weight of the recent batch in the learning procedure is increased. The effect of learning can be nearly avoided, if negative values for δ are chosen. If we let $\delta = -\alpha$, there is no chance to react on trends. Then the only way of learning is to completely relearn the rule from recent data. We also want to emphasize the special case $\alpha = C$ and $\delta = -C$ with a high value for C (for example 100 or 1000). That leads to a case of no learning at all (see section 5).

4.2 Experiments with Quadratic Discriminant Rule

In comparison to the previous section we used the quadratic discriminant rule for the data learning approach, where we tried to model the learning process by manipulation of the data, available for establishing the discriminant function(see section 3.2). First of all the learning set was used for a 10-fold cross-validation in order to get an unbiased estimation of the average error rate. This time, however, we made use of a third data set, the template set, from which we relearned the rule after each new batch of data coming in. We want to call that algorithm a combined one, since we used a data selection procedure as well as the error handling to have some kind of shift detection and drift observation. Figure 4 and the description will explain more.

Initially we executed a k-fold cross-validation (we selected $k = 10$) in order to find out the error behaviour (mean error rate $\bar{e}$ and standard deviation s). As mentioned in section 4.1 we wanted to mimic the learning process. Therefore we learned the discriminant function from the entire training data and applied it to compute the posterior probabilities of each observation. We discarded the example, if the posterior probability for the true class was low

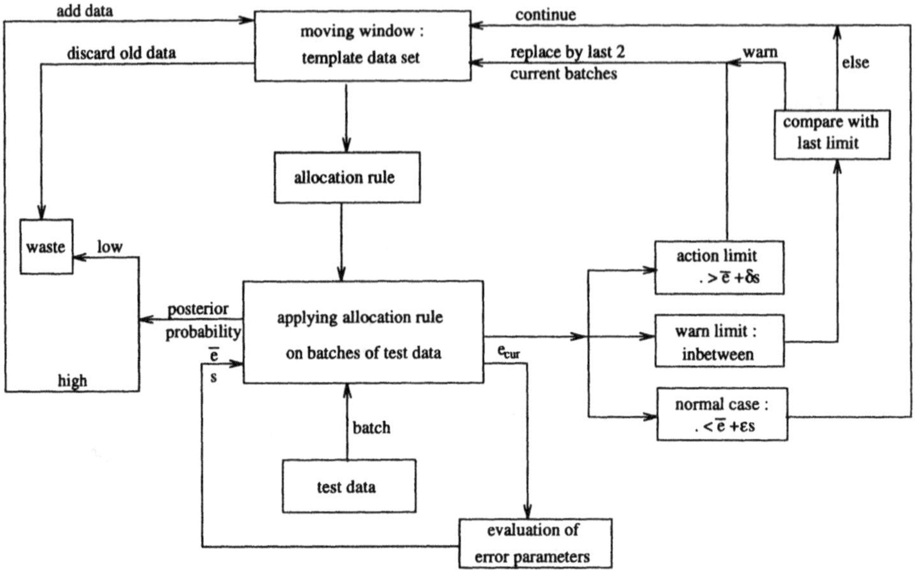

Figure 4: Control flow chart of combined approach (quadratic discrimination)

and added it to the initially empty template set otherwise. From then on we sequentially tested batches of new data, again (as in the previous section) to mimic the flow of time. As soon as this happened the new batch was allocated according to the most up-to-date rule. As a result we can get the current error rate e_{cur} that we can use to find a general trend of the data (also for monitoring purposes).

If the error rate exceeded the **action limit** ($\bar{e} + \delta s,\ \delta > 0$), a change in the population was supposed to have been occurred, so that the template data was no longer valid, but had to be replaced by the most recent data (we always took the last two batches for reasons that have been discussed before).

If the error rate e_{cur} was smaller that the **warn limit** ($\bar{e} + \epsilon s,\ \epsilon < \delta$), we supposed that there was at most little change (slight drift) in the data. For the reason that the allocation rule was re-established anyway we saw no need for any reaction.

In the other case, $\bar{e} + \delta s > e_{cur} > \bar{e} + \epsilon s$, we called it **warn status**, a more detailed observation seemed to be necessary. We looked at the previous status. If that one had also been warn status, the trend seemed to be obvious. Therefore we treated it exactly as surpassing the action limit in the knowledge that the last two recent batches were used to reestablish the discriminant function.

In the case that we replaced the template data in the previous step (last status: action) we saw no need for any special treatment assuming that this might be the result of down-going oscillations from relearning. Besides the weight of the current batch when relearning the rule after that step (about one third of the template set) could result in accepting that trend.

If the last error rate was below the warning limit, the reason could be either the beginning of a future trend that would be detected in the next batch or an unusual batch that should not be paid much attention, so that we wait for the next batch to arrive in order to make the right decision.

After getting and allocating each batch the current error rate e_{cur} was added and the new error rule was computed.

Our **moving window approach** for the selection of data was implemented in the following way: After a new batch arrived each data set was considered and the probabilities of class membership computed (that also had to be done to find the allocated class). If the posterior probability for the right class was high, i.e. bigger than γ/k, where k is the number of possible classes and γ is a parameter to control the area of the desired posteriors ($0 \leq \gamma \leq k$), the data was added to the template set; else it was discarded. As a special case we want to pick $\gamma = 0$, where all examples are added to the template set. After the batch is completely allocated the template set is reduced to the predefined size by using only the newest data. In that way older trends are lessened/eliminated and the newer data is given more weight, since only the template data is used for relearning the new discriminant rule (most up-to-date rule) after this step.

5. Results and Conclusions

In this section we want to describe some results of our adaptive updating ideas and compare them to conventional statistical classification methods in an example. Subsequently the same types of procedures (for benchmark purposes) where used, to simulate other more simple approaches (see special cases in sections 4.1 and 4.2):

- The simplest and nonadaptive approach is to use the classification rule, that was learned from the training data and apply that to all batches. So that there is neither an updating procedure using new data nor updating any kind of rule. We refer to this as **no-learn** approach.
- An alternative method is to completely re-learn the rule at each time point. So after the new batch arrived and was classified, the batch is added to the training data and the rule is replaced by a new one, that uses all up-to-that-time data (training data). Reference: **re-learn**.

5.1 Simulated Data

We tried out some of the above ideas on two simulated datasets (dat1 and dat2). At each of 1000 time points we generate an example with 3 variables (X_1, X_2, X_3) from 2 classes. We use the first 500 observations as the *training data* and the remaining 1500 as *test data.* The distributions of each class has 2 independent normal variables (with unit standard deviation) and a uniformly distributed (on $[0,1)$) "noise" variable. The mean of the noise variable $\mu_3 = 0.5$ was independent of time; the means of the normal variables vary with time as follows:

dat1: Class 1 has $\mu_1, \mu_2 = 0$ for $t \leq 750$ and $\mu_1, \mu_2 = t/1000$ for $751 \leq t \leq 1000$, whereas Class 2 differs in that $\mu_2 = 2$ for $t \leq 750$ and $\mu_2 = 2 + t/1000$ for $751 \leq t \leq 1000$, so that there is no change to the distributions until two thirds of the testing phase, when there is a sudden jump followed by a slow drift.

dat2: Class 1 has $\mu_1, \mu_2 = 0$ and Class 2 has $\mu_1 = 2t/1000, \mu_2 = 2 - 2t/1000$ for $1 \leq t \leq 1000$. In this case there is a gradual shift in the training and test phase of the second group in the mean of (X_1, X_2) from $(0,2)$ to $(2,0)$.

Since we split the testing data into batches of 50 observations (which always corresponds to 25 observations from each class) the change should happen in batch 21 when applying dat1 and should go through the whole training and testing phase when considering dat2. Table 1 shows the results of our calculations using the updating-rule approach. It can be seen that our updating routine achieves better performances than both conventional approaches. In order to see where the improvement occurred, we monitored the error rates of single batches.

data set	dat1		dat2	
parameters of / approach	linear	logistic	linear	logistic
$\delta = 0.5, \alpha = 2$	0.166	0.163	0.228	0.231
$\delta = 1, \alpha = 3$	0.160	0.168	0.229	0.223
no-learn	0.186	0.175	0.360	0.366
re-learn	0.166	0.163	0.271	0.273

Table 1: Average error rates of linear and logistic discrimination rule applying the updating approach (see section 4.1)

As expected due to the large training set, there was no great difference between our classification procedures until the drift happened (e.g. the no-learn routine achieved best results in that area) in the data dat1. But beginning with batch 21 our classifier was much better than conventional methods, with the no-learning rule worst. The error rates of the linear updating rule with parameters $\alpha = 3, \delta = 1$ after this batch were 5.2% (8.4%) smaller than the comparable rules, which results in an enhancement of 24% (33%) less misclassified observations.

data set	dat1	dat2
parameters of / approach	mean error rates	
$\gamma = 0.25$,ws= $300, \delta = 3, \epsilon = 2$	0.164	0.232
$\gamma = 0.25$,ws= $600, \delta = 3, \epsilon = 2$	0.168	0.241
$\gamma = 0.5$,ws= $300, \delta = 3, \epsilon = 2$	0.165	0.236
$\gamma = 0.5$,ws= $500, \delta = 2.5, \epsilon = 1.5$	0.164	0.229
$\gamma = 1$,ws= $300, \delta = 3, \epsilon = 2$	0.175	0.239
no-learn	0.190	0.353
re-learn	0.175	0.267

Table 2: Average error rates of quadratic rule applying the data learning approach (4.2) (ws =window size)

The second data set caused greater problems for our discriminant function. But here also our algorithm showed higher success rates than the other approaches. So that there were 65 (198) less misclassified observations when comparing the linear adaptive updating rule (parameters: $\delta = 1, \alpha = 3$) to the relearning (no-learning) rule, which is 3.2% (9.9%) of all observations. If we consider the quadratic (data-learn) approach, we get similar results (see Table 2). For the first 20 batches of dat1 all routines seem to have almost the same performance, although the relearning method is slightly more accurate than the other ones. But again after the shift and during the drift (dat1) the no-learn method loses performance, whereas the relearning rule works almost as well as our algorithms. But as soon as we apply the overall drift data set (dat2) to the different approaches the enhancement by our algorithm is obvious. The monitoring process shows almost constant error rates, while the error rates of the re-learn approach gradually increases (the no-learn rule completely ceases to work: three batches produced error rates of 50 percent).

We tried to apply values for our parameters from which the reader can observe the effects of changing them. So that we stepwise decreased only γ from 1 to 0.25 with other parameters constant. Then we increased the window size from 300 to 600.

5.2 Real Data

The data set concerns applications for credit; the class indicates the success of the application (*i.e.* two classes). There are 156 273 observations which cover a 2-year period. Originally there were many qualitative attributes which were then coded into indicator variables (since all the programs used here require real-valued data). Subsequently, 15 of these 0/1 variables were used for classification purposes. The initial 10-fold cross-validation was computed using the first 5000 observations. Subsequently, we used batch sizes of 1000. The error rates are shown in Table 3 for the linear and logistic

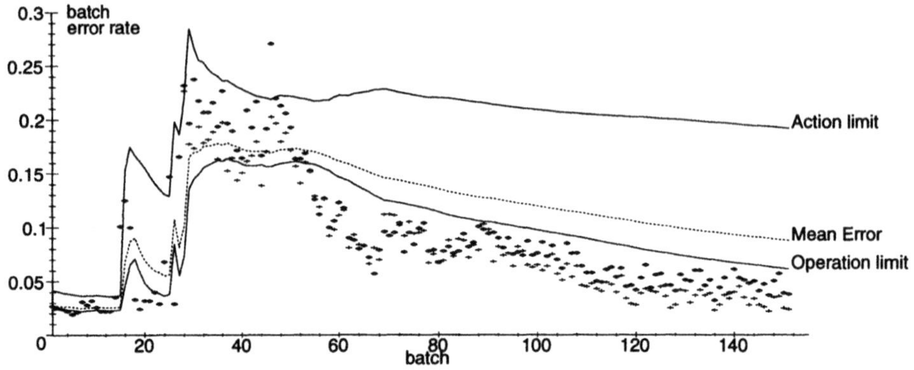

Figure 5: Typical output of monitoring process for linear and logistic discrimination applying the "errorlearn approach". This example uses the real data and compares linear adaptive discrimination (+) with parmeters $\delta = 0.5, \alpha = 2.0$ and re-learn method ($\diamond$).

approach and in Table 4 in the case of the quadratic method.

param. / approach	linear	logistic
$\delta = 0.25, \alpha = 2$	7.97	–
$\delta = 0.5, \alpha = 2$	7.96	7.45
$\delta = 0.5, \alpha = 1$	8.54	–
$\delta = 1, \alpha = 3$	8.49	7.46
$\delta = 1.5, \alpha = 2$	8.56	7.36
$\delta = 2, \alpha = 4$	–	7.33
$\delta = -1, \alpha = 1$	7.73	7.45
no-learn	8.97	8.26
re-learn	8.97	8.97

Table 3: Average error rates (%) of linear and logistic discrimination rule applying the updating approach (section 4.1) and using real data

param. of quadratic / approach	$\bar{e}$
$\gamma = 0$,ws$= 3000, \delta = 100, \epsilon = 100$	11.94
$\gamma = 1$,ws$= 3000, \delta = 7, \epsilon = 8$	8.789
$\gamma = 0.25$,ws$= 5000, \delta = 1, \epsilon = 3$	8.865
$\gamma = 0.7$,ws$= 7000, \delta = 1, \epsilon = 2$	8.861
$\gamma = 0.5$,ws$= 5000, \delta = 1.5, \epsilon = 3$	8.830
$\gamma = 0.3$,ws$= 8000, \delta = 1, \epsilon = 4$	8.863
no-learn	8.974
re-learn	8.974

Table 4: Average error rates (%) of quadratic discrimination rule applying the data-learn-approach (section 4.2) and using real data

Figure 5 suggests that there is a shift in the data in batches 30 – 50 which was where our algorithms showed the greatest difference in the success rates. For example the linear update process only differed from the conventional methods in this region. Unexpectedly the re-learn and no-learn approaches had the same success rates when using linear discriminant rule. As expected the logistic approach showed the best performance, but even here we could find parameters that improved the overall performance. Also astonishing were the results when applying the parameters ($\alpha = 1, \delta = 1$), which causes only relearning at certain steps but no updating at all.

As in previous studies the quadratic discriminant rule did not work as well as the other classifiers in the case that the data was binary.

5.3 Conclusions

It can be seen that as soon as trends seem to be in the observations, our algorithms improve the results in comparison to the conventional methods. But there are some difficulties to find the right parameter values for the data sets. Also we want to emphasize the importance of the monitoring process, without which our algorithm would not be possible. We think that further research is not only necessary but also worthwhile to find out more details in our approach and to build some new ideas to improve the old idea of classification. The interested reader is referred to the "Diplomarbeit" of Kunisch (1996), where more details can be found.

References:

KUNISCH, G. (1996, July). Anpassung und Evaluierung statistischeLernverfahren zur Behandlung dynamischer Aspekte in Data Mining. Diplomarbeit Universität Ulm.

MARDIA, K. V.; KENT, J. T. and BIBBY, J. M. (1979). *Multivariate Analysis.* London: Academic Press.

MClACHLAN, G. J. (1992). *Discriminant Analysis and Statistical Pattern Recognition.* New York: John Wiley.

MICHIE, D. M.; SPIEGELHALTER, D. J. and TAYLOR, C. C. (Eds.) (1994). *Machine Learning, Neural and Statistical Classification.* Chichester: Ellis Horwood.

NAKHAEIZADEH, G.; TAYLOR, C. C. and G. KUNISCH (1996). Dynamic aspects of statistical classification. *submitted.*

UTGOFF, P. E. (1989). Incremental learning of decision trees. *Machine Learning 4*, 161–186.

UTGOFF, P. E. (1994). An improved algorithm for incremental induction of decision trees. In *Proceedings of Eleventh Machine Learning Conference*, Rutgers University. Morgan Kaufmann.

A Hierarchical Neural Network Model for Pattern Recognition

Steffen Simon[1,2], Friedhelm Schwenker[2], Günther Palm[2]

[1] University of Ulm, Division of Nephrology, Dept. Medicine II, D-89081 Ulm

[2] University of Ulm, Dept. of Neural Information Processing D-89081 Ulm, Germany

Summary: A hierarchical neural network architecture for classification tasks is proposed. This algorithm has been evaluated in an OCR problem. The basic components of the network architecture are LVQ and RBF networks.

1. Introduction

Feedforward neural networks are used in different applications for classification tasks (KOHONEN (1995), BISHOP (1995)). Typically, in this kind of application the neural network performs a mapping from the feature space $\mathbb{R}^n$ into a finite set of classes $C = \{1, 2, \ldots, l\}$. During the training phase of a supervised learning algorithm the parameters of the network are adapted by presenting a finite set of training examples to the network. The training set $S := \{(x^\mu, t^\mu) \mid \mu = 1, \ldots, M\}$ in a supervised learning scenario consists of feature vectors $x^\mu \in \mathbb{R}^n$ each labeled with a class membership $t^\mu \in C$. After the training procedure – in the recall phase – unlabeled observations $x \in \mathbb{R}^n$ are presented to the trained network. This outputs a class $z \in C$ estimated for the input x.

Our application is the classification of gray scale images. The patterns are machine written symbols from a large alphabet containing the whole Latin and Greek and several mathematical symbols as subalphabets. Furthermore, the symbols of the Latin alphabet may appear from different fonts. Because the patterns are extracted from structured mathematical formulas the input patterns appear in different sizes.

We use a hierarchical network architecture, because in an application with many classes ($l = 135$ in section 5), a simple classifier is able to discriminate between patterns from a large subset of these classes. Furthermore, misclassifications of patterns typically take place between two or three classes not between all classes. Thus, only for a small fraction of patterns a complex classifier is necessary. This observation leads us to the construction of a hierachical neural network based on (1) an LVQ network for the rough classification of the patters and (2) a set of RBF networks each appropriate for the discrimination of patterns from the so-called confusion classes.

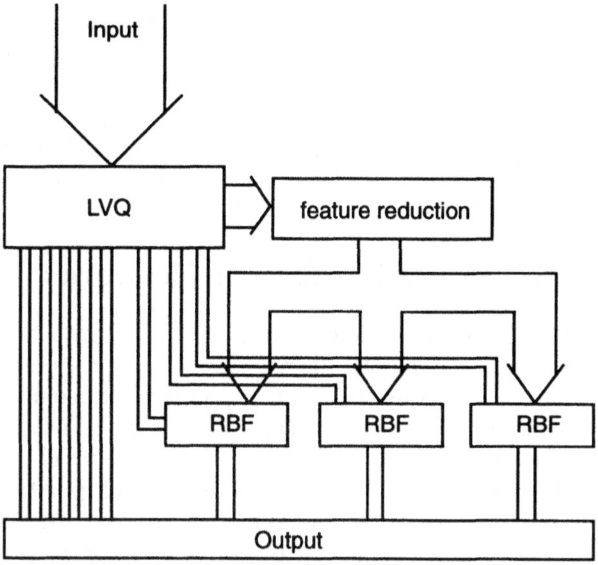

Figure 1: Overview of network architecture

2. Learning Vector Quantization (LVQ) Networks

An LVQ network is a competitive neural network consisting of a single layer of k neurons. The neurons have binary valued output where $y_j = 1$ stands for an active neuron and $y_j = 0$ for a silent one. The synaptic weight vectors $c_1, \ldots, c_k \in \mathbb{R}^n$ of an LVQ network divide the input space into k disjoint regions $R_1, \ldots, R_k \subset \mathbb{R}^n$, where each set R_j is defined by

$$R_j = \{x \in \mathbb{R}^n \mid \|x - c_j\| = \min_{i=1,\ldots,k} \|x - c_i\|\}. \tag{1}$$

Here $\|\cdot\|$ denotes the Euclidean norm. Such a partition is called a *Voronoi tesselation* of the input space. The weight vector c_j is also called a *prototype vector* of region R_j.

When presenting an input vector $x \in \mathbb{R}^n$ to the net, all neurons $j = 1, \ldots, k$ determine their Euclidean distance $d_j = \|x - c_j\|$ to input x. The competition between the neurons is realized by searching for the prototype with minimum distance: $d_{j^*} = \min_{j=1\ldots,k} d_j$. In LVQ-networks each prototype vector c_j is labeled with its class membership $y_j \in C$.

In LVQ1 training – the simplest version of LVQ-training – only the neuron j^* – the *winning neuron* – is adapted according to the learning rule:

$$\Delta c_{ij^*} = \eta(t)\delta_{y_{j^*}t^\mu}(x_i^\mu - c_{ij^*}). \tag{2}$$

Here is $\delta_{ij} = 1$ if $i = j$, $\delta_{ij} = 0$ otherwise, and $\eta(t) > 0$ is a decreasing learning rate.

This competitive learning algorithm has been suggested (KOHONEN (1990)) for vector quantization and classification tasks. From this algorithm Kohonen derived the OLVQ1, LVQ2 and LVQ3 network training procedures, which are useful algorithms for the fine-tuning of a pre-trained LVQ1 network. We use the OLVQ1 training procedure in our application. OLVQ1 realizes prototype depending learning rates which are adapted during training.

3. Radial Basis Function (RBF) Networks

In an RBF network the input vector $x \in \mathbb{R}^n$ is fed into a layer of k hidden neurons, each calculating the distance $d_j = \|x - c_j\|$ between its weight vector $c_j \in \mathbb{R}^n$ and the presented input x. This distance d_j is transformed by a nonlinear transfer function $h : \mathbb{R}_+ \rightarrow \mathbb{R}_+$ – the radial basis function – to the neuron's output $y_j = h(\|x - c_j\|)$. A popular choice for h is the Gaussian density function $h(s) = exp(-s^2/\sigma^2)$, with positive variance σ^2. Each neuron $p \in \{1, \ldots, l\}$ of the output layer receives inputs from all hidden neurons $j \in \{1, \ldots, k\}$ weighted by b_{jp}, such that its activity is given by the weighted sum:

$$z_p = \sum_{j=1}^{k} b_{jp} h(\|x - c_j\|). \tag{3}$$

In function interpolation, radial basis functions with fixed prototype vectors have been extensively studied (MICCHELLI (1986)). RBF networks with free prototype vectors are shown to be universal approximators which are able to approximate continous functions with arbitrary precision (PARK (1993)) – this property implies that RBF networks with adjustable prototypes are powerful enough for classification tasks (POGGIO (1990)).

Learning in an RBF network stands for (a) adapting the weight vectors $b_p \in \mathbb{R}^k$ between hidden and output layer and (b) moving the location of the prototype vectors $c_j \in \mathbb{R}^n$. This can be achieved by minimizing the least mean square error function:

$$E(c_i, b_p) := \sum_{\mu=1}^{M} \|t^\mu - z^\mu\|^2 = \sum_{\mu=1}^{M} \sum_{p=1}^{l} (t_p^\mu - \sum_{j=1}^{k} b_{jp} h(\|x^\mu - c_j\|))^2, \tag{4}$$

by a gradient descent optimization algorithm. The initial positions of the RBF neuron can be determined by LVQ training (SCHWENKER (1994)).

Determining the gradient direction leads to the learning rules for both types of parameters b_p and c_j. For a single example of the training set (x^μ, t^μ) these weight update rules are:

$$\Delta b_{jp} = \eta_1(t)(t_p^\mu - z_p^\mu)y_j^\mu \tag{5}$$

$$\Delta c_{ij} = \eta_2(t)(-h'(d_j^\mu))(x_i^\mu - c_{ij})\sum_p (t_p^\mu - z_p^\mu)b_{jp}. \tag{6}$$

These learning rules decrease the error function $E(c_i, b_p)$ for small enough learning rates at each step, and let successive steps adapt to the local gradient.

In the recall phase an input vector $x \in \mathbb{R}^n$ is presented to the network. This leads to an activation $z_p = \sum_{j=1}^k b_{jp}h(\|x - c_j\|)$, $p = 1, \ldots, l$, of the output neurons. The class of x is determined by the maximally activated output neuron.

4. Hierarchical Classifier

In this section we describe the hierarchical neural network classifier. The training phase of this network contains the following steps:

1. Training of the LVQ network
2. Calculation of confusion classes utilizing the trained LVQ classifier
3. Feature selection for each confusion class
4. Training of the RBF networks for each confusion class

In the first step of this training procedure an LVQ network is trained with the whole training set – we call this the training of the rough classifier. After this, the classification performance of this LVQ network is tested on the training set. The result of this classification is represented in a $l \times l$ confusion matrix V (l the number of classes), where $V_{ij} \in \mathbb{N}$ denotes the number of training patterns with target output i but LVQ network output$j$.

4.1 Calculation of the Confusion Classes

From the confusion matrix V we calculate a Boolean matrix V' by:

$$V'_{ij} = \begin{cases} 1 & : \frac{V_{ij}}{n_i} > \Theta \\ 0 & : \text{otherwise} \end{cases}$$

where $\Theta \in [0, 1)$ is a fixed threshold value and n_i is the number of training patterns of class i. Thus, a matrix entry $V'_{ij} = 1$ indicates a confusion from target class j to class i.

The following conditions on V' define an equivalence relation on the set of confusions:

Figure 2: Typical distribution of variance σ^2 inside a confusion class

1. $V'_{ij} = 1$ implies $V'_{ji} = 1$ (symmetric relation)
2. $V'_{ij} = 1$ implies $V'_{ii} = 1$ and $V'_{jj} = 1$ (reflexive relation)
3. $V'_{ij} = 1$ and $V'_{jk} = 1$ implies $V'_{ik} = 1$ (transitive relation)

A completion of the Boolean matrix V' by these 3 conditions results into a disjoint partion of the classes, the set of confusion classes $\mathcal{V}_i$.

In applications where the number of classes is small or where confusions appear between many different classes, the transitivity condition leads to the trivial solution with a single confusion class containing all classes. This effect can be controlled by setting Θ to a larger positive value.

4.2 Feature Selection

Misclassifications by the LVQ network take place if vectors from different classes are to close together measured by the Eucldean distance defined on the feature space. In particular, this happened for data points which are assigned to a certain confusion class $\mathcal{V}$. In other words, for each confusion class $\mathcal{V}$ there may exist a large subset of features where the input patterns are very similar. Therefore, the feature variances $\sigma^2 = (\sigma_1^2, \ldots, \sigma_n^2)$ for each confusion class $\mathcal{V}$ are small in there components. In Fig. 2 a typical variance distribution is shown.

The procedure to determine the relevant features for each confusion class $\mathcal{V}$ is the following:

1. Calculate $Q = (n-1)\sigma^2$
2. Calculate Q_I the sum of the prototype within variances
3. Calculate $Q_Z = Q - Q_I$ the sum the prototype between variances
4. Select the $N < n$ features with largest Q_Z values

1234567890
ABCDEFGHIJKLMNOPQRSTUVWXYZ
ÄÖÜ
abcdefghijklmnopqrstuvwxyz
äöü
≠<>~+-.,:;*/∞ø∫()[]{}|
ΔΦΓΛΠΘΣΩΞΨ
αβχδεφγηιϕκλμνπθρστυϖωξψζ

Figure 3: Patterns

The prototype vectors of the hidden layer RBF neuron for each network can be initialized by the calculated means and variances of the confusion class.

5. Application and Results

In our application we use a data set of artifically generated gray scale images showing symbols of the Latin and Greek alphabet and mathematical symbols. These patterns have been generated in the range from 6pt to 12pt in a solution of 300 dpi. Finally, these images have been scaled to an 8 bit gray scale image of size 16x16 pixel (see Fig. 3). From these images 98 features have been extracted by parallel projection and filtering (cf. chapter 8 in ROSENFELD (1982)). These 98 dimensional feature vectors serve as input vectors to the LVQ network.

For this kind of application 4 different architectures have been realized:

LVQ	classification by a LVQ network
RBF	classification by a RBF network
LVQ/LVQ	LVQ rough classification and classification of the confusion classes
LVQ/RBF	rough classification by LVQ and RBF for the confusion classes

In each architecture 2700 patterns have been used for training and 1000 patterns for testing the classification performance. We use 198 neurons (135 classes in the data set) in the hidden layer of the LVQ and RBF architecture as well as for the LVQ network in the hierarchical architectures LVQ/LVQ and LVQ/RBF. All calculations have been carried out on a Power Macintosh 7500/100.

	LVQ/LVQ	LVQ/RBF	LVQ	RBF
training cycles	100/100	100/100	100	100
error rate on training set	5.1%	1.1%	7.9%	0.8%
error rate on test set	7.8%	2.1%	10.1%	1.7%
time for training	8 min	12 min	6 min	8h40

The LVQ/RBF architecture show classification rates which are close to those which were achieved by RBF networks, but with training times close to these of LVQ nets. This is possible because such a hierachical architecture is able to split the classification problem in several classification subproblems. Wheras nonhierachical classifiers like RBF nets have to deal with the worst case classification problem, which drastically effects the computational resources.

References:

BISHOP, C. M. (1995): Neural Networks for Pattern Recognition, Clarendon Press, Oxford

KOHONEN, T. (1990): The self-organizing map, *Proc. IEEE, 78(9), 1464–1480*

KOHONEN, T. (1995): Self-Organizing Maps, Springer, Berlin

MICCHELLI, C. A. (1986): Distance matrices and conditionally positive definite functions, *Constructive Approximation, 2, 11–22*

PARK, J. and SANDBERG, I. W. (1993): Approximation and Radial Basis Function Networks, *Neural Computation, 5, 305–316*

POGGIO, T. and GIROSI, F. (1990): Networks for Approximation and Learning, *Proc. IEEE, 78, 1491–1497*

ROSENFELD, A. and KAK, A., (1982): Digital Picture Processing, volume 1, Academic Press, Oxford

SCHWENKER, F.; KESTLER, H. A.; PALM, G. and HOEHER, M. (1994): A survey of learning rules and the application to the classification of signals from high-resolution electrocardiography, *Proc. IEEE, International Conference on Systems, Man and Cybernetics, 646–651*

Section 3

Statistical Models and Methods

Markov Random Field Models with Spatially Varying Coefficients

J. Dreesman

Fachgebiet Statistik u. Wirtschaftsmathematik, Fachbereich Informatik, TU Berlin, Franklinstr. 28/29, 10587 Berlin, Germany

Summary: Markov random field models are used to describe the distribution of spatially arranged random variables. When estimating the parameters of such models, usually second order stationarity is assumed, but often doubted in applications. We present an approach which allows for nonstationarities of the mean and the covariance. For this purpose spatially varying coefficients are introduced. In applications we obtained more appropriate models as well as further insight into the structure of spatial data. The Mercer and Hall wheat-yield data are presented as example.

1. Markov random field models

Various areas of statistical application deal with random variables that are observed at spatial locations, for example geostatistics, environmental sciences, epidemiology, biostatistics (field trials) and image analysis. In these applications it often occurs that random variables lying close together are likely to be correlated, a phenomenon also known from time series data. Markov random fields provide a class of models for analysing spatially dependent random variables of this kind.

In the simplest case, the random variables are located on a regular lattice $L = \{(i,j), 1 \leq i \leq I, 1 \leq j \leq J\}$. The whole set of variables $\{X_{i,j}, (i,j) \in L\}$ can also be considered as one multivariate variable $\mathbf{X}$. Usually the dependence structure of the components of $\mathbf{X}$ is modelled in terms of the full conditionals. These are the conditional distributions of each univariate variable, given the values of all the others. $\mathbf{X}$ is called a Markov random field if every single random variable is conditional independent on all the others, given those in its immediate neighbourhood:

$$\pi\left(x_{i,j} | \{x_{r,s}, (r,s) \neq (i,j)\}\right) = \pi\left(x_{i,j} | \{x_{r,s}, (r,s) \in D_{i,j}\}\right). \qquad (1)$$

Here $D_{i,j}$ denotes the neighbourhood at site (i,j).

Often a first order neighbourhood and a normal distribution are assumed, leading to the following auto-normal model (Besag (1974)):

$$\begin{aligned} &X_{i,j} | \{x_{r,s}, (r,s) \neq (i,j)\} \\ &\quad \sim \; N(\beta_0 + \beta_1 \cdot (x_{i+1,j} + x_{i-1,j}) + \beta_2 \cdot (x_{i,j+1} + x_{i,j-1}), \tau^2). \end{aligned} \qquad (2)$$

Objectives of the application of Markov random field models are the prediction of missing or unobservable values and the detection of outliers. The influence of external variables can be taken into consideration by including additional regressors. The analysis of spatially arranged random variables via Markov random field models can also be useful to get some insight into the spatial dependence structure.

2. Parameter estimation

Usually a field is observed only once. Therefore serious problems occur, if the parameters of a Markov random field model, e. g. the βs determining the full conditionals of an auto-normal model, have to be estimated. In this case no sample of independent, identically distributed random variables is available. For this reason some stationarity assumptions are made for the field's components, in order to treat the single observed field as a sample of identically distributed random variables. Usually second order stationarity is assumed:

$$E(X_{i,j}) = \mu \quad \text{for all } (i,j) \in L, \tag{3}$$

$$cov(X_{i,j}, X_{i+u,j+v}) = \gamma(u,v) \quad \text{for all } (i,j),(i+u,j+v) \in L. \tag{4}$$

In this case the mean and the covariance function do not depend on the position on the lattice and the covariance of two variables only depends on their distance vector. The auto-normal model (2) is stationary since the βs and τ^2 are the same for the full conditionals at any site.

Additional difficulties are caused by the dependence of the field's components, which makes maximum likelihood estimation very awkward, because the joint log-likelihood cannot be given as sum of single log-likelihoods and often contains terms that still cannot be calculated with a justifiable amount of computer time. As a solution Besag and Moran (1975) developed an approximation to the likelihood. Geyer (1991) suggested a kind of stochastic integration using Markov chain Monte Carlo to simulate the random field.

As more practicable likelihood based estimation methods Besag (1974, 1975) proposed the coding method and the pseudolikelihood method. The coding method uses only a restricted subset of the data to calculate the likelihood estimates. The subset is constructed in a way, that the required independence conditions are fulfilled for its elements. In contrast the pseudolikelihood method uses the whole set of single likelihoods for estimation but these likelihoods are treated as if the variables were independent. In many cases the pseudolikelihood method seems to be a good compromise between the higher efficiency of the maximum likelihood estimator and requirements of computability.

3. The Mercer and Hall wheat-yield data

The possibly best analysed lattice data are the wheat-yields of Mercer and Hall (1911), that were obtained in an uniformity trial, where the same treatment was applied to each plot of a 20 $\times$ 25 lattice. Figure 1 gives an impression of this data set, that was chosen by several researchers as an examplary application of their methods, see for example Whittle (1954), Besag (1974) and Cressie (1993).

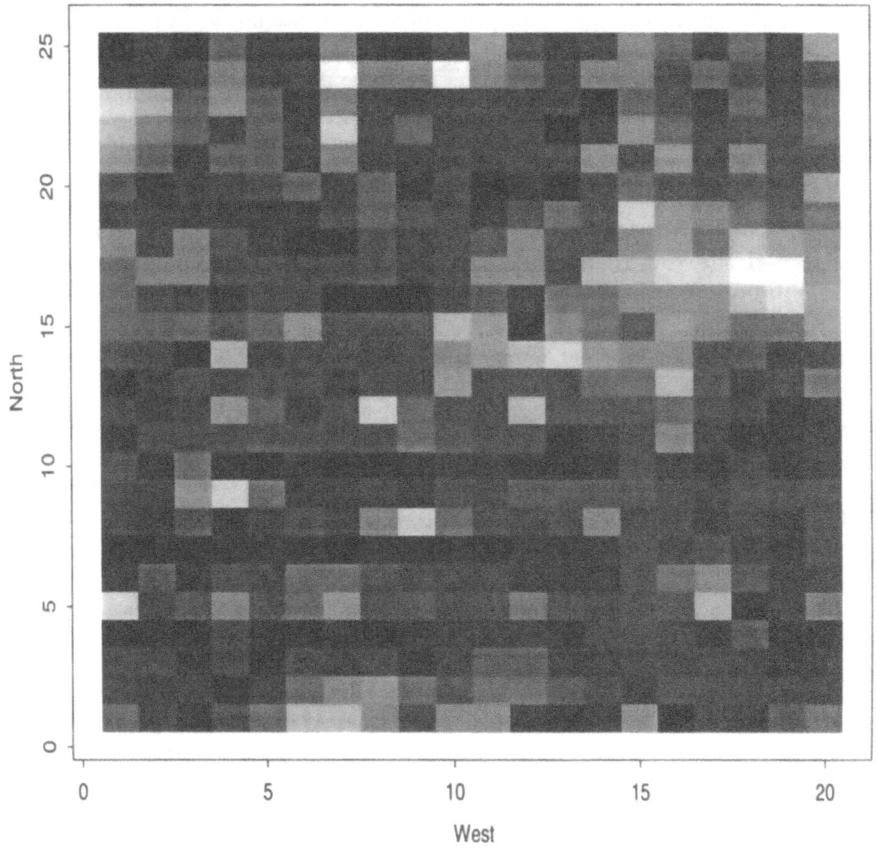

Figure 1: Wheat-yields on 20×25 lattice.

Besag (1974) has found the auto-normal model (2) as the most appropriate model for these data. He obtained the approximate maximum likelihood estimates $\hat{\beta}_1 = 0.368$ and $\hat{\beta}_2 = 0.107$, indicating that there is a stronger dependence between neighbours in the rows than between neighbours in the columns.

4. The problem of stationarity

In many applications the stationarity assumption for the mean is obviously not fulfilled. Deviations from the mean stationarity often can be detected by comparing the row and column means or medians. Such a comparison shows that this assumption is violated also for the wheat yields. In the case of a conditional specified Markov random field, mean nonstationarities can lead to parameter estimates that fall outside the parameter space. Therefore it is desirable to remove these nonstationarities. This can often be done by subtracting the row means. Other possibilities are median polishing, smoothing the mean or explaining the mean by covariates. For an overview, in particular for this data set, see Cressie (1993).

On the one hand nonstationarity of the mean is taken into consideration in many applications, on the other hand the covariance stationarity is hardly ever doubted.

Visual inspection of the wheat-yields leads to the impression that neighbours on the rows are stronger connected at the bottom of the field than at the top. Even if the row means are subtracted, this impression still remains. The neighbours on the columns appear to be stronger connected in the right of the field compared with the left. Therefore the covariance stationarity assumption also seems to be violated for the wheat-yield data.

5. Varying coefficient models

Covariance nonstationarity cannot be taken into account with standard methods. The task is to find a model that allows for nonstationarity of the covariance as well as of the mean. For this purpose spatially varying coefficients are introduced. Varying coefficient models (Hastie and Tibshirani (1993)) are generalized linear models, where the coefficients are not assumed to be fixed, but vary across some covariate called effect modifier.

For example, after introduction of varying coefficients, the standard linear model with response variable Y, covariates $X_1, \ldots, X_P$ and effect modifier R is

$$Y = f(X_1, \ldots, X_P, R) + \epsilon = \beta_0(R) + X_1\beta_1(R) + \ldots + X_P\beta_P(R) + \epsilon. \quad (5)$$

Furthermore the βs are assumed to be smooth functions of the effect modifier R, which can be considered as an additional explanatory variable. If $\beta_p(R) = \beta_p$ for $0 \leq p \leq P$, the ordinary linear model is obtained as a special case. A smooth regression of Y on R is obtained as another special case if no regressor X_p is given.

Hastie and Tibshirani (1993) suggested a penalized least squares approach in order to estimate the smooth functions $\beta_p(R)$, while other authors used the local likelihood method, earlier proposed by Tibshirani and Hastie (1987), which is related to kernel methods (see e. g. Staniswalis (1989)).

If kernel based methods are used, each function $\beta_p(R)$ is estimated as a nonparametric curve, composed of estimates at many target points, distributed over the range of the effect modifier R.

For each estimate, weights are given to all the observations. Let r be the target point and $d(r, r_n)$ the distance of the nth observation's effect modifier from the target point. Then the general formula of a weight function based on a kernel is

$$w_\lambda(r, r_n) = cK\left(\frac{d(r, r_n)}{\lambda}\right). \qquad (6)$$

The weight function can be adjusted through the constant factor c and the parameter λ. Usually the weights decrease as the distance increases. An example for a weight function is the tricube function of Cleveland (1979):

$$w_\lambda(r, r_n) = \begin{cases} (1 - (\frac{|r_n - r|}{\lambda})^3)^3, & \text{if } 0 \leq |r_n - r| < \lambda; \\ 0, & \text{otherwise.} \end{cases} \qquad (7)$$

Here the bandwidth λ gives the maximum distance of the effect modifier from the target point, for which a positive weight is received. The range of effect modifier values, for which observations receive positive weights, is called window.

Since the bandwidth controls the smoothness of the curve, its choice is governing the trade-off between the bias and the variance of the estimates. Increasing λ decreases the variance but tends to increase the bias, because the estimate at target point r might involve more terms with β-values different from $\beta(r)$. If λ tends to ∞, all the observations are receiving equal weights and a fixed parameter model is obtained. On the other hand decreasing λ leads to an increasing variance and a decreasing bias. If λ tends to zero, the number of observations involved in an estimate at any point becomes small and the estimated curves are quite jagged.

To consider both, bias and variance, often the cross-validation sum of squares $CV(\lambda) := \frac{1}{N}\sum_{n=1}^{N}(y_n - \hat{f}_{\lambda,n}^{-n})^2$ is minimized with respect to λ in order to find the bandwidth that leads to the smallest average mean squared error. In the case of dependent errors, van der Linde (1994) has shown that instead of $CV(\lambda)$ the cross-validation sum of squares for dependent observations $DCV(\lambda) := \frac{1}{N}\sum_{n=1}^{N}(y_n - (\hat{f}_{\lambda,n}^{-n} + \hat{\epsilon}_{\lambda,n}^{-n}))^2$ should be minimized. In these expressions $\hat{f}_{\lambda,n}^{-n}$ denotes the predictor at $(x_{1n}, \ldots, x_{Pn}, r_n)$ and $\hat{\epsilon}_{\lambda,n}^{-n}$ the corresponding predicted residual, which is obtained using the residuals' correlation structure. Both values are calculated with coefficients that were estimated leaving out the nth observation.

6. Markov random field models with spatially varying coefficients

If the covariances of variables on a lattice are assumed to be not fixed but not abruptly changing either, a Markov random field model with coefficients that vary across the location might be appropriate:

$$\begin{aligned} X_{i,j}|\{x_{r,s}, (r,s) \neq (i,j)\} \sim \; & N(\beta_0(i,j) + \beta_1(i,j) \cdot (x_{i+1,j} + x_{i-1,j}) \\ & + \beta_2(i,j) \cdot (x_{i,j+1} + x_{i,j-1}), \tau^2(i,j)). \end{aligned} \qquad (8)$$

The pseudolikelihood method can easily be adapted to this estimation problem. If maximum pseudolikelihood is applied to an auto-normal model in a fixed coefficients setting, the method leads to a least squares estimation problem, in which each observation once is the value of the response variable with all its neighbours being explanatory variables and is also explanatory for all its neighbours being responses.

In matrix notation the estimate of the parameter vector $\beta = (\beta_0, \ldots, \beta_P)'$ is

$$\hat{\beta} = (\mathbf{Z}'\mathbf{Z})^{-1}(\mathbf{Z}'\mathbf{x}), \tag{9}$$

where $\mathbf{Z} = (\mathbf{1}, \mathbf{H}_1\mathbf{x}, \ldots, \mathbf{H}_P\mathbf{x})$ is the design matrix and every $\mathbf{H}_p$ is a $(I{\cdot}J) \times (I{\cdot}J)$ matrix with entries 1, where two variables are neighbours with respect to β_p, and entries 0 elsewhere. In the auto-normal model (2), for example, $x_{i,j+1}$ is a neighbour of $x_{i,j}$ with respect to β_2 and therefore $\mathbf{H}_2$ has entry 1, where the line corresponds to the position of $x_{i,j}$ in the $\mathbf{x}$ vector and the column corresponds to the position of $x_{i,j+1}$.

If now the coefficients are assumed to vary, parameters are estimated at each point of the lattice. After introducing weights, the local-pseudolikelihood estimator is

$$\hat{\beta}(i,j) = (\mathbf{Z}'\mathbf{W}^{\lambda}_{(i,j)}\mathbf{Z})^{-1}(\mathbf{Z}'\mathbf{W}^{\lambda}_{(i,j)}\mathbf{x}), \tag{10}$$

where $\mathbf{W}^{\lambda}_{(i,j)} = \text{diag}\{w_\lambda((i,j),(1,1)), w_\lambda((i,j),(1,2)), \ldots, w_\lambda((i,j),(I,J))\}$. Since the effect modifier is measured in $\mathbb{R}^2$, an appropriate distance function is needed, which will usually be the Euclidean distance.

7. Application to the wheat-yield data

Model (8) was fitted to the wheat-yield data using the tricube weight function with $\lambda = 15$. Fitting an auto-normal model to a spatial data set leads to correlated residuals and therefore the value for the bandwidth was selected by minimizing an approximation to $DCV(\lambda)$.

In figure 2, for each coefficient β_0, β_1 and β_2 a separate contour plot shows, how the estimate is varying across the field. Since the variation of $\hat{\beta}_1$ and $\hat{\beta}_2$ is of most interest, their distributions are additionally presented through box plots.

From the contour plots it is obvious that the values of $\hat{\beta}_1$ are highest at the bottom in the middle and decrease to the top. For $\hat{\beta}_2$ an increase from the bottom left to the top right can be observed. This confirms the impression about a change in the covariance structure that was mentioned above.

The values of $\hat{\beta}_0$ should be interpreted carefully. They involve the nonstationarity of the mean, but they also depend heavily on the values of $\hat{\beta}_1$ and $\hat{\beta}_2$.

The gradients of the varying coefficient surface show, how strong the coefficients are varying in the corresponding directions, but they cannot exceed an upper limit determined by the bandwidth.

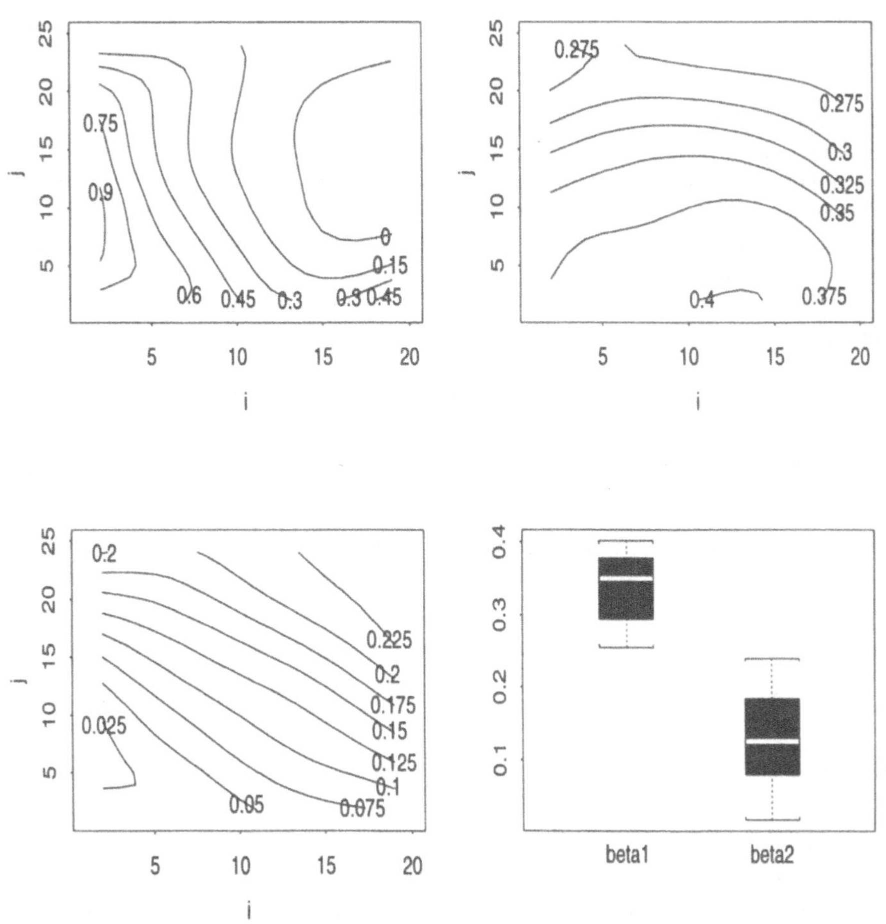

Figure 2: Spatial variation of $\hat{\beta}_0$ (top left), $\hat{\beta}_1$ (top right) and $\hat{\beta}_2$ (bottom left), boxplots for $\hat{\beta}_1$ and $\hat{\beta}_2$ (bottom right).

8. Summary

Markov random fields with spatially varying coefficients provide a class of models, which combine a parametric approach with smoothing ideas and therefore belong to the semiparametric model-family.

With these models the stationarity assumptions, that are often doubted in practice, can be relaxed. The process of removing nonstationarities of mean in order to prepare a data set before fitting a Markov random field model becomes unnecessary. Nonstationarity of the covariance function can also be taken into account, as long as the covariance function is smoothly changing across the field. So models are obtained which can lead to more appropriate fits and better prediction results.

Since the nonparametric component of these models is included in a parametric structure, the models allow for an interpretation of the coefficients. Graphical tools like contour or image plots are available, that give a visual impression of the varying coefficients. Therefore this technique can also be considered as an explorative method, which can be helpful for obtaining more insight into spatial structures and dependencies. This could be especially valuable for application areas such as epidemiology or environmental sciences, where one is interested in the direction of a spatial spread of diseases or environmental risks.

References:

BESAG, J. E. (1974): Spatial interaction and the statistical analysis of lattice systems. *J. R. Statist. Soc. B, 36, 192 - 225.*

BESAG, J. E. (1975): Statistical analysis of non-lattice data. *The Statistician, 24, 179 - 195.*

BESAG, J. E. and MORAN, P. A. P. (1975): On the estimation and testing of spatial interaction in Gaussian lattice processes. *Biometrika, 62, 552 - 562.*

CLEVELAND, W. S. (1979): Robust locally weighted regression and smoothing scatterplots. *J. Amer. Statist. Assoc., 74, 829 - 836.*

CRESSIE, N. (1993): Statistics for spatial data. Revised edition, Wiley, New York.

GEYER, C. (1991): Markov chain Monte Carlo maximum likelihood. In: E. M. Keramidas (ed.): *Comput. Sci. Statist. Proc. 23rd Symp. Interface, 156 - 163.*

HASTIE, T. and TIBSHIRANI, R. (1993): Varying-coefficient models. *J. R. Statist. Soc. B, 55, 757 - 796.*

MERCER, W. B. and HALL, A. D. (1911): The experimental error of field trials. *J. Agric. Sci., 4, 107 - 132.*

STANISWALIS, J. (1989): The kernel estimate of a regression function in likelihood-based models. *J. Amer. Statist. Assoc., 84, 276 - 283.*

TIBSHIRANI, R. and HASTIE, T. (1987): Local likelihood estimation. *J. Amer. Statist. Assoc., 82, 559 - 567.*

VAN DER LINDE, A. (1994): On cross-validation for smoothing splines in the case of dependent observations. *Austral. J. Statist., 36(1), 67 - 73.*

WHITTLE, P. (1954): On stationary processes in the plane. *Biometrika, 41, 434 - 449.*

Distribution of a Likelihood Ratio Statistic for Spatial Disease Clusters

J. Krauth

Department of Psychology, University of Düsseldorf
Universitätsstr. 1, D-40225 Düsseldorf, F.R.G.

Summary: As pointed out by Upton and Fingleton (1985), there are two major approaches commonly used for the analysis of spatial point patterns. These are, on the one hand, the so-called distance methods, where the test statistic is based on the distances between the disease cases; on the other hand, these are the quadrat methods, where the variability of the case-counts in certain subsets of the study region is studied. A quadrat-based test for the detection of disease clusters was proposed by Choynowski (1959). This procedure was improved by Openshaw et al. (1987) and Turnbull et al. (1990). Further improvement was given by Kulldorff and Nagarwalla (1995) who proposed a likelihood ratio statistic. The corresponding test was based on the Monte Carlo method. We study the exact distribution of this statistic and derive bounds for the upper P-values. In particular, we are interested in focused tests. These may be used, if the study region contains some putative health hazard, such as, for example, a nuclear power station.

1. Some Strategies for Detecting Spatial Disease Clusters

Many different approaches have been adopted for detecting spatial disease clusters and almost every year several new procedures are introduced. Following a classification used by Upton and Fingleton (1985, p. 27) two major approaches can be distinguished. For the *distance methods* the test statistic is based on the distances between the disease cases, while for the *quadrat methods* the variability of case-counts in certain subsets of the study region is studied. Another classification due to Kulldorff and Nagarwalla (1995) differentiates between *tests for overall clustering*, which only ascertain the existence of clusters in the data, and *tests for the detection of clusters*, which additionally allow to locate clusters.

A further classification distinguishes between *focused tests*, where the study region contains some known putative health hazard, and *general tests*, where no such hazard is known.

By Choynowski (1959) a quadrat method was proposed which was improved by Openshaw et al. (1987, 1988). This extension was in turn improved by Turnbull et al. (1990) (cf. also Waller et al. (1995)). The Openshaw et al. and Turnbull et al. approaches form the basis of the method proposed by Kulldorff and Nagarwalla (1995) which is described in the following with a slightly modified notation:

(1) Denote by N the total number of subjects in the study region and by C the total number of disease cases. Consider N and C with $1 \leq C < N$ as known constants.

(2) The study region is partitioned into K (with $K \geq 2$) geographic subdivisions called *cells*. For each cell k are known the co-ordinates of its geographical or population centroid, the number of subjects (n_k), and the number of cases (c_k), where

$$1 \leq n_k < N, \quad \max\{0, C - N + n_k\} \leq c_k \leq \min\{n_k, C\}, \quad 1 \leq k \leq K.$$

Cells without subjects are deleted or adjoined to non-empty cells. However, cells may include no cases. The cell centroids form a regular or irregular *lattice*. If each cell includes only one subject, its centroid is defined by the co-ordinates of this subject. In this case, the data are not *aggregated*.

(3) Beside the lattice of K cell centroids another regular or irregular lattice of I centers is considered with $I \geq 1$. Each of the I centers defines the centre point of a class of circles for which the radius varies continuously from zero upwards. Each circle defines a *zone* consisting of all subjects in those cells whose centroids lie inside the circle. Though there are infinitely many circles, they define only a finite number of zones.
The interpretation of a zone as a cluster will be against intuition, if the zone includes almost all of the study area, even if the incidence rate in that zone is higher than outside of it. Therefore, it is assumed for the radius (r) of each circle to be considered that it is bounded by a number r_ϵ which is chosen in such a way that no circle contains more than ϵ percent of the total number (N) of subjects. Kulldorff and Nagarwalla (1995) advocate $\epsilon = 50\%$ as a natural rule of thumb.

(4) Z denotes the set of all zones generated in the manner described above. Each zone $\vec{z} \in Z$ is defined by a three-dimensional vector $\vec{z} = (x_z, y_z, r_z)$, where (x_z, y_z) are the central co-ordinates and r_z is the radius of the corresponding circle.

(5) The probability of being a case is given by p for all subjects within the zone $\vec{z}$ and by q for all subjects outside $\vec{z}$ $(p, q \in [0,1])$.
The null hypothesis
$H_0 : p = q$ for all $\vec{z} \in Z$
is to be tested against the alternative hypothesis
$H_1 : p > q$ for at least one $\vec{z} \in Z$.

(6) For a fixed zone $\vec{z}$ denote the number of subjects by n_z and the number of cases by c_z, where c_z is the realization of a random variable C_z. The

likelihood function is given by

$$L(\vec{z}, p, q) = \binom{n_z}{c_z} p^{c_z} (1-p)^{n_z - c_z} \binom{N - n_z}{C - c_z} q^{C - c_z} (1-q)^{(N-n_z)-(C-c_z)}$$

and the likelihood ratio by

$$\frac{\sup\limits_{p>q} L(\vec{z}, p, q)}{\sup\limits_{p=q} L(\vec{z}, p, q)} = \frac{\sup\limits_{p>q, p, q \in [0,1]} p^{c_z} (1-p)^{n_z - c_z} q^{C-c_z} (1-q)^{(N-n_z)-(C-c_z)}}{\sup\limits_{p \in [0,1]} p^{c_z} (1-p)^{n_z - c_z} p^{C-c_z} (1-p)^{(N-n_z)-(C-c_z)}}.$$

The denominator is given by the constant value

$$L_0 = \sup_{p \in [0,1]} p^C (1-p)^{N-C} = \left(\frac{C}{N}\right)^C \left(1 - \frac{C}{N}\right)^{N-C}.$$

The nominator is calculated as

$$L(\vec{z}) = \left(\frac{c_z}{n_z}\right)^{c_z} \left(1 - \frac{c_z}{n_z}\right)^{n_z - c_z} \left(\frac{C - c_z}{N - n_z}\right)^{C - c_z} \times \left(1 - \frac{C - c_z}{N - n_z}\right)^{(N-n_z)-(C-c_z)}, \quad \text{for } \frac{c_z}{n_z} > \frac{C - c_z}{N - n_z}$$

and as

$$L(\vec{z}) = L_0, \text{ for } \frac{c_z}{n_z} \leq \frac{C - c_z}{N - n_z}.$$

(7) By Kulldorff and Nagarwalla (1995) it is proposed to consider that zone $\hat{\vec{z}} \in Z$ as the most likely cluster, for which $L(\hat{\vec{z}}) \geq L(\vec{z})$ for all $\vec{z} \in Z$. Obviously, more than one such zone may exist. The authors denote $\hat{\vec{z}}$ as the maximum likelihood estimate of $\vec{z}$. In our opinion this is not quite correct and this notation should be reserved for that value of $\vec{z}$ which maximizes

$$\binom{n_z}{c_z} \binom{N - n_z}{C - c_z} L(\vec{z}).$$

As a test statistic the authors consider

$$\lambda = \frac{\max\limits_{\vec{z} \in Z} L(\vec{z})}{L_0}.$$

The authors state that the exact distribution of λ can only be obtained for small values of N by enumerating all possible assignments of the C cases to the N subjects and by computing λ for each outcome.
For large values of N the authors propose a simulation test: A number (M) of replicates is fixed, where M should be "large". For each of

the M replicates C subjects are chosen at random from altogether N subjects and λ is calculated. The percentage of replicates with a value of λ larger than or equal to the value from the data yields an estimate of the P–value. If this estimate is not larger than the given level of significance (α), a significant result is obtained.

(8) The authors argue against taking the maximum number of cases among all zones instead of λ, because the zones may have different population sizes. It is also argued against taking the maximum of the incidence rates among all zones: First, because the variances of these rates may be unequal, and second, because, in many cases, the zone with the smallest number of subjects from among those zones with at least one case would be picked.

2. Distribution of the Likelihood Ratio Statistic

From

$$0 \leq \frac{c_z}{n_z} \leq 1, \quad 0 \leq \frac{C - c_z}{N - n_z} \leq 1$$

we conclude that

$$\frac{c_z}{n_z} > \frac{C - c_z}{N - n_z}$$

is equivalent with

$$1 + \left\lfloor \frac{C}{N} n_z \right\rfloor \leq c_z \leq \min\{n_z, C\}.$$

For

$$\min\{n_z, C\} < 1 + \left\lfloor \frac{C}{N} n_z \right\rfloor$$

the inequality above is satisfied for no value of c_z.

Further

$$\frac{c_z}{n_z} \leq \frac{C - c_z}{N - n_z}$$

is equivalent with

$$\max\{0, C - N + n_z\} \leq c_z \leq \left\lfloor \frac{C}{N} n_z \right\rfloor.$$

We consider now $L(\vec{z})$ as a function $f(c_z)$ of c_z and derive

$$\frac{d}{dc_z}\ln f(c_z) = \ln c_z - \ln(n_z - c_z) - \ln(C - c_z) + \ln((N - n_z) - (C - c_z)) > 0,$$

$$\frac{d^2}{dc_z^2}\ln f(c_z) = \frac{1}{c_z} + \frac{1}{n_z - c_z} + \frac{1}{C - c_z} + \frac{1}{(N - n_z) - (C - c_z)} > 0$$

for $1 + \left\lfloor \frac{C}{N} n_z \right\rfloor \leq c_z < \min\{n_z, C\}$.

Thus, $f(c_z)$ is increasing in c_z for these values of c_z. Since $f(c_z)$ is continuous in c_z for $c_z = \min\{n_z, C\}$, the function is maximum in this point. For

$$\max\{0, C - N + n_z\} \leq c_z \leq \left\lfloor \frac{C}{N} n_z \right\rfloor$$

$f(c_z)$ takes its minimum (L_0).

Now we consider

$$f_{\text{max}} = \max_{\vec{z} \in Z} L(\vec{z}) = \max_{\vec{z} \in Z} f(c_z).$$

Obviously, we have

$$P(F_{\text{max}} \geq x) = P\left(\bigcup_{\vec{z} \in Z} \{f(C_z) \geq x\}\right),$$

where F_{max} is a random variable with the realization f_{max} and x is an arbitrary real number. In particular, we find

$$P(F_{\text{max}} \geq L_0) = 1$$

and for $x \geq f\left(1 + \left\lfloor \frac{C}{N} n_z \right\rfloor\right)$ for at least one $\vec{z} \in Z$

$$P(F_{\text{max}} \geq x) < 1, \quad P(f(C_z) \geq x) < 1.$$

For

$$P(F_{\text{max}} \geq f_{\text{max}}) \leq \alpha,$$

where the probability is calculated under the null hypothesis, where f_{max} is the value found for the given data, and where α with $0 < \alpha < 1$ is a given level of significance, we assume the existence of clusters.

By considering all possible ways in which the C cases can be assigned to the N subjects, we can calculate the exact distribution of F_{max} under the null hypothesis. As an alternative we can calculate exact upper bounds for $P(F_{\text{max}} \geq f_{\text{max}})$ and compare these to α.

We assume in the following $f_{\text{max}} > L_0$, because otherwise we know that we certainly have no significant result.

For two different zones $\vec{z}$ and $\vec{z}'$ we define $n_{zz'}$ as the number of subjects and $c_{zz'}$ as the number of cases in the intersection of the two zones. Obviously, under the null hypothesis, the trivariate hypergeometric distribution (Jogdeo and Patil (1975)) yields

$$P\left(\{C_z = c_z\} \cap \{C'_z = c'_z\}\right) =$$

$$\frac{\binom{n_z - n_{zz'}}{c_z - c_{zz'}}\binom{n_{zz'}}{c_{zz'}}\binom{n_{z'} - n_{zz'}}{c_{z'} - c_{zz'}}\binom{N - n_z - n_{z'} + n_{zz'}}{C - c_z - c_{z'} + c_{zz'}}}{\binom{N}{C}},$$

$$\max\{0, C - N + n_z + n_{z'} - n_{zz'}\} \leq c_z + c_{z'} - c_{zz'} \leq \min\{n_z + n_{z'} - n_{zz'}, C\},$$
$$\max\{0, C - N + n_z - n_{zz'}\} \leq c_z - c_{zz'} \leq \min\{n_z - n_{zz'}, C\},$$
$$\max\{0, C - N + n_{z'} - n_{zz'}\} \leq c_{z'} - c_{zz'} \leq \min\{n_{z'} - n_{zz'}, C\},$$
$$\max\{0, C - N + n_{zz'}\} \leq c_{zz'} \leq \min\{n_{zz'}, C\}.$$

From this, in particular, the univariate hypergeometric probabilities

$$P(C_z = c_z) = \frac{\binom{n_z}{c_z}\binom{N - n_z}{C - c_z}}{\binom{N}{C}},$$

$$\max\{0, C - N + n_z\} \leq c_z \leq \min\{n_z, C\}$$

result, if we consider the intersection of identical events, because in this case we have $n_z = n_{z'} = n_{zz'}, \quad c_z = c_{z'} = c_{zz'}$.

For two disjoint zones $\vec{z}$ and $\vec{z}'$ the trivariate hypergeometric probabilities above are reduced to bivariate hypergeometric probabilities

$$P\left(\{C_z = c_z\} \cap \{C'_z = c'_z\}\right) = \frac{\binom{n_z}{c_z}\binom{n_{z'}}{c_{z'}}\binom{N - n_z - n_{z'}}{C - c_z - c_{z'}}}{\binom{N}{C}},$$

$$\max\{0, C - N + n_z + n_{z'}\} \leq c_z + c_{z'} \leq \min\{n_z + n_{z'}, C\},$$
$$\max\{0, C - N + n_z\} \leq c_z \leq \min\{n_z, C\}, \max\{0, C - N + n_{z'}\} \leq c_{z'} \leq \min\{n_{z'}, C\},$$

because of $n_{zz'} = 0, \quad c_{zz'} = 0$.

Bivariate hypergeometric probabilities also arise, if zone $\vec{z}$ is contained in zone $\vec{z}'$, in which case we have $n_z = n_{zz'}, \quad n_z < n_{z'}, \quad c_z = c_{zz'}, \quad c_z \leq c_{zz'}$:

$$P(\{C_z = c_z\} \cap \{C_{z'} = c_{z'}\}) = \frac{\binom{n_z}{c_z}\binom{n_{z'} - n_z}{c_{z'} - c_z}\binom{N - n_{z'}}{C - c_{z'}}}{\binom{N}{C}},$$

$\max\{0, C - N + n_{z'} - n_z\} \leq c_{z'} - c_z \leq \min\{n_{z'} - n_z, C\}$,
$\max\{0, C - N + n_z\} \leq c_z \leq \min\{n_z, C\}, \max\{0, C - N + n_{z'}\} \leq c_{z'} \leq \min\{n_{z'}, C\}$.

This last case of zones, in which one zone is contained in the other one, is typical for focused tests.

With the definition $A_{\vec{z}} = \{f(C_z) \geq x\}, \vec{z} \in Z$, we derive

$$P(F_{\max} \geq x) = P\left(\bigcup_{\vec{z} \in Z} A_{\vec{z}}\right).$$

An upper bound for this probability is given by

$$U = \min\left\{1, \sum_{\vec{z} \in Z} P(A_{\vec{z}}) - \max_{\vec{z} \in Z} \sum_{\substack{\vec{z}' \in Z \\ \vec{z} \neq \vec{z}'}} P(A_{\vec{z}} \cap A_{\vec{z}'})\right\}$$

(Kounias (1968)).

In order to evaluate the goodness of the approximation of the true probability by this upper bound we can use the lower bound

$$L = \frac{2}{w(w+1)}\left(w \sum_{\vec{z} \in Z} P(A_{\vec{z}}) - \frac{1}{2} \sum_{\substack{\vec{z}, \vec{z}' \in Z \\ \vec{z} \neq \vec{z}'}} P(A_{\vec{z}} \cap A_{\vec{z}'})\right),$$

$$w = 1 + \left\lfloor \sum_{\substack{\vec{z}, \vec{z}' \in Z \\ \vec{z} \neq \vec{z}'}} P(A_{\vec{z}} \cap A_{\vec{z}'}) / \sum_{\vec{z} \in Z} P(A_{\vec{z}}) \right\rfloor \qquad \text{(Kwerel, 1975).}$$

Since $f(c_z)$ is increasing in c_z, we have to consider

$$P(A_{\vec{z}}) = \sum_{c_z \geq c_z^\star} P(C_z = c_z), \quad c_z^\star = \min\{c_z : f(c_z) \geq x\}, \quad \vec{z} \in Z,$$

where $P(A_{\vec{z}})$ is set equal to zero if no such value of $c_z^\star$ exists. Further we consider

$$P(A_{\vec{z}} \cap A_{\vec{z}'}) = \sum_{c_z \geq c_z^\star} \sum_{c_{z'} \geq c_{z'}^\star} P(\{C_z = c_z\} \cap \{C_{z'} = c_{z'}\}),$$

$c_z^\star = \min\{c_z : f(c_z) \geq x\}$, $c_{z'}^\star = \min\{c_{z'} : f(c_{z'}) \geq x\}$, $\vec{z}, \vec{z}' \in Z, \vec{z} \neq \vec{z}'$, where $P(A_{\vec{z}} \cap A_{\vec{z}'})$ is set equal to zero if at least one of the values of $c_z^\star$ and $c_{z'}^\star$ does not exist.

3. Detection of Clusters

Kulldorff and Nagarwalla (1995) applied their approach to a data set from Upstate New York. They considered $N = 1,057,673$ subjects, $C = 592$ cases of leukaemia, $K = 790$ census tracts (or cells), $I = 790$ center points, and $\epsilon = 20$ (i. e. at most 20 percent of the total population could be contained within a circle). This yielded a total of approximately 124,820 distinct zones. Since, in this example, the calculation of an exact P–value would not be possible within a reasonable amount of time, the authors used a simulation procedure.

In a situation with such a large number of zones our method, most probably, would be of no practical use. If, however, the data is aggregated to a high degree, i. e. if only a small or medium number of zones have to be considered, our approach might be of interest. The number of zones can be reduced by pooling adjacent cells and by considering only a small number I of centers, e. g. $I = 1$ for a focused test.

We illustrate our procedure by two examples which we already used for demonstrating the efficiency of a maximum statistic in Krauth (1995). The first example concerns Figure 2 in Pilar, Landmesser and Burstein (1980), where a camera lucida reconstruction shows the distribution of labeled ciliary and unlabeled choroid cells in a ciliary ganglion from a white leghorn chick embryo. We considered rectangles of size 75.9 μm × 62.1 μm and counted the labeled and unlabeled cells within the $K = 16$ rectangles. These numbers are listed in Table 1 of Krauth (1995). The $I = 4$ centroids of the inner four rectangles were chosen as centers of the circles and the maximum percentage of the population to be considered was set to $\epsilon = 50$. For the total numbers of $N = 651$ cells and $C = 360$ labeled cells resulted 12 zones. Our procedure yielded the upper and lower bounds $U = L = 0.000000$ for the P–value, i. e. the existence of clusters may be assumed. The zone with the highest value of the test statistic is given by the second rectangle in the second row and its four direct neighbours. Therefore, we consider this zone as a presumable cluster.

Our second example concerns Figure 12C in Sun and Cassell (1993). Here, gamma–aminobutyric acid–immunoreactive (GABA–IR) neurons and horseradish peroxidase (HRP)–labeled neurons in the central amygdaloid nucleus of the rat are shown in a camera–lucida drawing. We considered $K = 16$ rectangles of size 235.3 μm × 211.8 μm and listed the corresponding numbers in Table 2 of Krauth (1995). The $I = 4$ centers of the circles and $\epsilon = 50$ were chosen as in the first example. With $N = 119$ and $C = 72$ we had to consider 9 zones and got $U = 0.041896$ and $L = 0.033688$. Because the value of U is smaller than $\alpha = 0.05$, we assume the existence of clusters. The zone with the highest value of the test statistic is given by the third rectangle in the second row and by its two neighbours above and below. Therefore, this zone is considered as a presumable cluster.

References:

CHOYNOWSKI, M. (1959): Maps based on probabilities. *Journal of the American Statistical Association, 54, 385–388.*

JOGDEO, K. and PATIL, G. P. (1975): Probability inequalities for certain multivariate discrete distribution. *Sankyā, Series B, 37, 158–164.*

KOUNIAS, E. G. (1968): Bounds for the probability of a union with applications. *Annals of Mathematical Statistics, 39, 2154–2158.*

KRAUTH, J. (1995): Spatial clustering of neurons by hypergeometric disjoint statistics. In: W. Gaul and D. Pfeifer (eds.): *From Data to Knowledge.* Springer, Heidelberg, 253–261.

KULLDORFF, M. and NAGARWALLA, N. (1995): Spatial disease clusters: detection and inference. *Statistics in Medicine, 14, 799-810.*

KWEREL, S. (1975): Most stringent bounds on aggregated probabilities of partially specified dependent probability systems. *Journal of the American Statistical Association, 70, 472–479.*

OPENSHAW, S.; CHARLTON, M.; WYMER, C. and CRAFT, A. (1987): A mark 1 geographical analysis machine for the automated analysis of point data sets. *International Journal of Geographical Information Systems, 1, 335–358.*

OPENSHAW, S.; CRAFT, A. W.; CHARLTON, M. and BURCH, J. M. (1988): Investigation of leukaemia clusters by use of a geographical analysis machine. *Lancet, 1, 272–273.*

PILAR, G.; LANDMESSER, L. and BURSTEIN, L. (1980): Competition for surviving among developing ciliary ganglion cells. *Journal of Neurophysiology, 43, 233-254.*

SUN, N. and CASSELL, M. D. (1993): Intrinsic GABAergic neurons in the rat central extended amygdala. *Journal of Comparative Neurology, 330, 381–404.*

TURNBULL, B. W.; IWANO, E. J.; BURNETT, W. S.; HOWE, H. L. and CLARK, L. C. (1990): Monitoring for clusters of disease: Application to leukaemia incidence in Upstate New York. *American Journal of Epidemiology, 132, S136–S143.*

UPTON, G. J. G. and FINGLETON, B. (1985): *Spatial Data Analysis by Example. Vol. I. Point Pattern and Quantitative Data.* John Wiley & Sons, Chichester etc.

WALLER, L. A.; TURNBULL, B. W.; GUSTAFSSON, G.; HJALMARS, U. and ANDERSSON, B. (1995): Detection and assessment of clusters of disease: An application to nuclear power plant facilities and childhood leukaemia in Sweden. *Statistics in Medicine, 14, 3–16.*

Biased Methods of Discrimination in High Dimensions: A Comparative Assessment[1]

B. J. A. Mertens[2], D. J. Hand[3]

[2] Department of Statistics,
Trinity College Dublin, Dublin 2, Ireland
[3] Department of Statistics,
The Open University, Milton Keynes MK7 6AA, United Kingdom

Summary: We discuss a project on the assessment of classification rules for high-dimensional data. We start the discussion with a typical example. Some of the existing approaches to classification with high-dimensional data are explained and we show how these methods have been implemented to allow for a comparison on unbiased estimates of error rate. We conclude with a brief discussion.

1. High-dimensional data

Due to improvements in instrumentation and computation, statisticians are increasingly confronted with classification problems on machine-generated data. A typical example is found in the abundance of classification problems which aim to extract information from spectroscopic curves, such as are generated in near-infrared and infrared spectroscopy. Similar examples are found in medicine where we may wish to exploit automated measurement techniques like encephalography, electro-cardiography, cat-scanning, and others. Further examples are in character or image recognition problems in engineering applications.

In statistics, data of this nature is often referred to as high-dimensional and experimental spectroscopy has become one of the standard examples in this field. Figure 1 shows near-infrared transmittance spectra for 18 samples of rice in an application to food analysis. Transmittance was recorded for the analysis of bulk grain first (measuring transmittance through whole grain), for an analysis of single grains (measuring transmittance for a single grain at a time and recording the average spectra obtained from a selection of grains) and finally, for the analysis of rice flour (following grinding of each sample). For each of these analyses, results are stored at a fixed grid of equidistant points across the measurement range. Thus, transmission is recorded at 100 wavelengths ranging from 850 to 1048 nm at equal intervals for the analysis of both bulk and single grains. The spectra for ground rice are recorded at 1050 wavelengths, ranging from 400 to 2500 nm.

The example gives an idea of the diversity in the appearance of spectroscopic curves which can be encountered in practice. In this application interest lies in the calibration of a discriminant rule for the prediction of the type of

[1]This work was supported by a grant from the Defence Research Agency.

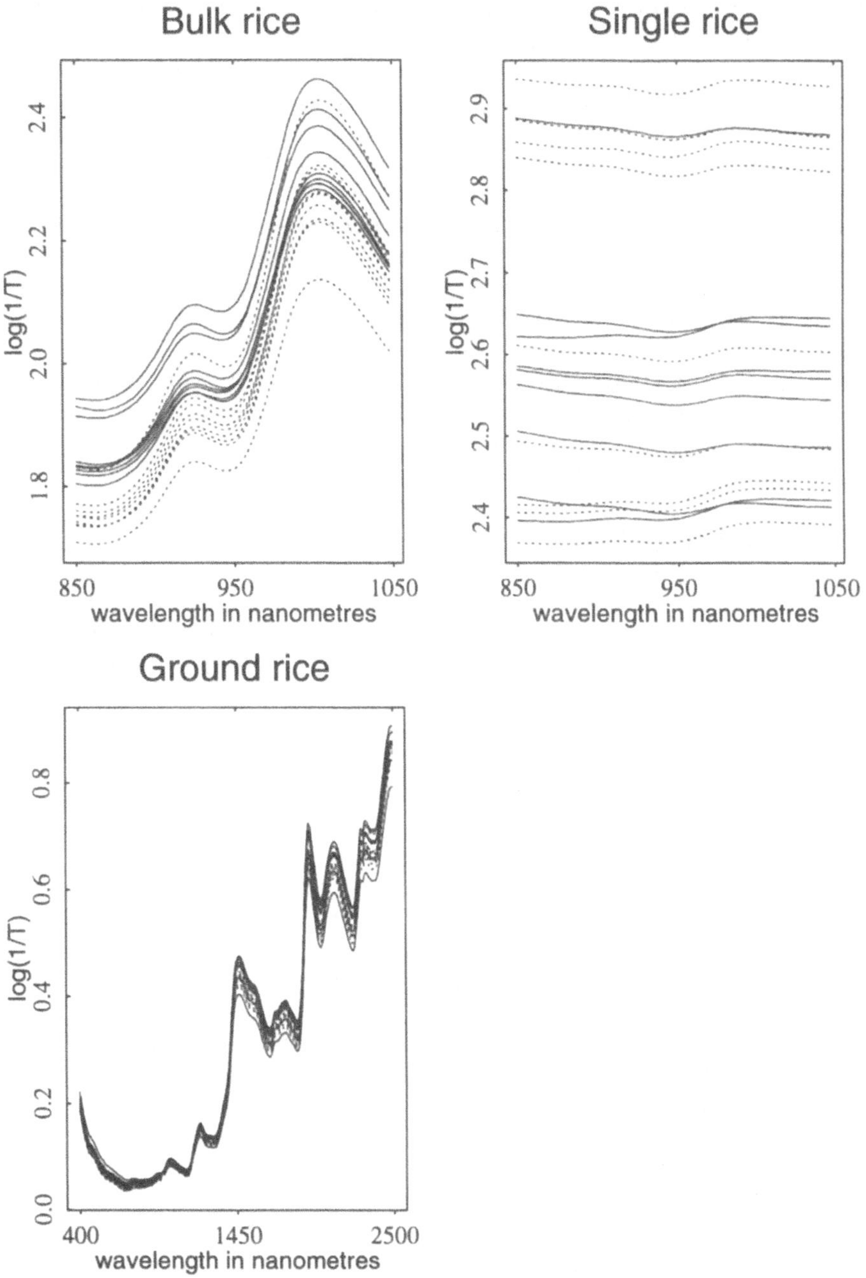

Figure 1: Near-infrared spectra for 18 samples of rice. Solid lines correspond to Basmati rice while the dotted lines are obtained from rice of Italian origin. The rice data was provided by B.G.Osborne of the Flour Milling and Baking Research Association.

rice from the observed near-infrared transmittance of a sample. The spectra obtained for bulk analysis appear to suggest at least a partial separation of Basmati and Italian rice at the lower end of the range of wavelengths. Such separation will typically be due to particle size effects. It is rare for high-dimensional problems in discrimination to be this simple, however. This applies to the present example as well, as only the spectra for samples of Basmati and Italian rice are shown here, for illustrative purposes. The actual problem is more complicated and involves several distinct classes of rice. There is some more information from a contrasting of the responses at the two ends of the wavelength spectrum. It is this type of information relating to changes in the shape of spectra which we must typically exploit in the calibration of discriminant rules for spectroscopic data. The analysis of single grains is a good example as the average transmission appears totally unrelated to the separation of interest. Visual analysis is cumbersome in this case, to say the least, a statement which applies even more strongly to the spectra of ground rice.

2. Methodologies

While statistical interest in the above data is fairly recent, there has been much success with *ad hoc* approaches in chemometrics. These consist, essentially, in leaving the discretized state of the stored spectra as is, without any attempt to exploit the continuous and smooth nature of the measured curves. As the dimensionality will usually swamp the sample size, an ill-specified problem in multivariate analysis emerges - hence the name high-dimensional analysis. Extreme correlations are induced as an artefact of the discretisation of continuous traces.

Many chemometric applications have tried to adapt linear or quadratic discrimination to the high-dimensional setting. Thus, for linear discrimination, we seek to classify a vector of observations $\mathbf{x}$ from the distances

$$(\mathbf{x}-\boldsymbol{\mu}_i)\mathbf{C}_p^{-1}(\mathbf{x}-\boldsymbol{\mu}_i)^T,$$

where $\mathbf{C}_p$ is the pooled within-group cross-products matrix and $\boldsymbol{\mu}_i$ the population mean vectors with $1 \leq i \leq I$. Similarly, for quadratic discrimination, we wish to calculate the quantities

$$(\mathbf{x}-\boldsymbol{\mu}_i)\mathbf{C}_i^{-1}(\mathbf{x}-\boldsymbol{\mu}_i)^T + \ln(\det(\mathbf{C}_i)),$$

where $\mathbf{C}_i$ is a within-group cross-products matrix. We will consistently work with the cross-products matrices in this article to simplify notations. In high-dimensional analysis these matrices will typically be singular and hence, the allocation rules corresponding to the above measures can only be applied through appeal to a generalised inverse, such as the Moore-Penrose. This may be achieved through reduction to principal components. As small-variance components are liable to increase the variability of the calibrated prediction rule, one will often find implementations on a suitably chosen subset of large-variance components. The number of components is usually chosen by cross-validation.

Many classification problems on spectroscopic data have in the past been analysed with the SIMCA method. This method, due to Wold (1976), is specific to chemometrics and is of an essentially algorithmic formulation. With relation to statistical discrimination, its features may be summarised in the following manner. For each group in the discriminant problem, the principal component decomposition $\mathbf{C}_i = \mathbf{Q}_i\mathbf{E}_i\mathbf{Q}_i^T$ of the within-group cross-products matrix $\mathbf{C}_i$ is computed, where $\mathbf{Q}_i$ is the matrix of principal component coefficients and $\mathbf{E}_i = \mathrm{diag}(e_1, \ldots, e_r)$ is the diagonal matrix of principal component eigenvalues in decreasing order on the diagonal. We will omit indices for eigenvalues and principal component weights from now on. r is the rank of $\mathbf{C}_i$. We then replace $\mathbf{E}_i$ with a diagonal matrix of weights $\mathbf{\Delta}_i = \mathrm{diag}(\delta_1, \ldots, \delta_r)$. The weights are chosen such that $\delta_1 = \ldots = \delta_k = 0$ for some integer k and the remaining weights are obtained from

$$\delta_{k+1} = \ldots = \delta_r = \frac{(n-k-1)}{\sum_{j=k+1}^{n} e_j}.$$

The dimension k is chosen through cross-validation. Thus, an allocation rule is implemented which is of quadratic nature. The scaling term from classical quadratic discrimination is discarded and all information on potential between-group differences in the space spanned by the first k principal components is lost. As a consequence, we obtain a distance-based allocation rule computed from small-variance components. The weights of these components are changed relative to the Mahalanobis distance by deriving the average weight of the selected components. Similar statistics have been discussed by Gnanadesikan and Kettenring (1972) in the context of outlier analysis (Mertens *et al.* 1994). With the above conventions, they define a group of principal component outlier statistics for $\delta_j = 1/e_j$ (for a Mahalanobis-type distance), $\delta_j = 1$ (for a distance of Euclidean nature) and finally $\delta_j = e_j$ for Gnanadesikan and Kettenring's distance, with $j = k+1, \ldots, r$. Their approach results in a relative reweighting of small- versus large-variance components in the distance. In contrast, Wold's implementation seeks to reduce the variability of the derived classification rule by averaging the weights.

Wold's implementation may be thought of as a form of regularisation which seeks to retain a measure of scale in the discriminant rule. The same approach is found in the novel method of regularised discriminant analysis which was proposed by Friedman (1989). While high-dimensionality of measurement may favour quadratic discrimination on purely geometric grounds, it is often found that linear discrimination will outperform quadratic discrimination even when the pooling assumption does not hold. Regularised discrimination seeks to exploit this phenomenon by optimising an allocation rule of quadratic type while allowing the procedure to partially pool the covariance matrices. The method proceeds in two stages. First, the within-group cross-products matrices are partially pooled by calculating the matrices

$$\mathbf{C}_i(\lambda) = (1-\lambda)\mathbf{C}_i + \lambda\mathbf{C}_p,$$

with $0 < \lambda < 1$. At the next stage, the partially pooled within-group cross-products matrices $\mathbf{C}_i(\lambda)$ are further regularized through mixing with a multiple of the identity matrix $\mathbf{I}$:

$$\mathbf{C}_i(\lambda,\gamma) = (1-\gamma)\mathbf{C}_i(\lambda) + \gamma \mathrm{tr}[\mathbf{C}_i(\lambda)]\mathbf{I}/d,$$

where d is the dimensionality of the data and $0 < \gamma < 1$. The second stage of regularisation is thus essentially identical to that employed by Wold as the identity matrices are multiplied with the average weight of components in the principal component decomposition of $\mathbf{C}_i(\lambda)$. The procedure is reminiscent of a "ridge-type" adjustment of the covariance matrices $\mathbf{C}_i(\lambda)$.

Finally, we conclude this section by mentioning the application of Moore-Penrose implementations of ordinary linear regression for dummy variables, as well as partial least squares and principal component regression. All these approaches are essentially shrinkage methods, which seek to reduce the weight of unhelpful measurements by either dimension reduction or some form of regularisation. More recent developments, following the increasing involvement of statisticians, have been to model the within-group covariance matrices explicitly. Thus, autoregressive structures have been pursued by Denham and Brown (1993). Similarly, antedepence models were proposed by Krzanowski *et al.* (1995). The model is specified through

$$X_{ij} = \mu_{ij} + f_j,$$

for the sequence of random variables $X_{i1}, ..., X_{ip}$, where j indexes the wavelength and the error terms f_j follow the structure

$$f_j = \sum_{k=1}^{s} f_{j-k}\alpha_{kj} + g_j.$$

The g_j are independent with mean zero and variance σ_j^2. s is the order of the model. With these assumptions, the inverse cross-products matrix has non-zero elements on the diagonal and s sub-diagonals only. Thus, the estimation of the pooled covariance matrix is effectively restricted, after which classical linear discrimination may be reapplied.

3. Implementation

While the number of methods applied in high-dimensional discrimination is large, no satisfactory methodology has yet emerged. At any rate, there is as yet no evidence to suggest that specific methods are superior to others. Moreover, there does not appear to be a method for which no data was found for which the method works well. This has also been noted by Stone and Jonathan (1994): ‘There is...such a plethora of unexplored possibilities that the only unifying idea seems to be that of realistic assessment...

Rice Data									
	Bulk			Single			Ground		
	B - I	B - U	I - U	B - I	B - U	I - U	B - I	B - U	I - U
Dimension	100	100	100	100	100	100	1050	1050	1050
Sample	18	36	36	18	36	36	18	36	36
Method									
LDA	0.000	0.056	0.111	0.000	0.194	0.417	0.222	0.028	0.056
LDA (L)	0.056	0.028	0.000	0.000	0.056	0.306	0.000	0.000	0.083
LDA (S)	0.000	0.083	0.111	0.056	0.194	0.444	0.222	0.028	0.083
QDA	0.056	0.111	0.250	0.333	0.250	0.250	0.111	0.000	0.250
QDA (L)	0.056	0.028	0.250	0.278	0.194	0.333	0.222	0.194	0.056
QDA (S)	0.111	0.111	0.222	0.222	0.306	0.250	0.000	0.028	0.250
RDA (P)	0.111	0.083	0.139	0.056	0.222	0.250	0.000	0.000	0.083
RDA (PR)	0.167	0.056	0.139	0.056	0.222	0.250	0.056	0.000	0.083
SIMCA (M)	0.111	0.083	0.194	0.222	0.139	0.389	0.000	0.028	0.028
SIMCA (E)	0.056	0.111	0.278	0.278	0.333	0.333	0.056	0.000	0.083
SIMCA (GK)	0.000	0.167	0.333	0.278	0.417	0.278	0.056	0.000	0.028
SIMCA (W)	0.333	0.139	0.278	0.333	0.222	0.361	0.167	0.083	0.139
OLS	0.167	0.056	0.028	0.056	0.083	0.333	0.000	0.028	0.056
PLS	0.056	0.028	0.083	0.056	0.056	0.306	0.000	0.028	0.056
PCR	0.056	0.028	0.056	0.000	0.111	0.306	0.000	0.000	0.056
NN	0.056	0.111	0.306	0.500	0.250	0.250	0.167	0.167	0.139
ATDP	0.056	0.000	0.333	0.222	0.361	0.500	0.056	0.028	0.083
Minimum	0.000	0.000	0.000	0.000	0.056	0.250	0.000	0.000	0.028

Table 1: The table shows cross-validated and double cross-validated error rates for the rice data. Results are shown for bulk analysis, the analysis of single grains and the analysis of ground rice. For each of these, the following two-group comparisons were investigated: B-I: Basmati versus Italian, B-U: Basmati versus American, I-U: Italian versus American. We must refer to the text for an explanation of the methods used.

What the area may need...is open-mindedness for the new - tempered by informed criticism.' For these reasons, we have resolved to compare a selection of methods across a large number of real data sets. The methods we have investigated at present are linear and quadratic discrimination using a Moore-Penrose inverse (LDA and QDA in the table). We have assessed these methods for classification on large-variance principal components (suffix L in the table). We chose to leave the order of components unchanged, partly to allow for a comparison with the SIMCA method. For the same reason, we also included implementations on small-variance components (suffix S). Regularised discrimination (RDA) was implemented, both with (PR) and without (P) a ridge-type adjustment in the estimation of the covariance matrices. Four versions of SIMCA were considered, among which Wold's original version (W) as well as the alternatives using Mahalanobis (M), Euclidean (E) and Gnanadesikan and Kettenring's (GK) distance measures. We also implemented ordinary least squares regression (Moore-Penrose)(OLS), partial least squares regression (PLS) and principal component regression (PCR). As for the above methods which use principal component reduction, we leave the order of components unchanged and select a subset from the first component onwards. Finally, we have implemented a k-nearest neighbour algorithm (NN) as well as a first-order implementation of Krzanowski's antedepence modelling (ATDP). All these implementations are for a two-group comparison. Cross-validation or double-cross validation were used to obtain unbiased estimates of misclassification rates as necessary. Thus, the implementations of LDA and QDA on principal components, SIMCA, PLS and PCR (choice of the number of components), RDA (choice of the pooling parameters) and finally, NN (choice of the number of neighbours) all require a double cross-validatory implementation. This is to allow for a joint cross-validatory choice of the tuning parameters and the unbiased estimation of error rates. No attempt was made at outlier analysis. Not only would this have made a comparison of results between methods intractable but also would have posed serious problems of implementation.

4. Results

In this article, we will only present results from an analysis of the rice data which we discussed earlier. We selected three two-group comparisons from the data: Basmati versus Italian rice, Basmati versus American rice and finally, American versus Italian rice. It is irrelevant that the classifications obtained in this way do not relate to the actual problem at hand, as we are only interested in an assessment of methods . Table 1 shows the validated error rates which are obtained from an application of the above described methodologies to each of these two-group classification problems. The dimensionality of each problem is shown at the top of the table. Sample sizes were 9 for both Basmati and Italian rice and 27 for American rice.

Interpretation of the results is not trivial, but a few points can perhaps be made at this stage. First and foremost among these is the apparent success of linear discrimination, across all of the two-group classification problems investigated and especially for the analysis on large-variance principal components. Not surprisingly, results from linear discrimination are comparable to those from regression for dummy variables In particular, cross-validated error rates from partial least squares and principal component regression are close to those from linear discrimination on large-variance components. There are no obvious benefits from using small-variance components instead, as is implemented in the SIMCA approach. As expected, error rates increase for quadratic classification as compared to those from linear discrimination, with a notable exception for the analysis of ground rice on small-variance principal components. Interestingly, the SIMCA methods produce respectable classifications for this same set of data, with the exception of Wold's approach, which does not appear to deliver successful classifiers for any of the two-group problems considered. A similar conclusion can be drawn from a comparison of the validated error rates between the two implementations of Friedman's regularised discriminant analysis (with and without ridge-type regularisation). There does not appear to be any significant decrease in the observed error rates from employing a secondary stage of regularisation in the estimation of the within-group covariance matrices. Krzanowski's antedepence implementation of linear discrimination disappoints for these data. Given the considerable added complexity of the approach, the results do not seem to warrant the effort involved.

5. Discussion

The methods assessed here ignore the continuous nature of the data. There is as yet no research on functional methods of classification for spectroscopic data. In prediction (of continuous data), some work may be found in the literature (Martens and Naes (1989))(Denham and Brown (1993)), but results have been ambiguous, at best. The reasons are due, essentially, to the fact that the information in the spectra which relates to the measure of interest in the prediction problem will typically form part of the more noisy contributions to the total variation observed in the predictor data.

Alternatively, we could of course argue that such sophisticated methods would be an unnecessary complication and irrelevant when only the calibration of a discriminant rule is at issue. Instead, we could focus on exploiting existing methods to the best of our ability. An extension of the work done by Breiman (1994) on the stacking of regressors to the discriminant setting may prove to be of interest here.

References:

BREIMAN, L. (1994): Stacked regression. Technical Report No. 367, Department of Statistics, University of California, Berkeley, California 94720, August 1992 (revised June 1994).

DENHAM, M. C. and BROWN, P. J. (1993): Calibration with many variables. *Applied Statistics, 42, 515-528.*

FRIEDMAN, J. H. (1989): Regularized discriminant analysis. *Journal of the American Statistical Association, 84, 165-175.*

GNANADESIKAN, R. and KETTENRING, J. R. (1972): Robust estimates, residuals and outlier detection with multiple response data. *Biometrics, 28, 81-124.*

KRZANOWSKI, W. J.; JONATHAN, P.; McCARTHY, W. V. and THOMAS, M. R. (1995): Discriminant analysis with singular covariance matrices: methods and applications to spectroscopic data. *Applied Statistics, 44, 101-115.*

MARTENS, H. and NAES, T. (1989): Multivariate calibration. Wiley, New York.

MERTENS, B.; FEARN, T. and THOMPSON, M. (1994): Principal component outlier detection and SIMCA: a synthesis. *The Analyst, 119, 2777-2784.*

STONE, M. and JONATHAN, P. (1994): Statistical thinking and technique for QSAR and related studies. Part II: Specific methods. *Journal of Chemometrics, 8, 1-20.*

WOLD, S. (1976): Pattern recognition by means of disjoint principal components models. *Pattern Recognition, 8, 127-139.*

A One-Step Optimization Procedure for ARFIMA Processes

J. S. Pai

Institut für Statistik und Ökonometrie, WWZ, Universität Basel, Holbeinstrasse 12, CH-4051 Basel, Switzerland

Summary: A one-step optimization procedure is suggested for parameter estimation in autoregressive fractionally integrated moving average (ARFIMA) processes. Comparisons are made between the proposed procedure and the well known two-step procedure of Haslett and Raftery (1989) using simulation results. The invertibility condition is handled by a convenient transformation. The optimization procedure is performed using the simulated annealing method.

1. Introduction

A time series $\{y_t\}$ is generated by an ARFIMA(p,d,q) process (Granger and Joyeaux (1980), Hosking (1981)) with shift μ if

$$\phi(B)y_t = \mu + \theta(B)(1-B)^{-d}\epsilon_t \tag{1}$$

where $\phi(B) = 1 - \phi_1 B - \cdots - \phi_p B^p$ and $\theta(B) = 1 - \theta_1 B - \cdots - \theta_q B^q$ are polynomials in B of degrees p and q respectively, p and q are known integers, B is the backward shift operator, $(1-B)^{-d}$ is the binomial series expansion and ϵ_t are independent and identically distributed as normal random variables with mean 0 and variance σ^2. In using ARFIMA processes to model time series, d is an additional parameter that is estimated simultaneously with the parameters of the ARMA part.

Haslett and Raftery (1989) approximate the log-likelihood function by the profile log-likelihood function of d, $\Phi = (\phi_1, \cdots, \phi_p)'$ and $\Theta = (\theta_1, \cdots, \theta_q)'$. They first obtain the maximum likelihood estimators of σ^2 and the mean of the process given values of d, Φ and Θ. They then separate the long memory parameter d and the short memory AR and MA parameters Φ and Θ in two groups. The optimization is carried out in two levels: an outer univariate unimodal optimization in d over the interval [0,.5] and an inner non-linear least-squares optimization in the AR and MA parameters to minimize white noise variance. For other two-stage estimation methods of the ARFIMA(p,d,q) process, see Geweke and Porter-Hudak (1983), Hosking (1984). Haslett and Raftery (1989) also approximate the partial linear regression coefficients $\phi^a_{k,j}$ of the ARFIMA(0,d,0) process by letting $j,k \to \infty$ with $j/k \to 0$, i.e.,

$$\phi^a_{k,j} \sim -j^{-d-1}/(-d-1)!. \tag{2}$$

Pai and Ravishanker (1996) suggest four closed form expressions for the exact likelihood function of a Gaussian ARFIMA process, which is useful in modeling time series with long memory and short memory behavior. Their fourth form enables the calculations of the profile log-likelihood function for ARFIMA(p, d, q) models. In this paper, the two-step optimization procedure suggested by Haslett and Raftery (1989) is modified to a one-step procedure. We first obtain the maximum likelihood estimators of μ, σ^2 and Φ given values of d and Θ. We then optimize the profile log-likelihood function of d and Θ by the simulated annealing method (Kirkpatrick (1984)).

Section 2 evaluates the exact likelihood function of the ARFIMA(0,d,q) process while the profile log-likelihood function is suggested in Section 3 for the ARFIMA(p, d, q) process. Section 4 expresses a convenient transformation to handle the invertibility restriction. Section 5 discusses the simulated annealing method. Simulation results are shown in Section 6. Section 7 presents the conclusion.

2. Likelihood function for ARFIMA$(0, d, q)$ models

Pai and Ravishanker (1996) suggest a factorization form for the likelihood function of an ARFIMA(p, d, q) process in terms of the partial linear regression coefficients. The partial linear regression coefficients are calculated from the autocovariances γ_k^y using Levinson's algorithm. The main problem is that the evaluation of the autocovariances of an ARFIMA(p, d, q) process is very complicated (Hosking (1981), Sowell (1992)). On the other hand, the evaluation of the autocovariances of an ARFIMA$(0, d, q)$ process is much easier (Hosking (1981)). Let $\alpha = (d, \theta_1, \cdots, \theta_q)'$ and $Y = (y_1, \cdots, y_n)'$; the exact likelihood function for an ARFIMA$(0, d, q)$ process can be written as

$$f(Y; \alpha, \sigma^2) = f_1(y_1|\alpha, \sigma^2) f_2(y_2|y_1, \alpha, \sigma^2) \cdots f_n(y_n|y_1, \cdots, y_{n-1}, \alpha, \sigma^2) \quad (3)$$

with

$$f_i \sim N(\eta_i, \tau_i^2)$$

where

$$\eta_1 = 0, \quad \tau_1^2 = \gamma_0^y,$$

and

$$\eta_i = \sum_{j=1}^{i-1} \phi_{i-1,j}^y y_{i-j}, \quad \tau_i^2 = \gamma_0^y \prod_{j=1}^{i-1} (1 - (\phi_{j,j}^y)^2), \quad \text{for} \quad i = 2, \cdots, n.$$

Simulating data from an ARFIMA(0,d,q) process is straightforward by using (3) recursively.

3. Profile log-likelihood for ARFIMA(p,d,q) models

Define $e_t = \theta(B)(1 - B)^{-d}\epsilon_t$, i.e., e_t follows an ARFIMA$(0, d, q)$ process. Given the initial values $(y_{1-p}, \cdots, y_0)$, equation (1) can be written as a regression equation

$$Y = X\beta + E \quad (4)$$

where

$$
\begin{aligned}
Y &= (y_1, \cdots, y_n)', \\
(X)_{i,1} &= 1, \quad \text{for} \quad i = 1, \cdots, n, \\
(X)_{i,j+1} &= y_{i-j}, \quad \text{for} \quad i = 1, \cdots, n, \quad j = 1, \cdots, p, \\
\beta &= (\mu, \phi_1, \cdots, \phi_n)', \\
E &= (e_1, \cdots, e_n)'.
\end{aligned}
$$

Consider the profile log-likelihood function

$$l^+(\alpha; Y) = \log f(\alpha, \beta^*, \sigma^{2^*}) = \sup_{\beta, \sigma^2} \log f(\alpha, \beta, \sigma^2; Y), \tag{5}$$

where

$$\beta^* = (X'\Lambda^{-1}X)^{-1}(X'\Lambda^{-1}Y), \tag{6}$$

and

$$\sigma^{2^*} = [(Y - X\beta^*)'\Lambda^{-1}(Y - X\beta^*)]/n. \tag{7}$$

The profile log-likelihood function can be written as

$$l^+(\alpha; Y) = \text{constant} - \frac{n}{2}\log \frac{(Y - X\beta^*)'\Lambda^{-1}(Y - X\beta^*)}{n}, \tag{8}$$

where Λ^{-1} is calculated as in equation (6) of Pai and Ravishanker (1996).

4. Reparametrization

The ARFIMA(p, d, q) process is invertible if $d > -0.5$ (Hosking (1981)) and the roots of $\theta(z) = 0$ lie outside the unit circle, i.e., the vector of parameters Θ is constrained to lie in region C_q corresponding to the polynomial operator root conditions. The constrained parameter region C_q is analytically intractable for $q > 4$.

To circumvent the problem of dealing with C_q, we consider successive transformations from C_q to a q-dimensional hypercube. We reparametrize Θ to Θ_r (Barndorff-Nielsen and Schou (1973)) by

$$\theta_i^k = \theta_i^{k-1} - \theta_k^k \theta_{k-i}^{k-1}, \quad for \quad k = 2, \cdots, q \quad i = 1, \cdots, k-1, \tag{9}$$

where $\Theta = (\theta_1^q, \theta_2^q, \cdots, \theta_q^q)'$ and $\Theta_r = (\theta_1^1, \theta_2^2, \cdots, \theta_q^q)'$. The inverse transformation (Monahan (1984)) is

$$\theta_i^{k-1} = (\theta_i^k + \theta_k^k \theta_{k-i}^k)/(1 - (\theta_k^k)^2), \quad k = q, \cdots, 2 \text{ for } i = 1, \cdots, k-1 \tag{10}$$

with Jacobian

$$J = \Pi_{k=1}^q (1 - \theta_k^{k^2})^{[(k-1)/2]} \Pi_{j=1}^{[q/2]} (1 - \theta_{2j}^{2j}).$$

The condition that $\Theta \in C_q$ now becomes $|\theta_k^k| < 1$, $k = 1, \cdots, q$. All function evaluations are made in the space of the Θ_r.

In this paper, we assume invertibility of the process and restrict d in $(-0.5, 0.5]$. We relax the stationary condition on the Φ parameters to allow our approach to capture models with possible random walk or trend.

5. Optimization by simulated annealing method

The simulated annealing method (Kirkpatrick (1984)) has attracted significant attention in finding a desired global extremum among many local extrema (Press et al. (1993). The method is an analogy with thermodynamics in the way that liquids freeze and crystallize, or metals cool and anneal. For slowly cooled systems, nature is able to find the minimum energy state to form crystals. For most of the minimization algorithms, efforts are made to look for the quick, nearby solution, which leads to a local, but not necessarily a global minimum.

Metropolis et al. (1953) suggest a numerical approach to allow the system to sometimes go uphill as well as downhill. The four elements required by the Metropolis algorithm applied to our problem are as follows: The objective function is the profile log-likelihood function of an ARFIMA(p,d,q) model. The system state is the point α. T is the control parameter with an annealing schedule by which it is gradually reduced. The fourth element is a uniform random increment with length controlled by T. The Metropolis algorithm is then carried out as follows. We first draw $\Delta\alpha$ from a uniform distribution on $(-T, T)$. We then move α to the new state $\alpha^* = \alpha + \Delta\alpha$ with probability $\min\{1, l^+(\alpha^*; Y)/l^+(\alpha; Y)\}$.

6. Simulation results

The procedure suggested in this paper is compared with the one by Haslett and Raftery (1989). Comparisons are made for ARFIMA(0,d,0) models and ARFIMA(0,d,1) models. In each case 100 simulations of the series $y_1, y_2, \cdots, y_{200}$ with $\mu = 0$ and $\sigma^2 = 1$ were performed and in each simulation the MLE's based on the proposed procedure (d^* and θ_1^*) and the MLE's obtained from the Haslett and Raftery's procedure (d^0 and θ_1^0) were calculated. In each simulation, the optimization was carried out using multiple chains with the same random increment. For example, in the case of the ARFIMA(0,d,0) models, we ran 4 chains with starting values $d = 0.1, 0.2, 0.3, 0.4$. For each chain we ran 100 iterations with the initial control parameter $T = 0.25$ which was reduced by 50% for every 5 iterations. The MLE d^* was then obtained from the average of the 100th iterations of the 4 chains. Figure 1 shows 10 such simulations with $d = 0.25$. The procedure had no problem in finding the MLE and the convergence was achieved at an earlier stage for all simulations. The CPU time (DEC station) to evaluate the 4000 Metropolis steps was 1.5 minutes.

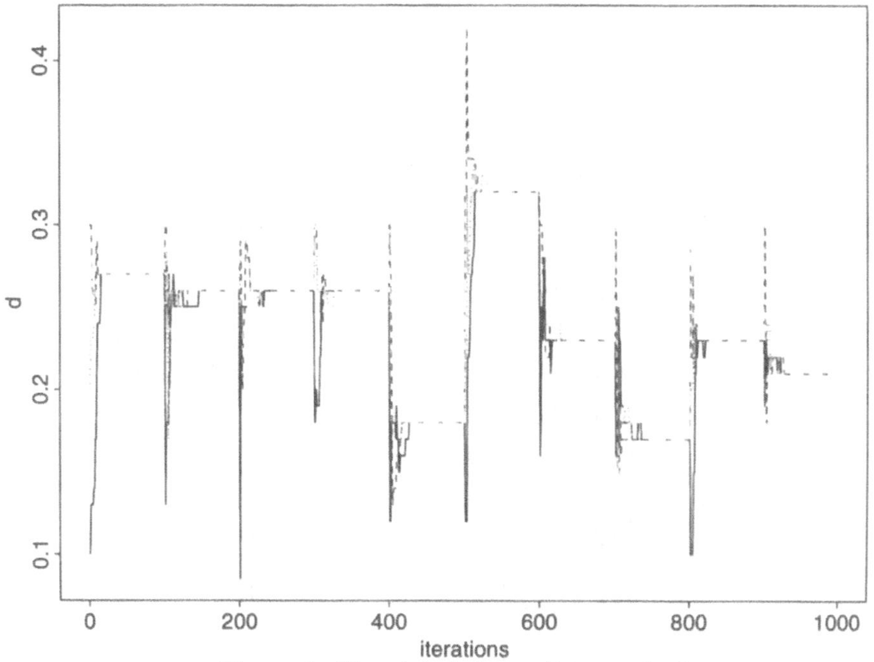

Figure 1: Simulated Annealing method

d	d^*	d^0
0.05	0.0502 (0.0407)	0.0445 (0.0450)
0.15	0.1479 (0.0567)	0.1153 (0.0614)
0.25	0.2444 (0.0475)	0.2288 (0.0505)
0.35	0.3470 (0.0520)	0.3321 (0.0532)
0.45	0.4328 (0.0402)	0.4121 (0.0472)

Table 1: Simulations for the ARFIMA(0,d,0) models

Table 1 and Table 2 show the simulation results for the ARFIMA(0,d,0) models and the ARFIMA(0,d,1) models respectively. The proposed procedure is better than Haslett and Raftery's procedure in the ARFIMA(0,d,0) models in terms of the estimated biases ($|d - d^*|$ and $|d - d^0|$). In the case of the ARFIMA(0,d,1) models, Haslett and Raftery's procedure is as good as the proposed procedure only when $\theta_1 \to -1$. In the situation where $\theta_1 \to 1$, Haslett and Raftery's procedure becomes worse especially when $d \to 0.5$. In this case, the likehood surface is a ridge from $(d = 0, \theta_1 = 0)$ to $(d = 0.5, \theta_1 = 1)$ and Haslett and Raftery's procedure always underestimates the true parameters. For example, for the parameter set $(d = 0.3, \theta_1 = 0.8)$, Haslett and Raftery's procedure provides estimates

d	θ_1	d^*	θ_1^*	d^0	θ_1^0
0.15	-0.8	0.1486 (0.0657)	-0.7965 (0.0485)	0.1248 (0.0656)	-0.8040 (0.0491)
0.15	-0.4	0.1322 (0.0590)	-0.4178 (0.0816)	0.1211 (0.0761)	-0.4213 (0.0754)
0.15	0.4	0.1359 (0.1041)	0.3949 (0.1352)	0.0694 (0.0843)	0.3096 (0.1105)
0.15	0.8	0.1540 (0.0891)	0.8048 (0.0757)	0.0465 (0.0743)	0.7204 (0.0851)
0.3	-0.8	0.2829 (0.0529)	-0.7986 (0.0520)	0.2729 (0.0586)	-0.8004 (0.0465)
0.3	-0.4	0.2803 (0.0621)	-0.4069 (0.0805)	0.2710 (0.0701)	-0.4116 (0.0661)
0.3	0.4	0.2708 (0.1000)	0.3621 (0.1239)	0.1561 (0.1032)	0.2677 (0.1192)
0.3	0.8	0.2419 (0.1288)	0.7647 (0.1330)	0.0963 (0.1271)	0.6241 (0.1343)
0.45	-0.8	0.4102 (0.0476)	-0.7910 (0.0767)	0.4079 (0.0547)	-0.8081 (0.0455)
0.45	-0.4	0.4127 (0.0469)	-0.4217 (0.0614)	0.3780 (0.0587)	-0.4430 (0.0789)
0.45	0.4	0.4213 (0.0698)	0.3743 (0.1004)	0.2900 (0.1065)	0.2566 (0.1133)
0.45	0.8	0.3523 (0.1237)	0.8467 (0.2182)	0.1080 (0.1369)	0.4856 (0.1625)

Table 2: Simulations for the ARFIMA(0,d,1) models

($d^0 = 0.0963, \theta_1^0 = 0.6241$) while for the parameter set ($d = 0.45, \theta_1 = 0.8$), the procedure gives ($d^0 = 0.108, \theta_1^0 = 0.4856$).

7. Conclusion

From simulation results, we have shown that the proposed one-step optimization procedure is better then the two-step procedure of Haslett and Raftery (1989).The proposed procedure is of order at most n^2 and is computationally feasible for maximum likelihood estimation. It can be extended to the multivariate case with the same approach suggested by Haslett and

Raftery (1989). The conditional means and variances are as same as in (3) for each series. The conditional correlation coefficient between any two series can be estimated from a regression approach (see page 9 of Haslett and Raftery (1989)). The simulated annealing optimization can be used with the downhill simplex method to improve the efficiency (Press et al. (1992)).

References:

BARNDORFF-NIELSEN, O. E. and SCHOU, G. (1973): On the reparametrization of autoregressive models by partial autocorrelations. *Journal Multivar. Analysis, 3, 408-419.*

GEWEKE, J. F. and PORTER-HUDAK, S. (1983): The estimation and application of long memory time series models. *Journal of Time Series Analysis, 4, 221-238.*

GRANGER, C. W. J. and JOYEAUX, R. (1980): An introduction to long-memory time series models and fractional differencing. *Journal of Time Series Anal., 1, 15-29.*

HASLETT, J. and RAFTERY, A. (1989): Space-time modeling with long-memory dependence: Assessing Ireland's wind power resource (with discussion). *Journal of the Royal Statistical Society, C, 38, 1-50.*

HOSKING, J. R. M. (1981): Fractional differencing. *Biometrika, 68, 165-176.*

HOSKING, J. R. M. (1984): Modeling persistence in hydrological time series using fractional differencing. *Water Resources Research, 20(12), 1898-1908.*

KIRKPATRICK, S. (1984): Optimization by simulated annealing: Quantitative studies. *Journal of Statistical Physics, 34, 975-986.*

METROPOLIS, N.; ROSENBLUTH, A. W.; ROSENBLUTH, M. N.; TELLER, A. H. and TELLER, E. (1953): Equations of state calculations by fast computing machines, *Journal Chemical Physics, 21, 1087-1091.*

MONAHAN, J. F. (1984): A note on enforcing stationarity in autoregressive moving models, *Biometrika, 71, 403-404.*

PAI, J. S. and RAVISHANKER, N. (1996): Exact likelihood function forms for an ARFIMA process. In: H.H. Bock and W. Polasek (eds.): *Data Analysis and Information System*. Springer-Verlag, Heidelberg, 323-331.

PRESS, W. H.; TEUKOLSKY, S. A.; VETTERLING, W. T. and FLANNERY, B. P. (1992): Numerical recipes. Cambridge University Press, Cambridge.

SOWELL, F. B. (1992): Maximum likelihood estimation of stationary univariate fractionally integrated time series models. *Journal of Econometrics, 53, 165-188.*

GARCH Models with Outliers

W. Polasek, S. Jin

University of Basel, CH-4051 Basel, Switzerland

Summary: This paper discusses the outliers in Bayesian Autoregressive and Conditional Heteroskedasticity (ARCH) model with AR structure in the mean (AR-ARCH model). A 'location shift' regression model is used to estimate the outliers. As an example, the model is applied to monthly exchange rate data, Japanese Yen to U.S. Dollars.

1. Introduction

Outliers can disturb the dynamic structure of a time series considerably. Since the estimation of an outlier model requires a large number of parameters, it is difficult to estimate such model from a classical point of view. It was shown by Verdinelli and Wasserman (1991) that the location shift outlier model can be estimated in a straightforward way by the Gibbs sampler which was developed by Gelfand and Smith (1990) for a wide class of Bayesian models. Bayesian time series models have been developed by Chib (1993), Albert and Chib (1993), and McCulloch and Tsay (1994).

In this paper we show how the Bayesian outlier model can be combined with the general autoregressive heteroskedasticity (GARCH) model. ARCH and GARCH models are used in financial economics to model volatile time series. The variances in GARCH model are time dependent, i.e. are a function of past squared observations or errors (ARCH) or includes past variances as well (GARCH). We will use a Metropolis step within Gibbs sampling for the distribution of the ARCH parameters. An example involving the USD/YEN exchange rate will demonstrate our approach.

2. The AR-GARCH(p,q) model

The univariate GARCH model with AR structure in the mean is written as

$$y_t \sim N[\mathbf{x}_t'\beta, h_t = \mathbf{z}_t'\gamma]$$

where $\mathbf{x}_t' = (y_{t-1}, y_{t-2}, \ldots, y_{t-p})$ contains p lagged observations and the prior information is given by

$$\begin{aligned} \beta &\sim N[\mathbf{b}_*, \mathbf{H}_*], \\ \gamma &\sim N[\gamma_*, \mathbf{R}_*], \end{aligned}$$

where the ARCH parameters are $\gamma = (\gamma_0, \gamma_1, \ldots, \gamma_q)'$, and $\mathbf{z}_t = (1, h_{t-1}, \cdots, h_{t-q})'$.

3. The AR model with outliers

The univariate location shift regression AR(p) model is given by the following decomposition of a time series observation with possible outliers

$$\mathbf{y} = \mathbf{X}\mathbf{b} + \mathbf{D}_\delta \mathbf{a} + \varepsilon$$

or

$$\mathbf{y} \sim N[\mathbf{X}\mathbf{b} + \mathbf{D}_\delta \mathbf{a}, \sigma^2 \mathbf{I}_n],$$

where $\mathbf{D}_\delta = diag(\delta_1, \delta_2, \cdots, \delta_T)$ is a diagonal matrix with location indicators δ_t on its diagonal, $\mathbf{a}$ is the the $n \times 1$ vector of shift parameter.
The location shift regression model has the density

$$f(y_t|\theta) = (1-\varepsilon)N[y_t|\mathbf{x}_t'\mathbf{b}, \sigma^2] + \varepsilon N[y_t|\mathbf{x}_t'\mathbf{b} + a_t, \sigma^2],$$

and the parameter vector θ is given by

$$\theta = (\mathbf{b}, \sigma^2, \varepsilon, \mathbf{a}).$$

Each observation has an ε chance of being an outlier. The location indicator ϑ_t follows an Bernoulli distribution with probability ε: $\vartheta_t \sim Ber(\varepsilon)$. When ϑ_t is 1, y_t is an outlier, and when ϑ_t is 0, y_t is not an outlier.

4. The AR-GARCH model with outliers

The location shift AR-GARCH outliers model is given by the following switching model for outliers

$$y_t \sim \begin{cases} N\left[\mathbf{x}_t'\beta, h_t = \mathbf{z}_t'\gamma\right] & \text{with prob. } \epsilon, \\ N\left[\mathbf{x}_t'\beta + a_t, h_t = \mathbf{z}_t'\gamma\right] & \text{with prob. } 1-\epsilon. \end{cases}$$

The location shift AR-GARCH model has the density

$$\begin{aligned} p(\theta, \mathbf{y}) &= \prod_{t=1}^{T} N\left[y_t|\mathbf{x}_t'\beta + \delta_t a_t, h_t\right] \cdot Ber\left[\delta_t|\epsilon\right] \cdot N\left[\beta|\mathbf{b}_*, \mathbf{H}_*\right] \\ & \quad \cdot N\left[\mathbf{a}|\mathbf{a}_*, \mathbf{G}_*\right] \cdot N\left[\gamma|\gamma_*, \mathbf{r}_*\right]. \end{aligned}$$

The full conditional distributions (f.c.d.'s) for θ are the following:

4.1 The f.c.d. for the AR coefficients β

Let θ^c denote all parameters accept β, then the f.c.d. of the posterior distribution is given by

$$p(\beta|\theta^c, \mathbf{y}) = N\left[\mathbf{b}_{**}, \mathbf{H}_{**}\right],$$

is a normal distribution with the parameters

$$\begin{aligned} \mathbf{H}_{**}^{-1} &= \mathbf{H}_{*}^{-1} + \mathbf{X}'\mathbf{D}_h^{-1}\mathbf{X}, \\ \mathbf{D}_h &= diag(h_1, \ldots, h_T), \\ \mathbf{b}_{**} &= \mathbf{H}_{**}\left[\mathbf{H}_{*}\mathbf{b}_{*} + \mathbf{X}'\mathbf{D}_h^{-1}(\mathbf{y} - \mathbf{D}_\delta \mathbf{a})\right]. \end{aligned}$$

4.2 The f.c.d. for the shift parameter a

$$p(\mathbf{a}|\theta^c, \mathbf{y}) = N\left[\mathbf{a}_{**}, \mathbf{G}_{**}\right],$$

is a normal distribution with the parameters

$$\begin{aligned} \mathbf{G}_{**}^{-1} &= \mathbf{G}_{*}^{-1} + \mathbf{D}_\delta'\mathbf{D}_h^{-1}\mathbf{D}_\delta, \\ \mathbf{a}_{**} &= \mathbf{G}_{**}\left[\mathbf{G}_{*}\mathbf{a}_{*} + \mathbf{D}_\delta'\mathbf{D}_h^{-1}(\mathbf{y} - \mathbf{X}\mathbf{b})\right]. \end{aligned}$$

4.3 The f.c.d. for the indicator variables δ_t

$$p(\delta_t|\theta^c, \mathbf{y}) = Ber\left[\epsilon_t = \frac{c_t}{c_t + d_c}\right],$$

is a Bernoulli distribution with the components

$$\begin{aligned} c_t &= \epsilon_* \Phi\left(\frac{y_t - \mathbf{x}_t'\beta - a_t}{h_t}\right), \\ d_t &= (1 - \epsilon_*)\Phi\left(\frac{y_t - \mathbf{x}_t'\beta}{h_t}\right). \end{aligned}$$

4.4 The f.c.d. for the ARCH parameter γ

$$\begin{aligned} p(\gamma|\theta^c, \mathbf{y}) \propto & \prod_{t=1}^{T}\{\frac{1}{2\sqrt{\mathbf{z}_t'\gamma}}\exp\{-\frac{(y_t - \mathbf{x}_t'\beta - \delta_t a_t)^2}{2\mathbf{z}_t'\gamma}\}\} \\ & \cdot \exp\{-\frac{(\gamma - \gamma_*)'\mathbf{R}^{-1}(\gamma - \gamma_*)}{2}\}. \end{aligned}$$

In the simulation we use for the proposal density in the Metropolis step

$$\gamma \sim N\left[\mathbf{1}/(q+1), \mathbf{I}/q\right].$$

5. Example

We have analyzed the monthly exchange rate of the Japanese YEN to the US Dollar from January 1974 to February 1989 for ARCH effects and outliers.

The estimation result for the AR outlier model of order 1 are shown in Figure 1 to 4. In the first panel of Figure 1, we have displayed the time series, in the second the series of boxplots of the shift parameters a_i and in the last panel the probability that the i-th observation is an outlier. We can see 8 observations which have a posterior probability (the Bernoulli parameter ϵ_{i**}) larger then 1/2 of being an outlier. Figure 2 shows the same type of information for the AR-GARCH outlier model. Now we can see that the number of outliers is reduced to 4 with a 5th observation just hitting the 1/2-line. This reduction can be partly expected, since the GARCH outlier model can produce volatility clusters, a phenomenon which is similar to an outlier effect. A location shift in an outlier model can be explained by a data generating process which has a large variance at this location.

Figure 3 shows the unconditional residual variances for 3 models: the AR outlier model, the AR-GARCH, and the AR-GARCH outlier model. The variance of the GARCH model can be derived from the Gibbs output by $\sigma^2 = \gamma_0/(1-\gamma_1-\cdots-\gamma_q)$. We see the largest residual variance for the AR-outlier model and about equal behaviour of the residual variance distribution for the AR-GARCH model with or without outliers. The mean and variance of the residual variance distribution is $(0.0035, 0.028^2)$ for the AR outlier model, while it is $(0.0026, 0.014^2)$ for the AR-GARCH model. The outlier AR-GARCH model is $(0.0028, 0.013^2)$ distributed, i.e. an increase in the mean and a decrease in the variance. This behaviour is confirmed by the behaviour of the predictive distributions which are shown in Figure 4. The predictive distribution for the AR-GARCH outlier model makes the tightest impression while the model without outliers has a small second mode at the the upper end of the distribution. The effect of the outlier model can be also seen in the estimated distribution of the GARCH coefficients which are less spread out.

A second example for the Deutsche Mark and US Dollar exchange rate has produced similar results.

6. Conclusion

This paper has shown that outliers should be considered additionally in GARCH models since the estimation performance can be improved quite considerably. The Gibbs sampler with a Metropolis step is a convenient (small sample) framework for estimation. The approach might be extended to multivariate GARCH models (see Polasek and Jin (1994)) or semi-parametric ARCH models (see Kozumi and Polasek 1996).

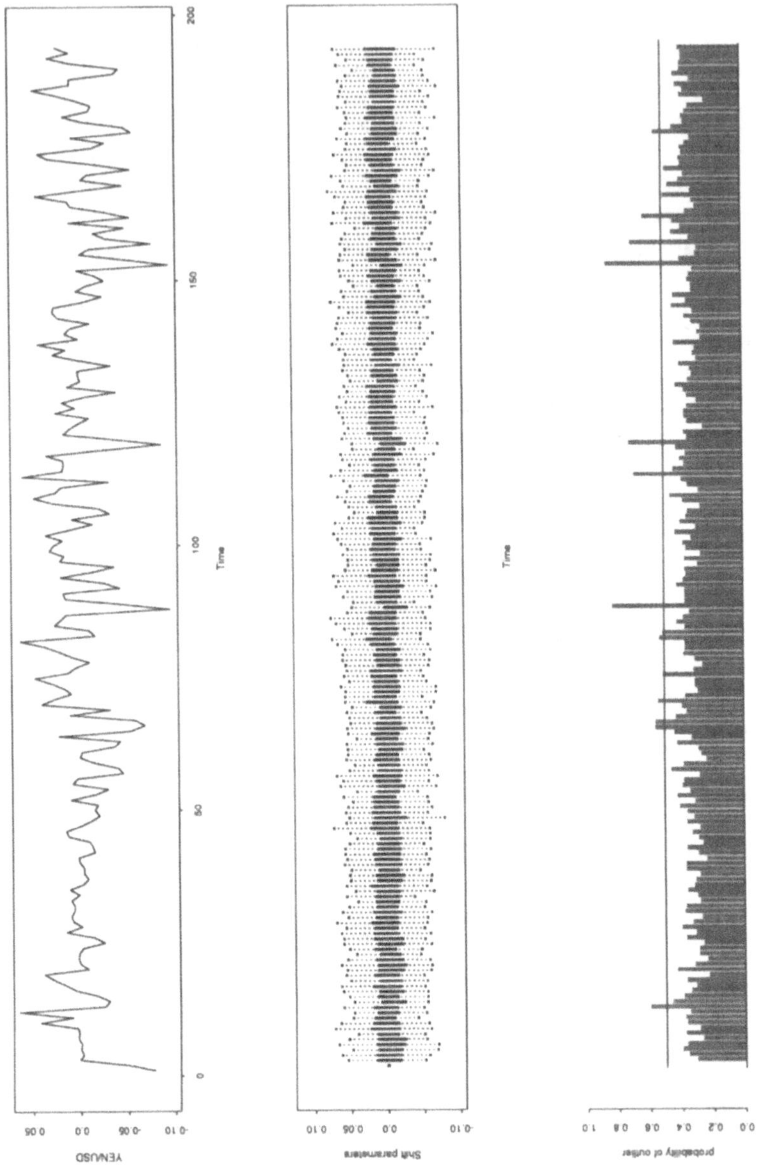

Figure 1: AR-Outlier model with YEN/USD

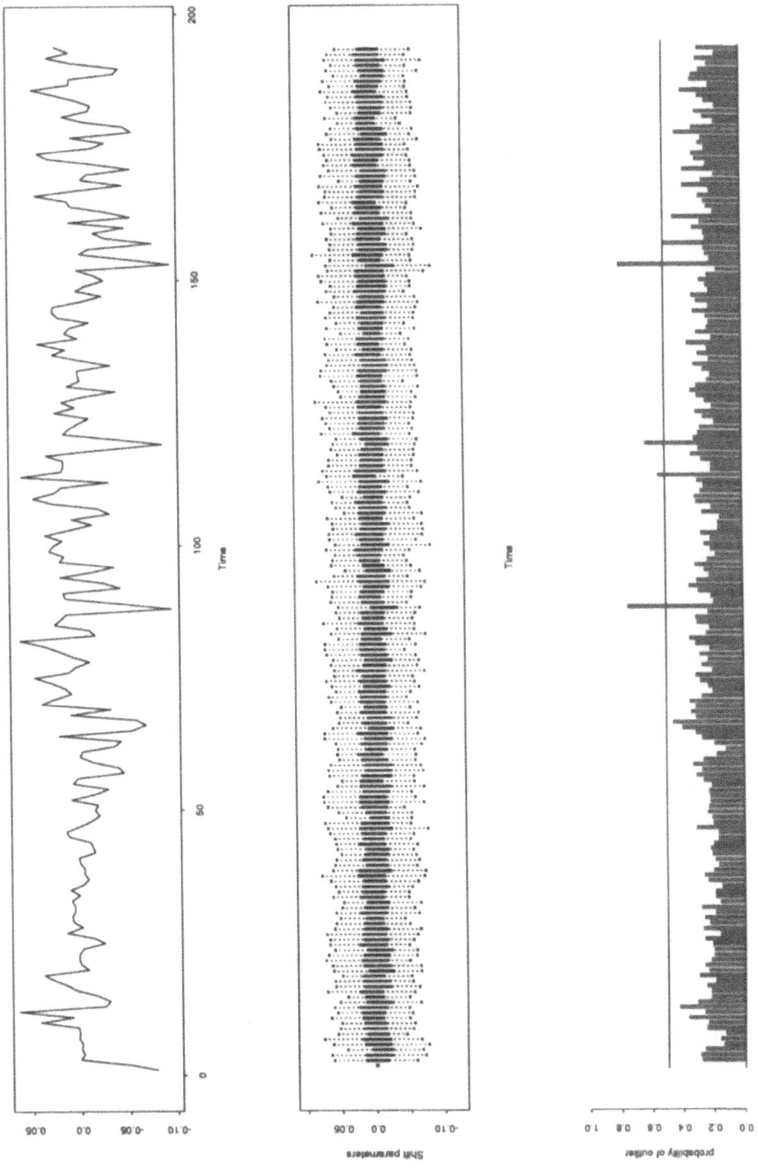

Figure 2: AR-ARCH-Outlier model with YEN/USD

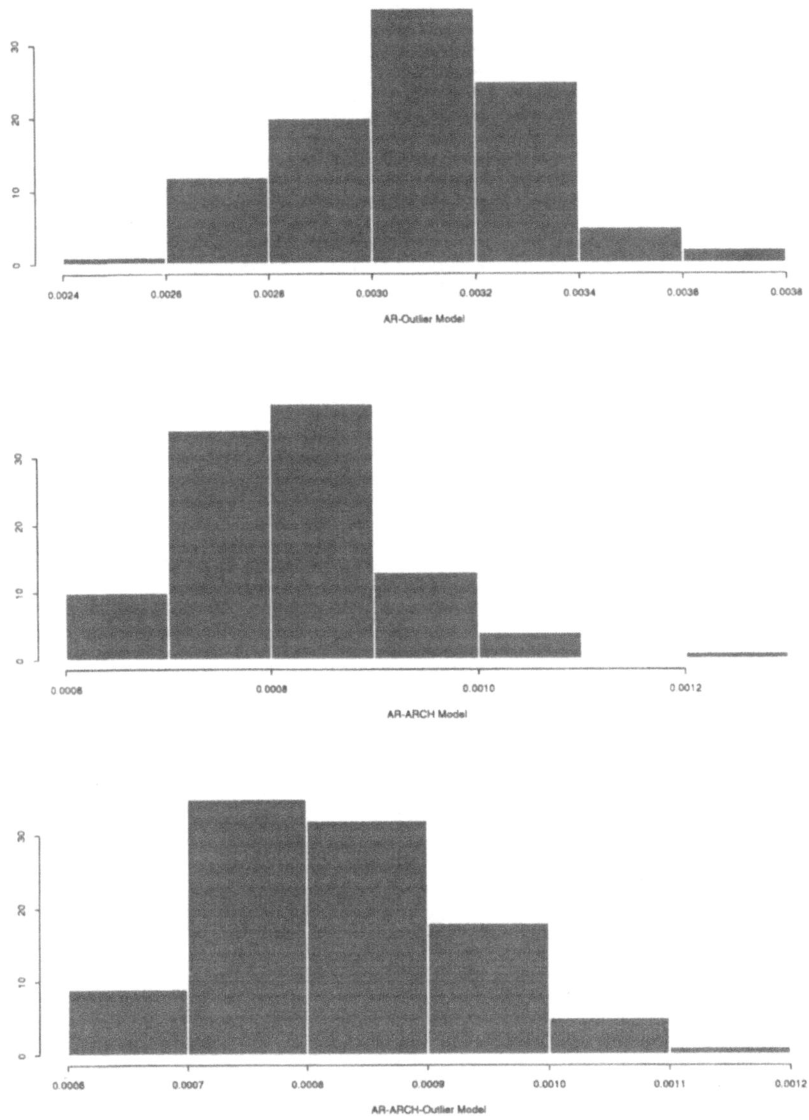

Figure 3: σ^2 for three models

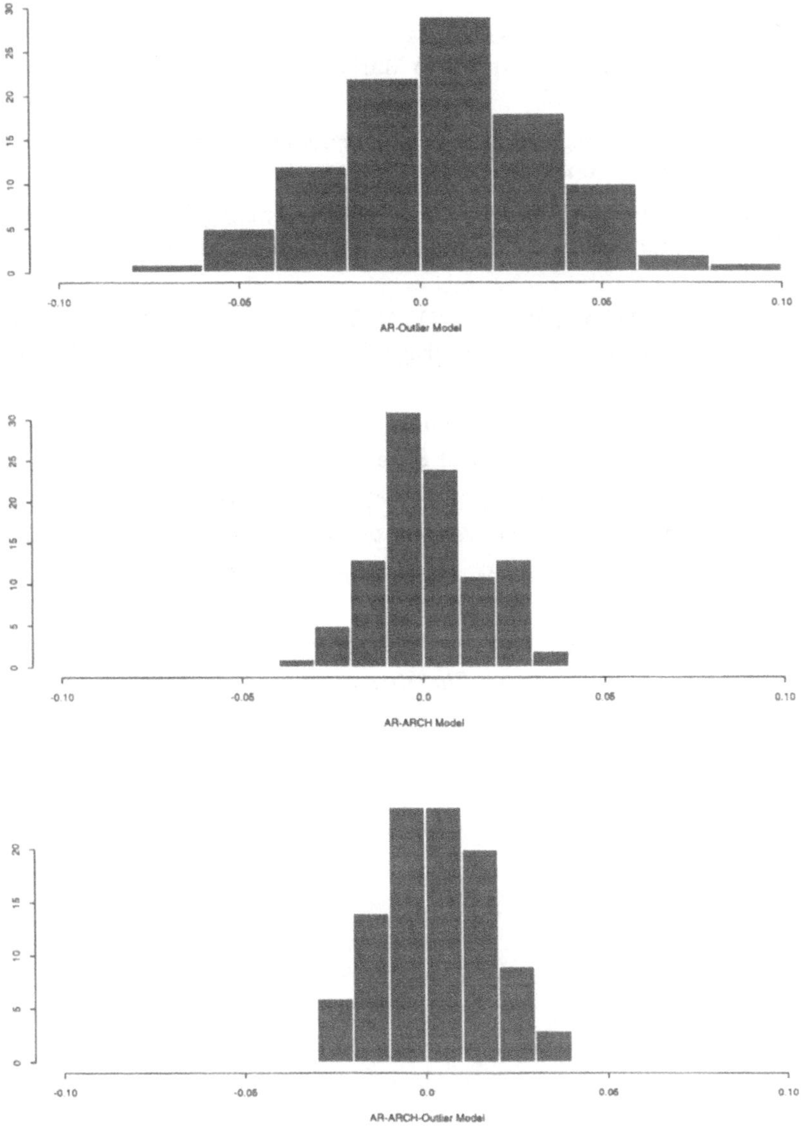

Figure 4: Predictive distribution

References:

ALBERT, J. and CHIB, S. (1993): Bayesian inference of autoregressive time series with mean and variance subject to Markov jumps, *Journal of Business and Economic Statistics, 11, 1-15.*

CHIB, S. (1993): Bayes regression with autoregressive errors: A Gibbs sampling approach. *Journal of Econometrics, 58, 275-294.*

GELFAND A. E. and SMITH A. F. M. (1990): Sampling based approaches to calculating marginal densities. *Journal of American Statistical Association, 85, 398-409.*

JIN, S.; PAI, J.; KOZUMI, H. and POLASEK, W. (1995): Bayesian Sampling Envirorment & Language: BASEL package, User's Guide Version 1.0, University of Basel.

KOZUMI, H. and POLASEK, W. (1996): A Bayesian Semiparametric Analysis of ARCH model, mimeo, University of Basel.

McCULLOCH, R. E. and TSAY, R. S. (1994): Bayesian analysis of autoregressive time series via the Gibbs sampler, *Journal of Time Series Analysis 15, 235-250.*

POLASEK, W. and JIN, S. (1994): Gibbs sampling in AR models with random walk prior, mimeo, University of Basel.

VERDINELLI, I. and WASSERMAN, L. (1991): Bayesian analysis of outlier problems using the Gibbs sampler, *Statistics and Computing 1991-1, 105-117.*

Minimum Volume Sets in Statistics: Recent Developments

Wolfgang Polonik

Institut für Angewandte Mathematik
Universität Heidelberg
D-69120 Heidelberg, Germany

Summary: We present a survey of recent developments in statistics connected to *minimum volume (MV)-sets*. MV-sets are sets which carry high mass concentration. Such sets are of interest in cluster analysis and can for example be used for estimating level sets of densities. Further statistical applications are prediction, testing for multimodality, goodness of fit, data analysis, regression problems, and density estimation.

1. Introduction

A classical example of a MV-set is the so-called *shorth*. Given a (one-dimensional) sample, the shorth is the shortest interval containing at least half of the sample. In other words, the shorth is an empirical MV-interval at level 1/2. It has first been used in statistics independently by Dalenius (1965) and Venter (1967). Of course also other fractions α than half can be used. The corresponding shortest interval is then called α-shorth. Andrews et al. (1972) defined the shorth to be the mean of the data inside the MV-interval at level 1/2, so that there is some confusion about this notion. In the present paper "shorth" always names the interval itself.

MV-sets are defined as follows: Let $X_1, X_2, \ldots$ denote observations drawn from a probability measure F. Let $\hat{F}_n$ denote a random probability measure based on the first n observations, as for example the empirical measure F_n which puts mass $1/n$ on each of the n observations. Further let $\mathcal{C}$ be a class of measurable sets.

Definition 1.1 *Let ν be a function defined on $\mathcal{C}$. For $\alpha \in [0,1]$ a set $C_{\mathcal{C}}(\alpha) \in \mathcal{C}$ is called* MV-set *for F in $\mathcal{C}$ at level α with respect to ν if*

$$C_{\mathcal{C}}(\alpha) \in \operatorname{argmin}_{C \in \mathcal{C}}\{\nu(C) : F(C) \geq \alpha\}.$$

Replacing F by $\hat{F}_n$ gives an empirical *MV-set $C_{n,\mathcal{C}}(\alpha)$.*

The phrase *minimum volume* stems from the case $\nu(C) = Leb(C)$ where Leb denotes Lebesgue measure. Unless stated otherwise, ν denotes a dominating measure for F as for example Lebesgue measure or counting measure. Important special cases of MV-sets are level sets of densities. Let f be a ν-density of F. Level sets of f are sets of the form $\Gamma_f(\lambda) = \{x : f(x) \geq \lambda\}$. $\Gamma_f(0)$ is defined to be the support of f. If $\Gamma_f(\lambda) \in \mathcal{C}$ then $\Gamma_f(\lambda)$ is a MV-set (see Lemma 2.1). This shows that $\mathcal{C}$ can in general be thought of as a shape/order restriction, or, as a statistical model. If for example the aim is to estimate f, then choosing $\mathcal{C}$ corresponds to choosing the model class of all ν-densities with level sets in $\mathcal{C}$. Standard choices for classes $\mathcal{C}$ in $\mathbf{R}^d$ are the classes of closed rectangles, balls, ellipsoids, or convex sets, denoted by $\mathcal{R}^d, \mathcal{B}^d, \mathcal{E}^d$, and $\mathcal{C}^d$, respectively. Empirical MV-sets exist in all these classes.

As surveyed in the present paper, MV-sets and other, closely related sets (cf. 2.2 and 2.3 below), have recently been used in various areas of statistics. The proposed procedures have an intuitive appeal, and they can in principle be computed on fast work stations, although in higher dimensions there is still a large gap to bridge to make most of the procedures fast enough to really use them in practice. In order to choose an appropriate model class $\mathcal{C}$ one has to balance between richness of the model (richness of $\mathcal{C}$), time needed for computation, and desireable statistical properties.

We briefly mention some related, but not so recent work. Grübel (1988) studied the length of the α-shorth (see also Beirlant and Einmahl (1995)) and applied it to robust scale estimation. Rousseeuw (1986) proposed the center of the minimum volume ellipsoid (MVE) at level 1/2 as an estimator of location and the matrix determing the shape of the MVE as a dispersion estimator. These are multivariate affine equivariant estimators of location and dispersion with highest possible breakdown points (see also Davies (1992)).

2. Level set estimation

Methods for level set estimation as discussed here are called direct methods. They don't require a preliminary density estimation step. It is worth noting, that for defining these estimators the existence of a density need not be assumed at all. And even in that case they still might contain useful information about mass concentration of the underlying distribution. A key lemma is the following (see Fig. 2.1 below):

Lemma 2.1 *Let f be the ν-density of F. If $\Gamma_f(\lambda) \in \mathcal{C}$ then for each $\lambda \geq 0$*

$$\Gamma_f(\lambda) \in \text{argmax}_{C \in \mathcal{C}}\{(F - \lambda\,\nu)(C)\}.$$

Hence, $\Gamma_f(\lambda)$ is a MV-set for F in $\mathcal{C}$ at the level $\alpha = \alpha_\lambda = F(\Gamma_f(\lambda))$. Moreover, with $\ell_\lambda = \nu(\Gamma_f(\lambda))$ we have $\Gamma_f(\lambda) \in \text{argmax}_{C \in \mathcal{C}}\{F(C) : \nu(C) \leq \ell_\lambda\}$.

2.1 Level set estimation by means of MV-sets

Lemma 2.1 says that if the assumption $\Gamma_f(\lambda) \in \mathcal{C}$ is true, then estimating MV-sets essentially means estimating level sets. One has the following general consistency result. For $C, D \in \mathcal{C}$ let $d_F(C, D) = F(C \Delta D)$ where Δ denotes set-theoretic symmetric difference.

Theorem 2.1.1 *(Polonik (1994)) Suppose that for each $\alpha \in [0,1]$ MV-sets $C_{n,\mathcal{C}}(\alpha)$ and $C_{\mathcal{C}}(\alpha)$ exist and that $C_{\mathcal{C}}(\alpha)$ is unique up to F-nullsets. Suppose further that there exists a distribution G with strictly positive ν-density such that $(\mathcal{C}, d_G)$ is quasi compact. Let $\eta \geq 0$. If*

(a) $\sup_{C \in \mathcal{C}} |(\hat{F}_n - F)(C)| \to 0$ *a.s. as* $n \to \infty$ *and*
(b) $\alpha \to \nu(C_{\mathcal{C}}(\alpha))$ *is continuous in* $[\eta, 1 - \eta]$,

then
$$sup_{\eta\leq\alpha\leq 1-\eta} d_F(C_{n,\mathcal{C}}(\alpha), C_{\mathcal{C}}(\alpha)) \to 0 \quad a.s. \ as \ n \to \infty.$$

If $\Gamma_f(\lambda) \in \mathcal{C} \ \forall \lambda \geq 0$ and if f has no flat parts, i.e. $F\{f = \lambda\} = 0 \ \forall \lambda \geq 0$, then (b) is satisfied for each $\eta > 0$. If in addition $\nu(\Gamma_f(0)) < \infty$ then (b) also holds for $\eta = 0$. The compactness assumption on $\mathcal{C}$ is satisfied for all standard classes $\mathcal{C}$ mentioned above. For $\hat{F}_n = F_n$ classes $\mathcal{C}$ satisfying (a) are known as Glivenko-Cantelli (GC) classes for F. The classes $\mathcal{R}^d, \mathcal{B}^d$, and $\mathcal{E}^d$ are GC-classes *for all F*. A sufficient condition for $\mathcal{C}^d$ to be a GC-class for F is that F has a finite Lebesgue density.

2.2 Modal sets

Definition 2.2.1 *Let $\ell \geq 0$. A set $M_{\mathcal{C}}(\ell) \in \mathcal{C}$ such that*
$$M_{\mathcal{C}}(\ell) \in \text{argmax}_{C\in\mathcal{C}}\{F(C) : \nu(C) \leq \ell\}$$
is called modal set (MO-set) *for F in $\mathcal{C}$ at level ℓ with respect to ν. Replacing F by $\hat{F}_n$ yields an* empirical MO-set $M_{n,\mathcal{C}}(\ell)$.

The modal set approach is inverse to the MV-approach. Chernoff (1964) was one of the first using MO-sets (intervals) in statistics. In the literature the notion modal set is also used for MV-sets or for level sets. Lemma 2.1 says that estimating modal sets means estimating level sets provided the model is correct. It can be shown under the same assumptions on $\hat{F}_n$ and $\mathcal{C}$ as in Theorem 2.1.1 that $M_{n,\mathcal{C}}(\ell)$ are consistent estimators for $M_{\mathcal{C}}(\ell)$.

2.3 The excess mass approach

The excess mass approach somehow lies in between the MV- and the MO-approach. It has first been considered independently by Hartigan (1987) and Müller and Sawitzki (1987).

Definition 2.3.1 *Let $\lambda \geq 0$. A set $\Gamma_{\mathcal{C}}(\lambda) \in \mathcal{C}$ is called* generalized λ-cluster *for F in $\mathcal{C}$ at level λ with respect to ν if*
$$\Gamma_{\mathcal{C}}(\lambda) \in \text{argmax}_{C\in\mathcal{C}}\{(F - \lambda\nu)(C)\}.$$
Replacing F by $\hat{F}_n$ gives an empirical *generalized λ-clusters* $\Gamma_{n,\mathcal{C}}(\lambda)$.

Hartigan (1975) defined λ-clusters as *connected* components of level sets. The sets $\Gamma_{\mathcal{C}}(\lambda)$ and $\Gamma_{n,\mathcal{C}}(\lambda)$ need not be connected. By definition MV-set as well as MO-sets are generalized λ-clusters. Polonik (1995c) gives relations between these classes of sets based on the least concave majorant of $\ell \to F_n(M_{n,\mathcal{C}}(\ell))$ (see also Section 4). Lemma 2.1 says that if $\Gamma_f(\lambda) \in \mathcal{C}$ then $\Gamma_f(\lambda)$ is a generalized λ-cluster in $\mathcal{C}$. Consistency results similar to the one for MO- and MV-sets also hold for $\Gamma_{n,\mathcal{C}}(\lambda)$ (cf. Polonik (1995a)). See Hartigan (1987), Müller and Sawitzki (1987, 1992), and Nolan (1991) for special cases and for algorithms.

The maximal value $E_{\mathcal{C}}(\lambda) = \sup_{C\in\mathcal{C}}\{(F - \lambda\nu)(C)\}$ is called *excess mass* over $\mathcal{C}$. Empirical versions are denoted by $E_{n,\mathcal{C}}(\lambda)$. The notion excess mass is motivated by the fact that if $\Gamma_f(\lambda) \in \mathcal{C}$ then $E_{\mathcal{C}}(\lambda)$ equals the volume of the area which lies below f and above the constant function λ (see Fig. 2.1).

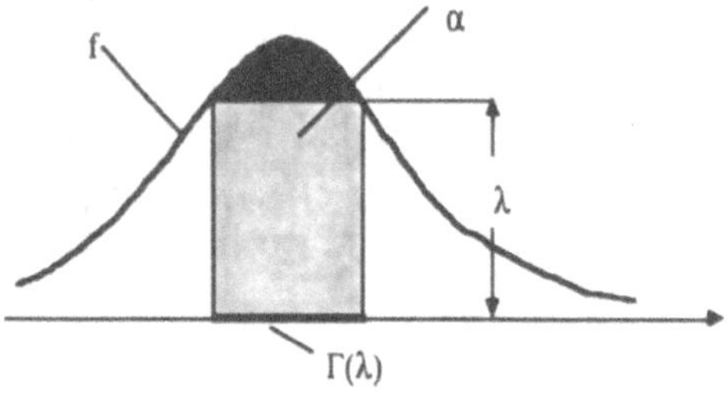

Fig. 2.1: The plot illustrates notation used above. The volume of the whole shaded area is $\alpha = F(\Gamma_f(\lambda))$. The volume of the area above the level λ equals the excess mass (cf. Section 2.3).

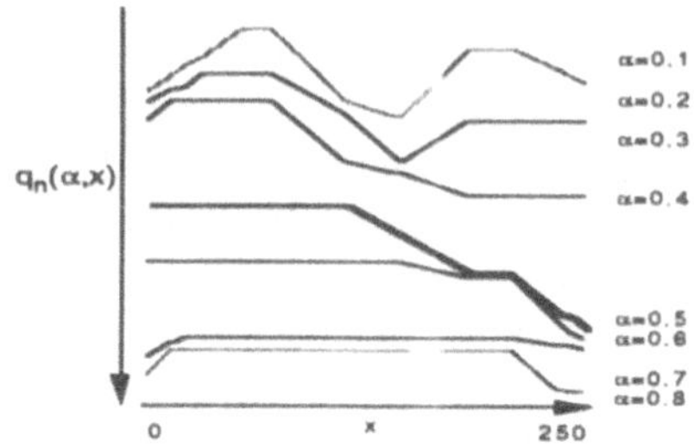

Fig. 2.2 (from Sawitzki (1994)): The plot shows an example of a shorth plot.

The above discussion might give the impression that all these estimators behave very similar. This however is not true in general. For example, there are qualitative differences in the asymptotic behaviour of the Lebesgue measures of empirical generalized λ-clusters on the one side and empirical MO- and MV-sets on the other side (compare Nolan (1989) and Polonik (1994)).

3. Recent work using the α-shorth

3.1 Data analysis: the shorth plot

The shorth plot is a data analytic tool proposed by Sawitzki (1994). It generalizes an idea of Lientz (1970). The shorth plot consists of a simultaneous plot of several localized α-shorth functionals $q_n(\alpha, x) = \inf_{a \le b}\{b - a : F_n([a,b]) \ge \alpha, x \in [a,b],\}$ (see Fig. 2.2). The shorth plot serves as a tool for inspecting mass concentration of the underlying distribution.

3.2 Regression problems

Beirlant and Einmahl (1995) considered the following regression problem. Let $(X,Y),(X_1,Y_1),\ldots,(X_n,Y_n)$ be i.i.d. random vectors, $X_i \in \mathbf{R}^d$, $Y_i \in \mathbf{R}$ where (X,Y) denotes a generic variable. Let further F_x denote the conditional distribution of Y given $X = x$ and let $U_x(\alpha) = inf\{b-a : F_x([a,b]) \ge \alpha\}$ be the length of the conditional α-shorth. The hypothesis of homoscedasticity can be expressed as $H_0 : U_x(\alpha) = \tilde{U}(\alpha)$ for some function $\tilde{U}$ independent of x. Instead of F_x for all x Beirlant and Einmahl actually consider $F_j, j = 1,\ldots,k_n$, where F_j denotes the conditional distribution of Y given $X \in A_{j,n}$, and $\{A_{j,n}\}$ defines a partition of the support of the distribution of X. Let $U_j(\alpha)$ denote the corresponding length of the α-shorth for F_j. Beirlant and Einmahl construct an estimator $\tilde{U}_n$ for $\tilde{U}$ based on estimates $U_{j,n}(\alpha)$ for $U_j(\alpha)$. Then $\tilde{U}_n(\alpha)$ is used to construct a test with asymptotic level $\beta > 0$ for H_0 of the form: reject H_0 if for some $j \in \{1,\ldots,k_n\}$ we have $\tilde{U}_n(\alpha) \notin [U_{j,n}(\alpha - k_{\beta,n}(\alpha)), U_{j,n}(\alpha + k_{\beta,n}(\alpha)))$. The choice of $k_{\beta,n}(\alpha))$ is based on asymptotic considerations.

3.3 Predictive regions for stochastic processes

Yao (1995) uses the MV-approach for constructing prediction regions for stochastic processes. Let $(X,Y),(X_1,Y_1),\ldots,(X_n,Y_n)$ be observations from a strictly stationary process where (X,Y) denotes some generic variable. One is interested in predicting Y from X. For example, given a (in general nonlinear) time series $\{Z_{t-k}, k \geq 1\}$ one might be interested in predicting Z_t from $\{Z_{t-k}, d \geq k \geq 1\}$. In contrast to classical prediction from linear time series with normal errors prediction accuracy for nonlinear time series depends on the current position in the state space (cf. Yao and Tong (1994)). Therefore Yao uses *conditional* predictive regions defined as MV-intervals for an estimator of the conditional distribution F_x of Y given X. One could instead use an empirical version of $I_\alpha = [k_{(1-\alpha)/2}(x), k_{(1+\alpha)/2}(x)]$, where $k_\alpha(x)$ denotes the α-quantile of F_x. However, in nonlinear cases the distributions F_x often are nonsymmetric, so that even in the unimodal case I_α is no MV-interval. And as is inherent in their definition, MV-intervals are the shortest intervals carrying a given portion of probability mass. The gain in relative size by using empirical MV-intervals instead of the empirical versions of I_α can be significant, especially if F_x is multimodal and one uses MV-sets in the class of unions of $k \geq 2$ intervals (see Yao (1995) for simulation studies).

4. Density estimation

Assume that $\Gamma_f(\lambda) \in \mathcal{C}\ \forall\, \lambda \geq 0$. Under this *shape restriction* level sets of f can be estimated as in Section 2. And once all level sets of f can be estimated it should also be possible to estimate f itself. In the context of the excess mass approach Müller and Sawitzki (1987, 1992) proposed a density estimator which they call silhouette. Here a key equality is $f(x) = \int \mathbf{1}_{\Gamma(\lambda)}(x)\, d\lambda$, where $\mathbf{1}_C$ denotes the indicator function of a set C. Plugging in $\Gamma_{n,\mathcal{C}}(\lambda)$ as an estimator for $\Gamma(\lambda)$ gives the definition of the silhouette (cf. Polonik (1995b)). If (for given data) the estimates $\Gamma_{n,\mathcal{C}}(\lambda)$ are monotone for inclusion (as is true for σ-lattices $\mathcal{C}$ (cf. Polonik (1995c)) then the silhouette can be visualized graphically as putting empirical generalized λ-clusters one on top of the other (as λ increases). For asymptotic results see Polonik (1995b,c).

It is well known (cf. Barlow et al. (1972)) that an order restriction for f, this is, $f(x) \leq f(y)$ if $x \preceq y$, where $\preceq$ is a (quasi) order, can be expressed as $\Gamma(\lambda) \in \mathcal{C}\ \forall\, \lambda \geq 0$ for an appropriate σ-lattice $\mathcal{C}$. It is shown in Polonik (1995c) that the silhouette is the ML-density estimator under order restrictions. A graphical representation of the silhouette can also be found in Polonik (1995c). It generalizes the well-known representation of the so-called Grenander density estimator as the slope of the least concave majorant to the empirical distribution function.

5. Goodness-of-fit techniques

The motivation for using MV-sets for constructing goodness-of-fit techniques is given by the fact that the ν-measures of the MV-sets "behave like" one-dimensional quantiles, and the latter are classically used in constructing goodness-of-fit techniques in one dimension.

Let $q(\alpha) = \nu(C_{\mathcal{C}}(\alpha))$ and $q_n(\alpha) = \nu(C_{n,\mathcal{C}}(\alpha))$ where $C_{n,\mathcal{C}}(\alpha)$ is a MV-set for F_n. Einmahl and Mason (1992) study the *generalized quantile process* $\beta_n(\alpha) = n^{1/2}(q'(\alpha))^{-1}(q_n(\alpha) - q(\alpha))$, where q' denotes the derivative of q. For $\mathcal{C} = \{(-\infty, x], x \in \mathbf{R}\}$ and $\nu((-\infty, x]) = x$ the process β_n reduces to the classical one-dimensional quantile process. Results in Einmahl and Mason show, that if the MV-sets in $\mathcal{C}$ are monotone for inclusion, then β_n converges in distribution to a standard Brownian Bridge on [0,1]. This is completely analogous to the classical quantile process. Einmahl and Mason proposed to use $q_n(\alpha)$ and $q(\alpha)$ to construct generalized Q-Q-plots.

Polonik (1994) derived further analogs of MV-sets to the classical quantiles. He showed that if $\Gamma_f(\lambda) \in \mathcal{C}$ then (under additional, more technical assumptions) the process β_n can be approximated asymptotically by $n^{1/2}(F_n - F)(C_{\mathcal{C}}(\alpha))$ which reduces to the classical empirical process in the case β_n is the quantile process (see above). In Polonik (1996) $F_n(C_{\mathcal{C}}(\alpha))$ and $F(C_{n,\mathcal{C}}(\alpha))$ are used for constructing generalized P-P-plots, and to construct tests for concentration and goodness-of-fit. For simple hypotheses these tests are (asymptotically) distribution free in arbitrary dimensions.

6. Testing for multimodality

Testing for multimodality is one of the applications of the excess mass approach (see Section 2.3). The underlying idea which goes back to Müller and Sawitzki (1987) and Hartigan (1987) is best explained in dimension one. Let $\mathcal{I}_1$ and $\mathcal{I}_2$ denote the class of closed intervals and the class consisting of (at most) two unions of closed intervals, respectively. Consider the excess mass difference $\Delta(\lambda) = E_{\mathcal{I}_2}(\lambda) - E_{\mathcal{I}_1}(\lambda)$. An estimator for $\Delta(\lambda)$ is given by $\Delta_n(\lambda) = E_{n,\mathcal{I}_2}(\lambda) - E_{n,\mathcal{I}_1}(\lambda)$. If f is unimodal, this is, the level sets of f are intervals, then $\Delta(\lambda) = 0 \; \forall \; \lambda \geq 0$. If however f is bimodal, such that f has level sets which lie in $\mathcal{I}_2 \setminus \mathcal{I}_1$, then, for such levels, $\Delta(\lambda) > 0$. This motivates to use $sup_{\lambda \geq 0}\Delta_n(\lambda)$ as a test statistic for unimodality. This maximal excess mass difference can be interpreted as the percentage of data not fitting the model of unimodality. For asymptotic properties of this kind of tests (in multivariate settings) see Polonik (1995a). Similar ideas have been used in Polonik (1994) and Barbe and Wei (1994) to construct tests for multimodality by comparing $\nu(C_{n,\mathcal{C}}(\alpha))$ and $F_n(M_{n,\mathcal{C}}(\ell))$, respectively, for different classes $\mathcal{C}$.

References:

ANDREWS, D. W.; BICKEL, P. J.; HAMPEL, F. R.; HUBER, P. J.; RODGERS, W. H. and TUKEY, J. W. (1972): Robust estimation of location: survey and advances. Princeton Univ. Press, Princeton, N.J.

BARBE, P. and WEI, S. (1994). A test for multimodality. Unpubl. Manuscript.

BARLOW, R. E.; BARTHOLOMEW, D. J.; BREMNER, J. M., and BRUNK, H. D.(1972): Statistical inference under order restrictions. Wiley, London.

BEIRLANT, J. and EINMAHL, J. H. J. (1995). Maximal type test statistics based on conditional processes. *Statist. Neerlandica, 49, 1–8.*

BEIRLANT, J. and EINMAHL, J. H. J. (1996). Asymptotic confidence intervals for the length of the shortt under random censoring. To appear in *J. Statist. Planning and Inference.*

CHERNOFF, H. (1964): Estimation of the mode. *Ann. Inst. Statist. Math. 16, 31–41.*

DAVIES, P. L. (1992): The asymptotics of Rousseeuws minimum volume ellipsoid estimator. *Ann. Statist., 20, 1828–1843.*

DALENIUS, T. (1965): The mode – a neglected statistical parameter. *J. Roy. Statist. Soc. A, 128, 110–117.*

EINMAHL, J. H. J. and MASON, D. M. (1992). Generalized quantile processes. *Ann. Statist., 20, 1062–1078.*

GRENANDER, U. (1956): On the theory of mortality measurement, Part II. *Skand. Akt., 39, 125–153.*

GRÜBEL, R. (1988): The length of the shorth. *Ann. Statist., 16, 619–628.*

HARTIGAN, J. A. (1975): Clustering algorithms. Wiley, New York.

HARTIGAN, J. A. (1987): Estimation of a convex density contour in two dimensions. *J. Amer. Statist. Assoc., 82, 267–270.*

LIENTZ, B. P. (1970): Results on non-parametric modal intervals. *SIAM J. Appl. Math., 19, 356-366.*

MÜLLER, D. W. and SAWITZKI, G. (1987): Using excess mass estimates to investigate the modality of a distribution. Preprint Nr. 398, SFB 123, Univ. Heidelberg.

MÜLLER, D. W. and SAWITZKI, G. (1992): Excess mass estimates and tests of multimodality. *J. Amer. Statist. Assoc., 86, 738–746.*

NOLAN, D. (1991): The excess mass ellipsoid. *J. Multivarite Anal., 39, 348–371.*

POLONIK, W. (1994): Minimum volume sets and generalized quantile processes. Beiträge zur Statistik Nr. 20, Institut für Angewandte Mathematik, Universität Heidelberg.

POLONIK, W. (1995a): Measuring mass concentrations and estimating density contour clusters - an excess mass approach. *Ann. Statist., 23, 855–881.*

POLONIK, W. (1995b): Density estimation under qualitative assumptions in higher dimensions. *J. Multivariate Anal., 55, 61–81.*

POLONIK, W. (1995c): The silhouette, concentration functions, and ML-density estimation under order restrictions. Technical Report No. 445, Department of Statistics, University of California, Berkeley.

POLONIK, W. (1996): Testing for concentration and goodness-of-fit in higher dimensions – (asymptotically) distribution-free methods. Beiträge zur Statistik Nr. 33, Institut für Angewandte Mathematik, Universität Heidelberg.

ROUSSEEUW, P. J. (1986): Multivariate estimation with high breakdown point. In: W. Grossmann et al. (eds): *Mathematical statistics and applications.* Reidel, Dordrecht, 283–297.

SAWITZKI, G. (1994). Diagnostic plots for one-dimensional data. In: Dirschedl P. and Ostermann, R. (eds.): *Papers collected at the occasion of the 25th Conference on Statistical Computing at Schloss Reisensburg.* Physica, Heidelberg.

VENTER, J. H. (1967). On estimation of the mode. *Ann. Math. Statist., 38, 1446–1455.*

YAO, Q. and TONG, H. (1994). Quantifiying the inference of initial values on nonlinear prediction. *J. Roy. Statist. Soc. B, 56, 701–725.*

YAO, Q. (1995). Conditional predictive regions for stochastic processes. Unpubl. manuscript.

Fusion of Data Sets in Multivariate Linear Regression with Errors-in-Variables

Albert Satorra [1]

Departament d'Economia i Empresa
Universitat Pompeu Fabra
Ramon Trias Fargas 23-25, 08005 BARCELONA, Spain

Summary: We consider the application of normal theory methods to the estimation and testing of a general type of multivariate regression models with errors-in-variables, in the case where various data sets are merged into a single analysis and the observable variables deviate possibly from normality. The various samples to be merged can differ on the set of observable variables available. We show that there is a convenient way to parametrize the model so that, despite the possible non-normality of the data, normal-theory methods yield correct inferences for the parameters of interest and for the goodness-of-fit test. The theory described encompasses both the functional and structural model cases, and can be implemented using standard software for structural equations models, such as LISREL, EQS, LISCOMP, among others. An illustration with Monte Carlo data is presented.

1. Introduction

Consider the classical regression equation

$$Y_i = \alpha + \beta x_i + v_i, \quad i = 1, \dots, n, \tag{1}$$

where, for individual i, Y_i and x_i are the values of the response and explanatory variables respectively, v_i is the value of the disturbance term, and α and β are respectively the intercept and slope parameters. Suppose we are unable to observe x_i. Instead we observe X_i which satisfy

$$X_i = x_i + u_i, \tag{2}$$

where u_i is an error term. We assume the v_i and u_i are iid with mean 0 and variances σ_v^2 and σ_u^2 respectively. Equations (1) and (2) with the associated assumptions define the classical simple regression model with errors-in-variables. When the x_i are a set of fixed values (across hypothetical sample replications) the model is called a functional model. When the x_i are random (i.e. varying across sample replications) we have the so-called structural model. See Fuller (1987) for a comprehensive overview of measurement error models in regression analysis. For recent work on the importance of assessing measurement reliability in multivariate linear regression, see Gleser (1992).

As it is well known, when $\sigma_u^2 > 0$ and X_i is used instead of x_i in (1), the

[1]Work supported by the Spanish DGICYT grant PB93-0403

usual least-squares estimate is not consistent for β; in fact, under normality, equations (1) and (2) with the associated assumptions fail to identify the parameters of the model. To obtain a consistent estimate of β, additional information related with the measurement error variance σ_u^2 is required. The parameter β is identified, for example, when we specify the value of σ_u^2, or the ratio of variances σ_u^2/σ_v^2, amid of other possibilities. In practice, however, it may be difficult to have such exact information on the value of error variances.

An alternative to specifying the size of measurement error variances is the multiple indicator approach. Consider for example two indicators X_{i1} and X_{i2} $(i = 1, \ldots, n)$ which satisfy

$$X_{i1} = x_i + u_{i1}, \quad X_{i2} = x_i + u_{i2}, \tag{3}$$

where the u_{i1} and u_{i2} are uncorrelated error terms. Equations (1) and (3) with the associated assumptions yield an identified model. Inferences for this type of models is usually carried out under normality and a single sample with complete data (see, e.g., Fuller, 1987).

In practice, however, the data may be composed of several subsamples with observable variables missing in the different subsamples. In the described regression model for example, we may have one sample with information only on Y_i and X_{i1} (X_{i2} missing), and a second sample with information only on X_{i1} and X_{i2} (Y_i missing). In the present paper we focus on the joint analysis of various samples of this type using Normal Theory (NT) methods while allowing the data to deviate from the normality assumption. NT methods are available in standard software for structural equation modeling, such as LISREL (Jöreskog & Sörbom, 1989), EQS (Bentler, 1989), LISCOMP (Muthén, 1987), the SAS Procedure CALIS (e.g., Hatcher, 1994). A clear advantage of the NT methods is that they require first- and second-order moments only.

The present paper relates with work of Arminger and Sobel (1990), where Pseudo-Maximum Likelihood (PML) analysis is applied to different data sets arising from missing data; Arminger and Sobel, however, do not address the issue of analyzing a functional model and the issue of correct NT inferences when data deviates from the normality assumption. The present work can also be seen as generalizing to multi-sample analysis and non-normality the results of Dham and Fuller (1986) that show the validity of NT-inferences in the case of a functional model. Our paper relates also with the asymptotic robustness results of Anderson and Amemiya (1988) and Amemiya and Anderson (1990); we extend such asymptotic robustness results to the case of multiple samples. Finally, we should mention that the following results are a specialization to the multivariate regression model set-up of more general results developed in Satorra (1995).

The structure of the paper is as follows. In Section 2 we present the general model set-up under consideration. Section 3 describes the NT generalized-least-squares analysis and provides a theorem which summarizes the major results of the paper. Finally, Section 4 presents a limited Monte Carlo study illustrating in finite samples and a specific model context the practical implications of the paper.

2. Multivariate Regression and Moment Structures

Let

$$\begin{cases} Y_i &= Bx_i + v_i \\ X_i &= \Lambda x_i + u_i, \end{cases} \quad i = 1, \ldots, n, \tag{4}$$

where for individual i, $Y_i\,(p \times 1)$, $X_i\,(q \times 1)$ and $x_i\,(m \times 1)$ are respectively the values of the response, indicator and true (latent) regressor variables; $v_i\,(p \times 1)$ and $u_i\,(q \times 1)$ are respectively the values of the disturbance and measurement error terms; $B\,(p \times m)$ is a matrix of (unknown) parameters, and $\Lambda\,(q \times m)$ is a matrix of known coefficients. Assume that the v_i and u_i are independent and both iid with mean 0 and variance matrices Φ_v and Ψ_u respectively; furthermore, assume

$$n^{-1} \sum_{i=1}^{n} x_i x_i{}' \to \Phi_x, \tag{5}$$

as $n \to \infty$, where Φ_x is a positive-definite matrix and the convergence is in probability in the case of a structural model. We further assume that for each i: (structural model) x_i, v_i and u_i are mutually independent; (functional model) v_i and u_i are independent with Φ_v and Ψ_u unrelated with the x_i. We consider that Λ is a 0-1 matrix of full column rank, [2] and $\Psi_u = H\psi_u$, where H is a 0-1 matrix and ψ_u is a vector of unrestricted parameters. Note that by allowing a component of X_i and x_i to be constant to 1, we can accommodate as parameters of the model the intercepts of the regression equations and the mean vector of x_i. In the model set-up of the present paper, we impose that the u_i are normally distributed.

Clearly, (4) can be written as

$$\begin{pmatrix} Y_i \\ X_i \end{pmatrix} = \begin{pmatrix} 0 \\ I \end{pmatrix} u_i + \begin{pmatrix} B \\ \Lambda \end{pmatrix} x_i + \begin{pmatrix} I \\ 0 \end{pmatrix} v_i; \tag{6}$$

or, alternatively, as

$$\begin{cases} z_i &= \Lambda^\star z_i^\star + u_i^\star \\ z_i^\star &= B^\star z_i^\star + v_i^\star, \end{cases} \tag{7}$$

where

$$z_i = \begin{pmatrix} Y_i \\ X_i \end{pmatrix}, \quad z_i^\star = \begin{pmatrix} Y_i \\ x_i \end{pmatrix},$$

[2]In the context of a multiple indicator model, Λ is of the form $1_K \otimes I_m$ where 1_K denotes a vector of ones and K is the number of "replicates".

$$\Lambda^\star = \text{diag}\,(I, \Lambda), \quad B^\star = \begin{pmatrix} 0 & B \\ 0 & 0 \end{pmatrix}$$

and

$$u_i^\star = \begin{pmatrix} 0 \\ u_i \end{pmatrix}, \quad v_i^\star = \begin{pmatrix} v_i \\ x_i \end{pmatrix}.$$

(Here 0 and I denote zero and identity matrices respectively of dimensions clearly determined by the context.). We denote by $\ell \equiv (p+q)$ the dimension of z_i. We should note that expression (7) enables the direct use of standard software for structural equation models for the purpose of model estimation and testing.

Now suppose that we have G different selection matrices T_g, $g = 1, \ldots, G$, where T_g is of dimension $\ell_g \times \ell$ with $\ell_g \leq \ell$. Suppose that instead of z_i we observe only

$$\tilde{z}_{ig} = T_g z_i,$$

where each i ($i = 1, \ldots, n$) is associated to one of the matrices T_g. Let the cases i associated with T_g form the group (or subsample) g, and denote by n_g the size of the gth group ($g = 1, \ldots, G$). For the validity of the asymptotic theory to be developed below, we require that G is "small" compared to n, so that $n_g/n \to c_g > 0$ as $n \to \infty$.

Define now the $(\ell_g \times \ell_g)$ uncentered sample cross-product moment matrix

$$S_g \equiv n_g^{-1} \sum_i \tilde{z}_{ig} \tilde{z}'_{ig},$$

where here $\sum_i$ denotes sum over the n_g cases in group g. Clearly, under the current model set-up,

$$S_g \xrightarrow{P} \Sigma_g,$$

where $\xrightarrow{P}$ denotes convergence in probability and Σ_g is a finite $\ell_g \times \ell_g$ matrix. From (6), and the definition of $\tilde{z}_{ig}$, we have

$$\tilde{z}_{ig} = \Pi_{g0} u_i^{(g)} + \Pi_{g1} x_i^{(g)} + \Pi_{g2} v_i^{(g)}, \tag{8}$$

where

$$\Pi_{g0} \equiv T_g \begin{pmatrix} 0 \\ I \end{pmatrix}, \quad \Pi_{g1} \equiv T_g \begin{pmatrix} B \\ \Lambda \end{pmatrix}, \quad \Pi_{g2} \equiv T_g \begin{pmatrix} I \\ 0 \end{pmatrix}, \tag{9}$$

and where $u_i^{(g)}$ and $v_i^{(g)}$ are subvectors of u_i and v_i respectively so that Π_{g0} and Π_{g2} are of full column rank. Note that the vector $x_i^{(g)}$ is just the whole vector x_i.

The following moment structure for Σ_g can be derived from (8) and the current assumptions:

$$\Sigma_g = \Pi_{g0}\Phi_{g0}\Pi_{g0}' + \Pi_{g1}\Phi_{g1}\Pi_{g1}' + \Pi_{g2}\Phi_{g2}\Pi_{g2}', \tag{10}$$

where $\Phi_{g0} \equiv \Phi_{u^{(g)}}$, $\Phi_{g1} \equiv \Phi_{x^{(g)}}$ and $\Phi_{g2} \equiv \Phi_{v^{(g)}}$ are respectively the "population" moment matrices associated to $u_i{}^{(g)}$, $x_i{}^{(g)}$ and $v_i{}^{(g)}$. Consider now the following (unconstrained) vector of parameters

$$\vartheta_1 \equiv (\mathrm{v}'(\Phi_{v^{(1)}}), \ldots, \mathrm{v}'(\Phi_{v^{(G)}}), \mathrm{v}'(\Phi_{x^{(1)}}), \ldots, \mathrm{v}'(\Phi_{x^{(G)}}))', \tag{11}$$

where the $\mathrm{v}(\Phi_{v^{(g)}})$ may be of varying dimensions (determined by T_g) while the $\mathrm{v}'(\Phi_{x^{(g)}})$ are all of dimension $m(m+1)/2$ (For a symmetric matrix A, $\mathrm{v}(A)$ denotes the vectorized matrix A with the redundant elements due to symmetry suppressed.). Furthermore, define

$$\tau \equiv \begin{pmatrix} \mathrm{vec}(B) \\ \psi_u \end{pmatrix} \quad \text{and} \quad \vartheta \equiv \begin{pmatrix} \tau \\ \vartheta_1 \end{pmatrix}. \tag{12}$$

Let denote by $t^\star$ and t the dimensions of τ and ϑ respectively.

In relation with (9), (10), (11) and (12), consider the (multiple-group) moment structure

$$\Sigma_g = \Sigma_g(\vartheta), \quad g = 1, \ldots, G, \tag{13}$$

where $\Sigma_g(.)$ is the matrix-valued function of ϑ associated with (10), with $\Pi_{g\ell} = \Pi_{g\ell}(\tau)$ and $\Phi_{g0} = H\psi_u$ ($\ell = 1, 2, 3$; $g = 1, \ldots, G$) as specified above. Note that in this model specification the non-redundant elements of Φ_x and of Φ_v are unconstrained parameters (even across-groups) of the model (this is in fact a key assumption for the theorem to be presented in Section 3). [3]

We now consider the estimation and testing of the multiple-group moment structure (13). We will do so by fitting simultaneously the S_g to the $\Sigma_g(\vartheta)$. We will use a fitting function that is optimal under the following NT assumption: the z_i are iid normally distributed. In fact, the major import of the present paper is a theorem that identifies conditions under which the NT-inferences are correct even though the assumption NT does not hold. In structural equation modeling, the validity of NT inferences when the NT does not hold has been called asymptotic robustness (Anderson, 1987).

To facilitate notation, we define

$$s \equiv \mathrm{vec}\{s_g;\, g = 1, \ldots, G\} \quad \text{and} \quad \sigma \equiv \mathrm{vec}\{\sigma_g;\, g = 1, \ldots, G\},$$

where $s_g \equiv \mathrm{v}\,(S_g)$ and $\sigma_g \equiv \mathrm{v}\,(\Sigma_g)$, and we write (13) as

$$\sigma = \sigma(\vartheta), \tag{14}$$

[3]We should note that the results to be described below apply also when B and Λ are continuously differentiable functions of the subvector of parameters τ.

where $\sigma(.)$ is a continuously differentiable vector-valued function of ϑ. Now we will fit s to $\sigma(\vartheta)$.

3. NT Generalized-Least-Squares

The NT Generalized-Least-Squares (NT-GLS) estimate of ϑ is defined as

$$\hat{\vartheta} \equiv \arg\min[s - \sigma(\vartheta)]'\hat{V}^{\star}[s - \sigma(\vartheta)], \tag{15}$$

where $\hat{V}^{\star}$ converges in probability to the following block-diagonal matrix:

$$V^{\star} \equiv \operatorname{diag}\left\{c_g V_g^{\star};\, g = 1, \ldots, G\right\},$$

where $\frac{n_g}{n} \to c_g$ when $n \to \infty$,

$$V_g^{\star} \equiv \frac{1}{2} D'({\Sigma_g}^{-1} \otimes {\Sigma_g}^{-1}) D, \tag{16}$$

and D is the duplication matrix for symmetry of Magnus and Neudecker (1988). It is well known that under the NT assumption this NT-GLS estimate is asymptotically optimal. For seminal work on GLS estimation of covariance structures in single-sample analysis, see Browne (1974); NT-GLS estimation in multi-sample analysis is treated in Lee and Tsui (1982).

We need to impose the following identification assumption: $\sigma(\vartheta^{\star}) = \sigma(\vartheta)$ implies (locally) that $\vartheta^{\star} = \vartheta$, and the Jacobian matrix $\Delta = \frac{\partial \sigma(\vartheta)}{\partial \vartheta'}$ is of full column rank in a neighborhood of ϑ. This assumption is needed to ensure uniqueness and consistency of the estimate $\hat{\vartheta}$.

An alternative to the NT-GLS estimate is the PML approach. Under the NT assumption,

$$F_{ML} \equiv \sum_{g=1}^{G} \frac{n_g}{n} [\log \mid \Sigma_g(\vartheta) \mid + \operatorname{trace}\{S_g \Sigma_g(\vartheta)^{-1}\} - \log \mid S_g \mid - \ell_g] \tag{17}$$

is an affine transformation of the log-likelihood function. Thus, under general distribution assumption for the z_i, maximization of $F_{ML} = F_{ML}(\vartheta)$ yields the so-called PML estimate of ϑ. This PML estimate is in fact asymptotically equivalent to the NT-GLS estimate described above. For a comprehensive overview of PML estimation in structural equation modeling, see Arminger and Sobel (1990). Multi-sample analysis of structural equation models using the maximum likelihood method was first proposed by Jöreskog (1971) in the context of factor analysis. Now we will review the asymptotic variance matrix of the NT-GLS and PML estimates.

Let Γ denote the asymptotic variance matrix of the scaled vector of sample moments $\sqrt{n}s$. Under the NT assumption, the asymptotic variance matrix of $\sqrt{n}s$ is (e.g., Satorra, 1993)

$$\Gamma^{\star} = \operatorname{diag}\left\{c_g^{-1}\Gamma_g^{\star};\, g = 1, \ldots, G\right\},$$

where

$$\Gamma_g^\star \equiv \Omega_g - \Upsilon_g,$$

$$\Omega_g \equiv 2D^+(\Sigma_g \otimes \Sigma_g)D^{+\prime}, \tag{18}$$

$$\Upsilon_g \equiv 2D^+(\mu_g\mu_g{}' \otimes \mu_g\mu_g{}')D^{+\prime}$$

and

$$\mu_g \equiv E(z_g),$$

where E denotes the expectation operator under the NT assumption. Furthermore, let

$$\Omega \equiv \operatorname{diag}\left\{c_g^{-1}\Omega_g;\ g = 1, \ldots, G\right\}, \tag{19}$$

and let $\hat{\Omega}$ denote the estimate of Ω obtained by substituting S_g for Σ_g and c_g^{-1} for $\frac{n}{n_g}$ in (18) and (19) respectively. Note that $V^\star = \Omega^{-1}$ (and that $\hat{V}^\star = \hat{\Omega}^{-1}$).

By using the standard delta-method, under general distribution assumption for the z_i, the asymptotic variance matrix of the NT-GLS and PML estimates is known to be (e.g., Satorra and Neudecker, 1994)

$$\operatorname{avar}(\hat{\vartheta}) = (\Delta' V^\star \Delta)^{-1}\Delta' V^\star \Gamma V^\star \Delta(\Delta' V^\star \Delta)^{-1}. \tag{20}$$

Under the NT assumption we have

$$\operatorname{avar}_{NT}(\hat{\vartheta}) = (\Delta' V^\star \Delta)^{-1}. \tag{21}$$

In addition to parameter estimation, we are interested in testing the goodness-of-fit of the model. The following goodness-of-fit test statistic will be used:

$$T^\star = n(s - \hat{\sigma})'(\hat{V}^\star - \hat{V}^\star\hat{\Delta}(\hat{\Delta}'\hat{V}^\star\hat{\Delta})^{-1}\hat{\Delta}'\hat{V}^\star)(s - \hat{\sigma}); \tag{22}$$

under NT, and when the model holds, $T^\star$ is asymptotically chi-square distributed with

$$r = \operatorname{rank}(P\Gamma^\star P') \tag{23}$$

degrees of freedom (df), where $P = I - \Delta(\Delta' V^\star \Delta)^{-1} V^\star$ (Satorra, 1993).

Direct specialization to the present model set-up of results of Satorra (1995) yield the following theorem.

THEOREM 1 (cf., Satorra, 1995, Theorem 1) *Under (6) and (13) and the current assumptions (NT is not included), the NT-GLS and PML estimates verify*

1. *$\hat{\vartheta}$ is a consistent estimate of ϑ.*
2. *The $t^\star \times t^\star$ leading principal submatrices of $avar(\hat{\vartheta})$ and $avar_{NT}(\hat{\vartheta})$ (recall (20) and (21)) coincide (that is, the asymptotic variance matrix $avar(\hat{\tau})$ is the same as under NT).*
3. *$\hat{\tau}$ is an efficient estimate within the class of GLS estimates of (15).*

4. *The asymptotic distribution of the goodness-of-fit test statistic $T_V^\star$ of (22) is chi-square with degrees of freedom r given by (23).*

Note that this theorem guarantees correct NT-inferences for the parameter vector τ, including the asymptotic efficiency, and also asymptotic chi-squaredness for the goodness-of-fit test, without the requirement for the z_i to be iid normally distributed. It is required, however, that for each i: (structural model) the x_i, v_i, u_i are mutually independent, not only uncorrelated; (functional model) the v_i, u_i are independent with the limit of the second-order moments of the x_i to be finite. With regard to the model specification, it is also required that the variances of the possibly non-normal constituents of the model, such as the x_i and v_i, are not constrained even across groups. See Satorra (1995) for full details of the proof of this theorem (in this reference, the theorem is formulated in a more general model context).

In the next section we present a limited Monte Carlo study to illustrate the performance in finite samples of the asymptotic results of the paper.

4. Monte Carlo Illustration

Consider the regression with errors-in-variables model set-up of equations (1) and (3) of Section 1, from which we simulate two-sample data of sizes $n_1 = 800$ and $n_2 = 400$, with X_{i2} missing in the first subsample and Y_i missing in the second subsample. The values of x_i and v_i are simulated as independent chi-square distributed of 1 degree of freedom (i.e., a highly skewed non-normal distribution) conveniently scaled; the u_{i1} and u_{i2} are simulated as uncorrelated iid normally distributed. Each Monte Carlo run consisted on replicating 1000 times the steps of (a) generating two-sample data, and (b) estimating (for each two-sample data) a specified model using the NT-GLS approach of Section 2.

The three models differ with respect to the restrictions imposed on equality across groups of means and variances of x_i. Two of the models considered are structural, one is a functional model.

Unrestricted structural model (USM): The mean of x_i is a parameter restricted to be equal across groups; the variance of x_i is a parameter different for each group.

Restricted structural model (RSM): The variance of x_i is a parameter restricted to be equal across groups; the mean of x_i is a parameter different for each group.

Unrestricted functional model (UFM): the (pseudo) mean and variance of x_i are (pseudo) parameters different for each group.

In all the three models, the variance of v_i in the first group and the variance of u_{i2} in the second group are unrestricted parameters of the model; the parameters α and β and the variance of u_{i1} are restricted to be equal across groups. With regard to the degrees of freedom of $T^\star$, the difference between the number of distinct moments (10 moments) and the number of free parameters in the model, yields 1 df in the case of models RSM and USM, and 2 df in the case of UFM. Tables 1-3 present the Monte Carlo results for models USM, UFM and RSM respectively.

Note that USM, UFM and RSM are in fact three alternative correctly specified models, since the restrictions of equality of parameters across groups imposed by the models are in fact true in the population. Only USM and UFM, however, satisfy the conditions for invoking Theorem 1; RSM does not satisfy such conditions, since it restricts across groups the variance of x_i, a non-normal constituent of the model.

In all the three models, we expect consistency of the parameter estimates. This is in fact corroborated by comparing the second and third columns of Tables 1-3. The second column corresponds to the population values of the parameters and the third column corresponds to the mean (across 1000 replications) of NT-GLS parameter estimates.

In the case of models USM and UFM, i.e. Tables 1 and 2 respectively, Theorem 1 guarantees asymptotic correctness of NT-inferences for the subvector of parameters τ: the parameters α, β, $\sigma^2_{u_1}$, $\sigma^2_{u_2}$ $\mu(x_i^{(1)})$ and $\mu(x_i^{(2)})$ in USM, and α, β, $\sigma^2_{u_1}$ and $\sigma^2_{u_2}$ in UFM. Thus, the rows of Tables 1-2 corresponding to the components of τ should show: (a) the Monte Carlo standard deviation of parameter estimates (column $sd(\hat{\vartheta})$) close to the Monte Carlo mean of NT standard errors (column $E(se)$); (b) the Monte Carlo variance of $d \equiv \frac{\hat{\vartheta}-\vartheta}{se(\hat{\vartheta})}$ (column var(d)) close to unity; and, (c) the Monte Carlo two-sided tails of d close to nominal values (columns 5%, 10% and 20%). Theorem 1 ensures also the asymptotic chi-squaredness of the goodness-of-fit test $T^\star$; thus, in Tables 1-2, the Monte Carlo mean (and variance) of $T^\star$ should be close to the degrees of freedom (twice the degrees of freedom), and the tails of the Monte Carlo distribution of $T^\star$ should be close to the nominal values. These expectations are clearly corroborated by the results shown in Tables 1 and 2. This correct performance of the NT-inferences is remarkable despite clear non-normality of the z_i.

In the case of model RSM, however, Table 3 shows that NT-inferences for parameters deviate dramatically from the correct performance (only NT-inferences for the means $\mu(x_i^1)$ and $\mu(x_i^2)$ seem to be appropriate), and the empirical distribution of $T^\star$ also deviates highly from the desired chi-square distribution. Note that this lack of robustness of NT-inferences resulted from the restriction of equality across groups of the variance of the x_i, a restriction which is in fact true in the population generating the data.

References:

AMEMIYA, Y. and ANDERSON, T. W. (1990): Asymptotic chi-square tests for a large class of factor analysis models. *The Annals of Statistics, 3, 1453-1463*

ANDERSON, T. W. (1987): Multivariate linear relations. In T. Pukkila & S. Puntanen (edits.) *Proceedings of the Second International Conference in Statis-*

tics, Pp. 9-36. Tampere, Finland: University of Tampere.

ANDERSON, T. W. and AMEMIYA, Y. (1988): The asymptotic normal distribution of estimates in factor analysis under general conditions." *The Annals of Statistics, 16, 759-771.*

ARMINGER, G. and SOBEL, M. E. (1990): Pseudo-maximum likelihood estimation of mean and covariance structures with missing data *Journal of the American Statistical Association, 85, 195- 203.*

BENTLER, P. M. (1989): EQS Structural Equations Program Manual. Los Angeles: BMDP Statistical Software, Inc.

BROWNE, M. W. (1974): Generalized least squares estimates in the analysis of covariance structures. *South African Statistical Journal, 8, 1-24.*

DHAM, P. F. and FULLER, W. A. (1986): Generalized Least Squares Estimation of the Functional Multivariate Linear Errors-in-variables Model. *Journal of Multivariate Analysis, 19, 132-141.*

FULLER, W. A. (1987): Measurement Error Models. New York: Wiley.

GLESER, L. J. (1992): The Importance of Assessing Measurement Reliability in Multivariate Regression. *Journal of the American Statistical Association, 87, 696-707.*

HATCHER, L. (1994): A step-by-step approach to using the SAS System for factor analysis and structural equation modeling, Cary, NC SAS Institute.

JÖRESKOG, K. (1971): Simultaneous factor analysis in several populations *Psychometrika, 57, 409-426.*

JÖRESKOG, K. and SÖRBOM, D. (1989): LISREL 7, A Guide to the Program and Applications (2nd ed.). Chicago: SPSS Inc.

LEE, S.-Y., and TSUI, K. L. (1982): Covariance structure analysis in several populations. *Psychometrika, 47, 297-308.*

MAGNUS, J. & NEUDECKER, H. (1988): Matrix differential calculus. New York: Wiley.

MUTHÉN, B. (1987): LISCOMP: Analysis of linear structural equations with a comprehensive measurement model (User's Guide). Mooresville, IN: Scientific Software.

SATORRA, A. (1993): Asymptotic robust inferences in multiple-group analysis of augmented-moment structures", in pp. 211-229 Multivariate Analysis: Future Directions 2, (edits. by C.M. Cuadras and C.R. Rao). Elsevier: Amsterdam.

SATORRA, A. (1995): Asymptotic robust inferences in Multi-Sample Analysis of Multivariate Linear Relations, Economics Working Paper 126, Universitat Pompeu Fabra, Barcelona

SATORRA, A. and NEUDECKER, H. (1994): On the asymptotic optimality of alternative minimum-distance estimates in linear latent-variable models. *Econometric Theory, 10, 867-883.*

		distribution of estimates						
para.	true val.	$E(\hat{\vartheta})^a$	$sd(\hat{\vartheta})^b$	$E(se)^c$	$\text{var}(d)^d$	5% [e]	10%	20%
$\sigma^2(v_i^{(1)})$	1.00	1.00	.25	.23	1.24	7.10	13.20	23.50
$\sigma^2(x_i^{(1)})$	1.00	1.00	.15	.08	3.43	29.40	37.10	49.50
$\sigma^2(x_i^{(2)})$	1.00	1.00	.20	.08	5.62	39.80	49.90	59.40
$\sigma^2(u_{2i}^{(j)})$	.30	.30	.05	.05	1.01	5.40	10.40	19.90
$\sigma^2(u_{2i}^{(2)})$	.40	.40	.05	.05	.89	3.70	8.20	16.60
β	2.00	2.00	.12	.12	1.04	4.90	9.20	20.20
α	1.00	1.00	.37	.37	1.03	4.40	9.00	20.50
$\mu(x^{(j)})$	3.00	3.00	.03	.03	1.05	4.50	10.10	20.90

		distribution of $T^\star$ (df=2)			
	mean	var	5%	10%	20%
	2.12	4.44	6.40	11.40	21.40

[a] Monte Carlo mean of estimates

[b] Monte Carlo standard deviation of estimates

[c] Monte Carlo mean of NT standard errors

[d] Monte Carlo variance of $d \equiv \frac{\hat{\vartheta}-\vartheta}{se(\hat{\vartheta})}$

[e] Monte Carlo nominal 5%, 10% and 20% two-sided tails for d

Table 1: Monte Carlo results for USM model, where $\sigma^2(x_i^{(1)})$ and $\sigma^2(x_i^{(2)})$ are unrestricted parameters and $\mu(x_i^{(1)}) = \mu(x_i^{(2)})$.

para.	true val.	$E(\hat{\vartheta})^a$	$sd(\hat{\vartheta})^b$	$E(se)^c$	var$(d)^d$	5% [e]	10%	20%
		distribution of estimates						
$\sigma^2(v_i^{(1)})$	1.00	1.00	.20	.16	1.41	8.70	16.80	26.60
$\sigma^2(u_{1i}^{(j)})$	.30	.30	.03	.03	1.08	6.20	10.90	21.80
$\sigma^2(u_{2i}^{(2)})$	.40	.40	.03	.04	.99	5.10	10.50	19.70
β	2.00	2.00	.08	.08	1.04	6.10	10.40	21.60
α	1.00	1.00	.17	.17	1.00	5.10	9.50	18.40

distribution of $T^\star$ (df=1)

mean	var	5%	10%	20%
1.05	2.27	5.40	10.00	19.90

[a] Monte Carlo mean of estimates

[b] Monte Carlo standard deviation of estimates

[c] Monte Carlo mean of NT standard errors

[d] Monte Carlo variance of $d \equiv \frac{\hat{\vartheta}-\vartheta}{se(\hat{\vartheta})}$

[e] Monte Carlo nominal 5%, 10% and 20% two-sided tails for d

Table 2: Monte Carlo results for UFM model, where $\sigma^2(x_i^{(1)})$, $\sigma^2(x_i^{(2)})$, $\mu(x_i^{(1)})$ and $\mu(x_i^{(2)})$ are unrestricted (pseudo) parameters.

		distribution of estimates						
para.	true val.	$E(\hat{\vartheta})^a$	$sd(\hat{\vartheta})^b$	$E(se)^c$	$\text{var}(d)^d$	5% [e]	10%	20%
$\sigma^2(v_i^{(1)})$	1.00	.99	.29	.22	1.72	13.50	20.80	32.80
$\sigma^2(x_i^{(j)})$	1.00	.98	.12	.06	4.37	34.90	43.60	55.30
$\sigma^2(u_{1i}^{(j)})$	.30	.30	.06	.05	1.49	10.80	17.30	29.50
$\sigma^2(u_{2i}^{(2)})$	.40	.40	.05	.05	1.12	5.50	10.90	22.70
β	2.00	2.01	.15	.11	1.85	14.80	23.20	35.00
α	1.00	.98	.45	.33	1.84	15.10	22.70	34.80
$\mu(x^{(1)})$	3.00	3.00	.04	.04	1.06	5.50	12.20	20.90
$\mu(x^{(2)})$	3.00	3.00	.05	.05	1.07	6.40	10.70	21.10

distribution of $T^\star$ (df=2)				
mean	var	5%	10%	20%
5.68	40.81	33.20	41.30	53.80

[a] Monte Carlo mean of estimates

[b] Monte Carlo standard deviation of estimates

[c] Monte Carlo mean of NT standard errors

[d] Monte Carlo variance of $d \equiv \frac{\hat{\vartheta}-\vartheta}{se(\hat{\vartheta})}$

[e] Monte Carlo nominal 5%, 10% and 20% two-sided tails for d

Table 3: Monte Carlo results for RSM model, where $\mu(x_i^{(1)})$ and $\mu(x_i^{(2)})$ are unrestricted parameters and $\sigma^2(x_i^{(1)}) = \sigma^2(x_i^{(2)})$.

Discriminant Analysis for Directional Data Exemplified in a Concrete Case

Franz Streit

Section de Mathématiques de l'Université de Genève
Case postale 240, CH-1211 Genève 24, Switzerland

Summary: Methods of identification analysis have been developed for directional statistics, i.e. for distributions on circles and spheres. For various situations optimal decision rules are known explicitely, see El Khattabi and Streit (1996). Howevever the practical application of these results is somewhat hindered by a lack of concrete examples which show the actual implementation of these methods. This contribution tries to alleviate this problem in working out the details of the computation involving a well-known data set published in one of the standard reference books on directional statistics, i.e. in Fisher et al. (1987).

1. Introduction

In El Khattabi and Streit (1996) the general methodology of identification analysis is used to derive an optimal discriminant rule, which permits to attribute an individual unit to one of two populations with different Dimroth-Watson distributions defined on the sphere S_3 in 3-dimensional Euclidean space R^3 . For the population G_i the density function of the direction in standard form specified by the colatitude ϑ and the longitude φ (measured with the principal axis chosen in vertical position) is given by

$$f_{(\Theta,\Phi)}(\vartheta,\varphi:\kappa_i) = (2\pi)^{-1}b(\kappa_i)exp[-\kappa_i cos^2(\vartheta)]sin(\vartheta)$$

$$[0 \leq \vartheta \leq \pi, -\pi \leq \varphi < \pi,$$
$$\kappa_i \in R, i = 1,2],$$

where

$b(\kappa) = (2\int_0^1 exp[-\kappa t^2]dt)^{-1}$ and
Θ and Φ are the random variables corresponding to the realized values ϑ and φ [see Mardia (1972), p.234].

In El Khattabi and Streit (1996) the individual classification of the units are indicated for the problem of discriminating between the values $\kappa_1 = 0$ and $\kappa_2 = 1$ respectively $\kappa_1 = -1$ and $\kappa_2 = 0$ for the concentration parameter (with equal a priori probabilities and equal costs caused by a misclassification of a unit to G_1 respectively to G_2). In the following we extend this analysis to the more realistic problem of discriminating the uniform distribution on S_3 from a Dimroth-Watson distribution which fits the data as closely as possible and we deduce optimal identification rules which permit to classify several units simultaneously.

i	ϑ	φ	i	ϑ	φ	i	ϑ	φ	i	ϑ	φ
1	90	41	26	73	49	51	85	-27	76	68	64
2	115	-13	27	80	67	52	101	24	77	51	-81
3	36	- 4	28	68	68	53	67	-66	78	74	-44
4	148	47	29	126	-56	54	131	28	79	141	13
5	108	-65	30	87	-43	55	125	62	80	113	-73
6	63	54	31	94	-68	56	92	69	81	108	-20
7	100	12	32	81	3	57	97	20	82	52	59
8	110	15	33	120	-86	58	60	44	83	158	-73
9	98	26	34	109	72	59	98	37	84	22	30
10	150	84	35	76	-28	60	141	14	85	63	-59
11	81	51	36	49	43	61	152	-18	86	156	-37
12	69	-26	37	38	77	62	78	49	87	155	-37
13	130	-23	38	123	2	63	83	88	88	147	- 1
14	95	-70	39	110	- 5	64	84	85	89	66	57
15	95	-54	40	95	16	65	94	33	90	125	-40
16	73	41	41	117	3	66	78	- 1	91	114	-12
17	91	4	42	88	71	67	59	-81	92	118	-14
18	109	-16	43	36	-75	68	126	-76	93	81	10
19	143	-57	44	123	65	69	60	-48	94	74	45
20	118	-10	45	116	70	70	85	31			
21	137	- 2	46	62	-24	71	114	-80			
22	108	-54	47	44	53	72	46	84			
23	74	-24	48	18	6	73	103	-10			
24	107	23	49	60	79	74	. 92	55			
25	163	14	50	82	6	75	41	38			

Figure 1: Coleman's cat data

Some indications on how to evaluate the probability of misclassification are given and a decision rule involving Bingham's distributions is presented.

We apply this methodology of discriminant analysis to Coleman's cat data, as described in Keilson et al.(1983) and in Fisher et al.(1987), p.295. The measurements were obtained for three groups of cats. Each group was subjected to a different visual stimulus and the orientation of the dendritic field was then measured at 15 or 16 sites in the retinas of the cats. In order to keep this article selfcontained we include a table of the data set (Table 1). It follows from the experimental set-up that the data are in axial form. Note that the units 1–15 and 16–31 respectively represent measurements taken from two cats exposed to horizontally polarized light, the units 32–47 and 48–63 represent measurements taken from two cats exposed to vertically polarized light and units 64–78 and 79–94 represent measurements from two cats exposed to normal light.

Besides the application of the identification rules we apply exploratory data analysis and statistical tests to these data in order to get guidance for the choice of the appropriate stochastic model.

2. Exploratory data analysis and statistical tests

For the data in Table 1 we have first calculated the Cartesian coordinates of the axes according to the transformation formulae

$$\begin{aligned} l_i &= \sin(\vartheta_i)\cos(\varphi_i) \\ m_i &= \sin(\vartheta_i)\sin(\varphi_i) \\ n_i &= \cos(\vartheta_i) \end{aligned}$$

and evaluated the orientation matrix (i.e. the matrix of sums of squares and products), which is needed in order to assess the situation and to find the maximum-likelihood estimates of the parameters.
We find

$$T = \begin{pmatrix} \sum l_i^2 & \sum l_i m_i & \sum l_i n_i \\ \sum l_i m_i & \sum m_i^2 & \sum m_i n_i \\ \sum l_i n_i & \sum m_i n_i & \sum n_i^2 \end{pmatrix}$$

$$= \begin{pmatrix} 37,82485 & 3,74173 & -4,10772 \\ 3,74173 & 32,29522 & 4,50266 \\ -4,10772 & 4,50266 & 23,87992 \end{pmatrix}.$$

Calculation of the eigenvalues of T yields
$\tau_1 = 20,3240, \tau_2 = 33,7474, \tau_3 = 39,9287$ (with $\tau_1 + \tau_2 + \tau_3 = 94$), and the normalized eigenvalues are $\overline{\tau_1} = \tau_1/94 = 0,2162$, $\overline{\tau_2} = \tau_2/94 = 0,3590$ and $\overline{\tau_3} = \tau_3/94 = 0,4248$. For the corresponding eigenvectors we obtain

$$\mathbf{e_1} = \begin{pmatrix} -0,30726 \\ 0,43825 \\ -0,90986 \end{pmatrix}, \mathbf{e_2} = \begin{pmatrix} -0,24394 \\ 0,75604 \\ 0,44654 \end{pmatrix}, \mathbf{e_3} = \begin{pmatrix} -0,95477 \\ -0,38808 \\ 0,13550 \end{pmatrix},$$

and the normalized eigenvectors are

$$\mathbf{e_1}/|\mathbf{e_1}| = \begin{pmatrix} -0,291 \\ 0,415 \\ -0,862 \end{pmatrix}, \mathbf{e_2}/|\mathbf{e_2}| = \begin{pmatrix} -0,268 \\ 0,830 \\ 0,490 \end{pmatrix}, \mathbf{e_3}/|\mathbf{e_3}| = \begin{pmatrix} -0,918 \\ -0,373 \\ 0,130 \end{pmatrix}.$$

Based on these indications we can determine in an explanatory analysis the likely type of distribution underlying the data and confirm this choice by means of statistical tests. The explanatory analysis may be based on the considerations in Mardia (1972), p.225 and in Fisher et al.(1987), pp. 46-49. Calculating $\zeta = \ln(\tau_3/\tau_1)$ and $\gamma = \ln(\tau_3/\tau_2)[\ln(\tau_2/\tau_1)]^{-1}$ we find $\zeta = 0,68$ and $\gamma = 0,33$. Following the instructions in the quoted literature this leads to the decision that a girdle distribution is the most likely type of distribution generating the data. This conclusion is confirmed by the application of statistical tests:

Anderson's and Stephen's test of uniformity against bimodal distributions explained in Fisher et al.(1987), pp. 160–161 gives a realized value of the test statistic 0,42 , which has an observed level of significance of 5 % . This indicates that a bimodal distribution is not overwhelmingly likely. In contrast Watson's test of uniformity against girdle distributions described in Fisher et al. (1987), p.179 yields as value of the test statistic 0,22 , which stays below the critical value 0,23 in the critical region of level $\alpha = 0,01$ and is thus to be considered as significative.

In the case of uniform distribution on the sphere all eigenvalues should be equal. Bingham's test statistic [see Mardia (1972), p.276]
$S_u = 15/(2n)\sum_{i=1}^{3}(\tau_i - n/3)^2$ may be used for a omnibus-test of uniformity. Asymptotically for large n $\chi_5^2(1-\alpha)$ with $P(\chi_5^2 > \chi_5^2(1-\alpha)) = 1-\alpha$ serves as critical value of size α. For our data $S_u = 16,03$ while $\chi_5^2(0,99) = 15,09$. Thus at a level of significance of $\alpha = 1\%$ we just barely accept the hypothesis of non-uniformity.

As noted in Fisher et al.(1987), p.193 the application of Ginés non-parametric test of uniformity leads to G= 0,918, a value of the test statistic with observed level of significance of 3,7%, so that also from this aspect of the situation there is only moderate evidence of significant deviation from the uniform distribution.

Since the Dimroth-Watson distribution is the most popular stochastic model for axial data, it seems reasonable to analyze the fit of such a model to the data.

Estimation of κ in the bipolar case, where κ is restricted to negative values leads to the equation

$-\hat{\kappa} = D^{-1}(\overline{\tau_3})$,

where $D(\kappa) = (\int_0^1 t^2 \exp[-\kappa t^2]dt)[\int_0^1 \exp[-\kappa t^2]dt]^{-1}$.

Appendix 3.4 in Mardia(1972), p.324 shows that the maximum-likelihood estimate of κ is -0,96 in this case. The principal axis is specified by the polar coordinates $\vartheta = 97,5^o$ and $\varphi = 22,1^o$.

However taking into account the reported testing results the choice of a girdle distribution from the class of Dimroth-Watson distributions seems to be more appropriate. We thus restrict κ to positive values and seek the solution of $\hat{\kappa} = D^{-1}(\overline{\tau_1})$. Application of Appendix 3.3 in Mardia (1972), p.323

yields $\hat{\kappa} = 1,6$ in this case and the study of the corresponding eigenvector reveals that the normal axis to the plane of maximal concentration of the data is specified by the indications $\vartheta = 30,5^o$ and $\varphi = -55^o$. Since the Dimroth-Watson distributions exhibit rotational symmetry it is indicated to check the hypothesis of rotational symmetry by a test. Application of the procedure explained in Mardia (1972), pp. 277,256,278 applied to our data yields

$S_g = 0,5\{0,4(39,9-33,7)\} = 1,24$

as compared to the asymptotic critical values $\chi_2^2(0,99) = 9,21$ and $\chi_2^2(0,95) = 5,99$. We therefore accept the null-hypothesis. The choice of a symmetric girdle distribution does not contradict the data.

3. Optimal identification rules for simultaneous attribution of several units

Extending the theoretical considerations in El Khattabi and Streit(1996) we derive now optimal identification rules which serve to attribute n units simultaneously to one of two given populations. In order to arrive at these decision rules we apply the methods of discriminant analysis described in Anderson(1984) to the *joint* density of the measurements $(\Theta_l, \Phi_l)(l = 1, \ldots, n)$ of the n units considered as a random sample from the population with density function $f_{(\Theta,\Phi)}(\vartheta, \varphi : \kappa)$. For the problem of discriminating a random sample from a population G_1 with the uniform distribution on S_3, i.e. for $\kappa = 0$, from a random sample taken from a population G_2 with the non-degenerate Dimroth-Watson distribution with fixed parameter $\kappa_2[\kappa_2 \neq 0]$, we are thus led to use as region of attribution to G_2

$$\{(\vartheta_i, \varphi_i), i = 1, ..., n : [\prod_{i=1}^{n} f_{(\Theta,\Phi)}(\vartheta_i, \varphi_i : 0)]^{-1}[\prod_{i=1}^{n} f_{(\Theta,,\Phi)}(\vartheta_i, \varphi_i : \kappa_2)] > 1\}$$

$$= \{\vartheta_i, i = 1, \ldots, n : (4\pi)^n([2\pi]^{-1}b(\kappa_2)^n \exp[-\kappa_2 \sum_{i=1}^{n} \cos^2(\vartheta_i)] > 1\}$$

$$= \{\vartheta_i; i = 1, \ldots, n : -\kappa_2 \sum_{i=1}^{n} \cos^2(\vartheta_i) > -n \ln(2b(\kappa_2))\}.$$

In the case of $\kappa_2 > 0$ (girdle distributions) we thus obtain as region of attribution to G_2 the set

$$\{\vartheta_i; i = 1, \ldots, n : \sum_{i=1}^{n} \cos^2(\vartheta_i < \kappa_2^{-1} n \ln(2b(\kappa_2))\},$$

and in the case of $\kappa_2 < 0$ (bipolar distributions) this set is given by

$$\{\vartheta_i; i = 1, \ldots, n : \sum_{i=1}^{n} \cos^2(\vartheta_i > \kappa_2^{-1} n \ln(2b(\kappa_2))\}.$$

For a standard Dimroth-Watson distribution with true parameter κ_0 the statistic $\sum_{i=1}^{n} \cos^2(\Theta_i)$ follows asymptotically for large n a normal distribution according to Bingham(1974). Thus putting $\kappa_0 = 0$ respectively $\kappa_0 = \kappa_2$ this fact can serve to evaluate the asymptotical misclassification probabilities by means of a table of the standard normal distribution. If κ_2 for G_2 is actually unknown and estimated from the data the described procedure leads to estimates of these probabilities.

We now apply the identification rule derived in this paper to the groupwise attribution of the 94 units to either the uniform distribution on S_3 or to the Dimroth-Watson distribution with parameter $\kappa = 1,6$ fitted to the data by the method of maximum-likelihood. In order to adjust the distribution to the data with respect to the orientation of the axes we replace the statistic

$\sum_{i=1}^{n} \cos^2(\Theta_i)$ by

$\sum_{i=1}^{n} [\sin(\Theta_i)\sin(\alpha)\cos(\Phi_i - \beta) + \cos(\Theta_i)\cos(\alpha)]^2$

where $\alpha = 30,5^o$ and $\beta = -55^o$.

The realized value of this sum is 20,324. Taking into account Abramowitz and Stegun (1970), pp.297,311 this value is to be compared to

$$94(1,6)^{-1}\ln(2b(1,6)) = 94(1,6)^{-1}\ln(2(1,6)^{0,5}[\pi^{0,5}\, erf((1,6)^{0,5})]^{-1})$$
$$= 58,75\ln(1,54084) = 25,40 \ .$$

Thus the identification rule prefers to attribute the observed units collectively to the optimally adjusted non-degenerate Dimroth-Watson distribution rather than to the uniform distribution on S_3.

The class of Dimroth-Watson distributions on the sphere S_3 is the analogon to the class of rotationally symmetric normal distributions in the plane. If we would like to admit also asymmetric distributions on the sphere we should consider the class of Bingham's distributions. In standard form (with axes coinciding with the Cartesian axes) these distributions are given by

$$f_{(\Theta,\Phi)}(\vartheta,\varphi : \kappa_1,\kappa_2) = [4\pi d(\kappa_1,\kappa_2)]^{-1} \cdot$$
$$\exp[(\kappa_1\cos^2(\varphi) + \kappa_2\sin^2(\varphi))\sin^2(\vartheta)]\sin(\vartheta).$$

with $d(\kappa_1,\kappa_2) = {}_1F_1(0,5,1,5,K)$ and $K = diag(\kappa_1,\kappa_2,0)$,

where *diag* indicates the diagonal matrix with the mentioned elements as entries in the main diagonal and ${}_1F_1$ designates the confluent hypergeometric function with matrix argument K. In our case we find

$${}_1F_1(0,5,1,5,K) =$$
$$(2\pi^{0,5})^{-1}\sum_{i,j=0}^{\infty}([\Gamma(i+j+1,5)i!j!]^{-1}\Gamma(i+0,5)\Gamma(j+0,5)\kappa_1^i\kappa_2^j).$$

Application of the methodology of discriminant analysis to the problem of distinguishing between the uniform distribution on S_3 (case $\kappa_1 = \kappa_2 = 0$) and a non-degenerate Bingham's distribution (under the usual conditions of equal a priori probabilities and equal costs of misclassification) results to collectively attribute a group of units with angles (ϑ_i, φ_i) as measurements

for $i = 1, \ldots, n$ to the non-degenerate Bingham's distribution if and only if
$\kappa_1 \sum_{i=1}^n \cos^2(\varphi_i)\sin^2(\vartheta_i) + \kappa_2 \sum_{i=1}^n \sin^2(\varphi_i)\sin^2(\vartheta_i) > n \ln({}_1F_1(0,5,1,5,K))$.
The application to our data of the approximative maximum likelihood method of estimation [Mardia (1972), pp. 256 & 278, f.(9.7.11)] leads to the choice

$\hat{\kappa_1} = -1,8$ and $\hat{\kappa_2} = -0,4$

as best fitting parameter values. The left hand side of the above inequality becomes -50,09 which is greater than -59,4, so that the non-degenerate Bingham's distribution with parameters and orientation optimally adjusted to the data is prefered to the uniform distribution.

References:

ABRAMOWITZ, M. and STEGUN, I. A. (1970): *Handbook of mathematical functions.* National Bureau of Standards, Washington.

ANDERSON, T. W. (1984) : *An introduction to multivariate statistical analysis.* John Wiley, New York. Chapter 6.

BINGHAM, C. (1974): An antipodically symmetric distribution on the sphere. *Annals of Statistics 2, 1201–1225.*

EL KHATTABI, S. and STREIT, F. (1996): Identification analysis in directional data. To appear in *Computational Statistics and Data Analysis* .

FISHER, N. I.; LEWIS, T. and EMBLETON, B. J. J. (1987):*Statistical analysis of spherical data.* Cambridge University Press, Cambridge.

KEILSON, J.; PETRONDAS, D.; SUMITA, U. and WELLNER, J. (1983): Significance points of some tests of uniformity on the sphere. *Journal of Statistical Computation and Simulation 17,195–218.*

MARDIA, K. V. (1972): *Statistics for directional data.* Academic Press, London.

Combination of Tests and Sequential Classification

G. Tusch

Medizinische Hochschule Hannover, D-30623 Hannover, Germany

Summary: The utility of a (dichotomous) test for a diagnostic or therapeutic decision can be improved by performing tests repeatedly, especially when the test results become available chronologically in time. If the tests are based on a quantitative (discriminant) score and the densities are known, partial classification is feasible (Tusch 1992). Conditional independence, the binormal model, and homoscedasticity are assumed for the assessment of the cut-off values of the single tests. When different tests are used, it is also assumed that the sequence of tests will be in chronological order or in ascending order of the Mahalonobis distances. The cost of the single tests should then be proportional or at least related to the Mahalonobis distances, i.e. tests with better discrimination assume higher costs. The aim is to find cut-off values for an average minimum number of tests. With slight adaptations group sequential strategies like the α- and β-spending approach can be applied in this context. Sequential strategies allow both for the control of the utility of the test in terms of sensitivity and specificity and the average total cost in terms of the average number of tests to be applied. A medical clinical example will be used for demonstration.

1. Introduction

When a classification procedure is applied primarily for routine use (e.g. implementation into a clinical environment), interest may be focused on one of the following three measures: 1. the average probability of misclassification (e.g. error rate), 2. the conditional probability of misclassification esp. if a-priori probabilities (prevalences) are very small for one class (e.g. sensitivity, predictive values), or 3. the costs of misclassification (from a Bayesian standpoint). If the measure in question does not give a sufficient value, the underlying scores might be improved. If this however is not possible the complete approach may be discarded and sequential classification attempted. This involves using the score as a first step and improving the result by repeated independent sampling or by using a different, more expensive score. Expense may be seen in monetary terms, but also in clinical terms (e.g. invasiveness of the related procedure, risk to health) or based on ethical considerations (decisions made as early as possible). We will restrict our discussion to dichotomous tests for two situations, other scenarios where the method could be applied may also be considered. We assume in the sequel that either the data are available in a chronological order and a decision should take place as carly as possible, or several tests are available and we start with a low cost, but minor effective test and finish up with a high cost and very effective test. In the latter case we assume that the costs are related to the separability of the test (the Mahalonobis distance). We believe that this is a reasonable assumption for practical purposes.

First of all we consider results from the classical approach to sequential classification using binary tests. In general, two procedures are applied, the 'unanimity rule' and the 'majority rule'. Applying the 'unanimity rule' means taking a positive decision only if all results are positive. For the 'majority rule' to give a positive result a minimum of k from N tests must be positive. Of course, these rules can also be defined for negative results. The unanimity (positive) rule results in an increase in specificity and decrease in sensitivity or vice versa (negative), while for the majority rule both sensitivity and specificity (predictive values or the likelihood quotient as well) may be improved compared to the single test.

In many clinical applications (e.g. screening) one test is applied repeatedly to improve sensitivity and specificity. Then it might be possible to select a pair of sensitivity and specificity values that fit the purpose of the clinical problem. However, it is not reasonable, to adapt the sequence of tests to given values of sensitivity and specificity, because the changes of sensitivity and specificity will occur in significantly large and discrete steps when only a few tests are in the series, and a smooth adaptation to the given values will only be possible by using a series with many tests. Here, tests based on quantitative scores and dichotomised by choosing a suitable cut-off value allow for a smooth adaptation. These tests will be considered in the following sections.

2. Sequential tests and sequential discrimination

Sequential classification for two classes 0 and 1 may best be described by the following general scheme. For this scheme we will assume conditional independence of the tests T_i, i.e. $P(T_2^+, T_1^+|G^+) = P(T_2^+|G^+) \cdot P(T_1^+|G^+)$ and $P(T_2^+, T_1^+|G^-) = P(T_2^+|G^-) \cdot P(T_1^+|G^-)$ where G means Gold standard. Furthermore we assume the a-priori probabilities p_0, p_1 (prevalences) and the densities to be known. Let $L_{10}^{(i)}(\mathbf{x_i})$ be the likelihood ratio of the conditional probabilities for class 1 and class 0. We further assume that a maximum number of tests K is available, and that for each test T_i two cut-offs $b_i^{(0)}$ and $b_i^{(1)}$ are determined.

Step 1: Sample $\mathbf{x_1}$ from a patient and apply the first test.
If $S_1 := p_1/p_0 \cdot L_{10}^{(1)}(\mathbf{x_1}) < b_1^{(0)}$ then allocate the patient to class 0, else if $S_1 > b_1^{(1)}$ then allocate to class 1, stop. Else: next step.

Step 2: If the score value is in between $b_2^{(0)}$ and $b_2^{(1)}$ sample again for the same patient, perform the second test with $\mathbf{x_2}$ and (modified) a-priori probability (including previous information).

If $S_2 := p_1/p_0 \cdot L_{10}^{(2)}(\mathbf{x_2}) = p_1/p_0 \cdot L_{10}^{(1)}(\mathbf{x_1}) \cdot L_{10}^{(2)}(\mathbf{x_2}) < b_2^{(0)}$ then allocate the patient to class 0, else if $S_2 > b_2^{(1)}$ then allocate to class 1, stop. Else: next step.

The scheme is continued until step K where it is useful to force $b_K^{(0)} = b_K^{(1)}$.

Every step in this scheme utilises partial classification (Tusch 1992). The cut-off points $b_i^{(0)}$ and $b_i^{(1)}$, i=1...K, are chosen to guarantee certain features of optimality, a given sensitivity of at least $1 - \alpha_i$, and a given specificity of at least $1 - \beta_i$ (see Tusch 1992). Then there exist constants q_i, such that $b_1^{(0)}$ and $b_i^{(1)}$ can be obtained for given costs of misclassification $C(i|j)$ from class i to j: $b_i^{(0)} = q_i \frac{C(1|0)}{C(0|1)}$, $b_i^{(1)} = \frac{1}{q_i} \cdot \frac{C(1|0)}{C(0|1)}$.

This is a generalisation of the Bayes procedure for forced classification. Notice also the strong relationship to the Neyman-Pearson theorem.

The separability of two groups by a score (test) can be measured by the Mahalonobis distance, i.e. the distance of the group centroids. There is also a strong relationship between the Mahalonobis distance and the error probability.

The underlying **minimisation problem** can now be stated as follows: Given K two-class- tests to be applied sequentially in a predefined order (e.g. time sequence or ordered in magnitude of Mahalonobis distances related to the costs) and based on a quantitative score (binormal distribution, homoscedasticity). Furthermore, we assume that the conditional errors are bound by given $\alpha = 1-$sensitivity and $\beta = 1-$ specificity. We now try to find cut-offs for the sequential discrimination tests consisting of all K discriminant scores that minimise the number of tests to be applied (in order to get the earliest decision or the presumable lowest overall costs) and meet the error bounds. The minimisation will be performed according to some technical restrictions mentioned below.

We will propose an approach for sequential classification that comprehends methods from classification, partial classification, and group sequential tests. Classification methods allow for the determination of the information value of a test, partial classification supplies means for the construction of cut-offs for given error rates α and β, and by group sequential methods an optimal distribution of the α_i and β_i may be found. We will now sketch the single topics briefly.

We will now show a relationship between sequential discrimination and sequential tests. Firstly the main differences of tests and discrimination: Sequential tests are based on univariate distributions, whereas discriminant scores are based on multivariate distributions, and the standardised distances δ for sequential tests (corresponding to Mahalonobis distances for sequential classification) are in general small in relation to those for practical clinical tests. We now want to apply methods developed for sequential tests in the setting of sequential discrimination. Sequential tests are directly applicable when the same test is used several times (one-sided one-sample tests). For an overview on sequential tests see Ghosh and Sen (1991).

We will restrict our considerations on a special class of group sequential tests, the *error probability spending function* approach (Lan-DeMets 1983). For sequential discrimination with different scores we will adopt the spending function approach, where the Mahalonobis distances will serve as accumulated information (group sizes). It is most promising because of its flexibility in group sequential designs. Then the minimisation problem is reduced to the adaptation to the given bounds for sensitivity and specificity. This can be achieved by use of the spending function family of the **power boundaries** of Wang and Tsiatis (1987). The procedure will be explained in the next section.

3. The spending function approach

The procedure of sequential testing is similar to sequential classification. On every step a decision is taken to accept or reject the null hypothesis, if the actual test score is lower (or higher) than a given lower (or upper) boundary value. The error probability spending function approach for group sequential tests is based on the specification of the error probabilities in advance. The boundaries for the sequential test are then constructed according to these error probabilities.

Let the test problem be testing $H_0 := 0$ against $H_1 := \delta_A$, where δ is the standardised mean difference, with errors of type I and II of α and β. Assume K normally distributed independent random variables $X_i \sim N(\delta v_i, v_i)$, v_i corresponding to the group size. $V_j = \sum_{i=1}^{j} v_i$ is Fisher's information after having observed groups 1...j, and $W_j = \sum_{i=1}^{j} X_i$, j=1,...,K, the efficient score for the parameter δ. The *information time* $t_j := \frac{V_j}{V_K}$, $t_j \in (0,1)$, is measured on a standardised time scale.

The error probability spending function $\alpha^*(t)$ is defined on the information time scale for an overall error α. It is strictly increasing in t with $\alpha^*(0) = 0$ and $\alpha^*(1) = \alpha$. At time t_1 the probability of crossing the boundary is set at $\alpha^*(t_1)$ and the boundary value b_1 at time t_1 can be calculated under the null hypothesis of no difference in treatment by $P_0(W(t_1) \geq b_1) = \alpha^*(t_1)$.

The remaining boundary constants b_j can be calculated by solving the equation of the probability to first cross the boundary at time t_j $P\{W(t_j) < b_j, j = 1, ..., i-1; W(t_i) \geq b_i\} = \alpha^*(t_i) - \alpha^*(t_{i-1})$.

This procedure guarantees that the overall level α is maintained. Whereas b_1 can be obtained directly, $b_2, ..., b_K$ have to be calculated by numerical integration (e.g. Arioan 1968).

We will now restrict the b_i's to the **power boundaries** of Wang and Tsiatis (1987) indexed by the power shape parameter ν of the form $W(t_j) > C(\nu, \alpha, K)t_j^{\nu}$, where $C(\nu, \alpha, K)$ is a suitable chosen constant depending on ν, α, and K. $\nu = 0$ yields the **O'Brien-Fleming** boundary, $\nu = 0.5$ the **Pocock** boundary, which are two 'extreme' boundaries corresponding to the Pocock and the O'Brien-Fleming spending function (see e.g. Lan and DeMets 1983). The first one almost achieves the minimum expected sample size with a large maximum sample size, the latter one achieves a maximum sample size close to the non-sequential test with a higher expected sample size. The maximum sample size corresponds to the maximum information available when all K tests are performed in classification (the sum of the Mahalonobis distances in case of conditional independence). The selection of α and β depends on the maximum available information, a common problem in classification. Therefore the O'Brien-Fleming spending function will be used when α and β shall be very small, the Pocock function, when the decision has to be very early. We assume in the sequel that suitable α and β are chosen.

The equivalent of sequential discrimination in the sequential testing setting are one-sided tests. They can be obtained for the α- and β-spending function by the following rule:

Reject H_0 if $W(t_j) > C_1(\nu, \alpha, \beta, K)t_j^\nu$ *(upper boundary)*

reject H_1 if $W(t_j) < \eta t_j - C_2(\nu, \alpha, \beta, K)t_j^\nu$ *(lower boundary)*

where the constants $C_i(\nu, \alpha, \beta, K)$, i=1,2, are positive and $\eta = \delta\sqrt{V_K}$. If the boundaries are finally forced to meet (at the 'last look' $t_K = 1$), i.e. $C_2(\nu, \alpha, \beta, K) = \eta - C_2(\nu, \alpha, \beta, K)$, then for any value of ν and K, there exist $C_i(\nu, \alpha, \beta, K)$, i=1,2, that satisfy α and β.

We now apply this theory to the setting of sequential discrimination for different scores. In the sequel we will restrict ourselves to K scores that follow a binormal distribution with homoscedasticity for each test separately. When the covariances Σ_i, i=1...K, are known, we obtain the K discriminant scores including Fisher's linear discriminant function $U_i(\mathbf{x})$, i=1...K. When $\mathbf{x}$ comes from class 1 and is normally distributed, then the U_i's are normally distributed with mean $\frac{1}{2}\Delta_i^2$ and variance Δ_i^2, where Δ_i^2 is the Mahalonobis distance. When $\mathbf{x}$ comes from class 0 and is normally distributed, then the U_i's are normally distributed with mean -$\frac{1}{2}\Delta_i^2$ and variance Δ_i^2.

Assume p_0, p_1, $C(1|0)$, and $C(0|1)$ are expressed by α and β as mentioned above. Then use $W_j(\mathbf{x}) = \sum_{i=1}^{j}(U_i(x) + \frac{1}{2}\Delta_i^2)$; U_i be the i-th LDF score.

Define the standardised $\delta := 1$ and $v_i := \Delta_i^2$. Then $V_K = \sum_{i=1}^{K} v_i = \sum_{i=1}^{K} \Delta_i^2$ and

$$W(t_j) = \sum_{i=1}^{j}(U_i(\mathbf{x}) + \tfrac{1}{2}\Delta_i^2)/\sqrt{\sum_{i=1}^{K}\Delta_i^2} \sim N(\eta t_j, t_j) \text{ for } \eta = \sqrt{\sum_{i=1}^{K}\Delta_i^2}.$$

We now can formulate the classification rule for α- and β -spending:

Allocate to class 1 if $W(t_j) > C_1(\nu, \alpha, \beta, K)t_j^\nu$ *(upper boundary)*,

allocate to class 0 if $W(t_j) < \eta t_j - C_2(\nu, \alpha, \beta, K)t_j^\nu$ *(lower boundary)*.

When the covariance is unknown, the U_i may be expressed by Anderson's W statistic (plug-in empirical covariance matrix S_i and use the resulting statistic) (Anderson 1984). The statements above remain true only asymptotically, i.e. only for scores developed on the basis of large samples. The cut-offs for the scores can be obtained approximately from the error rates of the spending function using the partial classification procedure of Anderson or it's modification by McLachlan (see Anderson 1984). The non-parametric procedure of Brofitt et al. (1976) might also be applied. However, we will show by an example in the next section, that reasonable results already may be obtained also for small samples, when an unbiased estimate of the Mahalonobis distance is used (Lachenbruch 1968, see also Anderson 1984).

4. Sample data set

The proposed procedure has been applied to a data set resulting from an $1\frac{1}{2}$-year period of consecutive and prospective accrual of all patients that underwent liver transplantation at Hannover Medical School (Gubernatis et al. 1989). Included were patient courses from the third to the 28th day after transplantation, because this period was assumed to be typical for acute rejection. If a patient was retransplanted, the course from the 2nd (or 3rd) transplantation was counted separately. Patient courses were only excluded from the study when organ survival was less than 4 days. This is the time of the transaminases peak and atypical for acute rejection.

Selected for the study were episodes starting with an increase in transaminases (SGOT, SGPT) and or an increase in bilirubin with or without fever. From the data set 44 episodes of 'acute rejection' and 43 episodes of 'no rejection' were obtained. 3 linear discriminant scores of transaminases and bilirubin were developed according to the changes on the first, second and third day of the episode. The data were transformed according to square root and reciprocal transformations and standardised according to a projection pursuit like method to obtain normality (see Huber 1985). Despite the small sample number of the study a split sample approach (2/3:1/3) was performed for demonstration purposes. The error rates were determined by reclassification, cross-validation, test sample, and the above mentioned parametric estimators.

The results are shown in tables 1 to 3. Table 1 demonstrates the partly extreme overestimation of the separability information of the scores for small Mahalonobis distances, and the marked difference of the Anderson and Lachenbruch estimate as well (it should be mentioned that the latter is a somewhat artificial construction, whereas the first is unbiased). We selected the latter one as a basis for the sequential procedure, because it showed the best fit to most of the sample information in terms of crossvali-

estimate	LDF1	LDF2	LDF3
sample	0.769	1.368	2.994
Anderson	0.520	1.079	2.594
Lachenbruch	0.351	0.849	2.244

LDFi = linear discriminant function U_i for day i=1,2,3 after first increase of transaminases

Table 1: Mahalonobis distances (learning sample)

error estimate	LDF1 no rej.	LDF1 reject.	LDF2 no rej.	LDF2 reject.	LDF3 no rej.	LDF3 reject.
reclassification (n=61)	29%	33%	23%	30%	23%	23%
crossvalidation (n=61)	35%	33%	26%	33%	23%	23%
test sample (n=26)	31%	38%	23%	31%	15%	46%
Anderson (param.)	36%	36%	30%	30%	21%	21%
Lachenbruch (param.)	38%	38%	32%	32%	23%	23%

(LDFi = linear discriminant function U_i for day i=1,2,3 after first increase of transaminases)

Table 2: Classification results

error estimate	no rej.	reject.
reclassification (n=61)	20%	20%
test sample (n=26)	8%	38%

Table 3: Classification results of the sequential rule ($\alpha = \beta = 20\%$)

dation error rates (table 2).

Table 3 shows reasonably good results for the misclassification error of the sequential rule for reclassification and test sample. Finally, to demonstrate the increase of the average number of tests when α and β decrease, we will give some figures for the sample test data set: For the case $\alpha = \beta = 20\%, 25\%$, and 30% the average number of tests were 2.8, 1.9, and 1.2 resp.

5. Discussion

Methods of sequential testing are applicable to sequential classification (dichotomous and normally distributed scores). For classification using different test scores the α- and β-spending function approach is the most flexible one, when Mahalonobis distance is used as accumulated information in the group sequential design. For practical applications an adequate estimation of the α- and β-spending function may be used (Anderson's W-statistic, McLachlan's statistic, Brofitt's non-parametric score). For the sequential test a 6-fold crossvalidation showed encouraging results. Of course, the example can give only a glimpse of the potential of the method because of the small sample size. This was selected only for demonstration purposes. The sample numbers are too small for a clinical application, but they can demonstrate the small sample behaviour of the method. Significant gain in costs compared to complete testing may be achieved in reasonably sized studies.

Acknowledgements

I would like to thank Prof. B. Schneider for helpful discussions, Prof. H. Hecker and an anonymous referee for many constructive comments to improve the presentation of the paper.

References:

ANDERSON, T. W. (1984): An introduction to multivariate statistical analysis. 2nd edition. New York, John Wiley.

AROIAN, L. A. (1968): Sequential analysis, direct method. *Technometrics, 10, 125-132.*

BROFITT, J. D.; RANDLES, R. H. and HOGG, R. V. (1976): Distribution-free partial discriminant analysis. *J. Amer. Statist., Assoc. 71(356), TMS 934-939.*

GHOSH, B. K. and SEN, P. K. (eds.) (1991): Handbook of Sequential Analysis. New York; Marcel Dekker, Inc.

GUBERNATIS, G.; KEMNITZ, J.; TUSCH, G.; RINGE, B.; BUNZENDAHL, H.; RIEDEL, T.; MÜLLER, R. and PICHLMAYR, R. (1989): Different features of acute liver allograft rejection, their outcome and possible relationship to HLA-compatibility. *Transplant. Proc. 21(1) 2213-2214.*

HUBER, P. J. (1985): Projection pursuit. *Ann. Statist. 13(2), 435-475.*

LACHENBRUCH, P. A. (1968): On expected probabilities of misclassification in discriminant analysis, necessary sample size, and a relation with the multiple correlation coefficient. *Biometrics 24, 823-834.*

LAN, K. K. G. and DEMETS, D. L. (1983): Discrete sequential boundaries for clinical trials. *Biometrika 70(3), 659-663.*

PAMPALONA, S. and TSIATIS, A. A. (1994): Group sequential designs for one-sided and two-sided hypothesis testing with provision for early stopping in favour of the null hypothesis. *J. Statist. Plann. Inference 42, 19-35* .

TUSCH, G. (1992): Leistungsbewertung bei medizinischer Entscheidungsunterstützung mit partieller Klassifikation. Ph.D. Thesis. Medizinischen Hochschule, Hannover (in German).

WANG, S. K. and TSIATIS, A. A. (1987): Approximately optimal one-parameter boundaries for group sequential trials. *Biometrics 43, 193-199.*

Classes of Influential Observations

R. Vonthein

ISO/WWZ/Uni-Basel
Holbeinstr. 12, CH 4051 Basel

Summary: "This observation does not fit in with the trend", "If the sample had been slightly different, we had reached another resul" – such statements shall be formalized. Regression diagnostics look for observations badly explained by the model. This is done by means of an influence measure that arises from a definition of influence. As influence has many aspects, there are many such diagnostics. Considering classical and Bayesian regression models, this work shall illustrate how to derive and use the diagnostics needed in specific cases.

1. Formal Introduction

An analysis is conducted to reach result E. That may be an estimate of parameter θ, transformation $\tau(\theta_1)$, or a forecast. E results from data D and model M (including prior information): $E = E(D; M)$.

For purpose of diagnosis $(D; M)$ is perturbed, e. g. one observation dropped, a case weight introduced. The perturbation is noted as mapping $\omega = \omega(D; M) = (D_\omega; M_\omega)$. The result is changed by the perturbation. One gets the following picture:

$$\left.\begin{array}{c} (D;M) \stackrel{E}{\longmapsto} E(D;M) \\ \omega\downarrow \\ (D_\omega;M_\omega) \stackrel{E}{\longmapsto} E(D_\omega;M_\omega) \end{array}\right\} \stackrel{\Delta}{\longmapsto} \Delta(E(D;M),E(D_\omega;M_\omega)) = \Delta(E,D,M,\omega)$$

Functional $\Delta = \Delta(E(D; M), E(D_\omega; M_\omega)) \in W$ is the diagnostic, with which the influence of perturbation ω on result E is assessed. The involved sets are

$$\left.\begin{array}{c} S_n \times \mathcal{M} \stackrel{\mathcal{E}}{\longrightarrow} \mathcal{E} \\ \omega\downarrow \\ S_n \times \mathcal{M} \stackrel{\mathcal{E}}{\longrightarrow} \mathcal{E} \end{array}\right\} \stackrel{\Delta}{\longrightarrow} W, \qquad \Delta : \mathcal{E} \times \mathcal{E} \longrightarrow \mathcal{W}.$$

Let Ω be the population and $S_n \subset \mathcal{P}(\Omega)$ the sample space of samples of n elements. Then $D \in S_n$. Denote by $\mathcal{M} = \{M_0, M_1, M_2, \ldots\}$ a set of models. Result E is an element of the set of possible results $\mathcal{E}$. Four important cases are: $\mathcal{E}_B = \{\mathrm{H}_0, \mathrm{H}_A\}$, $\mathcal{E}_R = \mathbb{R}^k$, $\mathcal{E} \subset \mathcal{P}(\mathbb{R}^k)$, $\mathcal{E}_M \subset \{p \mid p : (\Omega, \mathcal{P}(\Omega)) \to (\mathbb{R}^k, \mathbb{B}_k)\}$, as to E is a decision between nullhypothesis and alternative, a point estimate, a set estimate or a posterior distribution.

Set W and funktional Δ are chosen accordingly. Good choices of W are $[-1; 1]$ or $[0; 1]$, for Δ to be interpreted and compared easily. Δ should be dimensionless or a probability measure.

2. Normal-Gamma-Regression

The Normal-Gamma regression ($\mathcal{N}\Gamma$ regression) model for the relation between $\mathbf{y} \in \mathbb{R}^n$ and $\mathbf{X} \in \mathbb{R}^{n \times p}$ consists of the following prior distribution and likelihood:

$$\begin{aligned} \sigma^{-2} &\sim \Gamma[s_*, n_*] \\ \boldsymbol{\beta}|\sigma^2 &\sim \mathcal{N}_p[\boldsymbol{\beta}_*, \sigma^2\mathbf{H}_*] \\ \mathbf{y}|\boldsymbol{\beta}, \sigma^2 &\sim \mathcal{N}_n[\mathbf{X}\boldsymbol{\beta}, \sigma^2\mathbf{I}]. \end{aligned} \tag{1}$$

From this follows by the Theorem of BAYES that

$$\begin{aligned} \sigma^{-2}|\mathbf{y} &\sim \Gamma[s_{**}, n_{**}] \\ \boldsymbol{\beta}|\mathbf{y}, \sigma^2 &\sim \mathcal{N}_p[\beta^{**}, \mathbf{H}^{**}] \end{aligned}$$

with parameters

$$\begin{aligned} \mathbf{H}^{**} &= (\mathbf{X}^{\mathsf{T}}\mathbf{X} + \mathbf{H}_*^{-1})^{-1} \\ \beta^{**} &= \mathbf{H}^{**}(\mathbf{X}^{\mathsf{T}}\mathbf{y} + \mathbf{H}_*\boldsymbol{\beta}_*) \\ n_{**} &= n_* + n \\ n_{**}s_{**} &= n_*s_* + (\boldsymbol{\beta} - \boldsymbol{\beta}_*)^{\mathsf{T}}\mathbf{H}_*(\boldsymbol{\beta} - \boldsymbol{\beta}_*) + (\mathbf{y} - \mathbf{X}\boldsymbol{\beta})^{\mathsf{T}}(\mathbf{y} - \mathbf{X}\boldsymbol{\beta}). \end{aligned}$$

3. Measures For Different Aspects of Influence

Four different aspects of influence shall be illustrated. As a model consider normal gamma regression, where M_0 uses all data whereas M_1 leaves out the i-th observation. So the perturbation ω is perforation. $E = \hat{\boldsymbol{\beta}}$ is viewed as result. ω is denoted by subscript (i) for leaving out case i. In case the data are a column vector

$$\mathbf{y}_{(i)} = (y_1, \ldots, y_{i-1}, y_{i+1}, \ldots y_n)^{\mathsf{T}}$$

and a matrix

$$\mathbf{X}_{(i)} = (\mathbf{x}_1, \ldots, \mathbf{x}_{i-1}, \mathbf{x}_{i+1}, \ldots \mathbf{x}_n)^{\mathsf{T}}$$

with rows $\mathbf{x}_j$ respectively, the perturbed result is $E_\omega = \boldsymbol{\beta}^{**}_{(i)}$.

Influence may be exerted but does not need to be. The best example of a measure of potential influence is leverage. In simple linear regression the regression line playes the part of the lever hinged – not too firmly – at the centroid of the data. Movements of points outlying in regressor space may move this lever more easily. This measure of influence in regressor space is given by

$$h_i^{**} = \mathbf{x}_i\mathbf{H}^{**}\mathbf{x}_i^{\mathsf{T}},$$

An observation i may have high leverage but lie on the regression line whether it is computed with or without that observation. Such an observation has no influence on the mean posterior coefficient $\boldsymbol{\beta}^{**}$.

Influence may be called fixed or moving with regard to the system of reference. If influence is measured relative to the undisturbed data, the system of reference is fixed. It is moving, if influence is assessed with respect to perturbed data, e. g. all data but the i-th case. As one leaves out one case after the other, the system of reference moves. The advantage of a fixed reference is self-evident. The moving reference is justified in the cases of interest. Applying diagnostics to a really influential observation yields a perturbed result that comes closer to the desired result. In linear regression a high leverage outlier affects the slope that much that it is no longer the desired result. This fact is underlined by robust procedures that downweight such observations. In the case of studentized residuals

$$r_i^{**} = \frac{\varepsilon_i^{**}}{s_{(i)}^{**}},$$

$s_{(i)}^{**2} = [n_* s_*^2 + (\boldsymbol{\beta} - \boldsymbol{\beta}^*)^{\mathsf{T}} \mathbf{H}_* (\boldsymbol{\beta} - \boldsymbol{\beta}^*) + (\mathbf{y}_{(i)} - \mathbf{X}_{(i)}\boldsymbol{\beta}_{(i)})^{\mathsf{T}} (\mathbf{y}_{(i)} - \mathbf{X}_{(i)}\boldsymbol{\beta}_{(i)})]/(n - p - 1)$ has the additional advantage of being independent of residual ε_i^{**}, as Belsley, Kuh, Welsch (1980) pointed out for the classical case $\mathbf{H}_* = \mathbf{0}$, $n_* = 0$.

How does $\boldsymbol{\beta}^{**}$ change, when the i-th observation is dropped? The leave-one-out measure of global influence on the regression parameters is

$$\mathbf{DFBETA}_i^{**} = \boldsymbol{\beta}^{**} - \boldsymbol{\beta}_{(i)}^{**} = \frac{\mathbf{H}^{**}\mathbf{x}_i \varepsilon_i^{**}}{1 - h_i^{**}}. \tag{2}$$

Once the system of reference is chosen, relative influence can be assessed, e. g. $\beta_1^{**}/\beta_{(i)1}^{**}$, in contrast to absolute influence $\boldsymbol{\beta}^{**} - \boldsymbol{\beta}_{(i)}^{**}$ that usually is incommensurable. The scaled influence diagnostic is called $\mathbf{DFBETAS}_i^{**}$. It is defined as:

$$\mathbf{DFBETAS}_i^{**} = (\boldsymbol{\beta}^{**} - \boldsymbol{\beta}_{(i)}^{**})/\sqrt{H_{(i)ii}^{**}} = \frac{\mathbf{H}^{**}\mathbf{x}_i \varepsilon_i^{**}}{1 - h_i^{**}} / \sqrt{H_{(i)ii}^{**}}. \tag{3}$$

Another aspect of influence is the degree of perturbation. Consider a single case weight w assigned to observation i,

$$\mathbf{y}_{(i)}(w) = (y_1, \dots, y_{i-1}, w y_i, y_{i+1}, \dots y_n)^{\mathsf{T}}$$

and

$$\mathbf{X}_{(i)}(w) = (\mathbf{x}_1, \dots, \mathbf{x}_{i-1}, w\mathbf{x}_i, \mathbf{x}_{i+1}, \dots \mathbf{x}_n)^{\mathsf{T}}.$$

The result as a function of w is called influence curve. Although cases $w = 0$ and $w = 1$ deserve most attention, small deviations from these are of much interest. Differentiation with respect to w yields the local influence of the i-th observation as opposed to the global influence assessed by deletion. In linear regression $\partial\boldsymbol{\beta}^{**}(w)/\partial w$ taken in $w = 1$ is

$$\frac{\mathbf{H}_{**}\mathbf{x}_i \varepsilon_i^{**}}{(1 - (1 - w)h_i^{**})^2} = \mathbf{DFBETA}_i^{**}(1 - h_i^{**})$$

and is used to measure the local influence on the slope. $\partial\beta^{**}(w)^2/\partial^2 w$ gives the curvature of the influence curve.

4. Example: Swiss Consumption

The influence diagnostics derived so far shall be applied to the estimation of a consumption function. Private consumption **C** is the response. Regressors are $\mathbf{C}_{-1}$ the private consumption of the previous period and gross domestic product **BIP**. The quarterly data range from 1966.II to 1988.IV. Estimating the model

$$\mathbf{C} = \boldsymbol{\beta}_0 + \boldsymbol{\beta}_1\mathbf{C}_{-1} + \boldsymbol{\beta}_2\mathbf{BIP} + \boldsymbol{\varepsilon},\ \boldsymbol{\varepsilon} \sim \mathcal{N}[\mathbf{0}, \sigma^2\mathbf{I}]\ i.i.d. \tag{4}$$

with prior information $n_* = 0$, $\boldsymbol{\beta}^* = \mathbf{0}$ and covariance scaling matrix

$$\mathbf{H}_* = \begin{pmatrix} 3 & 40000 & 70000 \\ 4E+04 & 7E+08 & 1E+09 \\ 7E+04 & 1E+09 & 2E+09 \end{pmatrix},$$

one gets the posteriori covariance matrix

$$\mathbf{H}_{**} = \begin{pmatrix} 3.13E-01 & 2.35E-05 & -2.71E-05 \\ 2.35E-05 & 4.54E-08 & -2.88E-08 \\ -2.71E-05 & -2.88E-08 & 1.89E-08 \end{pmatrix}.$$

Parameters of the posterior distribution are

	a priori	a posteriori	standard deviation
constant	0	-3.08E+01	99.1
C_-1	0	8.66E-01	0.0377
BIP	0	8.70E-02	0.0243
s2	---	3.06E+04	4620
n	0	8.80E+01	

with a good fit:

stddv o res	multiple R^2	n	F.stat	d.f.num	d.f.denom
177	0.991	88	4610	2	85

Assuming $\mathbf{H}_*^{-1} = \mathbf{0}$ does not alter these results remarkably, so that a classical analysis was carried out.

Because the regressors are almost collinear, the observations form a cigar shaped cloud in regressor space. Leverage is high near the tips and low in the center, as is to be expected. Some high leverage points lie near the fringes of the cloud. These should be scrutinized.

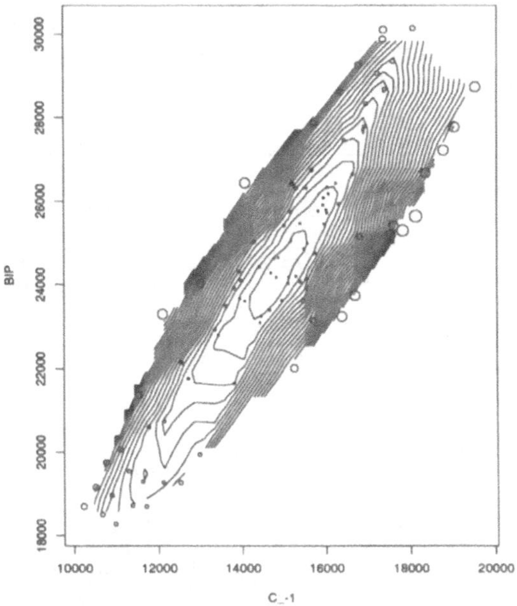

Figure 1: Leverage in the space spanned by regressors $\mathbf{C}_{-1}$ and **BIP**. Observations are marked by ellipses proportional in size to h_i^{**}. The largest values lie on the convex hull. Contours were drawn according to an interpolation with Splus.

Observations far from the regression plane have the highest residuals. This is illustrated in the $\mathbf{C}_{-1} - \mathbf{C}$ and $\mathbf{BIP} - \mathbf{C}$ planes.

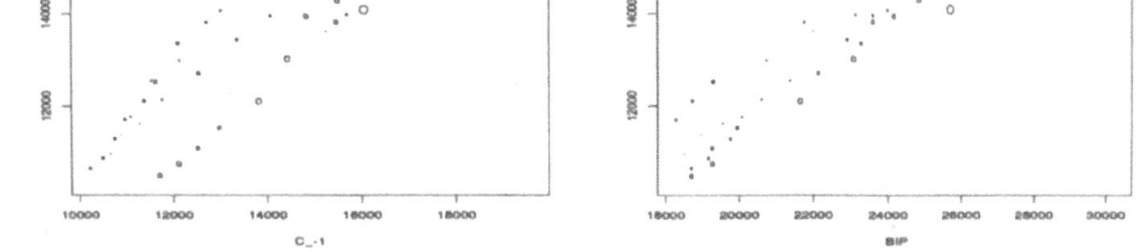

Figure 2: Studentized residuals of model (4) depicted as ellipses of proportional size. The largest values occur far from the regresion plane measured parallel to the ordinate.

For practical purposes other graphics will be employed. One should see outliers at once and check normality. The histogram of residuals is bimodal and near symmetric, as one mode is very small. In the QQ-plot of residuals a deviation from a normal is aparent. The points form two straight lines. Both graphics are not shown.

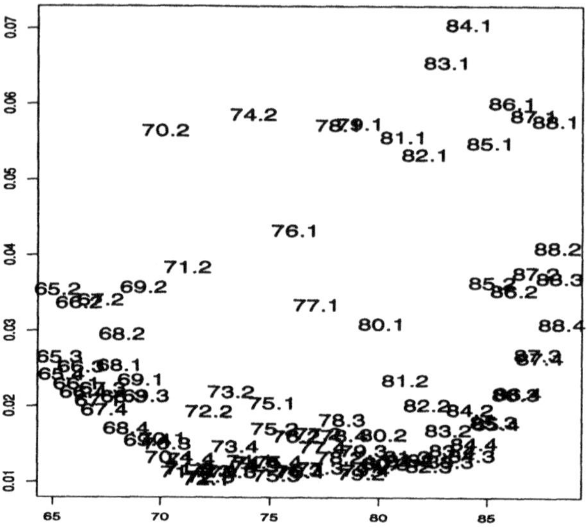

Figure 3: The index plot of leverage forms a parabola, if regressors change linear in time. Several observations in the seventies and eighties do not follow this rule.

Leverage is strongest in 1970.II and 1974.II and the first quarters of 1978, 1979, and 1976 to 1988. This is remarkable, for many of them lie near the middel of time scale. This potential influence is not equal to the real influence on the fit, as the respective studentized residuals are small. In the indexplot (not shown) of studentized residuals observations 1977.IV, 1975.IV, ..., 1988.IV form an outlying cluster having r_i^{**} of about 2. This masks the outlier 1974.I which has a studentized residual less than -2 even now.

Influence on the goodness of fit corresponds to the influence on the regression parameters which is assessed by DFBETAS$_i^{**}$.

The outlier 1974.I does not matter much, but the mentioned high leverage points and outlying cluster affect the intercept and the coefficients of the regressors respectively. In the estimation of intercept the first and last observations have high DFBETAS$_i^{**}$(intercept), due to the non zero slope. High leverage points have high DFBETAS$_i^{**}$($\mathbf{C}_{-1}$) and DFBETAS$_i^{**}$(**BIP**) as a rule, but there are more of that order. All β^{**} is more sensitive in the eighties.

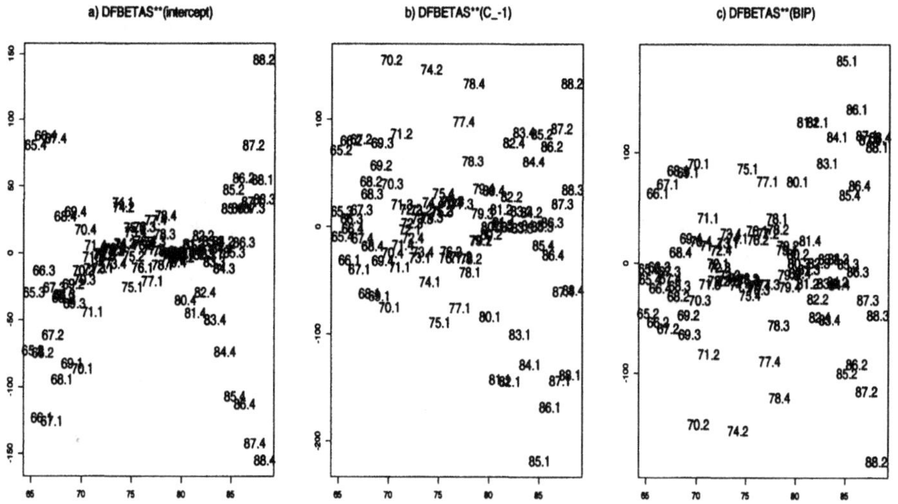

Figure 4: Indexplots of components of DFBETAS$_i^{**}$. a) Influence on the intercept b) Influence on the coefficient of $\mathbf{C}_{-1}$ c) Influence on the coefficient of **BIP**.

Obviously a seasonal variation was discovered first. It has become stronger in the eighties. But there has to be another reason for its high influence. One can conclude, that in the beginning of 1974 a macroeconomic shock occured and a different consumption function applied since the mid-seventies. This kink lets observations in the shorter and more volatile arm appear to be influential. They either determine more than others where the regression plane is – high leverage –, or they lie far from that plane – large studentized residuals.

One should try a segmented regression. Segments could be formed by the ranges 1966.II - 1973.IV, 1974.I - 1975.IV and 1976.I - 1988.I.

This segmentation makes sense. After the oil shock at the end of 1973 that drove the gross national product to a one time peak, private consumption fell. A new foreign-labor regime in 1974 caused migration and loss of productivity. Thus all components of $\boldsymbol{\beta}$ were affected.

The segmented regression can be estimated using the Gibbs sampler.

5. Conclusion

A cross-classification leverage by studentized residual finds the most influential points, where both are high. That is why special purpose diagnostics like **DFBETAS**$_i^{**}$ combine elements of both. Cases with either h_i^{**} or r_i^{**} high may form special subsets of the data, as in the example.

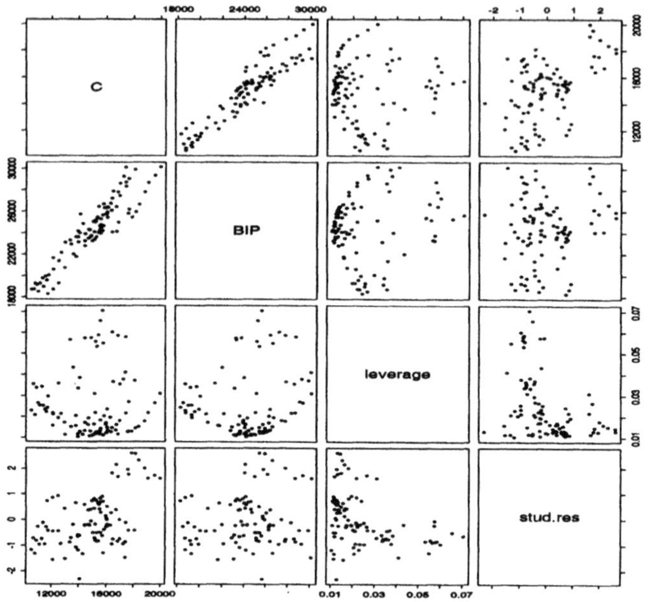

Figure 5: The scatterplot matrix of the time series **C** and **BIP** and two diagnostics h_i^{**} and r_i^{**} shows which data have high leverage or large studentized residuals and that no observation has both.

The relation of h_i^{**} and $|r_i^{**}|$ may be approximated by a hyperbola, in accord with the part of leverage in the formula of the studentized residuals. Thus h_i^{**} and $|r_i^{**}|$ tend to direct attention to different observations.

Other approaches, like influence curve and influence function (ω is perforation, Δ is FRECHET-differential with respect to sample size), lead to influence diagnostics that are proportional to $\mathbf{DFBETA}_i$.

So the dignostics employed above are regarded as the essential tools for diagnosis in the specified model, perturbation and result. Their posterior distributions remain to be computed.

References:

BELSLEY, D. A.; KUH, E. and WELSCH, R. E. (1980): Regression Diagnostics: Identifying Influential Data And Sources of Collinearity. Wiley, New York.

CHATTERJEE, S. and HADI, A. S. (1988): Sensitivity Analysis in Linear Regression. Wiley, New York.

COOK, R. D. and WEISBERG, S. (1982): Residuals and Influence in Regression. Chapman and Hall, New York.

POLASEK, W. (1993): Variance Diagnostics for Classical and Bayesian Linear Regression. *WWZ Discussion Papers Nr. 9302.* WWZ, Universität Basel, Basel.

WEISS, R. (1996): Sufficiency and Influence. *Journal of the Royal Statistical Society Series B,* to appear.

Bayesian Non-Linear Latent Variable Models

Jörg Wittenberg and Gerhard Arminger

Department of Economics,
Bergische Universität - GH Wuppertal, D-42097 Wuppertal, Germany

Summary: Non-linear latent variable models are specified that include squares and interactions of latent regressor variables, observed regressors and missing data as special cases. To estimate the parameters, the models are put in a Bayesian framework. The posterior distributions of the parameters and the latent variables are estimated using Markov chain Monte Carlo methods such as the Gibbs sampler and the Metropolis-Hastings algorithm. The proposed estimation methods are illustrated by two simulated examples.

1. Model specification

We introduce a non-linear version of the LISREL model (Jöreskog and Sörbom 1993) described in Arminger and Muthn (1996). Let $\boldsymbol{\xi}_i$ be a $\tilde{m} \times 1$ vector of random variables that is multivariate normal with $\boldsymbol{\xi}_i \sim \mathcal{N}(\mathbf{0}, \boldsymbol{\Phi})$. Let $\boldsymbol{\alpha}_i = g(\boldsymbol{\xi}_i)$ a $m \times 1$ deterministic function of $\boldsymbol{\xi}_i$ that is known. A $p \times 1$ random variable vector $\boldsymbol{\eta}_i$ is connected with $\boldsymbol{\alpha}_i$ through a linear regression model

$$\boldsymbol{\eta}_i = \boldsymbol{\gamma}_0 + \boldsymbol{\Gamma}\boldsymbol{\alpha}_i + \boldsymbol{\zeta}_i \ , \quad \forall i = 1, \ldots, n \ , \tag{1}$$

where the $p \times 1$ random variable $\boldsymbol{\zeta}_i \sim \mathcal{N}(\mathbf{0}, \boldsymbol{\Psi})$ is a disturbance, $\boldsymbol{\gamma}_0 \sim p \times 1$ is a vector of regression constants and $\boldsymbol{\Gamma} \sim p \times m$ is a matrix of regression coefficients. It should be noted that the model is linear in the parameters, but is non-linear in the components of $\boldsymbol{\xi}_i$.

The random variables $\boldsymbol{\xi}_i$ and $\boldsymbol{\eta}_i$ are connected to observed variables $\boldsymbol{x}_i \sim r \times 1$ and $\boldsymbol{y}_i \sim s \times 1$ with the usual factor analytic measurement models

$$\boldsymbol{x}_i = \boldsymbol{\nu}_x + \boldsymbol{\Lambda}_x\boldsymbol{\xi}_i + \boldsymbol{\delta}_i \ , \quad \boldsymbol{\delta}_i \sim \mathcal{N}(\mathbf{0}, \boldsymbol{\Theta}_\delta) \ , \tag{2}$$

$$\boldsymbol{y}_i = \boldsymbol{\nu}_y + \boldsymbol{\Lambda}_y\boldsymbol{\eta}_i + \boldsymbol{\varepsilon}_i \ , \quad \boldsymbol{\varepsilon}_i \sim \mathcal{N}(\mathbf{0}, \boldsymbol{\Theta}_\varepsilon) \ . \tag{3}$$

For simplicity, it is assumed that $\boldsymbol{\Theta}_\delta$ and $\boldsymbol{\Theta}_\varepsilon$ are diagonal matrices according to the classical factor analytic tradition (Lawley and Maxwell 1971).

An important special case is the model with a generalized second order polynomial of the vector $\boldsymbol{\xi}_i$ and a univariate dependent variable y_i

$$\boldsymbol{\alpha}_i = (\xi_{i1}, \ldots, \xi_{i\tilde{m}}, \xi_{i1}^2, \ldots, \xi_{i\tilde{m}}^2, \xi_{i1}\xi_{i2}, \ldots, \xi_{i,\tilde{m}-1}\xi_{i\tilde{m}})' \ , \tag{4}$$

$$\eta_i = \gamma_0 + \boldsymbol{\Gamma}\boldsymbol{\alpha}_i + \zeta_i \ , \quad \zeta_i \sim \mathcal{N}(0, \psi) \ , \tag{5}$$

$$\boldsymbol{x}_i = \boldsymbol{\nu}_x + \boldsymbol{\Lambda}_x\boldsymbol{\xi}_i + \boldsymbol{\delta}_i \ , \quad \boldsymbol{\delta}_i \sim \mathcal{N}(\mathbf{0}, \boldsymbol{\Theta}_\delta) \ , \tag{6}$$

$$y_i = \eta_i \ . \tag{7}$$

This model allows first order interactions of latent variables $\xi_{ij}, j = 1, \ldots, \tilde{m}$ in the regression model for y_i. Of course, higher order interactions may be formulated in the same way.

Important extensions to the framework of Arminger and Muthn (1996) are

- inclusion of observed regressors

$$\boldsymbol{\alpha}_i = \left(g(\boldsymbol{\xi}_i), z_{i1}, \ldots, z_{iq}\right) . \tag{8}$$

  These regressors are not restricted by distributional assumptions such as normality. In fact, the observed variables may be either metric and/or dummy variables. Therefore variables such as sex, region or occupation may be introduced as explanatory variables with dummy regressors.
- the possibility to model different regression coefficients for the latent variables $\xi_{i1}, \ldots, \xi_{i\tilde{m}}$ within categories of observed regressors by multiplying the latent variables or a subset thereof with a subset of dummy regressors. In this case, $\boldsymbol{\alpha}_i$ may be written as

$$\boldsymbol{\alpha}_i = (\xi_{i1}, \ldots, \xi_{i\tilde{m}}, z_{i1}, \ldots, z_{iq}, \xi_{i1}z_{i1}, \ldots \xi_{i1}z_{im}, \xi_{i2}z_{i1}, \ldots) . \tag{9}$$

- inclusion of missing data. This is done within the Bayesian framework discussed in section 2. and discussed in detail in subsection 3.4.

2. The Bayesian framework

The usual way to estimate the parameters $\boldsymbol{\vartheta} = \{\vec{\gamma}_0, \boldsymbol{\Gamma}, \boldsymbol{\Phi}, \boldsymbol{\Psi}, \boldsymbol{\nu}_x, \boldsymbol{\Lambda}_x, \boldsymbol{\Theta}_\delta, \boldsymbol{\nu}_y, \boldsymbol{\Lambda}_y, \boldsymbol{\Theta}_\varepsilon\}$ is a frequentist approach. Given the normal distribution of $\boldsymbol{\xi}_i, \boldsymbol{\zeta}_i, \boldsymbol{\delta}_i$ and $\boldsymbol{\varepsilon}_i$, the parameter vector $\boldsymbol{\vartheta}$ is estimated with the ML method. This approach works nicely if the function g in $\boldsymbol{\alpha}_i = g(\boldsymbol{\xi}_i)$ is the identity function. However, if g is a non-linear function, such as a polynomial, the unconditional distribution of $\boldsymbol{\eta}_i$, and therefore of $\boldsymbol{y}_i$ is unknown, but certainly not normal. To avoid these complications we switch from the frequentist to the Bayesian viewpoint and consider only the data $\{\boldsymbol{y}_i, \boldsymbol{x}_i\}, i = 1, \ldots, n$ as fixed and in - addition to the random variables $\boldsymbol{\xi}_i, \boldsymbol{\zeta}_i, \boldsymbol{\delta}_i$ and $\boldsymbol{\varepsilon}_i$ - the parameter vector $\boldsymbol{\vartheta}$ as a random variable. Given weak assumptions about the prior distribution of $\boldsymbol{\vartheta}$, we are interested in estimating the posterior distributions of $\boldsymbol{\xi}_i$, and $\boldsymbol{\eta}_i$ and the components of $\boldsymbol{\vartheta}$ given the data $\{\boldsymbol{y}_i, \boldsymbol{x}_i\}, i = 1, \ldots, n$.

As noted before, the a priori distributions of $\boldsymbol{\xi}_i, \boldsymbol{\zeta}_i, \boldsymbol{\delta}_i$ and $\boldsymbol{\varepsilon}_i$ are normal with $\boldsymbol{\xi}_i \sim \mathcal{N}(\mathbf{0}, \boldsymbol{\Phi}), \boldsymbol{\zeta}_i \sim \mathcal{N}(\mathbf{0}, \boldsymbol{\Psi}), \boldsymbol{\delta}_i \sim \mathcal{N}(\mathbf{0}, \boldsymbol{\Theta}_\delta)$ and $\boldsymbol{\varepsilon}_i \sim \mathcal{N}(\mathbf{0}, \boldsymbol{\Theta}_\varepsilon)$. $\boldsymbol{\xi}_i$, $\boldsymbol{\zeta}_i$, $\boldsymbol{\delta}_i$, and $\boldsymbol{\varepsilon}_i$ are assumed to be independent of each other. $\boldsymbol{\Theta}_\delta$ and $\boldsymbol{\Theta}_\varepsilon$ are assumed to be diagonal. The components of $\boldsymbol{\vartheta}$ are assumed to be stochastically independent and to have flat priors. Our aim is to describe the joint and marginal distribution of $\boldsymbol{\xi}$, $\boldsymbol{\eta}$, and $\boldsymbol{\vartheta}$ given $\boldsymbol{x}$ and $\boldsymbol{y}$. Using the notation of Gelfand and Smith (1990) for distributions, we are interested in $[\boldsymbol{\xi}, \boldsymbol{\eta}, \boldsymbol{\vartheta} | \boldsymbol{x}, \boldsymbol{y}]$.

To describe the joint and marginal posterior distributions of $\boldsymbol{\xi}$, $\boldsymbol{\eta}$ and $\boldsymbol{\vartheta}$ given $\boldsymbol{x}$ and $\boldsymbol{y}$ we use two different versions of the Markov chain Monte Carlo (MCMC) methods in combination (Müller 1994). The first method is the Gibbs sampler (Geman and Geman 1984, Gelfand and Smith 1990), the second is the Metropolis-Hastings (M-H) algorithm (Metropolis et al. 1953, Hastings 1970, Tanner 1993, Chib and Greenberg 1995b). Easily accessible descriptions of the Gibbs sampler are found in Arnold (1993) and Casella and George (1992). Statistical theory underlying the Gibbs sampler and the M-H algorithm is found in Chib and Greenberg (1995b) and Tierney (1994).

In the Gibbs sampler one draws iteratively from the conditional distributions

$$[\boldsymbol{\xi}^{(k+1)}, \boldsymbol{\eta}^{(k+1)} | \boldsymbol{x}, \boldsymbol{y}, \boldsymbol{\vartheta}^{(k)}] , \tag{10}$$

where $k+1$ denotes the present iteration and the distributions

$$[\boldsymbol{\vartheta}_q^{(k+1)} | \boldsymbol{x}, \boldsymbol{y}, \boldsymbol{\xi}^{(k)}, \boldsymbol{\eta}^{(k)}, \boldsymbol{\vartheta}_{l<q}^{(k+1)}, \boldsymbol{\vartheta}_{l>q}^{(k)}] , \tag{11}$$

where $\boldsymbol{\vartheta}_q^{(k+1)}$ is the qth subvector of the vector $\boldsymbol{\vartheta}$ at the present iteration. After a burn in phase, the conditional distributions converge to the posterior marginal distributions of the parameters of interest.

3. The MCMC method

We start by assuming that $\boldsymbol{\xi}_i$ and $\boldsymbol{\eta}_i$ are known for the kth iteration taking on the values $\boldsymbol{\xi}_i^{(k)}, \boldsymbol{\eta}_i^{(k)}$. Then we can compute

$$\boldsymbol{\alpha}_i^{(k)} = g(\boldsymbol{\xi}_i^{(k)}) . \tag{12}$$

3.1 Parameters of the regression model

The conditional distribution of the regression parameters in the model (1) - (3) is

$$[\gamma_0, \boldsymbol{\Gamma} | \boldsymbol{x}, \boldsymbol{y}, \boldsymbol{\xi}, \boldsymbol{\eta}, \boldsymbol{\vartheta} - \{\gamma_0, \boldsymbol{\Gamma}\}] . \tag{13}$$

We define the $p(m+1)$ dimensional column vector $\boldsymbol{\gamma}$ as the vectorized form of $\boldsymbol{\gamma}_0$ and $\boldsymbol{\Gamma}$:

$$\boldsymbol{\gamma} = (\gamma_{01}, \gamma_{11}, \ldots, \gamma_{1m}, \gamma_{02}, \gamma_{21}, \ldots, \gamma_{2m}, \ldots, \gamma_{pm})' . \tag{14}$$

From the multivariate regression equation (1), the posterior distribution of $\boldsymbol{\gamma}$ given $\boldsymbol{\eta}_i$ and $\boldsymbol{\alpha}_i$ may be derived. Because $\boldsymbol{\zeta}_i$ is multivariate normal, the vector $\boldsymbol{\gamma}$ is normally distributed with expected value

$$E(\boldsymbol{\gamma}) = (\sum_{i=1}^{n} \boldsymbol{A}_i' \boldsymbol{\Psi}^{-1} \boldsymbol{A}_i)^{-1} \sum_{i=1}^{n} \boldsymbol{A}_i' \boldsymbol{\Psi}^{-1} \boldsymbol{\eta}_i , \tag{15}$$

and variance covariance matrix

$$V(\boldsymbol{\gamma}) = (\sum_{i=1}^{n} \boldsymbol{A}_i' \boldsymbol{\Psi}^{-1} \boldsymbol{A}_i)^{-1} . \tag{16}$$

The regressor matrix $\boldsymbol{A}_i \sim p \times (p(m+1))$ is defined by

$$\boldsymbol{A}_i = \boldsymbol{I}_{p\times p} \otimes (1, \boldsymbol{\alpha}_i') \ , \tag{17}$$

where $\otimes$ denotes the Kronecker product. Hence $\boldsymbol{\gamma}^{(k+1)}$ is drawn from the posterior distribution $\mathcal{N}(E(\gamma), V(\gamma))$ where $\boldsymbol{\alpha}_i, \boldsymbol{\eta}_i$ and $\boldsymbol{\Psi}$ are replaced by $\boldsymbol{\alpha}_i^{(k)}, \boldsymbol{\eta}_i^{(k)}$ and $\boldsymbol{\Psi}^{(k)}$.

The conditional distribution of

$$[\boldsymbol{\Phi}|\boldsymbol{x}, \boldsymbol{y}, \boldsymbol{\xi}, \boldsymbol{\eta}, \boldsymbol{\vartheta} - \{\boldsymbol{\Phi}\}] \tag{18}$$

is found using the inverse Wishart distribution if $\boldsymbol{\xi}$ is multivariate normal. However, the Gibbs sampler yields realizations $\boldsymbol{\xi}_i^{(k+1)}$ not from the multivariate normal a priori density but from the a posteriori density which is discussed in section 3.3. Hence, $\boldsymbol{\Phi}^{(k+1)}$ should in principle be generated using the M-H algorithm. Details are found in section 4.4 of Arminger and Muthn (1996). In simulation studies we have found that the direct use of the inverse Wishart distribution gives satisfactory results. Because of the heavy computational burden of the Metropolis-Hastings algorithm within the Gibbs sampler, the direct use of the driving distribution for generating $\boldsymbol{\Phi}^{(l)}$ seems to be a feasible alternative.

The conditional distribution of

$$[\boldsymbol{\Psi}|\boldsymbol{x}, \boldsymbol{y}, \boldsymbol{\xi}, \boldsymbol{\eta}, \boldsymbol{\vartheta} - \{\boldsymbol{\Psi}\}] \tag{19}$$

is found using the inverse Wishart distribution because the error terms $\boldsymbol{\zeta}_i$ are multivariate normal. To generate $\boldsymbol{\Psi}^{(k+1)}$, we sample $(\boldsymbol{\Psi}^{(k+1)})^{-1}$ from a Wishart distribution with parameter matrix

$$\boldsymbol{B}_\Psi = \sum_{i=1}^{n} (\boldsymbol{\eta}_i^{(k)} - \boldsymbol{A}_i^{(k)} \boldsymbol{\gamma}^{(k)})(\boldsymbol{\eta}_i^{(k)} - \boldsymbol{A}_i^{(k)} \boldsymbol{\gamma}^{(k)})' \tag{20}$$

and $n-p-1$ degrees of freedom. If η_i is univariate, the variance $\psi^{(k+1)}$ is sampled from a χ^2 distribution with $n-2$ degrees of freedom as

$$\psi^{(k+1)} = \sum_{i=1}^{n} (\eta_i^{(k)} - \gamma_0^{(k)} - \boldsymbol{\Gamma}^{(k)} \boldsymbol{\alpha}_i^{(k)})^2 / \chi^2_{n-2} \ . \tag{21}$$

3.2 Parameters of the factor analytic models

Given $\boldsymbol{\eta}_i$ and the fact that $\boldsymbol{\Theta}_\varepsilon$ is diagonal, one finds the posterior distribution of the vector $\tilde{\boldsymbol{\lambda}}_{y,j} = (\nu_{y,j}, \lambda_{y,j,1}, \ldots \lambda_{y,j,p})'$ for the jth component of $\boldsymbol{y}_i$ from the factor analytic regression equation

$$y_{ij} = \nu_{y,j} + \lambda_{y,j,1}\eta_{i1} + \ldots + \lambda_{y,j,p}\eta_{ip} + \varepsilon_{ij} \ . \tag{22}$$

The conditional distribution

$$[\boldsymbol{\nu}_y, \boldsymbol{\Lambda}_y | \boldsymbol{x}, \boldsymbol{y}, \boldsymbol{\xi}, \boldsymbol{\eta}, \boldsymbol{\vartheta} - \{\boldsymbol{\nu}_y, \boldsymbol{\Lambda}_y\}] \tag{23}$$

is found by drawing for each vector $\tilde{\boldsymbol{\lambda}}_{y,j}$ from a $p+1$ dimensional normal distribution with expected value

$$E(\tilde{\boldsymbol{\lambda}}_{y,j}) = (\boldsymbol{Z}'\boldsymbol{Z})^{-1}\boldsymbol{Z}'\boldsymbol{y}_j \tag{24}$$

and variance covariance matrix

$$V(\tilde{\boldsymbol{\lambda}}_{y,j}) = \Theta_{\varepsilon,jj}(\boldsymbol{Z}'\boldsymbol{Z})^{-1} \; . \tag{25}$$

The $n \times (p+1)$ regressor matrix $\boldsymbol{Z}$ is defined as

$$\boldsymbol{Z} = \begin{pmatrix} \boldsymbol{z}_1 \\ \vdots \\ \boldsymbol{z}_n \end{pmatrix} \quad \text{with} \quad \boldsymbol{z}_i = (1, \eta_{i1}, \ldots, \eta_{ip}) \; . \tag{26}$$

The regressand vector $\boldsymbol{y}_j = (y_{1j}, y_{2j}, \ldots, y_{nj})'$.
The conditional distribution

$$[\boldsymbol{\Theta}_\varepsilon | \boldsymbol{x}, \boldsymbol{y}, \boldsymbol{\xi}, \boldsymbol{\eta}, \boldsymbol{\vartheta} - \{\boldsymbol{\Theta}_\varepsilon\}] \tag{27}$$

is found by drawing for each $\Theta_{\varepsilon,jj}, j = 1, \ldots, r$ from the inverse χ^2 distribution with $n-2$ degrees of freedom

$$\Theta_{\varepsilon,jj}^{(k+1)} = \sum_{i=1}^{n} (y_{ij} - \nu_{y,j}^{(k)} - \boldsymbol{\Lambda}_{y,j}^{(k)} \boldsymbol{\eta}_i^{(k)})^2 / \chi_{n-2}^2 \; , \tag{28}$$

where $\boldsymbol{\Lambda}_{y,j}$ is the jth row of $\boldsymbol{\Lambda}_y$.
Given $\boldsymbol{\xi}_i$ and the fact that $\boldsymbol{\Theta}_\delta$ is diagonal, similar calculations can be made to find the marginal posterior distributions of $(\boldsymbol{\nu}_x, \boldsymbol{\Lambda}_x)$ and $\boldsymbol{\Theta}_\delta$.

3.3 Drawing from the distribution of latent variables

The conditional distribution of $\boldsymbol{\xi}$ and $\boldsymbol{\eta}$ given $\boldsymbol{x}, \boldsymbol{y}$ and $\boldsymbol{\vartheta}$ is now considered. Since independence of $(\boldsymbol{x}_i, \boldsymbol{y}_i)$ and $\boldsymbol{\xi}_i, \boldsymbol{\eta}_i, i = 1, \ldots, n$ across units has been assumed, it suffices to look at the conditional distribution

$$[\boldsymbol{\xi}_i, \boldsymbol{\eta}_i | \boldsymbol{x}_i, \boldsymbol{y}_i, \boldsymbol{\vartheta}] \; . \tag{29}$$

Unlike for the conditional distributions considered before, we have not been able to derive a form of the conditional distribution from which values of $(\boldsymbol{\xi}_i, \boldsymbol{\eta}_i)$ given $\boldsymbol{x}_i, \boldsymbol{y}_i, \boldsymbol{\vartheta}$ can be sampled easily. Therefore, we use the M-H algorithm as described in Tanner (1993), and Chib and Greenberg (1995b) to generate values of $(\boldsymbol{\xi}_i, \boldsymbol{\eta}_i)$ that follow the conditional distribution of (29). Let $(\boldsymbol{\xi}^{(0)}, \boldsymbol{\eta}^{(0)})$ be the current value and and $(\boldsymbol{\xi}^{(1)}, \boldsymbol{\eta}^{(1)})$ denote the value generated by a distribution from which one can easily sample. This easily sampled distribution is called the driving distribution. Let $\pi^{(0)}$ and $\pi^{(1)}$ denote the density of (29) evaluated at $(\boldsymbol{\xi}^{(0)}, \boldsymbol{\eta}^{(0)})$ and $(\boldsymbol{\xi}^{(1)}, \boldsymbol{\eta}^{(1)})$. Then the value

$$c = \min\{\frac{\pi^{(1)} q^{(0)}}{\pi^{(0)} q^{(1)}}, \ 1\} \tag{30}$$

is calculated, where $q^{(i)}$ is the density of the driving distribution evaluated at $(\boldsymbol{\xi}^{(i)}, \boldsymbol{\eta}^{(i)})$. If $c = 1$, then the value $(\boldsymbol{\xi}^{(1)}, \boldsymbol{\eta}^{(1)})$ is accepted and becomes the current value. If $c < 1$, then the value $(\boldsymbol{\xi}^{(1)}, \boldsymbol{\eta}^{(1)})$ is accepted only with probability c, otherwise $(\boldsymbol{\xi}^{(0)}, \boldsymbol{\eta}^{(0)})$ stays the current value. This algorithm ensures that the accepted values form a sample from the distribution of (29). To use the M-H algorithm we have to compute $\pi^{(0)}$ and $\pi^{(1)}$ and to sample from a simple driving distribution. First, we deal with the computation of $\pi^{(0)}$ and $\pi^{(1)}$. We note that $[\boldsymbol{\xi}_i, \boldsymbol{\eta}_i | \boldsymbol{x}_i, \boldsymbol{y}_i, \boldsymbol{\vartheta}]$ may be written as

$$\begin{aligned} [\boldsymbol{\xi}_i, \boldsymbol{\eta}_i | \boldsymbol{x}_i, \boldsymbol{y}_i, \boldsymbol{\vartheta}] &\propto [\boldsymbol{\xi}_i, \boldsymbol{\eta}_i | \boldsymbol{\vartheta}][\boldsymbol{x}_i | \boldsymbol{\xi}_i, \boldsymbol{\eta}_i, \boldsymbol{\vartheta}][\boldsymbol{y}_i | \boldsymbol{x}_i, \boldsymbol{\xi}_i, \boldsymbol{\eta}_i, \boldsymbol{\vartheta}] \\ &= [\boldsymbol{\xi}_i, \boldsymbol{\eta}_i | \boldsymbol{\vartheta}][\boldsymbol{x}_i | \boldsymbol{\xi}_i, \boldsymbol{\vartheta}][\boldsymbol{y}_i | \boldsymbol{\eta}_i, \boldsymbol{\vartheta}] \,. \end{aligned} \tag{31}$$

The last equality holds because $\boldsymbol{x}_i$ depends only on $\boldsymbol{\xi}_i$ and not on $\boldsymbol{\eta}_i$ and because $\boldsymbol{y}_i$ given $\boldsymbol{\eta}_i$ does not depend on $\boldsymbol{\xi}_i$ or $\boldsymbol{x}_i$. Substituting

$$[\boldsymbol{\xi}_i, \boldsymbol{\eta}_i | \boldsymbol{\vartheta}] = [\boldsymbol{\xi}_i | \boldsymbol{\vartheta}][\boldsymbol{\eta}_i | \boldsymbol{\xi}_i, \boldsymbol{\vartheta}] \tag{32}$$

yields the expression:

$$[\boldsymbol{\xi}_i, \boldsymbol{\eta}_i | \boldsymbol{x}_i, \boldsymbol{y}_i, \boldsymbol{\vartheta}] \propto [\boldsymbol{\xi}_i | \boldsymbol{\vartheta}][\boldsymbol{\eta}_i | \boldsymbol{\xi}_i, \boldsymbol{\vartheta}][\boldsymbol{x}_i | \boldsymbol{\xi}_i, \boldsymbol{\vartheta}][\boldsymbol{y}_i | \boldsymbol{\eta}_i, \boldsymbol{\vartheta}] \,. \tag{33}$$

Since the ratio $\pi^{(1)}/\pi^{(0)}$ does not depend on the proportionality constant, it suffices to evaluate $[\boldsymbol{\xi}_i | \boldsymbol{\vartheta}][\boldsymbol{\eta}_i | \boldsymbol{\xi}_i, \boldsymbol{\vartheta}][\boldsymbol{x}_i | \boldsymbol{\xi}_i, \boldsymbol{\vartheta}][\boldsymbol{y}_i | \boldsymbol{\eta}_i, \boldsymbol{\vartheta}]$ at the points $(\boldsymbol{\xi}^{(0)}, \boldsymbol{\eta}^{(0)})$ and $(\boldsymbol{\xi}^{(1)}, \boldsymbol{\eta}^{(1)})$. The components of this joint density of $\boldsymbol{\xi}_i, \boldsymbol{\eta}_i, \boldsymbol{x}_i$ and $\boldsymbol{y}_i$ given $\boldsymbol{\vartheta}$ are implied by the model in (1) - (3).

An important special case occurs if $\boldsymbol{y} = \boldsymbol{\eta}$, that is, the measurement model for $\boldsymbol{y}$ is given by $\boldsymbol{y} = \boldsymbol{\eta}$. In this case, we consider the conditional distribution:

$$[\boldsymbol{\xi}_i | \boldsymbol{x}_i, \boldsymbol{y}_i, \boldsymbol{\vartheta}] \propto [\boldsymbol{\xi}_i | \boldsymbol{\vartheta}][\boldsymbol{x}_i | \boldsymbol{\xi}_i, \boldsymbol{\vartheta}][\boldsymbol{y}_i | \boldsymbol{\xi}_i, \boldsymbol{\vartheta}] \,. \tag{34}$$

For the driving distribution, we take the joint distribution $[\boldsymbol{\xi}_i, \boldsymbol{\eta}_i | \boldsymbol{\vartheta}] = [\boldsymbol{\xi}_i | \boldsymbol{\vartheta}][\boldsymbol{\eta}_i | \boldsymbol{\xi}_i, \boldsymbol{\vartheta}]$ by first sampling $\boldsymbol{\xi}$ from the a priori distribution $\mathcal{N}(\mathbf{0}, \boldsymbol{\Phi})$ and then sampling $\boldsymbol{\eta}$ from $\mathcal{N}(\gamma_0 + \boldsymbol{\Gamma}\boldsymbol{\alpha}(\boldsymbol{\xi}), \boldsymbol{\Psi})$ where $\boldsymbol{\Phi}, \gamma_0, \boldsymbol{\Gamma}$, and $\boldsymbol{\Psi}$ are evaluated at the kth iteration step. Alternatively, we sample $\boldsymbol{\xi}$ from the conditional distribution $[\boldsymbol{\xi} | \boldsymbol{x}, \boldsymbol{\vartheta}]$ which may be derived from the joint normal distribution of $\boldsymbol{x}$ and $\boldsymbol{\xi}$.

$$\begin{pmatrix} \boldsymbol{x} \\ \boldsymbol{\xi} \end{pmatrix} \sim \mathcal{N}\left(\begin{bmatrix} \boldsymbol{\nu}_x \\ \mathbf{0} \end{bmatrix}, \begin{bmatrix} \boldsymbol{\Lambda}_x \boldsymbol{\Phi} \boldsymbol{\Lambda}_x' + \boldsymbol{\Theta}, & \boldsymbol{\Lambda}_x \boldsymbol{\Phi} \\ \boldsymbol{\Phi} \boldsymbol{\Lambda}_x', & \boldsymbol{\Phi} \end{bmatrix} \right) . \tag{35}$$

Hence, the conditional distribution of $\boldsymbol{\xi}$ given $\boldsymbol{x}_i$ is multivariate normal with

$$E(\boldsymbol{\xi}|\boldsymbol{x}_i, \boldsymbol{\nu}_x, \boldsymbol{\Phi}, \boldsymbol{\Lambda}_x, \boldsymbol{\Theta}_\delta) = \boldsymbol{\Phi}\boldsymbol{\Lambda}_x'(\boldsymbol{\Lambda}_x\boldsymbol{\Phi}\boldsymbol{\Lambda}_x' + \boldsymbol{\Theta}_\delta)^{-1}(\boldsymbol{x}_i - \boldsymbol{\nu}_x) \tag{36}$$

and covariance matrix

$$\begin{aligned} V(\boldsymbol{\xi}|\boldsymbol{x}_i, \boldsymbol{\nu}_x, \boldsymbol{\Phi}, \boldsymbol{\Lambda}_x, \boldsymbol{\Theta}_\delta) &= \boldsymbol{\Phi} - \boldsymbol{\Phi}\boldsymbol{\Lambda}_x'(\boldsymbol{\Lambda}_x\boldsymbol{\Phi}\boldsymbol{\Lambda}_x' + \boldsymbol{\Theta}_\delta)^{-1}\boldsymbol{\Lambda}_x\boldsymbol{\Phi} \\ &= (\boldsymbol{\Phi}^{-1} + \boldsymbol{\Lambda}_x'\boldsymbol{\Theta}_\delta^{-1}\boldsymbol{\Lambda}_x)^{-1} . \end{aligned} \tag{37}$$

Since the driving distribution does not depend on $(\boldsymbol{\xi}^{(0)}, \boldsymbol{\eta}^{(0)})$, the M-H chain generated in this way is called an independence chain (Tierney 1994). Note that when the driving distribution is symmetric, (30) reduces to $c = \min\{\pi^{(1)}/\pi^{(0)}, 1\}$.

3.4 Missing data

We sample the missing data from the conditional distribution of $[\boldsymbol{y}_i|\, \boldsymbol{\xi}_i, \boldsymbol{\eta}_i, \boldsymbol{\vartheta}_i]$ and $[\boldsymbol{x}_i|\boldsymbol{\xi}_i, \boldsymbol{\eta}_i, \boldsymbol{\vartheta}_i]$. In the full model (eq. (1) – (3)) $\boldsymbol{x}_i$ and $\boldsymbol{y}_i$ are normally distributed with expected values

$$\mathbf{E}(\boldsymbol{x}_i) = \boldsymbol{\nu}_x + \boldsymbol{\Lambda}_x\boldsymbol{\xi}_i \text{ and } \mathbf{E}(\boldsymbol{y}_i) = \boldsymbol{\nu}_y + \boldsymbol{\Lambda}_y\boldsymbol{\eta}_i \tag{38}$$

and variance-covariance matrices

$$\mathrm{V}(\boldsymbol{x}_i) = \boldsymbol{\Theta}_\delta \text{ and } \mathrm{V}(\boldsymbol{y}_i) = \boldsymbol{\Theta}_\varepsilon . \tag{39}$$

In the case when the measurement model for $\boldsymbol{y}$ is given by $\boldsymbol{y} = \boldsymbol{\eta}$, the conditional distribution of $[\boldsymbol{y}_i|\, \boldsymbol{\xi}_i, \boldsymbol{\eta}_i, \boldsymbol{\vartheta}_i]$ is not known. Hence we use the M-H algorithm to generate values of $(\boldsymbol{\xi}_i, \boldsymbol{y}_i)$ that follow the conditional distribution

$$[\boldsymbol{\xi}_i, \boldsymbol{y}_i|\boldsymbol{x}_i, \boldsymbol{\vartheta}_i] \propto [\boldsymbol{\xi}_i|\boldsymbol{\vartheta}][\boldsymbol{y}_i|\boldsymbol{\xi}_i, \boldsymbol{\vartheta}][\boldsymbol{x}_i|\boldsymbol{\xi}_i, \boldsymbol{\vartheta}] \tag{40}$$

and sample the missing values for $\boldsymbol{x}_i$ afterwards from a normal distribution with expected value and variance-covariance matrix given in (38) and (39).

Another special case is given by the measurement model $\boldsymbol{x} = \boldsymbol{\xi}$. Here we sample the missing values for $\boldsymbol{x}_i$ from a normal distribution with empirical expected value and variance-covariance matrix of $\boldsymbol{x}$. After using the M-H algorithm to sample $\boldsymbol{\eta}_i$ the missing values for $\boldsymbol{y}_i$ are generated using the normal distribution with expected value and variance-covariance matrix given in (38) and (39).

4. Simulated examples

The above model and MCMC methods are demonstrated using two simulated data sets, the first coming from a model with two latent and one categorical regressor variables and interaction terms between the regressor variables, the second coming from a model with missing data and two latent regressor variables and an interaction term between the regressor variables.

Gibbs: Cycles = 2000, M-H: Cycles = 100

	True	Med.	Mean	S.D.		True	Med.	Mean	S.D.
γ_{01}	-0.91	-0.743	-0.751	0.266	γ_{02}	1.96	1.630	1.630	0.255
γ_{11}	3.38	3.238	3.240	0.236	γ_{12}	-2.96	-2.780	-2.807	0.263
γ_{21}	2.02	1.774	1.787	0.234	γ_{22}	1.70	1.923	1.933	0.250
γ_{31}	-1.63	-1.650	-1.662	0.187	γ_{32}	2.39	2.619	2.644	0.243
γ_{41}	0.51	0.412	0.401	0.448	γ_{42}	0.15	0.639	0.636	0.421
γ_{51}	0.32	0.410	0.424	0.307	γ_{52}	0.15	-0.025	-0.021	0.349
γ_{61}	-0.21	-0.189	-0.181	0.390	γ_{62}	-0.23	-0.449	-0.467	0.399
ϕ_{11}	2.07	2.179	2.211	0.349					
ϕ_{21}	-0.37	-0.367	-0.381	0.208	ϕ_{22}	1.47	1.523	1.549	0.230
ψ_{11}	1.41	2.063	2.126	0.559					
ψ_{21}	0.77	0.199	0.162	0.372	ψ_{22}	1.79	1.681	1.760	0.583

Table 1: Non-linear regression model with categorical regressors

Following Chib and Greenberg (1995a), we have implemented the single run method because it is less wasteful in the number of iterations needed. Following the arguments of McEachern and Berliner (1994) we use the full Gibbs sample rather than subsampling because subsampling is inefficient in comparison to full sampling. Methods of judging convergence from a single run are found in Ritter and Tanner (1993). We have settled for a burn in phase of 500 Gibbs cycles and additional 2000 Gibbs cycles from which parameters of the posterior distribution are estimated. The number of M-H cycles depends on the estimated model. As a rule of thumb the number of M-H cycles needs to be increased for more complicated models. We started with 20 cycles but up to 100 cycles are needed for satisfactory results.

For both examples the sample size is $n = 100$. The regression model is given by equation (1), the factor analytic model connecting $\boldsymbol{\xi}_i$ and the observed indicators $\boldsymbol{x}_i$ is

$$\boldsymbol{x}_i = \boldsymbol{\Lambda}_x \boldsymbol{\xi}_i + \boldsymbol{\delta}_i \ , \quad \boldsymbol{\delta}_i \sim \mathcal{N}(\boldsymbol{0}, \boldsymbol{\Theta}_\delta) \ . \tag{41}$$

and the factor analytic model connecting $\boldsymbol{\eta}_i$ and the observed indicators $\boldsymbol{y}_i$ is

$$\boldsymbol{y}_i = \boldsymbol{\Lambda}_y \boldsymbol{\eta}_i + \boldsymbol{\varepsilon}_i \ , \quad \boldsymbol{\varepsilon}_i \sim \mathcal{N}(\boldsymbol{0}, \boldsymbol{\Theta}_\varepsilon) \ . \tag{42}$$

4.1 Categorical regressors and interactions with latent variables

Here the deterministic function $\boldsymbol{\alpha}_i = g(\boldsymbol{\xi}_i)$ of (1) is

$$\boldsymbol{\alpha}_i = (\xi_{i1}, \xi_{i2}, \xi_{i1}\xi_{i2}, z_i, \xi_{i1}z_i, \xi_{i2}z_i) \ . \tag{43}$$

For (41) we use six indicator variables yielding the structure

$$\mathbf{\Lambda}_x = \begin{pmatrix} 1 & 0 \\ \lambda_{x21} & 0 \\ \lambda_{x31} & 0 \\ 0 & 1 \\ 0 & \lambda_{x52} \\ 0 & \lambda_{x62} \end{pmatrix}, \mathbf{\Theta}_\delta = \text{diag}\left\{\vartheta_{\delta 11}, \vartheta_{\delta 22}, \vartheta_{\delta 33}, \vartheta_{\delta 44}, \vartheta_{\delta 55}, \vartheta_{\delta 66}\right\}. \tag{44}$$

The values of λ_{x11} and λ_{x42} are set to 1 to fix the scale of ξ_{i1} and ξ_{i2}. For (42) we used four indicator variables yielding the structure

$$\mathbf{\Lambda}_y = \begin{pmatrix} 1 & 0 \\ \lambda_{y21} & 0 \\ 0 & 1 \\ 0 & \lambda_{y42} \end{pmatrix}, \mathbf{\Theta}_\varepsilon = \text{diag}\left\{\vartheta_{\varepsilon 11}, \vartheta_{\varepsilon 22}, \vartheta_{\varepsilon 33}, \vartheta_{\varepsilon 44}\right\}. \tag{45}$$

The values of λ_{y11} and λ_{y32} are set to 1 to fix the scale of η_{i1} and η_{i2}. Table 1 shows the results of the non-linear regression model. Most means and medians show close agreement to the true values. Exceptions are the parameters γ_{42}, γ_{52} and γ_{62} of the categorical regressors of η_{i2} which are not significant in the simulated data set and ψ_{11} and ψ_{21}. The parameters for the factor analytic models are not shown since they are very close to the true values. The results are satisfactory for the complex model and the small sample size $n = 100$.

4.2 Interactions between latent variables based on missing data

In this example the deterministic function of (1) is

$$\boldsymbol{\alpha}_i = (\xi_{i1}, \xi_{i2}, \xi_{i1}\xi_{i2}) \ . \tag{46}$$

For (41) we use five indicator variables yielding the structure

$$\mathbf{\Lambda}_x = \begin{pmatrix} 1 & 0 \\ \lambda_{x21} & 0 \\ \lambda_{x31} & 0 \\ 0 & 1 \\ 0 & \lambda_{x52} \end{pmatrix}, \mathbf{\Theta}_\delta = \text{diag}\left\{\vartheta_{\delta 11}, \vartheta_{\delta 22}, \vartheta_{\delta 33}, \vartheta_{\delta 44}, \vartheta_{\delta 55}\right\} \tag{47}$$

and for (42) four indicator variables yielding the structure

$$\mathbf{\Lambda}_y = \begin{pmatrix} 1 \\ \lambda_{y21} \\ \lambda_{y31} \end{pmatrix}, \mathbf{\Theta}_\varepsilon = \text{diag}\left\{\vartheta_{\varepsilon 11}, \vartheta_{\varepsilon 22}, \vartheta_{\varepsilon 33}\right\}. \tag{48}$$

Each variable contains 10% missing data. The estimation was carried out with 60 M-H cycles. Table 2 shows the results for the non-linear regression model. Again the means and medians show close agreement with the true

	Gibbs: Cycles = 2000, M-H: Cycles = 60								
	True	Med.	Mean	S.D.		True	Med.	Mean	S.D.
γ_0	-1.08	-1.185	-1.190	0.199	γ_2	1.74	1.593	1.618	0.362
γ_1	0.80	0.912	0.924	0.236	γ_3	0.69	0.663	0.683	0.382
φ_{11}	1.27	1.195	1.237	0.312					
φ_{21}	0.21	0.223	0.235	0.161	φ_{22}	0.74	0.836	0.867	0.306
ψ_{11}	0.65	0.914	1.019	0.454					

Table 2: Non-linear regression model based on missing data

paramaters except ψ_{11} which is slightly overestimated. The parameters for the factor analytic models are not shown, since they all are also very close to the true values. Nevertheless, the results are satisfactory keeping in mind that with standard listwise deletion of missing values the sample size would reduce to $n = 40$.

5. Conclusion

We have formulated a non-linear latent variable model including quadratic functions, interactions of latent regressors, observed regressors and missing data. Estimation of model parameters using conventional covariance structure analysis is almost impossible yielding unwieldy mean- and covariance structures and cumbersome numerical algorithms. Putting the non-linear latent variable model in a Bayesian framework allows the use of MCMC methods to estimate the posterior distributions of the parameters and the latent variables given the data. The application of the MCMC methods is complicated by the use of the M-H algorithm within iterations of the Gibbs sampler. The proposed model and the numerical algorithms are illustrated by two simulated examples, with a maximum of 65 minutes computing time on a 90 Mhz Pentium computer.

References:

ARMINGER, G. and MUTHÉN, B. O. (1996): A Bayesian Approach to Non-Linear Latent Variable Models using The Gibbs Sampler and the Metropolis-Hastings Algorithm. Unpublished manuscript, Universität Wuppertal, Germany.

ARMINGER, G. and SCHOENBERG, R. J. (1989): Pseudo Maximum Likelihood Estimation and a Test for Misspecification in Mean and Covariance Structure Models. *Psychometrika, 54, 409–425.*

ARNOLD, S. F. (1993): Gibbs Sampling. In Rao, C. R., editor, *Handbook of Statistics, Vol. 9, 599–625.* Amsterdam: North Holland.

BOX, G. E. P. and TIAO, G. C. (1973): Bayesian Inference in Statistical Analysis. Reading: Addison-Wesley.

BROWNE, M. W. (1984): Asymptotic Distribution-free Methods for the Analysis of Covariance Structures. *British Journal of Mathematical and Statistical Psychology, 37, 62–83.*

CASELLA, G. and GEORGE, E. I. (1992): Explaining the Gibbs Sampler. *The American Statistican, Vol. 46, 167–174.*

CHIB, S. and GREENBERG, E. (1995a): Markov Chain Monte Carlo Simulation Methods in Econometrics. Forthcoming in *Econometric Theory.*

CHIB, S. and GREENBERG, E. (1995b): Understanding the Metropolis Hastings Algorithm. *The American Statistican, Vol. 49, 327–335.*

GELFAND, A. E. and SMITH, A. F. M. (1990): Sampling-Based Approaches to Calculating Marginal Densities. *Journal of the American Statistical Association, Vol. 85, 398–409.*

GEMAN, S. and GEMAN, D. (1984): Stochastic Relaxation, Gibbs Distribution and the Bayesian Restoration of Images. *IEEE Transactions on Pattern Analysis and Machine Intelligence, 6, 721–741.*

HASTINGS, W. K. (1970): Monte Carlo Sampling Methods Using Markov Chains and Their Applications. *Biometrika, 57, 97–109.*

JÖRESKOG, K. G. and SÖRBOM, D. (1993): LISREL 8: Structural Equation Modeling With the SIMPLIS Command Language. Hillsdale, NJ, Lawrence Earlbaum Associates.

LAWLEY, D. N. and MAXWELL, A. E. (1971): Factor Analysis as a Statistical Method. London, Butterworth.

MACEACHERN, S. N. and BERLINER, M. L. (1994): Subsampling the Gibbs Sampler. *The American Statistican, 48, 188–190.*

METROPOLIS, N.; ROSENBLUTH, A. W.; ROSENBLUTH, M. N.; TELLER, A. H. and TELLER, E. (1953): Equations of state calculations by fast computing machines. *Journal of Chemical Physics, 21, 1087–1091.*

MÜLLER, P. (1994): Metroplis Based Posterior Integration Schemes. Unpublished manuscript, Duke University, Durham.

MUTHÉN, B. O. (1988): LISCOMP – Analysis of Linear Structural Equations With a Comprehensive Measurement Model. Mooresville.

MUTHÉN, B. O. and ARMINGER, G. (1996): Bayesian Latent Variable Regression for Binary and Continuous Response Variables Using the Gibbs Sampler. Unpublished manuscript, University of California at Los Angeles.

TANNER, M. A. (1993): Tools for Statistical Inference. 2nd edition, Heidelberg - New York, Springer Verlag.

TIERNEY, L. (1994): Markov Chains for Exploring Posterior Distributions (with Discussion). *Annals of Statistics, 22, 1701–1762.*

Section 4

Information Systems: Design and Inplementation

Case Based Software Engineering CBSE The Example of a Store Control System

H. Czap

Abteilung Wirtschaftsinformatik, Universität Trier,
Postfach 3825, D-54286 Trier, Email: cz@wiinfo.uni-trier.de

Summary: The development of software systems for specific customer needs may be substantially simplified by using past solutions. With respect to the more technical aspects of a store control system we show how the applicational situation can be modeled in order to build the needed case-base. Details are provided of how the customer situation will be represented and how access to stored cases works.

1. Introduction

Small software companies will have success only if they operate in specialized markets or if they have enough flexibility to develop customer specific applications. The tool CBSE, Case Based Software Engineering, being described, supports the development of customer specific applications, as far as they show same basic functionality.

As a typical example, store control systems (SCS) will be considered. They control the more technical aspects of the movement of goods in a store. Typical sub-functions consist of delivery point, the selection of suitable transport units, the possible repacking, the determination of store location, the deposition etc. and finally the commission and supply at the distribution counters.

In the following we present the typical characteristics of SCS in order to illustrate the variance of these applicational systems. Subsequent paragraphs show the knowledge representation of CBSE and concentrate on the structure of program-objects and the way to develop an application using CBSE.

2. Characteristics of Storage Control Systems

Basic functionality of all SCS is comparable, as far as headlines and macroprocesses are considered. For example, one finds functions like delivery point, quality assurance, deposition and so on in all SCS. But looking more closely to the customer specific situations, differentiation with respect to many aspects is necessary resulting in completely different SCS.

Differentiation of goods occurs with respect to size and kind of material (oversize, smaller respectively larger items, plates, packs, barrels, cans, bottles, bulk goods), with respect to storing temperature (normal tempered places, cool places, deep freeze places), danger (danger of explosion, toxicity etc.), production series and duration.

In addition, differentiation occurs with respect to shelves, their technical features and placements and if the store is run automatically or manually. Also, conveyance and containers used for storage differ from case to case (for details compare Jünemann (1989), Grochla (1978), Rupper(1988) and Rupper/Scheuchzer (1988)).

Major differences also occur with respect to functionality. Quality assurance will serve as an example.

> *In order to assure quality in some cases a sample survey is selected and carried out. In the case where this quality check takes some time, the remaining goods are labeled with a lock and put into store. In other cases the whole delivery is stored after labeling with a lock and the sample to do quality assurance will be taken from store at a later time. Depending on the quality check the locks are removed or the whole delivery will be send back. Since quality assurance may be destroying or not the sample itself needs some special treatment.*

As this example shows, some functions of a SCS are not independent of others. The function *'put into store'* consists of different variants corresponding to the *'quality assurance'*. The same holds for the function *'take from store'*. Also primitive parts of a SCS like the function *'find a store location to take goods off'* exist in different versions. In general, the store location will be sought to minimize the movement of vehicles to operate the store. But with chemicals and pharmaceuticals commissioning of only the same production series must be observed and in the case of perishable goods, the expiry date will be the determining characteristic.

These interdependencies of functions of a SCS contribute essentially to the high complexity. Interdependencies and the low market volume hinder the development of a standardized solution which can be customized by adjusting parameters.

Also, this marked segment is not very price sensitive. Cost of a SCS ranges between 300 to 500 thousand DEM while cost of the overall store investment ranges over 10 millions of DEM. Therefore, software houses doing individual software development can easily exist.

3. Knowledge Bases of CBSE: The Reference Model 'store-keeping'

3.1 Basic Structures: Object, Relationships 'part of' and 'variant of'

The knowledge base of CBSE is modeled by a hierarchical conceptual graph (Sowa (1984)), whose nodes consist of *objects* and the edges of the relationships *'part of'* and *'variant of'*. Additionally, the knowledge base consists of *production rules*:

- *Objects* represent organizational units, concepts or processes.
- The *'part of'*-relationship represents aggregation, i.e. the father-node is an aggregation of all its son-nodes. Therefore, with respect to infer-

ence the existence of the father node implies the existence of all the son nodes.

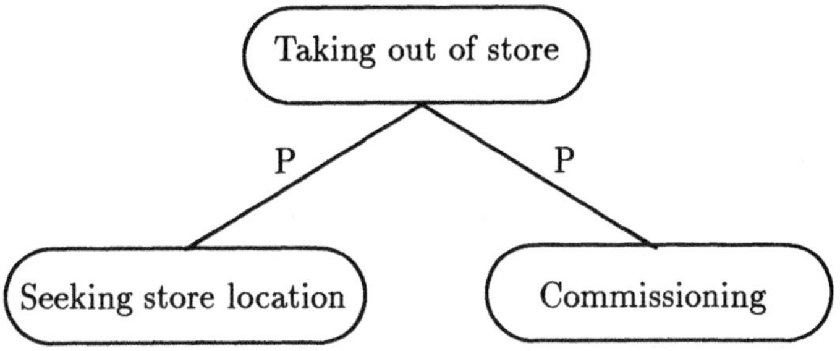

Figure 1: The relationship 'part of'

- The *'variant of'*-relationship represents specialization, i.e. alternative instances of the father node. With respect to inference, the existence of the father node implies the existence of at least one of the son nodes. Structural aspects of the reference model 'store-keeping' are described by *'part of'*- and *'variant of'*- relationships.

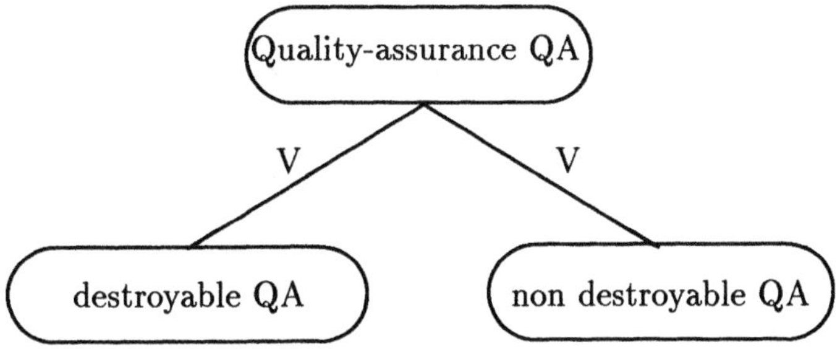

Figure 2: The relationship 'variant of'

- Production rules are used to represent non hierarchical relationships between objects. Example:

  IF *production-enterprise (as 'variant of' enterprise)*

  THEN *store-keeping of finished goods (as 'variant of' store)*

  AND *goods delivery from production (as 'variant of' goods delivery)*

  Production rules express generalized domain-knowledge. If during modeling of a concrete enterprise the objects *'production enterprise'* and *'store keeping of finished goods'* are recorded, then existence of the object (resp. process) *'goods delivery from production'* will be infered. Additionally, rules will be used to assure consistency of structural model with process model.

Expressions of the kind *'if A then B or C'* can be modeled by using the *'variant of'*-relationship taking A as father node and B and C as son-nodes.

3.2 Architecture of CBSE

The basic structures described in the previous paragraph are combined in order to represent knowledge within CBSE. On the top-level (figure 3) CBSE has the parts *'reference model'* and *'program objects'*. The reference model consists of domain specific knowledge needed to represent structural aspects of store organization (*structure*) and process related features to model work-flow (*process model*).

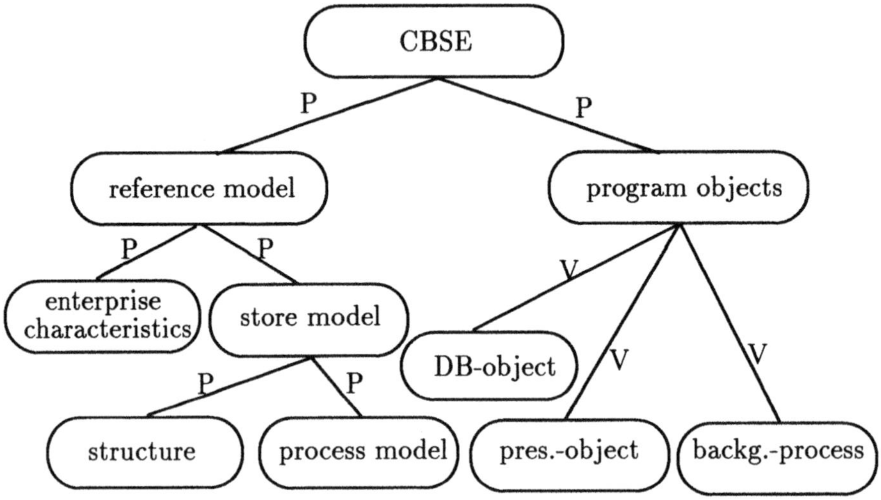

Figure 3: Main structure of CBSE

The submodel *'enterprise characteristics'* allows the differentiation of commercial enterprises and production enterprises as well as different distribution channels and allows formulation of important production rules: ***if*** *'commercial enterprise'* ***then not*** *'store for unfinished goods'* as well as ***if not*** *'store for unfinished goods'* ***then*** *'goods delivery from production'*. Production rules also very often link static aspects of the structural model *'structure'* to procedural ones of the process model: ***if*** *'definite type of material'* ***then*** *'destroyable quality assurance'*.

4. Programming objects

CBSE supports the development of applicational systems in administering and storing the separate programming objects, selecting suited ones based on recorded user needs and checking consistency.

4.1 Types

Just as with programming objects of any object oriented programming language, those of CBSE consist of methods and data. In addition, *'data base objects'* i.e. stored procedures or triggers, as well as the user oriented *'p-resentational objects'* are augmented by a section recording the events that result in a method call. A method of a CBSE-programming object may call a method of an other CBSE-programming object. In order to administer these references, each object has an identification consisting of a customer specific label and the function name. The function name corresponds to an object of the reference model.

CBSE-programming objects are differentiated in:

- *data base objects* to represent context independent methods. For example, if stock of an article in the commissioning store settles down beyond security level, supply is initiated.
- *presentational objects* realize the user interface.
- *background processes* represent the concrete flow of control during execution and realize batch processes.

4.2 Structure of program objects

Any CBSE programming object contains the following attributes and tables:

- *type of object* (as in paragraph 4.1 mentioned)
- *date of creation*
- *customer/project label.* Programming objects of similar functionality differ with respect to the customer/project label. Since this label is part of the object key, some default projects '$\boldsymbol{standard_1}$',..., '$\boldsymbol{standard_n}$' should be created in order to store different versions of a programming object, that is not related to a specific customer respectively project.
- *name of object.* There are two versions of the name of any object. The *extended version* contains an arbitrary string of up to 50 characters and should be chosen as descriptive as possible. The *shortcut name* consists of exactly 8 characters carrying signatures (each two bytes long) for the *function group* the object belongs to, the *function* within this group, the *object type* (as in 4.1) and finally a two byte *sequential number.* The shortcut name is used as primary key to reference the object within the actual project. The combination *(customer/project label, shortcut name)* is an identification throughout CBSE.
- *pseudo-code of object.* This is a *long field* describing the code of the object. Usually, before use in a new project any object needs some adjustments, that must be done by the programmers. Since as production environment for programming a Fourth-Generation Language (4GL) is used the step to move from pseudo-code to real programming code was not considered to be essential.
- *event-action table.* The event-action table relates possible events to methods which are executed in the case the event occurs. Possible

events in the case of *presentational objects* are mouse-clicks, key pressures etc. and in the case of *data base objects* these are trigger-events.

The action to be taken may be a method of the object in question. For that purpose, any method consists of the *method name*, which is identifying (up to 8 characters long) within the object, and a *long field* to define the method (in pseudo code).

Also, the action referenced by the *event-action table* can be a method of a different object. In that case, the reference consists of the *shortcut object name* and the *method name.*

4.3 Connectivity of program objects

The interdependencies of different program objects are represented by two different mechanisms. As described in the more general section above any program object corresponds to a node in the structural graph. Thus, the relationship of a father - son structure may be of type *'variant of'* or of type *'part of'*.

The *'variant-of'*-relationship simply allows reference to program objects of similar functionality. The *'part of'*-relationship of program objects within CBSE is interpreted as the well known *inheritance*-mechanism of object oriented programming, i.e. the son program object may reference any method of the father program object.

Besides this graph structure inside of any program object the *event-action table* allows the call of methods belonging to different program objects. That way arbitrary links between different program objects are realized.

5. Cases in CBSE

A case in CBSE corresponds to a programming object, which are stored as separate units. Because programming objects are identified by *(customer/project label, function name)*, where function name refers to an object of the *reference model*, there is a 1:n relationship between the objects of the reference model and the implementational objects of the *program store* (= *case store*). Indeed, the reference model acts as a thesaurus, describing the concept of any of its objects by the path to the object. Since, in general, there are more than one implementation of an object of the reference model additional key-words are stored for differentiation purpose. The links between key-words and program objects are stored in a separate table.

6. Working with CBSE

In order to develop a customer specific store control system SCS, the peculiarities of the customer must be recorded. To do so, the reference model serves as a checklist, where each entry must be labeled by TRUE or FALSE. In this way, the characteristics of the goods, the store technology and the wanted procedural details are registered. The general reference model is tailored to the *'customer model'*.

The customization process is supported by an inference mechanism which operates on a simple table having as attributes *(possible) facts* and *assertions*. The *(possible) facts* correlate by a 1:1 relationship to each node of the reference model, the *assertions* simple state if any *fact* is *true*, *false* or so far *unknown*. Initially any *fact* has value *unknown*.

During recording of the customer situation any *unknown fact*, i.e. the nodes of the reference model, are marked with *true* or *false*. The inference mechanism uses the dependencies of the reference model and the production rules to simplify the task of answering all unknown facts:

- The above mentioned production rules allow marking the conclusion node of the reference model if the condition is satisfied.
- If an object of the reference model has been marked with *true*, all nodes laying on its access path must also be marked with *true*.
- If the father - son relationship is of type *'variant of'* and the father node has been marked with *true*, at least one of its sons must be *true* too.
- If the father - son relationship is of type *'part of'* and the father node has been marked with *true*, all its sons must be *true* too.
- Analysis of customer situation at times results in an incomplete description. Missing facts cause expensive re-inquiries or even worse the follow-up treatment of delivered software. The reference model of CBSE implies soundness and completeness of recorded facts as long as any node has been marked, i.e. any *fact* has *assertion true* or *false*.

The *'customer model'* consists of the basic functionality of the needed programming objects (interface objects) and their interdependencies. Thus function and structure of the customer application is determined at that time. It remains to select the best suited implementational object for any interface object i.e. in CBSE-terminology the *'programming object'*. To support this task the property list of any of these programming objects will be used. Final selection and adaptation to the specific customer situation is done manually by code inspection.

Compared to the case based reasoning cycle, as described by Aamodt/Plaza (1994), the phase RETAIN needs some explanation. Customer specific peculiarities may result in a functional description of a programming object and subsequent implementation that should be stored as a 'new case' in the *program store* (case base).

In the simple case this new object is a *'variant of'* an existing object. In more complicated ones additional hierarchies must be added to the *'reference model'* resp. the *'program model'*.

7. Status of CBSE

CBSE is a prototype developed in close coordination with an software house specialized on SCS. But the applicational situation of CBSE is not limited to storage control systems. One of the design features was to have an application independent software development tool.

CBSE uses the relational data base management system ORACLE as the case store and the knowledge base. The system itself has been programmed using the development tools SQL*FORMS etc. of ORACLE supplemented by plain C-code. This is a serious draw back. An object oriented database system would certainly ease inheritance and inference. However, at time of the design-phase, a reliable oo-DBMS was not available.

References:

AAMODT, A. and PLAZA, E. (1994): Case-Based Reasoning: Foundational Issues, Methodological Variations, and System Approaches. *AI Communications, Vol. 7, Nr.1, 1994, p.39-59.*

CZAP, H. and GRIMM, C. (1996): CBSE: Ein Fallbasiertes CASE-Tool zur Entwicklung kundenspezifischer Anwendungssysteme am Beispiel von Lagerführungssystemen. *Wirtschaftsinformatik, Heft 1/96, 32-38.*

GROCHLA, E. (1978): Materialwirtschaft. Das materialwirtschaftliche Optimum im Betrieb. Wiesbaden 1978.

JÜNEMANN, R. (1989): Materialflu und Logistik - Systemtechnische Grundlagen mit Praxisbeispielen. Berlin u.a. 1989.

RUPPER, P. (1988): Logistische Systeme. Köln, 1988.

RUPPER, P. and SCHEUCHZER, R. (eds.) (1988): Lager- und Transportlogistik. Planung, Steuerung und Kontrolle im Transport und Lagerbereich. Zürich 1988.

SOWA, J. F. (1984): Conceptual Structures. Information Processing in Mind and Machine. Reading Massachusetts, Menlo Park California u.a. 1984.

Optimization in Probabilistic Logic for Decision Support Systems

F. Dellmann

FB Wirtschaftswissenschaften, Wirtschaftsinformatik und BWL,
Universität GH Essen, D-45117 Essen, Germany

Summary: Probabilistic logic (PL) is a well-suited framework for representing and processing partial information (PI) in Decision Support Systems (DSS) (Dellmann (1995)). This paper discusses the problems arising from conditional probabilities and the applicability of signomial programming algorithms to the nonlinear and nonconvex optimization problems resulting from nonlinear PI. An algorithm is proposed which converges to the global optimum of the resulting nonlinear optimization problem almost surely.

1. Introduction

In economic reality, the probability information the decision maker disposes of is rather partial than exact. It has been shown by Dellmann (1995) that DSS are well-suited to process PI. In this framework, PL is used to represent PI and to support the decision maker by determining bounds of unknown probabilities, which is achieved by solving optimization problems.

Kofler and Menges (1976) define PI as a proper subset of the distribution simplex containing more than one element where the distribution simplex is the set of all possible probability distributions. Following this definition, PI is given if the underlying probability information is not characterized by the knowledge of exact numerical probabilities nor by complete ignorance, i. e. the probability information is equal to the distribution simplex.

DSS are defined by providing "Computerbased support for management decisionmakers who are dealing with semistructured problem" (Keen and Scott Morton (1978), 97). As discussed by Dellmann (1995), uncertainty and especially PI is a characteristic feature of semistructured problems, which are the main feature of this definition of DSS.

This paper discusses the problems in PL arising from conditional probabilities and the applicability of signomial programming algorithms to the nonlinear and nonconvex optimization problems resulting from nonlinear PI. Finally, an algorithm is proposed which converges to the global optimum of the resulting nonlinear optimization problem almost surely.

2. Fundamentals of probabalistic logic

PL can be seen as a generalization of propositional logic, since the truth values in propositional logic are binary while they are real numbers in the interval $[0; 1]$ in PL. These are interpreted as probabilities of propositions.

By assigning binary truth values to the propositions, one gets possible worlds. A possible world is a particular assignment of truth values to the propositions $S_1, \ldots, S_n$. Representing n propositions one gets $N = 2^n$ possible worlds. This exponential growth can be reduced by exact knowledge about the propositions like $S_1 \wedge S_2 = \emptyset$ (Nilsson (1986)).

The assignment of truth values to the propositions yields an $n \times N$-matrix T. The elements t_{jk} of T denote the truth value of state S_j in the possible world W_k. For example, for two propositions the matrix T is given by

$$T = \begin{pmatrix} 0 & 0 & 1 & 1 \\ 0 & 1 & 0 & 1 \end{pmatrix}. \tag{1}$$

A probability vector $\pi = (\pi_1, \ldots, \pi_N)$ is assigned to the possible worlds with

$$\sum_{k=1}^{N} \pi_k = 1, \ \forall_k \, \pi_k \geq 0. \tag{2}$$

Then, the probabilities $p_1, \ldots, p_n$ of the propositions can be determined by

$$p = T \cdot \pi'. \tag{3}$$

3. Exact and linear partial probability information

3.1 Unconditional probabilities

Point probabilities and linear partial information (LPI) can be represented by using equation (3). A partial information is called linear if it can be expressed in terms of linear equations and inequalities (Kofler and Menges (1976)). Examples for LPI are comparative and interval probabilities. Generally, LPI can be represented in PL by M linear equations and inequalities

$$G \cdot \pi' \begin{Bmatrix} = \\ \geq \end{Bmatrix} g \tag{4}$$

where G: $M \times N$, g: $M \times 1$ (Rimscha (1990)).

(2) and (4) determine graphically a convex polyhedron in the N-dimensional space. Any non-redundant, additional probability information cuts off a part of the polyhedron. Inconsistent probability information leads to an empty polyhedron requiring a correction of the probability information.

Because the provided probability information needs not to be complete, probabilities of propositions might be unknown. To support the decision maker, bounds of these unknown probabilities can be computed by solving a linear programming problem given the case of a nonempty polyhedron

(Hailperin (1965)). The upper (lower) probability bound of a proposition S_j can be determined by maximizing (minimizing) p_j subject to the convex polyhedron of the underlying probability information. The linear programming problem

$$p_j = \sum_{k=1}^{N} t_{jk} \cdot \pi_k \rightarrow \begin{cases} \max \\ \min \end{cases} \quad \text{s.t. (2), (4)} \tag{5}$$

will be solved by the well-known simplex algorithm. The determination of the bounds of unknown probabilities is a remarkable support for the decision maker since he gets additional probability information.

3.2 Conditional probabilities

Conditional probabilities can be represented in PL by linear equations and inequalities using (3). For example, $P(S_1|S_2) \geq 0,5$ yields $\pi_4 - 0,5 \cdot (\pi_2 + \pi_4) \geq 0$ in the context of (1). Doing so, one must take into account explicitly that $P(S_2) > 0$, which leads to the strict inequality $\pi_2 + \pi_4 > 0$ in PL. This strict inequality has to be added to the constraint set of the linear programming problem (5) which has not been done in the literature of PL.

Since it is also possible to express LPI in strict inequalities, one has generally a system of M_0 strict inequalities

$$G_0 \cdot \pi' > g_0 \tag{6}$$

where G_0: $M_0 \times N$, g_0: $M_0 \times 1$. This leads to the problem that the simplex algorithm only handles inequalities like (4) but not like (6).

One might transform the strict inequalities into "$\geq$"-inequalities, i. e. use the convex hull of the constraint set, but this procedure leads to erroneous feasibility of contradictory restrictions like e.g. $p_1 < 0,5$ and $p_1 > 0,5$.

Thus it is necessary to check the feasibility of the system of strict inequalities which can be done by the theorem of Carver (1921/22) from linear algebra.

Theorem of Carver: *Let G_0 be a $M_0 \times N$-matrix and let g_0 be a M_0-dimensional vector. The system $G_0 \cdot \pi' > g_0$ is feasible, if and only if $z = 0$ is the only solution for the system*

$$z \cdot G_0 = 0, \quad z \cdot g_0 \geq 0, \quad z \geq 0. \tag{7}$$

A solution for the system (7) can be found by minimizing z using the big-M method or the two-phases method (see Murty (1983), for example).

Unfortunately, an optimal solution can not be reached because of the strict inequalities. But using parametric optimization, it is possible to find an ϵ-solution, i. e. a point in an ϵ-neighbourhood of the optimal solution where $\epsilon \in R^+$ is very small. Therefore, the strict inequalities have to be transformed into "$\geq$"-inequalities by subtracting ϵ from g_0 which yields

$$G_0 \cdot \pi' \geq g_0 - \epsilon. \tag{8}$$

The optimization problem (5) must be solved with the additional constraint (8) which can be done by the simplex algorithm for parametric right hand sides (see Murty (1983), for example). Doing so, the right hand side in the optimal tableau can become $\boldsymbol{g}_0^* - \epsilon > \mathbf{0}$ because ϵ can be as small as you like. Otherwise, the right hand side becomes $-\epsilon$, if $\boldsymbol{g}_0^* = \mathbf{0}$, i. e. the nonparametric optimization problem is degenerated. In the first case, the ϵ-solution is found, and in the second case, a dual pivot step is necessary.

4. Nonlinear partial probability information

4.1 Representation

In addition to the probability information of the last chapter there are some more types of information the decision maker may dispose of, like conditional comparative probabilities or (conditional) independence of propositions. It has been shown by Rimscha (1990), Andersen and Hooker (1994) and Dellmann (1995) that this information can be represented in PL by nonlinear equations and inequalities. In these cases, the decision maker disposes of nonlinear partial information (NLPI). Applying (3) to all these types of information can generally be represented in PL by

$$\sum_{h=1}^{q} \alpha_h \cdot \prod_{k=1}^{N} \pi_k^{\beta_{kh}} \begin{Bmatrix} < \\ \leq \\ = \end{Bmatrix} \gamma \tag{9}$$

where $\boldsymbol{\alpha_h}, \boldsymbol{\beta_{kh}}, \gamma \in \boldsymbol{R}$.

This type of PI can not be represented graphically as a convex polyhedron but it is a nonlinear and nonconvex subset of the distribution simplex. The nonconvexity results from the fact that $\boldsymbol{\alpha_h}, \boldsymbol{\beta_{kh}} \in \boldsymbol{R}$.

To determine the bounds of unknown probabilities one has to solve the nonlinear, nonconvex optimization problem

$$p_j = \sum_{k=1}^{N} t_{jk} \cdot \pi_k \rightarrow \begin{cases} \max \\ \min \end{cases} \quad \text{s.t. (2), (4), (6), (9).} \tag{10}$$

Andersen and Hooker (1994, 208) mention the possibility to solve this problem by a signomial programming (SP) algorithm but they do not check it.

4.2 Signomial Programming

To examine the applicability of SP algorithms to the nonlinear, nonconvex optimization problems yielded by NLPI, the resulting optimization problem (10) is compared with the standard form of SP (Dembo (1978)):

$$\sum_{h=1}^{q} \alpha_h \cdot \prod_{k=1}^{N} \pi_k^{\beta_{kh}} \rightarrow \min \tag{11}$$

$$\text{s.t.} \sum_{h=1}^{q} \alpha_h \cdot \prod_{k=1}^{N} \pi_k^{\beta_{kh}} \leq 1 \tag{12}$$

$$\forall_k \, \pi_k > 0. \tag{13}$$

It is easy to see that the linear objective function of the nonlinear, nonconvex optimization problem (10) is a special case of the objective function (11) of the standard SP problem. The same applies to the linear constraints (2), (4) and (6). The nonlinear constraints (9) have to be divided by γ to get the same right hand side as in (12). The "$<$"-restrictions of (9) can be transformed into "$\leq$"-restrictions by adding positive artificial variables. The "$=$"-restrictions must be splitted into "$\leq$"- and "$\geq$"-restrictions enhancing the optimization problem considerably. Finally, the variables in the SP problem must be positive while they could be zero in the problem (10). Since $\beta_{kh} < 0$ is incompatible with $\pi_k = 0$, it has to be taken into account explicitly that $\pi_k > 0$ for those k of which exponents β_{kh} are negative.

There exist several algorithms for solving SP problems (see Dembo (1978) and Rijckaert and Martens (1978)). But many algorithms need the declaration of positive lower and upper bounds for all variables. Since $\pi_k = 0$ for some k, it is impossible to set such bounds. Additionally, the problems have to be enhanced considerably by introducing slack variables in some algorithms and by splitting the "$=$"-restrictions into two inequalities.

One may conclude that SP algorithms are applicable to the resulting nonlinear, nonconvex optimization problems but there emerge difficulties with the requirement that all variables must be strictly positive. Furthermore, there does not exist any applicable algorithm for determining the global optimum.

4.3 An algorithm for nonlinear, nonconvex optimization problems

Because of these problems applying SP algorithms, an algorithm is suggested which combines the penalty function method of Fiacco and McCormick (1968), the unconstrained optimization algorithm of Hooke and Jeeves (1961) and the multistart method from global optimization (see Rinnooy Kan and Timmer (1987)). This algorithm converges to the global optimum of the resulting nonlinear optimization problem almost surely.

The penalty function method approximates a generally constrained, nonlinear optimization problem by an unconstrained minimization problem adding penalty terms, which enhance the objective function whenever the constraints are violated. Denoting the objective function with $g_0(\pi)$ and the constraints with $g_f(\pi)$, where f is an index variable for r "$\leq$"-constraints, $s - r$ "$=$"-constraints and $t - s$ "$<$"-constraints, one has to solve the unconstrained minimization problem

$$g_0(\pi)+\sum_{f=1}^{r} \rho_f \cdot \phi(g_f(\pi))+ \sum_{f=r+1}^{s} \rho_f \cdot \psi(g_f(\pi))+ \sum_{f=s+1}^{t} \rho_f \cdot \chi(g_f(\pi)) \rightarrow \min. \tag{14}$$

The penalty functions ϕ, ψ and χ are defined such that they take positive values, if their constraint is violated, and they become zero, if not. This unconstrained minimization problem is solved iteratively for growing values of ρ_f until the termination criterion

$$\max_f\{\phi(g_f(\pi)), \psi(g_f(\pi)), \chi(g_f(\pi))\} < \epsilon_1^P \tag{15}$$

comes true, where $\epsilon_1^P \in \boldsymbol{R}^+$ is a given small parameter. In each iteration, the values of ρ_f are increased only for the violated constraints. Obviously, this algorithm converges to the local optimum of the constrained optimization problem from outside the feasible region.

In addition to the termination criteria reported in Fiacco and McCormick (1968), another one is stated to reveal inconsistent probability information:

$$(1 + \max_f\{\phi(g_f(\pi)), \psi(g_f(\pi)), \chi(g_f(\pi))\}) \cdot (m - c) > \epsilon_2^P \tag{16}$$

where $c \in \boldsymbol{N}$ is a constant, $m \in \boldsymbol{N}$ counts the number of iterations and $\epsilon_2^P \gg 1$. In the case of contradictory restrictions, the algorithm determines the point which has the smallest distance to the restrictions. The maximal constraint violation has been added with one, because this distance could be smaller than one. The resulting term is multiplicated with $m - c$ instead of m to avoid that the criterion comes true erroneously in an early iteration.

If no assumptions shall be made regarding continuity or differentiability of the objective and the penalty functions, an unconstrained optimization algorithm is needed, which works without such assumptions, like the algorithm of Hooke and Jeeves (1961).

The Hooke/Jeeves algorithm searches for improving the objective function value alternately along parallel lines to the coordinate axes and along a line of which direction is given by the difference of the solution vectors of the last two iterations. In later iterations, the searching step size is reduced to search in smaller regions around the local optimum. The algorithm proceeds until the step size is less than a given small $\epsilon^{HJ} \in \boldsymbol{R}^+$.

Because the optimization problem (10) is nonconvex, a detected local optimum need not be global. In global optimization one differentiates between deterministic and stochastic methods. Deterministic methods find the global optimum warrantedly but demand specific features of the optimization problem which are not fulfilled by (10). Under mild conditions, stochastic methods, which are applicable to all optimization problems, converge to a point in the ϵ-neighbourhood of the global optimum with probability 1 as the sample size increases (Solis and Wets (1981)).

Following the study of Dixon and Jha (1993), the multistart method is a relatively good stochastic method. It intends to find all local optima by determining iteratively random points to start a local optimization algorithm until the solution meets the Bayesian stopping rules

$$\frac{\boldsymbol{w} \cdot (\boldsymbol{l}-1)}{\boldsymbol{l}-\boldsymbol{w}-2} < \boldsymbol{w}+0.5 \quad \wedge \quad 1-\frac{\boldsymbol{w} \cdot (\boldsymbol{w}+1)}{\boldsymbol{l} \cdot (\boldsymbol{l}-1)} \geq 0.95 \tag{17}$$

for $\boldsymbol{l} > \boldsymbol{w}+2$. Herein, $\boldsymbol{l}$ is the number of iterations and $\boldsymbol{w}$ is the number of the different local optima found. These stopping rules are derived from the assumption that the number of successes in finding a certain of the different local optima follows a multinomial distribution. The first rule estimates the posterior expectation of the total number of local optima. The second rule estimates the posterior expected relative size of the observed region of attraction of the local optima found to the feasible region. The observed region of attraction is the set of the points from which a local optimization algorithm converges to a local optimum.

The multistart method is now combined with the penalty function method and the Hooke/Jeeves algorithm to obtain an algorithm which converges to the global optimum of the resulting nonlinear optimization problem almost surely. This algorithm consists of the following steps:

1. The constrained optimization problem (10) is transformed into an unconstrained minimization problem (14) using penalty functions.
2. After initializing the parameters, a point $\boldsymbol{\pi}^{l,m}$ is chosen at random.
3. $\boldsymbol{\pi}^{l,m}$ is used as starting point for the Hooke/Jeeves algorithm to solve the unconstrained problem (14), which yields the solution point $\boldsymbol{\pi}^{l,m+1}$.
4. If $\boldsymbol{\pi}^{l,m+1}$ meets the criterion (15), $\boldsymbol{\pi}^{l,m+1}$ is a local optimum. If it is a new one, the multistart parameter $\boldsymbol{w}$ is incremented, and the algorithm continues with step 5. If the criterion (16) comes true, the optimization problem (10) has no feasible region indicating inconsistency of the probability information, and the algorithm stops. If none of these criteria comes true, the penalty parameters $\boldsymbol{\rho_f}$ are increased for the violated restrictions, the penalty iteration parameter $\boldsymbol{m}$ is replaced by $\boldsymbol{m}+1$, and the algorithm continues with step 3.
5. If $\boldsymbol{\pi}^{l,m+1}$ meets the Bayesian stopping rules (17), the global optimum is found, and the algorithm stops. Otherwise, the multistart iteration parameter $\boldsymbol{l}$ is incremented, the penalty parameters $\boldsymbol{\rho_f}$ and $\boldsymbol{m}$ are set to 1 and the algorithm continues with step 3 using a new random probability point.

This algorithm avoids the problems arising from the application of SP. It was implemented in a DSS for investment decisions and it has been proven reliable in finding the global optimum. With a suitable setting of the parameters, the running time of the algorithm is satisfactory.

5. Conclusion and suggestions for future research

The paper has shown the problems arising from the representation of conditional probabilities and NLPI in PL. The conditional probabilities yield strict inequalities which can be handled in linear optimization problems with the theorem of Carver and parametric optimization.

NLPI leads to nonlinear, nonconvex optimization problems. SP algorithms can be applied with the reservations as discussed. An algorithm, which converges to the global optimum of the resulting nonlinear optimization problem almost surely, is suggested.

Future research may be a comparison regarding efficiency and reliability of different possible algorithms of nonlinear and stochastic global optimization.

References:

ANDERSEN, K. A. and HOOKER, J. N. (1994): Bayesian Logic. *Decision Support Systems, 11, 191–210.*

CARVER, W. B. (1921/22): Systems of Linear Inequalities. *Annals of Mathematics, 23, 212–220.*

DELLMANN, F. (1995): Processing Partial Information in Decision Support Systems. In: W. Gaul and D. Pfeifer (Eds.): *From Data to Knowledge: Theoretical and Practical Aspects of Classification, Data Analysis and Knowledge Organization.* Springer, Berlin, 423–432.

DEMBO, R. S. (1978): Current State of the Art of Algorithms and Computer Software for Geometric Programming. *Journal of Optimization Theory and Applications, 26, 149–183.*

DIXON, L. C. W. and JHA, M. (1993): Parallel Algorithms for Global Optimization. *Journal of Optimization Theory and Applications, 79, 385–395.*

FIACCO, A. V. and McCORMICK, G. P. (1968): Nonlinear Programming: Sequential Unconstrained Minimization Techniques. Wiley, New York.

HAILPERIN, T. (1965): Best Possible Inequalities for the Probability of a Logical Function of Events. *The American Mathematical Monthly, 72, 343–359.*

HOOKE, R. and JEEVES, T. A. (1961): "Direct Search" Solution of Numerical and Statistical Problems. *Journal of the ACM, 8, 212–229.*

KEEN, P. G. W. and SCOTT MORTON, M. S. (1978): Decision Support Systems. Addison–Wesley, Reading.

KOFLER, E. and MENGES, G. (1976): Entscheidungen bei unvollständiger Information. Springer, Berlin.

MURTY, K. G. (1983): Linear Programming. Wiley, New York.

NILSSON, N. J. (1986): Probabilistic Logic. *Artificial Intelligence, 28, 71–87.*

RIJCKAERT, M. J. and MARTENS, X. M. (1978): Comparison of Generalized Geometric Programming Algorithms. *Journal of Optimization Theory and Algorithms, 26, 205–242.*

RIMSCHA, M. v. (1990): The Determination of Comparative and Lower Probability. In: Proceedings of the Workshop "Uncertainty in Knowledge-Based Systems", FAW-report 90025, 2, 344–376.

RINNOOY KAN, A. H. G. and TIMMER, G. T. (1987): Stochastic Global Optimization Methods. *Mathematical Programming, 39, 27–78.*

SOLIS, F. J. and WETS, R. J.-B. (1981): Minimization by Random Search Techniques. *Mathematics of Operations Research, 6, 19–30.*

Learning Strategies for Managing New and Innovative Products

A. Geyer-Schulz

Abteilung für Angewandte Informatik insbesondere Betriebsinformatik, Wirtschaftsuniversität Wien, A-1090 Vienna, Austria

Summary: Markets for new and innovative products are often characterized by a boom and bust dynamic. In such situations feedback effects as for example word of mouth, marketing, and learning curve effects can lead to rapid growth followed by overcapacity, price war, and bankruptcy. In the following we investigate the possibility of learning strategies for managing new and innovative products in such a setting with genetic programming methods. For this purpose we developed a simulator with a market sector based on a diffusion model, a firm, and its competition. The structure of the model is similar to the model in Paich and Sterman (1993). In the experiment strategies for managing new and innovative products are bred by applying genetic programming to the simulation. The genetic programming method is based on a simple genetic algorithm over context-free languages as presented in Geyer-Schulz (1995). The result of the experiment is compared with the findings of Paich and Sterman (1993).

1. Motivation

Markets for new and innovative products are often characterized by a boom and bust dynamic. Sales of a new product often boom, driven by word of mouth, advertising and – later in the life cycle – by price reductions. But eventually the market saturates and sales fall to the level of replacement purchases. During the transition to replacement demand producers often suffer grievous losses due to excess capacity and falling prices. A typical example is Atari, once leader in video games. The pervasiveness of boom and bust is well documented (e.g. snow mobiles, hand calculators, tennis courts, integrated circuits, consumer electronics, VCRs, PCs). For additional examples see Paich and Sterman (1993). The boom and bust dynamic is characteristic for dynamic decision making systems. The evolution of such systems strongly depends on the behavior of decision makers. Unfortunately, recent studies e.g. of Sterman (1989) have shown that the performance of human decision makers is poor when compared with normative standards or even with simple heuristics. Moreover, learning of human decision makers from outcome feedback with repeated experiments is slow and does not improve the situation considerably. See e.g. Paich and Sterman (1993).

In this paper we investigate the possibility of learning superior decision rules in nonlinear dynamic systems by applying a simple genetic algorithm over context-free languages (a generalization of *genetic programming*) to a simulation of a market for new and innovative products.

2. The Boom and Bust Game

For the task of managing a new product from launch to maturity we have developed a simulator based on a reconstruction of the model of Paich and Sterman (1993) which represents a market with a firm and its competitor. Subjects make price, capacity and marketing expenditure decisions each quarter for a ten year period.

2.1 The Market Model

The market model is based on well-known diffusion models in the tradition of Bass (1969). The basic idea of these models is the feedback structure shown in figure 1 which models the way in which potential customers become aware of the product and reach a purchase decision. Product adoption increases the customer base and generates a (positive) word of mouth effect and depletes the potential customers (a negative effect). The customer base typically follows an s-shaped pattern. Sales first rise exponentially, then peak and decline to the rate of repurchases as soon as the market saturates.

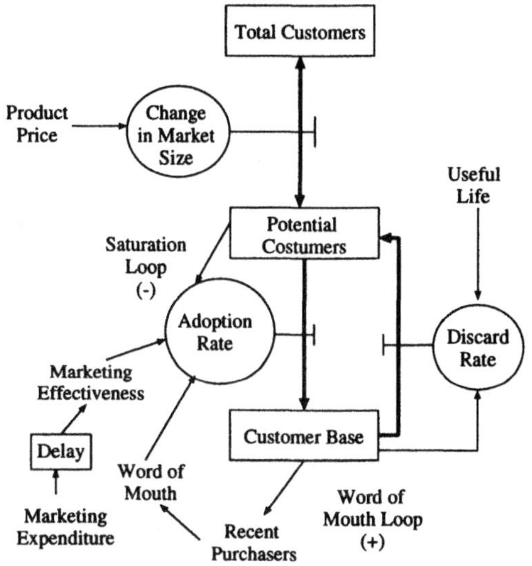

Figure 1: Structure of the Market Sector, Paich and Sterman (1993)

In addition to the diffusion model the following assumptions are modeled (see Paich and Sterman (1993)):

1. The number of potential customers depends on the product price. The price elasticity of demand is less than unity (**0.9**).
2. Industry marketing expenditures increase the fraction of potential customers who buy with diminishing returns.
3. Demand is also generated by word of mouth. The word of mouth effect is modeled as the average number of purchases generated per quarter by each recent purchaser.

4. A fraction of the customer base replaces worn or obsolete products each quarter. Repurchases start after the *useful life* of the product is exhausted (see Paich and Sterman (1993)).
5. The market share of the firm and its competitor is proportional to the attractiveness of the product. Attractiveness depends on price, delivery delay and effective marketing expenditure and is modeled with a standard multinomial logit model (see Lilien et al., (1992), p. 670).

2.2 The Model of the Firm

The supply side of the market, the firms, are explicitly modeled. The internal structure of the firms is based on the following key assumptions from Paich and Sterman (1993):

1. Product is built to order. The firm has **no inventory**. Inadequate capacity leads to order backlogs and delivery delays. A rise in delivery delays reduces the attractiveness of the firm's product and market share.
2. Target capacity is set each quarter. Adjustments of current capacity to target capacity take **4** quarters for capacity increases and **5** quarters for capacity decreases.
3. A standard organizational learning curve is assumed. This means that unit variable cost is reduced by **20** percent at each doubling of cumulative production.
4. Profit is revenue minus costs. Costs include fixed costs, variable costs, marketing expenditures and investment and decommissioning costs. Revenue is orders delivered in the quarter times prices in effect at ordering.
5. Fixed costs are proportional to current capacity. Fixed cost of a unit of capacity is constant. Variable costs are proportional to production, unit variable costs fall as production increases according to the organizational learning curve.
6. Investment costs include all costs resulting from an increase in capacity (e.g. planning, administrative, installation, training), decommissioning costs include all costs resulting from a decrease in capacity. Both are proportional to the size of capacity adjustment.
7. The firm has unlimited credit. The manager may loose as much money as he likes without bankruptcy.

2.3 The Competitor's Strategy

The manager's firm faces competition from another, artificial firm which has launched a similar product at the same time. Fairness of competition is granted, both companies start with identical structure and identical parameters. However, the artificial firm bases its price, capacity and marketig expenditure decisions on the following extremely simple heuristics which are in the spirit of the theory of bounded rationality (Simon (1969)):

1. The capacity decision rule aims at setting target capacity $\boldsymbol{t_t}$ to meet exepected orders and maintain normal capacity utilization. Extrapolation of expected growth rate of orders $(\boldsymbol{O})$ is tacitly assumed, the recent growth

rate is projected α periods ahead, with $\alpha = 2.88$ (see Paich and Sterman (1993)):

$$t_t = O_t \left(\frac{O_t - O_{t-1}}{O_{t-1}} \right)^{\alpha}$$

2. Price decisions are based on a simple markup rule. The price of a unit is set to unit costs at a reasonable employment level (**60** percent in the experiment) plus a fixed margin (**25** percent) to cover marketing and investment costs as well as a normal return when capacity is well utilized.
3. Marketing expenditures are held at a fixed percentage of revenue (**5** percent).

3. Genetic Programming

In this section we present simple genetic algorithms over context-free languages, a variant of the genetic programming approach of Koza (1992) with explicit representation of the grammar of the language (see figure 2). For an impressive collection of genetic programming applications we refer the reader to Koza (1992) and (1994). For a more theoretical treatment of simple genetic algorithms over context-free languages used in this paper see Geyer-Schulz (1995). Genetic algorithms simulate natural evolution processes in order to solve problems by breeding solutions. They are programs for an artificial evolution process. As natural evolution acts on populations of individuals, so do simple genetic algorithms over context-free languages with sets of derivation trees.

```
S := <tree> ;
<tree> := <priceexpr> <capacityexpr> <meexpr> ;
<priceexpr> := <const> | <pricevar> |
            '(' <priceexpr> <op> <priceexpr> ')' ;
<capacityexpr> := <const> | <capvar> |
          '(' <capacityexpr> <op> <capacityexpr> ')' ;
<meexpr> := <const> | <mevar> |
          '(' <meexpr> <op> <meexpr> ')' ;
<const> := '1'|'2'|'3'|'4'|'5'|'6'|'7'|'8'|'9'|'0' ;
<pricevar> := 'REV' | 'VC' | 'FC' | 'UVC' | 'PROFIT' ;
<pricevar> := 'CAP' | 'SHARE1' | 'P1' | 'SHARE2' | 'P2' ;
<capvar> := 'TBACK' | 'CAP' | 'SALES1' | 'SALES2' | 'P_CAP' ;
<mevar> := 'SALES' | 'SHARE2' ;
<mevar> := 'PRICE' | 'UVC' | 'DELAY' | 'MC' | 'SHARE1' ;
<op> := '+' | '-' | 'x' | '÷' | '*' ;
```

Figure 2: The Grammar BNF30

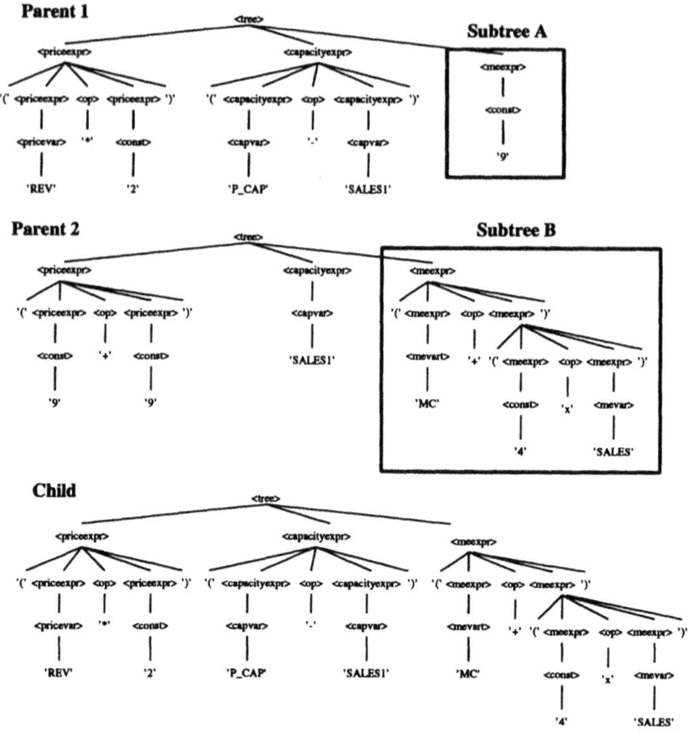

Figure 3: Crossover: Feasible Subtree Substitution of Subtree B by Subtree A

In a simple genetic algorithm derivation trees have a similar role like chromosomes. They encode a set of solutions (parameters, actions, rule, programs) which are evaluated by a fitness function. A simple genetic algorithm over context-free languages generates a new population of derivation trees with three simple genetic operators on derivation trees, namely reproduction, crossover, and mutation which taken together act as a probabilistic state transition function.

For example, in figure 3 we show how the crossover operation works on derivation trees. First, we select a subtree at random from parent 1. In the second parent we select the subtree which is going to be replaced with the restriction that feasible subtree substitution requires both subtrees to have the same root. Finally, subtree A of parent 1 replaces subtree B in parent 2 and we obtain the child shown in figure 3. Mutation works in a similar way. In the first step the subtree which should be replaced is chosen by random choice, a random subtree starting with the same root is generated and inserted into the derivation tree.

The key insight which guarantees that crossover and mutation work for all context-free languages is that, although no string manipulation operation which is closed over words of a context-free language seems to exist, a tree manipulation operation, namely feasible subtree substitution, which is closed over complete derivation trees exists. See Geyer-Schulz (1995), p. 248. Feasible subtree substitution means that a complete derivation tree remains complete if we substitute a subtree only by a subtree with the same root. The basic algorithms needed are an algorithm for decoding a complete derivation tree, an algorithm for generating a random complete derivation tree, an algorithm for extracting the n-th subtree of a complete derivation tree, and an algorithm for replacing the n-th subtree of a complete derivation tree with a complete derivation tree. See Geyer-Schulz (1995).

The Price Rule:
(((P2×REV)+P2)×((FC−8)÷(2×3)))
The Capacity Rule:
(SALES1⋆((((0⋆(TBACK÷(SALES1−P_CAP)))× P_CAP)−(1+((P_CAP⋆((((9×SALES1)÷SALES1) ÷SALES1)+TBACK))−SALES1)))⋆(SALES1−8)))
The Marketing Expenditure Rule:
MC
Cumulated Loss: −880

Figure 4: A Heuristic bred with BNF30

In our example, a derivation tree encodes the decision rules for making price, capacity and marketing expenditure decisions. The grammar of the language for these decision rules is shown in figure 2. In this example, decision rules are arithmetic expressions. Note, that the set of variables which can be used in each decision rule is kept as small as possible. For convenience, no general notation for lags is available in the language. Decisions at time t are always based on the state of variables at time $t - 1$. However, for a few variables additional states can be accessed. For example, for the variable *price*, P1 denotes the price at $t - 1$, the price at $t - 2$. An example of decision rules generated with this grammar is shown in figure 4. We compute the fitness value of the decision rules by decoding the derivation tree and running the simulation of the game with the decision rules. The fitness of the decision rules is the cumulative profit in a game.

4. Results

In figure 5 we show the best heuristic found in a small genetic programming experiment. Surprisingly, the heuristic is very simple. The price rule implies that the price is held constant throughout the simulation. The capacity rule indicates that target capacity is set to orders at $t - 2$. The rule for marketing expenditure suggests that the fraction of marketing expenditure from revenue should be set to the market share in the previous period. This heuristic achieves a profit of **16658.38** compared to a profit of **8821.12** of its competitor.

The Price Rule:	**P1**
The Capacity Rule:	**SALES2**
The Marketing Expenditure Rule:	**SHARE1**
Cumulated Profit (Genetic Programming):	**16658.38**
Cumulated Profit (Competitor):	**8821.12**

Figure 5: The Best Heuristic bred with BNF30

The comparison of the strategies shown in figure 6 shows that the profit reaped by the genetic programming heuristic can be attributed to the high marketing spending in the growth phase of the market and to a reasonable capacity adjustment.

5. Conclusion

It would be premature to assert any generalizations based on the computational experiment reported in this paper. Yet the experimental result shows that a simple and yet efficient heuristic could be bred by genetic programming. In the experiments of Paich and Sterman (1993) subjects performed very poorly. **87** percent of them were outperformed by a naive heuristic. It is encouraging that the heuristic bred by genetic programming could beat its competitor and still be understood by human decision makers in the simplified experimental setting of Paich and Sterman (1993).

References:

BASS, F. M. (1969): A new product growth model for consumer durables. *Management Science, 15, 215-227.*

GEYER-SCHULZ, A. (1995): Fuzzy Rule-Based Expert Systems and Genetic Machine Learning. Physica-Verlag, Heidelberg.

LILIEN, G. L.; KOTLER, P. and MOORTHY S. K. (1992): Marketing Models. Prentice Hall, Englewood Cliffs, New Jersey.

KOZA, J. R. (1992): Genetic Programming: On the Programming of Computers by Means of Natural Selection. MIT Press, Cambridge.

KOZA, J. R. (1994): Genetic Programming II: Automatic Discovery of Reusable Programs. The MIT Press, Cambridge.

PAICH, M. and STERMAN, J. D. (1993): Boom, bust, and failures to learn in experimental markets. *Management Science, 39, 1439-1458.*

SIMON, H. A. (1969): The Sciences of the Artificial. MIT Press, Cambridge.

STERMAN, J. D. (1989): Modeling managerial behavior: Misperceptions of feedback in a dynamic decision making experiment. *Management Science, 35, 321-339.*

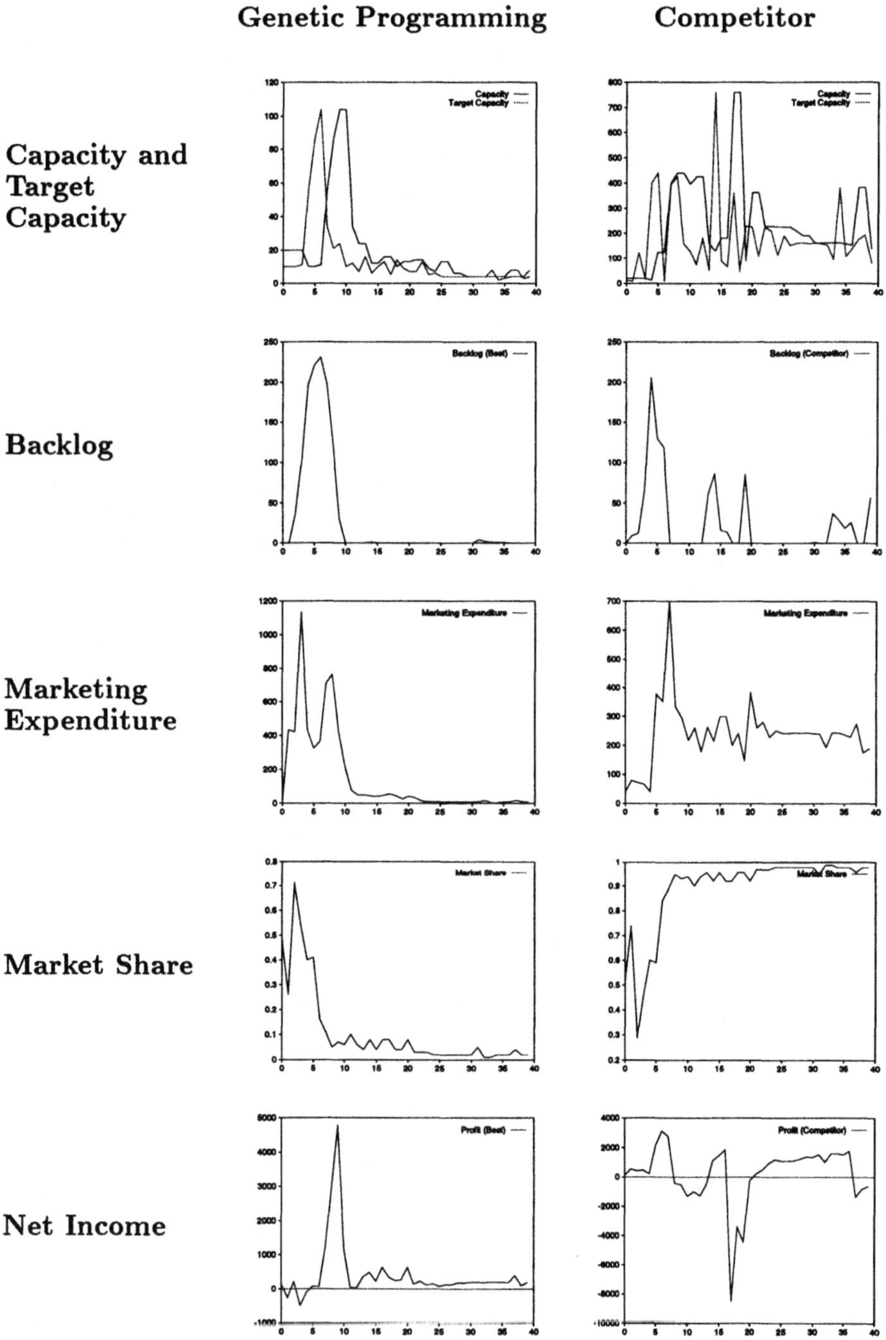

Figure 6: Behavior of the Competitors

Interoperable Database Systems

J. Hunstock[1], O. Rauh[2], E. Stickel[1]

[1] Universität Frankfurt/O., Postfach 776, D-15207 Frankfurt/O., Germany
[2] Fachhochschule Heilbronn, Daimlerstraße 35, D-74653 Künzelsau, Germany

Summary: Global access to databases is a key success issue for most of today's companies. Empirical evidence shows that a lot of data integration projects failed or did not provide the expected return. 'Technical' reasons are structural and semantic heterogeneity, as well as the complexity of the integration task, another major problem is that costs and benefits of data integration have never been seriously considered. A tool that addresses these problems and that supports all stages of the data integration process is introduced in this paper.

1. Introduction

Interoperability of information systems is an important issue. Data integration projects are carried out to gain global access to databases which is one aspect of interoperability. Such projects have to face various difficult problems, such as semantic and structural heterogeneity, very high complexity of the integration task and a lack of methods for cost/benefit analysis. Hence, it is not surprising that a lot of such projects fail as empirical research shows Goodhue et al. (1992).

It is necessary to develop a methodology for data integration that is driven by cost/benefit considerations. Before an integration project is launched it has to be decided on order and degree of the necessary integration. If more than two databases need to be integrated the order of integration is important. Also, it is not always necessary to totally integrate databases. In fact, it may be sufficient to allow global access to only parts of a database.

In this paper the focus lies on the early stages of generating interoperability. A methodology for cost/benefit analysis of data integration and a modified approach to view integration is presented. The approach sketched is part of a larger project that is sponsored by the German Science Foundation (DFG).

2. Basic facts and assumptions

There is a large number of literature on the technical aspects of database integration (see e.g. Sheth and Larson (1990), Chatterjee and Segev (1991)). Some approaches use view integration techniques. It is assumed that conceptual schemata (e.g. ER-diagrams) exist. These schemata are integrated into a global schema. After that it may be decided whether physical migration to a new database system is carried out or whether a federated database approach is chosen. The latter one is of particular interest for cooperations between different firms. A summary of various strategies and techniques for view integration is contained in Batini et al. (1986). More recent work on this subject may be found in Gotthard et al. (1992) and Reddy et al. (1994). These approaches use an object-oriented global conceptual data model.

Other approaches perform the actual integration during query time (Merz and King (1994)). These approaches seem to have cost advantages. Tedious determination of a global schema is no longer necessary. On the other hand, the efficiency of such techniques is significantly lower since the same heterogeneity issues may have to be resolved over and over again (whenever the same or similar queries are used). This is one of the reasons why coordination approaches have been suggested. Coordination approaches try to match objects of different databases by establishing suitable mappings. Usually, databases (as well as schemata) are not altered (see Zhao (1994)).

Our approach relies on the well-known ER-methodology. Local schemata are not altered. Hence local systems retain their autonomy. Cost/benefit analysis helps to determine order and degree of necessary data integration. To each local schema an export schema is constructed. The export schemata are then integrated into one or more global federated schemata. Global queries of the federated schema are mapped into local queries.

Throughout this paper we assume the existence of local ER-schemata. We use only the most basic constructs as presented in Chen (1976) together with specialization/generalization hierarchies. The graphical representation of cardinalities uses the 'crowfoot notation' as introduced in Martin (1990).

We use a binary view integration technique (see e.g. Batini et al. (1986), p. 345). The basic innovation of our view integration approach lies in the use of a clustering step. Clustering allows to simplify schemata and to reduce the complexity of the integration task at least at the beginning. To resolve naming conflicts a data dictionary approach is applied. At the beginning of a project the dictionary contains basic information about possible synonyms and homonyms. More domain specific information is added during the time of integration. The presented approach is part of a larger project. The prototype constructed consists of four modules (cf. Fig. 1). The pre-integration module addresses economic aspects and reduces the complexity of the integration task by means of schema clustering. If no conceptual schemata

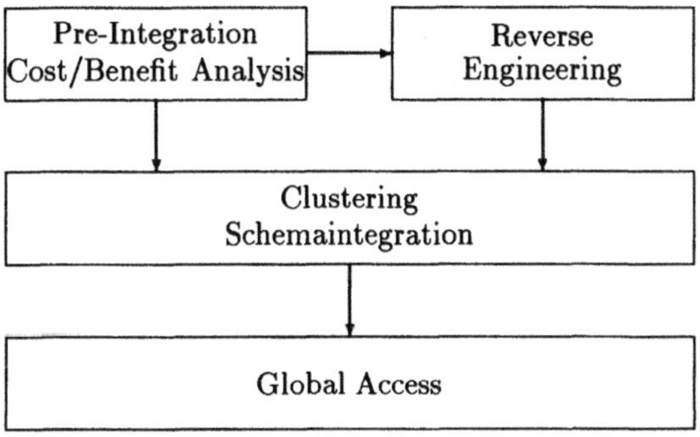

Figure 1: Project modules

are available they need to be generated from the existing databases. This is done with the help of the reengineering module. The view integration module generates global conceptual schemata. Finally the global access module allows global querying of the databases under consideration. Global queries are mapped into local ones. Also, instance integration is performed within this module. Refer to Lim and Prabhakar (1993) for details and for a problem description.

We will only concentrate on economic aspects of data integration (pre-integration module) and view integration. Hence, we do not address reengineering issues and issues related to global queries.

3. Data sharing economics

A decision support module (DSSDI = Decision Support System for Data Integration) is part of our integration tool. The purpose of this module is to aid the analyst in deciding whether and/or to which degree data sharing is necessary for the firm. The basic architecture of this system consists of three sub modules. Refer to Fig. 2 for a survey. Even in highly decentralized orga-

Figure 2: Architecture of DSSDI

nizations there is at least some necessity for data sharing. It is well-known from principal-agent-theory that managers of decentralized units may pursue goals that need not be beneficial or optimal for the firm as whole (Ross (1973)). To overcome this difficulty there is some need for coordination and close monitoring. This, however, calls for global access to at least some of the data of the decentralized unit. Benefits, as well as costs of data integration have been classified in Goodhue et al. (1992). If integration is carried out we may distinguish between costs for standardization, costs due to loss of flexibility of the local units, transformation costs resulting from necessary changes to local systems or from physical data migration if a new system is used in the future, reduced costs to access data and an improved decision quality. To summarize, there is some tradeoff between costs associated with the loss of local autonomy and benefits related to improved decision making. Of course, it is quite difficult to quantify costs and benefits. On the other hand, a lot of integration projects have failed or have not been beneficial (Goodhue et al. (1992)). Hence, there is some need for at least rough estimates. Within DSSDI we analyze these effects on an internal and external level.

It is first tried to determine the effects of data integration on internal business processes of the firm (e.g. efficiency of production). The next step analyzes possible market effects. The results are combined to form a profile of chances and risks. If this evaluation suggests that benefits are higher than costs it is determined in which order the integration is carried out (cf. Stickel et al. (1994)). It is also investigated if partial integration is sufficient.

4. Schema integration

4.1 Complexity reduction

We now turn to the actual schema integration task. Generally large ER-schemata are hard to analyze. The analyst is exposed to the full complexity of two schemata at once. This is why we try to first simplify schemata before they are actually integrated. We use a schema clustering algorithm that was presented in Rauh and Stickel (1992). Suppose we wish to integrate the schema contained in Fig. 3 with an other. This schema is one of 17 conceptual schemata of a large financial services group in Germany. It corresponds

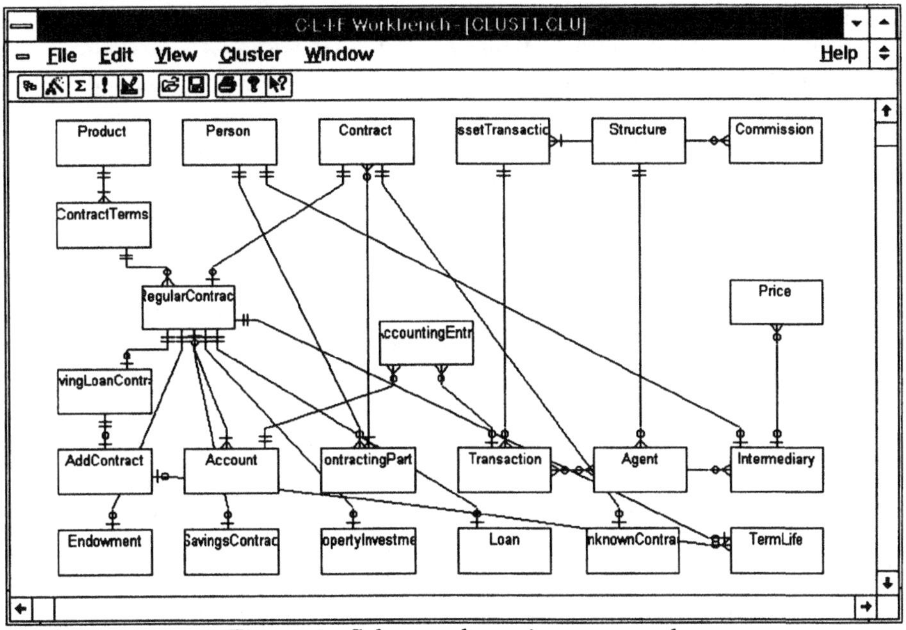

Figure 3: Schema clustering - example

to an IMS database. The clustering algorithm simplifies this schema such that in the first step we would only have to integrate a less complex schema (Fig. 4). Having done this the last clustering step is reversed. This adds more schema objects which may now be integrated. This process continues until all clustering steps have been reversed.

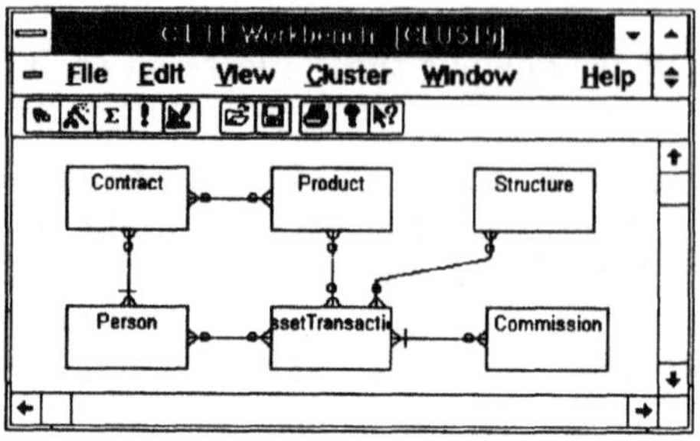

Figure 4: Schema clustering - final result

4.2 The process of view integration

The approach presented draws on the work of Gotthard et al. (1992) and Navathe et al. (1986). We also use the work of Spaccapietra and Parent (1994) in solving structural conflicts. Two conceptual schemata are integrated as follows: (1) Determination of order and degree. (2) If suitable and/or possible already known identical schema constructs may be identified, the same holds for already known homonyms and synonyms, as well as for structural conflicts. (3) Clustering of schemata, after some clustering steps the second step may be redone. (4) Analysis of schemata. Predicates are used to derive assertions about the similarity of schema constructs. (5) Integration is carried out, for detection of homonyms and synonyms a dictionary may be used. At the beginning standard terms are defined in this dictionary. During the integration process more knowledge is added. (6) Reversal of clustering process and return to step 5 unless clustering is already completely reversed. (7) Integration of relationship sets. (8) Restructuring of integrated schema, quality assurance.

These steps will now be commented on in detail by starting with the fourth one. Comparison of schemata is done to detect similarities, as well as differences between their objects. We use all information that is typically available in ER diagrams or ER description languages. Of course, this is not enough to completely automate the integration process and some amount of user interaction is necessary. This is supported by our prototype tool.

First we use the names of the entity sets in the two schemata. An assertion *ass_name* is derived. A specific weight between **0** and **1** is assigned to this assertion to take care of possible homonyms and synonyms. If two entity sets have the same name a weight of **1** is assigned unless there is some indication about possible homonyms in the dictionary. If this is the case a weight of **0.5** is assigned. If two entity sets with different names that are known to be candidates for synonyms are compared, a weight of **0.75** is assigned to that name assertion. In the next step we compare attributes of entity sets. The result is an assertion *ass_structure*. To determine the corresponding weight the ratio of the number of (weighted) common attributes (weighted because homonym and synonym information is again considered) and the square root of the product of the number of attributes of each of the two entity sets is computed. Experiments with different approaches have been carried out but no significantly different results have been found to evolve. The tool, however, supports the use of individually defined weights.

The two assertions already discussed are now combined to form an assertion *ass_similar* which is basic for subsequent steps. Within subsequent steps already derived assertions may be altered. If we have an assertion of type *ass_similar* for two entity sets we next consider the relationship sets they are engaged in. We try to estimate the number of similar relationship sets in the two schemata. Two relationship sets are considered to be similar if we have an assertion *ass_similar* for all participating entity sets and if the cardinalities are the same. If role names are available we may use this information too. An assertion *ass_relation* is derived. Its weight is computed in a similar way as the weight for the assertion *ass_structure*.

In general, more information is not available. The assertions *ass_similar* and *ass_relation* are now combined to derive an assertion *ass_overlap*. This assertion implies that two entity sets may represent the same real world object type. It is assumed that the domains are overlapping.

To derive this assertion one more condition is verified. It is checked if one of the two entity sets is part of a specialization or generalization hierarchy. If this holds, the 'inherited' relationship sets of this entity set need to be contained in the relationship sets of the other entity set (possibly including its inherited ones). If this does not hold all assertions referencing these pairs of entity sets are deleted. This condition implies that an entity set needs to have the same relationship sets as potential super classes of its partner.

The assertion *ass_overlap* is still quite a weak one. Using information on key attributes and domains of attributes (also constraints that further reduce these domains) it may be possible to derive assertions *ass_subset* or *ass_disjoint*. Furthermore, we derive an assertion *ass_is-a*. This assertion deals with an already introduced 'artificial' super class and an entity set that may be a sub class of it (in that case relationship sets and key attributes permit this generalization/specialization hierarchy). This terminates the process of deriving assertions.

It is possible, that more than one assertion for a pair of objects have been derived. Moreover, there may be contradictory assertions. To overcome this difficulty integration lists are associated with each entity set of both local schemata. For each of the lists the assertion with maximal weight is put on top of the list. This process continues until all assertions have been ranked. The integration lists then contain ranked assertions such that for each entity set the top ranked assertion has the highest weight and does not contradict equally ranked assertions about other entity sets of the schema. 'Top ranked' refers to a ranking that is as high as possible. Assertions of already ranked entity sets on the same level may not be contradicted, however (otherwise, a lower rank has to be assigned). After that step thc actual integration of entity sets takes place. With each assertion there is associated a set of transformation operators to form the global integrated schema. If an assertion *ass_overlap(A.X,B.Y)* is found in the first place of both integration lists for entity sets A.X of schema X and B.Y of schema Y, respectively, A and B are integrated by forming a common 'artificial' super class. Note, that in this step we do not really merge the entity sets since the domains are not identical but only overlapping. Later on the user may do the merging. Note, however, that he needs to define suitable mappings of the primary keys then. If an assertion *ass_is-a* is found instead, we introduce an is-a-relationship to the already existing 'artificial' super class. If the key attributes of sub class and super class do not coincide the user has to provide a mapping of key values. Similar transformations are defined for the other assertions.

If integration was carried out for two entity sets all assertions containing the name(s) of these entity sets are deleted. If there was no integration carried out (e.g. because of user interaction) only this particular assertion is deleted. This process continues until all integration lists are empty. The clustering strategy applied requires some special bookkeeping since entity sets that should be integrated may not be on the same clustering level.

Finally the relationship sets are integrated. Relationship sets are only integrated if the cardinalities coincide and if both partner entity sets have

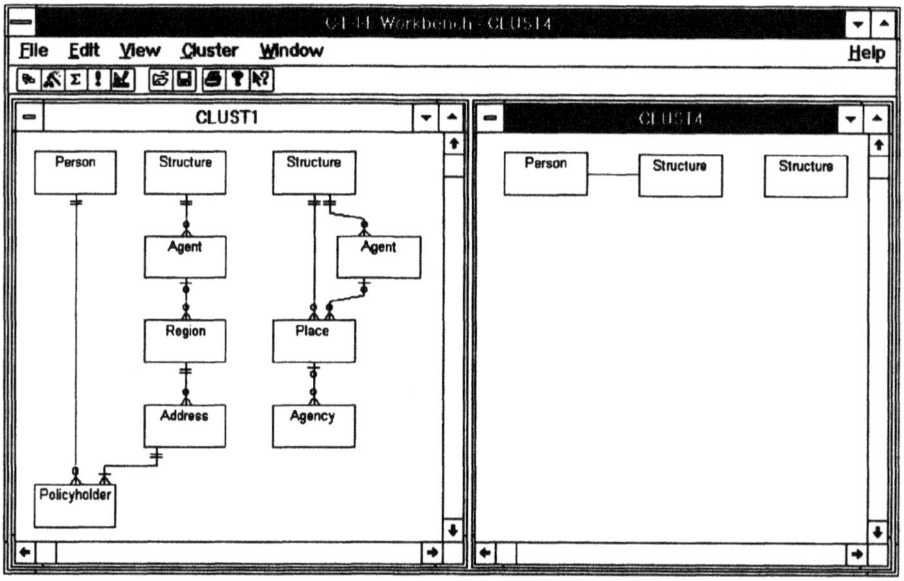

Figure 5: View integration - example

been integrated, say e.g. by forming an artificial super class (assertion *ass_overlap*). In that case the relationship set is established between the super classes. If role names have been assigned further checks are applied.

We now sketch the integration process by considering a simple example. The two schemata that should be integrated are on the left side of Fig. 5. Note, the existence of a synonym, namely the entity sets 'Region.X' in schema X (left schema) and 'Place.Y' in schema Y (right schema). Clustering leads to the schemata on the right side of Fig. 5. An assertion *ass_overlap(Structure.X, Structure.Y)* is derived. Integration is carried out by forming an artificial super class 'Structure'. The last clustering step is reversed introducing the entity sets 'Agent.X' and 'Agent.Y' Again this leads to an assertion *ass_overlap(Agent.X, Agent.Y)*. Fig. 6 shows the final result after relationship sets have been integrated too. User interaction is possible in every step of the integration process. On the other hand, integration may be carried out without interaction too. In that case possible changes to the global schema are deferred to the restructuring phase.

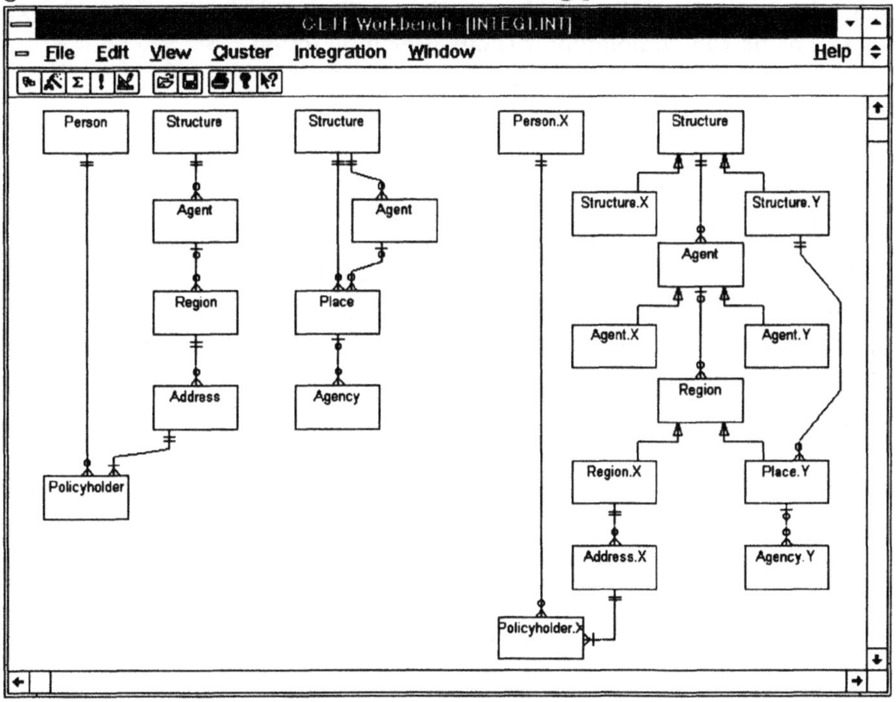

Figure 6: View integration - final result

4.3 Restructuring and quality assurance

After the integration process is finished the user may decide about restructuring the global schema. Most restructuring work may be due to (1) Markers set by the tool; (2) Entity sets that should be merged, super classes that have been introduced may be removed; (3) Different modelling constructs applied in the two schemata; (4) Restructuring of relationship sets.

5. Conclusion

A methodology that supports the generation of integrated conceptual schemata has been introduced. The integration process is driven by considering costs and benefits of data integration. For that purpose a decision support system was developed that allows to judge possible merits by modelling the internal structure and by considering possible effects of data integration on the ability of the firm to stay competitive. If integration is worthwhile the DSS supports the process of planning the integration process. If more than two databases or conceptual schemata have to be considered the order of integration is suggested. In any case the degree of integration is proposed. The method applied uses mappings of entity sets and databases to critical business processes of the firm. The presented algorithm for view integration uses a schema clustering technique to reduce complexity. The integration process uses all information that is available in ER designs to derive certain integration assertions. Local schemata are not modified. The process may or may not be run completely without user interaction.

References:

BATINI, C.; LENZERINI, M. and NAVATHE, S. (1986): A Comperative Analysis of Methodologies for Database Schema Integration. *ACM Computing Surveys, Vol.18, No.4, 1986 , 323–364.*

CHATTERJEE, A. and SEGEV, A. (1991): Data Manipulation in Heterogeneous Databases. *Sigmod Record Vol. 20, No. 4, 1991, pp. 64-68.*

ChHEN, P. (1976): The Entity-Relationship Model - Toward a Unified View of Data. *ACM Trans. on Database Systems Vol. 1, No. 1, 1976, pp. 9-36.*

GOODHUE, D. L.; WYBO, M. D. and KIRSCH, L. J. (1992): The Impact of Data Integration on the Costs and Benefits of Information Systems. *MIS Quarterly Vol. 16, No. 3, 1992, pp. 293-311.*

GOTTHARD, W.; LOCKEMANN, P. and NEUFELD, A. (1992): System-Guided View Integration for Object-Oriented Databases. *IEEE Transactions on Knowledge and Data Engineering, Vol. 4, No. 1, 1992, pp. 1-22.*

LIM, E. P. and PRABHAKAR, S. (1993): Entity Identification in Database Integration. Conference on Data Engineering. Vienna, 1993, pp. 294-301.

MARTIN, J. (1990): Information Engineering. Book II: Planning and Analysis. Englewood Cliffs, NJ.

MERT, U. and KING, R. (1994): DIRECT: A Query Facility for Multiple Databases. *ACM Transactions on Information Systems Vol. 12, No. 4, 1994, pp. 339-359.*

NAVATHE, S.; ELMASRI, R. and LARSON, J. (1986): Integrating User Views in Database Design. *IEEE Computer Vol. 19, No. 1, 1986, pp. 50-61.*

RAUH, O. and STICKEL, E. (1992): Entity Tree Clustering - A Method for Simplifying ER-Designs. Conference on Entity-Relationship Approach - ER '92, LNCS 645, 62–78, Berlin (1993).

REDDY, P. G. and PRASAD, B. E. (1994): A Methodology for Integration of Heterogeneous Databases. *IEEE Trans. on Knowledge and Data Engineering Vol. 6, No. 6, 1994, pp. 920-933.*

ROSS, S. (1973): The Economic Theory of Agency: The Principal's Problem. *American Economic Review Vol. 63, No. 2, 1973, pp. 134-139.*

SHETH, A. and LARSON, J. (1990): Federated Database Systems for Managing Distributed, Heterogeneous, and Autonomous Databases. *ACM Computing Surveys Vol. 22, No. 3, 1990, S. 183-236.*

SPACCAPIETRA, S. and PARENT, C. (1994): View Integration: A Step Forward in Solving Structural Conflicts. *IEEE Trans. on Knowledge and Data Engineering Vol. 6, No. 2, 1994, pp. 258-274.*

STICKEL, E.; HUNSTOCK, J.; ORTMANN, A. and ORTMANN, J. (1994): Data Sharing Economics and Requirements for Integration Tool Design. *Information Systems Vol. 19, No. 8, 1994, pp. 33-54.*

ZHAO, J. L. (1994): Schema Coordination in Federated Database Systems. Proc. 4th Annual Workshop on Information Technologies & Systems (WITS '94), Vancouver, 1994, pp. 29-38.

Consistent Completion of Incomplete Data Objects

R. Kiel, M. Schader

Lehrstuhl für Wirtschaftsinformatik III, Universität Mannheim,
Schloß, D-68131 Mannheim, Germany

Summary: To enable convenient application of a knowledge-based system it is necessary to give users the possibility to alter and re-evaluate previously handled inquiries. Thus, users are relieved of the task of entering a complete new query which possibly differs only slightly from earlier input. As a consequence of such a desirable system behavior, dialog control can no longer be managed by simple if/then rules. Instead, a more elaborated method controlling the system's dialog with its users must be developed. In this paper, we show how such a method can be implemented. The process is based on the evaluation of a finite automaton and refers to dialog-controlled rule systems. Our approach, however, is not restricted to queries on knowledge bases realized with that type of rule system.

1. Introduction

During consultation of any knowledge-based system the user has to enter a description of the problem which is to be analyzed by the system. (An example of a system supporting knowledge-based analysis of marketing data which implements our approach is described in Gaul and Schader (1994).) To enter the data of such a problem description into the system the user is typically required to answer several questions. Since the answers to the various questions are not independent of each other, the system presents each question together with a context-dependent set of admissible answers. The user is now requested to select an appropriate answer from this set and is thus prevented from entering unreasonable or inconsistent answer combinations. Typically, knowledge about the dependencies between the data elements of a problem description is given by a set of if/then rules.

For an efficient repeated use of a knowledge-based system its users must be enabled to interactively alter parts of a query already processed by the system in order to evaluate alternative scenarios. This has the consequence that the system's dialog with the user can no longer be conducted with a fixed given order of questions and answers. We are therefore confronted with the problem of possibly arising *pending inconsistencies* which we will discuss later by means of a simple example. In the following, we will use the term *data object* to denote the set of answers a user has entered during a consultation of the system.

2. Problems concerning incomplete data objects

Since the problem descriptions of real-world systems are too complex to be useful for illustrating the difficulties of consistent completion of incomplete data objects, we will use a small hypothetical example in order to discuss this subject. In this example different passenger cars are examined with respect to their ecological compatibility. The system comprises two rules, namely:

1. Install a catalyst if the car has a four-stroke engine, no catalyst, and doesn't need leaded gasoline.
2. Install a soot filter if the car has a diesel engine and no soot filter.

The set of all questions this system may ask during a consultation as well as the respective sets of possible answers are given in table 1. Because the answers which are admitted are not independent of previously given answers, we specify the dependency of answers by the following consistency rules: – the conditions of $\boldsymbol{r}_s$ are evaluated sequentially.

1. A four-stroke engine shall only consume gasoline.
2. A diesel engine shall only consume diesel oil.
3. A car with a four-stroke engine doesn't have a soot filter.
4. A car with a diesel engine doesn't have a catalyst.
5. A car with an engine needing leaded gasoline doesn't have a catalyst.

Questions		Possible Answers		
Full text	Abbr.	Full text	Abbr.	Set
Motor type?	q_1	4-stroke engine	a_{11}	A_1
		Diesel engine	a_{12}	
Fuel?	q_2	Diesel oil	a_{21}	A_2
		Normal unleaded	a_{22}	
		Super unleaded	a_{23}	
		Super leaded	a_{24}	
Soot filter?	q_3	Yes	a_{31}	A_3
		No	a_{32}	
Catalyst?	q_4	Yes	a_{41}	A_4
		No	a_{42}	

Table 1: Questions and Answers of the Example System

If our system asks its questions in the order given by their indexes (see table 1) these consistency rules can directly be used to compute the respective sets of admissible answers.

At the beginning of a consultation the system starts with asking which motor type our car has. It offers the alternatives *four-stroke engine* and *diesel engine* to choose from. If we select the first answer, the second question concerning the fuel our car needs is asked. Now, the admissible answers are *normal unleaded*, *super unleaded*, and *super leaded*. Due to consistency rule 1 the system doesn't offer the possible answer *diesel oil*. Suppose we answer *super unleaded*. Then the system answers the next question whether a soot filter is installed by itself – due to consistency rule 3 *no* is the only admissible answer. Finally, the question whether a catalyst is installed is posed and the answers *yes* and *no* are admitted. If we choose the first answer the system will not make any proposal because none of its rules can fire.

In such a situation it would be natural to change the query in order to test alternative scenarios. For example, we delete the first three answers and restart the system. We decide to answer the third question concerning the soot filter first. Now, the consistency rules cannot be directly used to compute the set of admissible answers to each question, because the last question *Catalyst installed?* was answered first. Since there is no consistency rule establishing a dependency between the fourth and third question we can answer the soot filter question with *yes* or *no*. If we select the first answer, we have established a pending inconsistency because now it is impossible to give correct answers to the remaining two questions.

Our goal is to enable the system to detect this problem early enough to prevent the occurence of pending inconsistencies. In the example the set of admissible answers to question three would have to be restricted to *no*. To develop a method that can handle dialog control in such a way we will first formalize the problem by use of dialog-controlled rule systems.

3. The DCRS approach

Dialog-controlled rule systems (DCRSs) are used to formally define the above illustrated problem and to develop a solution to it. First, we concisely recall the notion of a dialog-controlled rule system and list some results which are needed for the following derivations.

3.1 Dialog-controlled rule systems

Dialog-controlled rule systems were introduced in Kiel and Schader (1991) to study how a knowledge-based system can manage the dialog with its users. In Kiel and Schader (1994) DCRSs were used as a tool to define consistent data objects. And in Kiel and Schader (1996) the architecture of an interpreter for DCRSs was described. We will briefly recall the definition of a DCRS and present some selected results which form the basis of our solution to the problem of consistently completing incomplete data objects.

Definition 1

A dialog-controlled rule system (DCRS) is a quintuple (Q, A, R, F, C), *where:*

1. $Q = \{q_1, \ldots, q_n\}$ *contains all questions the user can be asked by the DCRS during a consultation;*
2. $A = \{A_1, \ldots, A_n\}$; *each* A_i *contains the (finite) set of possible answers to question* $q_i \in Q$;
3. $R = \{r_1, \ldots, r_k\}$ *is the set of rules. Notation of a single rule is given by* $r_s : [[q_{\pi_{s1}}, A_{s\pi_{s1}}, \alpha_{s\pi_{s1}}], \ldots, [q_{\pi_{sn}}, A_{s\pi_{sn}}, \alpha_{s\pi_{sn}}]] \rightarrow c_s$. $\pi_{s1}, \ldots, \pi_{sn}$ *is a permutation of* $1, \ldots, n$. *The subsets* $A_{s\pi_{si}}$ *of* $A_{\pi_{si}}$ *contain the answers to the question* q_i *satisfying condition* i *of rule* r_s;
4. *The finite set* F *contains the functions* $\alpha_{s\pi_{si}} : A_{\pi_{s1}} \times \cdots \times A_{\pi_{s(i-1)}} \rightarrow \mathcal{P}(A_{\pi_{si}})$ *which are used in the description of the rules in R.*
   These "dialog control functions" specify the answers to question $q_{\pi_{si}}$ *that are admissible if questions* $q_{\pi_{s1}}, \ldots, q_{\pi_{s(i-1)}}$ *have been answered with* $a_{\pi_{s1}} \in A_{\pi_{s1}}, \ldots, a_{\pi_{s(i-1)}} \in A_{\pi_{s(i-1)}}$. *For* $i = 1$ *we simply define* $\alpha_{s\pi_{s1}} = A_{\pi_{s1}}$;
5. $C = \{c_1, \ldots, c_k\}$ *is the set of conclusions.*

As usual, we assume that – while testing whether the antecedent of a rule r_s is true – the conditions of r_s are evaluated sequentially.

A DCRS corresponding to the example of section 2 comprises two rules. The five consistency rules are encoded by means of the system's dialog control functions. The rules are

r_1 : [['Motor type?', {'4-stroke engine'}, α_{11}],
['Fuel?', {'Normal unleaded','Super unleaded'}, α_{12}],
['Soot filter?', {'No'}, α_{13}],
['Catalyst', {'No'}, α_{14}]]
→ 'Install catalyst'

r_2 : [['Motor type?', {'Diesel engine'}, α_{21}],
['Fuel?', {'Diesel oil'}, α_{22}],
['Soot filter?', {'No'}, α_{23}],
['Catalyst', {'No'}, α_{44}]]
→ 'Install soot filter'

with

$\alpha_{11} := A_1$

$\alpha_{12} : A_1 \rightarrow \mathcal{P}(A_1)$, $\alpha_{12}(x) =$ {'Normal unleaded', 'Super unleaded', 'Super leaded'} for all $x \in A_1$

$$\alpha_{13} : A_1 \times A_2 \to \mathcal{P}(A_3),\ \alpha_{13}(x,y) = \{\texttt{'No'}\} \quad \text{for all } (x,y) \in A_1 \times A_2$$
$$\alpha_{14} : A_1 \times A_2 \times A_3 \to \mathcal{P}(A_4),\ \alpha_{14}(x,y,z) = A_4$$
$$\quad \text{for all } (x,y,z) \in A_1 \times A_2 \times A_3$$
$$\alpha_{21} := A_1$$
$$\alpha_{22} : A_1 \to \mathcal{P}(A_2),\ \alpha_{22}(x) = \{\texttt{'Diesel oil'}\} \quad \text{for all } x \in A_1$$
$$\alpha_{23} : A_1 \times A_2 \to \mathcal{P}(A_3),\ \alpha_{23}(x,y) = A_3 \quad \text{for all } (x,y) \in A_1 \times A_2$$
$$\alpha_{24} : A_1 \times A_2 \times A_3 \to \mathcal{P}(A_4),\ \alpha_{24}(x,y,z) = \{\texttt{'No'}\}$$
$$\quad \text{for all } (x,y,z) \in A_1 \times A_2 \times A_3$$

3.2 Dialog-controlled rule systems and formal languages

In Kiel and Schader (1991, 1994, 1996) we have defined several languages describing special properties of dialog-controlled rule systems or properties of individual rules of a DCRS. Here, we only provide the definitions of those languages that are essential to develop a solution to our problem.

Definition 2

a) For every rule $\mathbf{r}_s$ of a DCRS D we define the language Lf_s which describes all combinations of answers that fire $\mathbf{r}_s$

$$\begin{aligned} Lf_s := \{q_{\pi_{s1}} a_1 \cdots q_{\pi_{sn}} a_n \mid & \\ & a_1 \in A_{s\pi_{s1}} \\ \wedge\ & a_2 \in A_{s\pi_{s2}} \cap \alpha_{s\pi_{s2}}(a_1) \\ & \vdots \\ \wedge\ & a_n \in A_{s\pi_{sn}} \cap \alpha_{s\pi_{sn}}(a_1, \ldots, a_{n-1})\} \end{aligned}$$

b) The "normalized" version Lfn_s of Lf_s which handles the answers in the order corresponding to the indexes of Q is given by

$$Lfn_s := \{q_1 a_1 \cdots q_n a_n \mid q_{\pi_{s1}} a_{\pi_{s1}} \cdots q_{\pi_{sn}} a_{\pi_{sn}} \in Lf_s\}$$

c) The language $Lfn := \bigcup_{s=1}^{k} Lfn_s$ then describes all combinations of answers which are accepted by the DCRS D and fire at least one of its rules.

d) For every rule $\mathbf{r}_s$ of a DCRS D we denote by La_s the language of all combinations of answers that are accepted by $\mathbf{r}_s$ without firing this rule

$$\begin{aligned} La_s \ := \ & \{q_{\pi_{s1}} a_1 \mid a_1 \in A_{\pi_{s1}} - A_{s\pi_{s1}}\} \cup \\ & \{q_{\pi_{s1}} a_1 q_{\pi_{s2}} a_2 \mid a_1 \in A_{s\pi_{s1}} \wedge a_2 \in \alpha_{s\pi_{s2}}(a_1) - A_{s\pi_{si}}\} \cup \end{aligned}$$

$$\bigcup_{i=3}^{n} \{q_{\pi_{s1}} a_1 \cdots q_{\pi_{si}} a_i \mid$$

$$\begin{aligned} & a_1 \in A_{s\pi_{s1}} \\ \wedge \; & a_2 \in A_{s\pi_{s2}} \cap \alpha_{s\pi_{s2}}(a_1) \\ & \vdots \\ \wedge \; & a_{i-1} \in A_{s\pi_{s(i-1)}} \cap \alpha_{s\pi_{s(i-1)}}(a_1, \ldots, a_{i-2}) \\ \wedge \; & a_i \in \alpha_{s\pi_{si}}(a_1, \ldots, a_{i-1}) - A_{s\pi_{si}}\} \end{aligned}$$

Obviously, it is impossible to normalize La_s – this language contains words describing *incomplete* question/answer combinations. Such a word does not assign an answer to every question $q_1, \ldots, q_n$. In the next section we introduce a language which can be normalized and encodes the same question/answer combinations as La_s.

4. The Solution

We now define a language Lp expressing all question/answer combinations which can be directly entered into a DCRS or constructed from such a combination by deleting one or several answers. To obtain Lp, a normalized version of La_s is needed. Since it is not possible to normalize this language directly, we *pad* it with additional question/answer pairs denoting explicitly that certain questions are not associated with an answer. Therefore, we attach a marker symbol # to these questions. We denote the resulting language by $La\#_s$. Its normalized version is $Lan\#_s$.

Definition 3

a) $La\#_s := La_s \cdot (Q\#)^* \cap (q_{\pi_{s1}} \cdot (A_{\pi_{s1}} \cup \{\#\}) \cdots q_{\pi_{sn}} \cdot (A_{\pi_{sn}} \cup \{\#\}))$

b) $Lan\#_s := \{q_1 a_1 \cdots q_n a_n \mid q_{\pi_{s1}} a_{\pi_{s1}} \cdots q_{\pi_{sn}} a_{\pi_{sn}} \in La\#_s\}$

c) $Lan\# := \bigcup_{s=1}^{k} Lan\#_s$

Note that, as usual, S^* denotes the language which consists of the words constructed by concatenating any number of symbols from a set S.

Now we have completed all preparations for defining the above mentioned language Lp.

Definition 4

For a dialog-controlled rule system $D = (Q, A, R, F, C)$ *let* $\sigma : (Q \cup A_1 \cup \cdots \cup A_n \cup \{\#\}) \rightarrow (Q \cup A_1 \cup \cdots \cup A_n \cup \{\#\})^*$ *be a regular substitution with*

$$\begin{aligned} \sigma(\#) &:= \{\#\} \\ \sigma(a) &:= \{\#, a\} \quad \text{for all } a \in A_1 \cup \cdots \cup A_n \text{ and} \\ \sigma(q) &:= \{q\} \quad \text{for all } q \in Q \end{aligned}$$

With the languages Lfn and Lan# from definitions 2 and 3 we define

$$\boldsymbol{Lp} := \boldsymbol{\sigma}(Lfn \cup Lan\#)$$

The language *Lp* has two important properties which we state in the following theorem. (The proof is straightforward and can be omitted here.)

Theorem 1

1. *For every question/answer combination entered into D during an ordinary consultation followed by deleting one or more answers there exists a word $\boldsymbol{w} \in Lp$ which encodes the resulting incomplete combination.*
2. *Every word $\boldsymbol{w} \in Lp$ encodes an answer combination which can be entered into D using the above illustrated method.*

This theorem allows us to decide whether a specific possible answer is admissible in a certain situation during consulting a DCRS.

Suppose that the problem is to find out if an answer $\boldsymbol{x} \in \boldsymbol{A_i}$ to question $\boldsymbol{q_i}$ is admitted in a situation where the user input consists of the answers $\boldsymbol{a_1, \ldots, a_{i-1}, a_{i+1}, \ldots, a_n}$ given to questions $\boldsymbol{q_1, \ldots, q_{i-1}, q_{i+1}, \ldots, q_n}$. Here, we only have to check whether the corresponding word $\boldsymbol{w} = \boldsymbol{q_1 a_1} \cdots \boldsymbol{q_i x} \cdots \boldsymbol{q_n a_n}$ belongs to *Lp*: If $\boldsymbol{w} \in Lp$ then the answer $\boldsymbol{x}$ is admissible; otherwise $\boldsymbol{x}$ is not admissible.

In the case of incomplete input we simply attach the marker # to the questions that are not yet answered or that had their previous answers deleted.

Recurring to the example of section 2 we would have to examine the words $\boldsymbol{w} = \boldsymbol{q_1 \# q_2 \# q_3 x q_4 a_{41}}$ and find out for which $\boldsymbol{x} \in \boldsymbol{A_3}$ they belong to *Lp*. The result is that only $\boldsymbol{x} = \boldsymbol{a_{32}}$ (i.e. *no*) is an admissible answer.

In order to obtain the set of all admissible answers to question $\boldsymbol{q_i}$ in the above described situation, we have to construct $\boldsymbol{w}$ for every $\boldsymbol{x} \in \boldsymbol{A_i}$ and to check whether $\boldsymbol{w} \in Lp$.

The task to find out whether a word $\boldsymbol{w}$ is in *Lp* or not can be assigned to a finite deterministic automaton accepting *Lp*. Such an automaton exists because *Lp* is regular since *Lfn*∪*Lan#* is regular and $\boldsymbol{\sigma}$ is a regular substitution. (See Hopcroft and Ullman (1979) for details.) An automaton accepting our example system's language *Lp* is shown in figure 1.

A disadvantage of using a deterministic automaton to recognize the language *Lp* is its potentially extremely high number of states relative to the amount of states of an automaton accepting *Lfn* ∪ *Lan#*. It is therefore necessary to develop an alternative method that determines whether a word $\boldsymbol{w}$ is in *Lp* or not. It would be desirable to use the automaton accepting *Lfn* ∪ *Lan#* directly. This goal can be reached in the following manner. First, we construct a nondeterministic automaton accepting *Lp* by starting with the automaton accepting *Lfn*∪ *Lan#* and then including an additional transition $\boldsymbol{s} \xrightarrow{\#} \boldsymbol{t}$ into its transition graph for every existing transition

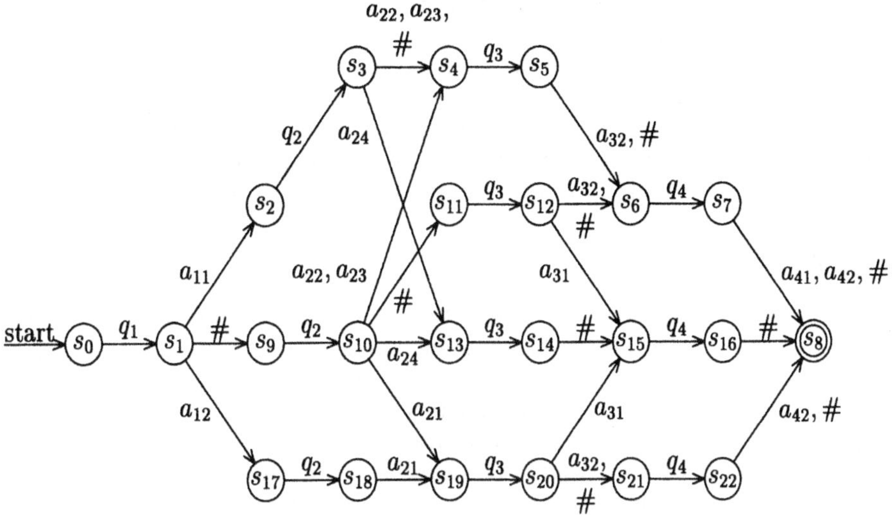

Figure 1: Automaton accepting *Lp* for the example DCRS

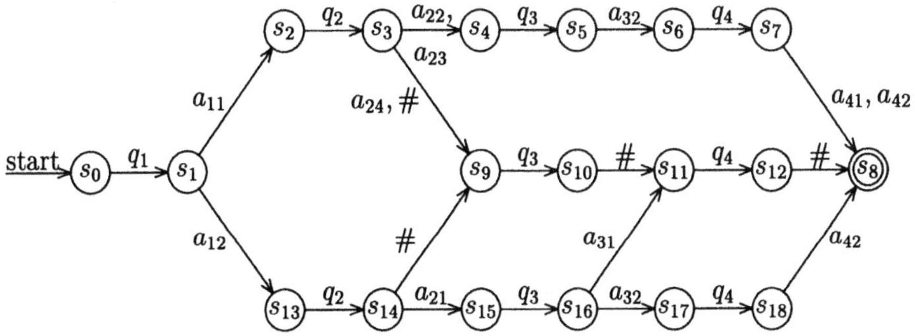

Figure 2: Automaton accepting $Lfn \cup Lan\#$ for the example DCRS

$s \xrightarrow{a} t$ with $a \in A_1 \cup \cdots \cup A_n$. Having introduced such a #-transition for every transition consuming one answer $a \in A_1 \cup \cdots \cup A_n$, we can integrate the handling of the #-transitions directly into a procedure evaluating a deterministic automaton accepting $Lfn \cup Lan\#$. For our example, the transition graph of this automaton is shown in figure 2.

Recurring, again, to the problem of section 2 one can see that each path from the initial state s_0 to the final state s_8 which accepts a_{41} (answer *yes* to the catalyst question) must transition through an arc labeled a_{32} (answer *no* to the soot filter question).

If we apply the nondeterministic approach to decide whether a certain answer is allowed in a specific situation, we must, generally, evaluate more states compared to the analysis of the deterministic automaton accepting Lp. In the second case we have to evaluate maximum $\mathbf{2n}$ states (at most one state for every symbol of $\boldsymbol{w}$). Following the nondeterministic approach, in the worst case we have to evaluate every state of the automaton accepting $Lfn \cup Lan\#$. This latter automaton, however, consists of considerably fewer states.

5. Outlook

There are two problems which might be of interest for future investigation. First, we have to analyze the number of states of an atomaton accepting Lp in comparison with the number of states of an automaton accepting $Lfn \cup Lan\#$. (Recall that Lp is constructed from $Lfn \cup Lan\#$.)
At the moment, we know that the number of states may maximally grow exponentially if we construct a deterministic automaton accepting Lp from a nondeterministic one accepting the same language – this nondeterministic automaton having the same number of states as the automaton accepting $Lfn \cup Lan\#$ (see, e.g., Hopcroft and Ullman (1979)).

A second goal would be to generalize the completion method, enabling it to to handle more complex data objects. For example, it would be interesting to complete a data object with list-valued attributes.

References:

GAUL, W. and SCHADER, M. (1994): Wissensbasierte Marketing-Datenanalyse. Peter Lang, Frankfurt am Main.

HOPCROFT, J. E. and ULLMAN, J. D. (1979): Introduction to Automata Theory, Languages and Computation. Addison-Wesley, Reading, Massachusetts.

KIEL, R. and SCHADER, M. (1991): Detecting Inconsistencies in Dialog-Controlled Rule Systems. In: G.H. Schildt and J. Retti (eds.): *Dependability of Artificial Intelligence Systems.* North Holland, Amsterdam, 141–152.

KIEL, R. and SCHADER, M. (1994): Using Dialog-Controlled Rule Systems in a Maintenance Module for Knowledge Bases. *Annals of Operations Research, 52, 171–180.*

KIEL, R. and SCHADER, M. (1996): The Design of an Interpreter for Dialog-Controlled Rule Systems. In: H. H. Bock and W. Polasek (eds.): *Data Analysis and Information Systems.* Springer, Heidelberg, 370–379.

Towards a Case–Based Assistant for the Conceptual Modelling of Information Systems

Dirk Krampe, Markus Lusti

Universität Basel, Institut für Informatik / WWZ
Petersgraben 51, CH–4051 Basel, Switzerland
email: krampe@ifi.wwz.unibas.ch

Summary: We present some aspects of a case–based assistant for the conceptual modelling of information systems. The *CBModeler* project aims to show, how the development of information systems can be supported by reusing requirements knowledge which has been acquired during former development projects.
The paper gives a short introduction into *case–based reasoning* and discusses some aspects of *CBModeler* in an analysis scenario. The following sections describe *CBModeler*'s general architecture, the representation of cases, the use of a *reference model* and some *indexing* and *matching* issues.

1. Introduction

The most expensive development mistakes are made during system analysis. One of the main tasks is therefore to get all user requirements and to formalize them in a conceptual model.

Most of the knowledge based tools concentrate on database design aspects. Only few support the conceptual modelling by reusing requirements knowledge (see Krampe et al. (1995/96) for further details).

CBModeler's case library contains conceptual models which may be reference models. In order to design a new conceptual model, *CBModeler* looks in its case library to match the given constraints for the new model. It then proposes the found cases or components to the system analyst who chooses appropriate ones. During the adaption of a conceptual model *CBModeler* gives hints. Then the conceptual model is indexed and stored as a new case in the library. The internal representation of the components supports specification reuse by combining logical and object–oriented programming concepts.

In this paper, we will not give an overview of the *conceptual modelling* of information systems itself and the tools that support this process. The interested reader may refer to Loucopoulos et al. (1992) and Krampe (1995).

2. Case–based design

Case–based reasoning (CBR) is a technology which proves successful in a wide range of application areas (see Kolodner (1993)). A case–based sys-

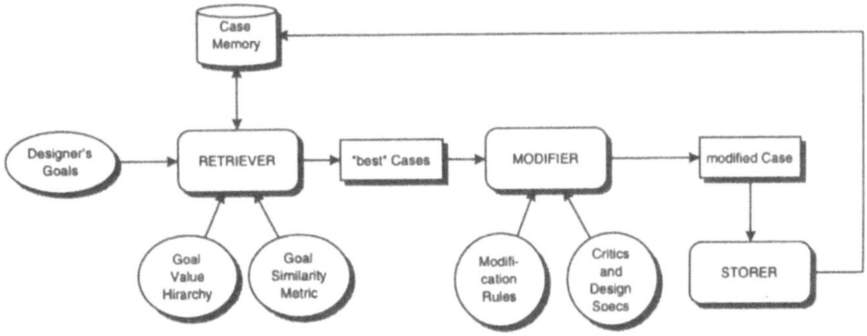

Figure 1: Architecture of a case–based designer (Hammond (1989))

tem "solves new problems by adapting solutions that were used to solve old problems" (Riesbeck and Schank (1989)). In the design domain the designer searches through a library of existing designs and chooses one which fits best in the given situation. The features which do not match the current requirements need to be adapted.

2.1 Case–based reasoning for design

Figure 1 illustrates the basic components of a case–based reasoner for design. The designer starts with the goals he wants to achieve. The problem is defined as a set of constraints and the *retriever* is asked to search through the *case memory* to find cases which solve the constraint problem. Usually the constraints underspecify the problem and the retriever will find many possible solutions. Therefore the retriever needs a *similarity metric* to evaluate which of the found cases match the constraints best. Then the retriever needs a *value hierarchy* that fulfills the most important constraints and compromises the others. After searching the case memory the retriever ranks the matching cases and presents the "best" ones. The designer selects one case or some components from several cases for further use.

During the next step, the selected design is adapted to the new situation. If done automatically the *modifier* needs adaption knowledge to control the adaption process. Finally, the modified case is stored in the case memory for later reuse.

In some architectures an *evaluation* and a *repair* component can be found (see Riesbeck et al. (1989), Hunt et al. (1995)). During evaluation the modified case is analysed to determine the applicability and usefulness of the design. The repairer is then used to suggest ways to avoid the problems identified during evaluation. Figure 1 shows the basic components.

3. A simple analysis scenario

The following analysis scenario illustrates how *CBModeler* reuses requirements knowledge to design a new conceptual model. Assume, the case library of *CBModeler* contains the conceptual model for an electronic equipment sales system. The goal is to design a conceptual model for a book sales system. Both companies deal with goods. Both buy their goods from suppliers and sell them to dealers.

Step 1: Describing the requirements and searching the case library

The system analyst describes the requirements for the new conceptual model using a *requirements form.* All requirements are described in terms of *business activities* and *events*, *data stores*, *external agents*, *organization units*, *entity types* and *attributes.* Every conceptual model contains these components (see 4.2). It is not necessary to fill in the requirements form completely. During the following example session no business events and attributes are specified[1]:

`Enter the name of the new case:` *book-wholesaler*
`Fill in the following requirements form for the`
`'book-wholesaler' system`
`Description:` *Selling system for a book wholesaler, who sells books to bookstores and buys books from publishing companies*
`Business Activities:` *add-order, process-order, add-application, process-selling-campaign*
`Business Events:`
`Data Stores:` *stock-of-bookstores, orders-on-hand, stock-of-available-books*
`External Units:` *bookstore, publishing-company, carrier*
`Organization Units:` *purchase, sale, incoming-goods*
`Entity Types:` *bookstore, book, order, publishing-company, stock-of-available-books*
`Attributes:`

After filling in the requirements form *CBModeler* searches through its case library and tries to find cases or components matching the given requirements. There are two different approaches to case library search. One is *searching components*: *CBModeler* looks for all components which may match the given requirements regardless to which cases they belong to. All found components are presented. The other one is *searching cases*: *CBModeler* presents all cases that partially match the given requirements. The system analyst selects the case or components he wants to work with. The following dialogue shows the search result:

```
I found the following components which match
the 'book-wholesaler' system:
Business Activities: add-order, process-order, add-application
Data Stores: orders-on-hand
```

[1] user input in italics

```
External Units: carrier
Organization Units: purchase, sale, incoming-goods

The following components for the 'book-wholesaler' system
are unknown:
Business Activities: process-selling-campaign
Data Stores: stock-of-bookstores, stock-of-available-books
External Units: bookstore, publishing-company
Entity Types:  bookstore, book, order, publishing-company,
stock-of-available-books
Can you give me some semantic links to these components?
```

stock-of-bookstores is_a stock-of-customers, stock-of-books is_a stock-of-goods, bookstore is_a customer, publishing-company is_a supplier, book is_a good, order is_a (order-heading,order-position), stock-of-available-books is_a stock-of-available-goods

All found components are automatically added to the conceptual model. For all unknown components the system analyst may define *semantic links* to components known by *CBModeler*. In the given situation *CBModeler* knows the conceptual model of the wholesaler for electronic equipment which contains customers, suppliers and goods. Looking from the book wholesaler's perspective the customers are bookstores, the suppliers are publishing companies and the goods are books. Semantic links make it possible to reuse components *CBModeler* could not find. Figure 2 shows the situation for suppliers and publishing companies after defining the semantic link: publishing companies are special suppliers. They inherit all attributes defined for suppliers. The inherited attributes `dyn.Class` and `count` are redefined during step 3 (validation and adaption).

Finally, the remaining unidentified components represent new concepts which have to be designed for the new conceptual model:

```
OK, then the following components for the 'book-wholesaler'
system are really new components:
business activities: process-selling-campaign
```

Step 2: Adding components which logically belong to the model

Step 1 produces a conceptual model which contains only the components the system analyst specified in the requirements form. Analysing these components *CBModeler* is able to complete the new model by adding components which logically belong to the model. Some examples illustrate this: The business activity *process-order* has several sub activities in the case library i.e. *add–order*, *reject–order* and *modify–order*. The activity *add–application* uses the data store *customer–data*, is triggered by the event *incoming–application* and is followed by several other activities i.e. *evaluate–credit-rating* and *modify–customer–data.* All these components can be added to the new conceptual model.

During completion *CBModeler* will use the following strategies: add all pre-

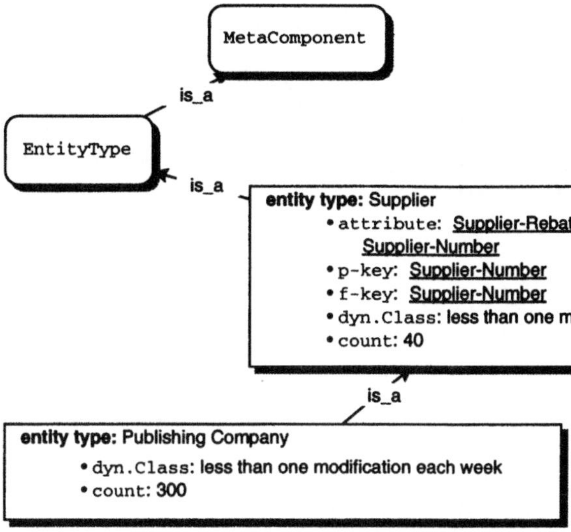

Figure 2: Specializing a model component

ceding, following, super and sub activities and events. Then add all data flows and data stores, external agents, organization units, entity types, attributes and relations as far as they are not added yet and a business activity uses them. Finally, note for each component why it is included into the new model. This makes it possible to generate explanations on why a component is part of the model.

Step 3: Validation and Adaption

The system analyst validates the new conceptual model to check wether it matches the given situation. Figure 3 shows a possible approach for the validation process: the conceptual model is presented using hypertext documents. The system analyst navigates through the conceptual model and inspects it using the hyperlinks attached to the components. The figure shows the data flow view of the example model[2].

The adaption of the model is done by hand because the design process is a creative task. We do not see how to capture the appropriate knowledge required to perform all the necessary modifications automatically. For a further discussion of this problem see Hunt et al. (1995).

[2]It is accessible in the Internet using the following WWW-address: **http://www.ifi.unibas.ch/grulu/projects/modeler/modeler.html**

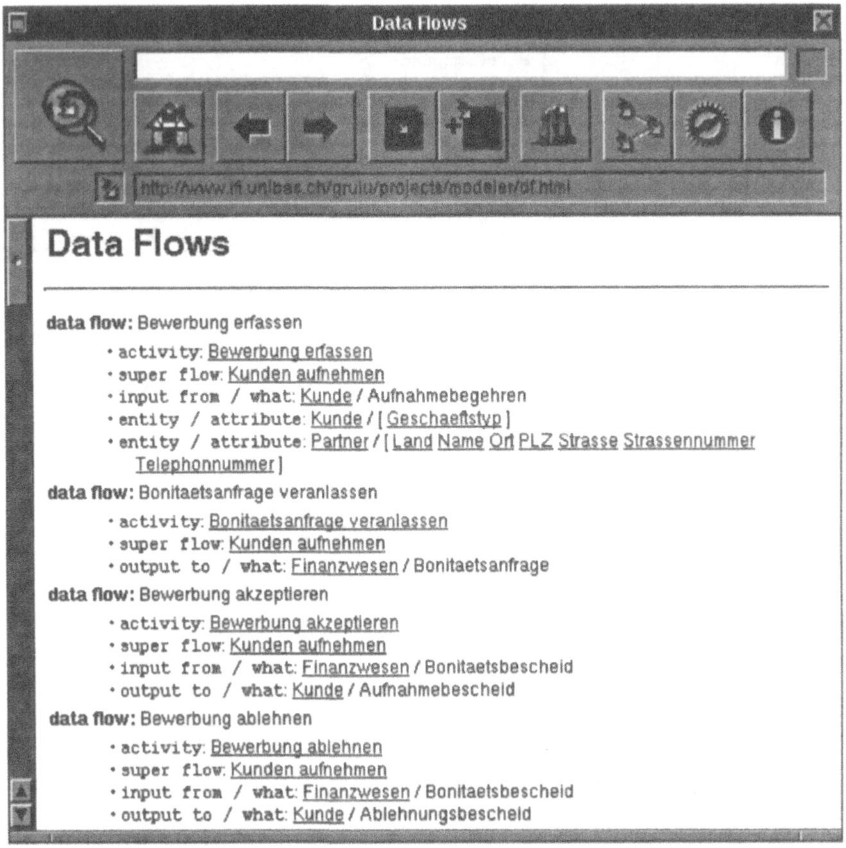

Figure 3: Hypertext view of data flows

Step 4: Consistency checking

During this step, *CBModeler* checks if all definitions have been made. It applies consistency rules such as "Each activity has to use at least one entity type" or "Each entity type is used by at least one activity". *CBModeler* uses the consistency rules defined by the CC RIM–Metamodel for the analysis (Gutzwiller et al. (1992)).

Step 5: Indexing the model and storing it in the case library

Before storing the new conceptual model in the case library it has to be indexed. Section 4.4 gives some notes on indexing issues.

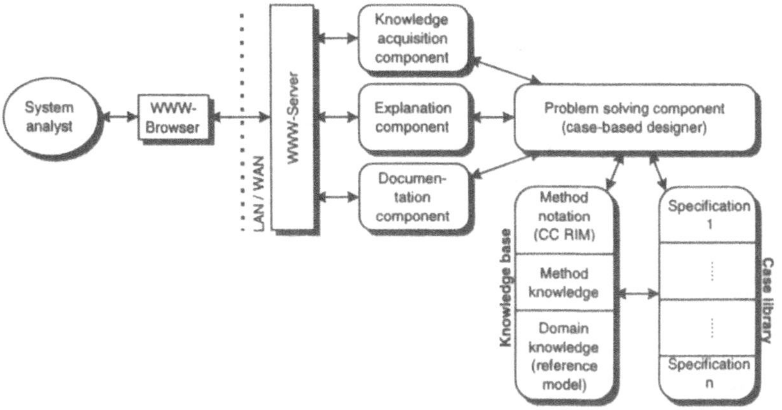

Figure 4: *CBModeler*'s architecture

4. The case–based assistant *CBModeler*

The following sections describe *CBModeler*'s architecture, some aspects of the knowledge base (method notation and domain knowledge) and the representation of conceptual models in the case library.

4.1 The architecture

Figure 4 describes *CBModeler*'s architecture. In order to be platform independent and to discuss *CBModeler*'s concepts in a team the communication interface is realised in World Wide Web (WWW) technology. The system analyst formulates his design goals using a *WWW browser*. The *WWW server* controls the components and functions *CBModeler* offers.

The *knowledge acquisition component* is used to enter and adapt cases as described in the scenario of section 3. Other components involve the modification of consistency rules and domain knowledge (section 4.3). The *explanation component* generates explanations such as why components have been added to a case. The *documentation component* generates different views on cases (see for example figure 3). Finally, the *problem solving component* realizes the case–based designer (see figure 1 and the scenario in section 3).

4.2 The representation of conceptual models

One part of the knowledge base is the *method notation* and the *method knowledge* used to describe the specifications stored in the case library. The method knowledge controls the adaption of cases and components and defines consistency rules. Figure 5 shows the different components of this metamodel and the relations between them.

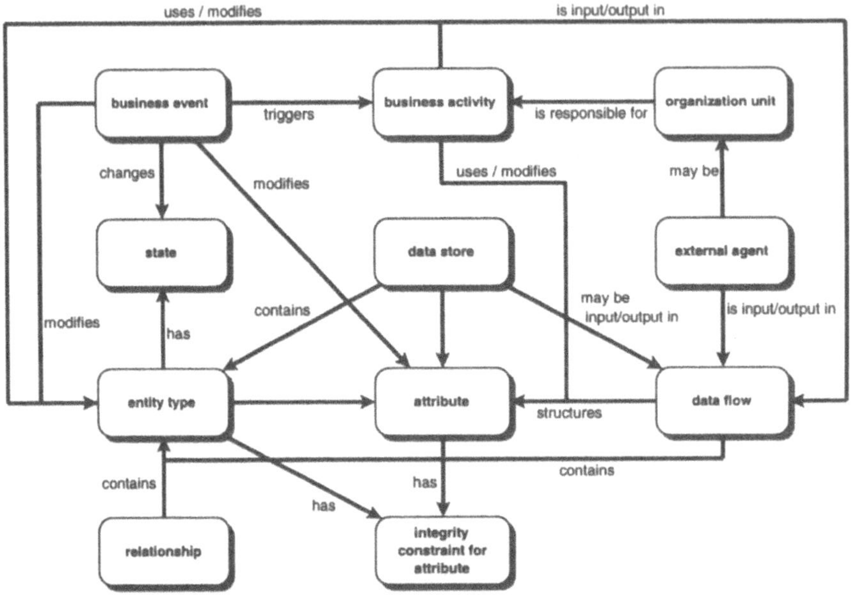

Figure 5: The CC RIM–Metamodel for the analysis (Gutzwiller et al. (1992))

A conceptual model is described in terms of ***business events***, ***business activities***, ***organization units***, ***states***, ***data stores***, ***external agents***, ***entity types***, ***attributes***, ***data flows***, ***relationships*** and ***integrity constraints***. The realization of conceptual models within *CBModeler* uses object–oriented knowledge representation techniques. Each component may be specialized from other components and inherits the concepts defined in the general components (see figure 2). This supports knowledge reuse.

4.3 The domain knowledge — reference models

Design problems of conceptual models tend to be large. A problem decomposition is hardly applicable because of the strong interactions between the subproblems. A better strategy to deal with large design problems is to use some cases as a basic framework and to take missing details from the case library (see Kolodner (1993)). The general frameworks *CBModeler* uses are reference models which generalize important aspects of business structures (see for example Scheer (1995), Mertens (1995)). The reference models are comparable to class or function libraries. Class libraries try to offer general functionality which may be reused in software development projects. Some classes/objects may be used as they are while others are adapted to match the given situation.

When using *CBModeler* for the first time the system analyst has to define a reference model of the domain the system should deal with. Entering such a reference model makes *CBModeler* domain dependent. For new design problems of the same domain the system offers a somewhat complete conceptual model and fills in the missing details using the case library.

4.4 Some notes about indexing and matching

As one may see in the scenario of section 3 we only use a very simple indexing and matching scheme: each case or component is identified by its symbolic name. This strategy is not satisfying. Therefore, we suggest some extensions. The first one is extending the indexing vocabulary. Each case or component should be indexed using its most important features. For example, each business activity may have data stores it reads from and data stores it writes to. Business events trigger several activities, entity types have key attributes and so on.

More important is the introduction of *abstraction hierarchies* for data stores, business activities, external agents and organization units. Abstraction hierarchies categorize components according to the role they play in a conceptual model and determine their semantics.

Data stores, for example, may be subdivided into *master data* and *transaction data.* Master data distinguish between *business data* (customers, suppliers, goods) and *technical data* (qualitiy of materials, compositions). Transaction data are subdivided into *notation data* (accounts payable, accounts receivable, outstanding offers), *transfer data* (journals, monthly accountings) and *archive data* (sales figures, stock turnover statistics)[3].

Business activities may be categorized using two abstraction hierarchies: the *kind of data access* and *kind of business function.* The first hierarchy looks on the data the business activity accesses. There are typical activities that process on master data (add, modify, delete) only. Others read master data and create notation data, for example creating an outstanding offer (read customers and goods, create an offer). The second hierarchy for business activities categorizes each activitiy according to its type. For example, a company may have a purchase department and within this an activity that calculates order quantities or a shipping department and an activity that processes the commission of goods.

Using abstraction hierarchies has several advantages. First, it is much easier for the system analyst to select the components he wants to include in his new conceptual model. Instead of naming a component he specifies nodes in the abstraction hierarchies and lets the system present suitable cases or components. Second, *CBModeler* is able to compute the degree of similarity of corresponding components. Using an abstraction hierarchy and relative ranking, one pair of corresponding values match better than another if the

[3]The abstraction hierarchies we introduce here are closely related to the categorizations reference models use to distinguish data and functions (see for example Mertens (1995), Scheer (1995)).

most specific common abstraction (MSCA) of the first is more specific then the MSCA of the second. For more details on matching and ranking using the MSCA strategy see Kolodner (1993).

5. Discussion and outlook

We presented some aspects of a case–based assistant for designing conceptual models for information systems. We believe that reuse has to start during the earliest part of the software development process — requirements acquisition and analysis. From the perspective of artificial intelligence, this is a good domain to study fundamental issues of knowledge acquisition. From the perspective of software engineering, it is obvious that errors in requirements are more costly than any other kind of error (see Rich (1990)). This is the reason why we suggest not only to reuse programming code in class libraries but also to reuse requirements knowledge.

We gave an analysis scenario, discussed the representation of conceptual models and sketched the architecture of *CBModeler*. Furthermore, we described the role of reference models in designing new conceptual models.

Until now the representation of conceptual models and some parts of the documentation component are realized. An example case is accessible through the Internet (see "Validation" in section 3). It illustrates the chosen structure for conceptual models. The current prototype is implemented using SICStus Prolog under UNIX. The case representation uses the opportunities the object–oriented extension of SICStus Prolog offers. SICStus Objects support logic programming in large applications, flexible sharing and reuse of knowledge and the combination of logic programming in Prolog style and object–oriented programming (see SICStus (1995)).

The main future task is applying case–based techniques to the domain of conceptual modelling. This means that we have to realize better indexing, matching and ranking mechanisms (see section 4.4). The communication interface should better support the definition of requirements used to identify cases and components in the case library. Finally, a reference model has to be included to show how it can support the conceptual modelling process.

References:

GUTZWILLER, T. and ÖSTERLE, H. (1992): Konzepte angewandter Analyse– und Design–Methoden; Band 1: Ein Referenz–Metamodell für die Analyse und das System–Design; Band 2: Ein Beispiel für die Analyse und das System–Design. *AIT Angewandte Informations Technik Verlags GmbH*

GUTZWILLER, T. (1994): Das CC RIM–Referenzmodell für den Entwurf von betrieblichen, transaktionsorientierten Informationssystemen. *Physica Verlag*

HAMMOND, K. J. (1989): Case–based planning: viewing planning as a memory task. *Academic Press, Inc.*

HUNT, J. and MILES, R. (1995): Toward an intelligent architectural design aid.

Expert Systems, 8/95, Vol. 12, No. 3: 209–218

KOLODNER, J. (1993): Case–Based Reasoning. *Morgan Kaufmann Publishers, Inc.*

KRAMPE, D. (1995): Wissensbasierter, konzeptioneller Entwurf betrieblicher Datenbanken — Forschungsstand. *Universität Basel, Institut für Informatik/WWZ, Interner Forschungsbericht*

KRAMPE, D. and LUSTI, M. (1996): Ein wissensbasiertes Werkzeug zur Unterstützung des konzeptionellen Entwurfs betrieblicher Informationssysteme. *Discussion Paper Nr. 9604 des Wirtschaftswissenschaftlichen Zentrums der Universität Basel*

KRAMPE, D. (1996): Conceptual Modelling via System Analysis. *Artificial Intelligence in Design 1996, Stanford University, USA; Workshop Notes on 'New Directions in Case–Based Design Systems'*

Loucopoulos, P. and Zicari, R. (ed.) (1992): Conceptual Modelling, Databases, and Case. *John Wiley & Sons, Inc.*

LUSTI, M. (1990): Wissensbasierte Systeme: Algorithmen, Datenstrukturen und Werkzeuge. *BI–Wissenschaftsverlag*

MERTENS, P. (1995): Integrierte Informationsverarbeitung I: Administrations- und Dispositionssysteme in der Industrie. *Gabler Verlag*

RICH, C. and WATERS, R. C. (1990): The Programmer's Apprentice. *ACM Press*

RIESVECK, C. K. and SCHANK, R. C. (1989): Inside Case–Based Reasoning. *Lawrence Erlbaum Ass.*

SCHEER, A. W. (1995): Business Process Engineering — Reference Models for Industrial Enterprises, 2nd Ed. *Springer Verlag*

SICStus (1995): SICStus v3.0: Advanced Prolog Technology; User's Manual *Swedish Institute of Computer Science*

Foundational Aspects of Knowledge-Based Information Systems in Scientific Domains

Wolfgang Lenski, Elisabeth Wette-Roch

Fachbereich Informatik, Universität Kaiserslautern,
Postfach 3049, D-67653 Kaiserslautern, Germany

Summary: This paper deals with foundational aspects for the design of information systems. We show in the concrete setting of an information system on mathematical logic that the classical philosophical distinction between syntax and semantics is insufficient to capture the main issues as the contribution of human interpretation acts for the (conceptual) representations have to be taken into consideration as well. We present an approach based on the philosophical position of the pragmaticism (Ch.S. Peirce) where in the triadic relation all the relevant issues are captured. We show that this approach is not only foundational but may well contribute to improvements of the retrieval quality of information systems.

1. Introduction

The aim of an information system is to assist users in finding material ('documents') that satisfy their information needs even if the formulation of their search requests contains some vagueness. To serve this purpose the data is represented in a way that effectively supports the retrieval process. In the long tradition of library organization a variety of methods to make the material accessible ranging from purely bibliographical references over topics up to classifications on the basis of a classification system may be identified. These are still the basic methods underlying modern information systems.

In this paper we discuss some principal problems we came across in the design of the LOGIC INFORMATION SYSTEM LIS based on the *Bibliography of Mathematical Logic*, an enterprise of the *Heidelberger Akademie der Wissenschaften.* The Logic Bibliography is not compiled for documentation purposes only that is to document for future generations the collection of material published so far in the field of logic. Though this may also be considered as a cultural impact, the aims of the Bibliography project are some more ambitious: LIS is designed to support research interests of users looking for specific information satisfying their information needs. This requires that contents information is accessible through LIS. Besides the standard bibliographical references LIS uses different ways to indicate to its users what the publications are about. As it will be explained in the following these reflect three different approaches which lead to principle considerations about their status and the way to handle these in an information system. In this paper we will present a philosophical foundation which will at the same time provide the guidelines for the general design of scientific information systems.

2. Fundamentals for the Design of Knowledge-Based Information Systems

The aim is to design an information system on a reseach area which in our case is mathematical logic. An information system is based on individual documents describing research results. Search requests are primarily scientific results presented through publications that may be retrieved from the system. In order to be able to perform this step the system must be enabled to access research results. In the absence of methodologies understanding full texts and being able to deduce consequences from these, techniques to represent the single documents are needed. So we have research communicated in the scientific community on one side and representation and communication media of the results on the other side. Our starting position is described by the following statements:

1. Research is manifested (and thus only accessible for an information system) through its traces in the process of scientific communication. In the current stage these traces are the scientific papers. Hence an information system must be enabled to handle publications.
2. A scientific document contributing to research is written in the vocabulary shared by a research group.
3. A research group shares research interests under assumptions which the group has agreed upon.
4. Research results are formulated and communicated in a language based on the special vocabulary of the research field they are related to. They are addressed to a scientific community and contribute to research lines.

These items will constitute the guideline for our exposition in the following. These points will be discussed in three dimensions which are not three separate topics but are closely interrelated and rather constitute three different aspects of the same phenomenon than three independent parts which may easily be isolated from each other:

- *What are the objects we are dealing with?*
- *In which way are these modeled?*
- *How can we handle these (semi-)automatically?*

These are the basic questions every information system is concerned with and (more or less implicitly) uses some pre-assumptions about these. We will make these explicit and present a holistic view ranging from philosophical considerations to models of data processing including representation models. Accordingly, we will isolate three aspects in our discussion.

Special emphasis is put on the fact that in the presence of a classification system research *interests* i.e. assumptions without claiming intersubjective validity have to be taken into consideration. Consequently, methods for dealing with these have to be established in a well-grounded way.

2.1 The Philosophical Background

Our approach is committed to the pragmatic philosophy in the succession of C. S. Peirce (see e.g. Peirce (1935)). *Pragmatism* (or *pragmaticism* as it was called later on) tries to react on deficiencies identified in the philosophical tradition where especially the gap between pure epistemology and practice could not be bridged in a philosophically sufficient way. The fundamental features of this position especially include a (re-)interpretation of the notion of a *concept* which has shown much influence on the philosophical discussion since then. While even a rough outline of the philosophical insights would certainly be out of the scope of this paper, we will instead briefly mention some features that may illustrate the influences this position has shown for the design of something carrying the name of an *information system.*

According to the philosophical tradition originating from the (idealistic) position of Kant, knowledge is composed via perception and subsumption of concepts under categories. While sharing this general approach (at least in principle) so far, Peirce emphasizes at this point that the actual establishment of knowledge is performed by individuals and in addition results in an orientation in the 'world'. This gives rise to the idea that some even more fundamental structure or categories prior to the application of categories in the classical sense must be present to enable subsumption process. These are the Peircean categories *firstness, secondness,* and *thirdness.* They are thought of as *universal* categories, i.e. categories that are always present in every attempt to establish knowledge (which are to be distinguished from particular ones that constitute a set or series of alternatives from which only one applies in each actual situation).

Because of their general epistemic status, they are also the basis of the semiotical relationships. According to the so-called *pragmatic semiotics* we have to deal with *signs, concepts*, and *objects* together with their relations. Concepts on the other hand must not be separated from the process of *establishing* knowledge (resp. orientation) and thus may not be thought without possible consequences or actions associated with them in our imagination.[1]

In the sequel a purely analytical reinterpretation of the semiotical relationships on the basis of the three categories by Morris (cf. Morris (1946)) leads to the distinction of *syntax, semantics*, and *pragmatics.* In the following we will adopt this analytical reduction for our considerations.

Philosophical ideas pre-structure the domain of things that may be dealt with. So these structures are all phenomena implicitly underlying and thus have to be taken into consideration in order to establish a cognitively adequate (information) model. But since the main focus of this paper is not on philosophical issues, we will interrupt the discussion of the philosophical background at this point. Instead, figure 1 is meant to serve as an 'icon' to represent the underlying ideas.

[1] "Considérer quels sont les effects pratiques, que nous pensons pouvoir être produits par l'object de notre conception. La conception de tous ces effets est la conception complète de l'object." Peirce (1935, p. 18; original in French)

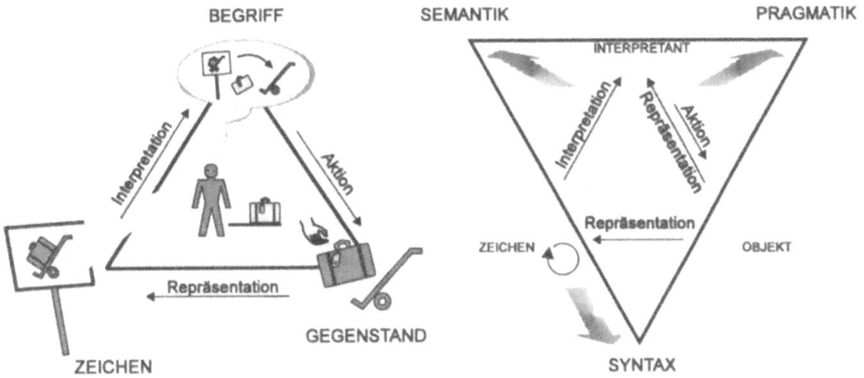

Figure 1: Dimensions of knowledge

From a purely philosophical point of view this certainly may be viewed as a shortening of the categorical status. However, we wish to emphasize at this place that a third dimension of human (inter-)action is added to the syntactical-semantical relationship which has been widely considered as the basis of the 'Logic of Science'; for a more detailed exposition of our approach see (Lenski and Wette-Roch (1996)). In the following, we will rather enter into a discussion of possible consequences for the design of an information system.

3. Consequences for the Design of an Information System for a Scientific Field

In this section we will show that these philosophical considerations are not only artificial or merely provide a foundational background but influence the design of an information system in a direct way. The attempt is to represent a scientific field in all of its aspects. According to the list of statements given in the previous paragraph we distinguish several levels of representation.

3.1 An Information System on Mathematical Logic

In this section we first describe the research enterprise *Mathematical Logic* of the *Heidelberger Akademie der Wissenschaften.* This is thought of as a general paradigm and at the same time constitutes a concrete setting for the questions under discussion. Moreover, we will then show how the theoretical considerations will again find their concrete expression.

3.1.1 The Bibliography: The *Mathematical Logic* enterprise maintains a comprehensive bibliography of mathematical logic which is meant to provide a complete overview over research activities in the field. The bibliographical entries are classified according to the *Omega Classification Scheme of Mathematical Logic* under the supervision of an international group of experts for specific subfields. Moreover, the Bibliography cites reviews in the leading reviewing journals (cf. Lenski, Richter, Wette-Roch (1995) for a more detailed description).

In addition, a structured thesaurus of the concepts of mathematical logic is currently being built which is meant to supplement the collection of publications by a nomenclature of the concepts of mathematical logic.

3.1.2 Data Organization: In this context it is sufficient to assume that the standard structure of bibliographical records is imposed onto the Bibliography. A frame-based representation of the thesaurus reflects the contents of the literature on mathematical logic (Wette-Roch, E. (1994)). A link to the *Zentralblatt für Mathematische Logik und ihre Grenzgebiete/Mathematics Abstracts* (*Zentralblatt* or simply *Zbl* for short) will make reviews of the *Zentralblatt* accessible. These are referred to as *Abstracts ABS* in the data model of figure 2.

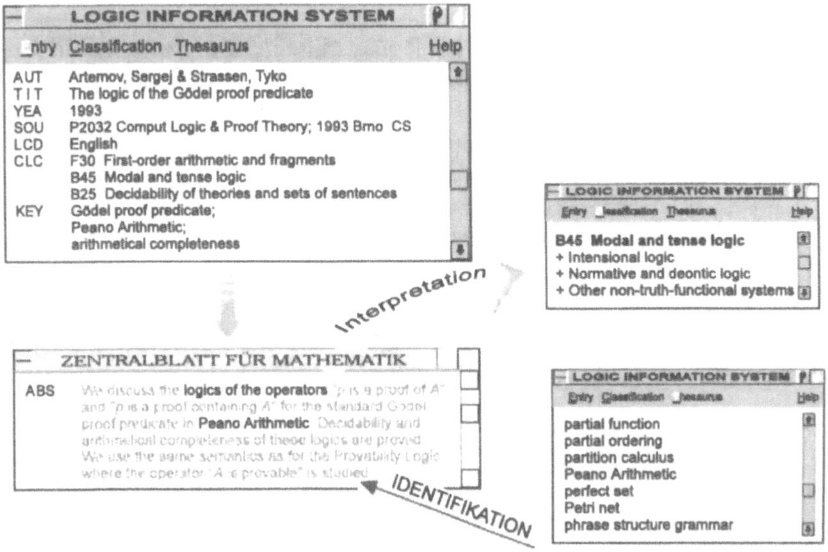

Figure 2: Data organization in the context of the Bibliography project

3.1.3 The Logic Information System LIS: The LOGIC INFORMATION SYSTEM LIS basically offers the standard retrieval techniques of a bibliographical information system. As usual the bibliographical records are structured via tags. The thesaurus of mathematical logic is designed to add further functionalities to LIS which will be explained in the sequel. Especially, the thesaurus will provide the information system with background knowledge needed to access and process the mathematical contents. Through pointers to the reviews of the *Zentralblatt* which is electronically available via the *Fachinformationszentrum (FIZ) Karlsruhe* a sort of distributed data organization may be assumed.

3.2 Conceptual Contents Representation in LIS

Even a structured bibliography only admits a very coarse view of a scientific discipline. A need for more intrinsic description techniques for scientific activities is obvious. Now the most adequate representation of a publication is the publication itself combined with methodologies of text understanding! But since respective data processing is certainly out of the focus at the moment though the near future will certainly bring considerable progress, we have to be somewhat modest. In this situation, statement 2 hints the direction in which we may go. Accordingly, a contribution to scientific discussion (which is manifested through a publication) is expressed in the *vocabulary* of the scientific community the contribution is addressed to. Thus the aim is to find methodologies to deal with this vocabulary.

3.2.1 Concepts: The terminology of a scientific field is bound to (more or less) well-defined concepts or concepts a scientific community has agreed upon. The concepts in a field are in general not isolated units but are interrelated through definitions or explanations.

3.2.2 Descriptors: The concepts dealt with in a scientific discourse are syntactically characterized as *descriptors*. The set of these represents a controlled vocabulary for the field. Moreover, the definitions in a field induce a structure on the set of descriptors which reflects the *conceptual dependencies* and interrelations. Hence an adequate representation of the concepts of a scientific field may only be provided by a *structured thesaurus*.

On the other hand this thesaurus should merely reflect dependencies which may be generally agreed upon, i.e. no counterarguments may be raised against. This means that they must be guaranteed by procedures not depending on individual opinions or actions. According to the philosophical background this means that the thesaurus relations shall only be *semantically* based.

On the other hand this requires respective methods being able to guarantee the wanted independence from individual verification procedures. In other words methods that must not depend on individual settings or actions but must be based instead on grounds of intersubjective validity. Such methods are especially provided by purely logical deductions based on definitions. Thus a structured thesaurus may only be established through defined concepts and solely include concept relations based on (logical) dependencies. (However, this does not exclude that additional relationships may be overlayed as we will see in the sequel.)

3.2.3 Indexing and Data Processing: One of the most prominent approaches to represent documents is known as *indexing*. The underlying paradigm is that the contents of documents may sufficiently be represented by the keywords a text is dealing with.

This approach has the advantage that there exist elaborate methods to handle contents representations this way. In the most prominent retrieval model, the *vector space model*, concepts are considered as isolated items independent from each other. Thus the related knowledge representation part is relatively poor but this is — from the viewpoint of an information system — compensated by highly efficient and elaborated retrieval techniques which may well compensate the lack of conceptual dependencies not being captured by the techniques.[2]

However, it may be expected that increased specifity (resp. precision) could be gained in case background knowledge is available, i.e. if also semantical relationships are used. Such background knowledge is especially provided by a structured thesaurus. Advantages are then more specific description facilities. However, in a specific field it has to be minded that the vocabulary the scientific community has agreed upon is indeed used. This requires that indexing and query formulation must be based on the standard terminology which in turn has to be accessible to users in an adequate way.

Hence in case indexing is based on a controlled vocabulary, the data processing techniques have to be supplemented with additional functionalities which actually make use of the structure of the thesaurus such as concept browsing or thesaurus based query (re-)formulation (Bates (1991)). However, since the independence of the individual concepts is not overcome this way, there is only a restricted way to reflect the complexity needed to describe research results.

3.3 Scientific Background of LIS

Whereas a thesaurus reflects the (interrelations of) concepts of a scientific field, research is certainly not described in an adequate way by just mentioning the concepts it is dealing with, i.e. by the set of descriptors. Instead the results are communicated in theorems, propositions, theses, explications, arguments, etc. Formally these are expressed as structured combinations of descriptors. These combinations may be natural language expressions, formal relationships or semi-formal statements as in mathematics or in the sciences.

3.3.1 Research Activities: Research activities in a scientific area are not isolated results. Rather groups of researchers share research *interests*. Accordingly, a scientific field is spread in the course of its development into *lines* of research or interestingness. Hence a characterization of research activities has to take this into account and outline the main streams and directions of (current) research.

3.3.2 Classification Systems: Whilst indexing via keywords may be viewed as a bottom-up approach to represent the contents of a document there exists another paradigm which may rather be characterized as a top-down approach. An adequate representation of the whole field of scientific activities may only be given by a system of research topics denoting the

[2]It should be mentioned that there have been attempts to enrich the vector space model with more background knowledge but these approaches have not been greatly acknowledged and have certainly not been considered as a break-through for this kind of approach; see e.g. (Giger (1988)) and (Meghini et al. (1993)).

landmarks in the landscape of activities in the area under consideration. A system of this kind is called an (annotated) *classification system.* It tries to capture these lines and describe it in the terminology of the field. This necessarily involves a process of abstraction of research activities and thus relies on human interpretation. Consequently, a classification system reflects *pragmatic* relationships.

3.3.3 The Classifying Process: While indexing techniques work independently of the scope (however the specific application may well determine the choice of key-expressions), it is quite obvious that scientific classification depends on a (more or less) complete survey of research activities, because it is designed to locate each individual research contribution within the network of general reserach interests.

Accordingly, the establishment of a classification system in a first step amounts to the awareness of research activities and then in a second step to the identification of research lines in which scientific activities may be bundled. These are often influenced by outstanding research results or methods which constitute a break-through in a scientific field. Since in general an act of interpretation is involved, there do not exist explicit criteria for the wanted coordination of individual contributions within the classification system. This again demonstrates the pragmatic character of a classification system. Moreover, this approach is very sensitive to scientific developments or changes of scientific paradigms which may result in revisions of the underlying paradigms (and thus cause reclassifications of all the documents). The advantage of this approach, however, is the chance to index documents by complex concept relations which are controlled by a fixed system of research activities.

3.4 Individual Research Contributions

Now we turn back to the single documents. In the sequel we will discuss ways in which research results are presented and describe methods to handle these in an information system.

3.4.1 Abstracts: The first way to represent the contents of papers besides the full text is an *abstract.* This is mostly written as a more or less condensed description of the contents oftenly combined with a scientific relevance judgement. As indicated by the notion *abstract* it is designed as an annotated abstraction of the contents surveying the results in a rather colloquial way i.e. not in the usual formal scientific language. As such it is primarily addressed to human recipients and not designed for computer based evaluation although it may well constitute the basic for computer based methodologies to extract, represent, and process information on the publication.

3.4.2 Phrases: Formally, abstracts may best be viewed as a sequence of phrases, i.e. natural language expressions formulated in the scientific vocabulary. It is based on a semi-formal language definitely different from a purely formal scientific language. The syntactical structure of the description of single research results and classification items is essentially the same.

It is obvious that research results are *pragmatic* by relating concepts to form propositions which do not merely reflect definitorial dependencies. On the other hand, the scientific paradigms in mathematics and the sciences imply an aspect of intersubjectivity in so far as mathematics and the sciences have established verification techniques which are independent of personal judgements (although in the history of science there exist remarkable exceptions where the question whether a proposed verification procedure may indeed be named a 'proof' was not totally agreed upon in the scientific community). However, it should be noted that such propositions may well establish semantical relationships via proofs. For example, the theorem stating that regular grammars are exactly those which are recognized by finite automata constitutes through its proof a relationship between the concepts 'regular grammar' and 'finite automata' which is by no means determined by purely conceptual relationship but by *structural* relationships.

3.5 Relating Research Results and Classifications

While representation models based on natural language processing seem out of focus at present, keywords and classifications may well be subject to automated procedures. Now classifications and single research results may both syntactically be viewed as (linguistic) combinations of keywords. So in one respect we have overcome the isolation of keywords as assumed in the vector space model. However, one principle key problem remains to be solved: the interplay of keywords with classifications.

To relate single research results to research lines involves an aspect of abstraction. We have seen that classification items are *abstractions* standing for *lines* of research rather than for individual research results. We will show how this process of abstraction can be modelled by manipulation techniques in the thesaurus. Given the thesaurus we have available all its concept relations. The basic relationship ('is-a') now implies a natural subsumption asymmetry. The same is true for concepts relations given by proofs. However, the identification of research lines also includes 'abstraction' phenomena which are definitely *not* grounded in the concept hierarchy (nor are given by proofs) but which are due to human interpretation claiming, however, that they are shared by the majority of the scientific community. An example is provided by the fact that (effective) quantifier elimination is almost exclusively used in order to proof the decidability of theories involving model theoretic methods. Consequently, this proof procedure should be rather subsumed into model theory instead of recursion theory where the concept 'decidability' belongs to by definition.

So in order to make this abstraction process leading to classifications operational we have to extend the concept relationships of the thesaurus by links which are merely required for this purpose. As such those links are clearly *pragmatic* links, because they provide the necessary paths to end the concept network propagation in the concepts appearing in the classification item. This is related to the usual abstraction process where the concepts higher up in the concept hierarchy are the more abstract ones. Now this exactly constitutes the guideline to relate individual research results to classification items: It may be interpreted as an abstraction process which can be procedurally described as subsumption in the extended concept hierarchy of the thesaurus including *pragmatic* links; for technical details cf. (Lenski, Richter, and Wette-Roch (1995)).

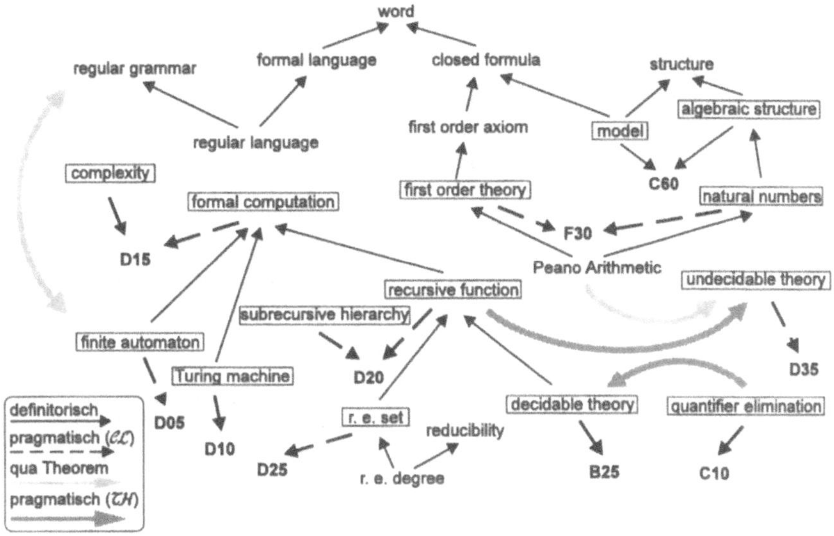

Figure 3: Extended thesaurus with additional pragmatic links

4. Summary

In this paper we have presented an attempt for a foundation of the different issues which appeared in the course of the development of an information system on mathematical logic. It has been demonstrated that philosophical aspects may well be taken into account for the design of an information system and are certainly needed to clarify the epistemological status of the basic concepts. We have given evidence that philosophical ideas are not only needed for foundational issue but rather influence the design of an information system in a direct way. In this context, we have argued that the pragmatic philosophy provides the theoretical grounds for activities in this area. More concretely, it has been shown that besides syntactical relationships also semantical and pragmatical aspects both have to observed and clearly distinguished from each other. We have thus identified the basis, described the principal positions and outlined the directions to be observed for automated procedures.

Based on this approach we have built up a prototype which has proven to show promising results for the data that had already been entered. Future work will be to put more material into this prototype in order to have better grounds for a judgement how powerful the system LIS/Thesaurus could be.

References:

APEL, K.-O. (1967): Der Denkweg von Charles S. Peirce. Frankfurt.

GIGER, H. P. (1988): Concept Based Retrieval in Classical IR Systems. In: *Proceedings of the Eleventh Annual International ACM SIGIR Conference on Research and Development in Information Retrieval.* ACM, New York, 275–289.

LENSKI, W.; RICHTER, M. M. and WETTE-ROCH, E. (1995): Consistency Conditions for the Classification in LIS/Cl. In: W. Gaul, D. Pfeifer (eds.): *From Data to Knowledge. Proceedings of the 18th Annual Conference of the Gesellschaft für Klassifikation.* Springer-Verlag, Heidelberg, 433–441.

LENSKI, W. and WETTE-ROCH, E. (1996): Terminologie und Wissensrepräsentation in pragmatischer Sichweise. In: H. Czap, P. Jaenecke, H.P. Ohly (eds.): *Analogie in der Wissensrepräsentation. Case-Based Reasoning und Räumliche Modelle.* Fortschritte in der Wissensorganisation, Bd. 4. Indeks Verlag, Frankfurt, 147–159.

MEGHINI, C. et al. (1993): A Model of Information Retrieval based on a Terminological Logic. In: *SIGIR '93: 16th Annual International ACM/SIGIR Conference on Research and Development in Information Retrieval*, ACM, New York, 298–307.

MORRIS, C. W. (1946): Signs, Language, and Behaviour. New York.

MÜLLER, G. H.; LENSKI, W. & BLASS; A. R., EBBINGHAUS, H.-D.; DALEN van; D.; HINMAN, P. G.; KISTER, J.; RAUTENBERG, W. and TROELSTRA, A. S. (1987): **Ω**-Bibliography of Mathematical Logic, Vols. I-VI, Springer-Verlag, Heidelberg, New York.

PEIRCE, C. S. (1931-1935): Collected Papers, Vol. I–VI, ed. by Ch. Hartshorne and P. Weiss. Harvard University Press, Cambridge.

WETTE-ROCH, E. (1994): Conceptual Structures in Mathematical Logic and Their Formal Representation, in: H.-H. Bock, W. Lenski, M. M. Richter (eds.): *Information Systems and Data Analysis.* Springer-Verlag Heidelberg, 204–214.

Section 5

Text Analysis and Information Retrieval

A Planning-Based Approach to Intelligent Information Retrieval in Text Databases

César Carranza, Wolfgang Lenski

Fachbereich Informatik, Universität Kaiserslautern,
Postfach 3049, D-67663 Kaiserslautern, Germany

Summary: Seeking information in text databases is a complex activity. It requires a variety of knowledge from the user. With the application of techniques from Artificial Intelligence we try to improve the retrieval effectiveness of traditional information retrieval systems (IRS). In particular, in the CoBRA system we study the use of planning techniques for the control of the knowledge-based information retrieval process, the transformation of the search terms into the descriptors known to the system and their conceptual reformulation, and the generation of the search plan for the IRS. We define the approach in CoBRA as cooperative planning for the information seeking process.

1. Introduction

The goal of an IRS is to retrieve documents that could be relevant to a particular user's information need. The decision to retrieve a document cannot be established in general by system's simple string-comparison of the index terms with the search terms or statistical retrieval techniques[1]. However, the natural human process of judgment whether a document is relevant to his information need involves a process of inference. For example, the content of the document's abstract together with the user's knowledge of the subject area is used to infer that the document will be a valuable source of information. In this situation methods from Artificial Intelligence (AI) are applied in the field of Information Retrieval (IR) to represent in declarative form the knowledge about the subject, to improve the dialog with the user, and to provide inference mechanisms. The field dealing with the application of AI methods in IR is known as *Intelligent Information Retrieval* (Croft (1987)). Hence traditional IR retrieval techniques and AI need not be seen as rivals. IR techniques can be applied independently of the domain and deal with very large text databases while AI techniques can be used to improve the quality of the retrieval process (Croft (1993)).

It is generally agreed upon that effective searches in bibliographic or other highly structured databases require elaborate expertise. On the other hand it is certainly desirable to have categorized data structures underlying because of the advanced retrieval facilities they offer. Moreover, tendencies

[1]A retrieval technique is defined as a technique for comparing the representation of a query with the document's representation (for a classification of retrieval techniques see Belkin and Croft (1987)).

(that are certainly pushed ahead by the WWW) may be identified which are characterized by hiding detailed background structures through the design of the user interface. The user often does not even know the type of machine he is dealing with not to speak of the underlying retrieval language. In this type of situation AI planning techniques may be applied to balance this seeminly diverging tendencies.

2. An Approach to IIR: the CoBRA System

With the aim of supporting a user in the search for documents in bibliographic databases we are currently implementing the CoBRA[2] system. The idea is to develop a flexible knowledge-based information retrieval assistant that based on a (pre-)categorization of the user's information need (see below) adapts its different ressources to find the possibly relevant documents. Especially, the system is designed to provide support for problems an unexperienced user of an IRS may find himself confronted with in the course of an interaction process to retrieve the relevant documents:

- *Which of the system's functionalities (Browsing, Direct Query Formulation with or without Inferences, etc.) may be used to formulate my information need?* A *search strategy* depending on the type of the user's information need is required!
- *What are the terms to be used to express the information need?* An identification of the appropriate descriptors that are most suitable to describe the user's information need and the relationship between them should be supported!
- *What are the query formulations recognized by the system's functionalities?* A *search plan* that the IRS has to follow must be generated!
- *May I have a chance to improve the query results through a reformulation of my query?* Conceptual assistance for query (re)formulation in case the user is not satisfied with the results should be accessible!

The approach in CoBRA uses multiple IR techniques (e.g. boolean, vector-based, cluster-based retrieval techniques), a frame-based representation to model the concepts and objects in the domain, and AI planning techniques to control the knowledge-based information retrieval process, the transformation of the search terms into the system-known descriptors and the generation of the search plan for the IRS. In all these subprocesses performed by the system, the user has an active role as we will explain below.

2.1 The Knowledge-based Information Retrieval Process

The search for information is not a one-step process. It involves an iterative process with the following parts which will just be sketched in the following figure.

---

[2]CoBRA = Cooperative knOwledge-Based Information Retrieval Assistant

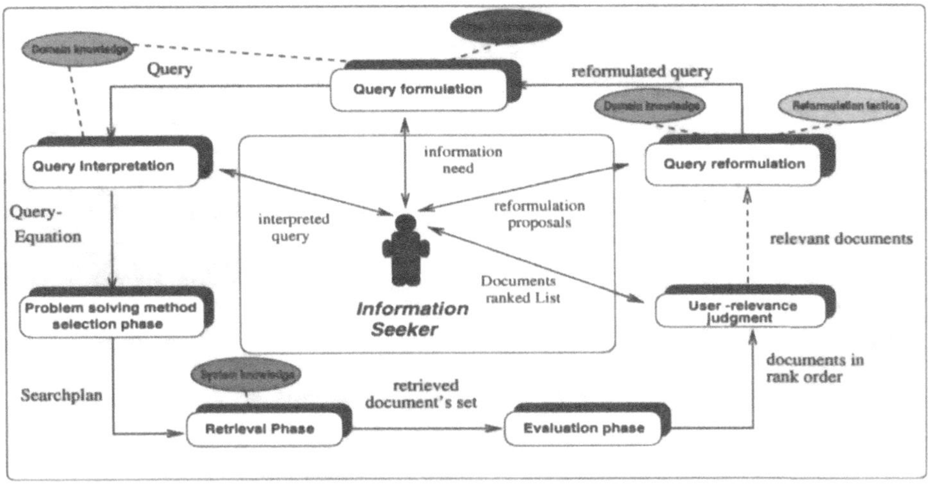

Figure 1: Knowledge-based IR Process

1. **Query formulation**. The information seeker expresses his request in pre-structured templates offering several levels of expertise.
2. **Query interpretation**. The request passed to CoBRA is analyzed and transformed to the system's internal representation: the query equation. Standard linguistic treatment of the user's search terms for the recognition of the index terms is applied. Moreover, modifications based on concept relations provided by a thesaurus are offered and transformations into (a set of) requests understood by the system's functionalities is performed.
3. **Problem solving method selection phase**. In this phase, the query equation is analyzed to capture aspects which permit the selection of an appropriate retrieval technique (boolean retrieval, vector matching, cluster-based retrieval, concept navigation, etc.) Also it can be decided whether and how to apply several retrieval techniques to the same request. As a result a *search plan* is generated.
4. **Retrieval phase**. The *search plan* generated in the above phase is converted into commands of the query language. The IRS then retrieves the appertaining documents.
5. **Evaluation phase**. In this phase the retrieved documents are ranked e.g. in decreasing order based on a query-document similarity measure.
6. **User-relevance judgment**. The system displays the retrieved documents to be judged by the information seeker. He expresses relevance statements according to his information need.
7. **Query reformulation**. Normally a query must be reformulated to improve precision. Extensive use of the information offerings of the system may be applied here, e.g. browsing in the thesaurus.

Figure 2: Domain representation in CobRA

2.2 Domain Representation in CoBRA

The domain representation is a frame-based network in which the relations betweeen the objects of a bibliographical database are represented (figure 2). It consists of objects that describe the formal attributes of the documents (such as document type, title, author, language, etc.) and on the other side of objects that are used to represent the content of the documents. In the latter ones we distinguish between classification codes, subject descriptors, and keywords. While the first two belong to a controlled vocabulary, the keywords are terms given by the author or are generated by automatic indexing procedures. However, it should be noted that not all items (objects) in the domain are index terms (i.e. terms which actually have a reference to bibliographic items) and that the items in the domain model are not mapped one-to-one to the database items.

3. Planning in CoBRA

In our point of view, information seeking in (partly) structured databases may be considered as a planning task where decisions in the course of the retrieval process (e.g. which descriptors may be appropriate to (re)formulate the query, or how to implements the query, etc.) can be supported by AI planning techniques. Before we explain the use of planning in this field, the AI planning problem will be sketched.

3.1 The AI Planning Problem

A planning setting consists of a formal description of a class of objects in some formal language, a set of operators that can be applied to transform states of the objects, and two special states: the initial state and the goal. The planning problem is to find (effectively) a sequence of operators turning the initial state into the goal. The standard example in the traditional planning domain is known as "Blocks World" where we have a table and three blocks on them. An initial state is some configuration of the blocks (for an example see fig. 3a) and the goal is the configuration of figure 3b. Each state is described by the collection of *On* and *Clear* predicates.

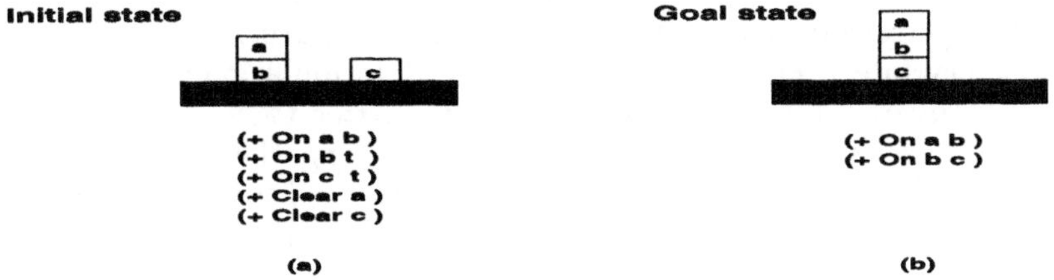

Figure 3: Problem in the Blocks World

Two operators are allowed to use to derive the desired state from the initial state: the operators *ClearTop* and *PutOn*. *ClearTop(x,y,z)* moves block y from the top of x to the top of z ; *PutOn(x,y,z)* puts the block x being on z on y. Each operator is defined using the precondition-effects representation where the preconditions are the propositions that must be valid in order to apply the operator and the effects are the changes caused by the application of the operator. For the given problem (fig. 3) the generated plan is as follows:

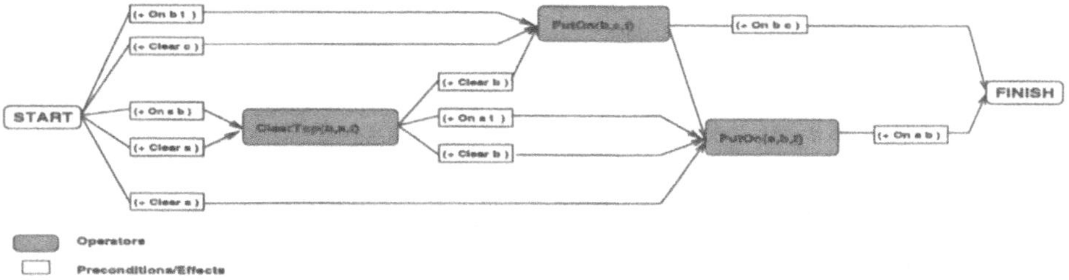

Figure 4: Solution of the above problem

According to figure 4 we first apply the *ClearTop* with the parameters *ClearTop(b,a,t)* (*t* for Table) then *PutOn(b,c,t)* and finally *PutOn(a,b,t)*. The important aspect here is that the generation of the plan is not obtained through trial and error but by goal directed search (for other ways to find plans in AI see Hendler et al. (1990)).

3.2 The Information Seeking (IS) Planning Problem

Within the framework of a multi-strategy knowledge-based information retrieval system, there are many ways to generate the *search plan* to retrieve the documents for the user's information need. In the IS planning problem the goal is to retrieve possibly relevant documents, the initial state is defined by the search-profile of the user and the type of his information need. To generate the *search plan*, planning activities occur at three levels:

1. **Planning at the strategic level.** For each of the subprocesses of figure 1 appropriate methods are selected to perform the task in question. At this level a rough plan containing the sequence for the overall IR process (selection of all methods for each subprocess) is generated.
2. **Planning at the conceptual level.** At this level the query is conceptually transformed partly based on the thesaurus. This occurs in the Query Interpretation as well as in the Reformulation phase of figure 1.
3. **Planning at the operational level.** After CoBRA obtains an acceptable internal representation of the query, the system selects the operations for processing the data and the order in which to perform them. This planning activity occurs in the problem solving method selection phase.

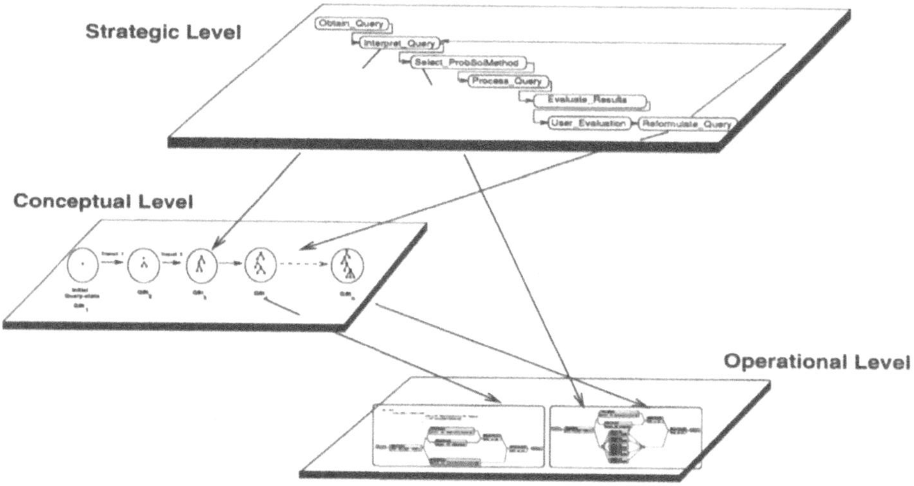

Figure 5: Interaction between the three levels

3.2.1 Planning at the strategic level

At the beginning of the planning process the user is faced with a few categories to express his information needs. These categories are independent of the structure of the underlying database(s). At the top level the goal of the system is to solve the information seeking problem. This goal is split into subgoals (each subprocess of figure 1 is a subgoal). Each subgoal determines tasks that solve this goal. Associated with each task are one or more methods that perform it. Based on the type of the information need the system configurates its methods for a session search. A brief overview of some typical situations will at the same time act as a guide and demonstrate the standard planning situations at the strategic level.

1. *Precise need:* the user nearly knows which documents are needed. He enters the identification numbers of the documents or the search terms he knows will retrieve the documents searched for.
2. *Exploratory need:* the user does not have a definite request for specific documents. What he wants is an overview of the database content. The appropriate *search strategy* is rather concept browsing in a thesaurus guided by a classification system.
3. *Thematic need:* the user is motivated by a more or less broad theme. As an adequate *search strategy* the system should offer concept navigation in a thesaurus.
4. *Connotative need:* the user has a definite need for information but he doesn't known the descriptors to find the desired documents. These are the more complicated user's queries which require more elaborate procedures: He makes an indirect reference, for example using circumscriptions for the descriptors. The system first tries to identify the specialized (sub-)databases being concerned with the user's request. It then handles the specific query language or functionalities according to the data that has been entered.

An example will illustrate the use of the system's search strategy: A person has recently read a paper on a certain subject. He then wants to know which papers have an association with the subject in question and addresses the problem to the system. This situation will then be analysed as a multi-step retrieval problem combining strategies. At first, the system will offer the possibility to input directly the title of the paper he has just read (*precise need*). The system will then find an exact match for this title and the index terms of the associated document will be used to find more documents about the same subject. (The *search strategy* applied so far is called "Known Item Instantiation" and is often used by human expert searchers.) If the user is interested in more information on related subjects, the system will then enter into planning strategies at the conceptual level (*thematic need*) (Bates (1990)).

3.2.2 Planning at the conceptual level

Planning at the conceptual level is primarily understood as a contents-driven process of query reformulation guided by conceptual dependencies or relationships. Planning in this context is *not* understood as a query revision process driven by relevance feedback provided by users (although users may well be involved into the query reformulation process). Instead, it is understood that a concept-relations-driven query reformulation may be produced by the system prior to relevance-feedback interaction.

To perform this task background knowledge of concept relationships must be accessible through the system. This implies that a thesaurus must be present covering the concepts which may be involved in the retrieval process. Whilst conceptual knowledge about 'the world' is out of focus at present, planning techniques may well be applied for information systems that are based on a controlled vocabulary in a specific (scientific) field.[3]

Now the setting for planning has to identifiy a class of objects to be dealt with as well as to specify operators on these objects. The class of objects for the planning procedures is clearly given by the concepts of the domain thesaurus. The operators may then be defined by 'adding new descriptors' to the query or by 'replacing descriptors by others'. Hence it remains to describe the selection process of new descriptors. It is obvious that this selection process has to be based on concept relations provided by the thesaurus. However, decisions have to be made concerning the general design of those relations in the domain knowledge base which may be subject to planning strategies. Here, in principal, different approaches may be assumed which in turn depend on philosophical considerations concerning the status of possible (concept) relationships (Lenski and Wette-Roch (1996)). Browsing through the thesaurus should especially be enabled in the planning process when no descriptors in the thesaurus may be identified to match the search terms in a request and thus descriptors the system will be able to deal with cannot be generated automatically.

First one could assume that only concept relations definable through the generic concept hierarchy may be allowed for use in the planning procedures and thus for query reformulations. These are essentially the sub- and superconcept-relations together with relations definable through these e.g. sibling concepts (of various stages) which are defined as the subconcepts of the superconcept(s) (Shute and Smith(1993)).

However, there may well situations be anticipated which make it highly desirable to include further kinds of relationship which may also be present in the thesaurus but which do not necessarily result in the basic 'adding/replacing'-operations in the Query Reformulation and Interpretation phase. Because of their epistemological status user-driven navigation should be involved to cover these additional features. These aspects require further discussion which can not be carried out in this paper (Lenski and Wette-Roch (1996)).

3.2.3 Planning at the operational level

As mentioned above, the user should not be concerned with specific retrieval languages. Since multiple databases with different retrieval languages may be involved in the retrieval process, a user otherwise must have detailed knowledge about these in the classical approach. Moreover, adding further functionalities to the IRS would require awareness of these as well as expertise in the underlying languages before the system's retrieval capabilities could be fully exhausted by the users. We draw the consequence that the organization of the operational process should be rather put on the system's side. Once CoBRA obtains an acceptable declarative representation of the query — which is a collection of expressions — it selects the adequate data manipulation operators for each expression and the order in which to per-

[3]An information system of this kind is currently being developed for mathematical logic by the LIS/Thesaurus project (Lenski, Richter, and Wette-Roch (1995)).

form these. These operations constitute the *search plan* that is executed by the IRS. At this level we have defined the following operators:

Operator	**Purpose**
Retrieve class :	gets a set of document identifiers (*Doc_Ids*) of a certain class.
Filter concept :	selects a subset of *Doc_Ids* out of some class.
Generate relation :	obtains the values of a relation for a set of *Doc_Ids*.
Filter relation :	puts constraints on the values of a relation.
Compare :	compares the values of a relation with a term/ list of terms.
Join :	combines two sets of *Doc_Ids*.
Compute :	performs additional processing, e.g. computes the differences of two numbers.

Suppose a user is interested in *documents that are articles in journals whose date of publication are greater than 1990 and have as their main theme the use of thesaurus and planning techniques in information retrieval.* At the conceptual level the system identifies *articles in journals* as a subclass of the concept Document Type (*DT_ArticleInJournal*, see figure 2), the *date of publication* is an attribute of this concept, the search term *information retrieval* is identified as a controlled term in the thesaurus (in short *CT_IR*) and for the other search terms of the query (*thesaurus* and *planning*) the system cannot find associated descriptors. But it finds out that these are keywords and can be processed using the keyword relation. Moreover, for the term *planning* it decides to truncate it to *plan** using the wildcard *, since the system finds different words such planning, planners, etc. . So the query equation is:

```
(:in
   (:in
      (DT_ArticleInJournal ?C_1)
      (greater_date_of_publication ?C_1 "1990") )
   (:and
      (:and
         (keyword_rel ?C_1 "plan*")
         (keyword_rel ?C_1 "thesaurus") )
      (CT_IR ?C_1) ) )
```

Figure 6: Query Equation

The above query equation is the goal for the planning problem at the opera tional level.

Some remarks before we explain the *search plan* of figure 7: first, the upper part of the boxes are the operators applied to satisfy the expressions (lower part of the boxes) which act as the effect for the operators (the preconditions are not shown). Second, we distinguish between some types of variables: those denoted by *C_number* reference to a type of class; *R_number* stands for relations; *L_number* for a list of document identifiers for a certain class and *D_number* for the values of a certain relation.

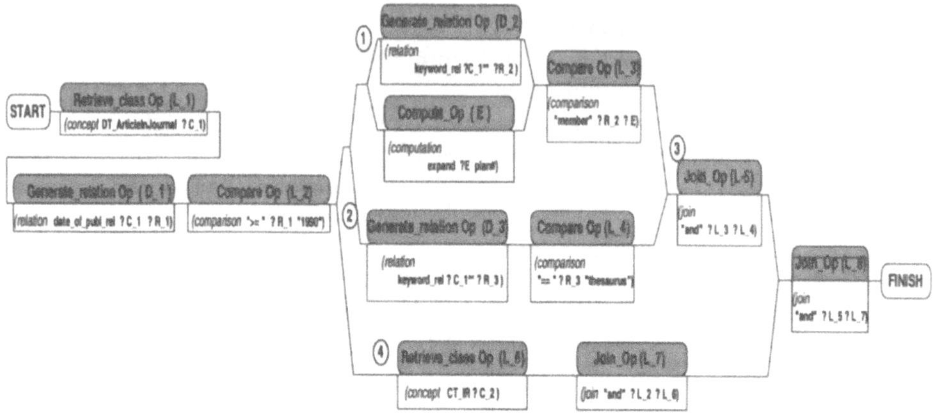

Figure 7: Search plan

So the *search plan* could be interpreted as: first the system retrieves documents of the class *DT_ArticleInJournal* (list of identifiers stored in *L_1*), then generates for these the values of the attribute (relation) *date of publication* (stored in *D_1*), uses the comparison $\geq$ to restrict the values to those greater than *1990* (stores the filtered *Doc_Ids* in *L_2*). Then various actions could be performed in parallel (to speed up the processing): the branch (1) generates values for the keyword relation and compares them with the root word *plan**, part (2) corresponds to the search for the *Doc_Ids* which have the keyword *thesaurus.* In step (3) the partial results are joined by the *and* connector which in this case is understood as the set intersection of *Doc_Ids* *L_3* and *L_4*, etc.

4. Summary

Existing approaches for IRS are inflexible in the sense that the solving process for satisfying an information need (i.e. a user's query) is fixed. If there are multiple retrieval techniques availables, several structured databases underlying, or the user has little expertise in query formulation, planning techniques may be used to hide technical details from the user and to improve search results. With the help of a planner we hope to gain more flexibility and user interaction already in the *process* of seeking for information. It synthesizes and executes plans dynamically to achieve the goal to find the possibly relevant documents according to a particular user's information need.

The planning approach will especially be useful under the following conditions: Highly structured documents with enough contents representation constitute the fundament of the database. Further background knowledge including conceptual interrelations is available. Multiple search techniques compete. And finally, query formulation differ from database to database and requires some advanced expertise even for standard retrieval situations. However, in this situation CoBRA is expected to demonstrate high quality retrieval results — higher precision combined with more user interaction. Respective evaluations will be carried out in the near future.

References:

BATES, M. (1990): Where should the person stop and the information search interface start? *Information Processing and Management, Vol. 26, No. 5, 575–591.*

BELKIN, N. and CROFT, B. W. (1987): Retrieval Techniques. *Annual Review of Information Science and Technology, Vol. 22, 109–145.*

CROFT, B. W. (1987): Approaches to Intelligent Information Retrieval. *Information Processing and Management, Vol. 23, No. 4, 249–254.*

CROFT, B. W. (1993): Knowledge-Based and Statistical Approaches to Text Retrieval. *IEEE Expert, April, 8–12.*

HENDLER, J.; TATE, A. and DRUMMON, M. (1990): AI Planning : Systems and Techniques. *AI Magazine, Vol. 11, No. 3, 61–77.*

LENSKI, W.; RICHTER, M. and WETTE-ROCH, E. (1995): Consistency Conditions for the Classification in LIS/Cl. In: W. Gaul, and D. Pfeiffer (eds.): From Data to Knowledge. Springer-Verlag, Heidelberg, 433–441.

LENSKI, W. and WETTE-ROCH, E. (1996): Foundational Aspects of Knowledge-Based Information Systems in Scientific Domains. *This volume.*

ROBERTSON (1977): The Probability Ranking Principle in Information Retrieval. *Journal of Documentation, Vol. 33, No. 4, 294–304.*

SALTON, G. and McGILL, M. (1983): Introduction to Modern Information Retrieval. McGraw-Hill Inc., New York.

SHUTE, S. and SMITH, P. (1993): Knowledge-Based Search Tactics. *Information Processing and Management, Vol. 29, No. 1, 29–45.*

3D-Navigation in Virtual Information Spaces: From Text-Based Retrieval to Cognitive User Interaction

R. Däßler, A. Otto

Institut für Geowissenschaften, Universität Potsdam,
D-14473 Potsdam, Germany
Daten- und RechenZentrum GeoForschungsZentrum Potsdam, D-14473
Potsdam, Germany

Summary: The information management of increasing masses of information has become a key problem of today's communication and information technologies. Due to the complexity of data and an information overload in databases, the access to information, namely the design of user interfaces, becomes an important factor. A new quality in the Human Machine Interaction is achieved by the modeling of abstract information spaces. In contrast to the traditional information retrieval by exact-matching and Boolean retrieval operations, Virtual Reality Interfaces allow for an interactive exploration of the information space by navigation and browsing (information access). This quasi-heuristic search for information simplifies the retrieval process as a result of the reformulation of the text-oriented retrieval dialogue with a cognitive user interaction. The *Daten- und RechenZentrum* at the *GeoForschungsZentrum* Potsdam has developed *GEOLIS*^{GFZ}, a metainformation system, which handles information on all geoscientific object types. In order to achieve a wide acceptance the information retrieval is implemented on the basis of WWW-browsers. On the application side there are the *Information Broker* and the *Navigator* available for the information retrieval. The scientist is supported by the *Information Broker* when exploring and making use of information resources, whereas in case of indistinct search criteria the *Navigator* supports the traditional retrieval process. The *Knowledge Browser* in GEOLIS^{GFZ} has been designed as a visual 3D-User Interface for the support of the text-oriented information retrieval; high demands were made on the *Knowledge Browser* referring to the user's needs. The *Knowledge Browser* consists of two main functions, the *Information Processor* that classifies the retrieval results and the *VRML-Converter* that dynamically creates a virtual information space based on the classification structure. This 3D-Retrieval Interface is platform- independent and is available within the standard WWW-browser technology.

1. Visual User Interfaces in Information Retrieval

In the information highway environment we have access to an explosive amount of data and information. The question how to benefit from these new technologies leads to the question how to find specific information. One powerful concept is to use the visualization paradigm for navigation, re-

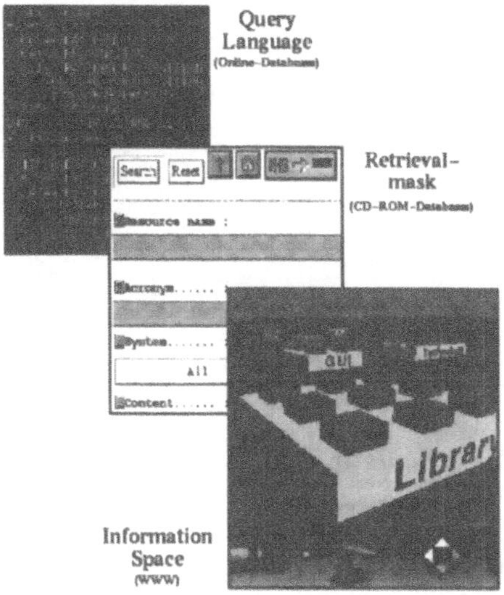

Figure 1 Evolution of Retrieval User Interfaces: Text-GUI-Information Space

trieval and access of information (Robertson et al., 1993). Current advances in computer graphics hardware and software have created the base for a new generation of user interfaces as well as for information retrieval systems. The evolution of the Human- Computer-Interface is strongly coupled with advances in computer technology. In a text- oriented environ-ment the user communicates by a command language and in retrieval systems by a query language. Most of the recent on-line databases or digital libraries are still working in this environment. On the other hand Graphical User Interfaces (GUI's) allow for advanced user interactions. Windows-Icons-Menus-Pointing interfaces are today available on nearly all platforms. With Graphical Retrieval Masks the knowledge of a special retrieval language is no longer necessary, however the retrieval dialogue is still text-based via search queries and a similarity analysis. In contrast virtual space concepts opens additional options as visualization of large data structures, navigation in data space and interaction with objects (Figure 1). Most of the recent progress in visualization techniques has occurred in the field of scientific visualization. Information spaces are abstract and differ from physical data spaces that typically have a spatial mapping. The representation of abstract data needs to create a virtual reality. The widest acceptance will probably find a virtual space with a real-worlds analog that mostly support our cognitive communication abilities. Furthermore the creation of an information space can be more difficult than a scientific visualization because the information space can be multidimensional and the kind of data representation depends much more on the specific nature of the information. However, one of the most sophisticated concept is undoubtedly the application of the Vir-

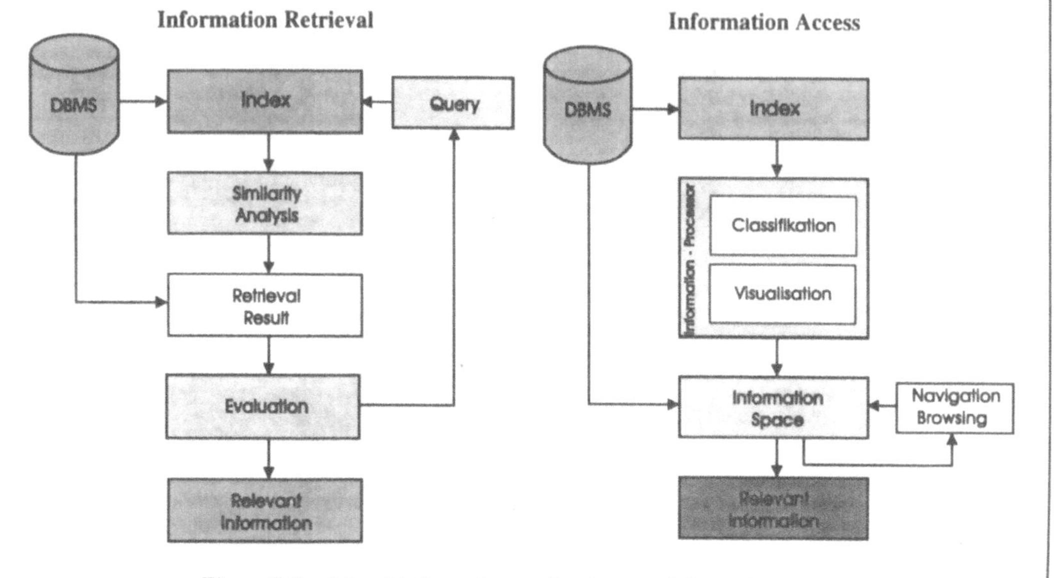

Figure 2 Traditional information retrieval versus information access

tual Reality (VR) paradigm to create and explore the abstract information space.

2. Information Access versus Information Retrieval

Information retrieval is concerned with the representation, storage, organization and accessing of information items. The advantages in high speed storage devices and networking have led to an imbalance in information sciences. Because information produces more information the problem is that we have an increasing amount of data and information to which accurate and speedy access is becoming more and more difficult. The basic retrieval process is quite simple. Information is stored and later retrieved when it is needed. On-line retrieval systems typically consist of a large document database. Terms that describe the document contents (index) are selected from manual or automatic indexing. The index terms are descriptors of the represented document. Queries are requests to process information and a search query consists of different terms combined in a structured query language. The traditional information retrieval paradigm is being a matching process according to the similarity between the keyword index entries and the search query. The problem is to find all and only the relevant documents. To evaluate the retrieval results two paradigms are used: *recall*, the percentage of all relevant documents found and *precision*, the percentage of the documents found that are relevant. From the evaluation of the retrieval results one can formulate a new query search.

The traditional model of information retrieval shown in Figure 2, however, has some strong limitations. The retrieval process might be very inefficient if the user vocabulary does not match the index vocabulary and the user can not explore parts of the document space by a context-oriented search path. The information visualization paradigm could help to overcome these problems. Virtual information spaces allow for an interactive exploration on the basis of both cognitive and motoric elements, such as navigating and browsing (information access). As shown in Figure 2 the abstract data model represented by the index is visualized as an information space. The user now has different ways to communicate. Navigation inside the information space is helpful following a context-oriented search path to find certain domains of interest. On the other hand a query leads to a certain position in the information space and a context-oriented exploration of related information is possible. The information space simplifies the user's dialogue, whereas the methodological effort with regard to information modeling and visualization does grow and additional information processing algorithms (*Information Processor*) need to be implemented.

3. The Knowledge Browser UI

The *Daten- und RechenZentrum* at the *GeoForschungsZentrum* Potsdam has developed GEOLIS^{GFZ}, a metainformation system[1] that handles information on all geoscientific object types. GEOLIS^{GFZ} has been designed to be an extendible set of software tools which have access to a common integration platform (GEOBASE^{GFZ}). There are four groups of tools available for the administration, the maintenance, and the use of the system, namely the *Operator* and the *GeoObject-Modeler* on the administrative side and the *Information Broker* and *Navigator* on the user's side. The information retrieval was implemented on the basis of WWW-Browsers in order to fulfill the user's needs. The tool *Information Broker* supports the scientist when exploring and using information resources by traditional retrieval methods, whereas the tool *Navigator* provides for advanced procedures that support the traditional retrieval dialogue, such as navigation in information spaces, relevance feedback, or natural- language user interfaces. The Visual User Interface *Knowledge Browser* is implemented for the retrieval support in GEOLIS^{GFZ}. This new VRML-based approach (Virtual Reality Modeling Language) to the modeling of the information space is to match the demands of the Internet mass market, since browsing in the information space is made possible by the use of standard WWW-Browsers (Pesce, 1995).

3.1 UI Architecture

The *Knowledge Browser* consists of two main functions, firstly the *Information Processor* that analyses the information and puts a structure to the retrieval results, and secondly the *VRML-Converter* that creates a virtual information space from the classification structure in the form of a VRML-Script. The *Knowledge Browser* may additionally be activated to a text-

[1]URL: http://www.gfz-potsdam.de/drz/geolis/

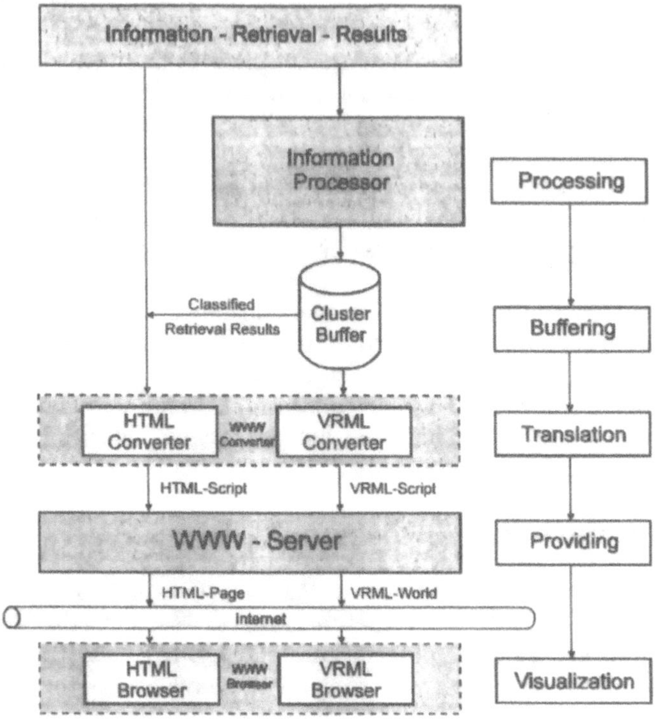

Figure 3 Knowledge Browser Architecture

oriented search query in GEOLIS^{GFZ}. It supports the retrieval when a high number of results has been found. After classifying the retrieval results the *Knowledge Browser* creates a structured information space. In this way the user is given a semantic support for the search for information; the *Information Processor*, a complex module, enables this function. The visual UI is a platform-independent, VRML-based information space which may be explored by the function of standard 3D-WWW-browsers. In this way we overcome the recent problems concerning the development of end-user tools such as VR-viewer. The *Knowledge Browser* makes use of the existent system architecture of the *Thematic-Mapper*[2], an interactive GUI which allows for the exploration of WWW information resources within a geographical context. An interactive retrieval function is achieved by the coupling of a WWW-server to a Relational Database Management System. The system architecture of the *Knowledge Browser* is shown in Figure 3. A search query via a text-oriented input mask in GEOLIS^{GFZ} is translated in the Standard Query Language (SQL) and the retrieval results are usually presented in a hypertext list which can be accessed via the WWW-Server. The *Knowledge Browser* may optionally be activated in the case the retrieval results exceed

[2]URL: http://www.gfz-potsdam.de/cgi-bin/thematic_mapper/

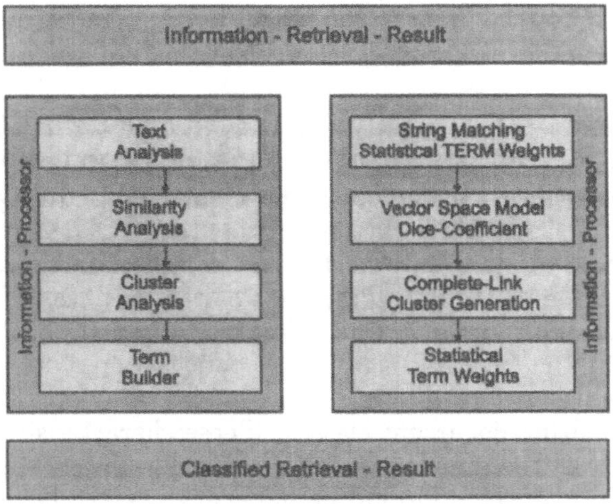

***Figure 4** Information Processor architecture*

a certain amount. A special processing module, the Information Processor is analyzing in a first step the retrieval results in real time. In a second step a VRML-scene, the virtual information space is dynamically generated and stored on the WWW-server. The processing steps which are necessary to build an information space are described in detail in the following:

1. Text analysis and information classification (*Information Processor*)
2. Storage of the classified information (*Clusterbuffer*)
3. Translation of the classification structure and link structure into a VRML-Script (information space) or HTML-Script (hypertext-list)
4. Information Access to the WWW-Scripts via an WWW-Server (Providing)
5. Presentation (Visualization) of the retrieval results in the form of a 3D-information landscape (VRML-browser) or a WWW-hypertext document (HTML-browser).

All functions of the *Knowledge Browser* are carried out on the basis of CGI-scripts (Common Gateway Interface) on the WWW-Server.

3.2 Information Processor

The *Information Processor* includes algorithms for text analysis, similarity analysis and cluster analysis as well as a term generator for the description of the document classes. The different processing steps and related standard algorithms are shown in Figure 4. In the following we present a condensed description of the algorithms for the automatic text analysis and automatic document classification. A detailed discussion of these standard procedures in the information retrieval is given elsewhere (Salton and McGill, 1983; Salton, 1989). The information processing is based on a statistical text analysis approach. The Vector Space Model is used to compute global similarities between the documents. The vector model assumes that an available index-term set is used to identify the stored information records. The term weighting of the index vocabulary is given by the following empirical approach:

$$\mathbf{TERMWEIGTHING}_k \sim \mathbf{FREQ}_{ik} \ / \ \mathbf{DOCFREQ}_k$$

Figure 4 Information Processor architecture where $\mathbf{FREQ}_{ik}$ is the document frequency of index term i in the document k and $\mathbf{DOCFREQ}_k$ is the total number of documents which include the index term i. The automatic document classification results from the standard complete-link algorithm which generates a document tree with different abstraction levels. The resulting dendrite-structure includes as a rule redundant information. An optimizing procedure results therefore in a significant condensation of the classification. The term generator determines in the last step characteristic terms which describe the resulting document classes. These characteristic terms determine the quality of the information retrieval inside the virtual information space. A summary of the information processing is given below:

1. Automatic text analysis: computation of term frequencies and term weights
2. Computation of the pairwise document similarity coefficients using the Dice coefficient
3. Determination of the hierarchical cluster structure: complete-link cluster generation
4. Condensation of the hierarchical classification structure
5. Generation of characteristic terms for the description of the document classes

3.3 Advanced Procedures for the Text Analysis

The quality of the visualization depends on the quality of the basic structure of the information space. At this time the statistical procedure, which was described above, is responsible for the structuring of the information space. In this context it is problematic to classify the documents by term frequency and additionally by term significance. This in particular holds true when documents that semantically belong to each other have to be found and to be placed into document classes. Due to this, we aim at successively improving the statistical procedure by implementing advanced procedures for text analysis such as computational-linguistical and knowledge-based procedures (Chen, 1995). The use of these advanced methods aims at concentrating firstly on the automatic indexing of full documents by controlled terms and secondly on the generation of characteristic terms that describe the classification structure. The architecture of the *Information Processor* may be developed further by advanced semantic procedures, such as semantic networks for knowledge representation, genetic algorithms for the generation of term hierarchies or the implementation of metathesauri.

3.4 Information Retrieval Process

Starting point of the visual retrieval support via *Knowledge Browser* is a

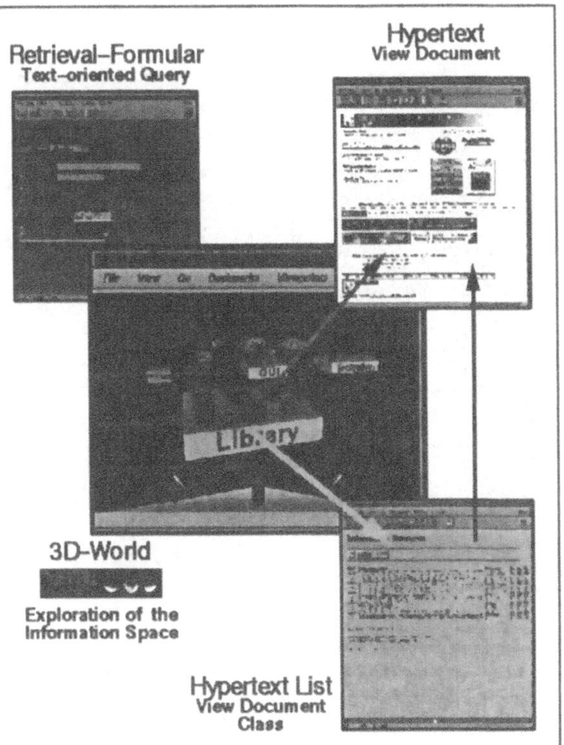

Figure 5 Information Retrieval Process

search query within the common retrieval dialog with a standard input mask in GEOLIS^{GFZ}. In case the retrieval result exceeds the defined number of retrieved information units (e.g. 30), the user may choose the retrieval support of navigating in a virtual information landscape. The 3D- information space may be navigated by common WWW-browsers and allows for a direct access to both the single information resource (e.g. a HTML-page) and the infor-mation cluster (hypertext list). The different levels of the information retrieval process are shown in Figure 5. The arrows represent the user-controlled information access via hyper-links. A new search query dynamically generates a new information space. The *Knowledge Browser* has been tested with the bibliographic reference database, an essential part of the metainformation system GEOLIS^{GFZ}.

4. The Virtual Information Space

The Virtual Reality Modeling Language has been established as a graphic standard for modeling 3-dimensional objects and virtual 3D-worlds in the Internet. Future generations of WWW-Browsers will be influenced by this standard. Now there are external VRML-Browsers (e.g. Webspace) and

Figure 6 Design of the virtual information space

WWW-Browsers that already include the 3D- capability (e.g. Quarterdeck or Harmony) or that (e.g. Life3D) may be made capable of the 3D-function by the help of Plug-In Software (Frey, 1996). VRML offers decisive advantages for the development of visual user interfaces for information retrieval systems. The VRML standard offers the following essential advantages in the design of 3D-UIs. VRML is a platform-independent graphic standard which is already supported by many WWW-Browsers. The development of the VRML-software is confined to the semantic and graphic design of the information space.

4.1 The Design of the Information Space

The algorithm which has been implemented in the *Knowledge Browser* generates a 3D hierarchic information landscape (Figure 6). Tree structures give the best opportunity for a structural search following a context-oriented search path. The classification of objects and the hierarchical data structure make it easy to browse the information space and finally to find information domains of special interests. One of the most efficient concepts of VR-based UI's is the paradigm of the information landscape; for the first time presented as the prototype File System Navigator[3] (Tesler and Strasnick, 1992). Fundamental aspects of this design concept for abstract information spaces have been implemented in the *Knowledge Browser* in a modified form. The information landscape consists of blocks (information classes), cells (final information resources within the information class and single documents) and connectors (relations). Each block and each cell is linked to a certain infor-

[3]URL: http://www.sgi.com/Fun/free/cool_sw_01.html

mation source. In this way any final information resource may be reached from the inside of the 3D-information space. The yellow blocks mark possible context-based searchpaths in this hierarchy. Apart from the interactive navigation in the information space it is possible to define various viewpoints on and entry points into this virtual world.

5. Summary

Increasing amounts of data and information and the unlimited access to information in advanced network environments such as the Internet have created the need for a new generation of UI's for querying, accessing and retrieving information. Major commercial on- line information services and network engines use keyword matching or text-oriented query languages for information managing. On the other hand, advances in graphics technology have created new possibilities for GUIs and VR-concepts.

Metainformation systems of this new generation ask for new paradigms with regard to the user's dialogue and the concept of a retrieval interface. *The Knowledge Browser*, a VRML- based retrieval interface, was therefore implemented in GEOLIS^{GFZ}, a geoscientific metainformation system. Thus the navigation in the virtual information space substitutes the traditional text-oriented retrieval dialogue. In the first place the *Knowledge Browser* supports the retrieval in bibliographic reference databases; in the future its support for the retrieval in OPACs and full document databases will be added to its range of capacity. The visualization of abstract information poses the problem of the basic classification. On the basis of statistics it will be necessary to include computational-linguistical and knowledge-based methods into the information processing.

VR-based prototypes of GUIs, such as the Knowledge Browser, are still on an experimental level. The experiences drawn from these prototypes will strongly influence the design and the efficiency of future information systems.

The user will decide whether the VR-based information space will be accepted as an UI of the next generation. The graphic design of the user interface and the classification of the information will determine the level of acceptance.

References:

CHEN, H. (1995): Machine Learning for Information Retrieval: Neural Networks, Symbolic Learning and Genetic Algorithms. *Journal Am. Soc. Inform. Sci. 46(3) 194-216.*

FREY, J. (1996): Durchblick-Die neue Generation der WWW-Browser. *c't, 2, 126-130.*

PESCE, M.(1995): VRML Browsing & Building Cyberspace, New Riders Publishing.

ROBERTSON, G. G.; CARD, S. K. and MACKINLEY, J. D. (1993): Information Visualization using 3D interactive animation. *Communications of the ACM, 36, 57-71.*

SALTON, G. and McGILL, M. J. (1983): Introduction to modern information retrieval, McGraw-Hill, New York.

SALTON, G. (1989): Automatic Text Processing: The transformation, analysis, and retrieval of information by computer. Addison Wesley, New York.

TESLER, J. and STRASNICK, S. (1992): FSN – 3D Information Landscapes. Man pages entry for an unsupported but publicity released system from Silicon Graphics. Mountain View.

A Note on Intelligent Information Retrieval Tools in the World Wide Web

Alexander Kaiser

Abteilung Angewandte Informatik,
Institut für Informationsverarbeitung und -wirtschaft
Wirtschaftsuniversität Wien, Augasse 2-6, A-1090 Wien
E-Mail: alexander.kaiser@wu-wien.ac.at

Summary: The worldwide network Internet and especially the World Wide Web is growing faster and faster. The available information has increased consequently and it is hard for the user to find relevant information. As a consequence there is a great need of Intelligent Information Retrieval Tools for finding relevant information. The following paper presents demands on Intelligent Information Retrieval Tools. Based on these demands we describe two protoypes of Intelligent Information Retrieval Tools (Letizia and the concept of Guided Information Exploration). At the end of the paper we show an approach of a case based retrieval tool which will fulfill most of the formulated demands.

1. Introduction

The worldwide network Internet is growing faster and faster. In July 1994 there were 3,212,000 hosts at the Internet. One year later, in July 1995 there were 6,642,00 hosts at the Internet (Lottor (1995)). During a year the number of hosts has doubled. In the same time the available information in the internet has increased rapidly. Table 1 shows the expansion of the Internet. However, there is a lack of adequate search engines and browsers.

Austria	92%
Germany	77%
Switzerland	40%
Europe	90%

Table 1: Growth of the Internet in 1994

As a consequence finding relevant information can be a challenge, because 99% of the available information is of no particular interest to 99% of Internet users (Bowman et al. (1994)). The World Wide Web (WWW) is decentralized, dynamic and diverse. The first WWW-hosts were installed in 1993. Since March 1994 the number of WWW-Hosts has been greater than the number of Gopher-Servers (Rutkowski (1994)). The WWW is a worldwide hypermedia system. So the great advantage of the WWW is the ease of use and the integration of multimedia like audio, video and graphics or pictures. As a consequence the number of WWW-hosts has increased

Time	Web-Server	commercial ratio	Host per Web-Server
June 1993	130	1,5%	13.000
December 1993	623	4,6%	3.475
June 1994	2.738	13,5%	1.095
December 1994	10.022	18,3%	451
June 1995	23.500	31,3%	270

Table 2: Growth of the World Wide Web (Gray (1995))

dramatically. Table 2 shows the expansion of WWW-hosts, the ratio of commercial WWW-hosts and the ratio of WWW-hosts to Internet hosts as a whole. This paper mainly concentrates on the topic of how to manage the lot of information in the World Wide Web.

2. Information Retrieval and the WWW

2.1 Status Quo

There are (at least) two possibilities to find information in the WWW:

1. The user can navigate from link to link in the hyperspace with a browser like Mosaic or Netscape. Browsers support bookmarks and hotlists. So the user can save interesting links for a later recherche.
2. The user can search for terms with search engines like Lycos, Wais, Alta Vista, etc. The result is a couple of links, which may satisfy the information need of the user.

There are a lot of problems involved with these two techniques:

- Hotlists and bookmarks typically save the title of each document only. Later the user often does not remember the content of the documents.
- Bookmarks are only useful if users consciously save the desired links. If a user has forgotten to save an interesting link there is often no chance to return to this link.
- Browser programs lack the ability to perform content-based search across multiple hypertext documents (Sun and Mayfield (1995)).
- The internet-address of bookmarks may change. So choosing a link after some time may result in error-messages like 'file not found'.
- There are different concepts of search engines. Often the user does not know the concept of a search engine.
- There are different search spaces of search engines. Often it is not clear in which document space the search engine is searching.

2.2 Demands on Intelligent Information Retrieval Tools

As a consequence of the problems shown in 2.1 we will formulate some demands on Intelligent Information Retrieval Tools (IIRT) in the WWW. We will understand IIRT as a combination of browsers and search engines.

- The need for "personal retrieval tools"
  - IIRT should help the user to formulate his information need.
  - IIRT should consider the "history" of the user. This means that IIRT should consider the personal knowledge of the user and his information need.
  - As a consequence it must be possible that two identical searches for information of two different users A and B give two different results at the same time t .
  - IIRT should perform some kind of relevance feedback of the user. There are two possibilities: Either explicit feedback as it is realized in Wais, or implicit feedback which means the interpretation of the user's actions.
  - IIRT should work in the background and assist the user on demand. That means IIRT should work while the user is searching or browsing himself.
- The need for "actual retrieval tools"
  - IIRT should verify the addresses of saved links (e.g. bookmarks). If a saved link does not have a valid address no longer, the address has to be corrected. If a correction is not possible, the saved link has to be deleted.
  - It is possible that two identical searches for information of a user A at time t and $t+1$ give two different results.
- The need for "efficient retrieval tools"
  - IIRT should make use of the users "idle-time". This is the time while a user is reading a document.
  - Similar links should be ranked according to the availability of the links.

Some of these demands are realized in search engines even now, for example the relevance feedback in Wais or the content of a link in Lycos or Alta Vista. But there are no available tools which fulfill more than two or three of these demands. In the literature there are some projects of IIRT, that fulfill many of the formulated demands above.

2.3 Letizia: An Agent That Assists Web Browsing

The following description of Letizia is based on Lieberman (1995). Letizia is a user interface agent that assists a user browsing the World Wide Web. The agent tracks user behavior and attempts to anticipate items of interest by doing concurrent, autonomous exploration of links from the user's current position. It uses a simple set of heuristics to model what the user's browsing behavior may be. So the search for information is a cooperative venture between the human user and an intelligent software agent. Letitia adopts a strategy that is midway between the conventional perspectives of information retrieval and information filtering. The goal of the Letizia agent is to automatically perform some of the exploration that the user would have done while the user is browsing these or other documents, and to evaluate the results from what it can determine to be the user's perspective. Letizia's leverage comes from overlapping search and evaluation with the "idle time" during which the user is reading a document. The best use of Letizia's recommendations is when the user is unsure of what to do next.

2.4 Guided Information Exploration

The following description of Guided Information Exploration (GIE) is based on Favela (1995). GIE works by monitoring user actions and formulating a hypothesis of her information needs. This hypothesis is used to locate documents similar to those which the information seeker has found to be interesting. In this case the computer has to act as an assistant that "looks over our shoulders" at what we do and the type of information we focus on and discard. So the formulation of hypothesis of user information needs is based on the monitoring of the user manipulation of electronic documents. The system that supports GIE must be ubiquitous and integrated with the application that the user is using to document support his decision making process. Instead of explicitly giving feedback about the relevance of the information that is presented to the user, implicit methods are used to extract this information. Such an implicit method is the interpretation of users actions. For the implementation of the GIE-process neural networks are used.

2.5 An Approach for a Case Based IIRT

An important fact formulated in the demands in 2.1 and realized with Letizia and GIE is the orientation of IIRT on the knowledge of the user and the user's bahaviour during the search- and browsing-process. By searching for literature in conventional libraries there is the possibility to solve an information need by remembering a book or article having red before, that contains similar information. We can adopt this "search strategy" for IIRT in the World Wide Web. It may be possible to make use of documents in the WWW that the user has visited in the history to solve an actual information need. A similar idea is behind the concept of Case based reasoning.

2.5.1 A Short Introduction in Case Based Reasoning

Case based reasoning means to solve a new problem by remembering a previous similar situation and by reusing information and knowledge of that situation (Aamodt and Plaza (1994)). So the main hypothesis behind Case based reasoning is quite simply "similar problems have similar solutions" - or put the other way round "you can reuse the solution of a similar problem in order to solve your actual problem" (Althoff and Bartsch-Spörl (1996)). One of the key issues in building a CBR system is to represent the cases in a way that is storable in a computer memory, accessible by a software programme and capable of capturing the true meaning of a case (Althoff and Bartsch-Spörl (1996)). There are several ways of representing a case. A case can be represented as a vector of a problem that describes the state of the world when the case occured, a solution which states the derived solution to that problem and an outcome which describes the state of the world after the case occured (Watson (1994)). So we can write for a case C_i:

$$C_i = (P_i, S_i, O_i)$$

A case basis is a set of all solved cases, so that we can write:

$$CB = \{C_1, C_2, ..., C_n\}$$

So the actual problem, which is the problem we would like to solve, can be written as

$$C' = (P', S', O')$$

Now there is the question if there is a case C_i of our case basis CB which is similar to our actual case C'.

A detailed description of the concepts of Case based reasoning and the problems is given for example in Kolodner (1993), Watson (1994), and Aamodt and Plaza (1994).

2.5.2 Case Based Reasoning and the WWW

The problem P_i in the conventional concept of Case based reasoning would correspond to the information need of a user searching for information in the WWW. The outcome O_i would correspond to the Uniform Ressource Locator (URL) of the link that satisfies this information need. The solution S_i would correspond for example to the content of a link. The solved cases are similar to the bookmarks in conventional browsers. But there are some advantages compared with the bookmark-concept:

- The URLs that have been found are based on problems. We can also represent these problems.
- It is possible to define a search space for an actual problem by the URLs of the solved cases. This search space represents a part of the "personal knowledge" of the user.

2.5.3 Problems With Case Based Reasoning in the WWW

There are some problems occuring in Case based reasoning in the WWW:

- The representation of cases
- The measure of similarity between cases
- The combination of the case based IIRT with conventional browsers and search engines

A case can be represented as a vector as shown in 2.5.1. The S_i of a vector may be either the "path" from an actual link in the WWW to the link which gives us the solution or it can be something else, for example the content of the link. It may also be possible to enlarge a vector C_i with an attribut

T_i which will represent the content of a link. Working with the content of a link has the advantage that we can compare the P' of our actual case C' with the content T_i and the problems P_i of the solved cases of our case basis CB.

2.5.4 Some Implementation Aspects of Case Based IIRT

Implementing (case based) IIRT we have to include knowledge of the field of Information Retrieval, aspects of the field of (relational) database systems and knowledge of the field of Case based reasoning. It may be possible to administrate the cases described above with database systems. We need knowledge of the field of Information Retrieval in order to index the information and searching. The case based IIRT has to make use of existing search tools in the Internet and has to be integrated in existing browsers like Netscape or Mosaic. These implementation aspects of IIRT are shown in figure 1.

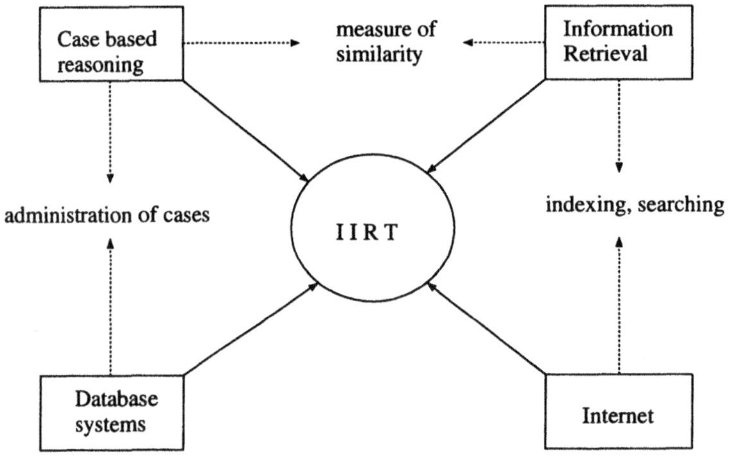

Figure 1: Implementation aspects of IIRT

3. Conclusion and Further Works

This paper has shown that there is a great need of Intelligent Information Retrieval Tools for finding relevant information in the World Wide Web. There are many problems with available browsers and search engines. The formulated demands on Intelligent Information Retrieval Tools are like a vision for a comfortable and efficient working in the World Wide Web. Some prototypes of IIRT like Letizia and the concept of Guided Information Exploration fulfill many of the formulated demands but they lack in the possibility to cover previous search results. As a consequence we presented an approach for a case based IIRT in the World Wide Web. This approach would be able to consider the personal knowledge of the user by covering previous search results. This paper has only shown the idea of a case based IIRT. Detailed implementation aspects and problems, for example the question of how to measure the similarity between two cases, will be the subject of further works.

References:

AAMODT, A. and PLAZA, E. (1994): Case-Based Reasoning: Foundational Issues, Methodological Variations, and System Approaches. In: AICom-Artificial Intelligence Communications, Vol. 7,No. 1. http://www.iiia.csic.es/People/enric/AICom.html

ALTHOFF, K. and BARTSCH-SPÖRL, B. (1996): Decision support for case-based applications. In: Wirtschaftsinformatik 38 (1996) 1, S. 8-16.

BOWMAN, C.; DANZIG, P.; MANBER, U. and SCHWARTZ, M. (1994): Scalable Internet Resource Discovery: Research Problems and Approaches. In: Communications of the ACM Vol. 37, No.8, p. 98-107

FAVELA, J. (1995): Guided Information Exploration. Technical Report. http://www.cs.uchicago.edu/~burke/aiakn/wnotes/favela.ps.

GRAY, M. (1995): WWW Server Growth. Technical report. ftp://ftp.isoc.org//isoc/charts/hosts-gifs/www.gif

KOLODNER, J. (1993): Case-Based Reasoning, Morgan Kaufman Publishers, 1993.

LIEBERMAN, H. (1995): Letizia: An Agent That Assists Web Browsing. Technical Report. http://www.cs.uchicago.edu/~burke/aiakn/wnotes/lieberman.ps

LOTTOR, M. (1995): Internet Domain Survey, Network Wizards. Technical report, January 1995. ftp://ftp.nw.com/zone/report-9501.doc

RUTKOWSKI, A. (1994): Growth of Internet Browsing Services. Technical report. ftp://ftp.isoc.org/isoc/charts/traffic-gifs/www-byte.gif

SUN, J. and MAYFIELD, J. (1995): Searching the World-Wide Web Using Signature Files. Technical Report. http://www.cs.umbc.edu/~mayfield/pubs/swiss.ps

WATSON, I. (1994): The Case for Case-Based Reasoning. Technical Report. http://www.salford.ac.uk/docs/depts/survey/staff/IWatson/ita01.htm

Computer-Aided Methods for Typification in Qualitative Social Research

U. Kelle

Sonderforschungsbereich 186,
University of Bremen, 28359 Bremen, Germany

Summary: In the past decade a variety of computer-aided techniques have been developed to aid the qualitative analysis of unstructured textual material in interpretive sociology and ethnography. This contribution gives an overview of these techniques, focusing especially on the building of typologies. To start, some fundamentals of qualitative or interpretive research will be outlined and thereafter the specific kind of typification will be discussed that can be regarded as a cornerstone of qualitative data analysis: the "data-driven" or empirically grounded construction of descriptive typologies. In the third part the general principles of *non-formatted textual database systems* that can be used to support this process of data-driven typification are described. In the fourth part some of the more complex techniques of data administration fostered by this software and how these techniques relate to typification are discussed. The final part investigates to what extent computer-aided methods can stimulate progress in the field of qualitative methodology.

1. The interpretive paradigm and its methodological requirements

In this paper the terms "qualitative research" and "qualitative data analysis" are not intended to refer to the statistical analysis of categorical data that are referred to as qualitative data in statistics, but to those traditions in social research that are rooted in hermeneutic, phenomenological or interactionist approaches and which see as their main task the interpretation and understanding of text. Although there is a great diversity of such approaches (cf. Tesch (1990: 55f)), they have in common a specific concept of human action which has been called *Interpretive Sociology* (Giddens (1976)), *Interpretive Interactionism* (Denzin (1989)) or the *Interpretive Paradigm* (Wilson (1970)).

Within the interpretive paradigm the importance of a shared system of communication, based on linguistic and other social knowledge for successful communication, is strongly emphasised. Like other actors, social scientists also have to draw upon a shared stock of culturally produced common-sense knowledge to understand the meaning of significant gestures, symbols and actions within a given social setting. Or in other words: social researchers will only be able to (adequately) discern the meanings of given empirical acts if they take into account the interpretations of the actors. This does not mean that researchers cannot develop more abstract notions than those that the actors use – but it does mean that the starting point of social research has to be the interpretations, the common-sense knowledge and the worldviews provided by the actors in the field under investigation.

If the researcher is not very familiar with the investigated social world, he or she must, at the start of their research, learn, to a certain extent, to perceive the world in the same way as the actors they are investigating do. This requirement often results in many qualitative researchers being strictly opposed to what is often seen as the methodological standard view in the social sciences, the hypothetico-deductive (H-D) model of research. It is of course debatable whether a general disapproval of H-D research can be justified, but at least if empirical research is conducted within unknown cultures or subcultures, the limitations of an H-D approach become obvious. If the aim is to discern the specific perspectives and 'local knowledge' (Geertz (1973)) held by members of a specific but unknown social world, the researcher would hardly meet with success if he or she formulated precise hypotheses and constructed survey questionnaires. Instead, interpretive research usually starts with observation, recording, listening etc., that means by collecting empirical data through interviewing and observation. Hypotheses and theories are then developed on the basis of unstructured textual data material consisting of lengthy field protocols or interview transcripts.

The necessity of starting empirical research with the collection of data and then developing theories has seduced many qualitative methodologists into adopting a rather naive inductivist model of the research process – according to this view theoretical categories will simply emerge from the empirical material (cf. Glaser and Strauss (1967), Glaser (1992)), if the researchers free their minds from useless theoretical preconceptions. However, one of the most crucial insights of the modern philosophy of science and cognitive psychology is the fact that "there are and can be no sensations unimpregnated by expectations" (Lakatos (1982: 15)). Therefore, if one does not want to follow a naive inductivist model of the research process, an important reservation has to be made here. The idea that hypotheses have to be developed after the researcher has gained an insight into the actors' perspectives must not lead to thinking of the researcher's mind as a tabula rasa. An open mind does not mean an empty head: social researchers always bring their theoretical preconceptions with them when entering a certain empirical domain. But such preconceptions often cannot be regarded as 'hypotheses' in the sense this word has in the context of an H-D research strategy (Kelle (1994)): at the start of the research process qualitative researchers' theoretical preconceptions do not represent explicit and precise propositions about measurable empirical facts. Rather, they should be referred to as *heuristic concepts* which serve as 'lenses' for the perception of the empirical world and which can be used to formulate *orientation hypotheses* (cf. Merton (1968)). At the start of qualitative field studies such heuristic concepts and orientation hypotheses are often implicit. If one looks at qualitative research projects and systematically tries to explicate heuristic concepts one can distinguish two types (the following examples are taken from research projects in which the author participated as a methodological consultant):

1. The first type of heuristic concepts used by qualitative researchers are *general topics of interest.* To take an example from social research: a project investigating social assistance started its empirical study by looking at the *trajectories of welfare recipients* and differentiated between *long-term recipients* and *short-term recipients* (Brückner (forthcoming)).
2. The second type of heuristic concepts frequently employed are concepts derived from a *general theory* of action. A research project which investigates the transition of young adults from the educational system to the labour market began the analysis of the qualitative data material by differentiating between the *aspirations, realisations* and *evaluations*[1] of the young adults to describe their decision processes during their status passage from education to work (Witzel and Mönnich (1994)).

Heuristic concepts are usually either rather *trivial* or highly *abstract* and thus have something in common: they fit many different kinds of empirical phenomena, and only very little information about the concrete domain under investigation is needed in order to use these concepts. In other words: heuristic concepts cannot be used to construct empirically contentful propositions without additional information about empirical facts. This makes them rather useless in the context of an H-D strategy, but is their strength in the context of explorative, interpretive research. Concepts of this kind can serve as a theoretical axis or "skeleton" to which the flesh of empirically contentful information is added (Strauss and Corbin (1990: 96ff.), Kelle (1994: 351ff.)). This empirically contentful information is drawn from the empirical data material which consists of unstructured verbal data and can be structured according to the heuristic concepts. The empirically contentful concepts which are developed on the basis of (1) such a theoretical axis and (2) empirical material usually take the form of typologies. To take the research project mentioned above, that conducted open interviews with young adults about their transition from vocational training to the labour market, as an example: the development of the typology started with structuring the material according to the general categories *aspirations, realisations* and *evaluations.* The project ended up discovering eight different types of biographical construction that lead to the formation of specific aspirations, for example the type "*delegation*" – young adults who try to delegate the responsibility for their occupational career to the gate keepers of the labour market, i.e. to the managers of their companies or to the officials at the employment office.

[1] *Aspirations* are rooted in individual preferences and interests that are used to account for occupational options. They relate to the experienced and anticipated conditions and situations of job-directed behaviour.
Realisations consist of actual steps of action to fulfil aspirations and the individual solutions to tasks defined by the status passage.
Evaluations focus on the results of decisions and activities. They are assessments of the relations between aspirations, conditions and the consequences of status passage results.

2. The process of typification in interpretive research

Typologies in qualitative research are constructed by integrating empirical observation into the empirically contentless skeleton of heuristic concepts. In the following this process is described in a step-by-step manner to clarify its underlying logic.

The researcher starts his or her investigation with some heuristic concepts in mind. Empirical material is collected on this basis without, however, neglecting the actors' perspectives. The balance between the researcher's heuristic concepts on the one hand and the actors' perspective on the other is best kept by semi-structured interviews with questions focusing on a specific area of interest, but simultaneously allowing the interviewees to freely unfold their relevancies (Patton (1990: 277 ff.)). If this interviewing strategy is successful, the interview will contain long narrative passages, resulting in sometimes large masses of unstructured verbal data (thirty to forty interviews with a transcribed length of sometimes more than fifty pages are quite normal for a qualitative research project).

The next step of qualitative analysis would be the structuring of this material. This is usually done by *coding*: project members look through the material, mark relevant text passages and assign codes to these passages. The codes are derived from the previously mentioned framework of heuristic concepts with which the researchers started their investigation.

This process of structuring the data is the preliminary for the actual analysis in which the analyst tries to discern relevant structures in the data. This is achieved by comparing the coded text passages in order to find commonalities or differences between them. This comparison helps to combine the pieces of information in a new way in order to develop an empirically grounded typology.

3. The use of non-formatted textual data base technology for qualitative typification

The construction of an empirically grounded typology by means of comparing text passages is a highly creative process. Nevertheless, there are certain mechanical tasks involved that have been applied as "folkloristic techniques" in the humanities or *Geisteswissenschaften* for hundreds of years. Textual material can be structured with the help of *concordances* or *indexes*. These can then be used to employ techniques for the systematic comparison of text passages which play an enormous role in different hermeneutic sciences. Take for example the technique of *synopsis* in biblical exegesis. In particular, the application of synoptical techniques to the investigation of the four gospels led to a major breakthrough in 18th and 19th century theology since it helped to overcome the dominance of dogmatic literal interpretations of the scriptures.

Qualitative researchers in the social sciences who - like theologians, historians or philologians - have to cope with large amounts of textual data, draw on similar techniques. Before the advent of computers, '*cut-and-paste*' techniques were the most widely used methods of organising the data material - the researcher had to literally 'cut up field notes, transcripts and other materials and place data relating to each coding category in a separate file folder or manila envelope' (Taylor and Bogdan (1984: 136)). However, a certain methodological problem was always prevalent with this technique: since one is forced to remove the cut-out text passage from its original context it could easily happen that, when reading the different text segments relating to one code, one finds most of them meaningless. Furthermore: "cards and file folders are reasonably workable if the number of sites is small and the data collection not extended. But they are increasingly difficult and very time-consuming as the database gets larger" (Miles and Huberman (1984: 106)).

The first attempts to computerise such manual storage-and-retrieval procedures in the early 1980s employed standard software like word processing programs or standard database systems. However, neither of these strategies proved satisfactory. Using a word processing program to copy text segments from one file to another has hardly any advantage over manual methods. And standard database systems like dBase do not offer standard features or modules that can easily be employed for a flexible definition of text segments in an already stored text corpus. Standard database applications follow a certain logic which often contradicts qualitative analysis: the researcher is forced to define the record structure for the storage of the text segments *before* entering the data. This made it very difficult to change the coding scheme *during* the ongoing process of analysis. But the style of empirically grounded theory building intimately related to qualitative research makes it necessary to have a very flexible code scheme which can be modified easily.

These problems could only be solved with the help of *non-formatted textual database technology* (Kelle (1995: 7)). The basic principle of this technology is to connect codes to *pointers*, containing the address of certain text segments (which are represented, for example, in terms of line numbers). Thus, codes are stored separately from the raw data and it is very easy to modify the code scheme. Non-formatted textual database technology also helped to solve the great methodological problem of all cut-and-paste techniques: when using computers, the text segments did not have to be removed from their original context.

Since the mid-1980s a variety of software programs have been developed (mostly by qualitative researchers) based on the principle of non-formatted databases. The first of these software packages[2] allowed only for simple *coding* and *retrieval* of text segments where

1. *coding* means attaching codes to text segments, and

[2]These were programs like TAP, QUALPRO, Textbase Alpha and the early versions of THE ETHNOGRAPH (Tesch (1990)). For more details see Kelle (1995: 120) or Weitzman and Miles (1995)

2. *retrieval* means searching for and displaying all text segments from a defined set of documents to which the same code has been assigned.

Straightforward code-and-retrieve techniques were the first computer-aided methods that were used in qualitative studies to assist the construction of typologies from unstructured textual material. To illustrate this process with an example: In a research project about men's attitudes towards their spouses' labour market participation (Braemer (1994)) lengthy open interviews with 60 to 70 year-old men were conducted. In these interviews a variety of normative aspects of marriage and family were stressed. The evaluation of the behaviour of their own children with respect to cohabitation and marriage turned out to be an important issue for the interviewees. As a first step of analysis the textual material was structured according to the topic (among others) "*cohabitation among people of the younger generation.*" Thereafter all text passages where this issue was mentioned were compared to look for similarities and differences. Take these three passages as an example:

> I mean, that was ... how we found us, how we lived together after that, the marriage, ... somehow, I liked it more. As it is today, I mean, I mean that this is not good today. It is not ideal, how they are together or not ... I don't know, that's nothing for me. (Case 60)
>
> Well, I mean, if one moves together in one apartment, (...) one should marry ... That has perhaps to do something with morality, since we were educated that way ... (Case 98)
>
> Well, I like it that one does obviously not live, so to speak, under that strong pressure (to get married) today, and the children can do it today in different ways ... But there is also some regret, that it is a little bit too loose today ... (Case 46)

All three cases show a certain similarity: the interviewees regret declining moral standards in the younger generation. But there are also differences which would be hard to capture with a standardised questionnaire. Case 46 and Case 98 both show general disapproval of the kind of cohabitation they think the younger generation prefer, but Case 46 is more or less ambivalent. These three passages can be regarded as specimens of two different types which can also be found in the utterances of other respondents. One could name these two types "ambivalence" (Case 46) and "disapproval" (Cases 60 and 98). Within the type "disapproval" one can differentiate between respondents with a more offensive position (Case 98), and a more defensive position (Case 60) (using phrases like, "perhaps", or "I don't know" ...). This example of course shows only the start of the process of typology building. Types cannot be constructed solely on the basis of a few cases, but have to be validated with further data material. In fact, the qualitative data material in this project showed a variety of moral arguments which could be grouped in a variety of types.

4. Complex coding-and-retrieval techniques

The first code-and-retrieve programs were nothing more than electronic cut-and-paste facilities. However, it rapidly became apparent to both developers and users that with non-textual database systems a whole variety of *enhanced coding and retrieval techniques* could be implemented which offered possibilities for qualitative typology construction that greatly exceed manual cut-and-paste (Kelle (1995: 11 ff.)). In the following section of this contribution those complex computer-aided techniques that are especially useful for typology construction will be described, namely

1. *complex coding techniques* that support the construction of *structured code schemes*
2. *complex retrieval techniques* that help with the building of *multi- dimensional typologies*

4.1 Complex coding techniques for the construction of structured code schemes

As has been said above the first non-formatted textual database systems that were applied in qualitative research helped the researcher to define linkages between codes and text segments (usually with the help of pointers that contained the addresses of text segments). With this simple code-and-retrieve technology the structure of the code scheme is not taken into account. All codes are treated equally, and the whole code scheme is simply regarded as a set of strings (=codes) which are attached to certain text passages whereby the single strings have no relation to each other.

But such a concept is hardly sufficient to capture the complexity of most of the code schemes which are used in the qualitative analysis of textual data. Although the code scheme is usually very simple and not highly structured in the beginning, it will gain complexity and structure in the ongoing process of analysis.

In the above mentioned panel study about young adults' transitions from education to the labour market, the research team started by looking for text passages where the respondents talked about their intention to marry and simply differentiated between those who wished to marry and those who did not. In terms of classical survey research, such a simple "typology" would be regarded as a categorical variable ("*wish to marry the partner*") where the categories of the variable represent the two types ("*Yes*" and "*No*"). But this typology formed only the basis of further analysis that helped (through the careful hermeneutic analysis and comparison of text passages) to discover three different types of orientations towards marriage (which could hardly have been discovered with the help of a survey questionnaire):

1. marriage is seen as the only acceptable form of cohabitation,
2. marriage is viewed as the *prerequisite for child-centred family formation*,
3. marriage is regarded as a kind of *safeguard*. Respondents with this last orientation advanced three different arguments: marriage was seen

as (1) offering financial protection, (2) providing a support for the bonding between the partners, or (3) a means to fulfil the expectations of their social environment (parents, relatives etc.).

Thus, three typologies were developed in the analysis process: *firstly* the typology of those who wish to marry vs. those who do not; *secondly*, among those who wish to marry, three types of orientations towards marriage can be found; and, *thirdly*, among those who see marriage as a safeguard, three different types of arguments for why marriage is a safeguard. The hierarchical relation between these three typologies is shown in Figure 1.

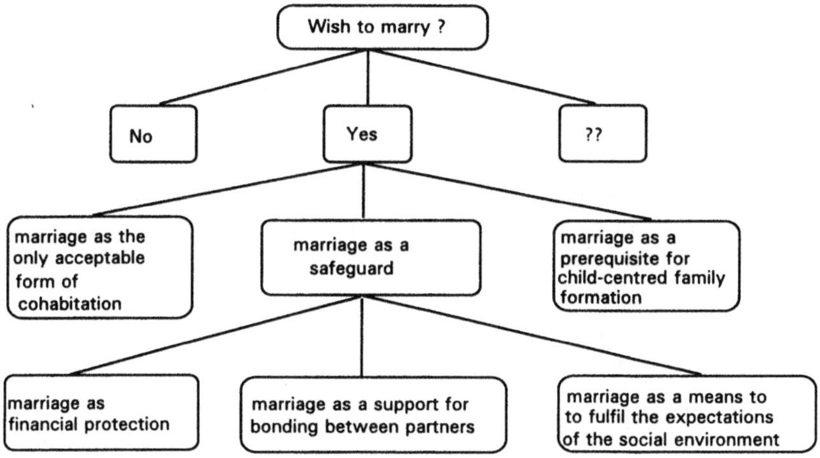

Figure 1: A hierarchical relation between typologies

In the terms of information science the tree structure displayed in Figure 1 can be formally described as a *network* or *graph* in which the types (=codes) represent the *nodes* of the graph and the lines between them the *edges*. Using such a network approach it is possible to expand the basic principle of the first code-and-retrieve programs in which codes were connected with pointers to text segments (Muhr (1991), (1992)). With such a network approach each code and text segment (and also other objects like short comments or "memos" written by the analyst) can be treated as a "node" which can be linked to all other nodes. Thus, it is possible to electronically store the whole structure of the hierarchical typology or graph shown in Figure 1 (which can then also be used as a hierarchical code scheme). Such a graph can not only be used to give an account of the emerging network of types and concepts, but it also allows for the construction of algorithms for rather complex retrieval procedures, algorithms that would be able, for example to follow a long path from one node at one end of the network or graph to a node at the other end.

It should be noted that graphs may be structured in quite diverse ways: either all possible linkages between the nodes can be allowed (permitting the researcher to define "cycles" and "loops")[3] or certain restrictions can be imposed, restricting the researcher, for example, to the construction of more tightly structured networks (e.g. hierarchical trees).[4]

4.2 Enhanced retrieval techniques for the construction of multi-dimensional typologies

To construct empirically grounded typologies means to proceed from one typology to another: on the basis of rather straightforward one-dimensional typologies (consisting of one categorical variable with a certain number of categories) more and more complex typologies are built. These typologies are often multi-dimensional structures: different categorical variables are combined to define a property space. The different types comprising the multi-dimensional typology are located within this property space, as in the following example.

We have already mentioned two one-dimensional typologies developed in the longitudinal project about work and family biographies of young adults in Bremen:

1. the typology of biographical constructions concerning occupational careers which lead to different work aspirations
2. the typology of orientations towards marriage.

To investigate the different combinations between work and family orientations one can now combine these two one-dimensional typologies in order to lay the groundwork for the formation of a two-dimensional typology.

	traditional orientation towards marriage	marriage as the prerequisite for a child-centred family formation	marriage as a safeguard
Delegation	Type 1	Type 2	Type 3
Search for orientation	Type 4	Type 5	Type 6
Self-maintenance	Type 7	Type 8	Type 9

Figure 2: Combining two one-dimensional typologies

As one can see in Figure 2, the combination of the two variables ("biographical constructions" and "family orientations"), each of which has three categories, would lead to a contingency table with $3^2 = 9$ cells. These cells contain all the types of the new two-dimensional typology which are *logically possible.* The next step in the ongoing process of analysis would be to analyse the empirical data (that is, the interview texts) to investigate the *empirical significance* of the logically possible combinations of categories. For this pro-

[3]ATLAS/ti and HYPERSOFT are programs that allow for the construction of different kinds of networks.

[4]NUD.IST is a software package especially designed for the construction of hierarchical trees of code categories.

cess two complex computer-aided retrieval strategies are of enormous value: *selective retrieval* and the *retrieval of co-occurring codes.*[5]

1. *Selective retrieval* means the retrieval of text segments according to document-specific variables such as the age, sex or profession of an interviewee. With selective retrievals the researcher can, for example, systematically compare the work orientations of men and women or the work orientations of members of certain professions.
2. A second useful complex retrieval technique utilises information on whether text segments coded with certain codes co-occur in a given document. Co-occurrences can be defined in various ways: indicated by the *overlapping* or *nesting of text segments* to which the codes under investigation are attached, or indicated by a certain specified *maximum distance* between the different text segments. With certain software packages[6] it is also possible to retrieve all text segments that follow each other in a certain *sequence.*

5. Methodological aspects of computer use in qualitative research

Since the advent of the first methods for computer-aided text retrieval there has been great concern that qualitative researchers can become alienated from their data by a machine which had shifted from being an aid in doing qualitative analysis to its definition (Agar (1991: 182), Seidel (1991)). But recent empirical investigations among users also show that the fear of the machine 'taking over the analysis' is clearly overemphasised (Lee and Fielding (1995)): in practice researchers tend to restrict themselves to using computer programs as mere electronic storage and retrieval facilities and not as "analytic tools". Given these experiences one can raise serious doubts about whether the computer will ever play a role in qualitative research that is comparable to that in statistical analysis. One has to keep in mind that the central analytical procedure of interpretive data analysis is the "operation called Verstehen" (Abel (1948)). Interpretive data analysis is mainly a hermeneutic process. Given the poor success of two decades of intensive efforts in the domains of machine translation and voice recognition, the idea of a hermeneutic machine seems far away. Furthermore, as critical computer scientists and linguists have pointed out (see Dreyfus (1972), Dreyfus and Dreyfus (1986), Winograd and Flores (1986)), hermeneutic understanding fundamentally contradicts the general principles of a Turing machine. Ordinary language is not strictly rule governed and unambigous: most of its symbols, i.e. the words and phrases we use in everyday talk, do not have precise lexical definitions (Giddens (1984)). Unlike a Turing machine, a human interpretor will be able to handle this ambiguity and context relatedness. This is a necessary prerequisite for the comparison of two different text passages in order to decide whether these passages mean "the same" or

[5]Such complex retrievals can be realised with almost all computer programs for qualitative research. For further details see Weitzman and Miles 1995 or Kelle 1995.

[6]THE ETHNOGRAPH and NUD.IST

text passages in order to decide whether these passages mean "the same" or "something different".

Since typology building in interpretive research cannot be regarded as a strictly rule governed process, the role of electronic data processing machines is therefore limited to a mere device for data management. But this does not mean that they do not have an important methodological function. Although the techniques of data management described above may seem trivial on first sight, their methodological significance for hermeneutic research can hardly be overrated. A good example for this would be the already mentioned technique of synopsis which provided 18th and 19th critical biblical hermeneuticians with good arguments against fundamentalist interpretations of biblical texts.

Intelligent data administration strategies are also of enormous importance for interpretive social research. Qualitative data analysis often means the collection of sometimes enormous masses of textual data. The danger of drowning or getting lost in this material is extremely high, and as the time pressure increases towards the end of a research project, researchers are often seduced into developing their typology on the basis of memory, supported by some hastily chosen quotations.

In contrast, the use of computer-aided coding and retrieval techniques offers two major advantages: (1) easy and quick retrievals help to explore the data more fully than would have been possible by manual means, and (2) having easy access to all of the collected data increases the confidence that typologies are not based on one or two highly untypical cases.

Facilities for enhanced coding and retrieval are extremely helpful for systematically developing typologies from the data. In particular, complex retrieval techniques, i.e.selective retrieval and retrieval of co-occurring codes, offer fascinating new possibilities for analysts to 'play' with their data and thereby help to open up new perspectives and to stimulate new insights. At the same time they will help to make qualitative data analysis more rigorous and systematic, by offering facilities that ensure that all relevant information is retrieved from the data, and by assisting the management of large qualitative samples. Thus, computer-aided methods can ensure that the typologies developed are really grounded in the data and not based on single and highly untypical incidents. In this way it will be possible to add trustworthiness to a methodology which had often suffered (not always without justification) from the reputation of seducing the researcher into employing an unsystematic, subjective and journalistic style of inquiry.

References:

ABEL, T. (1948): The Operation called *Verstehen. American Journal of Sociology, 54, 211–218.*

AGAR, M. (1991): The Professional Stranger: An Informal Introduction to Ethnography. Academic Press, New York.

BRAEMER, G. (1994): Wandel im Selbstbild des Familienernährers? Reflexionen über vierzig Jahre Ehe-, Erwerbs- und Familienleben. Working Paper No. 29, Sonderforschungsbereich 186, Bremen.

BRÜCKNER, H. (forthcoming): "Times of Poverty": Lessons from the Bremen Longitudinal Social Assistance Sample. In: D. Chekki (ed.): *Urban Poverty in Affluent Nations.* DENZIN, N. (1989): Interpretive Interactionism. Sage, Newbury Park/London/ New Delhi.

DREYFUS, H. (1972): What Computers Can't Do - A Critique of Artificial Reason. Harper and Row, New York.

DREYFUS, S. and DREYFUS, H. (1986): Mind Over Machine - The Power of Human Intuition and Expertise in the Era of the Computer. Blackwell, Oxford.

GEERTZ, C. (1973): The Interpretation of Cultures. Hutchinson, London.

GIDDENS, A. (1976): New Rules of Sociological Method. A Positive Critique of Interpretive Sociologies. Hutchinson, London.

GIDDENS, A. (1984): The Constitution of Society. Outline of the Theory of Structuration. Polity Press, Cambridge.

GLASER, B. (1992): Emergence vs. Forcing: Basics of Grounded Theory. Sociology Press, Mill Valley, CA.

GLASER, B. and STRAUSS, A. (1967): The Discovery of Grounded Theory: Strategies for Qualitative Research. Aldine, Chicago.

LEE, R. and FIELDING, N. (1995): User Experiences of Qualitative Data Analysis Software. In: U. Kelle (ed.), *Computer-Aided Qualitative Data Analysis.* Sage, London, 29–40.

MERTON, R. (1968): Social Theory and Social Structure. The Free Press, New York.

MUHR, T. (1991): ATLAS/ti: A Prototype for the Support of Text Interpretation. *Qualitative Sociology, 14, 349–371.*

MUHR, T. (1992): Catching Bugs and Butterflies in Networks. Paper presented at the conference 'The Qualitative Research Process and Computing', 7–9 October, Bremen.

KELLE, U. (1994): Empirisch begründete Theoriebildung. Zur Logik und Methodologie interpretativer Sozialforschung, Deutscher Studienverlag, Weinheim.

KELLE, U. (1995): An Overview of Computer-aided Methods in Qualitative Research. In: U. Kelle (ed.): *Computer-Aided Qualitative Data Analysis.* Sage, London, 1–17.

LAKATOS, I. (1982): The Methodology of Scientific Research Programs. Philosophical Papers, Vol. 1. Cambridge University Press, Cambridge.

MILES, M. and HUBERMAN, M. (1984): Qualitative Data Analysis: A Sourcebook of New Methods. Sage, Newbury Park.

PATTON, M. (1990): Qualitative Evaluation and Research Methods. Sage, Newbury Park.

SEIDEL, J. (1991): Method and Madness in the Application of Computer Technology to Qualititative Data Analysis. In: N. Fielding, R. Lee (eds): *Using Computers in Qualitative Research.* Sage, London.

STRAUSS, A. and CORBIN, J. (1990): Basics of Qualitative Research: Grounded Theory Procedures and Techniques. Sage, Newbury Park.

TAYLOR, S. and BOGDAN, R. (1984): Introduction to Qualitative Research Methods: The Search for Meanings. Wiley and Sons, New York.

TESCH, R. (1990): Qualitative Analysis: Analysis Types and Software Tools. Falmer Press, London and Philiadelphia.

WILSON, T. (1970): Conceptions of Interaction and Forms of Sociological Explanation. *American Sociological Review, 35, 697-710.*

WINOGRAD, T.; FLORES, F. (1986): Understanding Computers and Cognition. Ablex Publishing Corp., Chicago.

WEITZMAN, E.; Miles, M. (1995): Computer Programs for Qualitative Data Analysis. Sage, Thousand Oaks.

WITZEL, A. and MÖNNICH (1994): Developing a Typology of Young Adults' Occupational Biographies. Paper presented at the conference "Empirically Based Theory Construction and Qualitative Life Course Research", 17- 19 November, Bremen.

Classification of Text Analysis Software

Harald Klein

University of Jena, Institute for Sociology, D-07740 Jena, Germany

Summary: This paper introduces a classification of text analysis software. At first the definition of terms show the different meanings of the same terms used by different approaches. Especially the implications of the qualiative and quantitative approach in the research process are discussed. The software currently available is classified using the classification of Weitzman and Miles (1995).

1. Overview on classification of text analysis software

The purpose of this paper is first to give an overview on text analysis and definitions of terms, their paradigms and their implications. The criteria for classification by Tesch (1990), Weitzman/Miles (1995), and Popping (1996) and the classification of the current packages form the last part. Information on text analysis is hard to find, the sources I used are the SIBYL data bank of the Erasmus University of Rotterdam/Netherlands, the ASK-SISY data bank of the University of Karlsruhe/Germany, and the CAQDAS-site of the University of Surrey/UK. Also discussion lists like *qual-soft, qualnet*, and *content* were monitored. Fig.1 shows the different types of text analysis software:

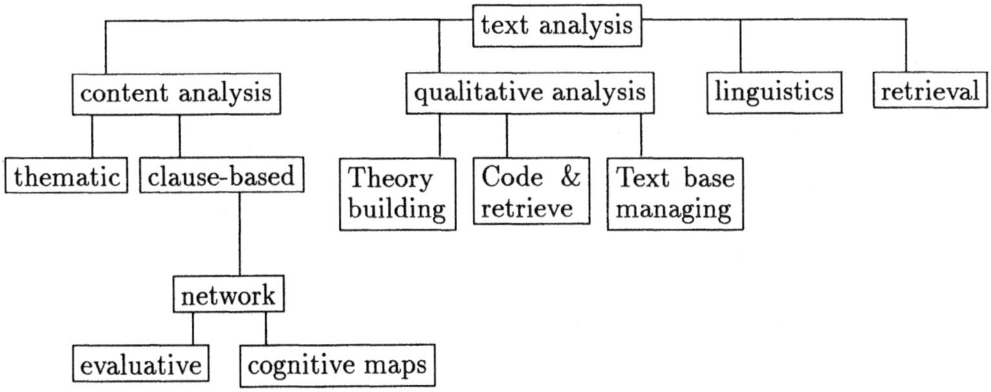

Figure 1: Overview on text analysis

2. Definition of terms

The definition of terms is necessary because these are often used in different contexts. The most common classification of text analysis is whether it is qualitative or quantitative. These terms are established, although I prefer the terms finding hypotheses for the qualitative approach and testing hypotheses for the quantitative approach are much more appropriate. During the discussion of the terms the differences between both approaches will become clearer.

2.1 Segmentation

A text is seldomly analysed as a whole, often the hypotheses or the use of theoretical implications (e.g. grounded theory) make a segmentation of the text necessary. The segmentation defines the structure of the text. Often external variables are attached to the segments for later analysis and for drawing statistical inferences on groups. The quantitative approaches often uses text units as text segmentation, and external variables are often called identifiers. These are used for later statistical analysis as control variables, depending on the hypotheses to be tested. Using the qualitative approach text segments are often – but not always – limited to lines, and for each text segment external variables can be defined to allow text retrieval and to generate subgroups.

2.2 Coding

Analysing texts often means that a kind of coding is performed. The coding process is used for the classification of phenomena of the text and does mean something different in the quantitive and the qualitative approach. The quantitative approach works with a category system that is derived from the hypotheses which have to be formulated before the analysis. The operationalisation is one of the key processes of this approach. The rationale is that the categories are the variables mentioned in the hypothesis. Indicators for these variables are – especially in computer aided content analysis – search patterns (mostly words, but also word sequences or co-occurences of words). The frequencies of these search patterns are the variables used for the statistical analysis. In computer aided content analysis the category system is organised as a file. The program reads a text unit and searches for the search patterns. If one is found, the appropriate code is written to the output file. The coding process is completly done by the software with no interactive elements although some problems like ambiguity and negation are mostly not detected and therefore can bias the results (see Klein 1992 for a detailed discussion). One program can also work interactive (Klein 1994). The qualitative approach assigns codes to parts of the texts, often limited to whole text segments. The researcher reads the text and attaches the codes at appropriate places for later retrieval. The work is organised interactive, the program itself does not code. Often one has the possiblity of writing memos on the reasoning for assigning the code, or to document regularities of the text or any other interesting things.

2.3 Files

Working with software implies using files, and the way the text analysis software works with them is totally different. Using the quantitative approache mostly one file is used, it is called the system file, whereas using the qualitative approach each text segment is often one file so that a project can consist of hundreds of files. One file has the advantage that it takes less disk space than a lot files. The advantage that the text segments are organised as files makes accessing them much faster, but the disadvantage is that the files take up more disk space. Often codes and memos are organised in separate files.

Files can be grouped by their contents:

- raw text: the file that is needed for all operations. Often the researcher has to insert control sequences for specifying the values of the external variables used. The raw text can be a plain file, must most software requires some sort of control sequence.
- system file without codes: the internal file that contains the text segments and the identifiers. It is generated from the raw text by an appropriate module of the text analysis program. This file is the basis for all further analyses.
- system file with codes: the internal files that contains additional information on coded text segments, their start, their end, and their code. It is often possible to assign several codes to a text segments, and often the codes are administered in another file.
- file of codes: depending on the approach used this file can be the category system with search entries (quantitative) or the codes and their position in the hierarchy in a coding tree (qualitative). The file of codes using the quantitative approach can be regarded as a set of rules for coding, whereas using the qualitative approach this file contains the results of the coding. One has to have in mind that coding does have a totally different meaning depending on the approach applied to the text.
- memo file: also memo files have different meanings depending on the approach used. The variable labels for categories can be regarded as memos in the quantitative approach, where in the qualitative approach memos contain remarks on the reasoning process of the coding, or on the text segments as such. They can be rather long. Memos can contain any information in the qualitative approach, whereas in the quantitative approach memos are only available as variable labels, most software packages do not support them.

3. Implication of text analysis approaches

If one has to decide what approach fits best to one's own goals, the implications of the different methods must be taken into account. The classical thematic content analysis is a method for testing hypotheses to be able to make statistical inferences. The frequencies of categories are regarded as indicators for theoretical constructs. The hypotheses can be statements on the text, the author or the recipient of the text.

Clause-based content analysis makes inferences about how words are used in the text. Semantic content analysis works with variables that are parts of speech in a semantic grammar, so parsing of the text is required. The network approach requires semantic links among themes to give information on the frequency of concepts in each block of text.

Qualitative data analysis is a summary term for all methods to analyse text that are not used to make statistical inferences. The basic method is simple text base management, so that texts can be sorted, indexed, or filtered. Software usinmg this approach are often data base programs. The next operation is assigning codes to the text and retrieving text using these codes. Often memos are assigned to text segments to document the reasoning process or other phenomena of the text or the coding. Software supporting the building of theory are mostly based on grounded theory (Glaser and Strauss 1967). It is an inductive working technique with the goal of finding regularities that can be coded into categories. With links between codes and/or text segments it is possible to build a network of nodes and links; this is called conceptual network and is often displayed as a graph.

4. Classification criteria

The text analysis software currently on the market offers a variety of features. It is rather difficult to decide what programs fits which needs. The following wants to support this decision making process.

Weitzman and Miles (1995) classify computer programs for qualitative data analysis as text retrievers, textbase managers, code and retrieve programs, codebase theory builders, and conceptual network builders. No program belongs purely to one category, the categories overlap. This classification of qualitative text analysis does have disadvantages, but currently there is no other available. Popping (1996) alters the Weitzman and Miles classification, he skips text retrievers and codebase theory builders, he also classifies some programs into different categories than Weitzman and Miles, and he classifies software based on the quantitative approach that is not classified by Weitzman and Miles at all. Tesch (1990) developed a completly different classification:

- the characteristics of language
  - as communication: with regard to its content or its process
  - as its mirror culture: in terms of the cognitive structure or the interactive process
- the discovery of regularities
  - as the identification and categorisation of elements and the establishment of their connections
  - as the identification of patterns
- the comprehension of the meaning of text or action: through the discovery of themes or through interpretation
- reflection

Unfortunately this classification is outdated, but new software should be classified using this classification. The following tables give an (incomplete) overview of the software currently on the market. For each program the name or the acronym, the version, the date of its publication, the operating system (OS), and the program author or vendor are given. The abbrevations for OS mean: DOS is MS-DOS 3.3 or better, Mac is Apple Macintosh System 6 or better, Win is MS-Windows 3.0 or better, multi means is runs on many hardware platforms, mostly on MS-DOS/Windows, Macintosh, and Unix. The addresses of the authors/vendors can be obtained from Weitzman and Miles (1995) or from the WWW-sites listed below.

5. Software for text analysis

5.1 Thematic content analysis

name & version	date	OS	author/vendor
DISCAN 2.0	1989	DOS	P. Maranda
FlexText 1.0	1991	DOS	Albert L. Baker
GENCA		DOS, Mac	Wayne Danielson
INTEXT 3.0	1995	DOS	Harald Klein
MCCA	1995	DOS	Don McTavish
TEXTPACK V 5.0	1995	multi	ZUMA
VB-pro	1995	DOS	Mark M. Miller

5.2 Semantic content analysis

name & version	date	OS	author/vendor
PLCA 2.1	1994	DOS	C. W. Roberts
PC-ACE	1990	DOS	Roberto Franzosi
Gottschalk	1989	DOS, Win	Louis Gottschalk
KEDS 0.5 B2	1993	Mac	Philip M. Schrodt

5.3 Network content analysis

name & version	date	OS	author/vendor
CETA 2.1	1988	DOS	Jan A. de Ridder
MECA 1.0	1990	multi	Kathleen Carley

5.4 Theory builders

name & version	date	OS	author/vendor
Atlas/ti 1.1 E	1995	DOS	Thomas Muhr
AQUAD 4.01	1995	DOS	Günter Huber
HyperResearch 1.56	1994	DOS, Mac	A. Dupuis & E. Tornabene
NUDIST 3.04	1994	multi	Tom & Lyn Richards
QCA 3.0	1992	DOS	Kriss A. Drass & Charles C. Ragin

5.5 Text retrieval

name & version	date	OS	author/vendor
askSam 5.1	1993	DOS, Win	askSam Systems
Concord 1.0	1989	DOS, Win	Peter Klicmann, Heinzgünter Klein
Folio Views 3.1	1993	Win, Mac	Folio Corporation
ISYS 3.02	1992	DOS	AskSam Systems
MICRO-OCP	1988	DOS	Oxford University Computing
Metamorph 4.0	1991	multi	Thunderstone Software
Orbis 1.0	1993	DOS, Win	The Technology Group
Sonar 8.5	1993	Win, Mac	Virginia Systems Software
Strap 2.0	1989	DOS	Teresea Snelgrave & Lidio Presutti
Tabletop 1.0	1994	Win	Chris Hancock
Tact 2.0	1992	DOS	John Bradley & Lidio Presutti
TextCollector 1.7	1987	DOS	O'Neill Software
WordCruncher 4.5	1993	DOS, Win	Johnston & Co.
Zyindex 5.14	1993	multi	ZyCo International, Inc.

5.6 Code & retrieve

name & version	date	OS	author/vendor
Childes 2.0	1995	DOS, Mac	Lawrence Erlbaum Associates
Code-A-Text	1996	Win	Alan Cartwright
Ethnograph 4.0	1995	DOS	John Seidel
FocusReports 3.0	1995	Win	Jeffrey Henning
HyperQual2 1.0	1993	Mac	Raymond V. Padilla
Kwalitan 4.0	1994	DOS	Vincent Peters
Martin 2.0	1991	Win	N.L. Diekelmann, S. Lam, R.M. Schuster
(Win)Max 3.5	1995	DOS, Win	Udo Kuckartz & Andreas Maurer
Qualpro 4.0	1993	DOS	Bernard I. Blackman
Textbase α	1995	DOS	Bo Sommerlund
Textbase β	1995	DOS	Bo Sommerlund

6. List of software acronyms

Atlas/ti	Archive for technology, the lifeworld, and everyday language
AQUAD	Analysis of qualitative data
CETA	Computer-aided evaluative text analysis
DISCAN	Discourse analysis
GENCA	General content analyser
Gottschalk	Gottschalk-Gleser content analysis computerized scoring system
INTEXT	INhaltsanalyse von Texten (content analysis of texts)
KEDS	Kansas Events data system
MCCA	Minnesota contextual content analysis
MECA	Map extraction, comparison, and analysis
Micro-OCP	Micro Oxford Concordance Program
NUDIST	Non-numerical, unstructured data indexing, searching, and theorizing
PC-ACE	Program for computer assisted coding of events
PLCA	program for linguistic content analysis
QCA	Qualitative comparative analysis
TACT	Text analysis computing tools
TEXTPACK	Text package
VB-Pro	VerBatim protocols

7. Internet links to more information

Atlas/ti homepage: http://www.cs.tu-berlin.de/~muhr/atlasti.html
demo versions: http://www.soc.surrey.ac.uk/caqdas/
Nudist homepage: http://qsr.latrobe.edu.au/Public/WWW/resource.html
Sibyl data bank: http://www.gamma.rug.nl/
ASK data bank: http://www.ask.uni-karlsruhe.de/
VB-Pro homepage: http://excellent.com.utk.edu/~mmmiller/vbpro.html
Code-A-Text homepage: http://ourworld.compuserve.com/homepages/code_A_Text/
Kwalitan homepage: http://www.kun.nl/methoden/kwalitan/
Textbase α homepage: http://www.psy.aau.dk/#Utility Software

References:

GLASER, B. and STRAUSS, A. (1967): The discovery of grounded theory: Strategies for qualitative research. Chicago: Aldine.

KLEIN, H. (1992): Validity Problems and their Solutions in Computer-Aided Content Analysis with INTEXT/PC and Other New Features. In: Faulbaum, F.; R. Haux, and K.-H. Jöckel (eds.): Advances in Statistical Software 3. Stuttgart:

Gustav Fischer, 483-488.

KLEIN, H. (1994): Computerunterstützte Inhaltsanalyse mit INTEXT – dargestellt am Vergleich von Nachrichtenfaktoren des Fernsehens. Münster, dissertation.

POPPING, R. (1996): Computer programs for the analysis of texts and transcrips. In: Roberts, Carl W. (ed.): Text Analysis for the Social Sciences. Methods for Drawing Statistical Inferences from Texts and Transcripts. Hillsdale, N.Y.: Lawrence Erlbaum Associates.

TESCH, R. (1990): Qualitative Research: Analysis types and software tools. New York: Falmer.

WEITZMAN, E. A.; MILES, M. B. (1995): Computer Programs for Qualitative Data Analysis. A Software Sourcebook. Thousand Oaks: Sage.

Computer Aided Text Analysis and Typology Construction

U. Kuckartz

SE Information und Datenverarbeitung Fachbereich Erziehungswissenschaft, Psychologie, Sportwissenschaften Freie Universität Berlin Arnimallee 12, D-14195 Berlin, Germany

Summary: In the last decade the development of software for computer aided text analysis has become a booming area in social science methododology. Recent developments offer a variety of tools for text analysis and complex types of retrievals. The paper focuses the different tools of the text analysis package WINMAX: codewords, memos, case variables, weight variables and memo-codewords. The base technique of "code and retrieve" and the possibilities for classification and typification are discussed.

1. Coding in Quantitative and Qualitative Content Analysis

In the social science-oriented text analysis a traditional split between *quantitative* and *qualitative* methods can be found. *Computer Aided Content Analysis* came up in the Sixties with the software program *General Inquirer.* In the beginning *Computer Assisted Content Analysis* was expected to be one of the upcoming new methods, but since more then a decade nothing really important happened in this field: either concerning methodology, research techniques as well as software.

The development of the *qualitative* type of text analysis was completely different. Till the mid Eigthies qualitative researchers did not think about using computers and software as helpful tools for their analysis. But at the end of the Eighties and beginning of the Nineties a boom of software started – and since that time all developments and innovation happened in the *qualitative* section of content analysis. Different traditions often have the effect that people are not aware of each other and do not recognize the strength and weaknesses of the different approaches. My work in the qualitative content analysis and the development of software is oriented towards bringing together qualitative and quantitative procedures and it is focused upon the procedures of classification and typification. What are the main characteristics of the qualitative and the quantitative approach?

In *Quantitative Content Analysis*

- large data sets, e.g. from the mass media, are analyzed
- pre-defined categories are used
- units of analysis havc to bc defined before data is categorized
- coding is performed by different coders independently from each other
- reliability of coding, e.g. inter-coder-reliability is checked

- finally statistics of categories like correlation coefficients are computed

In *Qualititative Content Analysis*

- small data sets are analyzed like field notes or narrative interviews with rather few cases (about 20 to 70)
- texts are interpreted by the researchers in a holistic way (hermeneutics)
- codes are used in different steps of the analysis
- in the different modes of coding the codes have a quite different status (see for instance the different techniques described by Glaser/Strauss (1967): 'open coding', 'axial coding', 'selective coding')
- chunks of data are related to categories ('cut-and-paste')
- all text passages that have something in common (the same code) are drawn together ('code-and-retrieve')
- finally key categories are developed and data is interpreted on this background

In spite of the differences, it is obvious that there are also important commonalities between the two paradigms: categorizing has a strong position in both approaches. Berelsons statement that *content analysis stands and falls with its categories* could also be formulated by scholars of the qualitative approach like Anselm Strauss. All different modes of coding can be described as a types of classification. The main difference between the approaches is that coding in qualitative data analysis is always performed by human coders and not automatically. In the following chapters the different tools supporting classification are described.

2. An Example

To discuss how classification and typification in qualitative text analysis can be reached and how this is performed with the software WINMAX, image the following example coming from our clinical research:

"...for instance I did say, once a week I will have my 'rice-day'. But till today I did not a carry out a single 'rice-day' – and I'm back home again since nearly two months. I even can't say the reasons for. Yes ... I could say that there wasn't enough time. Yeah, really sometimes it is in deed a time problem, but in the same time I'm preparing something other in the kitchen I could also cook my rice. And my family supports me – they all say, allright do it. So, I can not find a true reason. I simply don't know."

(Woman, 36 years, married, one child, heart and circulatory disease, employed)

What does this segment mean to us? Obviously there is a person speaking about problems with a so-called 'rice day'. The person is now back home, has a family that gives her support. But the person is not successful with the 'rice-day' and does not know the true reason. The meaning of the segment becomes more clear, when some context information are given: the person is a woman, 36 years old, married, has one child, a heart and circulatory disease, employed etc.

How can we analyze the segment? The first point is, that there is no analysis in the social sciences without having questions towards the data. Nothing 'emerges' directly from the data. Of course, some operations can be performed with the data without any particular research question: It is possible to compute the frequencies of words or to set 'hyper links' from one text segment to the other using the same words, e.g. from 'rice-day' to 'rice day' or 'family' to 'family'. But this 'visualizes' only a formal structure of the text like the appearance of the same strings and their frequencies. Interpretation of the segment is not possible without ideas about the text.

3. Four Tools of Computer Aided Text Analysis

Quantitative content analysis has only one tool to analyze texts and classify text's characteristics: the *categories*. Qualitative content analysis, however, uses several tools like *codes, dimensions* and *memos*. The software WINMAX contains four different tools for text analysis:

1. codewords and subcodewords that can be attached to text segments
2. case-oriented variables
3. fuzzy variables that can be used to weight coded segments
4. codewords that can be assigned to memos

3.1 Codewords and subcodewords that can be attached to text segments

This is the standard tool of text analysis software. Text is segmented and categories are assigned by the researcher. These codes can be grouped to a hierarchical system of categories (see fig. 1). Several codewordwords can be attached to the same segment, overlapping and nesting of segment is allowed. It is very useful to arrange the screen as shown in figure 1. One window contains the categories, the other window contains the text to be coded. Regarding our example the following categories could be defined and attached:

'changes in way of life'
'role of the family'
'good intentions during rehabilitation'
'reasons for patients behavior after rehabilitation'

Imagine all interview-transcripts have been coded. Now the category 'changes in way of life' might hold all references to all text passages where people talk about changes of their way of life that happened after they left the rehabilitation hospital. This is the first step of classifying the data: identification of meaningful segments and definition of codes. In the second step simple retrievals are performed that bring together all text segments that belong to the same category. More complex types of retrievals might look for co-occurences of codes, e.g. where patients talk about their family support and talk about their good intentions. The basic scheme looks as follows. The table looks like a normal crosstabulation of two variables. But the problem is that the frequency count of cell d is not known. How often do people *not*

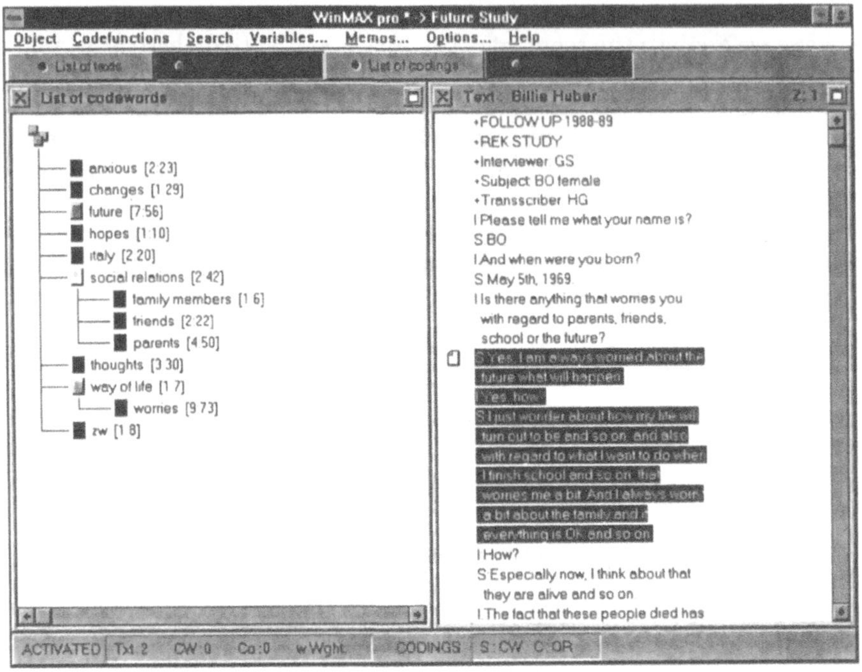

Figure 1: Attach Codewords to Text Segments

speak about their family support *and* do *not* talk about their good intentions at the same time. Cell d is only known, if fixed-size units of analysis have been defined and overlapping and nesting of segments are not permitted. This is normally not the case in the analysis of narrative material. So, working with tables like in figure 2 has a heuristic value, but hypothesis testing in the statistical sense is not possible.

Extended versions of the basic scheme of co-occurences introduce multiple codes and the concept of 'vicinity of codes' (with a distance measure in terms of lines). If the analysis also allows for distances between the attachement of Code A and Code B another problem arises: the results depend directly on the chosen distance. Increasing the distance results automatically in in-

	Code B attached	Code B not attached	Total
Code A attached	a	b	
Code A not attached	c	d ???	
Total			???

Figure 2: Basic Scheme for Co-occurence of Codes

creasing the frequency count of cell a.

Anyway, codewords are powerful heuristic tools for classification and typification. In WINMAX patterns of co-occurences can be searched for and all data about codings can be exported as a data base file in order to analyze it with a statistics program.

3.2 Case oriented variables

The method of case-oriented quantification (Kuckartz 1995a) uses case-oriented variables as a basis for constructing a typology. To give an example: all text segments belonging to 'role of family' might be examined to find out for each patient if the family supported him or not. The variable values for all the interviewees are coded - at least by two coders. These subjective ratings can be stored in a dichotomous case-variable 'Family Support' with the values 'Yes' and 'No'.

Often, the analysis of the coded segment belonging to a particular subject shows that different dimensions play a role. Analyzing the changes of patients' daily life three dimensions could be identified. The first one concerns the degree of changes that took place in the daily life. A variable 'change of life' with ordinal scaling from (1) 'strongly changed' thru (4) 'no changes' was created. A second dimension concerned the kinds of changes. We differentiate between (1) 'nutrition' (2) 'sports, fitness, jogging' (3) 'meditation' (4) 'job' and (5) 'others'.

The third dimension concerned the relation between the reported changes and the chronical illness. A variable 'pressure for changes' was created, which was set to 'forced' when the illness forced an issue (like giving up smoking or giving up the job after cardiac infarction) and to 'voluntary' when the changes were results of the patients own decision and were not directly forced by the illness.

During the research process, the case-oriented variables can be used for different purposes:

- computing frequencies gives a descriptive overview ('How often patients feel themselves supported by their families?')
- performing 'selective retrievals' allows to compare different subgroups of patients ('Is it easier for 'supported patients' to achieve changes in their daily life?')
- combining variables allows to construct typologies - for instance the variables mentioned above could be combined to achieve a typology of 'coping strategies'

To perform typification with WINMAX different strategies have to be used. Monothetic typologies can be easily constructed with WINMAX. The typology can then be used as criterion for 'selective retrievals'. Polythetic typologies can be computed, by exporting the data matrix of case oriented-variables in data base format to SPSS or another statistics program. After a cluster analysis has been computed the cluster codes - that means which person belongs to which cluster (=type of a typology) - can than be used in combination with codewords and the other classification tools of WINMAX.

3.3 Fuzzy variables that can be used to weight coded segments

The third tool for classification and typification of data are 'weight-variables' that can be attached to coded segments. Cut-and-paste, normally, works by choosing a codeword from the codeword tree and attaching it to a previously marked segment. WINMAX allows to assign a weight to such a coded segment. Sometimes qualitative researchers want to find text passages that are most characteristical for an argumentation or for a theoretical concept. In this case it might be very helpful to give each coding a score that indicates the degree of similarity to an ideal type (e.g. to an ideal type of argument or argumentation). A weight score is like an attribut of a coded segment. It indicates, for example, a text passage where the interviewer points out very*clear* why it was so difficult for her to change the daily life.

When codewords are used like this, they are like sign-posts that lead us to particular text segments. The weight scores helps us to find the most important segments later on when retrieving the text segments belonging to this particular category.

The weight scores are used like fuzzy variables. They are not intended as an exact value of similarity with an ideal type. The fuzzy variables can be used in simple retrievals as well as when searching for patterns of co-occring codes. For example, when searching for co-occurence of code 'role of the family' and 'good intentions during rehabilitation' introducing a fuzzy variable effects that only those segments are retrived that have anassociated score inside the defined range.

3.4 Codewords that can be assigned to memos

Memos play an important role in qualitative data analysis, particularly in the grounded theory approach. In his works Anselm Strauss differentiates between different types of memos like 'code notes', 'theoretical notes' and 'operational notes'. Memo writing is seen as an analytical operation for organiszing coded data.

In WINMAX memos can be attached to text lines. The number of attached memos are not limited they are symbolized by a yellow sheet before the text line. Memos do not consist only of text and informations on their origin, also codewords can be attached to a memo. And this offers new possibilities for classification and typification, that we us in a new type of text analysis, that we call in German 'Leitbild Analysis' ('analysis of guiding visions').

In a research project 'Leitbilder in youth literature on environmental issues (Schack 1994) the memo function of WINMAX was used as follows. The books - represented by their detailed descriptions and citations of characteristic parts - were defined as 'units of analysis'. According to previously defined categories of 'leitbild components' the books were analyzed and classified (Kuckartz 1995b). Doing this, one memo per book was needed. The attached codewords of the memo indicate the pattern of categories for any

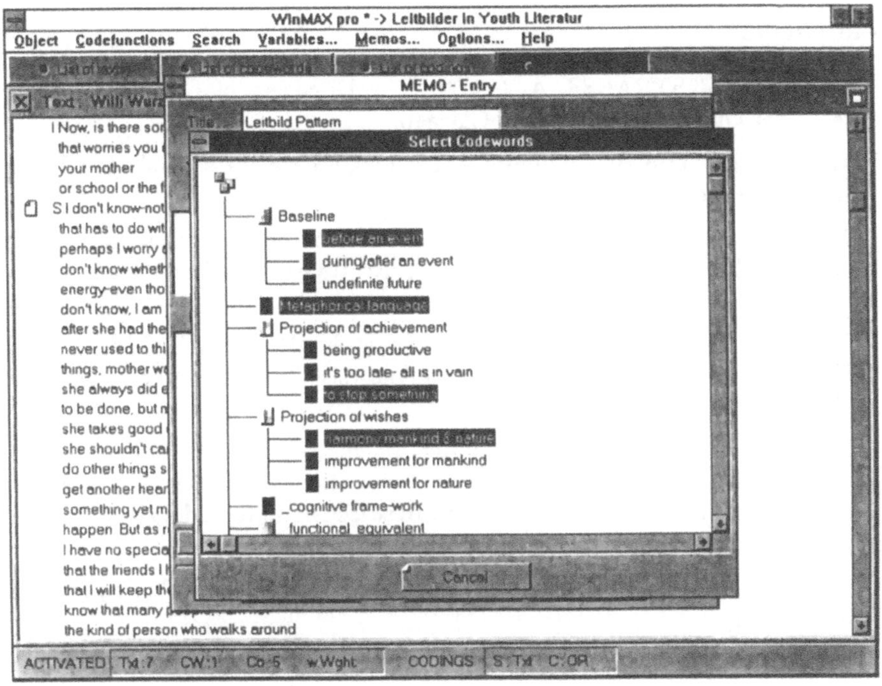

Figure 3: Attach a Codeword-Pattern to Memos

particular text (see figure 3). Theoretically thousands of combinations between the defined categories exists. But only a few configurations can be observed. Some are unique, some exist more or less numerous. Identifying similarities between configurations has a lot in common with finding homogeneous groups in cluster analysis. But the style of analysis is more heuristic - different patterns can be 'tested' to see if they are present in the examined literature. If they are, the memos lead us directly to the books.

4. Conclusions

It has been shown that there are many different opportunities to construct typlogies from textual data. A good software package should offer the whole variety of tools to construct typologies and not only one type.

The techniques of analysis, decribed above, are oriented towards summarizing and classification. This is quite different from hypertext techniques which try to make a text accessible and not to analyze in the sense of abstracting, categorizing, dimensionalising and typifying. Programs like WINMAX try to assist qualitative researchers in the process of classifying and typifying their data, they do not perform any 'automatic analysis' and are not substitutes for social scientists.

References:

GLASER, B. and STRAUSS, A. (1967): The Discovery of Grounded Theory: Strategies for Qualitative Research. Chicago.

KELLE, U. (ed.) (1995): Computer-Aided Qualitative Data Analysis. Theory Methods and Practice. London.

KUCKARTZ, U. 1992: Textanalysesysteme für die Sozialwissenschaften, Einführung in MAX und TEXTBASE ALPHA. Stuttgart/New York/Jena: Gustav Fischer

KUCKARTZ, U. (1995a): Case oriented Quantification. In: KELLE, U. (ed.) (1995): Computer-Aided Qualitative Data Analysis. Theory Methods and Practice. London.

KUCKARTZ, U. (1995b): The Computer-Aided Qualitative Analysis of Argumentations and "Leitbilder". Paper presented at the Conference "Text Analysis and Computers", ZUMA Center for Survey Research and Methodology, Mannheim, September 18-21, 1995

KUCKARTZ, U. (1995c): MAX for WINDOWS. Version 1.0 - User's guide and reference, Berlin

SCHACK, K. (1994): Das Thema Ökologie in der erzählenden Kinder- und Jugendliteratur. Eine Leitbildanalyse. Paper 94-114 der Forschungsgruppe Umweltbildung, Berlin.

Rotwang's Children: Information Ecology and the Internet

Bonnie A. Nardi[1], Vicki O'Day[2] and Edward J. Valauskas[3]

[1] Apple Computer, Inc., Advanced Technology Group, 1 Infinite Loop, Cupertino, CA 95014 USA
[2] Xerox Palo Alto Research Center, 3333 Coyote Hill Road, Palo Alto, CA 94304 USA
[3] Internet Mechanics, 5050 South Lake Shore Drive, Suite 3214, Chicago, IL 60615 USA 12

Summary: An analysis of the work of librarians in two corporate libraries was completed to find ideas for software agents. As the analysis proceeded, a larger message came to the fore: we developed a clear understanding of why the services of skilled librarians cannot be replaced by automated tools – why it is impossible to do as some software developers have suggested and create a so-called librarian in a box. Currently, the best software agents on the Internet only do a small portion of the navigational and interactive work of an experienced librarian. The increased digitization of information instead points to a new kind of information ecology, where librarians and software tools work together to gather information. Librarians can work with their patrons in the construction of this diverse information ecology in several ways. These collaborations of librarians, software, and patrons include:

a) designing software agents for clients and for other librarians;

b) cataloging, distributing, and customizing software agents;

c) helping software agents to evolve;

d) contributing to the quality of information databases, and

e) working collaboratively with software agents in the actual processing of information.

1. Introduction

In this time of terabytes of information at our fingertips, of Internet search engines and software agents, of digital tools electronically doing our bidding, of online analytical processing digesting gigabytes at the press of a button, you may wonder why I mention someone called Rotwang, let alone bring up the topic of his children. If you are not a familiar with movies created before, say, the midpoint of this century, you may wonder about the identity of Rotwang. And what does Rotwang have to with information or computers or the Internet in any case, you may ask. Let me explain.

2. Metropolis

First about Rotwang. Rotwang was a crucial character in the one of the earliest science fiction films, Fritz Lang's wondrous *Metropolis. Metropolis* debuted in Berlin in January, 1927. This movie provides a glimpse of life in the next century, seventy three years away from Berlin, where a vast city is powered by workers and their machines, hidden well underground. This large enterprise is managed by a character simply known as the Master, with his son at his side patiently learning the trade of being a big boss. Rotwang is an inventor working for the Master, sort of the equivalent of Chief of Advanced Technology Development.

Early in the movie, Rotwang shows off to the Master his latest invention, a robot. All Rotwang needs to make it work is a human model. The Master readily suggests Maria, a daughter of one of theworkers. Maria is a social worker among the downtrodden masses, preaching an optimistic view that someday a mediator will come to bridge the distance between the mind and hands with heart. The Master has tired of Maria's gospel and welcomes Rotwang's invention as a way of ridding himself of Maria and her influence over the workers. With Rotwang's robot in the form of Maria, the Master will urge the workers to violence through her. These destructive acts on the part of the workers will give the Master the opportunity to replace all of them with robots, thus giving him absolute control over the infrastructure of Metropolis once and for all.

The Master's use of technology – in the form of the robot – is ultimately evil. On the other hand, Rotwang's invention of the robot is based on his own attempt to recreate a lost love. Rotwang's original purpose is simple-minded and good. Rotwang's robot as Maria indeed leads the workers to riot, urging a destruction of the very technology that keeps Metropolis alive. The workers realize their error in destroying their own city's infrastructure, and turn on the robot and its inventor. The Master's son, quietly and meekly observing everything, finally pleads with his father to cooperate with the workers and with technology. In the end, the Master's son is the very mediator that Maria predicted, bringing labor and intelligence together with compassion.

What are the lessons from this movie for designing the human interfaces of technology? According to *Metropolis*, technology should be more than just a collection of dumb tools. In *Metropolis*, rigid and simplistic applications of machines and tools tormented workers, so much so that they were quite ready to listen to the ravings of a robot. Technology instead should be well integrated and adaptive in human activities. Technology that ignores humanity, that ignores the mind and heart, is self-defeating and even self destructive. Creative technology that demonstrates the importance of a very human interface is far more successful, and it makes for more complex and richer interactions with the world.

3. Metropolis' and Rotwang's legacy

Where are Rotwang's children today? When we look at information technology, and especially at tools designed to manipulate Internet based information, how are they being created? Do today's search engines and knowbots start out as well-intentioned and misguided as Rotwang's original robot? In their ultimate application, do these search engines become like tools of the

Master, frustrating the very people that they were suppose to help? More importantly, what kinds of background assumptions or evidence are being used in the design of these agents, avatars, and search engines? Are search engines being created because programmers themselves find tinkering with computer interfaces easier than talking about search problems with people in libraries? Are these designs relying on controlled experiments with people in information mazes? Or on real-life experiences in natural settings?

If you look at the results so far, many popular, information-seeking programs are based on simple or poorly defined models and assumptions. Take any group of Internet search engines. Different search engines, different agents give different results for the searches constructed in the same way. These software solutions are frankly stubborn and very pigheaded, because they are search with a narrow set of assumptions and options about electronic information (Bringsjord, 1995). They repeated hunt across the Internet by sheer computational force, often very stupidly, consuming vast amounts of bandwidth. One Internet server found itself dealing with over 600,000 queries a week from bands of roving knowbots. Other servers have simply banned knowbots and agents because of the way that they greedily use resources. Postings are appearing on more and more servers telling users not to run knowbots or simply bots on specific machines, under penalty of loss of access to an entire site or domain (Markoff, 1994). These warnings wouldn't appear if some intelligence - based on the ways in which humans use and seek information with other humans - was used in their design.

Most of these agents are a dangerous combination of ignorance and arrogance in computer code. Most of the programmers of these agents know that they can ask a machine to predict the eighth member of an infinite sequence, beginning 2, 4, 6, 8 . . . . The program will reply 16 and all seems well to machine and programmer. But if they ask the same machine to tell them the eighth member in the sequence of numbers and words, beginning "2, 4, 6, 8, who, do, we . . ." the program will bomb (Kugel, 1990). Even though a small child with certain experiences in the English language will know that the eighth member of that sequence is "appreciate". Simple human experience is an invaluable resource for handling many basic information problems.

4. New methodologies

Software and machines can only begin to show signs of understanding the complexity of information, of comprehending the human need for a huge daily diet for information, by interacting and learning with human information experts. This sort of software and computer design calls for a radically different way of approaching machines and people. Efforts and methodologies based on activity theory are helping to create new views of human computer interactions.

What is activity theory? Activity theory was born in Soviet psychology in the 1920s, as a practical approach to define human consciousness. In this view of the world, there are no complex undercurrents of motives, there's a lack of subtexts. Forget stratified reasons for behavior. With activity theory, everything is on the surface in daily acts and choices. You are what you do (Nardi, 1996).

Activity theory, in its application to human computer interactions, calls for actual field studies of how people use technology in real life, of how technology works in specific and very real social and physical contexts. Laboratory experiments in human computer relations may produce one kind of view of reality, but activity theory begs for a more holistic perspective, based on actual and real experiences of humans with their machines. Activity theory has inspired alternative human-computer interface research that fundamentally describes the ways humans, computers, and information interact (Nardi, O'-Day, and Valauskas, in preparation).

Of particular note is research into the invention of a digital library services. With the principles of activity theory firmly in mind, efforts were made to study the ways of human information hunters and gatherers, and their interactions with technology and people (Nardi and O'Day, in press). Let me provide you with a brief overview of this investigation and discuss its implications for information retrieval, librarians, and future software solutions.

5. Libraries, librarians and intelligent agents

Librarians, from the perspective of activity theory, are perfect models for understanding human and computer interactions with information. Many librarians, as we well know, have been using networked computers and online resources almost longer than anyone else in our current digital community. Many libraries were among the earliest users of large computers, along with banks, government agencies, and universities. In fact, many libraries and librarians were the earliest users of new computing technologies, from minicomputers to personal computers to personal digital assistants. Librarians created some of the earliest international standards for computerized information, so that bibliographic data could be transferred around the world without a glitch. Librarians indeed have been information gatekeepers as long as records have been gathered and preserved. Clay tablets, papyrus, parchment, paper, magnetic tape, diskettes, and the Internet have all served librarians as media.

What could be learned from this analysis of librarians? Would it be possible to actually create software programs that could do what librarians do?

The heart and soul of a librarian's job is helping clients understand what they know, and more importantly what they don't know and need right now. Librarians are information therapists. It's surprising how many people think they know what they are looking for when they really don't. Information requests that librarians receive clearly indicate this lack of knowledge about a lack of knowledge, with queries that are fuzzy or awkward, too narrow or too broad. Questions appear on a daily basis that are stated "I want books with connectors, contacts and reliability in the title" or "I need right now all technical articles about the UNIX operating system." Could a software program be also an information therapist?

Questions that sound straightforward to an information-starved library patron are not very straightforward, when you think about the exacting requirements to actually search a database or the Internet. For example, one client wanted to be enlightened about "the presence of Hewlett Packard in Japan and in Europe."What does that really mean? As one of the librarians in the study explained: "First of all, what does person mean by Japan vs. Europe? Are they thinking about market share or the number of units? Do they mean plant size or relative presence? Do they need something that is more economic, which might be conversion ratios? And when you finally find some articles and reports that you think they need, you maybe take a couple others." The librarian plays a role in focusing a search, but also in providing some elbow room by adding in a couple of other articles and citations of possible areas of interest to the client. Can a search engine understand the concept of informational elbow room?

This ability to focus and anticipate is an art. The librarian, unlike a simple program, provides a human-based interpretation of the needs of an information-deficient person.

Librarians are masters at helping fuzzy-thinking information seekers create crisp statements of real needs. This translation is accomplished through artful conversations that are called reference interviews. These conversations are often gems of strategy and tact. In them, librarians maneuver to get their clients to pare down the search space or open it up so the search won't come up completely dry. And this conversational process occurs without ever making the client feel that the original request was ill-formed or just plain ridiculous. Has your favorite Internet search engine made you feel dumb recently?

Librarians adjust quickly to the rapidly changing landscape of Internet resources, online databases, and printed materials. They understand when a database is returning lousy results because it has not been updated or the index terms have been changed. They know how to save a client money by doing pre-searches, by using the most cost-effective commercial databases for a given request, and using the right combination of keywords to focus but not over constrain a search.

Librarians recognize when their information hungry clients are frustratedand want a helping hand, as much as they need a certain fact. They know when someone is really in a crunch, really depending on the librarian to weed out irrelevant materials from a search. Librarians are tuned to non-verbal signals, changes in voice and posture, emotional cues. They respond with intelligence, sympathy, and a sense of humor, unlike any program.

Indeed unlike software programs, librarians find and broker proprietary material because they are always quietly sniffing around, finding out what their colleagues are up to. Since they usually know the information interests of many people, they don't need to scan hard disks or networks for new facts and ideas. In this process, librarians are the souls of discretion. They distribute proprietary material by putting the right people in touch, and then leave the parties themselves to decide if they want to share secrets. It seems that its part of a librarian's culture to be discreet and nosy at the same time. How would you write a program to do thesame?

Librarians understand paper sources. Huge, stunning amounts of information are not online and never will be. Librarians know how to navigate through journals, books, newsletters, conference reports, with a trained eye and a sense for the sublime.

They know that information wears all sorts of disguises, as official releases of financial data, scientific articles, analyst reports, news, product reviews, statistical data, government reports, and patents, to name a few. They can judge the reliability of sources, estimate costs, and find material with a particular slant or perspective.

Librarians understand the frustrations of exploratory searches, when someone needs to learn about a new area. Exploratory searches are vaguely expressed because users cannot think of the right words at the moment. Better search interfaces will not make these searches less vague. Users actually learn about what they want by going ahead with the search. Librarians are willing to figure out a sensible way to execute a search even if it is poorly defined, if that is what the client needs to move ahead.

Librarians personalize searches for clients. Does someone want historical background, because she is new to the topic, or only the latest developments? Should the retrieved articles be for experts or for novices, for scientists or for administrators? What about regional, national, or international perspectives? Is the client seeking articles that are centrally focused on a topic, or articles that may have only peripheral references? Does she want complete, detailed coverage, or a sneak preview? Can she wait till next week, or is it a rush job?

Clients often become attached to a personal librarian because these ongoing personal relationships form the basis for an extraordinary level of on-target search performance. With a client history to call upon, librarians can aim their search spotlight very exactly, without having to ask the client a lot of questions.

Not surprisingly, librarians retain much of what they have read. Computers are a long way from the kind of information processing skills of librarians. It is unclear from this study exactly how it is that librarians mentally catalogue what they read to recall it later for clients, but they assimilate startling amounts of information for later recall.

Not only do librarians remember substantive material, but they track the way tables of contents are set up, remember indexing schemes, and store away other means by which a search can be organized. They might remember the kind of material that appears in a quarterly column of a particular journal, or the fact that a major article in a certain magazine runs only once a year. This mental cataloguing enables librarians to perform searches that do not fall into convenient keywords or phrases. These searches depend on a grasp of the kind of information that might appear in a particular format in a publication. When presented with a request, librarians perform incredibly fast searches – by using their own neural storage to remember what they have recently read.

Perhaps librarians are born with special neural data quality sensors. In this study, librarians were sometimes heard to mutter that a particular commercial database was "lame," or "pathetic." These descriptions are based on an analysis of the quality of data in a particular information database and depend on several factors. Large commercial databases contain millions of items and, because of their sheer size, they are not always perfectly up to date or complete. The indexing scheme in a database may change, providing better or worse access depending on the kinds of alterations. A company controlling a specific database can change hands, meaning that differing policies will be introduced that may affect the database and the quality of data. Databases themselves are sometimes set up to receive new information in a "first load" containing only citations and no full text, with a "second load" including full text done later. A skilled librarian is well aware of these subtleties. Librarians understand if the search results are too sparse, too dated, or too off the mark. A client has no way of understanding these nuances, without the daily contact to these files.

Finally, it seems librarians offer a human touch to information gathering. This factor will become more important as electronic access to information grows and tools for information access proliferate.

6. Future scenarios for intelligent agents

Conclusions? Based on an analysis of librarians and their interactions with information, technology, and patrons, it would be impossible to replace much of librarians' highly-skilled services with software tools. Proposals for a librarian-in-a-box completely misconstrue what it is that librarians do and the value they provide. The best-written software agent on the Internet can only do a small portion of the navigational and interactive work of an experienced librarian.

But where does the increased digitization of information leave librarians? What are the roles of librarians as software tools proliferate that claim to be search agents for the masses? The future points to a new kind of information ecology, one where humans, software agents, and librarians work together to hunt and gather information. This information ecology – which we can define as a living, evolving system of people, things, ideas, and relationships, a phenomenon of neurons, electrons, wires and social networks – means new approaches to software agents.

Designs for intelligent search agents at the moment concentrate on such things as search algorithms, keywords, and relevance rankings. These are fine and useful, but we imagine more personalized software search tools for specific clients, designed with the help of librarians. We see at least six ways that librarians will contribute to a diverse information ecology:

6.1 Designing software agents for clients

A client and a librarian will collaboratively design intelligent search agents for the client, for searches that need to be run on a daily, weekly, or monthly schedule. Rather than asking the client – who is not an expert in searching – to design his own agent, the client will work with a librarian who understands search techniques, cost constraints, and all the other technical issues. Designing agents would be a new role for librarians but one that grows naturally out of work they are already doing. This kind of collaboration would work to the librarian's advantage while at the same time educate the client and allow the client to get his work done. The librarian could continue the relationship with the client for those times when changes need to be made in the software agent that are beyond the capability of the client. Such collaborations would leverage the special skills of librarians while taking advantage of the potential of software agents. This sort of collaborative programming is typical in offices where spreadsheets and CAD systems are successfully being used (Nardi, 1993). This kind of collaboration will occur in libraries withsoftware agents.

6.2 Designing specialized software agents for librarians

Since librarians know their work best, they will take part in efforts to design "agents for agents," for their own use in libraries.

6.3 Cataloging, distributing and customizing software agents

Librarians will be critical to a future diverse information ecology in cataloguing and distributing intelligent search agents. With new architectures and languages that enable the development of small software programs that can be easily transported across systems, there will be a need to access software agents developed by others. Knowing where these agents are, which ones will be useful for a particular client, and how to customize them for specific needs will be a yet another role.

6.4 Encouraging the evolution of agents

Librarians adjust their search strategies in response to changes in the information landscape. New information appears, terminology evolves, old sources become unreliable, cost structures change. Even the best-designed software agents will gradually become out of date and less useful if they are not tuned up regularly. Librarians will become mechanics for software agents. Software agents will be built with the expectation that they will change and grow, with internal hooks built in for easy modification.

6.5 Contributing to information bases

Intelligent software agents should have deep technical and practical knowledge about searching. Descriptions of information sources, guidelines for cost optimization, focus and defocus strategies, and other elements of expert search knowledge can be shared among a group of software search agents. Librarians will supply this knowledge.

6.6 Working cooperatively with software agents

We should not assume that a particular search task must be carried out from start to finish by either a software agent or a librarian. Sometimes the best approach will be to use both. Librarians can help get searches started by working with paper sources when that is necessary, assisting with source selection, or supplying initial guidance on terminology, allowing clients to follow through with the help of software agents. Conversely, a librarian might take over an unsatisfactory search performed by a software agent, using the partial results generated by the agent as a starting point. For such cooperative efforts to work, the designers of software agent environments must educate users about their choices and make it technically feasible for agents to start or end at various points "in the middle" of a search.

7. Designing the diverse information ecology

The potential of software agents to enable much larger populations to access information requires that librarians be right in the loop in creating this diverse future information ecology. Librarians can draw upon their experience of serving real users, their search expertise, and their culture to forward goals of broad information access. Both librarians and software agents mediate the vast but sometimes unfathomable information resources made available through networking. Often, librarians will instigate and cultivate software agent-human client relationships.

An information ecology should provide diverse communication channels between librarians and clients. In the libraries at Apple and Hewlett-Packard, telephone and electronic mail were and still are used effectively to provide services for employees all over the world from an information base in California. Future communication channels will no doubt be extended to include video conferencing so that the expertise of librarians can be shared with people on a global scale. With reasonably good resolution for video, clients in far-flung locales with network access could enjoy the services of expert librarians and well-tuned software tools.

Of course this diverse information ecology also includes the kinds of interactions between clients and librarians that currently take place in libraries. The value of these interactions will persist and grow as more information becomes available through networked sources.

8. Conclusion

We envision a rich and complex future, where print, electronic data, and humans all interact, taking advantage of the best aspects of ink, neurons, and electrons. For this future, we should be designing ecologies that knit together people, print and machines in brilliant new designs. These diverse information ecologies will be the triumph of a human network, a synthesis of computer code and human memory, and even emotion and processing speed.

References:

BRINGSJORD, S. (1995): Could, How Could We Tell If, and Why Should – Androids Have Inner Lives? In: K. M. Ford, C. Glymour and P. J. Hayes (eds.): Android Epistemology. *MIT Press, Cambridge, 93-121.*

KUGEL, P. (1990): Is it time to replace Turing's test? In: *Artificial Intelligence: Emerging Science or Dying Art Form.* SUNY Binghamption Dept. of Philosophy's Program in Philosophy and Computer & Systems Sciences, Workshop, quoted in BRINGSJORD, S. (1995): Could, How Could We Tell If, and Why Should – Androids Have Inner Lives? In: K. M. Ford, C. Glymour and P. J. Hayes (eds.): Android Epistemology. MIT Press, Cambridge, 114.

MARKOFF, J. (1994): The Fourth Law of Robotics. *Educom Review, 29, 45-46.*

NARDI, B. A. (1993): A Small Matter of Programming: Perspectives on End User Programming. MIT Press, Cambridge.

NARDI, B. A. (1996): Activity Theory and Human-Computer Interactions In: B. A. Nardi (ed.): Context and Consciousness: Activity Theory and Human-Computer Interaction. *MIT Press, Cambridge, 7-16.*

NARDI, B. A. and O'DAY, V. (in press): Intelligent Agents: What We Learned in the Library. *Libri, 46.*

NARDI, B. A.; O'DAY, V. and VALAUSKAS, E. J. (in preparation): Rotwang's Children: Technology, Ecology and People.

Network Approaches in Text Analysis

R. Popping[1], C. W. Roberts[2]

[1] Department of Social Science Information Technology and Department of Sociology, University of Groningen, NL 9712 TG 15

[2] Department of Statistics and Department of Sociology, Iowa State University, Ames IA 50010, USA

Summary: In the last decade a broad spectrum of text analysis methods have been developed. One class of these involves the building of network- depictions of texts. At present there are four primary approaches within this class: network evaluation, map analysis, knowledge graphing, and functional depiction. The first two of these approaches are quantitative text analysis methods, in that they were developed to yield statistical inferences about text populations. The latter two approaches are qualitative, in that they offer methods whereby texts can be "simplified," not into a concise statistical model, but into a generalized graphic representation. This paper provides brief introductions to the four methods, as well as to computer programs developed for each.

1. Introduction

Despite some disagreement on the importance of explicit versus implied (i.e., manifest versus latent) nature of their subject matter, classic texts on content analysis have been relatively permissive on the types of qualitative data (e.g., words, gestures, art forms, etc.) that are appropriate for scientific analysis (Berelson, 1952; Stone, et al., 1966; Gerbner, et al., 1969; Holsti, 1969; Krippendorff, 1980). Moreover, these texts are also consistent in that each treats content analysis as a *quantitative* enterprise in which qualitative data are mapped into a data matrix for the purpose of drawing statistical inferences about a relatively well-defined population of symbolic phenomena. This quantitative approach experienced a revival in the late 1980s via numerous articles (and related software) on methods for quantifying texts (Carley, 1986; Van Cuilenburg, et al., 1988; Franzosi, 1989; Roberts, 1989).

What distinguishes the new methods from classical ones, is that with few exceptions (e.g., Osgood, et al. (1956) and Gottschalk & Gleser (1969)), the earlier methods quantified texts as "counts" of words and phrases that were classified according to a set of general themes. In contrast to these thematic text analysis methods, the new quantitative text analysis methods required not only the identification of themes, but also the encoding of relations among themes in texts. These researchers' relational methods for encoding texts are strikingly similar. In each case, a Subject–Valence– Verb–Object (or S–V–V–O) syntax is applied during the encoding process. Unlike thematic content analysis, such "clause-based content analysis" affords inferences about how texts' sources use words in their speech or writings. The new text analysis methods differ primarily according to the research purposes to which each's relationally encoded texts can be applied (Roberts, forthcoming).

The early 1990s saw the publication of a torrent of books and articles on *qualitative* text analysis methodologies and related computer programs. (See Tesch (1990) and Weitzman & Miles (1995) for overviews.) In these methodologies, the goal is not to obtain statistical inferences, but may be to identify structure in texts, to build theories, and/or to derive hypotheses. In general, these methods seek to summarize large quantities of text in terms of some simplifying user- (or software-) imposed framework or text-based structure. Current software applications typically use path algebra for this purpose (Carr, 1979: 84–85), and may use fuzzy logic in the near future (Yuen & Richards, 1994). The focus of this paper is on network approaches to quantitative and qualitative text analysis. In general, these are methods in which pairs of words/phrases are encoded as related themes, and then that these theme–relation–theme tuples (more often than not in S–V–V–O form) are combined into networks of interrelated concepts. The following sections first discuss characteristics that quantitative and qualitative network text analysis methods have in common. As it turns out, both types of methods have similar procedures for encoding linguistic data, and similar types of relations that related software make available to the user. Next, overviews are provided of the distinguishing characteristics of four network text analysis methodologies: network evaluation, map analysis, knowledge graphing, and functional depiction. The paper concludes with a brief discussion of the sorts of research applications to which each type of network text analysis is appropriate.

2. Encoding of texts into networks

There are two important steps required when one encodes networks from texts. First, one must specify the themes, or concepts, that are to be related, or linked, within networks. These themes may originate with one's theory or with the texts themselves (i.e., with themes that are empirically recurrent). Encoding then involves the classification of texts' words/phrases as occurrences of these themes. Yet this encoding process is likely to generate invalid data if contextual information is not taken into consideration. Thus it is crucial that native speakers determine when specific words/phrases are instances of one theme, and, given their idiomatic usage, when they are instances of another.

The second step in network encoding involves the assignment of links between pairs of concepts. After removing idiomatic ambiguity during the classification of theme occurrences, most interrelations among words/phrases can be unambiguously identified according to the grammar of the language in which the texts were written. Yet even identical nonidiomatic, but grammatically correct statements can have different meanings, depending upon the meaning intended by the statement's source. (For example, the statement, "Joe was abandoned," could mean either that Joe was alone or that others departed?) All relational text analysis methods must assure that such illocutionary ambiguity is systematically disambiguated (Roberts, 1989).

Within any single network, no theme appears more than once. As a direct consequence, graphical displays of networks become complex when specific pairs of themes are allowed multiple types of links. For this reason, network analysis software developers have tended to gravitate toward six general types of links–types that we pretend neither to be exhaustive nor exclusive:

- Similarity – records that one theme is identical with another. The relation is symmetric (abbreviations: ALIke, SIMilarity, EQUal, EQuiValent);
- Causal – denotes a cause-effect relation. The relation is asymmetric and transitive. In all methods using networks based on text the causal relation is read as "might" cause (CAUses; is Caused BY);
- Relation – indicates an ASSociation, an ORDering, an EVAluation, or a REAlization. In the first case the relation is symmetric, in the other three it is asymmetric;
- Classification – indicates transitive (is A Kind Of, Has As Kind), asymmetric (Is Property Of), or symmetric (INConsistent with or contradicts, DIStinct) classification;
- Structure – indicates a structuring. The relation is transitive (is PARt of, Has As Part);
- Affective – establishes a judgment of the subject about the object (AFFective, WILL).

Several software applications also allow the user to weight the degree of similarity, causal determinance, etc. of links, usually on a 0 to 1 scale. Evaluative links can often be encoded on a minus–neutral–positive scale. Moreover, statements may themselves be conditionally encoded. For example, they may be encoded as "*if* statement 1 *then* statement 2" or as "statement 1 *or* statement 2 is true." Finally, users may be allowed to encode statements according to their respective sources.

3. Network text analysis methodologies

Network text analysis originated with the observation that after one has encoded semantic links among concepts or themes, one can proceed to construct networks of semantically linked concepts. When concepts are depicted as points in networks, one is afforded more information than the frequency at which specific concepts are linked in each block of text; one is also able to characterize concepts and/or linkages according to their position within the network. For example, consider the following pairs of causal relations among the themes A, B, and C:

$$\begin{aligned} A &\longrightarrow B \\ A &\longrightarrow C \\ B &\longrightarrow C \end{aligned}$$

Note that by combining these three theme-causes-theme tuples into a network, the following indirect relation results:

$$A \longrightarrow B \longrightarrow C$$

Thus, the information in a network is greater than the sum of its S-V-V-O components. It is precisely this added information that specific network text analysis applications seek to exploit. In particular, computer programs for network text analysis typically allow the user to infer such added information by tracing transitive relations among networks' concepts. In fact, the premise that meaning can be derived from statements' logical interrelations is basic to all artificial intelligence endeavors (Van Cuilenburg, et al., 1988).

In the following subsections, descriptions are given of four network text analysis methods and related computer programs. The first two methods are quantitative; the second two are qualitative. The two quantitative methods differ primarily in that network evaluation infers transitive information based entirely on the texts themselves, whereas map analysis may derive such information after text-based networks are supplemented by user-supplied expert knowledge. The principal difference between the two qualitative methods is that knowledge graphing simplifies texts by grouping networks' concepts into ever-larger conceptual frames, whereas functional depiction simplifies texts by characterizing networks according to their respective discursive functions.

3.1 Network evaluation

The network evaluation approach has its roots in the work on evaluative assertion analysis by Osgood, et al. (1956). The fundamental premise of this methodology is that every language has three kinds of words:

- "Common meaning terms", words that have a common evaluative meaning among "reasonable sophisticated users of the language;"
- "Attitude objects", words that have no discernable evaluative meaning;
- "Verbal connectors", words that indicate the association ("it is...") or dissociation ("it is not...") of attitude objects with common meaning terms or with other attitude objects.

By investigating how attitude objects are associated or dissociated, one can investigate how these attitude objects are valued in a text. In brief, the method requires that texts be parsed into clauses (or nuclear sentences), that instances of the three above-mentioned word-types be located within these clauses, and that the clauses be recombined in a way that reveals structure in the text.

The NET content analysis method goes beyond this approach to reveal the logically implied structure of texts as well. CETA (Computer-assisted Evaluative Text Analysis) is the computer program developed specifically for this method (De Ridder, 1994). The program facilitates the parsing of text into nuclear sentences, as well as the assignment of words and phrases to meaning categories.

The program uses four classes of links (SIM, CAU, AFF, ASS), which are further subdivided into 15 types of links that may be associated with a positive (did) or negative (did not) valence. Two special-purpose concepts are hard-wired into the software. By linking it to the abstract concept, "Ideal," the user can encode a statement as a positive (is good) or negative (is bad) evaluation of a concept. By linking it to the concept, "Real," the user can encode a statement as an affirmation that a concept's referent exists (is) or does not exist (is not).

3.2 Map comparison

Cognitive mapping involves extracting linguistic relations from texts or transcripts and then representing the 'mental models' or 'cognitive maps' that individual sources had in their memory at the time the relations were expressed (Johnson-Laird, 1983). The following are three key theoretical assumptions that tie this literature into cognitive mapping methodology:

- Mental models are representations of internal (i.e., subjective) states;
- A source's linguistic expressions reveal its mental model;
- Conceptual networks provide valid representations of mental models.

Within a cognitive map, the meaning of a concept is the aggregate set of relations it has to all other concepts that make up a conceptual network. Mental models are dynamic structures that are constructed and expanded as individuals make inferences and gather information. They contain both specific information about particular items, and general (or social) knowledge. A transcript of an individual's speech is a reflection of the individual's mental model at a particular point in time. Accordingly, such texts may be thought of as a sampling of information from the individual's memory.

The map comparison method (Carley, 1986) affords not only graphic descriptions of individuals' mental models, but also comparisons among models maintained by various social groups. MECA (Map Extraction Comparison and Analysis) is a set of programs developed precisely for this type of text analysis. Like CETA, the programs facilitate the parsing of text into nuclear sentences and the assignment of words and phrases to meaning categories. The investigator begins by specifying the concepts and types (or meanings) of links that may exist between pairs of these concepts. Links have four formal properties: meaning, directionality, strength, and sign. Once texts have been encoded in terms of these concepts and links, they can be depicted graphically as discrete mental models or as "unions" or "intersections" of mental models among or between groups of individuals. The sort of data matrix yielded in a map comparison analysis lists propositional information (i.e., S–V–V–O relations aggregated across blocks of text supplemented with frequency data).

Given the quantitative nature of their methodologies, both network evaluation and map analysis require the development of a data matrix in anticipation of statistical analysis. When based on networks, the columns of this data matrix will typically contain information about where concepts or concept-pairs are located in each network listed as a unit of analysis in the matrix's rows. (See Roberts & Popping (1993) for more detail on these and other quantitative text analysis methodologies.)

3.3 Knowledge graphing

Knowledge graphing methodology was originally developed to describe the state of the art in some scientific field. It begins with a researcher's explicit choice of a relatively few types of relations (James, 1992: 98). The method then proceeds to the *text analysis* step, in which a list of concepts is generated. An "author graph" is then constructed with each node corresponding to a specific concept, and with each link of one of the chosen types.

In the next, *concept identification* step, the author graphs are compiled into a single graph by combining their corresponding nodes. During this process linguistic ambiguities are eliminated as synonyms across author graphs are combined, and as homonyms are not. The resulting "compiled graph" is further refined by alternating between *concept integration* and *link integration* steps. In the first refinement step, theoretically interesting substructures are identified; in the second, new links are inferred from those already encoded. This "integrated graph" can then be further refined (i.e., further "integrated") with the initiation of another concept integration step. By superimposing various frame relations on the graph, this step identifies further knowledge structure. Each frame relation generalizes several concepts and relations as a single concept. Additional link integration is then possible as inferences are made based on these concepts newly added to the graph.

Bakker's (1987) KISS program allows seven relations (PAR, CAU, ORD, ASS, EQU, DIS) between pairs of concepts. Restrictions can be placed on the types of concepts that may be related according to each of these relations. Yet the program also allows concepts to be related as tokens and types. By definition, tokens are instances of types (and are thus always related as ALIke). At the most primitive level, a word (e.g., "Pluto") might be an instance of a more general concept (e.g., "dog"). Within the context of knowledge graphing, types are always concepts. And so as integrated graphs become refined, concept types may capture global abstractions generalized from a large number of author graphs.

COPE is another program for aggregating texts into a unified network representation (Cropper, et al., 1990). Unlike KISS, COPE only permits CAUsal relations between concepts. The program is designed to assist in the delineation of causally related classes of phenomena. It accomplishes this goal by making it easy for the user to redefine concepts as additional causal data arise.

3.4 Functional depiction

Most qualitative network text analysis software was developed (as were KISS and COPE) to generate an aggregated network representation of a large corpus of texts. Once obtained, this representation's broad overview provides a resource for the user to construct a grounded theory based on these aggregated observations. In contrast to such programs, ATLAS\ti (Weitzman &

Miles, 1995: 217 ff.) does not aggregate structures within networks. Instead, the program *distinguishes* structures among texts according to the functions they might have within a larger discursive context. ATLAS\ti allows text-text links to be encoded in five ways. One linkage (CRITicizes) is asymmetric; the others (SUPports, DISCusses, JUSTifies, EXPLains) are transitive. More fundamental concept-concept links can either be user-defined, or selected from among six built-in links (ASS, PAR, CAU, INC, AKO, IPO).

4. Conclusion

One leaves the above brief descriptions with a long list of distinctions among the programs CETA, MECA, KISS, COPE, and ATLAS\ti. Only CETA and MECA generate data matrixes. COPE only allows the encoding of CAUsal relations among concepts. Unlike the other four programs, MECA does not derive transitive S-V-V-O relations from encoded texts, but generates them based on an expert's knowledge. KISS has powerful procedures for concept and link integration. This list could go on.

The most important information that a review article such as this can provide is guidance on what sort of research agenda each program is "good for." One must begin, of course, by considering whether one's research question requires the complexity of information afforded via the time-consuming task of encoding one's data as networks. If one need only know which themes were prominent in some texts, a more traditional (and less laborious) thematic text analysis may be called for. If one wishes to know how concepts are related in texts, but not how they fit into a larger network of concepts, a semantic text analysis may be most appropriate (Roberts & Popping, 1993).

Given that one is interested in understanding how concepts and concept-pairs are located in network depictions of texts, the next question is whether one wishes to perform an analysis from which probabilistic inferences can be drawn from a sample to a population of texts. For such analyses, one will require software (such as CETA or MECA) that download data matrices with concepts and/or concept-pair characteristics appropriate to randomly sampled text blocks. If one's networks are to be characterized as mental models, MECA may be more useful than CETA. If one wishes to derive concept linkages based on concept–relations manifest in one's texts, CETA may be the preferred research tool.

On the other hand, if one's research objectives are qualitative (that is, if one seeks to describe structures underlying a corpus of texts), KISS, COPE, or ATLAS\ti may be most appropriate. The qualitative network text analyst must next consider whether the data are to be aggregated into a single abstract rendering of text structure, or divided according to the various discourse functions each subdivision serves. KISS and COPE are more appropriate for the former, and ATLAS\ti more appropriate for the latter type of analysis. If the links in one's networks are restricted to CAUsal ones (as would be the case if one wished to synthesize causal implications within a large body of scientific research), COPE might be the best computer aide for text aggregation. Yet if the goal is aggregation of networks comprised (at least in part) of noncausal links, KISS would be more appropriate.

Thus the choice of text analysis methodology depends fundamentally on the research question being asked. Despite its platitudinous ring, this statement is far from obvious to many a zealous champion of a single (usually self-created) text analysis methodology. Its repetition here is to highlight that the current plurality of text analysis methods can only serve to improve the overall quality of the research in which they are applied. The contribution of this paper is to help researchers decide which (here, network) text analysis method will be most legitimate for their purposes.

References:

BAKKER, R. R. (1987): Knowledge Graphs: Representation and Structuring of Scientific Knowledge. PhD thesis University of Twente.

BERELSON, B. (1952): Content Analysis in Communication Research. The Free Press, New York.

CARLEY, K. (1986): An Approach for Relating Social Structure to Cognitive Structure. *Journal of Mathematical Sociology 12: 137–89.*

CARRÉ, B. (1979): Graphs and networks. Clarendon Press, Oxford.

CROPPER, S.; EDEN, C. and ACKERMAN, F. (1990): Keeping sense of accounts using computer–based cognitive maps. *Social Science Computer Review 8: 345–366.*

DE RIDDER, J. A. (1994): Van tekst naar informatie [From Text to Information]. PhD thesis, University of Amsterdam.

FRANZOSI, R. (1989): From words to numbers: A generalized and linguistics–based coding procedure for collecting event–data from newspapers. In: C. Clogg (ed.),: *Sociological methodology, 1989.* Basil Blackwell, Oxford, 263– 298.

GERBNER, G.; HOLSTI, O. R.; KRIPPENDORFF, K.; PAISLEY, W. J. and STONE, P. J. (Eds.) (1969): The analysis of communication content: Developments in scientific theories and computer techniques. Wiley, New York.

GOTTSCHALK, L. A. and GLESER, G. C. (1969): The measurement of psychological states through the content analysis of verbal behavior. University of California Press, Los Angeles.

HOLSTI, O. R. (1969): Content analysis for the social sciences and humanities. Addison-Wesley, Reading, MA.

JAMES, P. (1992): Knowledge graphs. In: R. P. Van de Riet and R.A. Meersman (eds.): *Linguistic instruments in knowledge engineering.* Elsevier, Amsterdam, 97–117.

JOHNSON–LAIRD, P. N. (1983): Mental models: Toward a cognitive science of language, inference, and consciousness. Harvard University Press, Cambridge, MA.

KRIPPENDORFF, K. (1980): Content analysis: An introduction to its methodology. Beverly Hills, Sage.

OSGOOD, C. E.; SAPORTA, S. and NUNNALLY, J. C. (1956): Evaluative assertion analysis. *Litera 3: 47–102.*

ROBERTS, C. W. (1989): Other Than Counting Words: A Linguistic Approach to Content Analysis. *Social Forces 68: 147–77.*

ROBERTS, C. W. and POPPING, R. (1993): Computer Supported Content Analysis. Some Recent Developments. *Social Science Computer Review 11: 283–291.*

STONE, P. J.; DUNPHY, D. C.; SMITH, M. S. and OGILVIE, D. M. (Eds.) (1966): The General Inquirer: A computer approach to content analysis. MIT Press, Cambridge, MA.

TESCH, R. (1990): Qualitative research: Analysis types and software tools. Falmer Press, New York.

VAN CUILENBURG, J. J.; KLEINNIJENHUIS, J. and DE RIDDER, J. A. (1988): Artificial intelligence and content analysis. *Quality and Quantity 22: 65–97.*

WEITZMAN, E. A. and MILES, M. B. (1995): Computer programs for qualitative data analysis. Thousand Oaks: Sage.

YUEN, H. K. and RICHARDS, T. J. (1994): Knowledge representation for grounded theory construction in qualitative data analysis. *Journal of Mathematical Sociology 19: 279–298.*

Qualitative Software and Analysis Structures: Solving Problems or Creating Them?

Tom Richards[1], Lyn Richards[2]

[1]Qualitative Solutions & Research Pty Ltd, Melbourne, Australia 3083
[2]School of Sociology & Anthropology, La Trobe University, Melbourne, Australia 3083

Summary: In our paper in the Denzin & Lincoln volume (Richards & Richards (1994)), we made a distinction between the conceptual and textual levels of work in qualitative data analysis (QDA), and argued that future work in the computerisation of QDA would involve finding ways of crossing the gap from the textual to the conceptual levels. In this paper we will discuss more closely the relation between the two levels; will argue that bridges from textual to conceptual levels are still poorly supported; and will discuss requirements for making the bridge.

1. Textual and conceptual levels of QDA work

Much of the current QDA software is based on the code-and-retrieve model. The software is used to mark text with codes, such as "Mary", "angry", "feeling peer group pressure", etc. The analysis is primarily supported by retrieving all text marked with some code or codes so it can be seen together, which assists the researcher in contemplating particular themes and developing ideas about them.

More sophisticated software will allow boolean and other combinations of retrievals. Thus one might look for where Mary is angry by intersecting text coded with "Mary" and "angry". This can be useful in comparative analyses: what does Mary get angry about, is it what makes John angry?

There is no doubt that this is a most useful procedure; certainly it is extremely widely practised and underlies a large number of research publications and PhDs. Equally there is no doubt that software aids this process. Its particular benefits when done computationally are:

- Clerical efficiency of management: codes don't get lost and a retrieval is guaranteed complete.
- Coding is minimum effort so thoroughness of coding is encouraged.
- It's easy to use many different codes, so finer discriminations in retrievals can be made.
- Speed: far more material can be handled at a useful level of coding.
- If the software supports specifying retrieval combinations, this can lead to complex and subtle probing of the text that would be impossible manually.

In sum, greater sensitivity to the text is possible, and deeper explorations are encouraged.

1.1 Label coding and the textual level

However, code-and-retrieve methods remain essentially about the passage of text they are coding and retrieving. The codes are usually labels to mark what people are saying in the text and perhaps how they are saying it. The code "mothering" labels text where people are talking about mothering, the code "anger" is labels text where people are talking angrily (or perhaps about anger), the code "feeling peer group pressure" labels text where the speaker's words evince (explicitly or subtly) subjection to peers. Let us call coding used to label what the text is about, *label coding* to remind of labels on jam jars.

Label coding is nearly always done on the basis of what is read in the text, hence marks features of the content or style of the text. This level of working is an example of what we called *textual level* work in our Denzin/Lincoln article, because what the terms, the words, the codes are about is the text. They are in effect labels for the text, noting what the text says or expresses.

1.2 Base data coding and the conceptual level

But most researchers also use codes for a quite different purpose, to record information, typically about interviewees. A code, for example "female", is often used to code all speeches by female interviewees, and similarly "Mary" to code all speeches by Mary. Codes used in this way we call *base data* codes. This technique can be exploited for retrieving coded text sorted by base data codes, for example female interviewees feeling peer group pressure, by searching for all text coded with the label code "feeling peer group pressure" which is also base data coded "female" – another example of searching for intersection of codes.

Q.S.R. NUD•IST supports both forms of coding with a hierarchical index system. This allows users to handle the relation between general and specific labels, or variables and values. For example to record information about interviewees the user can set up a code "interviewee" (not actually used to code text) under which go as subsidiary codes all the demographic categories one wishes to have for interviewees, such as "gender" and "age-group". Under these demographic variables in turn go the values of each variable, in this example "male" and "female" under "interviewee/gender"; and for age decades one could use "interviewee/age-group/30s". Then these values are used to code text in the manner described. The higher codes: "interviewee/gender..."are not used for actual text coding but to give meaning to the concrete ones: ".../female". In this way they together show the *propositional structure* of the information that is being recorded about: the system is recording quitc cxplicitly thc proposition that the age-group of the interviewee is 30s, not merely recording a code "30s" that the user must remember is cryptic for saying that proposition.

These hierarchies also provide a useful organisational structure for the system of codes. This supports locating and disambiguating codes, and can be exploited in analyses.

Base data coding can usually be distinguished from label coding by what it codes. If a project has one document per interviewee for example, all the demographic facts about the interviewee will code the entire document. In the case of focus group interviews, base data codes about interviewees will code the document speech by speech. Label codes, on the other hand, usually pick up much more specific portions of text, such as where someone is talking angrily or is showing peer group pressure or is talking about the neighbours. This difference arises because base data coding is about the world of the project, not about the text. We attach base data coding to text *only* as a convenient device for computational recording of information about aspects of the project – that the interviewee Mary is a woman in her 30s. Because we have text by Mary, we can computationally record the above information by labelling her speeches with "interviewee/Mary", "interviewee/gender/female" and "interviewee/age- group/30s".

Base data coding is not only distinct from label coding, it is not textual-level coding either. Its primary purpose is to record *propositions* (facts, true/false statements, information) about aspects of the project or the project's world. Our example has been that Mary is a female interviewee in her 30s. Such propositions get recorded, at least in Q.S.R. NUD•IST, by the "gimmick" of labelling all Mary's text accordingly. The virtue of the gimmick is that we can then retrieve her text, text by females, text by people in their 30s. It also means that if we are looking at a piece of text spoken by Mary, inspection of the cross- references for that passage will show us the base-data facts about Mary.

Base data information mostly belongs at what we have, in the Denzin & Lincoln article, perhaps misleadingly, called the *conceptual level* of work, as opposed to the textual level. Most of a project operates at the conceptual level, for there belong the thinking, the writing, the knowledge, the conjectures and their tests, and the theories and categories we use to explain and understand our project material. The conceptual level is about things in the world, people and their behaviour and their places, plus the concepts we use for discussing them. This is to be contrasted to the textual level of work, which is about the text and what it says. Even apparently conceptual-level label codes applied to text, such as "neighbours" or "power" are really saying "text about neighbours" or "talk about power". They are there just to help the researcher to get to such text to read what the interviewees say about neighbours or power; in themselves they don't say anything about power or express any description of power wielding or hold any theory about power acquisition. All of that, which is where the researcher's work lies, is unsupported by code-and-retrieve in computational QDA.

2. Retrievals

Effectively all QDA programs support compound retrievals, that is, the ability to make a selection of the project's text based on a combination of two or more codes. Usually this is confined to *boolean retrievals*, eg the ability to find passages of text coded by all of two or more codes (*intersection retrieval*) or text coded by one code but not another (*relative negation retrieval*). Label coding (and what goes along with it, textual-level coding) and base-data coding (and what goes along with it, conceptual-level coding) behave very differently under such retrievals.

2.1 Base data retrievals

Consider project text coded with the sort of base data codes described above, such as "interviewee/age-group/30s" and similar. If we ask the question "Are any interviewees women in their 30's?" we can get back a definite yes/no answer by doing an intersection retrieval on "interviewee/gender/female" and "interviewee/age-group/30s". If any text at all is retrieved, we have an unambiguous Yes answer, and otherwise No. This is a good example of one sort of conceptual-level work: a propositional question about the world of the project is posed, and the computer is able to compute the correct answer.

Note the computer delivers the correct answer because each passage of interview text comes from just one interviewee, so all interviewee base data codes for that passage have to be about the same person.

An extension of this sort of conceptual-level question can also be supported by an extension of base-data coding. Suppose we wish to record propositions concerning people talked about in the project, such as the children or workmates of the interviewees. For an example suppose we know that Annie is aged 5, female, a child of Mary and John, and is healthy. Then we can create base data codes "subject/child/Annie", "subject/child/child-of/Mary" and similar for this information about Annie, and use them to code any text talking about Annie. This enables us to find base data facts about Annie by looking at the cross-references for a piece of text where Annie is talked about – just as we do for base data facts about speakers.

Boolean retrievals will of course allow us to find passages where healthy 5- year-olds are being talked about (intersection retrieval) and unhealthy ones (relative negation). So far this still parallels base-data retrievals about speakers.

But speakers and subjects can be mixed in a boolean retrieval. We can retrieve wherever a woman is talking about a 5-year-old, wherever Mary is talking about her children, wherever women in their 30s are talking about healthy 5- year-old girls, etc.

This can be very valuable and powerful. But note what can not be retrieved here is where women are talking about *their own* children, only individual cases where Mary is talking about Mary's children, where Maria is talking about Maria's children, and so on. To carry out this sort of retrieval essentially requires some capability of the program to process logic. (We won't go into the formalities of it here.)

Note too that in base data retrievals about subjects, and unlike for interviewees, there is room for error, for a given passage of text, such as "John and Mary are ..." can be about two or more subjects with different base data profiles, permitting mixups in retrievals. If John is in his 40s but Mary isn't, this passage will still be retrieved by the intersection of "subject/age-group/40s" and "subject/gender/female", in an attempt to find text about females in their 40s.

2.2 Mixed retrievals

A useful boolean retrieval arises by combining any number of base data codes with one label code. This is where the researcher wants to find out what some group of interviewees have to say on a particular topic, eg. "What do women in their 30s have to say about divorce?". These can be essential for comparative analyses, eg. "Now, what do men in their 30s have to say about divorce?".

The next step with mixed retrievals was the ability of Q.S.R. NUD•IST to store the results of a compound retrieval as a new code, coding all the text in the retrieval. We call this process *System Closure*, an important distinguishing principle of the social sciences in which results and analyses become more data (or put another way one cannot distinguish absolutely between "raw data" and "theory"). For example, we can store as a new code the retrieval from "What do women in their 30s have to say about divorce?". Then, in a project with one interview per document, we can in Q.S.R. NUD•IST easily find what the women in their 30s who have discussed divorce have to say about adultery, by retrieving text label-coded "adultery" restricted to the documents coded by the new node. A parallel sequence of retrievals on men will find the views of the corresponding male group on adultery. Minor variations on such a retrieval sequence can find what women or men who have not talked about divorce have to say on adultery, or women or men with other demographic characteristics such as their religion or marital status.

2.3 Label coding retrievals and the Retrieval Fallacy

Compound retrievals of label codes do not fare so elegantly as with base data codes. An important type of label coding retrieval request is for text about several things related propositionally, eg. "Where is someone talking angrily about their neighbour?" or "Who thinks adultery is unimportant?" The obvious way to answer these is to use boolean intersection searches on the label codes, e.g. the intersection of a label code "being angry", and a label code "about neighbour".

But such a method of retrieval can be both too narrow and too wide. Too narrow, because where they are talking about their neighbour, speakers may never be showing or speaking of their anger. They may for example say their neighbour is extremely noisy (code that "about neighbour") and much later say in a more general way, without reference to any situation at all, that intrusive noise makes them angry (code that "being angry" if you like, it won't help). Equally the intersection method can be too wide, because the interviewee may be talking angrily about something else and say "it annoys my neighbour too" – which gets coded "about neighbour".

Are there other ways we could retrieve where speakers are angry about their neighbours? If intersection is too wide and too narrow, it's hard to see what other sort of textual retrieval based on those two codes could do it.

We will term the *Retrieval Fallacy* as the assumption that by label coding passages of text by what's in them, or even by what you know about that passage, the retrievals answer real-world questions about the subject matter of the research project. It's the fallacy of moving from "Where in the text are interviewees speaking angrily and about their neighbours?" to "Which interviewees are angry with their neighbours?", and expecting the retrieval from the first to answer the second. True, the text retrieval may be some help to the researcher, but it does not answer the question. It can often be downright misleading. We are not getting correct and unambiguous answers to our questions the way we do with compound retrievals on base data codes for interviewees. And when compound retrievals on label codes are used, as they often are, as a basis for some statistical analysis, we are in real trouble.

The Retrieval Fallacy is even more obvious when we do context searches, a type of non-boolean compound retrieval. Q.S.R. NUD•IST for example will find all passages marked by code A if they're inside passages marked by the contextual code B, or within a certain textual distance of them, or before or after it to within a certain textual distance. An example is "Find me all cases of what people have to say about divorce in a religious context". You cannot translate such a "context" to a textual context, even where the textual context is the answer to a direct question "In terms of your religious beliefs, what are your views on divorce?" What is said in such a textual context may have nothing to do with one's religious beliefs (too wide) and religious beliefs may be relevant to views on divorce recorded elsewhere and even without any textual indication of religious relevance (too narrow).

Methodologically, if every retrieval from text were accompanied by a careful Retrieval Fallacy check of what coded text will be returned versus what questions are being asked and what if anything is the relation of one to the other, then in the authors' experience a lot of wishful but invalid analyses would be exposed.

The source of the Retrieval Fallacy is simply assuming that codes about text represent information about the world being studied, plus the compounded felony that combinations of retrievals of codes about text represent propositional information constructed, like the grammar of a sentence, from the component codes. That holds un-fallaciously for base data codes for interviewees. But with label codes, finding passages where people are talking at once about their neighbours and about noise and expressing anger, does not mean they are angry about their neighbours' noise (a proposition); nor does the lack of such retrievals mean they are not.

In other words, label codes, operating at the textual level, cannot (except by lucky chance) be combined to express propositions (operating at the conceptual level) and retrieve the text to which that proposition properly applies. Compound retrievals of textual level codes do not bootstrap us up to the conceptual level.

3. Propositional codes and the conceptual level again

Part of the problem with the Retrieval Fallacy is that it's *structured propositional information* we want: we want to store it, and we want to do retrievals that get it.

Let us explore this with a simple example. Suppose we wanted to label a passage where it is said that John is afraid of Mary; and we want to label it with that proposition – that John is afraid of Mary. One option is to label it with "John" and "Mary" and "about fear". But then, as we have seen, an intersection retrieval is far too wide: it can retrieve where Mary is afraid of John, where John is saying Mary is afraid of spiders, and many other things.

Another option is to label the passage with a sentence "John is afraid of Mary". That however doesn't help if we want to get back all passages about John, or fear, or Mary, because we have not labelled it accordingly. To be thorough we would have to label it all four ways, which is becoming a mighty task for all the propositional information we might want to store.

Q.S.R. NUD•IST offers another approach here. We could devise some way of storing the proposition in the index hierarchy. Suppose we are interested in fear, so we have a code "emotion/fear". We could record subjects and objects of fear under that, like "emotion/fear/subject/John" and "emotion/fear/object/Mary". But that still doesn't record the proposition we want because John and Mary are unrelated by this hierarchy. Of course we could overcome this is by listing cases of fear, as "fear-case-1" etc. Then you can have "emotion/fear/fear-case-1/subject/John" and "emotion/fear/fear-case- 1/object/Mary". The proposition is expressed, although clumsily; and as a way of recording many propositions this leads to huge proliferation of nodes in an index tree with consequent storage and speed problems, let alone consumption of large quantities of valuable researcher time in carrying out such complex coding. And still, at the end of the day, this makes many simple retrievals impossible. Even "Get me everything about John" is problematic, for now John is a code cropping up in as many places in the index system as there are propositions involving John, instead of in one place.

In our experience of qualitative coding procedures, researchers are helped by the valuable *One Place Principle*, to code everything about one thing in one place, ie. on one code. The above sort of fudge to code propositions in a way that codes their structure too, violates that.

The trouble is, that it's far from obvious computationally how to store structured propositions as textual codes. You're damned if you store them just by coding for the terms in them, and you're damned if you make a sentence as the code. Combining both, or using structural fudges, condemns the researcher to a lifetime of massive coding and unworkable index systems. But work at the conceptual level demands propositions.

One approach to solving this problem is in the program Atlas/ti, which supports the creating of relationships between textual label codes. By taking the code "John" and joining it to the code "Mary" with a link "afraid of", we can represent "John is afraid of Mary" in a way that allows the system also to recognise that this is about John, and about Mary, and about being afraid. And, they are joined together in the right way to make a proposition, not just a label. Thus the system, not just the researcher, knows this is about John, and Mary, and fear. If we want to find out about John, the system will deliver this relationship to us. If we want to find out about fear, the system can tell us John is one who fears. And if we retrieve on the code "John" in the text, we can also see that he's afraid of Mary, even if no text is so coded.

However, in Atlas/ti, that propositional information, represented by semantic nets, does not code the text. The terms in the relationship link, "John" and "Mary", may do so, but their relationship doesn't. So the problem of supporting propositional coding remains.

It is important to acknowledge that this problem is compounded, even created, by computational QDA. Researchers using paper and pencil, who eschew computational QDA packages, are under no such disadvantage. They can, and commonly do, put propositional remarks in the margins of their document transcripts, and "retrieve" on those remarks and their component terms. This reflects as much the useful informality of free marginal notes and serendipitous coding and retrieval "methods" as the superior intelligence of humans who can read and understand sentences.

4. Conclusion

So, the present state of computational QDA programs seems to be that powerful coding and retrieval engines have enabled all the thoroughness and rigour we described at the start of this paper, and in so doing have directed work away from the conceptual level and the exploration of propositions. Computationally we are, with the exception of base data coding, mainly working at the level of terms labelling bits of text, and not the wide world of the project, populated with ideas, people, places, attitudes, and arguments related in webs by propositions and anchored in the textual evidence.

A partial exception, as has been pointed out, is the facility of Q.S.R. NUD•IST to provide propositional structuring through its hierarchical index system, and the complex relating of propositions in retrieval through its system closure facility.

It remains then, to build software that really supports conceptual-level propositional coding of text, in a way that provides powerful compound retrievals on propositions and their parts. This article explores the current deficiencies, and while a deficiency doesn't supply its own remedy, at least we know what needs to be remedied, if QDA projects, and not a corner of them, are to be supported computationally. We are still only at the beginning.

References:

RICHARDS, T. J. and RICHARDS, L. (1994): Using computers in qualitative research. In: N. K. Denzin and Y. S. Lincoln (eds.): Handbook of Qualitative Research. Sage, Thousand Oaks, 445-462.

Computer Tools for Grounded Theory: Introducing ATLAS/ti for Windows 95

J. Strübing

Institut für Soziologie der Freien Universität Berlin D-10715 Berlin, Germany

Summary: Grounded theory as a somewhat "paradigmatic" methodology for the development of qualitative data analysis (QDA) software is discussed in its basic statements, emphasizing its strict work process orientation. Requirements for good QDA-software are deduced from these statements. Finally, the forthcoming Windows version of ATLAS/ti is introduced, using central issues of grounded theory as touchstones for the usability of the program.

1. Introduction

While in the 80th advances in software support for qualitative analysis were rare, during the last years there has been a remarkable development in this domain. Qualitative researchers by and by adapted advanced computer tools to their daily work. By way of practicing computer-aided qualitative analysis (CAQA) a process of refining and extending requirements for supportive computer tools began and is still ongoing.

First of all a main question arises: Why talking about grounded theory at all? Instead of talking about, for instance, "hermeneutic circles" or "qualitative content analysis"? One answer is because it is said that grounded theory plays a somewhat "paradigmatic" role in the discussion around CAQA (Lonkilla 1995: 41). I think, there are certain reasons for this dominance. I will discuss them before introducing the key concepts of grounded theory.

2. Grounded theory: basic statements

Grounded theory is by no means what its label suggests: it is not a *theory.* Instead it is both an epistemologically grounded *style of research* and a well-suited *set of single research techniques* (see Legewie in Strauss and Corbin (1996)). The research style of grounded theory derived from decades of practical work in qualitative social research. It reflects the learning processes Strauss, Glaser and other contributors experienced while tracing their sociological research questions in various empirical projects. It is presumably because of this experience-based approach to methods and methodology that grounded theory is explicitly *work process oriented.* Grounding in the American tradition of pragmatic philosophy it avoids to simply deduce research consequences from a largely believed or accepted epistemology. Instead, grounded theory is a systematization of those everyday research practices which turned out to produce sound and interesting results. In this sense, grounded theory makes explicit, what I would call the hidden "quick and dirty practices" of everyday research. In my opinion, this is the main reason why grounded theory is sort of a "paradigmatic model" for the design of software tools meant to be useful to qualitative researchers. If designing software means to develop an environment for an existing work process, than it is essential to understand how this work process is performed – not in theory but in *reality*.

2.1 No restrictive rules but "guidelines" and "rules of thumb"

Grounded theory is not meant to be a set of prescriptive rules but experience-proven suggestions of how a good research process should be organized and performed in order to produce relevant theories. Researchers are encouraged to develop their own work style – strongly referring to their personal abilities and the specifics of the ongoing analytical task (Strauss 1987: 7). If we translate this statement into requirements for QDA-software, openness and flexibility seem to be highly desirable: The software should allow for divergent and changing work styles. It is not enough to simply model a research process from a methodological concept.

2.2 Work process and cooperation

A second basic claim of grounded theory is the indispensability of teamwork: Though subjective achievements are essential in interpretative analysis, cooperation is needed to (1) enrich the scope of ideas and possible interpretations, and (2) to minimize idiosyncrasies. In terms of QDA-software this underlines the need for an integrated co-authorship administration including, for instance, author-stamped outputs or features to display individual work results of team members. But moreover, to really support teamwork processes, we need tools to integrate work results produced by authors in different places. We also could make good use of tools allowing members of a research team to work simultaneously at the same analytical unit or project.

2.3 Inductive/abductive logic

Grounded theory is often – even by its authors; see Strauss and Corbin (1990: 23) – said to follow an inductive logic, inferring from empirical data to theoretical concepts. However, I would like to stress that on the whole it is more likely to correspond to an abductive type of logic. Following a concept of Charles S. Peirce (1976: 400pp.), abductive logic is kind of a *third way* of logical inference, integrating steps of deductive and inductive logic. Translated into practical empirical analysis this means a permanent "oscillation" between (1) conclusions drawn from the collection and comparative analysis of empirical data, (2) the testing of preliminary working hypothesis deduced from the emerging theory, and (3) the invention of a new theoretical concept if the preliminary hypothesis does not explain all related data.

In grounded theory we find this "oscillation" on a second, more technical level of research work: There is also a constant movement back and forth between *open, axial* and *selective* coding strategies – where the first is strongly data-driven, whereas the second and the third are more theory-driven. Again we may translate these properties of grounded theory into requirements for QDA-software: A good tool supporting abductive processes should allow for a flexible process organization of research tasks. That means, it should be possible to introduce new data all along the analytical process, it should allow for flexible changes of code-labels, comments, links and all other elements administrated by the program, and it should finally give us some transparency with respect to the "history" of our analytical process.

2.4 Coding paradigm

Coding is not always the same. Especially the grounded theory concept of coding is quite different from a mere technical definition of coding like "assigning labels to pieces of data". For grounded theorists, in contrast, coding is a basic step in generating empirically grounded theories. Therefore it is important to build codes not on the level of descriptive labels but as constituents of the theory under construction. Strauss and Corbin insist that coding a phenomena means to seriously seeking for (1) conditions, (2) interaction among actors, (3) strategies and tactics, and (4) consequences related to this phenomena. They call it a "coding paradigm" (Strauss 1987: 27f), which should underlie all actions of open, axial and selective coding in order to align the codes to the emergent theory[1]. But how to support the use of a coding paradigm? Certainly, following a coding guideline like this one, first of all would mean to have it in mind while reading carefully through the data. But, in addition to this, a good code comment editor right at hand and a set of – probably graphical – tools for defining relations among codes would mean a substantial support.

2.5 Concept – indicator model

A central element in the methodological concept of grounded theory is the so-called "concept – indicator model" of developing theoretical concepts when analyzing empirical data. Following this model, the analytical process *"is based on constant comparing of (1) indicator to indicator, and then when a conceptual code is generated (2) also comparing indicator to the emerging concept"* (Glaser 1978: 62). Obviously, this is the basic claim underlying the trinity of open, axial, and selective coding techniques. But moreover, the model stresses the aspect of *comparison*. Building valid concepts means to carefully compare empirical instances with each other in order to figure out the variety of properties and dimensions belonging to a certain concept. Once a conceptual code is generated, further comparison with new empirical instances serve to sharpen its outlines and structure. Besides the technical features needed for the three different coding techniques, a good program to

[1]However, Barney Glaser, another leading contributor to grounded theory, prefers a more open approach to coding. He suggests to work with what he calls "coding families" of either substantive or theoretical character, instead of tying the analysis to a symbolic interactionist action model right from the outset. See the comprehensive discussion of Glasers fundamental critique to the type of grounded theory currently represented by Strauss and Corbin in Kelle (1994: 333pp.).

work with this type of concept-indicator model should provide a high degree of transparency with respect to quotations and codes. Practically spoken it is like that: The more complicated the quick retrieval of related quotations is, the less likely we are to perform a painstaking comparison. And, if the access to code comments is difficult we are less likely to carefully note the changing characteristics of the code under question.

2.6 Saturation of concepts and theories

Not just by its name, grounded theory is explicitly devoted to *generate* theory. In fact, abductive logic as mentioned above means more than just coding back and forth. On a second level it also means to check first concepts and tentative hypothesis, elaborated on the base of a small, homogenous sample considering contrasting cases in a sample expanded by means of theoretical sampling. If these concepts are also discovered in the new data it is considered as a proof for its enhanced validity. If they cannot be discovered there, a new conceptual frame has to be elaborated – integrating both the 'old' concepts, valid in the 'old' homogenous sample and new ones valid in the new data. Thus, we need to be able to introduce additional data, to code them in a similar manner and then to check its dissemination in all related cases. Therefore our QDA software should have strong elaborated search and retrieval features. We should be able to link codes and other elements of our analytical unit to complex concepts. And we should be able to use these concepts as elements in search requests to check for structural relations among them.

2.7 Relation between previous theories and actual data

Different from early stages in the development of grounded theory, the current approach considers previous theories as a useful means of empirical research work-at least if these are properly explicated[2]. Thus, Strauss' and Corbins' position is as follows: *"if existing (grounded) theories seem appropriate (...), then these may be elaborated and modified as incoming data are meticulously played against them (...) Researchers can also usefully carry into current studies any theory based on their previous research, providing it seems relevant to these-but again the matching of theory against data must be rigorously carried out"* (Strauss/Corbin 1994 :273; emphasize by the authors).

This comprises a set of requirements for QDA-software: We should be allowed to import concepts from theory and from previous research projects to compare our recent data with them. Technically spoken this means to be able (1) to build codes apart from related quotations, (2) to import complete code lists and relations between these codes as they have been developed in other analytical units, and (3) to merge various analytical units into a new one. Of course, this does not release us from the need to carefully examine the relation between 'old' concepts and new data. But, at least it would

[2]This position is not shared by Glaser (see Kelle 1994, Glaser 1992).

provide a suitable technical support to this task.

Regarding all these 'essentials' of the grounded theory method, it is hard to believe that most of the basic grounded theory oriented empirical studies have been conducted without supporting QDA software. Anyhow, in the next section I will examine one of the few approaches to technically support grounded theory based analytical work.

3. ATLAS/ti

The windows version of ATLAS/ti is still under construction. This paper refers to a very early Beta version. Therefore, the final version will differ in some aspects from the one introduced here. As there is enough published information about the DOS version of ATLAS/ti available (see Weitzmann and Miles 1995) I will concentrate on the new user interface and some of the new features.

3.1 Readiness at hand: The user front-end

Considering that ATLAS/ti has been developed in order to serve the goal of "readiness at-hand" (Muhr 1991), a first step of inspection should concentrate on the user interface. Compared to the old version it is interesting to see, that the fixed-windows screen is now split up: There is a main screen centered around the primary text (PT) window and giving access to all other features related to the analytical work. The old small list windows for primary texts, quotations, codes, and memos have been modified into pull down windows to be opened with a mouse click. For intensive work with the objects administrated in these boxes the user is encouraged to open an extra, always-on-top list window showing, for instance, the list of codes. With network editors, memo editors and family browser its the same: They will be presented in extra windows on top of the main screen. All functions are available either via the main menu or via context menus opened by a mouse click to the related window or object. Together with enhanced 'drag & drop' capabilities, this is just the adaptation of ATLAS/ti to the advanced windows standard. An important new feature is the narrow annotation panel alongside the PT window. It marks the quotations in the active PT and the codes assigned to them. At the moment it is just a visual support – like annotations in the conventional paper-and-pencil technique. In the next version the annotation panel will have interactive properties: A click on a code will activate the related quotations or give direct access to code comments.

3.2 Cooperation

Considering the importance of teamwork abilities for QDA software, the new ATLAS/ti version shows up with two additional features: First of all, there is a procedure to merge two or more analytic or "hermeneutic" units (HUs). This allows for the integration of work results of different team members into one final unit. Together with a co-authorship administration, this is a quite suitable feature for teamwork purposes. An improved cooperation support is also provided by an HTML-conversion tool. It enables the user to translate a HU or parts of it into an HTML file easily exchangeable via WWW. The HTML file can be re-imported to ATLAS/ti. Especially research teams cooperating separated by long distances might benefit from this tool. However, the possibility of working simultaneously at the same HU is still missing.

3.3 Modeling concepts – Building theory

There are a lot of QDA programs designed for coding activities. But only a remarkably small number of them is devoted to the goal of theory-building (Weitzman and Miles 1995). ATLAS/ti is among those few. "Traditionally" ATLAS/ti comes along with features like (1) user-definable hyperlinks between quotations; (2) user-definable relations between codes; (3) a code family browser. A fourth feature relevant for theory building and maybe the heart of ATLAS/ti is its network editor. It allows for visualizations of relations between quotations, codes and memos. It does not only visually represent what we defined elsewhere but it offers the opportunity to model complex concepts graphically by manipulating the relations between objects.

Two enhancements in the network editor are worth to be mentioned: (1) More graphical support to mark groups of nodes as having something in common: Now it is possible to assign bitmaps to nodes and chose special fonts and colors, same with the links indicating the various types of relations. (2) The logical properties of a relation are now used to display the network in a "semantic layout" which is user-definable. For instance, we can define a transitive relation like *"is cause of"* to always appear in a left-to-right layout.

3.4 Exploring Structures: The new Query tool

Developing grounded theories always consists of exploring our data thoroughly from different points of view. We are continuously demanded to examine whether a certain phenomena always occurs under determinable conditions or if a code is spread out over all cases, respectively: whether it is limited to a certain subset of cases with identifiable properties. The old quotation retrieval tool of ATLAS/ti has been somewhat limited in its capacities. Thus, one of the important innovations of the new ATLAS/ti version is its completely renewed Query tool. It offers additional search operators and the construction of search expressions is based on what is called the "Reversed Polish Notation" (RPN).

While in the DOS version one could only apply the Boolean AND resp. the OR operator to all search terms included, the new tool provides access to all *Boolean operators* including the NOT and an XOR, i.e. an exclusive OR. In addition to this one might use these operators selectively related to single search terms in a more complex search request. Additional so-called *thesaurus operators* are dedicated to retrieve quotations coded with codes of certain types of relationships: With three operators users are able to selectively retrieve all quotations connected (1) to code X or any of its subterms (or "children"), (2) to code X or any of its direct ancestors ("parents"), or (3) to code X or any of its "sibling" codes, i.e., all codes that are subcodes of one of the ancestors of code X. Finally, there is a third group of operators realized in the new query tool. They are called *proximity operators* and they are dedicated to detect the spatial relations between quoted text segments. Thus, it is an approach to the co- occurrence problem in qualitative data analysis. There are the various operators to retrieve constellations of (1) inclusion, (2) intersection, and (3) spatial nearness.

Concerning the employment of the RPN for constructing queries, the first notion is that in principle it represents nothing but the consequence of windows techniques. Composing a search process that way means to first click at the one or two objects a certain operator shall work on and then chose the related operator and get a first result. For next search steps the already existing expression represents a complex object which can be related to another object by a new operator. The advances of this technique compared with the conventional arithmetic method are the following: *First of all*, every expression is syntactically correct, while in arithmetic methods we are most likely to confuse the number and position of parenthesis if queries become more complex. *Secondly*, all queries are performed step by step. That is, we can see the results of every added object and operator immediately and we may step by step expand our query. Together with "undo" and "redo"-buttons this is very handy for all these tentative, trial-and-error-based retrievals we often conduct while exploring our material. *Finally*, a further aspects of this tool should be mentioned: It is possible to transform every query into a *supercode*. That is, the query expression (not the result) is labeled as a code and appears in the code list.

There has been a bit of arguing about pros and cons of RPN. Some contributors worried if it is the right choice to confront qualitative researchers without much mathematical background with a notation so completely different from the common arithmetic notation of Boolean expressions. It is right at hand, that it takes some time to get accustomed to this type of notation. In the beginning, it is simply a problem of how to read the query expressions. Especially, it is difficult to figure out how the selected operator works on which of the chosen objects. This seems due to the fact that average users are used to the arithmetic notation, as used, for instance, in database programs. Therefore RPN seems counter-intuitive at first sight. However, if it comes down to it, it reveals itself to be an easy to handle and valuable heuristic tool.

4. Concluding remarks

More and enhanced features will again raise the question of usability and methodological appropriateness: Is a tool as complex as this one usable to "normal" non-expert researchers? Is it helpful, respectively: does it expand the scope of analytical possibilities? And, does it offer proper procedures in terms of methodological rules? These questions have been discussed intensively and controversially between developers and users. Especially the use of sophisticated query tools was concerned. In my opinion, *first of all*, we should not confuse the number and complexity of features with the way how researchers make use of them. At the beginning I mentioned that grounded theory encourages people to develop their own work style and that a supporting software should be open and flexible. In this respect a software which offers a large variety of features and alternative procedures is most likely to let users find their appropriate way of how to get things accomplished. *Secondly*, it is not necessary that every retrieval operation is explicitly covered by methodological principles and formal procedures. In fact, researchers do a lot more than strictly following precise step by step rules in qualitative data analysis.

As a more general result I would like to emphasize that QDA programs must not only comply with the conditions of methods for qualitative data analysis but with the whole research act as a work process. That is, they have to deal with the linkage between data collection, coding sessions, and the modeling of theory and they have to support the various tasks during the research process flexible enough to assist researchers in performing their own work style.

References:

GLASER, B. G. (1978): Theoretical Sensitivity. Advances in the Methodology of Grounded Theory. Sociology Press, Mill Valley, CA.

GLASER, B. G. (1992): Emergence vs. Forcing: Basics of Grounded Theory. Sociology Press, Mill Valley, CA.

KELLE, U. (1994): Empirisch begründete Theoriebildung. Deutscher Studienverlag, Weinheim.

KELLE, U. (1995b): Computer-aided Qualitative Data Analysis: An Overview. Paper presented at the "Text Analysis and Computers"-Conference, September, 18-21 1995. Mannheim (ZUMA).

LONKILLA, M. (1995): Grounded Theory as an emerging paradigm for computer-assisted qualitative data analysis. In: Kelle, U. (ed.): Computer-Aided Qualitative Data Analysis. Sage, London, S. 41-51.

MUHR, T. (1991): ATLAS/ti-A Prototype for the Support of Test Interpretation. Qualitative Sociology (Special Issue: Computers and Qualitative Data II, Part 1) 14: 349-371.

PEIRCE, C. S. (1976 (1903)): Schriften zum Pragmatismus und zum Pragmatizismus; edited by Karl-Otto Apel, Frankfurt a. M. 1976, 2. Ed.

STRAUSS, A. L. (1987): Qualitative Analysis for Social Scientists. Cambridge UP, Cambridge.

STRAUSS, A. L.; CORBIN, J. (1990): Basics of Qualitative Research. Grounded Theory Procedures and Techniques. Sage, London. (German: (1996): Grounded Theory: Grundlagen qualitativer Sozialforschung. Beltz/PVU, Weinheim)

STRAUSS, A. and CORBIN, J. (1994) Grounded Theory Methodology. An Overview. In: Denzin, N. (Ed.): Handbook of qualitative Research. Sage, London, 273-285.

WEITZMAN, E. and MILES, M. B. (1995): Computer Programs for Qualitative Analysis. A Software Sourcebook. Sage, London.

Section 6

Applications in Medicine

Some Issues in the Comparison of Diagnostic Tests from a Paired Experiment

Colin B. Begg and E. S. Venkatraman

Memorial Sloan-Kettering Cancer Center
New York, New York 10021

Summary: Data from a study to compare two diagnostic approaches for differentiating melanoma from atypical melanocytic nevi are analyzed. The study involved the paired-sample design, commonly used in diagnostic test evaluations, and involved the comparison of two diagnostic scoring systems. The scoring systems are based on clinical evaluation of the lesion, and dermoscopic evaluation. The study motivated the development of a new distribution-free procedure for comparing the receiver operating characteristic (ROC) curves generated by the two test scores. The new method permits an evaluation of the hypothesis that the two curves are identical for all operating points, unlike conventional methods that compare the areas under the curves. The method has been shown by simulation to have very similar operating characteristics to the conventional method based on comparisons of the areas under the ROC curves when the curves are parallel, but markedly superior power when the curves cross, that is when the curves are different but have similar areas. Analysis of the melanoma data using this test, and generalized versions of the test, demonstrates that use of the dermoscope adds little if any diagnostic information in the classification of a suspicious nevus.

1. Introduction

Considerable research effort in recent years has been focused on the development of new technologies and medical instrumentation designed to facilitate the non-invasive diagnosis of serious illnesses. These new diagnostic tools require evaluation in valid clinical research studies. Commonly the new technique will be compared with the standard diagnostic test in an experiment in which all patients are evaluated by both techniques, and the results correlated with a "gold standard" definitive diagnosis to determine whether the new procedure is more accurate. These studies must be evaluated with caution since both the design (Begg, 1987) and the analysis (Metz, 1986) present opportunities for misleading interpretation of the data. In this article we analyze in detail the results of such a study, in which a new device, dermoscopy, is evaluated for its utility in the differential diagnosis of cutaneous malignant melanoma and atypical melanocytic nevi. In the process, we highlight the conceptual problems of comparing diagnostic tests, and describe some recently developed methods for illustrating the critical aspects of the analysis.

The discrimination of atypical melanocytic nevi (benign pigmented moles) from malignant melanoma is an important clinical problem, in view of the fact that malignant melanoma is a potentially fatal disease. Although most individual lesions can be removed and biopsied relatively easily, this is impractical and disfiguring for individuals with many nevi. A reliable method for the non-invasive diagnosis of these lesions is highly desirable. Dermatologists make this diagnosis on the basis of the visible appearance of the lesion, paying particular attention to the ABCD features of melanoma, namely asymmetry, border irregularity, color variegation, and diameter (Friedman et al., 1985). Recently, dermoscopy has been advocated as a non-invasive technique for improving diagnostic accuracy. In this procedure oil is applied, and the lesion is examined through a dermatoscope, which permits magnification by a factor of ten, allowing identification of structural features of the lesion believed to be relevant to the diagnosis (Stolz et al., 1994). Stolz et al. developed a dermoscopic scoring system based on a slightly different ABCD set of features, in which the D refers to different structural components. These features were identified as the most important ones in a multivariate analysis of 31 features, and in a subsequent study the scoring rule was shown to have higher sensitivity and specificity than use of the conventional ABCD features without any formal scoring (Nachbar et al., 1994).

The dataset at our disposal was collected by a different group of investigators in an attempt to provide confirmation of the results of the original study (Rao et al. 1996). In this article we analyze the results of the study by Rao et al. in detail. The basic question addressed in their study was the following: does dermoscopy improve the detectability of melanoma? Our analysis is directed at the question of how to address this issue scientifically. The investigators, in their primary analysis, considered the attainment of increased sensitivity as evidence supporting the hypothesis that dermoscopy increases detectability. [Note: the sensitivity is the proportion of melanomas diagnosed as melanoma by the diagnostic test.] However, they acknowledged that the increased sensitivity was achieved at the expense of a larger false positive rate.

Our thesis is that a valid comparison of diagnostic tests requires calibration of the scales of measurement, and that the concepts of ROC (receiver operating characteristic) analysis are essential for this purpose (Metz, 1986). By developing continuous diagnostic scores based on the ABCD features for both clinical and dermoscopic evaluations, we can accomplish the appropriate calibration. In practice dermatologists employ judgment in classifying patients, and indeed this is ultimately a binary decision, i.e., perform a biopsy or do not perform a biopsy. Therefore evaluation of the sensitivities and specificities at the implicit decision thresholds actually employed in practice is also important, and this may reflect the phenomenon whereby a new test encourages a more "conservative " strategy where more patients receive a biopsy, and the detection rate is increased correspondingly, without necessarily any real increase in the true "accuracy" of the strategy, as represented by the ROC curves.

2. Study Design and Basic Results

The study consisted of a series of 63 patients who received a biopsy evaluation of at least one lesion (total of 72 lesions). Each lesion was classified histologically as either a malignant melanoma or a benign melanocytic nevus by an experienced dermatopathologist. Photographs were taken of each lesion, first without oil and then with oil and a dermoscopic camera (10X magnification), and these were evaluated retrospectively by three readers, one experienced dermatologist (observer 1) and two fellows in training (observers 2 and 3). Each observer rated the lesion clinically, giving ratings for asymmetry, border irregularity, color, and diameter, and then provided a subjective clinical diagnosis of either positive or negative for melanoma. In addition, the lesions were rated independently on the basis of the four dermoscopic features, asymmetry, border irregularity, color, and dermoscopic structures, without knowledge of the clinical ratings. The observer provided an analogous subjective dermoscopic diagnosis of positive or negative, and then rated the lesion formally using the Stolz scoring criteria (Stolz et al. 1994), a linear function of the results of the four features, and diagnosed the lesion according to whether the score was greater than (positive) or less than or equal to 5.45. Further details of the study are described in Rao et al. (1996).

In all of the analyses presented we have considered the lesion as the unit of analysis. While some patients contributed more than one lesion to the study, there were too few to evaluate independence empirically. Furthermore, since the evaluation of the lesions was accomplished independently and retrospectively by photographs, without knowledge of the patient or the histologic diagnosis, the assumption of statistical independence of the lesions is plausible.

The basic results of the study are presented in Table 1, for the subjective clinical and dermoscopic diagnoses, and for the formal scoring method of Stolz et al. (1994). The results are presented separately for each observer. Use of dermoscopy increases the sensitivity for the experienced observer. The numbers of melanomas detected, of the 21 malignant lesions in the study, increases from 16 to 19. Specificity is decreased to compensate for this increased detectability. The results for the inexperienced observers (2 and 3) are less clear cut, especially for observer 2 who demonstrates an apparent degradation of accuracy with the use of the Stolz system, i.e. lower sensitivity and specificity.

Rao et al. concluded from these data that when used by experienced dermatologists, dermoscopy would increase the detectability of melanoma at the cost of an increased number of biopsies. Although the statistical significance of this conclusion is eroded by the small number of melanomas in the study, the pattern of the results certainly supports the inference. However, the fact that the increased sensitivity is achieved at the expense of decreased specificity presents a challenge to the implicit conclusion that dermoscopy truly adds information of relevance to the differential diagnosis, as opposed to merely encouraging a more conservative diagnostic classification. Put in more formal terms, does dermoscopy lead to a more accurate, more informative diagnosis, or does it simply cause observers to change their implicit classification thresholds without contributing real diagnostic information? [Note: the threshold is explicit in the case of the Stolz scoring sytem.] The intent of the detailed analysis in the following section is to address this question formally.

3. Detailed Analysis of the Data

The primary goal of our analysis is to reconcile whether the altered diagnostic performance using dermoscopy truly reflects the contribution of additional diagnostic information, or whether it merely reflects an arbitrary change in the classification threshold in the absence of a real contribution of information. ROC analysis is the tool for accomplishing this task. The ROC curve is a plot of the sensitivity against the false positive rate for all possible classifications on this continuous scale, and thus it characterizes in a complete way the diagnostic accuracy of the procedure. By evaluating the ROC curves for the two comparative diagnostic strategies, we can evaluate the incremental diagnostic information in the dermoscopic procedure at classification points that are genuinely calibrated, unlike the single classification points embedded in the subjective classifications in Table 1.

It transpires that the "correct" way to calibrate the two diagnostic procedures in a paired experiment is by equating the percentiles of the marginal distributions of the measures representing the two strategies. Let X and Y represent these two measures for the clinical evaluation and the dermoscopic evaluation, respectively. We will describe later in this section how we determined these two measures for each patient. Let D be a binary indicator of the definitive histologic diagnosis where 1 represents melanoma and 0 represents a benign lesion. Setting

$$F_x(x) = pr(X \le x|D = 1) \qquad G_x(x) = pr(X \le x|D = 0)$$
$$\text{and} \qquad F_y(y) = pr(Y \le y|D = 1) \qquad G_y(y) = pr(Y \le y|D = 0)$$

then the sensitivity and specificity of X at cut-off value x are $1 - F_x(x)$ and $G_x(x)$ respectively, with corresponding definitions for Y. The two ROC curves are equivalent if, for every cut-off value x of X, there exists a corresponding cut-off value y of Y at which $F_x(x) = F_y(y)$ and $G_x(x) = G_y(y)$ and vice versa. Denoting the unconditional distributions of the markers by $M_x(x)$ and $M_y(y)$, we see that

$$\begin{aligned} M_x(x) = \quad & pr(X \le x) & = \theta F_x(x) + (1-\theta)G_x(x) \\ M_y(y) = \quad & pr(Y \le y) & = \theta F_y(y) + (1-\theta)G_y(y), \end{aligned} \tag{1}$$

where θ is the sampling prevalence of melanoma. Let $x_z = M_x^{-1}(z)$ and $y_z = M_y^{-1}(z)$. Then x_z and y_z represent the cut-off values on the respective measures corresponding to the zth quantile of the marginal distributions. That is $M_x(x_z) = M_y(y_z), \forall z$. The two ROC curves are equivalent at the zth quantile if both the sensitivities of the two tests are equal and the specificities are equal, i.e. $1 - F_x(x_z) = 1 - F_y(y_z)$ and $G_x(x_z) = G_y(y_z)$. It is easily seen from (1) that $F_x(x_z) > F_y(y_z) \Leftrightarrow 1 - G_x(x_z) > 1 - G_y(y_z)$, and so the difference in the total number of errors for each test at the zth quantile is a measure of the discrepancy in diagnostic accuracy at the zth quantile since the overall error rates are $F_x(x_z) + 1 - G_x(x_z)$ and $F_y(y_z) + 1 - G_y(y_z)$, respectively. That is, if the sensitivity of test X is larger than the sensitivity of test Y, then the specificity of test X is also larger than the specificity of test Y. Later we shall demonstrate how this result can be used to develop a permutation test of the hypothesis that the two ROC curves are equivalent at all operating points. Calibration of the marginal quantiles in this manner has been used previously by Jung et al. (1995) for the purpose of comparing prognostic markers. Examination of the subjective classifications of the

	Clinical		Dermoscopy		Scored Dermoscopy	
Observer	Sensitivity	Specificity	Sensitivity	Specificity	Sensitivity	Specificity
1	76%	82%	81%	76%	90%	57%
	(16/21)	(42/51)	(17/21)	(39/51)	(19/21)	(29/51)
2	90%	80%	61%	82%	76%	71%
	(19/21)	(41/51)	(13/21)	(42/51)	(16/21)	(36/51)
3	85%	74%	85%	76%	90%	59%
	(18/21)	(38/51)	(18/21)	(39/51)	(19/21)	(30/51)

Table 1: Diagnostic Classifications

four observers shows clearly that they are poorly calibrated, both within observers and between observers (Table 1). Observer 1, for example, classifies more subjects as positive for melanoma using dermoscopy than using the clinical evaluation (40% (29/72) versus 35% (25/72)). In general scored dermoscopy leads to more patients being classified as positive (57% (41/72) for Observer 1). It is especially surprising that the thresholds for scored dermoscopy vary considerably (and significantly) between observers, since a major purpose of the scoring scheme is to reduce such subjectivity and homogenize the classifications.

The fact that the different tests are not calibrated means that it is neither valid to compare the observed sensitivities, nor valid to compare the observed specificities. In fact it is clear from Table 1 that the high sensitivity achieved by scored dermoscopy is at least in part due to the high thresholds employed by this scheme. Valid comparison of the testing strategies is only possible if we can find a way to compare them at equivalent, calibrated thresholds.

A more satisfactory and complete way to evaluate the incremental diagnostic yield of dermoscopy is to create a continuous clinical scale, analogous to the Stolz scale, and compare the two scales at all possible classification points, i.e. compare the entire ROC curves. Since the four components of the ABCD rule were evaluated by each observer for each lesion as part of the study, this is relatively easy to accomplish.

In our principal analysis we elected to aggregate the results of the three observers. We created our clinical score, denoted S_c, by using simple logistic regression to relate the histologic diagnosis to the four variables, c_1, c_2, c_3 and c_4, representing the four components of the clinical ABCD scheme. [Note: The four components are asymmetry (c_1), measured on a three-point scale $(0, 1, 2)$; border irregularity (c_2), measured on a three-point scale $(1, 2, 3)$; number of colors (c_3); and the largest in-situ diameter (c_4), measured in millimeters.] For each patient the values of the four variables were obtained by averaging the ratings of the four observers. This analysis led to the following linear predictor:

$$S_c = -8.19 + 1.91c_1 + 0.71c_2 + 0.83c_3 + 0.12c_4.$$

The probability that a lesion will be classified histologically as a melanoma is thus $\exp(S_c)/(1 + \exp(S_c))$. We developed an analogous predictor based on the four dermoscopy ratings. These were asymmetry (d_1), measured similarly to the clinical scale; border (d_2), measured on a scale 0 to 8 on the basis of the number of abrupt boundaries; number of colors (d_3), up to six as for the clinical scale; and differential dermoscopic structures (d_4), up to five distinct structural features. This led to the following predictor:

$$S_d = -7.58 + 2.54d_1 + 0.19d_2 + 1.11d_3 - 0.53d_4.$$

We employed S_d in our primary comparisons with S_c, rather than the Stolz scheme, since it seemed fairer to compare predictors that were derived from the same dataset, although our analyses demonstrate that the operating characteristics of all three systems are very similar. Note: The Stolz scheme has no constant term, and is calculated using $1.3d_1 + 0.1d_2 + 0.5d_3 + 0.5d_4$.

The scores were then ranked to facilitate the required calibrated comparisons at all classification points (Table 2). At each classification point (rank) the total number of errors (false positive + false negative) for each scoring system can be compared directly. It is clear from this table that the performances of the two systems are very similar at all operating points, differing by at most 4 errors. Figure 1 demonstrates that these results produce similar ROC curves. This great similarity occurs despite the fact that the actual rankings of the individual lesions differ considerably. This can be ascertained by matching the lesion identifiers. For example, although lesion number 33 is ranked 1 by both scores, lesion number 5 is ranked 2nd lowest on the dermoscopy score, but 12th on the clinical score. Lesion 44 is ranked 15th on dermoscopy and 25th on clinical. Nonetheless, the classification abilities are remarkably similar.

A conventional comparison of the two ROC curves involves a test of the hypothesis that the areas under the curves are equivalent. Taking account of the pairings in the results, and building on the ideas of Hanley and McNeil

(1983), DeLong et al (1988) developed a nonparametric test of this hypothesis. This test exploits the fact that the nonparametric areas under the two ROC curves form a generalized U-statistic, and a consistent nonparametric estimator of the two variances and the covariance can be obtained for such U-statistics. We have developed a permutation test that addresses the null hypothesis that the entire ROC curves are identical at all (calibrated) operating points (Venkatraman and Begg, 1996). The test statistic is the total of the error differences, in the last column of Table 2, and a reference distribution is obtained by permuting the ranks of the observations within pairs. This procedure will be described in more detail, with a discussion of its properties, in Section 4.

	Clinical (X)				Dermoscopic (Y)				
Rank	*ID*	S_c	*Status*	*Error*	*ID*	S_d	*Status*	*Error*	$e_{\cdot k}$
1	33	-5.239	0	50	33	-5.947	0	50	0
2	62	-5.206	0	49	5	-5.493	0	49	0
3	22	-4.456	0	48	62	-4.893	0	48	0
4	53	-4.335	0	47	6	-4.877	0	47	0
5	6	-4.275	0	46	53	-4.557	0	46	0
6	24	-4.125	0	45	10	-4.443	0	45	0
7	39	-4.047	0	44	57	-4.407	0	44	0
8	41	-3.993	0	43	27	-4.380	0	43	0
9	11	-3.966	0	42	2	-4.357	0	42	0
10	27	-3.846	0	41	9	-4.300	0	41	0
11	19	-3.821	0	40	19	-4.157	0	40	0
12	5	-3.819	0	39	39	-4.153	0	39	0
13	25	-3.637	0	38	59	-4.150	0	38	0
14	59	-3.593	0	37	68	-4.103	0	37	0
15	68	-3.438	0	36	44	-4.080	0	36	0
16	9	-3.413	0	35	64	-4.050	0	35	0
17	2	-3.135	0	34	28	-3.953	0	34	0
18	18	-2.984	0	33	42	-3.910	0	33	0
19	57	-2.977	0	32	22	-3.837	0	32	0
20	1	-2.877	1	33	25	-3.667	0	31	2
21	28	-2.793	0	32	41	-3.607	0	30	2
22	17	-2.775	0	31	11	-3.483	0	29	2
23	8	-2.677	1	32	18	-3.387	0	28	4
24	21	-2.624	0	31	46	-3.330	0	27	4
25	44	-2.558	0	30	43	-3.317	0	26	4
26	43	-2.483	0	29	52	-2.690	0	25	4
27	66	-2.382	0	28	4	-2.497	0	24	4
28	32	-2.347	0	27	66	-2.373	0	23	4
29	49	-2.205	0	26	49	-2.310	0	22	4
30	55	-2.101	0	25	37	-2.073	0	21	4

Table 2: Overall Comparision of Clinical and Dermoscopic Scores. Column label *ID* indicates the lesion identifier, *Status* the disease status, *Error* the total number of errors by that decision rule and $e_{\cdot k}$ is as defined in the text.

Table 2 continued

	Clinical (X)				Dermoscopic (Y)				
Rank	*ID*	S_c	*Status*	*Error*	*ID*	S_d	*Status*	*Error*	$e_{\cdot k}$
31	15	-1.916	0	24	21	-2.053	0	20	4
32	46	-1.806	0	23	1	-2.037	1	21	2
33	64	-1.765	0	22	32	-2.007	0	20	2
34	13	-1.713	0	21	55	-1.943	0	19	2
35	67	-1.656	1	22	60	-1.737	0	18	4
36	7	-1.512	0	21	50	-1.727	0	17	4
37	10	-1.498	0	20	7	-1.290	0	16	4
38	58	-1.483	0	19	67	-1.227	1	17	2
39	42	-1.412	0	18	63	-1.217	1	18	0
40	35	-1.395	0	17	61	-1.197	1	19	-2
41	50	-1.306	0	16	23	-1.150	0	18	-2
42	4	-1.238	0	15	35	-1.017	0	17	-2
43	52	-1.188	0	14	36	-0.933	0	16	-2
44	70	-1.167	0	13	58	-0.900	0	15	-2
45	36	-1.117	0	12	17	-0.843	0	14	-2
46	65	-1.113	1	13	15	-0.830	0	13	0
47	26	-0.894	0	12	40	-0.707	1	14	-2
48	37	-0.807	0	11	24	-0.697	0	13	-2
49	31	-0.749	1	12	29	-0.557	1	14	-2
50	29	-0.683	1	13	65	-0.410	1	15	-2
51	61	-0.558	1	14	14	-0.240	0	14	0
52	60	-0.120	0	13	47	-0.230	1	15	-2
53	16	-0.068	0	12	16	-0.007	0	14	-2
54	23	-0.058	0	11	54	-0.003	1	15	-4
55	71	-0.042	1	12	72	0.220	1	16	-4
56	12	-0.028	1	13	20	0.330	0	15	-2
57	20	0.032	0	12	70	0.357	0	14	-2
58	38	0.192	1	13	8	0.410	1	15	-2
59	14	0.511	0	12	38	0.437	1	16	-4
60	47	0.612	1	13	56	0.513	0	15	-2
61	63	0.749	1	14	31	0.683	1	16	-2
62	56	0.888	0	13	48	0.753	0	15	-2
63	30	1.051	1	14	13	0.753	0	14	0
64	51	1.170	1	15	51	1.433	1	15	0
65	45	1.182	1	16	3	1.440	1	16	0
66	72	1.317	1	17	26	1.443	0	15	2
67	48	1.463	0	16	12	1.477	1	16	0
68	40	1.639	1	17	45	1.523	1	17	0
69	54	1.641	1	18	69	1.913	1	18	0
70	3	1.777	1	19	71	1.967	1	19	0
71	34	1.945	1	20	30	2.993	1	20	0
72	69	2.101	1	21	34	3.123	1	21	0

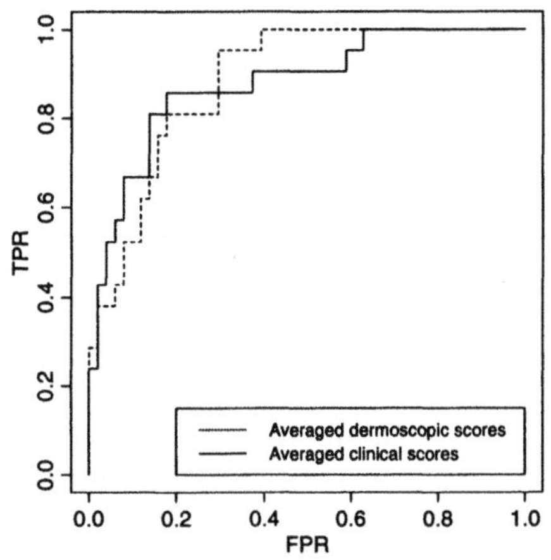

Figure 1

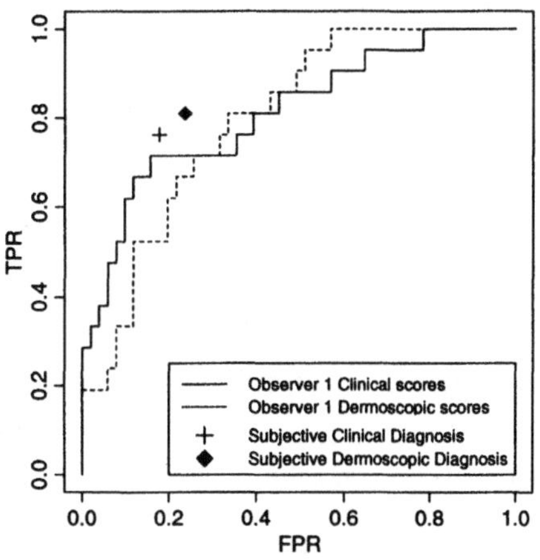

Figure 2

The test has a p-value of 0.37 (two-sided), demonstrating the absence of any apparent incremental information in the dermoscopic scores. The conventional test of the equivalence of the two areas leads to a p-value of 0.88.

In the preceding analysis we averaged the individual features recorded by the three observers prior to obtaining the scores. However it is clear from Table 1 that the observers differ substantially in their choices of thresholds, and possibly also in their inherent accuracies. Moreover, one of the observers is experienced (1) while two are not (2 and 3). We have repeated the preceding analysis for each reader individually, first by obtaining reader-specific scores for clinical and dermoscopic procedures, and then by testing the equivalence of the ROC curves. These analyses lead to the following results:- observer 1 ($p = 0.43$); observer 2 ($p = 0.44$); observer 3 ($p = 0.74$). These data provide no evidence that the accuracy of any of the observers is influenced by dermoscopy.

To evaluate the hypothesis that the subjective evaluation of the lesions contains more information than the quantitative content of the ABCD ratings and their derived scores, S_c and S_d, one can plot the points corresponding to these subjective classifications, and compare them with the ROC curves for the corresponding scores. As an example this is shown for Observer 1 in Figure 2. For the other observers these points did not lie consistently above the curves. It is of interest to consider whether an individual point is significantly superior to the corresponding curve in terms of accuracy. It is possible to simplify the procedure for comparing diagnostic rules when one of the rules involves only a single cut-off value, and the comparison is with a continuous test. First, the classification point for the continuous test that is calibrated with the binary rule is selected. Second, the analysis is restricted to subjects (lesions) with "discordant" classifications, i.e. where the scored rule is positive and the subjective rule is negative, and vice versa. The crosstabulation of these two discordant possibilities and true (histologic) disease status is a 2×2 table with fixed margins, and can be analyzed using Fisher's exact test. [Note: one margin is fixed on the basis of true disease status, and the other is fixed by the calibration of the marginal distributions of the test results.] Of the six comparisons performed, clinical and dermoscopy for each of the 3 observers, Fisher's test is significant at the 5% level for one of these comparisons, but the overall pattern of results does not lend support to the hypothesis that the subjective diagnoses are inherently more accurate.

4. Theoretical Issues

4.1 Details of the Permutation Test

In Section 3 it was shown that equality of the total numbers of errors of the two testing strategies (X and Y) at the classification thresholds calibrated on the basis of zth quantile of the marginal distributions is compatible with equality of the two ROC curves at that threshold. More formally, consider the function

$$\mathcal{E}(z) = \begin{cases} 1 & \text{if } (X \le x_z, Y > y_z, D = 0) \text{ or } (X > x_z, Y \le y_z, D = 1) \\ -1 & \text{if } (X > x_z, Y \le y_z, D = 0) \text{ or } (X \le x_z, Y > y_z, D = 1) \\ 0 & \text{otherwise} \end{cases}$$

Our test statistic, the total of the unsigned differences in total errors across all ranks is designed as means of estimating $\mathcal{E}(\cdot)$, averaged over all classification points. In fact, estimation of $\mathcal{E}(\cdot)$ directly is hindered by the fact one needs knowledge of exact form of $M_x(\cdot)$ and $M_y(\cdot)$, or more succinctly of $M_x^{-1}(M_y(\cdot))$, in order to calibrate X and Y precisely. This would be necessary if we were to exchange the actual scores when evaluating the permutation distribution of the test statistic. To circumvent this problem we permute the ranks rather than the actual scores. The entire procedure is constructed as follows. Denote the entire dataset by $\{(X_i, Y_i, D_i); i = 1 \ldots, n\}$, where $D_i = 1$ if the lesion is melanoma and $D_i = 0$ if the lesion is benign, and let $\{R_i\}$ and $\{S_i\}$ denote the corresponding ranks of $\{X_i\}$ and $\{Y_i\}$, respectively. Then, setting $k = 1, \ldots, n-1$, we define an empirical error matrix by

$$e_{ik} = \begin{cases} 1 & \text{if } (R_i \le k, S_i > k, D_i = 0) \text{ or } (R_i > k, S_i \le k, D_i = 1) \\ -1 & \text{if } (R_i > k, S_i \le k, D_i = 0) \text{ or } (R_i \le k, S_i > k, D_i = 1) \\ 0 & \text{otherwise} \end{cases}$$

The statistic $e_{\cdot k} = e_{1k} + \ldots + e_{nk}$ is the measure of the "closeness" of the two ROC curves at the kth order statistic, and the overall test statistic is

$$E = \sum_{k=1}^{n-1} |e_{\cdot k}|.$$

Under the null hypothesis we can randomly exchange the ranks of X and Y for each subject to generate the permutation distribution. Let $(q_1 \ldots, q_n)$ represent a sequence of 0's and 1's. Then a permuted dataset $\{R_i^*, S_i^*\}$ indexed by that sequence is given by

$$\left.\begin{array}{rcl} R_i^* & = & q_i R_i + (1-q_i) S_i \\ S_i^* & = & q_i S_i + (1-q_i) R_i \end{array}\right\} \quad i = 1, \ldots, n.$$

This process will invariably introduce numerous ties, so it is necessary to have a second randomization step to break the ties. That is, we generate $\{R_i^{**}, S_i^{**}\}$, where

$$\left.\begin{array}{rcl} R_i^{**} & = & J(R_i^*) \\ S_i^{**} & = & J(S_i^*) \end{array}\right\} \quad i = 1, \ldots, n.$$

where $J(\cdot)$ represents the process by which ties are broken by randomization. The permutation distribution is the distribution which assigns a uniform mass to each value of E^* given by all the 2^n sequences of 0's and 1's. Since this may be a very large number in practice we have used a sampling scheme where $(q_1 \ldots, q_n)$ is a random permutation generated by n fair coin tosses and the process is repeated a sufficiently large number of times to obtain a stable p-value.

The operating characteristics of this procedure have been evaluated using simulations, and detailed results are presented in Venkatraman and Begg (1996). Briefly the test is shown to possess very similar power to the non-parametric area test of DeLong et al (1988) in circumstances where one of the ROC curves is uniformly higher than the other, but markedly superior power when the curves "cross", i.e. when the curves are different but have similar areas. For example, if nondiseased subjects have test results from a $N(0,1)$ distribution, and diseased subjects have test results generated from $N(\mu, \sigma^2)$, then the diagnostic tests with distributions ($\mu = 0.74$, $\sigma^2 = 1$) and ($\mu = 1.17$, $\sigma^2 = 2$) both produce ROC curves with an area of 0.7. In this setting the power of the DeLong et al. test is always in the region of 5%, regardless of sample size. The power of our procedure increases from 30% if the sample size is 80 (40 cases, 40 controls) to 65% if the sample size is 160. It is also demonstrated that the test is valid under an assumption of exchangeability of the transformed test scores, and that the test is consistent.

4.2 Issues of Cross-Validation

The ROC curves presented in Section 3 are based on the linear scores derived from logistic regression analyses of all 72 lesions in the study. Since the maximum likelihood procedure used to estimate the coefficients of the scores optimizes the classifications for the dataset, the estimated ROC curves will tend to overestimate the true classification performance. If the classification rule was applied to an independent (test) dataset, it would lead to an

ROC curve with smaller area with high probability. That is, the approach employed leads to upwardly biased estimates of the ROC curve.

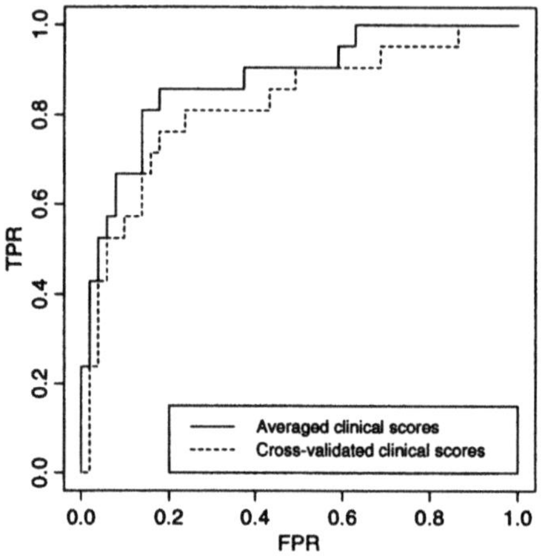

Figure 3

This bias can be reduced substantially by the use of cross-validatory techniques (Stone (1974), Copas (1983)). We have repeated our analyses by using the following procedure. The clinical score for the ith lesion, for example, was obtained by determining the score function using a logistic regression analysis of the remaining 71 lesions, excluding the ith lesion, and then applying the function to the four components of the ABCD scheme for the jth lesion. This entire procedure was repeated for each lesion, and the ROC curve was evaluated from the rankings of these new scores. This is plotted in Figure 3, along with the original curve calculated without cross-validation. The Figure shows that there is a small but noticeable upward bias in the uncorrected curves. Similar results are obtained for cross-validated dermoscopic scores (data not shown).

5. Discussion

We have demonstrated by example a detailed analytic method for evaluating the incremental diagnostic information in a new technology, the use of the dermoscope for diagnosing suspicious nevi as melanomas. This study employed the most common design for comparing diagnostic tests, the retrospective paired design. This design ensures that the sample prevalence of disease is identical for the two tests, and thus validates the use of quantiles of the marginal distributions of the test results for calibrating the comparisons of sensitivities and specificities. This is important since use of the new test will often encourage readers to classify more patients as positive, as was the case in this study, and may thus give a spurious impression of improved accuracy. In fact the substantive conclusion of our analysis, that dermoscopy does not appear to contribute diagnostic information, runs counter to the impression of investigators involved in using the technology (Rao et al. 1996).

Calibration of the marginal quantiles for the purpose of comparing accuracies in an appropriate way is not always possible, especially in studies where only binary or ordinal classifications have been ascertained from the readers (Venkatraman and Begg, 1996). We were fortunate in this study to have data rated on several factors, the ABCD factors, both for the standard clinical evaluation and for the experimental dermoscopic evaluation. As a result it was possible to create linear score functions based on these factors with no tied data elements, permitting the comparison of the test strategies at every value of the rank order statistics. This allowed us to take advantage of a recently developed permutation test that provides a powerful distribution-free test of the hypothesis that the two ROC curves are identical at all classification thresholds. However, it is possible to use this methodology even when some of the scores are tied, as long as there exist some values of the marginal quantiles at which calibration is achievable. The minimum degree of calibration is where the comparison can only be accomplished at a single cut-off point. This is the case for comparisons of subjective diagnoses with scored diagnoses, and we have indicated that it is appropriate to use Fisher's exact test in this setting.

By evaluating the null hypothesis that the ROC curves are identical at every operating point the test provides leverage to distinguish diagnostic tests that may have dissimilar ROC curves with similar areas, in contrast to conventional tests that simply compare the areas. To some extent this merely provides a more satisfying theoretical benchmark of comparability. However, it also means that, for example, if it is important clinically to employ a cut-off with a high specificity in order to minimize medical interventions on healthy subjects, then one of the two tests will be clearly superior (on the basis of higher sensitivity at the calibrated specificity). Conversely, if high sensitivity is important, e.g. in a screening context, different specificities can be distinguished.

There are a few additional qualifications or observations pertinent to the interpretation of these data. First, we used throughout the lesion as the unit of analysis. In fact the data consisted of 72 lesions derived from 63 patients. While the evaluation of two lesions from the same patient may be dependent in practice when a dermatologist evaluates the patient, this study was conducted retrospectively based on photographs, and so the observers were not aware of the details of the patients from whom the individual lesions were derived. So the assumption of independence implicit in our analyses is probably reasonable. In any case the small number of patients with multiple lesions prohibits the realistic evaluation of dependence.

There are two issues that bear upon the validity of the estimates of accuracy, that is the estimates of the sensitivities, specificities and ROC curves. First, we have assumed that the histopathologic diagnosis is an accurate gold standard. In fact, there exists considerable controversy among pathologists regarding the diagnosis of melanoma, especially lesions restricted to the epidermis (Rosai (1996)). It is unclear how such misclassifications would affect the estimates of accuracy, although it is probable that they would lead to a positive bias, in view of the fact that the pathologist would be privy to the clinical and dermoscopic results when making the diagnosis (Begg (1987)). A second issue is verification bias, which occurs when the study is restricted to those lesions that were removed for pathological evaluation, as was the case in this study. Since the larger population of evaluated lesions would be likely to consist of a preponderance of lesions classified as benign the study sample inevitably consists of a group of lesions with a higher prevalence of melanoma classifications. This phenomenon results, typically, in overestimation of the sensitivities and underestimation of the specificities, and the biases can only be corrected if detailed data (excluding pathology) are available on a random sample of evaluated lesions in addition to the study sample (Begg and Greenes (1983)).

Acknowledgements

We are grateful to Dr. Alfred Kopf and colleagues for permission to use their data. The research was supported by the National Library of Medicine, Award LM 05227.

References:

BEGG, C. B. (1987). Biases in the assessment of diagnostic tests. *Statistics in Medicine, 2, 411–423.*

BEGG, C. B. and GREENES, R. A. (1983). Assessment of diagnostic tests when disease verification is subject to selection bias. *Biometrics, 39, 207–215.*

COPAS, J. B. (1983). Regression, prediction and shrinkage (with discussion). *Journal of the Royal Statistical Society B, 45, 311–354.*

DELONG, E. R.; DELONG, D. M. and CLARKE-PEARSON, D. L. (1988). Comparing the areas under two or more correlated receiver operating characteristic curves: a nonparametric approach. *Biometrics, 44, 837–846.*

FRIEDMAN, R. J.; RIGEL, D. S. and KOPF, A. W. (1985). Early detection of malignant melanoma: the role of physical examination and self-examination of the skin. *CA, 35, 130–151.*

HANLEY, J. A. and MCNEIL, B. J. (1983). A method of comparing the area under two ROC curves derived from the same cases. *Radiology, 148, 839–843.*

JUNG, S. H.; WIEAND, S. and CHA, S. S. (1995). A statistic for comparing two correlated markers which are prognostic for time to an event. *Statistics in Medicine, 14, 2217–2225.*

METZ, C. E. (1986). ROC Methodology in radiologic imaging. *Investigative Radiology, 21, 720–733.*

NACHBAR, F.; STOLZ, W.; MERKLE, T. et al. (1994). The ABCD rule of dematoscopy. *American Academy of Dermatology, 30, 551–559.*

RAO, B. K.; MARGOOB, A. A.; STOLZ, W. et al. (1996). Can early malignant melanoma be differentiated from atypical melanocytic nevi by in-vivo techniques? Part 1: clinical and dermoscopic characteristics. Unpublished manuscript.

ROSAI, J. (1996). Ackerman's Surgical Pathology. C.V. Mosby Co., St. Louis, 8th Edition, 1996, pp 138-156.

STOLZ, W.; REIMANN, A. COGNELLA, A. B. et al. (1994). ABCD rule of dermoscopy: a new practical method for early recognition of malignant melanoma. *European Journal of Dermatology, 7, 521–528.*

STONE, M. (1974). Cross-validatory choice and assessment of statistical predictions. *Journal of the Royal Statistical Society B, 36, 111–147.*

VENKATRAMAN, E. S. and BEGG, C. B. (1996). A distribution-free procedure for comparing receiver operating characteristic curves from a paired experiment. *Biometrika.* (in press).

Formal Modeling of Medical Concept Systems Considering Part-Whole Relations

Jochen Bernauer[1] , Mareike Schoop[2], Dominik Schoop[2]

[1]Institute for Medical Informatics, University of Hildesheim, 31141 Hildesheim, Germany

[2]Medical Informatics Group, Department for Computer Science, University of Manchester, Manchester M13 9PL, U.K.

Summary: Conventional medical classification systems are monolithic and are lacking multiple views on concepts through the strictly mono-hierarchical organisation imposed by hierarchical coding. The only conventional way for reuse and multiple classification is by a multi-axial design. Formal concept representation and classification in medicine is a strategic approach for overcoming these problems. This paper identifies the major principles of subordination in conventional medical classification systems and outlines how these principles can be modeled through formal concept representation considering both subsumption and part-whole relation.

1. Introduction

Classification of concepts with respect to different types of characteristic is a common requirement for medical information processing either for the purpose of health care statistics and information retrieval, or in the field of knowledge based systems. A common problem for instance is the ordering of diseases with respect to the type of the underlying nosological process or the affected body parts. Conventional medical classification and terminology systems approach multiple classification by different principles: The only way for mono-hierarchical systems to support ordering of concepts under multiple aspects is through multiple assignment of codes (e.g. ICD-10 (1992), MeSH), for multi-axial systems (e.g. SNOMED (1993)) through juxta-position codes consisting of hierarchical code segments each organised in a mono-hierarchical way on the particular axes. As a relative weakness, in both cases the potential views on concepts are defined in an a priori manner by hierarchical codes which represent predefined levels of abstraction. As an additional obstacle, conventional medical classification systems are constrained in their hierarchical organisation if the applied coding scheme restricts the hierarchies in breadth and depth. A weakness specific to multiaxial systems is the lack of compositionality on the particular axes. This is a problem for instance for anatomical descriptions since the representation of body structures strongly calls for a compositional mechanism.

2. Formal concept representation

The rigid ordering inherent to conventional medical concept systems can be overcome by formal approaches for concept representation and classification, as proposed for instance by terminological logics. Formal concept representation languages have gained attractiveness in Medical Informatics primarily for the language-independent representation of findings (Campbell et al. (1994)), as target structures for medical text analysis (Baud et al. (1992)), as intermediate language for text generation and report writing (Bernauer et al. (1996b)) in structured data entry systems. In recent time also formal reconstruction and coordinate development of medical classification systems are gaining interest. The European Fourth Framework project GALEN-IN-USE develops methods for the coordination and harmonisation of national classifications for medical procedures through a principled way for representing rubrics (GALEN-IN-USE (1996)).

Typically formal concept representation systems comprise a language for the definition of concepts through necessary and sufficient characteristics, and a formal criterion for subsumption, which can be used for classifying concept descriptions automatically. Concept definitions are based on elementary concepts and roles. Elementary concepts represent conceptual primitives in a granularity fine enough to support classification with respect to particular views. Roles represent the semantic relations between constituent concepts. It has been debated in the past about the trade-off between expressiveness of the representation language and tractability of the classifier, and whether the classifier should consider definitorial knowledge only or also assertional knowledge (Doyle et al. (1991)). The actual discussion about formal concept representation in medicine concerns the following aspects: which constructs to be provided by the concept description language – i.e. operators for negation, conjunction or disjunction, number restrictions, statements for disjointness – how to represent assertional knowledge, and how to use it for classification, how to represent temporal or spatial relations, and to what degree compositional restrictions for controlling sensible concept formation are required. An additional aspect concerns a commonly agreed modeling style, i.e. a uniform way medical language expressions are represented by conceptual descriptions.

In the following the most important intuitive principles of subordination in conventional medical classification systems are analysed and it is outlined how they are modeled in BERNWARD, a graph oriented concept representation system. Its classifier considers both subsumption and part-whole relation (Bernauer et al. (1996c)). A comparison between GRAIL, the representation system used in GALEN (Rector et al. (1995)), and BERNWARD with a discussion of desirable constructs and features for the purpose of classification modelling is given in this volume (Schoop et al. (1996)).

3. Subordination in medical classification systems

In conventional medical classification systems subordination of rubrics is almost equivalent to generic or partitive specialisation. In most cases – for instance in disease classifications – the subordinate rubric is kind of the superordinate which functions as an abstraction of the subordinate. Mainly in anatomical or topographical systems (cf. anatomical axis of SNOMED (1993)) subordination represents a partitive aspect: the superordinate rubric is more comprehensive than the subordinate. Typically both types of subordination are not explicitly separated (cf. for instance MeSH). Moreover, in many cases generic subordination of rubrics follows from the partitive relationship between some characteristics. This is for instance the case if a disease at body location A subsumes a disease at body location B, if B can be seen as a part of A: colon tumor is kind of tumor of the digestive system.

In conventional medical classification systems subordination of rubrics is intuitive and syntactically expressed by the assignment of hierarchical codes. The identification of these principles provides some clues for terminological modeling of rubrics and for formal classification of concept descriptions. The following list covers the most important principles: introduction of a modifying criterion, generic refinement, introduction of a partitive criterion, partitive refinement, and coordination (Bernauer (1994)). The examples are taken from the German version of ICPM (International Classification for Procedures in Medicine) (ICPM (1994)). Often several principles are concurrently applied in a single step of subordination.

3.1 Introduction of a modifying criterion

The subordinate rubric is more specific than the superordinate, since an additional modifying criterion is introduced:

5-433 Excision of abnormal tissue of the stomach
5-433.1 – by endoscopy

In this example the criterion which adds detail ("by endoscopy") modifies the base concept ("excision") which is the primary category of the rubric. A slightly different situation occurs if a modifying criterion is introduced for a criterion of a rubric. This is the case in the following example:

5-401 Excision of a lymph node
5-401.1 – axillary

Here the criterion "axillary" modifies the criterion "lymph node" as the target of excision. For both cases – modification of the base concept, and modification of a criterion – subordination follows intuitively for terminological reasons only, i.e. the principle is independent from domain knowledge. Abstractly speaking, the subsumption has the structure: (A *mod* B) is kind of A, and (A *mod* (B *mod* C)) is kind of (A *mod* B), where *mod* expresses some modificational type of characteristic.

3.2 Generic refinement

Subordination by generic refinement can be illustrated by the following case:

5-12 Operation at the cornea
5-12.0 Removal of a foreign body at the cornea

In this example the base concept "operation" is refined by "removal", which is kind of "operation". The replacement of "cornea" by "foreign body at the cornea" which concurrently adds detail follows the principle of introduction of a partitive criterion (see below). Like introduction of a modifying criterion generic refinement can also regard the concept of a criterion:

5-783 Collection of a bone transplant
5-783.1 - Spongiosa

Here subordination is caused by spongiosa being kind of a bone transplant.

Generic refinement is not restricted to "primitive" concepts only. A more complex case is given in the following example which shows the relevance of generic refinement for composite constituents of rubrics:

5-787 Removal of osteosynthetic material
5-787.1 - Screw

In this case screw is seen as kind of osteosynthetic material where the latter concept can be considered as a composite concept description in the sense of "material having an osteosynthetic function". Regarding the examples given *removal being kind of operation, or spongiosa being kind of bone transplant* are axioms from the surgical domain. In contrast to the principle of introduction of a modificational criterion, subordination by generic refinement follows from assertional grounds: it depends on knowledge about the application domain.

A more complex variant of generic refinement is given in the following example, where an entire description is genericly refined:

5-434 Atypical partial resection of the stomach
5-434.2 Antrectomy

On the one hand it is true that antrum is kind of a part of the stomach and ectomy is kind of resection. In so far it follows by already known principles that antrectomy is kind of a partial resection of the stomach. However, antrectomy being an atypical type of resection of the stomach is an aspect based on additional knowledge from the surgical domain. This case of subordination can be modeled as an instance of generic refinement which regards the entire concept representing the superordinate rubric.

3.3 Introduction of a partitive criterion

This principle resembles structurally the principle of introduction of a modifying criterion, but in contrast adds a partitive concept to the concept of a criterion. An instance occurs in the following example:

5-502 Liver resection

5-502.1 Left hemihepatektomy

In the subordinate rubric "liver" as the target of "resection" is further qualified by the partitive concept "left half segment". The partitive relation between "left half segment of the liver" and "liver" follows for terminological grounds, i.e. it can be derived from the structure of the concept description only. This is in analogy to the principle of introduction of a modifying criterion, where the generic relationship of the subordinate to the superordinate rubric follows also from structure solely.

3.4 Partitive refinement

Partitive refinement resembles structurally generic refinement, but regards the replacement of a concept of a criterion by one of its parts. As for generic refinement, this principle refers to domain knowledge, since the part-whole relation cannot be simply derived from the structure of the partitive concept:

5-448 Reconstruction of the stomach
5-448.4 Fundoplication

In this example the subordinate rubric is intuitively subsumed by the superordinate, because fundus is an anatomical part of the stomach (partitive refinement) and "plicatio" is kind of "reconstruction" (generic refinement).

The principles of introduction of a partitive criterion and partitive refinement as defined make no difference between the type of part-whole relation involved. Instead, the idea of partitive relation is used in a broad sense, covering for instance structural or functional parts, segments, or members of a collection. For a classification of types of part-whole relations see for instance Winston et al. (1987), for an analysis in the medical domain see Bernauer (1996a).

3.5 Coordination

A common principle for creating rubrics and for "balancing" mono-hierarchical classifications is coordination of concepts or characteristics. In this context an often unsystematic usage of "and" and "or" in the language expressions can be observed. In the case of the following example coordination is used in a disjunctive sense:

5-45 Incision and excision on the small and large intestine
5-450 Excision on the small intestine

Here, the superordinate rubric comprises instances of incision or excision on the small or large intestine. Therefore, it subsumes excision on the small intestine. In the superordinate rubric of this example both the base concept as well as a criterion are coordinated logically by "or".

The inverse case in which coordination is applied in the sense of a logical "and" is given in the following example. It leads to subordination of the rubric containing the coordination:

5-803 Suture of the articular capsule
5-803.1 Suture of the anterior crucial ligament
5-803.2 Suture of the posterior crucial ligament
5-803.8 Combined reconstruction

Here, *5-803.8* would be properly represented as subordinate to *5-803.1* and *5-803.2* however such an ordering is prevented by the mono-hierarchical structure of the classification system. This is a typical case in conventional medical concept organisation where logical principles are sacrificed to coding principles.

4. Modelling subsumption and part-whole relation in BERNWARD

BERNWARD supports classification of concept descriptions by several principles. Firstly, primitive or composite concept descriptions can be explicitely set into a generic or partitive relationship. The respective relations $<_{es}$ (explicit subsumption) and $<_{ep}$ (explicit part-of) are reflexive, transitive and antisymmetric and induce each a partial ordering on concept descriptions which are called "taxonomy" resp. "partonomy". Both are to represent hierarchical relationships that hold for assertional grounds. $<_{es}$ corresponds to the generic ordering relation of the type lattice in conceptual graphs and to partially defined concept description of terminological logics (Sowa (1991)). The provision of an explicit partonomy is motivated by the specific import of the part-whole relation in medical concept representation and the effect it has on subsumption.

Secondly, BERNWARD provides formal criteria for subsumption and part-whole. Both criteria are based on the respective explicit ordering and on the structure of concept descriptions. A brief outline of these criteria is given in the following. For an abstract notation and formal definitions and of these criteria see Bernauer et al. (1996c). Formal subsumption ($<_{fs}$) realizes the above mentioned principles introduction of a modifying criterion, generic refinement and coordination:

$A <_{fs} (A\ mod\ B)$, and
$(A\ mod\ B) <_{fs} (C\ mod\ D)$, if $A <_{es} C$ and $B <_{es} D$
$(A\ and\ B) <_{fs} A, (A\ and\ B) <_{fs} B$
$(A\ mod\ and\ BC) <_{fs} (A\ mod\ B), (A\ mod\ and\ B\ C) <_{fs} (A\ mod\ C)$

Theidea of formal part-whole ($<_{fp}$) is like follows. $<_{fp}$ entails explicit part-whole:

$A <_{fp} B$, if $A <_{ep} B$

and additionally derives partitive relations from the structure of concept descriptions. For that reason a particular semantic role denoting some kind of partitive relation, called *part-of*, is used:

$(A\ part-of\ B) <_{fp} B$

A generalisation of formal subsumption is a formal criterion for part-sensitive subsumption ($<_{ps}$). $<_{ps}$ entails formal subsumption, but additionally allows for formal part-whole relations between criteria of concept descriptions. Part- sensitive subsumption corresponds to the principle of introduction of a partitive criterion and partitive refinement. Its idea is the following:

$(A\ mod\ B) <_{ps} (A\ mod\ C)$, if $B <_{fs} C$ or $B <_{fp} C$

Part-sensitive subsumption provides flexible means for deriving subsumption across part-whole relations, as in the following example: osteosynthesis at the upper extremity subsumes part-sensitively osteosynthesis at the humerus, which subsumes part-sensitively osteosynthesis at the humerus metaphysis. For a discussion how this problem is approached in contrast by terminological logics and GRAIL see Bernauer (1996a).

5. Discussion

The described principles of subordination in conventional classification systems are the most common ones. It could be shown that they can be modeled using a compositional concept representation formalism in combination with axioms for expressing assertions about generic and partitive relationships. Besides, there are additional intuitive mechanisms for the formation of rubrics that need concern with respect to formal reconstruction. Among them are the notion of exclusion which is used for separating out certain concepts from a rubric, the explicit negation of characteristics (*ICPM: 5-063.0 Thyreoidectomy without neck dissection*), and relative complements (*5-063.x Other kind of thyreoidectomy*).

A different class of questions to be dealt with regards the style of modeling, i.e. how to transform language information to conceptual, language independent representations. Part of this problem is the question what kind of information to put into concepts and what into semantic relations. In this context there is a clear need for standardisation of medical terminology and semantics and for merging existing medical terminology systems with formal approaches.

References:

BAUD R. H.; RASSINOUX A. M. and SCHERRER J. R. (1992). Natural Language Processing and Semantic Representation of Medical Text, *Meth. of Information in Medicine, 31:117-25.*

BERNAUER J. (1994). Subsumption principles underlying medical concept systems and their formal reconstruction. In: Ozbolt J.G. (ed.), Proceedings 18th Annual Symposium on Computer Applications in Medical Care (SCAMC94), Washington DC, pp 140-144

BERNAUER, J. (1996a), Analysis of part-whole relation and subsumption in the medical domain, to appear in: Data and Knowledge Engineering

BERNAUER J.; BENNEKE A.; FUESESCHI A. and URBAN M. (1996b), Structured data entry for medical records and reports. To appear in: van de Riet (ed.), NLDB96 Second international workshop on applications of natural language to information systems, Amsterdam, IOS Press

BERNAUER J. et al. (1996c), A Model for Representation, Classification and Composition of Medical Concept Descriptions Based on Conceptual Graphs, to be published

CAMPBELL K. E.; DAS A. K. and MUSEN M. A. (1994), A Logical Foundation for Representation of Clinical Data, *J Am Med Informatics Assoc. 1:218-232.*

DOYLE J. and PATIL R. (1991). Two theses of knowledge representation: language restrictions, taxonomic classification, and the utility of representation services. *Artificial Intelligence 48; 261-297.*

GALEN-IN-USE (1996). Telematics Application Programme – HC 1018 HC

ICD-10 (1992). International Statistical Classification of Diseases and Related Health Problems. Tenth Revision, World Health Organization Geneva

ICPM (1994), Operationenschluessel nach Paragraph 301 SGB V – Internationale Klassifikation der Prozeduren in der Medizin herausgegeben vom Deutschen Institut fuer medizinische Dokumentation und Information DIMDI im Auftrag des Bundesministeriums fuer Gesundheit, Version 1.0, 1994

MeSH, Medical Subject Headings, National Library of Medicine, Bethesda (quarterly updated).

RECTOR A. L.; GLOWINSKI A. J.; NOWLAN W. A. and ROSSI-MORI A. (1995), Medical- concept Models and Medical Records: Am Approach Based on GALEN and PEN&PAD, *J Am Med Informatics Assoc. 2:19-35.*

SCHOOP M. and SCHOOP D.(1996). Differences of representing a conventional classification by the compositional concept representation languages BERNWARD and GRAIL, GfKL96

SNOMED III (1993), American College of Pathology, Skokie, IL

SOWA J. F. (ED.) (1991). Principles of Semantic Networks. Morgan Kaufmann Publishers, San Mateo CA

WINSTON M. E.; CHAFFIN R. and HERRMANN D. (1987). A Taxonomy of Part-Whole Relations. *Cognitive Science 11: 417-444.*

Classification of Oligodendrogliomas Using Neural Networks

A. Brodbeck[1] , J. R. Iglesias-Rozas[2], A. Zell[1]

[1] Institut für Parallele und Verteilte Höchstleistungsrechner (IPVR) Breitwiesenstr. 20-22, D-70565 Stuttgart, Germany

[2] Pathologisches Institut (Neuropathologie), Katharinenhospital, Kriegsbergstr. 60, D-70174 Stuttgart, Germany

Summary: Oligodendrogliomas are brain tumors which, according to the WHO guidelines, can be classified into three groups: typical oligodendro-gliomas, oligoastrocytomas, and malignant oligodendrogliomas. For the following paper, 108 oligodendrogliomas, 168 oligoastrocytomas, and 131 malignant oligodendrogliomas were examined. Self-organising cards and feed-foward networks of various topologies were created to process this visually collected data. Discriminant analysis was then applied to the same data sets. For classification purposes, it was found that neural networks could improve on the results of discriminant analysis.

1. Introduction

Brain tumours are normally histologically diagnosed and classified by pathologists. How correct the diagnosis is depends on the experience of the specialist. Furthermore, histological diagnosis has meant that the pathological and anatomical nomenclature embody the interpretation of visually collected findings. Transitional states between the different tumor types are not unusual in general tumor pathology. It is exactly this distinction, between the nature of the tumor and its biological dignity, which needs to be optimisied. Above and beyond this, the total structure of the tumor, its growth and spreading characteristics, the morphology of the cell and nucleus, and the vascular provision all need to be taken into account. Oligodendrogliomas are tumors in which different cell types can also lie beside each other, thus creating a clinical- pathological entity, the so-called mixed glioma or oligoastrocytoma. Statistical tools for the diagnosis and evaluation of the malignancy of brain tumors have been described before (see Iglesias et al. (1986a), (1986b), (1988), (1991). As an alternative to these, it is intended to examine whether a suitable solution can be found concerning the ability to reproduce the classification of brain tumours with the assistance of neural networks, as Iglesias-Rozas et al. (1996) have shown with Meningiomas.

2. Material and Methods

For this research paper 108 typical oligodendrogliomas, 168 oligoastrocytomas, and 131 malignant oligodendrogliomas were surveyed (WHO guidelines.) From each group, about three quarters were available for the training set and one quarter for the testing set. For the purposes of diagnosis, 50 histological characteristics were examined , and evaluated on a scale of four (0 = not present, 1 = present to a small degree, 3 = middling presence, and 4 = abundant), which together with the age and sex of the patient made up the basis of the classification. Self Organising Cards (SOC's) and feed-foward networks of different topologies using various learning methods such as Backpropagation (BP), BP with Momentum, Quickprop, trained Rprop, and Pruning techniques such as Skeletonisation, Optimal Brain Damage (OBD) and Magnitude Based Pruning (MBP) were then applied to the visually captured data (Zell 1994). The Stuttgart Neural Network Simulator (SNNS) running on SUN-Workstations in versions 3.3 and 4.0 (Zell (1994)) was used for the simulation. Discriminant analysis (DA) was applied to the same datasets as a comparison to the Neural Network technique using the SPSS program (SPSS Inc software package, Michigan)

3. Results

To train the oligodendrogliomas two training sets and two testing sets were created from the available datasets. One training set was given training vectors from all three classes, as was its associated testing set. The same vectors were applied to the second training set as in the first case, but with the difference that all the training vectors that were initially described as oligoastrocytomas were put together with the oligodendrogliomas. The aim was to limit the classification to a differentiation of malignant tumors from benign ones.

3.1 Classification of Oligodendrogliomas Using SOC's

These unsupervised learning techniques prove to be very suitable to examine the similarities within the classes. Improved results were achieved for the examination of all three classes as well as for the classification of malignant and benign oligodendrogliomas using the smaller networks of 10 x 10 neurons (Table 1).

Class	25 x 25 neurons		10 x 10 neurons	
	Training data correct	Testing data correct	Training data correct	Testing data correct
2	97.5	78.3	84.0	80.0
3	95.8	61.5	78.5	63.2

Table 1: Classification Results (in %) for Oligodendrogliomas using SOC's

3.2 Classification of the Oligodendrogliomas Using Feed-foward Networks

In comparison to self-organising cards feed-foward networks showed markedly improved results. Using networks for the classification of benign and malig-

Learning technique	50 hidden neurons Training data correct	Testing data correct	20 hidden neurons Training data correct	Testing data correct
BP	89.9	87.1	90.2	87.1
BP with Momentum	87.6	88.1	89.2	86.1
Quickprop	89.2	90.1	68.6	60.6
Rprop	99.0	90.1	98.0	90.1

Table 2: Classification Results (in %) using 52-50-2 and 52-20-2 Networks (2- Class problem)

nant oligodendrogliomas, the results could be improved from 80 % to 90 % as Table 2 shows.

With networks of three classes the results in comparison to SOCs could be improved from 63 % to 79 % as Table 3 shows. In this task the results from the larger networks returned markedly better results than the smaller networks.

Learning technique	50 hidden neurons Training data correct	Testing data correct	20 hidden neurons Training data correct	Testing data correct
BP	82.0	79.2	82.3	73.3
BP with Momentum	81.1	76.2	79.1	71.3
Quickprop	82.0	78.2	51.6	39.6
Rprop	86.6	76.2	83.3	70.3

Table 3: Classification Results (in %) using 52-50-3 and 52-20-3 Networks (3- Class problem)

3.3 Reduction of the Input Parameters for Oligodendrogliomas (2- Class Problem)

The number of input parameters for the classification of oligodendrogliomas is, at 52 neurons, comparatively large. For certain, not all of these inputs are relevant for the classification of the three classes. With the aid of pruning techniques it was intended to examine which characterisitics are essential for the classification problem. To this end three different pruning processes each with four learning techniques were employed on networks with 50 and 20 hidden neurons.

A recognition rate of 90 % for the testing set and 89 % for the training set was acheived for networks with fixed topology. These classification results relate to a network with 20 hidden neurons trained with backpropagation and thinned out using skeletonisation (Table 4). 12 characteristics were not

	50 hidden neurons		20 hidden neurons	
Pruning technique	Training data correct	Testing data correct	Training data correct	Testing data correct
Skeletonisation	88.6	86.1	90.2	90.1
OBD	96.4	87.1	95.1	89.1
MBP	94.8	91.1	96.1	88.1

Table 4: Classification Results (in %) for 52-50-2 and 52-20-2 Networks

removed in the skeletionisation process (2, 6, 10, 11, 13, 20, 27, 28, 29, 34, 36, 40). Only the characteristics 6, 11, 13, 34, and 40 have any relevance for the conventional classification between benign and malignant oligodendrogliomas. Sixteen input neurons remained after the OBD process and were not removed during any training cycle; these are also the most important characteristics to distinguish between malign and benign tumors: nodular architecture (1), reticular arrangement (8), necrosis (13), hemorrhages (14), vascularisation quantity (17), anomalous vessels (18), polymorphic cells (21), polymorphic nuclei (23), spherical nuclei (25), quantitiy of chromatin (30), typical mitosis (31), lympho-granular-plasma-cellular infiltrates (35), presence of astrocytes (38), presence of oligodendrocytes (42), dedifferentiated cells (44), and the sex (52). The only characteristic not deleted by MBP was (31), typical mitosis. The following characteristics were each deleted once: vascularisation quantity (17), polymorphic nuclei (23), quantity of chromatin (30), and presence of glioblasts (40).

3.4 Reduction of the Input Parameters for Oligodendrogliomas (3- Class Problem)

Also when training with all three classes, no improvement could be seen from applying pruning techniques to the networks.

MBP classified 79 % of the Testing set and 90 % of the Training set correctly and needed only 20 hidden neurons in the initial topology for this improvement (Table 5).

MBP left only seven input neurons undeleted from training: Necrosis (13), vascularisation quantitiy (17), polymorphic nuclei (23), typical mitosis (31), atypical mitosis (32), glial fibres (33), and presence of astrocytes (38). These characteristics are the most important histological parameters for the neuropathologist when differentiating between the three tumor types.

3.5 Classification of the Oligodendrogliomas using DA

Table 6 shows a comparison between the results of DA and those of neural

Pruning technique	50 hidden neurons Training data correct	Testing data correct	20 hidden neurons Training data correct	Testing data correct
Skeletonisation	85.6	74.3	83.7	74.3
OBD	87.6	74.3	90.2	79.2
MBP	88.2	70.3	92.2	77.2

Table 5: Classification Results (in %) for 52-50-3 and 52-20-3 pruned networks

networks. In each case the best neural network has been used for the comparison. As far as training with two classes is concerned, the most successful network was trained with Rprop using 20 hidden neurons. For three classes the most successful was a network using BP, OBD and 20 hidden neurons as the initial value. In both cases neural networks bettered the results acheived by discriminant analysis.

Class	Discriminant Analysis Training data correct	Testing data correct	Neural Networks Training data correct	Testing data correct
2	94.1	88.1	98.0	90.1
3	73.2	70.3	90.2	79.2

Table 6: Classification Results (in %) for Oligodendromas

4. Discussion

Classification systems of human tumors based on statistical methods have already been successfully implemented as a semi-automatic diagnostic aid. The most used statistical methods are cluster analysis and DA (Späht (1977), Iglesias (1986a and 1986b), Boulay et al. (1987)). Furthermore, research has been done into experimental tumors in mice and rats (Iglesias et al (1991)) In Pathology, histological diagnosis is a complicated process. Not only does the experience of the pathologist in dealing with the specific tumor class play a role during observation and identification of the characteristics and defintion of a possible diagnosis, but also the subjectivity of the descriptive terminology and the exhaustion-factor of the pathologist during the examination. These factors vary amongst the pathologists, neuropathologists, and for the individual himself. The preparation of the diagnosis is mainly based on two processes: the identification of the presence – and degree thereof – of morphological anomalies, and the evaluation of the significance of these alterations for the possible diagnosis. Classification using neural networks should minimise these problems. The results acheived by this research show that classification of brain tumors using neural networks is possible. Moreover, better results have been acheived than with statistical methods. Similar results have been shown concerning the diagnosis of Meningiomas (Iglesias et al. (1996)) The influence on the diagnosis itself could also be examined. With the assistance of pruning methods the input parameters could be reduced to only a few characteristics. As is also to be seen from DA and cluster analysis, the results presented in this paper show that in many cases only a few characteristics are of relevance for correct classification. These characteristics represent the qualities of the respective class. As concerns reduction of the input parameters, OBD and MBP are the most appropriate pruning techniques. The undeleted characteristics are, according to our experience as neuropathologists, the most important parameters needed to diagnose the malignancy of the process (two-class problem) and also the tumor type (three-class problem). Malignant and benign oligodendrogliomas are relatively easy to differentiate from one another with conventional histological methods. However the specific differentiation of oligoastrocytomas from typical oligodendrogliomas is difficult even with current dying techniques and modern immuno-histological methods (Burger and Scheithauer (1994)). The characterisation of oligoastrocytomas is of great importance as this tumor type behaves differently biologically to the typical oligodendrogliomas and malignant oligodendrogliomas.

The advantages of using neural networks for the diagnosis of brain tumours are: relatively simple computer-supported evaluation of the microscope pictures , and the possibility to establish the morphological evaluation (Aston and Wilding (1992), Christy et al. (1995)). The histological comparison between the diagnoses of various pathologists is thereby able to be more objective. Neural networks are objective analytical methods in which neither the independance nor the linear relationship of the variables or parameters is necessary. As tools, neural networks are more useful for the pathologist in deciding on a diagnosis than statistical methods. Mathematical differences between neural networks and statistical methods have been analysed by White (1989). Methods of classification with the aid of neural networks are relatively cheap and can be realised on a modern desktop computer. They require a minimum of statistical knowledge and personnel. In the future such methods are very promising for the histological diagnosis of other entities, in particular the relationship of clinical data to laboratory parameters.

References:

ASTION, M. L. and WILDING, P. (1992) Application of neural networks to the interpretation of laboratory data in cancer diagnosis. *Clin Chem, 38, 34-38.*

BOULAY, G. H.; TEATHER, D.; MORTON, B. A.; WILIS, K. M.; INOCENT, P. R. and PLUMMER, D. (1987) Brains a computer advisor system to aid in CT scan interpretation and cerebral disease diagnosis. *Neuroradiology, 29, 196-199.*

BURGER, P. C. and SCHEITHAUER, B. W. (1994) Atlas of Tumor pathology.

Tumors of the Central Nervous System. AFIP (eds.). Washington D.C.

CHRISTY, P. S.; TERNOVEN, O.; SCHEITHAUER, B. W. and FORBES, G. S. (1995) Use of a neural network and a multiple regression model to predict histologic grad of astrocytoma from MRI appearances. *Neuroradiology, 37, 89-93.*

IGLESIAS, J. R.; PFANNKUCH, F.; ARUFFO, C.; KAZNER, E. and CERVÓS-NAVARRO, J. (1986a) Histopathological diagnosis of brain tumours with the help of a computer: mathematical fundaments and practical application. *Acta Neuropathol (Berl), 71, 130-135.*

IGLESIAS, J. R.; KAZNER, E.; ARUFFO, C. and ESPARZA, J. (1986b) A model semiautomatical type-specific CT diagnosis of brain tumours. Mathematical fundaments and practical application. *Br.J.Radiology, 59, 895-900.*

IGLESIAS, J. R.; ESPARZA, J.; ARUFO, A. and MAIER-HAUFF, K. (1988) Diagnóstico diferencial de los neurinomas y meningiomas intrarraquídeos con ayuda de uns sistema bayesiano. *Arch de Neurobiol, 51, 333-341.*

IGLESIAS, J. R.; KROH, H.; MATYJA, A. and ARUFFO, C. (1991) Experimental CNS tumors in rats. Testing field for a computerized classification and tumor grading. *Exp Pathol, 43, 25-31.*

IGLESIAS-ROZAS, J. R.; GRIESHABER, F. A. and SOBOL, A. (1996) Superiority of neural networks over discriminant analysis methods for histological classification of human meningiomas. *EJPathol-Electronic Journal of Pathology and Histology. 2.1,961-05.txt.*

SPÄTH, H. (1977) Cluster-Anaylse-Algorithmen zur Objektklassifizierung und Datenreduktion. R. Oldenburg Verlag, München.

WHITE, H. (1989) Learning in Artificial Neural Networks: A Statistical Perspective. *Neural Computation, 1, 425-464.*

ZELL, A. (1994) Simulation Neuronaler Netze. Addison-Wesley Publishing Company. Bonn.

Statistical Methods to Estimate the Impact of Risk Factors on the Disease Load in a Population

O. Gefeller

Abteilung Medizinische Statistik,
Georg-August-Universität Göttingen, D-37073 Göttingen, Germany

Summary: In epidemiologic studies the impact of an exposure factor on the disease load in a population is quantified by the attributable risk. In this concept the observed cases of disease are classified into two (unobservable) subpopulations: those cases due to the exposure and the remaining part resulting from other factors. Statistical methods for estimating the parameter have been extensively developed during the last decade. Especially, the multifactorial situation in which several exposure variables have to be taken into account has been analyzed leading to the development of the concept of *adjusted attributable risks.* In the paper, the different statistical strategies to adjust for confounding and effect-modification in the estimation of attributable risk proposed in the literature are reviewed.

1. Introduction

In epidemiologic studies on the effect of a potentially harmful exposure factor on some disease variable the attributable risk serves as a useful methodologic tool by quantifying a meaningful aspect of the relationship under study. It addresses the public health importance of the exposure for the disease in the population under investigation by answering the question, "What proportion of the observed cases in the study population suffers from the disease due to the exposure factor of interest?". The observed cases of disease are thus classified into two — principally unobservable — subpopulations: those cases due to the exposure under study and the remaining part resulting from other factors. In providing this information the attributable risk places the concept of the relative risk, commonly used in epidemiologic studies for the evaluation of the etiologic relevance of a specific exposure, in a public health perspective. For example, a factor associated with a small relative risk can account for a substantial proportion of the total number of disease events provided that a large number of individuals is exposed to it. On the other hand, an exposure revealing a high relative risk may represent a negligible public health problem if the subpopulation exposed to the factor is small.

Attributable risk estimation has become an important subject of methodologic investigation in recent years. Statistical methods for estimating the parameter have been extensively developed during the last decade (Gefeller (1992a)). Especially, the multifactorial situation in which several exposure variables have to be taken into account has been analyzed leading to the development of the concept of *adjusted attributable risks*. This approach considers the case in which the impact of the exposure factor(s) of primary interest on the disease has to be controlled for the effects of other covariables when calculating the attributable risk. Parallel to the rising statistical treatment of the measure, a growing number of practical applications employing the attributable risk as the primary tool of the analysis can be found in the epidemiologic literature. For example, attributable risk calculations are especially useful in quantifying the maximal benefit that can be achieved by an intervention campaign aiming at the reduction of the disease load in a population (Browner (1986); Chyou et al. (1992)). An introductory review of the attributable risk methodology is provided here with special emphasis on the different adjustment methods.

2. Notation and Model

Consider a population of size n that is classified into groups according to a dichotomous, non-recurrent disease variable D and some harmful dichotomous exposure variable E. Let $P(D|E)$, $P(D|\overline{E})$, and $P(D)$ represent the probability of disease among the exposed, the unexposed, and the entire population, respectively. Analogously, denote $P(E|D)$, $P(E|\overline{D})$, and $P(E)$ the exposure probability among the diseased, the non-diseased, and the entire population, respectively. The relative risk RR is defined as the ratio of the disease probabilities among the exposed and the unexposed $(P(D|E)/P(D|\overline{E}))$ and is assumed to be greater than or equal to 1. The case of protective exposures, i.e. $RR < 1$, leads to the concept of prevented fractions which will not be considered here (Miettinen (1974)).

The observed data on the joint distribution of the disease and exposure variable can be arranged in the following 2×2 table:

		disease status		
		D	$\overline{D}$	Σ
exposure	E	X_{11}	X_{12}	$X_{1\cdot}$
status	$\overline{E}$	X_{21}	X_{22}	$X_{2\cdot}$
	Σ	$X_{\cdot 1}$	$X_{\cdot 2}$	n

The appropriate sampling model corresponding to these data depends on the design of the study. Throughout this paper, a multinomial model achieved by an unrestricted sampling of the n study subjects is assumed to hold. In epidemiology, this sampling model applies to cross-sectional and cohort studies in which the sampling is realized without prestratification on the exposure status.

The extension to the multivariable case to adjust for the effects of other covariables leads to the situation of a $2 \times 2 \times K$ table where the third dimension is represented by a stratum variable C with K levels which may be one observed variable (K refers to the number of different values of C) or a combination of two or more variables (K refers to the number of all possible combinations of observed values of these variables). In the second situation, C is a hypothetical construct and not an observable variable. In the case of a $2 \times 2 \times K$ table all quantities receive an additional subscript indicating the stratum of C. The distributional assumption in the situation of an unrestricted sampling of study subjects with post–sampling stratification remains that of one multinomial distribution, however, now with $4K-1$ cell probabilities as independent parameters.

3. Definiton of the Attributable Risk

The definition of the attributable risk dates back to 1953 when Levin introduced the parameter as follows (Levin (1953)):

$$AR = \frac{P(E) \cdot [RR-1]}{P(E) \cdot [RR-1]+1}.$$

This definition can be equivalently expressed as

$$AR = P(E \mid D) \cdot \frac{RR-1}{RR} \quad \text{or} \quad AR = 1 - \frac{P(D \mid \overline{E})}{P(D)}.$$

The attributable risk AR can be interpreted as the proportion of cases of disease D due to the exposure factor E among all observed disease events in the population, or as the percentage of disease events potentially preventable by a total elimination of the exposure in the entire population. Already Levin recognized that this "causal interpretation" is critically based on the assumption that exposed subjects, if they had not become exposed, would have had the same probability of developing a disease event as that found among unexposed subjects.

4. Estimation of the Attributable Risk in 2×2 Tabels

In the situation of a simple 2×2 table of disease versus exposure the maximum likelihood estimator (MLE) of AR and its asymptotic variance can be derived easily under all sampling models as shown first by Walter (1976). In the present case of a multinomial model the MLE of AR is given by the following expression:

$$\widehat{AR} = \frac{X_{11} \cdot X_{22} - X_{12} \cdot X_{21}}{X_{.1} \cdot X_{2.}}.$$

An estimator of the asymptotic variance of $\widehat{AR}$ can be obtained by plugging in the empirical quantities for the corresponding probability terms in the particular diagonal element of the inverse of the Fisher information matrix:

$$\widehat{Var\widehat{AR}} = \frac{n \cdot X_{21} \cdot [X_{11} \cdot X_{22} \cdot (n - X_{21}) + X_{12} \cdot X_{21}^2]}{X_{\cdot 1}^3 \cdot X_{2\cdot}^3}.$$

To construct asymptotic confidence intervals for AR based on the normality of the MLE $\widehat{AR}$, one should be aware of the results given by Leung and Kupper (1981). These authors proved that confidence intervals based on the logit transformation of $\widehat{AR}$ exhibit widths smaller than the symmetric intervals around the MLE for all $AR \in (0.21, 0.78)$. Simulation studies have further demonstrated that the finite confidence coefficients corresponding to the logit–based intervals are slightly conservative, whereas the symmetric confidence intervals are slightly anti–conservative in most situations.

5. Adjustment Methods for the Attributable Risk

Typically in observational epidemiologic studies the exposure–disease relationship is affected by other covariables leading to a biased attributable risk estimate when the estimation is based on the marginal 2×2 table. In the epidemiologic context this distortion of the estimation due to other covariables is referred to by the terms "confounding" and "effect–modification". In both (conceptually very different) situations the adjusted attributable risk taking the effects of the covariable(s) C into account can be expressed as

$$AR_{adj} = 1 - \frac{\sum_{i=1}^{K} P(C_i) \cdot P(D \mid \overline{E}, C_i)}{P(D)},$$

where $P(C_i)$ represents the proportion of the population in the i-th stratum of C, and $P(D|\overline{E}, C_i)$ denotes the stratum–specific probability of developing the disease among the unexposed in the i-th stratum of C.

Different strategies to estimate adjusted attributable risks from the data of epidemiologic studies have been proposed in the methodologic literature. These methods can be classified into three groups which are discussed separately in the following subsections.

5.1 The Weighted-Sum Approach

This method was informally introduced by Walter (1976) and further elaborated by Whittemore (1982). The basic idea is to rewrite AR_{adj} as follows

$$AR_{adj} = 1 - \frac{\sum_{i=1}^{K} P(C_i) \cdot P(D \mid \overline{E}, C_i)}{P(D)}$$

$$= \sum_{i=1}^{K} P(C_i \mid D) \cdot \frac{P(D \mid C_i) - P(D \mid \overline{E}, C_i]}{P(D \mid C_i)}$$

which shows that AR_{adj} can be viewed as a weighted sum

$$AR_{adj} = \sum_{i=1}^{K} w_i \cdot AR_i$$

of the stratum–specific attributable risks AR_i, $i = 1, \ldots, K$. The weights w_i, $i = 1, \ldots, K$, correspond to the proportion of the diseased subjects (the "cases") in the i-th stratum of C among all cases in the population. An estimator $\widehat{AR_{adj}}$ employing this weighted-sum approach can now be written as

$$\widehat{AR_{adj}} = \sum_{i=1}^{K} \frac{X_{\cdot 1i}}{X_{\cdot 1 \cdot}} \cdot \widehat{AR_i} = 1 - \sum_{i=1}^{K} \frac{X_{\cdot\cdot i} \cdot X_{21i}}{X_{\cdot 1 \cdot} \cdot X_{2 \cdot i}}$$

where $\widehat{AR_i}$ refers to the stratum–specific attributable risk estimate obtained from the data of the i-th stratum of C. Owing to the interpretation of the weights, the term "case load weighting" can be found in most papers referring to this estimator. Expressions for the asymptotic variance of this estimator can be found in Eide and Gefeller (1995).

Another weighting scheme was proposed by Ejigou (1979) who chose the weights proportional to the inverse of the estimated variance of the stratum-specific attributable risk estimates. Such a "precision weighting" implicitly assumes, however, identical attributable risks over the strata, a condition which rarely holds in practice and is even incompatible with the usual confounding situation (Gefeller (1992b)). Consequently, this proposal leads to an adjusted attributable risk estimator which is not consistent in most situations.

5.2 The Indirect Method

Using the functional relationship of RR and AR

$$AR = P(E \mid D) \cdot \frac{RR - 1}{RR}$$

an adjusted attributable risk estimate can be derived via adjustment of the relative risk (Miettinen (1974)). To this end, any consistent estimate of the adjusted relative risk in this multifactorial situation can be used to replace RR in the equation above yielding a consistent estimate of the adjusted attributable risk (Greenland (1984)). This approach is valid, however, only under the homogeneity assumption on the relative risks, which means that the stratum-specific relative risks have to be identical over the strata (otherwise a summary estimate of relative risk would be less meaningful). In this homogeneity model the indirect method, often referred to as "adjustment of the Mantel-Haenszel type" due to its similarity with the classic Mantel-Haenszel procedure for estimating a common odds ratio from stratified contingency tables, yields an asymptotically unbiased estimate not only for the usual "large strata" situation (K fixed, $n \to \infty$) but also in the "sparse data" situation (n fixed, $K \to \infty$) characterized by small stratum-specific counts for a large number of strata. This property is not shared by the MLE of AR introduced in the previous subsection which is only consistent in the "large strata" case (Greenland (1987)).

5.3 Model-Based Techniques

In the context of attributable risk estimation from case-control data Bruzzi et al. (1985) were the first to propose the incorporation of logistic regression models in the adjustment process. Based on rewriting AR_{adj} as

$$AR_{adj} = 1 - \sum_{i=1}^{K} \left(\frac{P(E, C_i \mid D)}{RR_i} + P(\overline{E}, C_i \mid D) \right)$$

they suggested to estimate the stratum-specific relative risks RR_i by $\widehat{OR}_i$ derived from a logistic regression model. The stratum-specific exposure prevalences also involved in this representation of AR_{adj} can be separately estimated from the case population.

The resulting estimator $\widehat{AR}_{adj}$ has been shown to be a flexible generalization of the previously introduced competitors (Benichou (1991)). Depending on which logistic regression model is specified the estimators resulting from "case load weighting" and the indirect method can be derived as special cases of this model-based approach. If the logistic model includes only main effects (exposure and covariables), then this procedure is analogous to the indirect method. If the completely saturated model (including all interaction terms between exposure and covariables) is specified, this technique results in the "case load weighting" estimator. The flexibilty of the regression-based method lies in the possibility to select any logistic model, which fits appropriately the data, between these two extremes.

The complex problems of asymptotic variance estimation for the model-based estimator $\widehat{AR}_{adj}$ taking the covariance between the stratum-specific relative risk estimates and the estimated exposure prevalences into account have been described by Benichou (1991) for the case-control design. For cohort studies the situation is slightly easier as the probability terms $P(D \mid \overline{E}, C_i)$, $i = 1, \ldots, K$, are directly estimable from the logistic model under this design. Thus, there is no need to use Bruzzi's roundabout way of approximating stratum-specific relative risks by model-derived odds ratio estimates; the estimator can simply be written as

$$\widehat{AR}_{adj} = 1 - \sum_{i=1}^{K} \widehat{P}^{LR}(D \mid \overline{E}, C_i) \cdot \frac{X_{\cdot\cdot i}}{X_{\cdot 1 \cdot}}$$

where $\widehat{P}^{LR}(\cdot)$ refers to the model-derived probability estimates (Eide and Gefeller (1995)). For the multinomial sampling framework Basu and Landis (1995) have recently provided the necessary variance formulae.

6. Discussion

The estimation of attributable risks from epidemiologic data forms an integral part of modern analytical approaches quantifying the relationship between some binary disease variable and exposure factors. Whereas the relative risk may be thought of as indicating the strength of the physiologic effects of an exposure to the disease, the attributable risk combines this information on the individual hazard of the exposure with its prevalence in the population under study. Thus, the attributable risk is the appropriate epidemiologic risk measure for evaluating the public health importance of an exposure factor for the disease load of a population and should guide health administrators to rational planning of prevention campaigns.

Statistical methods for estimating the attributable risk in a variety of design situations have been developed in the methodologic literature. Especially the multifactorial situation has attracted considerable research activities, since mostly the disease variable under study exhibits a multifactorial etiology. This paper has briefly reviewed different approaches for estimating adjusted attributable risks which account for potential confounding and/or effect-modification when quantifying the attributable risk of the exposure(s) of primary interest in a multifactorial design. The model-based techniques offer the greatest flexibility in practical applications and are thus the most promising candidates for further research. Alternative recent concepts for defining attributable risks in this case, such as the "sequential" and the "average (partial)" attributable risk, have not been considered here, but nevertheless they deserve attention in methodologic-oriented and applied epidemiology (Eide and Gefeller (1995)).

A final word of caution relating to the interpretation of attributable risk estimates in epidemiologic practice has to be added. The calculation of an attributable risk in a given study revealing some positive value for AR, say 50%, does by no means guarantee that one half of the observed cases in the study population can be causally attributed to the exposure of interest and are "saved" when the exposure is eliminated from the population, even if all obstacles of the statistical calculation have been properly taken. Unidentified confounding variables, measurement errors in the assessment of exposure and disease status, selective participation of eligible study subjects in the investigation, and a wealth of other problematic issues can render the calculated estimate rather useless. The particular problem of the interpretation of attributable risk estimates results from the hardly solvable linguistic dilemma that in each simple description of the estimates it cannot be avoided to implicitly anticipate a causal relationship by using such phrases as "attributable/due to". Consequently, statisticians have to be very cautious in communicating their results of attributable risk calculations to physicians and other public health professionals.

References:

BASU, S. and LANDIS, J. R. (1995): Model-based estimation of population attributable risk under cross-sectional sampling. *American Journal of Epidemiology, 142, 1338-1343.*

BENICHOU, J. (1991): Methods of adjustment for estimating the attributable risk in case–control studies: a review. *Statistics in Medicine, 10, 1753-1773.*

BROWNER, W. S. (1986): Estimating the impact of risk factor modification programs. *American Journal of Epidemiology, 123, 143–153.*

BRUZZI, P.; GREEN, S. B.; BYAR, D. P. et al. (1985): Estimating the population attributable risk for multiple risk factors using case–control data. *American Journal of Epidemiology, 122, 904–914.*

CHYOU, P. H.; NOMURA, A. M. Y. and STEMMERMANN, G. (1992): A prospective study of the attributable risk of cancer due to cigarette smoking. *American Journal of Public Health, 82, 37–40.*

EIDE, G. E. and GEFELLER, O. (1995): Sequential and average attributable fractions as aids in the selection of preventive strategies. *Journal of Clinical Epidemiology, 48, 645-655.*

EJIGOU, A. (1979): Estimation of attributable risk in the presence of confounding. *Biometrical Journal, 21, 155-165.*

GEFELLER, O. (1992a): An annotated bibliography on the attributable risk. *Biometrical Journal, 34, 1007-1012.*

GEFELLER, O. (1992b): Comparison of adjusted attributable risk estimators. *Statistics in Medicine, 11, 2083-2091.*

GREENLAND, S. (1984): Bias in methods for deriving standardized morbidity ratio and attributable fraction estimates. *Statistics in Medicine, 3, 131-141.*

GREENLAND, S. (1987): Variance estimators for attributable fractions estimates consistent in both large strata and sparse data. *Statistics in Medicine, 6, 701-708.*

LEUNG, H. M. and KUPPER, L. L. (1981): Comparisons of confidence intervals for attributable risk. *Biometrics, 37, 293–302.*

LEVIN, M. L. (1953): The occurrence of lung cancer in man. *Acta Unio Internationalis Contra Cancrum, 9, 531-541.*

MIETTINEN, O. S. (1974): Proportion of disease caused or prevented by a given exposure, trait or intervention. *American Journal of Epidemiology, 99, 325–332.*

WALTER, S. D. (1976): The estimation and interpretation of attributable risk in health research. *Biometrics, 32, 829–849.*

WHITTEMORE, A. S. (1982): Statistical methods for estimating attributable risk from retrospective data. *Statistics in Medicine, 1, 229–243.*

Neural Networks for Classification of Image Data in Quantitative Pathology

C. P. Janson, D. D. Komitowski

Abteilung Histodiagnostik und Pathomorphologische Dokumentation, Deutsches Krebsforschungszentrum, D-69120 Heidelberg, Germany

Summary: Computerized image analysis in tumor grading offers new possibilities for the decision support in diagnosis. But one of the major problems remains the selection of the optimal classifier for the quantitative data. From this point of view artificial neural networks were examined, particularly after architecture selection. Optimized nets were tested on several data sets and compared to statistical classification methods. Concerning generalization ability, neural nets did not surpass statistical procedures.

1. Introduction

The assessment of cancer malignancy is based primarily on TNM staging and histologic grading. The histologic diagnosis is routinely done by skilled pathologists in microscopical investigations of the tumor specimen. This method, however, depends on individual human experience and subjective judgements. Recently, systems for computerized analysis of microscopic images provide methods to support the diagnostic decision by quantitative reproducible measurements. The tumor grading based on those quantitative features can be regarded as a classification task. The selection of the most suitable classifier is an important part of it. As one is interested in correctly classifying future tumor cases, the optimum classifier is the one with highest generalization ability.

Artificial neural networks arise as a new class of flexible nonlinear classifiers. The objective of this study was to investigate the suitability of feedforward neural networks for classification, particularly in quantitative histodiagnostics.

Concerning the model equations of feedforward neural networks, there is strong relationship to nonlinear regression methods (Ripley (1994), Sarle (1994), Schumacher et al. (1994)). Empirical investigations comparing statistical methods with neural nets failed to prove the superiority of neural net classification concerning generalization ability (Dawson et al. (1991), Palcic et al. (1992), Reibnegger et al. (1991), Ripley (1994)). Up to now such comparative studies paid little attention to the selection of optimal network architecture. Architecture denotes the number and location of nodes and connections within a network and can be considered in the sense of model selection.

In this contribution we report on two ways of architecture selection, the sequential network construction (SNC) of Moody (1994), and the Cascade-

Correlation of Fahlman and Lebiere (1990). They are applied to three different data sets and the results compared to those of statistical classifiers.

2. Methods for Architecture Selection

2.1 Sequential Network Construction

In the sequential network construction by Moody (1994) a hierarchical sequence of fully connected networks of increasing number of hidden nodes is developed. Only nets with one hidden layer are considered.

Before starting the construction, a selection criterion has to be defined. As suggested by Utans and Moody (1991), Akaike's final prediction error FPE can be used as a measure of generalization ability (Akaike (1970)). It is defined by

$$FPE = MSE(\lambda)(\frac{1 + \frac{Q(\lambda)}{N}}{1 - \frac{Q(\lambda)}{N}}) \tag{1}$$

where λ denotes the (net-)model under consideration, MSE the mean squared error, $Q(\lambda)$ the number of free parameters in the model, and N the number of samples used for parameter estimation.

The algorithm for SNC is as follows:

Start Generation of the starting network, with one or few hidden units. As usual, all weights in the net are initialized by random values.

Iteration

Step 1 Training of the net, the learning function with corresponding parameters as well as the number of training cycles must have been determined previously. In our case, backpropagation learning with momentum term was used.

Step 2 The selection criterion is computed and the trained network stored.

Step 3 One (or more) new hidden units are added to the net. The weights of existing connections are taken on, new weights randomly initialized.

The iteration stops when a net of some predetermined size has been computed.

Out of the resulting sequence of networks, the one with best values of selection criterion is choosen, which means the net with lowest FPE.

Supplementary, weight decay was used to eliminate single connections: by adding a cost-term to the error function of the network, the weights tend to decrease during learning. Their connections are cut, if their values fall below a threshold (Werbos (1988)).

2.2 Cascade-Correlation

The Cascade-Correlation learning architecture (Fahlman and Lebiere (1990)) is believed to build the optimal network architecture during training. It works according to the following algorithm:

Start The starting network has no hidden layer. Input and output layer are fully connected, all weights initialized by random values.

Iteration

Step 1 Training of the net until no significant error reduction occurs within a certain number of training cycles.

Step 2 A candidate unit is created with trainable connections with all input nodes and with all pre-existing hidden nodes.

Step 3 The weights leading to the candidate unit are adapted by maximizing the correlation between the output of the candidate unit and the residual net output error.

Step 4 Finally the candidate unit is included as a new hidden unit and is connected with the output units. The connections leading to the candidate unit are 'frozen', that means their weights are fixed.

The algorithm stops if the net error falls below a predetermined threshold.

Cascade-Correlation is characterized by its cascade-like architecture where each new hidden unit represents a hidden layer, fully connected with input and output, and by the use of linear correlation. As the connections to the hidden units are frozen after training, no error backpropagation is necessary and training in step 1 can be done quickly.

The backpropagation learning with momentum term, computation of cascade correlation, and weight decay was done using the SNNS software (Zell et al. (1995)).

3. Empirical Studies

The investigations of architecture-optimized neural nets and statistical classifiers were performed with three different data sets. The first two sets were intensively studied by B. Ripley (1994), such that his reliable classification results could be used for comparisons. In order to be consistent with those results we took the same training and testing sets.

3.1 Sonar Data

Originally generated by Gorman and Sejnowski (1988), the sonar data consist of 208 sets from a sonar study. As there were 60 measurements for discrimination of two classes of sonar targets, the neural nets had 60 input nodes and 2 output nodes.

SNC was repeated 10 times for 1 to 10 hidden nodes and FPE computed. The weights were randomly initialized by values $x \in (-0.25; 0.25)$ before each run. Figure 1 shows the 10 FPE-curves as results of SNC.

The differences between the curves demonstrate the variability of SNC. The reason for this must be the random initialization of the weights: depending

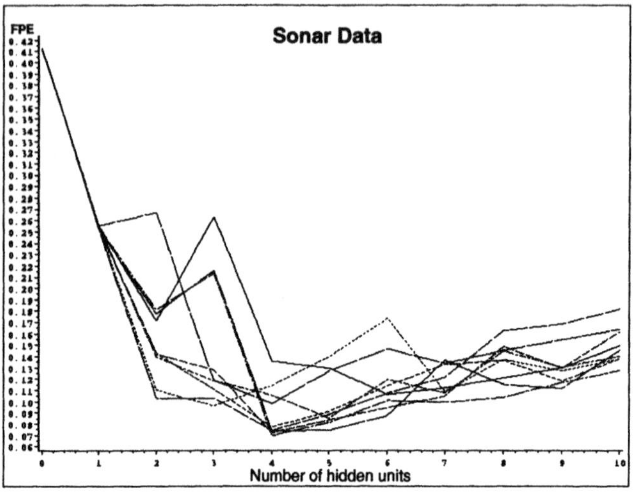

Figure 1: Results of SNC for sonar data: curves of FPE as function of number of hidden units for 10 repeats

on these initial weights, the learning algorithm starts from different points on hypersurface of the error function. Thus the gradient descent might converge to different local minima of the error function.

In most curves, FPE was minimal with 4 hidden units. Thus a net architecture with 4 hidden units was further investigated in the crossvalidation. Different net models were studied, for each the crossvalidation was repeated 10 times.

The mean results of classification are shown in table 1, additionally Ripley's neural net results and best results of statistical classifiers are given.

The lowest errors were reached by the nearest neighbor classification method. The 'optimized' nets with 4 hidden units did not yield the expected generalization ability, nor did the cascade correlation nets. Better results were obtained by arbitrarily choosen 11 hidden units.

3.2 Forensic Glas Data

For classification of forensic glass data (original source: B. German, Central Research Establishment, Home Office Forensic Science Service, Aldermaston, Reading, Berkshire RG7 4PN) the nets had 9 input and 4 output units, according to the number of independents and dependents. The optimization and classification procedure were the same as described above (section 3.1). There was also some variability in repeated runs of SNC. In contrast to the SNC results of 3.1, no clear minimum for FPE could be identified (Fig. 2). Because the curves approached minimal values at 18 hidden units, neural nets with that hidden layer size were considered in the further investigation.

Classifier	No. falsely classified
NN, 4 HU	15.4
NN, 4 HU, Weight Decay	17.2
NN, 11 HU	11.2
NN, 11 HU, Weight Decay	16.9
Cascade-Correlation	20.7
Results of Ripley:	
NN, 2 HU	17
NN, 6 HU	12
NN, 12 HU	11
Nearest Neighbor after rescaling	4
Nearest Neighbor	9

Table 1: Results of crossvalidations, sonar data, mean number of falsely classified test observations over 10 runs, 104 training-, 102 test-sets, NN-Neural Network, HU - No. hidden units

Crossvalidation results are given in tab. 2. The mean number of falsely classified cases was lowest for the nearest neighbor method and for classification and regression trees. Neural Nets with 18 hidden units and cascade correlation nets did not yield much worse classification of test data.

3.3 Image Data of Quantitative Pathology

For the quantitative assessment of histologic grading in pathology, methods of image analysis have been developped. They include recording, segmentation, and feature extraction of microscopic images of tissue sections. The measurements yielded 40 quantitative features concerning structural characteristics of tumor tissue such as growth patterns, glandular organization, and spatial distribution of cell nuclei (Janson (1996)).

The methods were applied to 50 prostatic adenocarcinoma. The classification task consisted in discriminating the tumor cases on the basis of that image data according to their histologic grade.

The results of SNC for 10 trials (fig. 3) again showed high variability. A number of 4 hidden units was considered in the further investigation.

Because of the small sample size of 50 cases, instead of the learning set/test set method the 'leave-one-out' crossvalidation (Lachenbruch and Mickey (1968)) was used to estimate the generalization ability of different classifiers (tab. 3). The nearest neighbor classifier yielded the lowest crossvalidation

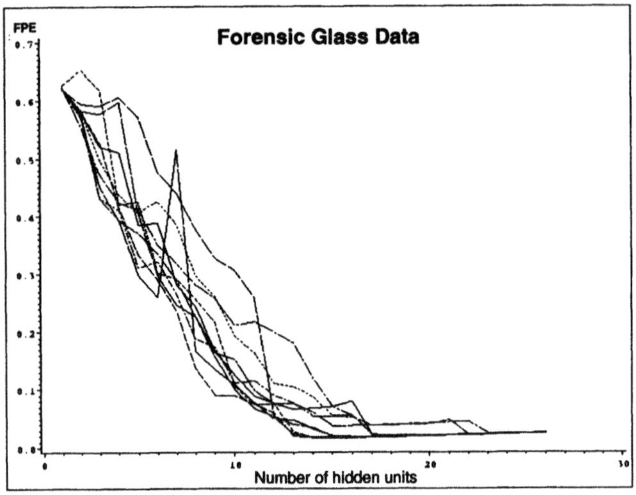

Figure 2: Results of SNC for glass data, curves of FPE as function of hidden units

Classifier	no. falsely classified
NN, 18 HU	30.2
NN Weight Decay, 18 HU	44.8
Cascade-Correlation	30.0
Results of Ripley:	
NN, 2 HU	38
NN, 6 HU	33
Nearest Neighbor	26
Classification and Regression Tree	28

Table 2: Results of crossvalidations, forensic glass data, mean number of falsely classified test observations over 10 runs, 89 training-, 96 test-sets, NN- Neural Network, HU - No. hidden units

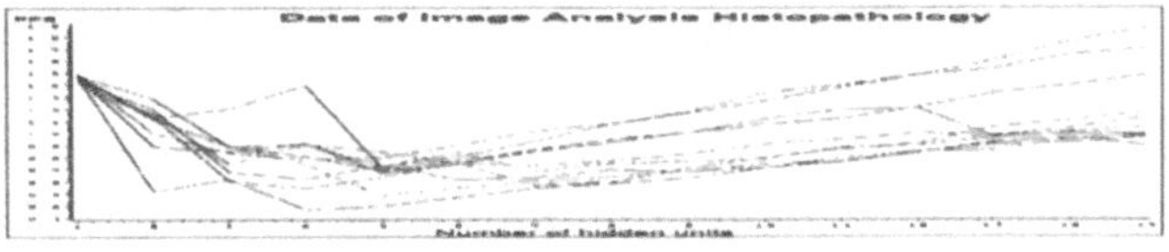

Figure 3: Results of SNC for histologic image data: curves of FPE as function of number of hidden units for 10 repeats

Method	% falsely classified
Linear discriminant analysis	47.0
Quadratic discriminant analysis	62.0
Nearest neighbor	12.5
NN, 20 HU	57.14
NN, 4 HU	62.55
Cascade-Correlation	61.70

Table 3: Results of leave-one-out crossvalidations, image data of histodiagnostics, NN- Neural Network, HU - No. hidden units

error of 12.5%. Linear and quadratic discriminant analysis as well as all neural net models resulted in high error rates of more than 50%.

4. Discussion

Despite architecture selection of neural networks, their generalization ability could not be increased in our studies. The performance of optimized nets was not better than of those with arbitrarily choosen architecture.

Applying neural nets for classification purposes one is faced with a number of questions. These concern variability due to the random initialization of weights, local minima of the error functions, choice of error function and learning algorithm as well as the values of parameters, such as learning rate, momentum term, weight decay term, and number of learning cycles in connection with overtraining (Schumacher et al. (1994), Ripley (1994)).

Problems of estimating the true classification error and of sample size affect both neural nets and statistical procedures. For error estimation, we used Akaike's FPE, learning-set/test-set and leave-one-out crossvalidation. Strictly speaking, FPE is unbiased asymptotically, so it was only used in SNC for its ease of computation.

The required sample size for neural nets is a multiple of the number of free parameters, which corresponds to the number of weights. As Ripley (1994) pointed out, one will e.g. need a multiple of ten, if the difference between true classification error and estimated error should be about 10%. These requirements can hardly be fulfilled in many studies and were a problem in our studies too.

This investigation concentrated solely on feedforward nets with one hidden layer or cascade architecture. There is a number of other network models whose classification performance might be investigated.

Apart from classification tasks, special neural networks are successfully used in robot control, speech recognition and in the field of image analysis for image recognition, restoration or segmentation (Rojas (1993)).

The nearest neighbor classifier, a classic and rather simple method, revealed the lowest error rates in all examples. These results depend on the data structure. In the two-spiral problem for instance (Fahlman and Lebiere (1990)), that classifier would presumably not be successful. But the question arises whether such 'artificial' problems exist in medical applications.

References:

AKAIKE, H. (1970): Statistical Predictor Identification. *Ann Inst Statist Math, 22, 203-217*

DAWSON, A. E.; AUSTIN, R. E. Jr. and WEINBERG, D. S. (1991): Nuclear Grading of Breast Carcinoma by Image Analysis - Classification by Multivariate and Neural Network Analysis. *Am J Clin Pathol, 95 (Suppl 1), 29-37*

FAHLMAN, S. and LEBIERE, C. (1990): The Cascade Correlation Learning Architecture. In: D.S. Touretzky (ed.): *Advances in Neural Processing Systems 2* Morgan Kaufman, San Mateo, 524-532

GORMAN, R. and SEJNOWSKI, T. (1988): Analysis of Hidden Units in a Layered Network Trained to Classify Sonar Targets. *Neural Networks, 1, 75-89*

JANSON, C. P. (1996): Bildanalysemethoden und mathematische Klassifikationsverfahren zum histologischen Grading des Prostatakarzinoms. Doctor Thesis, Universität Heidelberg

LACHENBRUCH, P. and MICKEY, T. (1968): Estimation of Error Rates in Discriminant Analysis. *Technometrics, 10, 1-11*

MOODY, J. (1994): Prediction Risk and Architecture Selection for Neural Networks. In: V. Cherkassky, J.H. Friedman, and H. Wechsler (eds.): *From Statistics to Neural Networks - Theory and Pattern Recognition Application.* Springer; Berlin, Heidelberg, New York, 147-165

PALCIC, B.; MACAULAY, C.; SHLIEN, S.; TREURNIET, W.; TEZCAN, H. and ANDERSON, G. (1992): Comparison of Three Different Methods for Automated Classification of Cervical Cells. *Analytical Cellular Pathology, 4, 429-441*

REIBNEGGER, G.; WEISS, G.; WERNER-FELMAYER, G.; JUDMAIER, G. and WACHTER, H. (1991): Neural Networks as a Tool for Utilizimg Laboratoty

Information: Comparison with Linear Discriminant Analysis and with Classification and Regression Trees. Proc. Natl. Acad. Sci. USA (Medical Sciences), 88, 11426-11430

RIPLEY, B. (1994): Neural Networks and Related Methods for Classification. *Journal of Royal Statistical Society B, 56 (3), 409-456*

ROJAS, R. (1993): Theorie der neuronalen Netze - eine systematische Einführung. Berlin, Heidelberg, New York : Springer

SARLE, W. (1994): Neural Networks and Statistical Methods. Proc. 19th Annual SAS Users Group Int. Conference

SCHUMACHER, M.; ROSSNER, R. and VACH, W. (1994): Neural Networks and Logistic Regression. Part I and Part II. Institut f. Medizinische Biometrie und Informatik, Universität Freiburg i. Breisgau, Germany

UTANS, J. and MOODY, J. (1991): Selecting Neural Network Architectures via the Prediction Risk: Application to Corporate Bond Prediction. First International Conference on Artificial Intelligence Applications on Wall Street, Los Alamitos, Ca, IEEE Computer Society Press

WERBOS, P. (1988): Backpropagation: Past and Future. IEEE Int. Conf. on Neural Networks, 1. New York : IEEE Press, 343-353

ZELL, A.; MAMIER, G.; VOGT, M. et al. (1995): SNNS Stuttgart Neural Network Simulator, User Manual, Version 4.0. Report 6/95, Universität Stuttgart

Variations on the Shapley Solution for Partitioning Risks in Epidemiology

M. Land, O. Gefeller

Abteilung Medizinische Statistik
Georg-August-Universität, D-37073 Göttingen, Germany

Summary: The attributable risk is an epidemiologic measure that quantifies the percentage of cases of disease that can be explained by an exposure under study. In the case predominantly found in epidemiologic practice when several interrelated exposures affect a disease it is of particular interest to apportion the joint attributable risk among the factors. This problem is formally analogous to the division of the profit that several players gain by acting together in a grand coalition. It is shown how game theoretic results on "fair" allocation rules can be used to develop a "reasonable" procedure of partitioning joint risks in epidemiology.

1. Introduction

In many epidemiologic situations it is of particular interest to quantify the proportion of cases of disease attributable to an exposure under study. Levin (1953) approached this problem by introducing a new epidemiologic measure, nowadays often referred to as *attributable risk* (AR), to quantify the "proportion of lung cancer attributable to smoking". It is defined as

$$AR = \frac{P(D) - P(D|\overline{E})}{P(D)}, \tag{1}$$

where $P(D)$ and $P(D|\overline{E})$ denote the probability of disease among the entire population and the unexposed subpopulation, respectively. The attributable risk can be interpreted as the percentage of cases of disease hypothetically preventable by the total elimination of the exposure in the population.

For practical purposes, however, it is particularly interesting to consider this risk concept in multifactorial situations as only they realistically describe the complex interrelations of several exposures affecting the disease. In order to assess the public health effect of one exposure of prime interest in the presence of confounding and/or effect-modification the attributable risk has to be adjusted as

$$AR_{adjusted} = \frac{P(D) - \sum_{i=1}^{K} P(C_i)P(D|C_i \cap \overline{E})}{P(D)}, \tag{2}$$

(Benichou (1991); Gefeller (1992)) where the stratum variables $C_1, \ldots, C_K$ denote all combinations of any possible values of the confounding or effect modifying factors, $P(C_i)$ represents the proportion of the population in the i-th stratum C_i, $P(D|C_i \cap \overline{E})$ refers to the stratum-related probability of disease among the unexposed. The adjusted attributable risk can again be interpreted as the proportion of cases of disease hypothetically preventable by the total elimination of exposure E, where it is assumed that the distribution of risk factor combinations $C_1, \ldots, C_K$ is not changed by erasing the exposure E from the population under study.

This approach is thus based on the implicit assumption of a hierarchy among the risk factors that affect the disease. Actually, the risk factors that determine the strata $C_1, \ldots, C_K$ are relevant only for adjustment. In many epidemiologic situations the adjustment of the attributable risk constitutes an appropriate procedure of measuring the public health effect of one exposure of particular interest. In those multifactorial designs, however, that require equal-ranking treatment of several exposures, say the exposures $E_1, \ldots, E_n$, none of the established methods of estimating the adjusted attributable risks $AR(E_i)$ of any exposure E_i leads to a suitable procedure of quantifying the contribution of a single factor to the joint attributable risk $AR(E_1, \ldots, E_n)$ of all variables.

One possibility of evaluating the effect of one factor on the joint attributable risk $AR(E_1, \ldots, E_n)$ is to partition it additively into components $(r_1, \ldots, r_n)$. For example, the definition

$$\begin{aligned} r_1 &= AR(E_1)\,, \\ r_i &= AR(E_1, \ldots, E_i) - AR(E_1, \ldots, E_{i-1})\,, \quad \text{if } i > 1, \end{aligned} \tag{3}$$

where $AR(E_1, \ldots, E_j)$ denotes the joint attributable risk of the exposures $E_1, \ldots, E_j$ (possibly adjusted for $E_{j+1}, \ldots, E_n$), yields an additive partition of $AR(E_1, \ldots, E_n)$. Suppose, for instance, the attributable risks of two exposures A and B are $AR(A) = 0.2$, $AR(B) = 0.4$ and $AR(A, B) = 0.5$, respectively. Whenever E_1 denotes exposure A the vector $(r_1, r_2) \hat{=} (r_A, r_B)$ is $(0.2, 0.3)$. In the case of E_1 denoting exposure B, however, the vector $(r_1, r_2) \hat{=} (r_B, r_A)$ changes to $(0.4, 0.1)$. Hence, equation (3) defines a partition that unfortunately depends on the arbitrarily selected numbering of exposures.

This paper shows how mathematic game theory can be used to find a suitable procedure of additively dividing joint attributable risks into components for each factor that is independent of the order of factors and fulfils other reasonable conditions. In preparation of this result a transfer of the game theoretic terminology to epidemiologic needs will be given in the following section.

2. Risk Allocation Functionals

2.1 The Linear Space of Risk Functions

Let a *risk function* be any mapping R from $\{E | E \subset \{E_1, \ldots, E_n\}\}$ into the real line $\mathbb{R}$ that assigns to every subset E of a set $\mathcal{E} = \{E_1, \ldots, E_n\}$ of exposure variables under study its real-valued "risk" $R(E)$ and satisfies

$R(\emptyset) = 0$. For example, the risk function might be defined so that $R(E)$ denotes the joint attributable risk of all exposures E_i that are included in the set E, possibly adjusted for those exposures that are not included. In that case, the condition $R(\emptyset) = 0$ simply expresses that the percentage of cases of disease cannot be changed if no exposures are eliminated.

As one can prove easily, the set $\mathcal{R}_\mathcal{E}$ of all risk functions, given the set $\mathcal{E}$ of exposures under study, forms a linear space of dimension $2^n - 1$, and the *simple functions*, defined as

$$S_H(E) = \begin{cases} 1 & \text{if} \quad H \subset E \\ 0 & \text{if} \quad H \not\subset E \end{cases}, \quad E \subset \{E_1, \dots, E_n\}, \tag{4}$$

where H is any nonempty subset of $\{E_1, \dots, E_n\}$, form a basis of $\mathcal{R}_\mathcal{E}$ that in the following will be called the *simple basis* of $\mathcal{R}_\mathcal{E}$.

2.2 Risk Allocation Functionals

A *partition of the joint risk* $R(E_1, \dots, E_n)$, given a risk function R and a set $\{E_1, \dots, E_n\}$ of exposure variables, is any n-dimensional vector $(r_1, \dots, r_n)$ with real-valued components that sum up to $R(E_1, \dots, E_n)$. Furthermore, let a *risk allocation functional* be any mapping Φ defined on the space $\mathcal{R}_\mathcal{E}$ of all risk functions that assigns to each risk function R a partition $\Phi(R) = (\Phi_1(R), \dots, \Phi_n(R))$ of the joint risk $R(E_1, \dots, E_n)$.

Several examples of risk allocation functionals can be obtained by a generalization of the increments in risk as defined in equation (3). Suppose π is any permutation of the integers $1, \dots, n$. Then the functional Φ^π defined as $\Phi_i^\pi(R) = R\left(\{E_{\pi(1)}, E_{\pi(2)}, \dots, E_i\}\right) - R\left(\{E_{\pi(1)}, E_{\pi(2)}, \dots, E_i\} \setminus \{E_i\}\right)$, where R is any risk function, serves as an example of a risk allocation functional. In the special case of Φ^π being used to partition adjusted attributable risk functions $\Phi_i^\pi(AR)$ was termed *sequential attributable risk* (Eide, Gefeller (1995)).

2.3 Symmetric Risk Allocation Functionals

The functionals Φ^π as defined above are not independent of the order of exposures. Therefore, it is particularly interesting to search for risk allocation functionals that are *symmetric* in the following sense.

Definition 1: A risk allocation functional $\Phi : \mathcal{R}_\mathcal{E} \longrightarrow \mathbb{R}^n$ is *symmetric* if $\Phi_{\pi(i)}(R_\pi) = \Phi_i(R)$ holds for any permutation π of the set $\{1, \dots, n\}$ and any risk function R, whenever R_π is defined for all subsets E of $\{E_1, \dots, E_n\}$ as

$$R_\pi(E) = R(\{E_i | E_{\pi(i)} \in E\}). \tag{5}$$

Whenever the numbering of exposures changes, symmetry demands of a risk allocation functional to exchange its components accordingly without affecting their explicit numerical representation. In that sense, a symmetric risk allocation is actually independent of the order of exposures.

2.4 The Principle of Marginal Rationality

In the following, the increments in risk, defined as

$$\Delta_{i,E}(R) = \begin{cases} R(E) - R(E \setminus \{E_i\}) & \text{if} \quad E_i \in E \\ R(E \cup \{E_i\}) - R(E) & \text{if} \quad E_i \notin E \end{cases} , E \subset \mathcal{E}, \tag{6}$$

are called the *marginal risks of factor* E_i. They describe the additive change in risk if an arbitrary collection of exposures that does not include factor E_i is extended by the additional consideration of this formerly excluded exposure. With regard to this nomenclature, the r_i as defined in equation (3) simply turn out to be marginal attributable risks.

Definition 2: A risk allocation functional $\Phi : \mathcal{R}_{\mathcal{E}} \longrightarrow \mathbb{R}^n$ is *marginally rational* if with regard to all risk functions $R_a \in \mathcal{R}_{\mathcal{E}}$ and $R_b \in \mathcal{R}_{\mathcal{E}}$ as well as any index $i \in \{1, \ldots, n\}$, the following implication is valid:

$$\Delta_{i,E}(R_a) \leq \Delta_{i;E}(R_b) \text{ for all } E \subset \mathcal{E} \quad \Longrightarrow \quad \Phi_i(R_a) \leq \Phi_i(R_b). \tag{7}$$

In order to understand the definition above, let R_a and R_b denote risk functions that describe two different strata A and B of a population under study. Suppose that the increase in risk is never smaller in stratum A than in stratum B whenever an arbitrary collection E of exposure variables is extended by the additional exposure E_i, then no smaller risk should be allocated to factor E_i in stratum A than in stratum B. A marginally rational risk allocation functional guarantees a "rational" quantification of the influence of the same exposure factor for different strata in a population whenever "rationality" is understood in a way as described above.

3. Partial Risks

The risk allocation functionals Φ^π are marginally rational but they are not symmetric. Averaging these functionals among all possible permutations π of the indices $1, \ldots, n$, however, leads to a symmetric and marginally rational solution. Moreover, the following theorem (Young (1985)) states that it is the only solution with these properties.

Theorem 1: The *partial risk* $P(R) = (P_1(R), \ldots, P_n(R))$ whose i-th component is defined as

$$P_i(R) = \frac{1}{n!} \sum_{\substack{\pi \text{ Permutation} \\ \text{der Zahlen } 1,\ldots,n}} \Phi_i^\pi(R) \quad , R \in \mathcal{R}_{\mathcal{E}}, \tag{8}$$

is the only n-dimensional, symmetric, marginally rational risk allocation functional on the space $\mathcal{R}_{\mathcal{E}}$, given a set $\mathcal{E} = \{E_1, \ldots, E_n\}$ of exposures. An equivalent representation of the partial risk is

$$P_i(R) = \sum_{E \subset \{E_1,\ldots,E_n\}: E_i \in E} \frac{(|E|-1)!\,(n-|E|)!}{n!} \Delta_{i,E}(R), \tag{9}$$

where $|E|$ denotes the cardinality of the set E.

Proof: The fact that P is symmetric and that the components of $P(R)$ sum up to $R(\{E_1, \ldots, E_n\})$ was already pointed out by Shapley (1953). Furthermore, P is obviously marginally rational. In order to prove the uniqueness of P the statement

$$\Phi_i(R) = P_i(R) \quad \text{for all } R \in \mathcal{R}_\mathcal{E} \tag{10}$$

has to be proved with regard to any symmetric and marginally rational risk allocation functional Φ and to any exposure E_i.

Let $c_H(R)$, $\emptyset \neq H \subset \mathcal{E}$, be the coefficients of a risk function $R \in \mathcal{R}_\mathcal{E}$ that are uniquely determined by the representation $R = \sum_{\emptyset \neq H \subset \mathcal{E}} c_H(R) S_H$ of R with regard to the simple basis of $\mathcal{R}_\mathcal{E}$ (see equation (4)) and let the *complexity of* R (in signs $\|R\|$) be the number of non-zero coefficients $c_H(R)$. Then equation (10) can be proved by induction with regard to $\|R\|$.

For symmetry, $\Phi_i(R) = 0 = P_i(R)$,$\forall i \in \{1, \ldots, n\}$, holds whenever R has complexity 0. Now suppose that equation (10) has already been proved for all risk functions with complexity $c < 2^n - 1$ and let R be any risk function with complexity $c+1$. With respect to every exposure E_i that is not included in the set $H_0(R) = \bigcap_{H \subset \mathcal{E}: c_H(R) \neq 0} H$ the risk function $R_i = \sum_{E_i \in H \subset \mathcal{E}} c_H S_H$ has complexity $\|R_i\| \leq c$ and satisfies $\Delta_{i,E}(R_i) = \Delta_{i,E}(R)$ for all $E \subset \mathcal{E}$. Marginal rationality of Φ and P implies $\Phi_i(R_i) = \Phi_i(R)$ and $P_i(R_i) = P_i(R)$. Finally, by induction hypothesis $\Phi_i(R) = \Phi_i(R_i) = P_i(R_i) = P_i(R)$.

Assume that $H_0(R)$ includes two different exposures, say E_j and E_k. It is easily proved that the permutation π_{jk} which exchanges j for k and leaves any other index unchanged induces a risk function $R_{\pi_{jk}}$ (compare equation (5)) that satisfies $R = R_{\pi_{jk}}$. Hence, $\Phi_j(R) = \Phi_{\pi_{jk}(k)}(R_{\pi_{jk}}) = \Phi_k(R)$ and $P_j(R) = P_{\pi_{jk}(k)}(R_{\pi_{jk}}) = P_k(R)$ because of symmetry.

Furthermore, the components of Φ and P sum up to the same value and are equal with respect to any exposure that is not included in $H_0(R)$. Therefore, equation (10) holds with respect to every index $i \in \{1, \ldots, n\}$ and every risk function $R \in \mathcal{R}_\mathcal{E}$. □

Remark: The preceding proof only requires the assumption that the partial risk of exposure E_i depends exclusively on the marginal risks of factor E_i, which means that implication (7) is valid when both inequalities were replaced by equality as Young pointed out in a game theoretic terminology. Hence, it is proved simultaneously that the partial risk is the only risk allocation functional that is symmetric and whose i-th component P_i depends only on the marginal risks of factor E_i.

The remark implies also that partial risks should only be applied if the marginal risks are meaningful in the present epidemiologic context.

Based on the "translation" of a game theoretic theorem proved by Owen (1972) the following interpretation of partial risks can be derived as a further step in elucidating the nature of the process of partitioning risks in epidemiology:

Theorem 2: Let $i \in \{1, \ldots, n\}$ be an index and let $R \in \mathcal{R}_\mathcal{E}$ be a risk function. With regard to each $\alpha \in [0;1]$ let $P_\alpha(E) = \alpha^{|E|}(1-\alpha)^{n-1-|E|}$ define a probability distribution on the system of all subsets of $\{E_1, \ldots, E_n\} \setminus \{E_i\}$. Furthermore, let $E_\alpha(\Delta_{i,E}(R))$ denote the expected marginal risk with regard to P_α. Then $P_i(R) = \int_0^1 E_\alpha(\Delta_{i,E}(R))\ d\alpha$.

Because of Theorem 2 the partial risks simply turn out to be the expected additional risk caused by the entering of exposure E_i to a random collection E of exposures, where the exposures that are different from E_i are assumed to join the random collection independently and with the same but completely unknown entering probability α.

3.1 The Shapley Value

In a game theoretic context Shapley (1953) established the functional P as a unique method of dividing the profit that several players are guaranteed to gain by acting together in a grand coalition. Following the game theorists von Neumann and Morgenstern (1944) he described an n-person game completely by a *characteristic function*, which is any positive and real-valued function defined on the system of all subsets of $\{E_1, \ldots, E_n\}$ that fulfils $v(E \cup F) \geq v(E) + v(F)$ whenever E and F are disjoint subsets of $\{E_1, \ldots, E_n\}$ (superadditivity) and $v(\emptyset) = 0$. Within this concept a subset E of the set of all players $\{E_1, \ldots, E_n\}$ can be interpreted as a coalition of players that is able to gain the (minimal) profit $v(E)$. Although Shapley derived his solution by regarding grand coalitions of variable cardinality which are part of an abstract set that he named the "universe of players", his derivation still stays valid whenever the set of players is assumed to be fixed with cardinality n. Shapley proved that the functional P, which in game theory is called *Shapley value*, is the only functional that maps the space of all characteristic functions that are defined on the system of all subsets of $\{E_1, \ldots, E_n\}$ into $\mathbb{R}^n$ and satisfies the following axioms.

(1) *Symmetry axiom:* P is symmetric as defined in Definition 1.

(2) *Additivity axiom:* $P(v_a + v_b) = P(v_a) + P(v_b)$
for all characteristic functions v_a and v_b.

(3) *Dummy axiom:* $P_i(v) = 0$ holds whenever E_i is a dummy-player, which means $v(E \cup \{E_i\}) = v(E)$ for all $E \subset \{E_1, \ldots, E_n\}$.

3.2 Comparison of Axiomatic Approaches

The applicability of game theoretic results to epidemiologic problems was first described by Cox (1985) who compared game theoretic terms such as *player, coalition, characteristic function, imputation, value, dummy player* with epidemiologic terms such as *exposure, collection of exposures, risk function, partition of the joint risk, risk allocation functional, irrelevant exposure.* Cox pointed out that P is the unique risk allocation functional that satisfies the axioms of symmetry, additivity, positivity (which means that P_i never

	Shapley	Cox	Young
Assumptions on risk functions	$R(\emptyset) = 0$ Superadditivity[1)]	$R \geq 0$ Monotonicity [2)] Superadditivity	$R(\emptyset) = 0$
Axioms that lead to P as a unique solution	Symmetry Additivity Dummy-Axiom	Symmetry Additivity Dummy-Axiom Positivity	Symmetry Marginal Rationality
[1)] This assumption can be dropped while the result stays valid [2)] These are the assumptions mentioned by Cox. Note that monotonicity as well as $R(\emptyset) = 0$ follows from superadditivity and $R \geq 0$			

Table 1: Shapley's, Cox' and Young's axioms that lead to P as the only risk

assigns negative values) and the dummy-axiom. This set of axioms, however, demands of the risk functions to be monotone (i.e. $R(E) \leq R(F)$ if $E \subset F$), positive and superadditive. Table 1 summarizes the sets of axioms used by Shapley, Cox and Young that lead to the uniqueness of P and reviews their assumptions on risk functions.

In order to oppose the criticism of arbitrariness the uniqueness of P is desirable, since without this uniqueness the partial risk would only be one of several risk allocation functionals. Therefore, the axioms as well as the conditions for risk functions have to be convincing in the actual epidemiologic situation. Cox's compelling assumptions on risk functions are very restrictive and cannot be applied in every epidemiologic context, whereas Shapley's assumption of superadditivity could even be dropped.

In game theoretic applications the additivity axiom was motivated by the "natural" additivity of profits from successive games. Transferring this axiom to epidemiologic applications, however, is a particularly problematic matter. Summing up attributable risks, for example, is sensible as long as they are calculated with regard to disjoint subpopulations but are based on the entire population as reference. Nevertheless, the sum of attributable risks can not be interpreted in general. In contrast, the comparison of marginal risks as used in the axiom of marginal rationality might be explainable in a wider spectrum of situations.

In order to justify the use of partial risks in a specific epidemiologic context, an appropriate set of axioms combined with well explainable assumptions on risk functions has to be chosen from the table above.

4. Partial Attributable Risks

Eide and Gefeller (1995) pointed out that Cox's derivation can easily be transferred to the problem of apportioning the joint attributable risk among a set of explanatory variables. If $R(E)$ denotes the joint attributable risk of all those exposures that are included in $E \subset \{E_1, \ldots, E_n\}$ (possibly adjusted for those that are not), the marginal risks of any factor can be interpreted well. In accordance with Theorem 2 the partial attributable risk of an exposure E_i can therefore be interpreted as the expected proportion of cases of disease preventable by the additional elimination of exposure E_i after a random collection $\tilde{E} \subset \{E_1, \ldots, E_n\}$ of exposures had already been eliminated in the population.

5. Conclusions

The partial attributable risk turned out to be a useful and well comprehensible tool for assessing the responsibility of one exposure for the joint public health effect of several exposures and might therefore be an interesting source of information for planning intervention strategies. Using partial risks as a procedure of partitioning can be justified by different sets of axioms that have to be chosen according to the actual context. In a wide spectrum of epidemiologic situations, however, symmetry as well as marginal rationality are particularly important features of this method of risk partitioning.

Acknowledgements

This work has been supported by a grant of the Deutsche Forschungsgemeinschaft (grant no. Ge 637/3-1).

References:

BENICHOU, J. (1991): Methods of adjustment for estimating the attributable risk in case-control studies: a review. *Statistics in Medicine, 10, 1753-1773.*

COX, L. A. Jr. (1985): A new measure of attributable risk for public health applications. *Management Science, 31, 800-813.*

EIDE, G. E. and GEFELLER, O. (1995): Sequential and average attributable fractions as aids in the selection of preventive strategies. *Journal of Clinical Epidemiology, 48, 645-655.*

GEFELLER, O. (1992): Comparison of adjusted attributable risk estimators. *Statistics in Medicine, 11, 2083-2091.*

LEVIN, M. L. (1953): The occurrence of lung cancer in man. *Acta Unio Internationalis Contra Cancrum, 9, 531-541.*

OWEN, G. (1972): Multilinear extensions of games. *Management Science, 18, 64-79.*

SHAPLEY, L. S. (1953): A value for n-person games. In: H. Kuhn, A. Tucker (ed.): Contributions to the theory of games, II. *Annals of Mathematics Studies, 28, 307-317.*

VON NEUMANN, J.; MORGENSTERN, O. (1944): Theory of games and economic Behavior. Princeton University Press, Princeton.

YOUNG, H. P. (1985): Monotonic solutions of cooperative games. *International Journal of Game Theory, 14, 65-72.*

Generalized Regression Trees Applied to Longitudinal Nutritional Survey Data

B. Lausen

Forschungsinstitut für Kinderernährung (FKE) Dortmund, Heinstück 11, D-44225 Dortmund, Germany

Summary: The method of classification and regression trees (CART) is one approach to model the relationship between a classification, response or dependent variable to factors or independent variables possibly measured on different scales. The generalization of regression trees is defined via the additive combination of tree models and generalized linear models (cf. Loh, 1991). The response variable Y is p-dimensional and the factor variable X is k-dimensional. I illustrate the approach by an exploratory analysis of calcium excretion data on 24 hour urine samples of a longitudinal study and other (longitudinal) factors, e.g. sex, age, anthropometric measurements, nutritional intake data.

1. Tree based models and prognostic factors

The method of classification and regression trees (CART) (Breiman et al., 1984 or Clark and Pregibon, 1992) can be used as a nonparametric regression model with a tree-like structural component (e.g. Lausen, Sauerbrei and Schumacher, 1994; Lausen and Dirschedl, 1994). The approach is an important model and data analysis tool for prognostic factors in clinical research and epidemiology (e.g. Ulm et al., 1989; Dirschedl, 1991; Sauerbrei et al., 1991; Wernecke, Possinger and Kalb, 1996).

Due to the different numbers of considered cutpoints in each factor, CART tends to a biased variable selection in the situation of factors measured on different scales. Lausen, Sauerbrei and Schumacher (1994) manage this problem by applying results on maximally selected rank statistics (cf. Miller and Siegmund, 1982; Lausen and Schumacher, 1992) and an improved Bonferroni inequality (cf. also Alman et al., 1994; Hilsenbeck and Clark, 1996).

Tree based models are especially of interest in case of longitudinal data (Ciampi, 1991 or Segal, 1992, 1994). Moreover, the approach allows in a simple and straightforward way the incooperation of "expert" knowledge. The expert's opinion and the split criterions numerical values may be used in an interactive way at each node of the tree (e.g. Zhang, Holford and Bracken, 1996).

Lausen, Kersting and Schöch (1995) apply CART in nutritional epidemiology and discuss the advantages and disadvantages of the tree model and two generalized linear models in this application. The combination of tree based models and generalized regression models provides a methodological framework for these alternative data representation. Furthermore, such an additive combination should be a powerful data analysis tool in many applications. Using a subset of a prospective nutritional survey I illustrate this argument in section 3.

2. Residual space tree model (REMODEL)

The generalization of regression trees is defined via the additive combination of tree models and generalized linear models. The response variable Y is p-dimensional and the factor variable X is k-dimensional. Therefore, I state the model in terms of the conditional expectation of Y given $X = x$:

$$E(Y|X = x) = h(x\beta) + tree(x),$$

where h denotes the inverse link function of the generalized linear model, (β the unknown k-dimensional parameter vector and *tree* the hierarchical partition of the k dimensional space defined by the unknown splits (cutpoints) in the factor variables (cf. also Loh, 1991). The implementation of this model is straightforward. I suggest a two step procedure:

1. *Fitting of the generalized linear model;*
2. *Regression tree analysis of the residuals.*

Obviously, this suggestion is a generalization of CART. Recently, one finds many variants, generalizations or improvements in the literature (e.g. Ciampi, 1991; Hecker and Wübbelt, 1992; Lausen, Sauerbrei and Schumacher, 1994; Breiman, 1994). Consequently, I suggest to call the generalization of the tree based model given above: *Residual space tree model (acronym: REMODEL).*

3. Calcium excretion data

The necessity of such a REMODEL approach is illustrated by a preliminary exploratory analysis of calcium excretion data on 24 hour urine samples of a longitudinal study and other (longitudinal) factors, e.g. sex, age, anthropometric measurements, nutritional intake data. The data is a subset of a prospective nutritional survey started 1985 at the FKE on *nutrition, metabolism, growth and development of healthy children.* The data set consists of 1862 calcium excretion measurements in 24 hour urine samples (cf. Fig. 1). The sample were collected from August 1985 to January 1996. For example I analyse in the following a longitudinal sub data set of 175 longitudinal measurements with more than 4 observations. The degree of heterogenity regarding the longitudinal information is shown in Figure 2. These heterogenity implies that standard multivariate techniques or modelings for longitudinal data are not feasible. The heterogenity itself may cause artefacts or biases in a tree based analysis. Consequently, it should be analysed and fitted as possible confounder. I measure the heterogenity in the preliminary data analysis by the variables age at first urine sample and number of the longitudinal measurements of each individual. The univariate distributions of the considered prognostic and confounding factors are shown in Figure 3. The longitudinal information of the considered factors is used via means as preliminary ad hoc approach, which can be interpreted as scaled response integrals.

Using as split criterion the two-sample Wilcoxon statistic and as stop criterion $P_{stop} = 0.01$, the improved CART procedure considers 45 rank statistics processes (cf. Figure 4). The resulting P-CART representation is shown in Figure 5. The P-CART representation shows the size of the subgroups (relative size of branches), the medians of the subgroups (x-axis) and the P-value of the Wilcoxon statistic (y-axis). Using $P_{stop} = 0.1$ I produce a bigger tree (N-CART representation Figure 5) and put more emphasis on the size of the subgroups (relative size of branches and y-axis). Moreover, the N-CART representation avoids problems caused by the possible nonmonotone behaviour of the P-values.

4. Discussion

Figure 1 and 5 suggest that a continuous modeling of a development factor (model term) should be done. The REMODEL approach allows as *first step* a fitting via a generalized linear model (GLM) including for example the factors age, body height, body weight and sex. Afterwards, these factors should be included in the *second step* of the REMODEL analysis. The tree based analysis of the residuals considers possible interaction terms, step function approximations of physiological models, search for homogeneous subgroups, etc.. But the recursive strategy of tree based models causes problems similar to the well known problems of forward selection procedures in variable subset selection of regression analysis (e.g. Breiman, 1995). Overinterpretation and instability of tree based results were recognized, discussed and solved to some extent regarding CART (e.g. Breiman et al., 1984; Ciampi, 1991; Clark and Pregibon, 1992; Lausen, Sauerbrei and Schumacher, 1994).

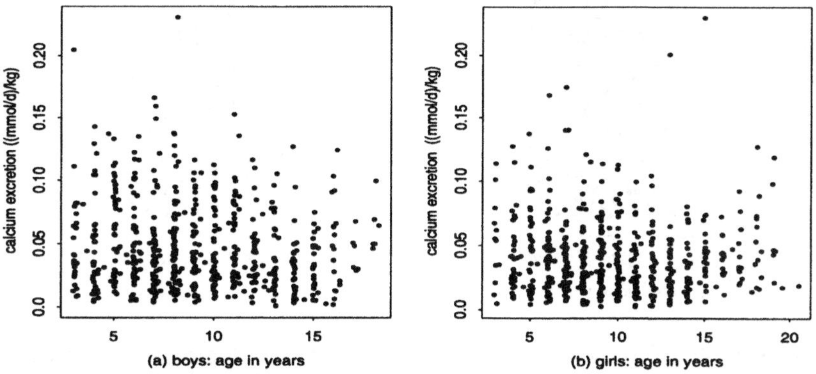

Figure 1: Age in years (Tanner decimal representation) vs. calcium excretion ((mmol/d)/kg) (boys (a), girls (b)).

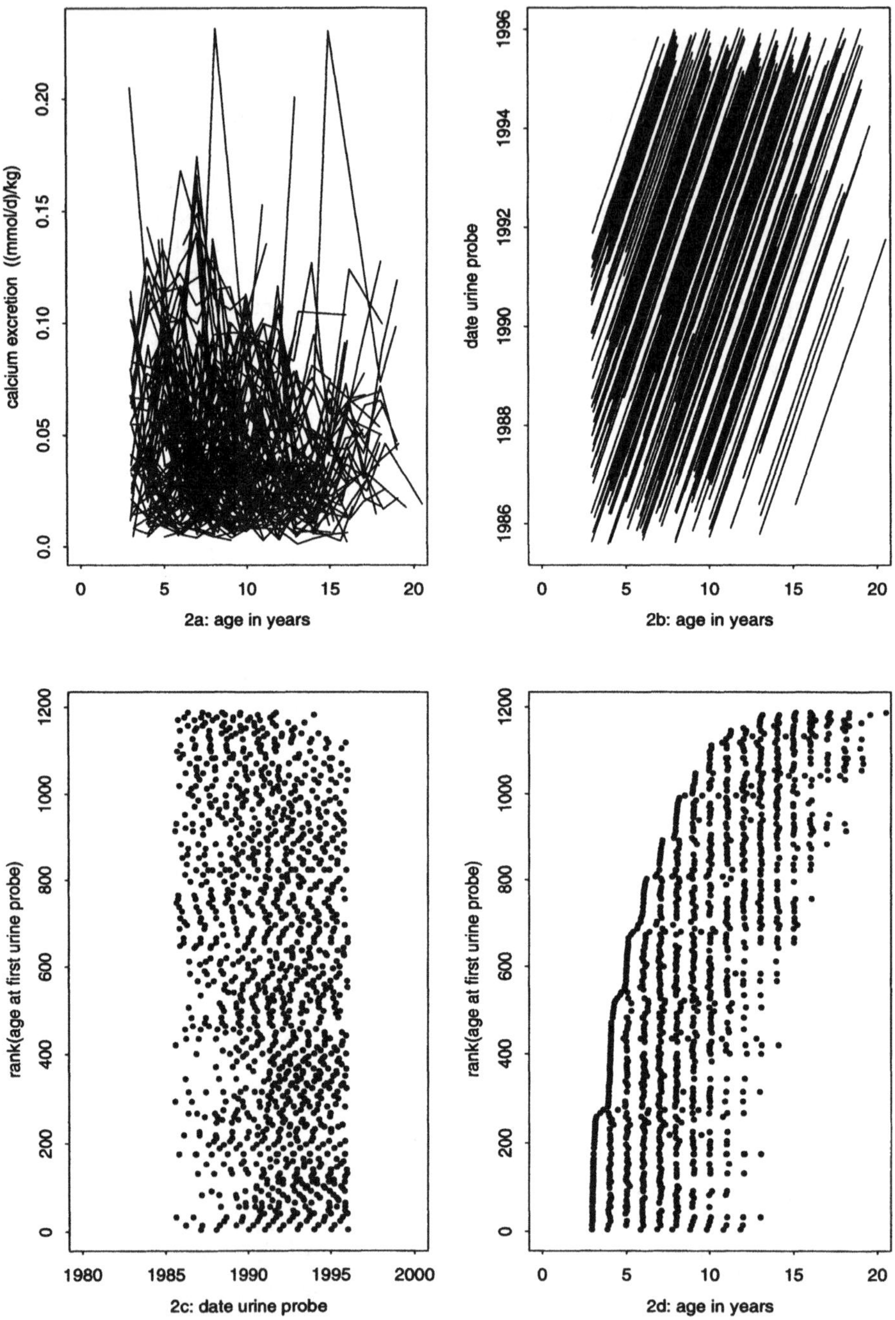

Figure 2: Graphical representation of the longitudinal measurements with more than 4 observations and survey structure.

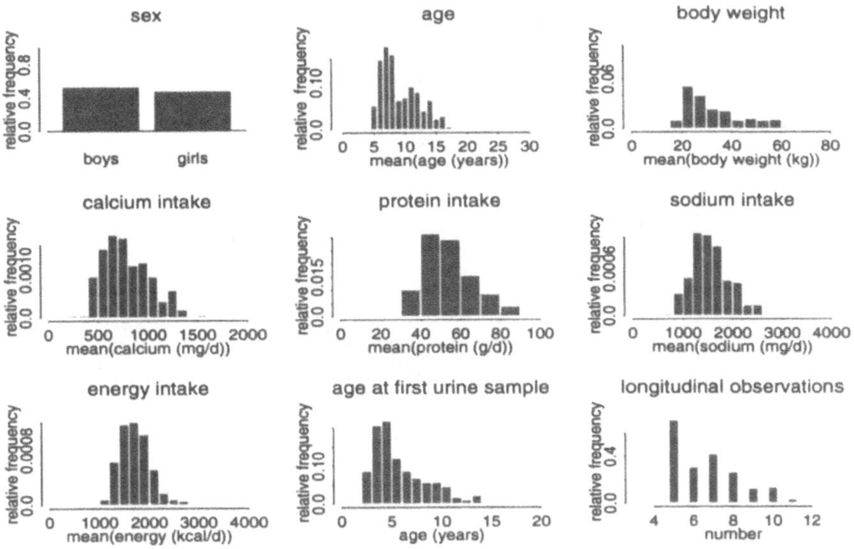

Figure 3: Univariate distributions of the response integrals of the considered influence factors

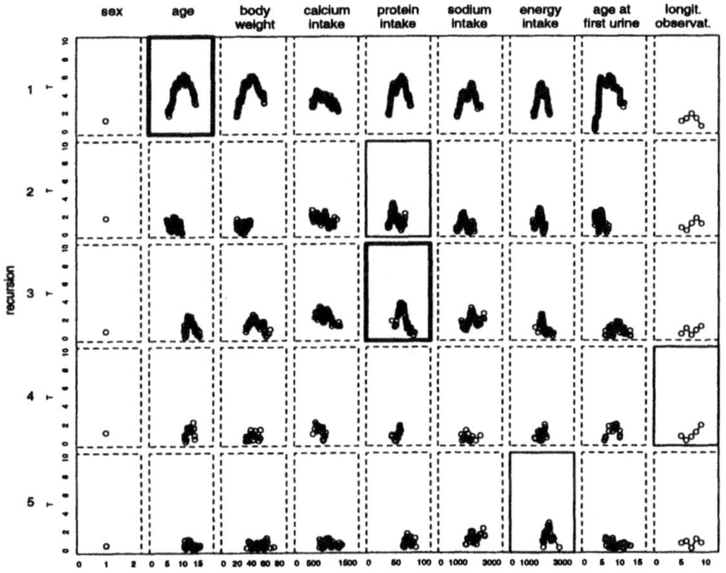

Figure 4: The processes of the considered rank statistics by CART (p_{stop} = 0.01).The thickness of the boxes indicate for each recursion (row) the process with the minimum P-value for the considered factors (column).

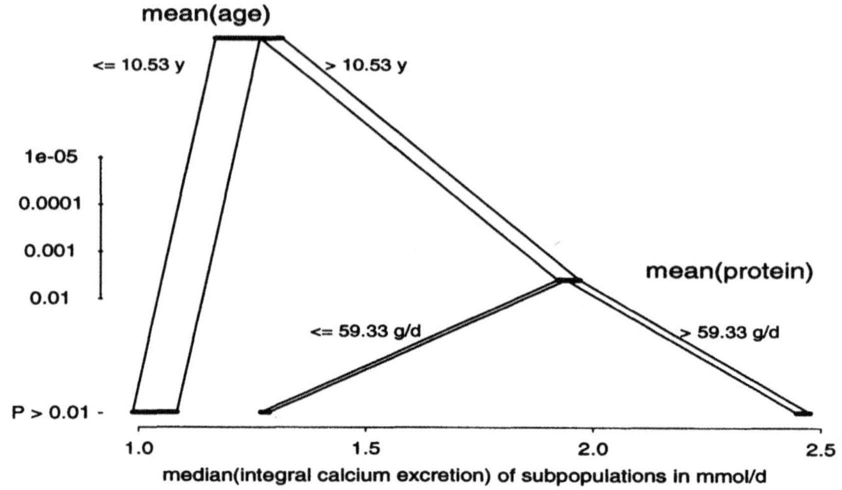

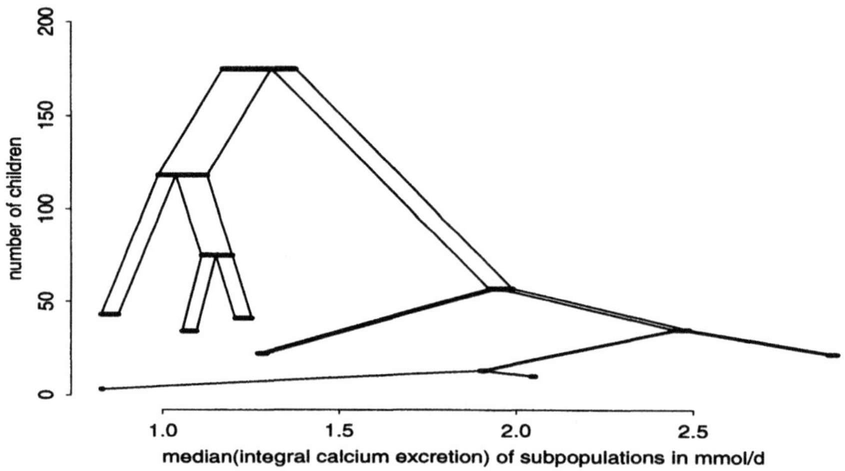

Figure 5: Computed regression tree representations (cf. Dirschedl, 1991; Lausen, Kersting and Schöch, 1995); top: P-CART with $p_{stop} = 0.01$; bottom: N-CART with $p_{stop} = 0.1$.

Bootstrapping CART is a straightforward approach to assess the stability of the results. In the case of classification and prediction the bootstrap idea is promising (e.g. Breiman, 1994, 1995; Vach, 1995; Wernecke, Possinger and Kalb, 1996).

But the quest and assessment of prognostic factors is a harder problem, there are some results in the bivariate situation (e.g. Siegmund, 1988; Verweij and van Houwelingen, 1993; Altman et al., 1994; Lausen and Schumacher, 1995; Schumacher, Holländer and Sauerbrei, 1995). Consequently, I suggest to use the results of Lausen and Schumacher (1995) on confidence regions and effect estimation for tree based models. Finally, I hope that the arguments above demonstrate that tree based models and REMODEL are important tools especially for applications in clinical research and epidemiology.

References:

ALTMAN, D.; LAUSEN, B.; SAUERBREI, W. and SCHUMACHER, M. (1994): Dangers of using "optimal" cutpoints in the evaluation of prognostic factors. *Journal of the National Cancer Institute, 86, 829-835.*

BREIMAN, L. (1994): Bagging predictors. Technical report No. 421, Department of Statistics, University of California, Berkeley.

BREIMAN, L. (1995): Better subset regression using the nonnegative carrote. *Technometrics, 37, 4, 373-384.*

BREIMAN, L.; FRIEDMAN, L.; OLSHEN, R. A. and STONE, C. J. (1984): Classification and regression trees, Wadsworth, Monterey.

CIAMPI, A. (1991): Generalized regression trees, Computational Statistics and Data Analysis, 12, 57-78.

CLARK, L. A. and PREGIBON, D. (1992): Tree-based models. In: J. M. Chambers and T.J. Hastie (eds.): *Statistical models in S.* Wadsworth and Brooks Cole Computer Science Series, Pacific Grove, 377-419.

DIRSCHEDL, P. (1991): Klassifikationsbäume - Grundlagen und Neuerungen. In: W. Fleischer, M. Nagel and R. Ostermann (eds.): *Interaktive Datenanalyse mit ISP.* Westarp Verlag, Essen, 15–30.

HECKER, H. and WÜBBELT, P. (1992): Klassifikation nach Mustern von Prognosefaktoren und Ausprägungen von Responsevariablen. In: S. Schach and G. Trenkler (eds.): *Data analysis and statistical inference* (Festschrift in honour of Prof. Dr. Friedhelm Eicker). Verlag Josef Eul, Bergisch Gladbach, 259-275.

HILSENBECK, S. G. and CLARK, G. M. (1996): Practical p-value adjustment for optimally selected cutpoints. *Statistics in Medicine, 15, 103-112.*

LAUSEN, B.; KERSTING, M. and SCHÖCH, G. (1995): The regression tree method and its application in nutritional epidemiology (submitted).

LAUSEN, B.; SAUERBREI, W. and SCHUMACHER, M. (1994): Classification and regression trees used for the exploration of prognostic factors measured on different scales. In: P. Dirschedl and R. Ostermann (eds.): *Computational Statistics.* Physica-Verlag, Heidelberg, 483–496.

LAUSEN, B. and SCHUMACHER, M. (1992): Maximally selected rank statistics. *Biometrics, 48, 73-85.*

LAUSEN, B. and SCHUMACHER, M. (1995): Evaluating the effect of optimized cutoff values in the assessment of prognostic factors. *Computational Statistics and Data Analysis.* (in press)

LOH, W.-Y. (1991): Survival modeling through recursive stratification. *Compu-*

tational Statistics and Data Analysis, 12, 295-313.

MILLER and SIEGMUND (1982): Maximally selected chi-square statistics. *Biometrics, 38, 1011-1016.*

SAUERBREI, W.; ZAISS, A.; LAUSEN, B. and SCHUMACHER, M. (1991): Evaluierung prognostisch homogener Subpopulationen mit Hilfe der Methode der Klassifikationsbäume. In: Guggenmoos-Holzmann, I. (ed.): *Quantitative Methoden in der Epidemiologie,* Medizinische Informatik und Statistik, 72, Springer-Verlag, Berlin, 229-234.

SCHULGEN, G.; LAUSEN, B.; OLSEN, J. H. and SCHUMACHER, M. (1994): Outcome-oriented cutpoints in analysis of quantitative exposures. *American Journal of Epidemiology, 140, 172- 184.*

SCHUMACHER, M.; HOLLÄNDER, N. and SAUERBREI, W. (1995): Resampling and cross-validation techniques: A tool to reduce bias caused by model building. Preprint.

SEGAL, M.R. (1992): Tree-structured methods for longitudinal data. *Journal of the American Statistical Association 87, 407-418.*

SEGAL, M. R. (1994): Representative curves for longitudinal data via regression trees. *Journal of Computational and Graphical Statistics 3, 2, 214-233.*

SIEGMUND, D. (1988): Confidence sets in change-point problems. *International Statistical Review, 56, 31-48.*

ULM, K.; SCHMOOR, C.; SAUERBREI, W.; KEMMLER, G.; AYDEMIR, Ü.; MÜLLER, B. and SCHUMACHER, M. (1989): Strategien zur Auswertung einer Therapiestudie mit derÜberlebenszeit als Zielkriterium. *Biometrie und Informatik in Medizin und Biologie, 20, 4, 171-205.*

VACH, W. (1995): Classification trees. *Computational Statistics (Workshop on modern methods of classification), 10, 1, 9-14.*

WERNECKE, K.-D.; POSSINGER, K. and KALB, G. (1996): On the validation of classification trees. In: H.-H. Bock and W. Polasek (eds.): *Data analysis and information systems - Statistical and conceptual approaches.* Springer, Berlin.

ZHANG, H.; HOLFORD, T. and BRACKEN, M. B. (1996): A tree-based method of analysis for prospective studies. *Statistics in Medicine, 15, 37-49.*

Conceptual Complexity in Biomedical Terminologies: The UMLS Approach

A. T. McCray

U.S. National Library of Medicine, Bethesda, Maryland 20894, U.S.A.

Summary: The U.S. National Library of Medicine's Unified Medical Language System (UMLS) project is a research effort to develop knowledge-based tools and resources to compensate for differences in the way concepts are expressed in the field of biomedicine. Since 1990, a set of UMLS Knowledge Sources has been released annually to the research community. This paper gives an overview of the current knowledge sources and highlights the nature of the concepts and their interrelationships in the most recent release of the UMLS. The paper illustrates the complexity of the biomedical terminologies that have been represented in the UMLS as well as the methods that have been used to harness this complexity.

1. Introduction

The U.S. National Library of Medicine's Unified Medical Language System (UMLS) project is a research effort to develop knowledge-based tools and resources to compensate for differences in the way concepts are expressed in the field of biomedicine, particularly in electronic sources that provide automated access to biomedical knowledge (Lindberg et al. (1993)). This knowledge may be stored in any of a variety of database types and may involve factual information, such as that found in clinical records, diagnostic expert systems, or toxicology databases, or it may involve bibliographic information, such as that found in the MEDLINE database. Importantly, a goal of the project is to provide tools and methods whereby users can easily navigate from one system to another without undue concern for the way in which concepts are expressed in these various systems.

The UMLS effort has resulted in a number of research products over the past several years. Since 1990, a set of UMLS Knowledge Sources has been released annually to the research community for experimentation and comment. The early development and use of the knowledge sources has proven to be a successful strategy for gaining valuable feedback from interested researchers and has significantly influenced not only the further development of the knowledge sources, but also the UMLS project itself. In the summer of 1995, the Internet-based UMLS Knowledge Source Server was released (McCray and Razi (1995)). The server provides access to the data contained in the knowledge sources in a convenient format through several client interfaces, including the World Wide Web.

The UMLS research products currently comprise four Knowledge Sources. The Metathesaurus and Semantic Network together express and classify a significant portion of the biomedical vocabulary. The Metathesaurus is composed of terminology from some thirty thesauri whose terms have been interlinked in the Metathesaurus conceptual structure (McCray and Nelson (1995)). The Semantic Network encodes high level semantic categories and important relationships between these categories in the biomedical domain. Each Metathesaurus concept is assigned to one or more of the 135 semantic types currently available in the Network. The SPECIALIST Lexicon and the Information Sources Map complete the current set of knowledge sources and are more fully described in McCray et al. (1994) and Rodgers (1995).

In the following sections an overview of the UMLS Knowledge Sources is presented, followed by a description of the Internet-based UMLS Knowledge Source Server. The server allows individuals to browse the knowledge sources, as well as to extract information from them. The current content of the Metathesaurus is discussed, and finally, the early results of experiments that explore the nature of the concepts and their interrelationships in the most recent release of the Metathesaurus are presented. The results point to further methods that may be brought to bear to increase the semantic coherence of the Metathesaurus, in particular by taking full advantage of the knowledge encoded in the UMLS Semantic Network. The results have general implications for the design and validation of classification systems.

2. Overview of the UMLS Knowledge Sources

2.1 Metathesaurus

The largest of the four UMLS knowledge sources is the Metathesaurus, which may be thought of as an associative network of biomedical terminology. It currently contains over 250,000 concepts, representing some 580,000 terms. The Metathesaurus draws its terminology from approximately thirty biomedical vocabularies, including broad coverage thesauri, such as NLM's Medical Subject Headings (MeSH), which is used to index the biomedical literature, and the Systematized Nomenclature of Medicine (SNOMED), a clinically oriented thesaurus. It also includes thesauri which have been designed for much smaller and more specific domains, e.g., the Physician Data Query (PDQ) system, which involves terminology in the domain of oncology, and the Universal Medical Device Nomenclature, which is a product category thesaurus. Some thesauri are included in their entirety, e.g., MeSH, DSM-III-R and DSM-IV (the Diagnostic and Statistical Manual of Mental Disorders), the Classification of Nursing Diagnoses, and WHOART (World Health Organization Adverse Drug Reaction Terminology). Others are only partially represented, but in most cases the remainder of each of these vocabularies will be added over time.

The Metathesaurus is updated and distributed annually. Each year since 1992 new vocabularies as well as additional terms from existing vocabularies have been added. The addition of terminology to the Methathesaurus is a process that involves not only precomputing as much information as possible as a concept record is being built, but also involves the participation of many domain experts who review and add information to the emerging concept records. A concept in the Metathesaurus may be thought of as a cluster of synonyms. For example, the Metathesaurus unique identifier "C0007134" refers to the following set of synonyms: "renal cell carcinoma", "hypernephroma", "renal cell adenocarcinoma", "Grawitz tumor", "hypernephroid carcinoma", "nephroid carcinoma", and "renal cell cancer". Additionally, translations of this concept currently exist for four other languages, French ("nephrocarcinome"), German ("Nierenzellkarzinom"), Portuguese ("carcinoma de celulas renais"), and Spanish ("carcinoma de celulas renales"). The synonyms are in many cases drawn from multiple vocabularies. The synonyms for this concept are derived from MeSH, the International Classification of Diseases (ICD-9 CM), SNOMED, PDQ, DxPlain, an expert diagnostic system developed by Massachusetts General Hospital, COSTART, terms from the Computer Stored Ambulatory Record system, also of Massachusetts General Hospital, and the Crisp thesaurus, a thesaurus used for indexing research grants awarded by the U.S. National Institutes of Health.

Synonyms from a constituent thesaurus are not always considered to be synonyms in the Metathesaurus. This is because each thesaurus has been developed for a particular purpose. If that purpose is information retrieval, for example, then some terms that are listed as synonyms may, in fact, not be synonyms, but, rather, may be more specific in meaning. This is entirely appropriate for this purpose, since the thesaurus will group the more specific terms together with the general term, thereby providing the user with direct access to all related relevant data. In the MeSH vocabulary, for example, entry terms (which are called 'entry terms", since they are not always synonymous) are sometimes narrower in meaning than the main heading and sometimes they are synonyms. As part of the process of building the Metathesaurus these distinctions have been made. Those entry terms that are synonyms are treated as such and are added as synonyms in the Metathesaurus concept record. Those entry terms that are not synonyms become part of a different Metathesaurus concept and cluster with other terms that share their meaning. For example, the MeSH heading "environmental pollution" has the entry term "soil degradation". When searching the MEDLINE database, entry terms are automatically mapped to main headings. Thus, a search for "soil degradation" will retrieve all citation records for articles relating both to environmental pollution in general and more specifically to soil degradation. In the Metathesaurus, however, since these terms have different meanings, they appear in two separate concept records.

Definitions are included in the Metathesaurus when these are available from one of the constituent thesauri. "Carcinoma", for example, includes definitions from two sources, the MeSH vocabulary, and the Dorland's Illustrated Medical Dictionary, 27th edition, where it is defined as follows: "A malignant new growth made up of epithelial cells tending to infiltrate the surrounding tissues and give rise to metastases." Semantic types from the 135 that are available in the UMLS Semantic Network are assigned to Methesaurus concepts. The semantic types are intended to give a high level computable meaning to the concepts. The semantic type assigned to "carcinoma" is "Neoplastic Process". In the hierarchical structure of the Semantic Network, a neoplastic process is a child of disease or syndrome, which is a child of pathologic function, which in turn is a type of biologic function. The semantic types may be related by a variety of relationships. Thus, for example, pathologic functions may be treated by therapeutic procedures or drugs; they are located in particular body sites; and they may co-occur with other pathologic functions.

As Metathesaurus concepts are added, related concepts are added as well. An interesting set of related concepts derives from the so-called co-occurrence information. Co-occurring MeSH terms in MEDLINE citation records give a particularly rich set of related terminology. Each MEDLINE citation record includes between ten to fifteen MeSH headings that characterize the referenced journal article. The Metathesaurus reports the frequency of pairs of co-occurring headings that have appeared in the last ten years, thus giving a very good idea of the recent topics of interest in the scientific literature. Some of the headings that co-occur most frequently with "renal cell carcinoma" are "nephrectomy", "kidney", "combined antineoplastic agents", and "lung neoplasms". These concepts have the semantic types, "Therapeutic or Preventive Procedure", "Body Part, Organ, or Organ Component", "Pharmacologic Substance", and "Neoplastic Process", respectively. The semantic relations among these concepts mirror the relationships that were alluded to above. Thus, renal cell carcinoma sometimes involves the procedure nephrectomy; the location of the disease is the kidney; it is often treated by a combined drug therapy; and it tends to co-occur with other types of neoplastic processes, e.g. lung neoplasms.

Most of the vocabularies that are included in the Methesaurus organize their terminologies in context hierarchies. These hierarchies are maintained in the Metathesaurus, since the context of a term contributes to an understanding of its meaning. To continue with the previous example, "renal cell carcinoma" appears in seven vocabularies. Figure 1 illustrates some of these hierarchies. The thesaurus imposes a certain point of view on the terms that it organizes. In PDQ the term is grouped together with other urinary tract disorders as well as with other solid tumors, while in MeSH it appears in a subtree that includes neoplasms organized by histologic type.

Other related concepts are added as concept records are built or augmented. As new concepts enter the system, domain experts evaluate potentially related terms, assigning these as synonyms of an existing concept, as narrower or broader in meaning than an existing concept, or as otherwise saliently related to a particular concept. Some non-synonymous terms that are related to "renal cell carcinoma" include "kidney cell dysplasia", "urinary system disorders", and "malignant neoplasm of kidney, except pelvis".

The Metathesaurus represents extensive information about its constituent thesauri. It maintains the, in some cases quite detailed, information that each vocabulary provides, and, importantly, it integrates the vocabulary

PDQ

- cancer
  - body system/site cancer
    - kidney/urinary cancer
      - kidney tumor
        - renal cell cancer

PDQ

- cancer
  - solid tumor
    - adult solid tumor
      - renal cell cancer

MeSH

- Diseases
  - Neoplasms
    - Neoplasms by Histologic Type
      - Neoplasms, Glandular and Epithelial
        - Carcinoma
          - Adenocarcinoma
            - Carcinoma, Renal Cell

Fig. 1: Some context hierarchies for "renal cell carcinoma"

with the larger Metathesaurus structure. This integration is at the concept level where synonyms, definitions, semantic types, and a range of lexical information are added. The recognition and expression of semantic links among concepts results in a highly interrelated conceptual structure for the biomedical domain.

2.2 SPECIALIST Lexicon and Lexical Programs

The SPECIALIST lexicon has been developed for natural language processing applications, and it serves as the basis for the lexical programs that are included with the UMLS. The lexicon contains grammatical information for a large number of biomedical and general English terms. (See a grammar such as Quirk et al. (1985) for discussion and elucidation of many of the grammatical characteristics of the English language.) Figure 2 shows some examples of lexical items in the SPECIALIST lexicon. The lexicon currently contains some 80,000 lexical records of the type shown below. The verb "treat" can be used in variety of ways, transitively (tran) as in "treat the patient" where it takes one nounphrase (np) object, but also quite commonly ditransitively (ditran), where it takes two objects, one of which may be a prepositional phrase (pphr), as in "treat the patient for a disease". Note that the nominalization ("treatment") for the verb is also encoded. Verbs may nominalized in a variety of ways, so it necessary to store this information explicitly, rather than being able to do it by rule. Many verbs allow nominalizations in "-ment", e.g. "movement ", "abatement" and "advancement"; some in "-tion", e.g. "exacerbation", "incubation", and "medication"; and still others in "-er", e.g. "reporter", "teacher", and "conveyer", and some allow multiple nominalizations with different meanings, e.g., "employer" and "employment".

Noun entries encode spelling variation, if there is any, as in the entry for

```
{base=treat
entry=E0061964
    cat=verb
    variants=reg
    intran
    tran=np
    tran=pphr(with,np)
    tran=pphr(of,np)
    ditran=np,pphr(to,np)
    ditran=np,pphr(with,np)
    ditran=np,pphr(for,np)
    cplxtran=np,advbl
    nominalization=treatment
entry=E0061965
    cat=noun
    variants=reg
}

{base=canula
spelling_variant=cannula
entry=E0014988
    cat=noun
    variants=glreg
    variants=reg
}
```

Fig. 2: Sample entries from the SPECIALIST Lexicon

"canula", and they encode the noun's inflectional variation type. The noun may form its plural in the native English way (reg), e.g. "canulas", but it may also form its plural using the rules of Greco-Latin variation (glreg), e.g. "canulae".

The importance of the SPECIALIST lexicon in the context of the UMLS is that it underlies the lexical programs that provide flexible access to UMLS terminology (McCray et al. (1994)). For example, when searching the Metathesaurus indexes for "gamma knife", "gamma knife radiosurgeries" is found. When invoking the lexical tools, "gamma knives" is found, which is simply the plural form of the term originally sought. A standard look up for "nasal cannula" finds nothing. However, when the lexical tools are invoked, several concepts are returned, including "nasal cannulae", and "nasal oxygen cannulae". As noted above, the SPECIALIST lexicon has encoded all these variations. The source code for the lexical tools is distributed together with the UMLS Knowledge Sources. Additionally, two normalized indexes for the Metathesaurus terminology have been created using the lexical programs. These indexes contain normalized, or canonical, forms that abstract away from inflectional variation, word order, punctuation, and other orthographic details. The complete lexical tool set also includes routines for linking derivationally related terms to each other. For example, adjectives such as "aphasic", "surgical", and "diabetic" will be matched to their noun forms, "aphasia", "surgery", and "diabetes", respectively, and nouns such as "behavior", "belief", and "lubrication" will be related to their respective verb forms, "behave", "believe", and "lubricate".

2.3 Semantic Network

The UMLS Semantic Network may be thought of as an ontology for the domain of biomedicine. As such, it shares many of the characteristics of such systems including nodes (semantic types), links (relationships), and inheritance properties. (See Lehman (1992) and Sowa (1984) for fuller discussion of some of the properties of semantic networks.) The network currently has 135 semantic types and 51 relationships. The network has been designed to include two semantic type hierarchies, one for entities and another for events. Types have been established for physical objects such as organisms, anatomical structures, manufactured objects, and substances. Conceptual objects such as temporal, qualitative, and quantitative concepts are also included. Figure 3 below illustrates a portion of the network for the semantic type "Group". Note that a grouping of individuals based on some

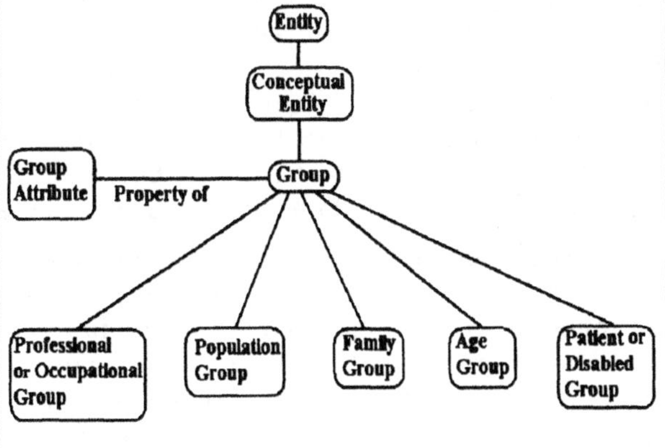

Fig.3: Portion of the UMLS Semantic Network

shared characteristic is a conceptual notion. The partitioning of groups in the UMLS Semantic Network is based on those that are important in the biomedical domain. Professional groups such as physicians, researchers, and educators, population groups such as Asian Americans, indigents, and blood donors, age groups such as infants, adolescents, and the aged are all salient in health care management and research.

The network also categorizes events such as behaviors, occupational activities, and biologic functions. This latter includes both physiologic functions, such as organism and tissue functions, and also pathologic functions, such as diseases and neoplastic processes. Since types are organized in type hierarchies, each type inherits properties from its parent and ancestors. Thus, for example as shown in figure 3, since a group attribute is a property of a "Group", by inheritance it is also a property of all the children of this type, i.e., "Professional or Occupational Group", "Population Group", "Family Group", "Age Group", and "Patient or Disabled Group".

Types in the network may be related to each other by multiple relationships. The primary relationship is "isa", and this links the types in the two major type hierarchies. The isa link allows a meaning to be computed for any given type. For example, the leaf node "Family Group" is a "Group", which is itself a "Conceptual Entity", which is an "Entity". Therefore, by transitivity, it is possible to calculate that a family group is an entity. There are five link hierarchies in addition to the "isa" link. These are physical (e.g. "part of", "contains"), spatial (e.g. "location of", "surrounds"), temporal (e.g. "precedes", "co-occurs with"), functional (e.g. "performs", "indicates"), and conceptual (e.g. "measures", "property of"). The links are binary, linking two types and asserting a relationship between them. The arguments in the relation are ordered and are generally asymmetric, i.e., "a drug prevents a disease", but not "a disease prevents a drug". Importantly, the relationships link semantic types and not the concepts that have been assigned to those types. While it is true that a relation is only established if at least some members of a set map onto at least one member of another set linked by the relationship, the relation will not hold for all members of the sets. For example, the relation "Pharmacologic Substance treats Disease or Syndrome" is expressed in the Semantic Network. It is clear that not all members of the set of drugs treat all members of the set of diseases. Thus, aspirin may treat rheumatoid arthritis, digoxin may treat congestive heart failure, and bronchodilator agents treat asthma, but aspirin does not treat asthma, and bronchodilator agents do not treat rheumatoid arthritis.

The Semantic Network contributes crucially to the conceptual structure expressed by the UMLS. By virtue of the semantic types assigned to all Metathesaurus concepts, concepts participate in the full network structure where high-level knowledge about the biomedical domain is encoded. The knowledge that is represented there, while not being highly detailed, does express the important interrelationships between concepts in the biomedical domain. Further, the network gives a computable semantics to the concepts represented in the Metathesaurus. The semantic types themselves can be used to partition the Metathesaurus into smaller domains. For example, it is possible to extract anatomical terminology by searching for all those concepts that have been assigned anatomical semantic types. Thus, cells, tissues, organs, and body locations can all be easily found. Similarly, all chemicals, all medical devices, or all diseases can be quickly and accurately identified.

2.4 Information Sources Map

The Information Sources Map characterizes existing biomedical databases. It currently contains descriptions for some eighty databases, but that number is expected to increase significantly in the next few years. Each database description, or record, includes narrative descriptions of the content of the database, contact information, intended audience, probability of use by discipline, and update frequency of the database. For example, NLM's Hazardous Substances Data Bank (HSDB) contains records for hazardous chemicals. For each chemical, information on manufacturing and use, toxicological effects, environmental exposure, and safety and handling is listed.

Interestingly, each database is also indexed by sets of semantic relations. For the Hazardous Substances Data Bank relevant relations include "Hazardous or Poisonous Substance causes Pathologic Function" and "Biologically Active Substance causes Congenital Abnormality". These semantic relations are intended to assist retrieval systems in locating appropriate databases given a user query.

3. Exploring UMLS Data Through the UMLS Knowledge Source Server

Since 1990, the UMLS data files have been distributed annually on CD-ROM. The data are stored in several formats including ASCII relational tables and in Abstract Syntax Notation 1 (ASN.1) format. In 1995, the Internet-based UMLS Knowledge Source Server was made available in order to allow users to access the data included in the UMLS knowledge sources without needing to be concerned about exactly how the data are represented on the CD-ROM's (McCray and Razi (1995)). The design of the system is client-server, with the server residing on a SparcCenter 2000 machine at the NLM. Three types of client interfaces have been developed, a command-line interface, an Application Programming Interface (API), and a World Wide Web interface. The Knowledge Source Server allows the user to retrieve individual concept and term data, for example, the synonyms, sources, semantic types, definitions, lexical variants and co-occurring terms for any given Metathesaurus concept. It also allows the user to retrieve on UMLS attributes, e.g. retrieving all concepts that share a semantic type, or all terms with a certain lexical tag. Additionally, if the user is interested in browsing the Metathesaurus for information on a particular constituent vocabulary, then the search can be restricted to that vocabulary. Figure 4 shows the initial Metathesaurus screen, where the user has decided to restrict the search to just that information that appears in the ICD-9-CM thesaurus. Access to the Semantic Network is illustrated in a sample screen shown in Figure 5. The user is interested in the semantic type "Alga". Clicking through the "Entity" hierarchy eventually brings the user to this screen. The definition is displayed and the user can click to see a list of all Metathesaurus concepts having the semantic type "Alga". Choosing this option displays terms such as "Acetabularia mediterranea", "Anabaena", and "Seaweed". Choosing the option to display types related to "Alga" indicates that, among others, "Alga has process Biologic Function", and "Alga has part Anatomical Structure".

The Knowledge Source Server frees developers from having to write programs to extract the data from the CD-ROM. The server is centrally maintained at NLM and any changes to the underlying data files are transparent to the developer when using the available API. Users can easily browse the extensive UMLS data through the Web interface, and they can also conduct experiments using the command-line interface, which allows them to flexibly combine the many options that are provided there.

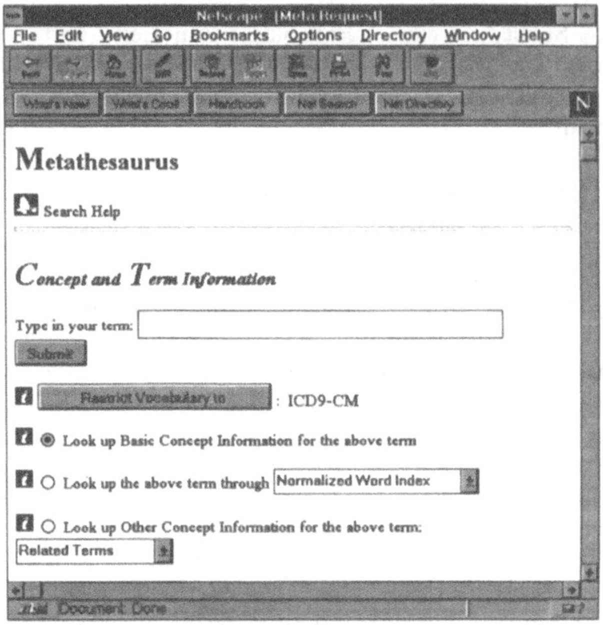

Fig 4: Initial UMLS Metathesaurus Screen

4. Current Content of the Metathesaurus

The 1996 edition of the Metathesaurus contains over 250,000 concepts from over thirty source vocabularies. Figure 6 lists the number of "preferred" terms from several of the thesauri included in the UMLS. Many thesauri include synonyms for the preferred names of terms in their systems. Thus, the total number of terms from each thesaurus is actually much larger than the numbers shown here. The 1996 edition of the Metathesaurus contains a large number of expressed relationships between concepts, including synonymy, plesionymy (similar in meaning), and hierarchical relationships, such as broader than, narrower than, child of, and sibling of. In some cases these relationships are further specified by relationships from the Semantic Network. Thus, a concept may be narrower than another concept and it may, more specifically, be in the "part of" relationship to that other concept. Relationships between approximately 9,000 child and parent MeSH terms have been made explicit in the Metathesaurus structure. The next section will discuss recent experiments involving these relations.

The National Library of Medicine served the Agency for Health Care Policy and Research (AHCPR) are sponsoring a test during 1996 that will evaluate the ability of the current Metathesaurus, together with specific additions, to represent the clinical vocabulary that will be required in a number of health care and research applications, including emerging patient record systems (Humphreys et al. (1996)). The additions will be that part of SNOMED In-

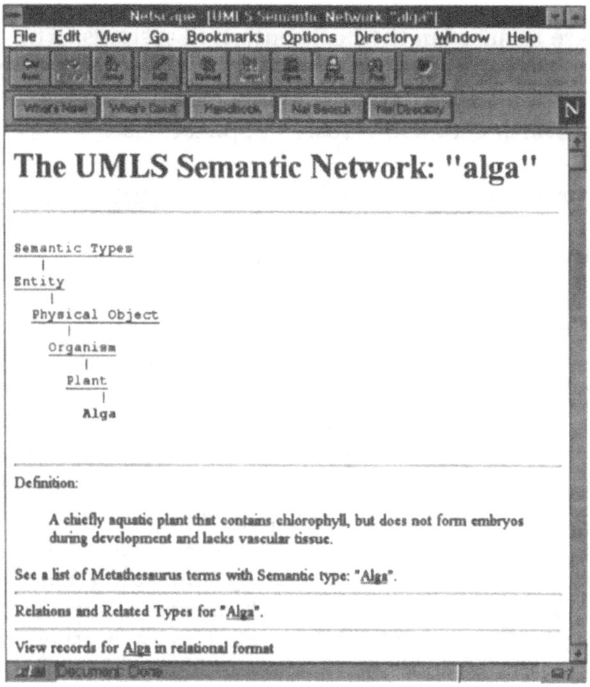

Fig 5: UMLS Semantic Network Screen

55,777	SNOMED International
19,651	SNOMED II
18,636	ICD-9-CM terms
18,023	MeSH
6,547	DXPlain
6,461	CRISP
6,292	Library of Congress Subject Headings
4,561	Universal Medical Device Nomenclature System
4,051	Physician Data Query System
3,461	COSTAR
3,084	Thesaurus of Psychological Index Terms
1,292	WHO Adverse Drug Reaction Terminology
1,182	Coding Symbols for Thesaurus of Adverse Reaction Terms
681	AI/RHEUM
666	Neuronames
494	DSM-IV R
468	Omaha System:Applications for Community Health Nursing
433	Nursing Interventions Classification
162,945	MeSH supplementary chemicals
17,997	German MeSH translations
17,996	Spanish MeSH translations
17,996	Portuguese MeSH translations
17,993	French MeSH translations

Fig. 6: Profile of portion of 1996 Metathesaurus Terminology

ternational that is not yet represented in the Metathesaurus, all of the Read Clinical Classification, and a set of codes for clinical laboratory test results (LOINC). Through a special interface to the UMLS Knowledge Source Server that is currently under development, participants will search the UMLS for the terms that are required in their computerized systems. The results of the test are expected to elucidate what clinical terminology is already covered by UMLS and what terminology may still be needed in order to accommodate a wide-range of health care and research applications.

5. Exploring Concepts and Relationships in the UMLS

5.1 Background and Methods

In an effort to increase the number of inter-concept relationships available in the UMLS, the author has recently conducted experiments to see if some of these relationships might be automatically inferred by taking advantage of the existing knowledge that the UMLS provides about the concepts it includes. In particular, the hypothesis is as follows. Since all concepts in the Metathesaurus have been assigned to semantic types, and since semantic types are related to each other in explicit ways, it should be possible to computationally infer the exact relationship between any given pair of concepts that have been related through the Metathesaurus structure. Since, however, semantic types can be related by multiple (though generally a small number of) relationships, the hypothesis was refined to consider only those concepts that are hierarchically related to each other in an existing thesaurus. Thus, if a term is related to another term in the child/parent relationship, and if the semantic types of the two terms are in the "isa" relationship, then the exact relationship between the child and the parent term should be "isa". For example, if a child term has been assigned the semantic type "Immunologic Factor", and if the parent term has been assigned a semantic type of any of the ancestors of this type, e.g., "Biologically Active Substance", or "Chemical", then the child and parent can be inferred to be in the "isa" relationship.

The initial experiment was conducted using the MeSH vocabulary, since many of the relationships between MeSH child/parent pairs had already been assigned by domain experts. These assignments then served as the gold standard for the experiment. The UMLS Knowledge Source Server, through its command-line interface, was queried for the semantic types of the descendants of the high-level nodes of interest in the MeSH thesaurus. The two nodes chosen were the disease and organism hierarchies. Then, based on the semantic types that had been assigned to the terms, the allowable semantic relationships for each child/parent pair were calculated. The "isa" relationship was algorithmically inserted if the semantic types of the child/parent pair were in the "isa" relationship to each other. The algorithmically inserted relations were compared with those originally assigned by domain experts.

5.2 Preliminary Results

There are 7,224 child/parent pairs in the MeSH disease hierarchy. Of these, according to the hypothesis, 6,110 child/parent pairs should have been inferrable as being in the "isa" relation. In fact, only 4,793 were correctly inferred when compared with the gold standard. The initial hypothesis, thus, predicted that 85inferrable. Of this 85(4793/6110) were correctly inferred. The organism hierarchy has 2,328 child/parent pairs. Of these, 2,293 (98algorithmically inferrable as being in the "isa" relation. In this case, there was no gold standard for comparison, but analysis of a random sample of the results by a domain expert indicates that the inferred relations are correct.

5.3 Discussion

A small portion of the disease hierarchy will illustrate the issues involved. Figure 7 below shows a sample with semantic type assignments. All but

C Diseases (MeSH Category) <Classification>

C01 Bacterial and Fungal Diseases <Disease or Syndrome>

C01.252 Bacterial Infections <Disease or Syndrome; Finding>

C01.252.100 Bacteremia <Disease or Syndrome>

C01.252.100.375 Hemorrhagic Bacteremia <Disease or Syndrome>

C01.252.300 Endocarditis, Bacterial <Disease or Syndrome>

C01.252.300.407 Endocarditis, Subacute Bacterial <Disease or Syndrome>

C01.252.354 Eye Infections, Bacterial <Disease or Syndrome>

C01.252.354.225 Conjunctivitis, Bacterial <Disease or Syndrome>

C01.252.354.225.250 Conjunctivitis, Inclusion <Disease or Syndrome>

Fig. 7: Portion of MeSH hierarchy with semantic type assignments

the top term have the appropriate semantic type, thus licensing the "isa" relationship according to the hypothesis. In fact, however, the relationship between "bacteremia" and "bacterial infections" was labelled as "occurs in" by a domain expert. All other links were labelled by the domain expert as "isa", as predicted by the hypothesis. When considering the meaning of "bacteremia" (the presence of viable bacteria circulating in the blood), it becomes clear that this is actually not a disease, and should more appropriately have been given the semantic type "Pathologic Function" (a disordered process, activity, or state of the organism). As such, its placement in the MeSH disease hierarchy might also be re-evaluated.

The results for the organism hierarchy were much more successful. A possible explanation is that since organism taxonomies have long existed and are quite well understood, it is not surprising that they are well formed, strict hierarchies. Thus the results of the experiment for this hierarchy were exactly as predicted by the hypothesis.

A preliminary conclusion that can be drawn from this experiment is that the Metathesaurus and Semantic Network can be used to acquire deeper knowledge about the thesauri that are included in the UMLS. When relations are correctly inferred, this indicates that the thesaurus hierarchy is well-formed. When relations cannot be correctly inferred, then either the items in the thesaurus are not correctly classified and the developers of the thesaurus might consider a modification, or the UMLS semantic type or semantic relation assignment is not correct. In this case, the UMLS need to be modified. Whether or not the relations are correctly inferred, the methodology reported on here provides a new analysis technique for the investigation of thesaurus well-formedness.

6. Conclusion

The UMLS knowledge sources represent a large, continually evolving source of biomedical knowledge. The knowledge sources are regularly made available to interested individuals for research and application. We have developed a set of research tools including the Internet-based Knowledge Source Server that should be useful in conducting this research. For example, the experiment reported above used the command-line client interface to the server, extracting and further manipulating the retrieved data. Other researchers have successfully used the Knowledge Source Server in their experiments and application programs, e.g., Lowe et al. (1995) and Ketchell et al. (1996). The UMLS together with its tools and application programs can be used to carry out a range of investigations, including studies that have implications for the design and validation of classification systems.

References:

HUMPHREYS, B. L.; HOLE, W. T.; McCRAY, A. T. and FITZMAURICE, J. M. (1996): Planned NLM/AHCPR Large-Scale Vocabulary Test: Using UMLS technology to determine how well controlled vocabularies cover terminology needed for health care and public health. *Journal of the America Medical Informatics Association, 3(4):281-287.*

KETCHELL D. S.; FREEDMAN M. M.; JORDAN, W. E., LIGHTFOOT E. M., HEYANO S. and LIBBEY, P. A. (1996): Willow: A uniform search interface. *Journal of the American Medical Informatics Association, 3(1):27-37.*

LEHMAN, F. (ed.) (1992): Semantic Networks in Artificial Intelligence. Pergamon Press, Oxford.

LOWE H. J.; BUCHANAN B. G.; COOPER G. F. and VRIES J. K. (1995): Building a medical multimedia database system to integrate clinical information: an application of high-performance computing and communications technology. *Bulletin of the Medical Library Association, Jan, 83(1):57-64.*

LINDENBERG D. A. B.; HUMPHREYS, B. L. and McCRAY, A. T. (1993): The Unified Medical Language System. *Methods of Information in Medicine, 32, No. 4, 281-291.*

McCRAY, A. T. and RAZI, A. (1995): The UMLS Knowledge Source Server. In: *R. A. Greenes, H. E. Peterson, D. J. Protti (eds.), Proceedings of the Eighth World Congress on Medical Informatics: Medinfo '95, 144-147.*

McCRAY, A. T. and NELSON, S. J. (1995): The representation of meaning in the UMLS. *Methods of Information in Medicine, 34, No. 1/2, 193-201.*

McCRAY, A. T.; SRINIVASAN, S. and BROWNE, A. C. (1994): Lexical methods for managing variation in biomedical terminologies. In: *J. G. Ozbolt (ed.): Proceedings of the 18th Annual Symposium on Computer Applications in Medical Care, 235-239.*

QUIRK, R.; GREENBAUM, S.; LEECH, G. and SVARTVIK, J (1985): A comprehensive grammar of the English Language. London: Longman Publishers.

RODGERS, R. P. C. (1995): Automated retrieval from multiple disparate information sources: The World Wide Web and the NLM's Sourcerer project. *JASIS, 46, No. 10, 755-764.*

SOWA, J. F. (1984): Conceptual Structures: Information Processing in Mind and Machine. Addison-Wesley, Reading, MA.

Sun Protection of Children: Changing Patterns of Preventive Knowledge and Behaviour

A. Pfahlberg[1,2], O. Gefeller[1], U. Cleffmann[2], K. F. Kölmel[2]

[1] Abteilung Medizinische Statistik,
Georg-August-Universität Göttingen, D-37073 Göttingen, Germany
[2] Universitäts-Hautklinik Göttingen,
Georg-August-Universität Göttingen, D-37075 Göttingen, Germany

Summary: Sufficient epidemiologic evidence exists that exposure to ultraviolet light in childhood is an important risk factor for developing malignant melanoma, a form of skin cancer rapidly increasing in incidence around the world. Public health campaigns focusing on sun protection of children have therefore been developed in several countries. In this paper, an evaluation of the effects of the German campaign launched during the summer of 1993 is described. The design, conduct and results of two successive cross-sectional studies devoted to evaluate changes in the patterns of preventive knowledge and behaviour among parents of children in nursery schools of Göttingen are delineated. In addition, the methodologic considerations when dealing with such an applied problem of epidemiologic classification are discussed.

1. Introduction

Malignant melanoma of the skin is increasing rapidly in incidence in fair-skinned populations around the world: in most countries with a 5 – 7% yearly increase in the rate of new occurrences more rapidly than any other cancer (Swerdlow (1990)). There are strong indications from birth cohort analyses that this rise in incidence will continue; time-trend data even suggest that malignant melanoma will soon become one of the most common malignancies in all age groups. Consequently, this type of cancer is today a major public health problem in developed countries.

Numerous epidemiologic studies addressing the relationship between exposure to ultraviolet light and the occurrence of malignant melanoma revealed a strong association between an excessive sunlight exposure and a subsequent development of skin cancer (Koh (1991)). Especially a skin damage due a high amount of ultraviolet light during childhood could be identified as an important risk factor for this disease (Osterlind et al. (1988)). The influence of the timing of exposure to sunlight during the individual life span has been further corroborated by several migration studies in Australia and Israel (Dobson and Leeder (1982), Movshovitz and Modan (1973)). When comparing the melanoma risk of immigrants with that of native-born people of the same ethnic origin and skin type it became apparent that immigrants had always a lower risk (with a pronounced gradient depending on their age at immigration even when controlled for the length of stay in that country). These epidemiologic results stimulated the development of preventive concepts for primary prevention of malignant melanoma starting early in life.

Public health campaigns directed towards educating parents of young children have therefore been created in several countries (Bolognia et al. (1991)). These activities focused on improving the preventive knowledge and behaviour of the parents with respect to sun protection of their children (Kölmel et al. (1993)). In Germany such a campaign was launched by the "Kommission für Früherkennung von Hautkrebs" of the German Dermatological Society prior to the summer holidays of 1993.

In this paper, a brief description of the design, conduct and results of an epidemiologic study devoted to evaluate potential changes in the patterns of preventive knowledge and behaviour among parents of young children in Göttingen is given. In addition, the methodologic considerations when dealing with such an applied problem of epidemiologic classification are delineated.

2. Study Design

The education campaign devoted to the hazards of unprotected exposure to ultraviolet light during childhood initiated and coordinated by the "Kommission für Früherkennung von Hautkrebs" took place prior to the summer holidays of 1993. Several information brochures were distributed in pediatricians' waiting–rooms and by health insurance companies. In addition, advertisements were placed in special-purpose magazines reaching parents of young children (e.g. "Eltern") and spots ran on TV and radio channels. Moreover, press conferences were organised to inform interested journalists about the topic and to initiate successfully press stories in a wide spectrum of magazines ranging from news magazines ("Der Spiegel", "Stern", "Focus") to yellow press gazettes ("Das Goldene Blatt", "Frau im Spiegel", "Die Aktuelle" etc.).

To evaluate the effects of these educational efforts on the preventive knowledge and behaviour of parents with young children as well as on the frequency of skin damages due to excessive sun exposure among their children, an epidemiologic study within the city of Göttingen, the capital of the southern part of Lower Saxony with about 130 000 inhabitants, was conducted. The target population of the study consisted of parents with children in the age of 3 – 6 years living in Göttingen. The sampling was realised by approaching all 56 nursery schools in Göttingen and distributing (with the help of the nursery–school teachers) a standardized questionnaire to all parents with children in these nursery schools. This procedure was implemented first in spring 1993, prior to the start of any activities of the educational campaign in Germany. A second identical approach was employed using the same questionnaire in autumn 1994, one year after the "hot phase" of the educational activities. No individual follow–up could be performed, the study was realised as two successive cross–sectional studies using the same methods in the same target population.

3. Classification of Knowledge and Behaviour

The questionnaire used in both cross-sectional studies contained three questions concerning the parents' knowledge about the dangers of exposure to ultraviolet light. First, the parents' opinion to several general statements on the topic (such as "Brown skin is healthy skin") was assessed on a four point scale ranging from "agree totally" to "disagree totally". Second, the parents were asked about the necessity of sun protection for their children in different situations (e.g."during hot hours at noon", "at cloudy days", "when the child's skin is dry"). Again, these answers were recorded on a four point scale. In a last question the parents had judge which factors increase the risk of developing melanoma. To this end, they could select risk factors from a prespecified list of nine exposures which included strong true risk factors (such as number of nevi, frequency of sunburns during childhood etc.) as well as exposures attracting public awareness, but which have not been identified as factors increasing melanoma risk (such as presence of allergies, rising air pollution, diet). From the parents' answers a score was built for each of the three questions separately.

It turned out, however, that only the score related to the melanoma risk factors revealed a sufficient variability. Therefore, further analyses on the parents' preventive knowledge were based on this score only. The score was further classified into three categories ("low", "medium", "high knowledge") based on a priori definition how many correct answers are necessary to qualify for the corresponding categories.

The parents' behaviour concerning sun protection of their children was assessed separately for two typical situations: playing at a beach and staying in a garden or a public playground. Questions were asked how the children are clothed in these situations, whether sun creams are used (and if yes, how often and which sun protection factor is employed) and whether the parents try to keep their children playing in the shade. The pattern of answers were used to classify the parents' behaviour into one of five predefined ordered groups with respect to the degree of sun protection translated into action separately for both situations. The "best" group I of this classification scheme "protected child" was defined by a sufficient textile sun protection (including a sun cap). If the parents did not keep their children wearing a sun cap (with otherwise sufficient textile sun protection), this behaviour was classified as group II. In the case of an insufficient textile sun protection but a frequent stay in the shade a classification into group III followed. When instead of textile sun protection and the avoidance of intensive sun exposure only a frequent use of sun creams with a high sun protection factor was reported, this response pattern led to a classification into group IV. All remaining types of behaviour were combined in group V.

The frequency of sunburns and skin reddenings were assessed in two questions related to the child's skin damages during the summer preceding the corresponding study (1992 and 1994, respectively). The categorization of answers to the two questions into five classes (frequency of sunburns/skin reddenings: 0 / 0, 0 / 1, 0 / ≥ 2, 1 / $\cdot$, ≥ 2 / $\cdot$) reflects the fact that sunburns are a major skin damage, whereas the deleterious effect of skin reddening is only of minor importance.

4. Methodologic Aspects

The design of the present investigation as two successive cross–sectional studies has a major limitation with respect of its ability to prove the efficacy of the public education campaign. Since we have no individual follow–up any changes in the response patterns cannot directly be attributed to the campaign. For example, the composition of the two study samples concerning relevant variables influencing the parents' answering behaviour could have changed. Thus, the analysis of the data has not been realized as a simple pre–post comparison where the statistical significance of differences between the studies is assessed by a univariate testing of the four variables of primary interest, the preventive knowledge on melanoma risk factors, sun–protective behaviour at the beach and the garden as well as sun–induced skin damages. We identified the child's sex as a potential confounder of the analysis as the sex distribution slightly differed between the studies and a remarkable relationship between the child's sex and the sun–protective behaviour was apparent. In both studies boys were much better protected than girls (at the beach and in the garden).

To evaluate potential changes in response patterns between the studies accounting for this confounding and the interrelations between the factors we therefore employed a log–linear modelling approach. We started with a saturated model including the five variables (study, child's sex, knowledge on melanoma risk factors, preventive behaviour, frequency of skin damages in the preceding summer) as well as all first, second, third and fourth order interaction terms between the variables. Two separate models were analysed, one for the preventive behaviour at the beach, and one for the situation in the garden. Their general form was:

$$\begin{aligned}
\log m_{ijklm} &= u + u_{1(i)} + u_{2(j)} + u_{3(k)} + u_{4(l)} + u_{5(m)} \\
&+ u_{12(ij)} + u_{13(ik)} + \ldots + u_{45(lm)} \\
&+ u_{123(ijk)} + u_{124(ijl)} + \ldots + u_{345(klm)} \\
&+ u_{1234(ijkl)} + u_{1235(ijkm)} + \ldots + u_{2345(jklm)} \\
&+ u_{12345(ijklm)} ,
\end{aligned}$$

where $m_{ijklm} := E(n_{ijklm})$ denotes the expected cell frequency, $i = 1,2$ (study), $j = 1,2$ (child's sex), $k = 1,2,3$ (knowledge), $l = 1,2,\ldots,5$ (behaviour) and $m = 1,2,\ldots,5$ (skin damage). The usual constraints on the model parameters (i.e. $\sum_i u_{1(i)} = 0, \sum_j u_{2(j)} = 0, \ldots, \sum_m u_{12345(ijklm)} = 0$) are assumed to hold to guarantee their estimability (Andersen (1994)). A backward–oriented elimination procedure was then implemented to delete unnecessary model parameters. This procedure, starting at the fourth order interaction term and then proceeding step by step to lower order interaction terms, took the "hierarchy principle" into account, i.e. if an interaction term stayed in the model then all factors involved in this interaction are automatically included in the model as main effects and in lower order interaction terms. The significance level defined for the backward elimination to delete automatically a model term was 0.1. In addition, for model terms associated with p–values in the range 0.05 – 0.1 a manual comparison of the goodness of fit between models with and without the corresponding parameter was performed. Such an semiautomatic procedure for model selection in log–linear analyses has been advocated as a good compromise between the objectivity of the final model and the adaptation of the analysis to context–specific needs (Bishop et al. (1975)).

5. Results

The response to the first study during the spring of 1993 ran up to 1341 completed questionnaires. Based on 2536 distributed questionnaires the proportion of participating parents was 52.9%. In the second study during the autumn of 1994 1150 completed questionnaires (out of 2302 distributed ones) were returned which led to a response of 50.0%. The information on the relative response has to be viewed, however, as a lower bound of actual response. The denominator of the calculation, the number of distributed questionnaires, is too high for the following reasons: (i) due to typical short–term fluctuations in the population of a nursery school all 56 nursery schools in Göttingen received more questionnaires than they had previously reported as their current capacity, (ii) some children did not show up in their nursery school during the study period (e.g. due to illness), (iii) not infrequently the same parents had more than one child in the nursery school which resulted in a double counting of these parents in the denominator of response calculations.

The distribution of the different classification schemes regarding preventive knowledge and behaviour in the two studies are compared in table 1. A consistent picture is apparent from these results: the remarkable shift towards a better knowledge on melanoma risk factors and a better sun protection of the children reported by the parents in the second study is accompanied by a pronounced reduction in the frequency of sun–induced skin damage. In the second study half of the children had no sun–induced skin damage at all and only 7% suffered from sunburns.

The methodologic problems of the interpretation of such successive cross–sectional studies leading to a log–linear analysis of the data have been delineated in the previous section. The two final models resulting from the application of our model selection procedure are shown in table 2. In both situations only main effects and first order interaction terms stayed in the model which simplifies the interpretation of the results. In the beach and the

	First study Spring 1993 (n = 1341) %	Second study Autumn 1994 (n = 1150) %
Score based on answers to risk factors for melanoma		
high	20.2	31.0
medium	63.8	59.1
low	15.9	9.9
Classification "protected child" Situation: beach		
I	21.0	33.7
II	4.0	3.9
III	28.3	32.1
IV	21.3	17.6
V	25.4	12.7
Classification "protected child" Situation: garden		
I	36.5	57.8
II	47.9	30.4
III	4.5	6.1
IV	4.2	3.4
V	7.0	2.3
Frequency of sunburns (S) and skin reddenings (R) S / R		
0 / 0	38.6	50.3
0 / 1	31.7	31.8
0 / ≥ 2	18.6	11.2
1 / ·	9.4	6.4
≥ 2 / ·	1.6	0.4

Table 1: Comparison of results concerning preventive knowledge, behaviour and skin exposure from two cross-sectional studies in Göttingen

Situation: beach
$\log m_{ijklm} = u + u_{1(i)} + u_{2(j)} + u_{3(k)} + u_{4(l)} + u_{5(m)}$ $+ u_{13(ik)} + u_{14(il)} + u_{15(im)}$ $+ u_{24(jl)}$ $+ u_{34(kl)} + u_{35(km)}$ $+ u_{45(lm)}$
Situation: garden
$\log m_{ijklm} = u + u_{1(i)} + u_{2(j)} + u_{3(k)} + u_{4(l)} + u_{5(m)}$ $+ u_{13(ik)} + u_{14(il)} + u_{15(im)}$ $+ u_{24(jl)}$

Table 2: Results of the log-linear analysis: Final models after application of the backward elimination

garden situation, a significant association between the study and all three factors of primary interest for the evaluation of the effect of the campaign (knowledge, behaviour and skin damages) could be verified. These results demonstrate thus that the remarkable differences between studies seen with respect to the univariate distributions of these factors in table 1 are still valid when analysed in multifactorial framework. Furthermore, a significant difference in the parents' preventive behaviour depending on the child's sex could be observed in both models. The relationships between preventive knowledge, sun–protective behaviour, and skin damages were only significant for the beach situation.

A similar trend (higher knowledge ⇒ better sun–protective behaviour and less skin damages, better sun–protective behaviour ⇒ less skin damages) could also be found – but only to a much lesser extent – for the garden situation. These interaction terms dropped thus out off this model during the selection process (the interaction term between preventive behaviour and skin damages as the last one with a p-value of 0.085, but with no substantial loss of model fit when omitted from the model).

6. Discussion

In the light of the rising incidence of malignant melanoma sun protection of children has become an important part of the preventive activities developed to counter this epidemic. The controversial discussion about the merits of public education campaigns revolve round the question: do these campaigns really increase the preventive knowledge and alter the preventive behaviour of the target population? In this paper we have described the results of two cross–sectional studies in Göttingen conducted prior and after an intensive public education campaign on the nationwide level in Germany to evaluate potential effects of these activities among parents of children in nursery schools. Based on large samples we found a consistent picture of improvement in preventive knowledge and behaviour after the campaign. In addition, the frequency of sun-induced skin damages dropped remarkably. Notwithstanding the methodologic problems of this analysis we feel confident to conclude that the German campaign had a pronounced beneficial effect on the level of sun protection for children achieved in Göttingen. The public education activities should thus be implemented as a continuing process to further improve the parents' preventive knowledge and behaviour on this topic and to apply successfully new methods in the primary prevention of malignant melanoma.

Acknowledgements

The study has been partially supported by a grant of the Niedersächsische Krebsgesellschaft. The authors thank Mrs B. Kropik, Ms H. Thölking and all nursery-school teachers in Göttingen for their help during the field phase of the study.

References:

ANDERSEN, E. B. (1994). The Statistical Analysis of Categorical Data – Third Edition. Springer-Verlag, Berlin.

BISHOP, M. M. Y.; FIENBERG, S. E. and HOLLAND, P. W. (1975): Discrete Multivariate Analysis: Theory and Practice. MIT–Press, Cambridge

BOLOGNIA, J. L.; BERWIG, M.; FINE, J. et al. (1991): Sun protection in newborns. *American Journal of Disease in Children, 145, 1125–1129.*

DOBSON, A. J. and LEEDER, S. R. (1982): Mortality from malignant melanoma in Australia: Effects due to country of birth. *International Journal of Epidemiology, 11, 207–211.*

KOH, H. K. (1991): Cutaneous melanoma. *New England Journal of Medicine, 325, 171–182.*

KÖLMEL, K. F.; ROSER, M.; CLEFFMANN, U. and BREITBART, W. (1993): Sonne, Kind und Melanome. Melanomprävention im Kindesalter. *Der Kinderarzt, 24, 470–481.*

MOVSHOVITZ, M. and MODAN, B. (1973): Role of sun exposure in the etiology of malignant melanoma: Epidemiologic Inference. *Journal of the National Cancer Institute, 51, 777–779.*

OSTERLIND, A.; TUCKER, M. A.; STONE, B. J. and JENSEN, O. M. (1988): The Danish case control study of cutaneous malignant melanoma. II. Importance of UV–light exposure. *International Journal of Cancer, 42, 319–324.*

SWERDLOW, A. J. (1990). International trend in cutaneous melanoma. *Annals of the New York Academy of Science, 609, 235–251.*

A Natural Language Understanding System for Knowledge-Based Analysis of Medical Texts

Martin Romacker[1], Klemens Schnattinger[1], Udo Hahn[1], Stefan Schulz[2], Rüdiger Klar[2]

[1]Freiburg University, CLIF Computational Linguistics Lab
Europaplatz 1, D-79085 Freiburg, Germany
[2]Freiburg University Hospital, Department of Medical Informatics
Stefan-Meier-Str. 26, D-79104 Freiburg, Germany
{romacker,schnattinger,hahn}@coling.uni-freiburg.de
{klar,schulz}@amil.ukl.uni-freiburg.de

Summary: An approach to knowledge-based text understanding of real-world texts from the medical domain (viz. gastro-intestinal findings) is presented. We survey major methodological features of an object-oriented, fully lexicalized, dependency-based grammar model which is tightly linked to domain knowledge representations based on description logics. The parser adheres to the principles of robustness, incrementality and concurrency. The substrate of automatic knowledge acquisition are text knowledge bases generated by the parser from medical narratives, which represent major portions of the content of these documents.

1. Introduction

Information contained in medical free texts (such as patient reports or discharge summaries) is relevant to a variety of different retrieval, coding and inference purposes. It should, for instance, provide support for medical decision making, for mapping data into medical coding systems or for quality assurance of medical treatment. With the growing availability of medical documents in machine-readable form, procedures for automatically analyzing and formatting textual data gain more and more importance, because hand-coding and manual indexing are time-consuming and usually error-prone. Therefore, automatic knowledge acquisition from medical free texts is highly desirable for hospital data management.

We currently port a German language text understanding system originally developed for the domain of information technology to the medical domain. In this paper, we will outline the basic methodological principles of our approach and illustrate them briefly with an example taken from a sample medical text.

2. Challenges of Medical Text Understanding

Acquiring knowledge from real-world, routinely produced medical texts at a reasonable level of detail and accuracy requires to meet at least two fundamental challenges – a methodological and an engineering one.

The *methodological* challenge consists of supplying an almost excessive amount of different kinds of knowledge, which are needed for proper text understanding. Of primary importance are grammar specifications capturing the linguistic knowledge – not only at the lexical (morphology), phrasal, clausal and sentential level (syntax, semantics), but also at the discourse level. By this, we mean phenomena of local as well as global coherence of texts as exemplified by (pro)nominal anaphora (Strube and Hahn, 1996) or functional anaphora (Hahn and Strube, 1996) and more global patterns of thematic progression in texts (Hahn, 1990). Unless these *discourse coherence* phenomena are adequately dealt with, referentially invalid and incohesive text knowledge bases are likely to emerge, thus seriously deteriorating the quality and subsequent usability of the acquired knowledge. Structural (i.e., syntax-based) linguistic descriptions, however, as sophisticated as they may be, are still not expressive enough for comprehensive inferencing and, thus, have to be backed up by conceptual knowledge of the underlying domain. This requires to specify a consistent, balanced *ontology*, with descriptions pertaining to major concepts (categories), their hierarchical and aggregational interrelations, and their incorporation into rule expressions to allow for various inference modes. Both levels are linked by mediating semantic representations at the linguistic level, which relate to conceptual specifications at the ontological (sometimes also called the encyclopedic) level.

The *engineering* challenge can be characterized by the observation that, at least in realistic text understanding scenarios, sufficiently complete and in-depth specifications at the linguistic and conceptual level are the exception rather than the rule. Hence, appropriate processing strategies have to be supplied to proceed with text understanding even in the presence of lacking or underspecified knowledge. Ideally, the degradation of the text understander's performance should be proportional to the degree of linguistic and conceptual underspecification encountered. Together with the observation that realistic text understanders also have to cope with ungrammatical (ill-formed) input, an inherent potential for *robustness* has to be built into these systems. Ignoring the quest for robust processing either leads to a huge knowledge engineering overhead (*viz.* advance manual coding of all lacking pieces of information) or fatal system behavior, *viz.* blockade at every occurrence of an unknown lexical item or an extra- or ungrammatical expression. Note also that this engineering challenge can easily be transferred into a methodological one by supplying text understanders with automatic learning facilities in order to actively account for novel input.

In the remainder of this paper, we elaborate in greater detail on our approach to text knowledge acquisition based on the above methodological principles. We present a model of automatic text understanding, with emphasis on the knowledge sources used and their underlying theoretical constructs. Procedural aspects of text parsing, e.g., robustness, preference and prediction strategies (cf. Neuhaus and Hahn, 1996), as well as issues relating to auto-

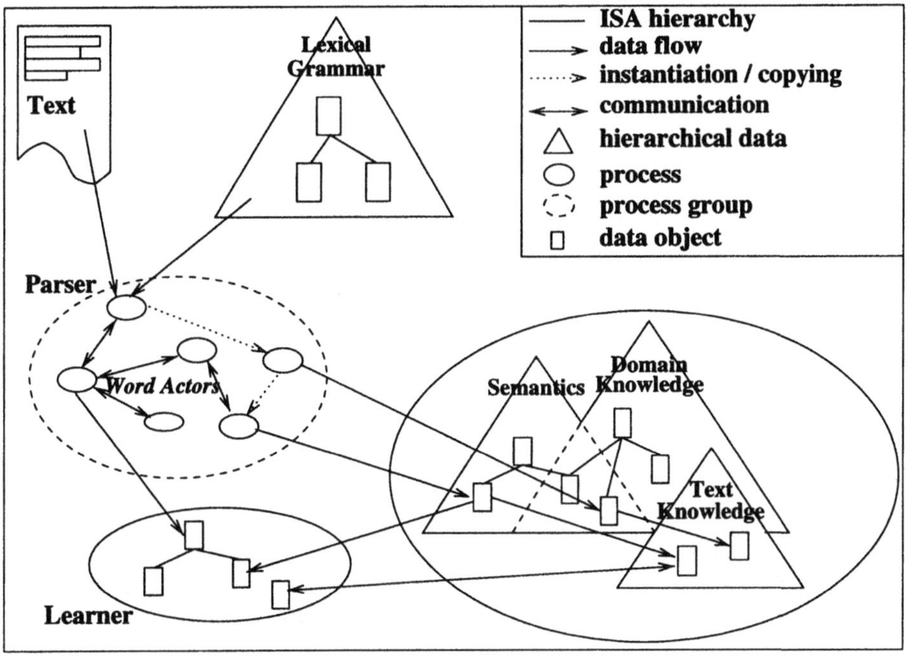

Figure 1: System Architecture

matic concept learning (cf. Hahn et al. 1996) are given almost no attention in this paper.

3. Architecture of the Text Understanding Kernel System

We here consider the basic knowledge sources and their organization in terms of the architecture (see Fig. 1) underlying the text understanding kernel system PARSETALK (for a survey, (cf. Hahn et al. 1994).

Linguistic knowledge comes with a completely lexicalized, head-oriented *grammar* system. Each lexical item is associated with specifications of its part of speech (e.g., noun, verb, preposition), morphosyntactic features (e.g., gender, case, mode, tense marking), word order constraints and a valency frame which specifies the potential modifiers the lexical item may govern as a head. Part of the valency description are semantic compatibility criteria (ako sortal constraints) between the head and each modifier.

The lexical distribution of declarative grammatical knowledge is complemented by the provision of lexicalized control knowledge. Knowledge of this sort is locally encoded in terms of scripts, i.e., collections of methods by which the behavior of a lexical parsing process is determined upon the reception of an incoming message. The object-oriented specification style of the lexical grammar accounts for the variety of control patterns between hetero-

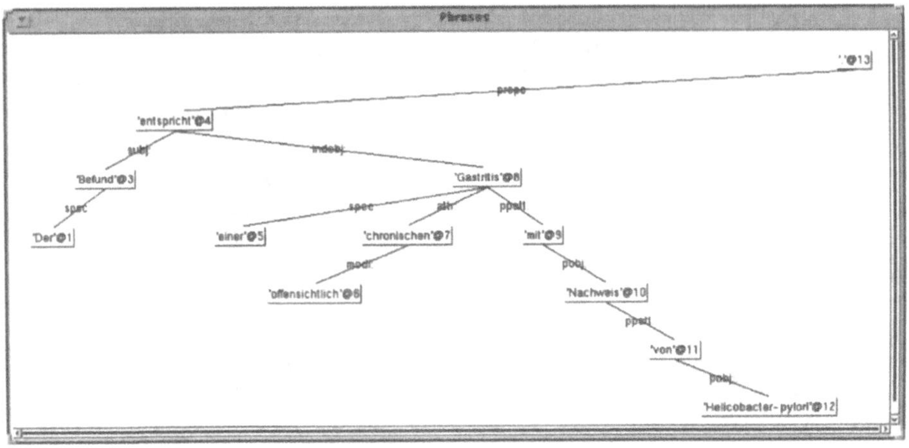

Figure 2: Sample Dependency Tree

geneous knowledge sources and the flexible switching of strategies to ensure efficient as well as robust system performance. An ubiquitous feature of the PARSETALK system are various forms of inheritance hierarchies, e.g., for parts of speech, morphosyntactic features, by which the size and complexity of the grammar specifications can be kept manageable and consistent. The lexical items form the leave nodes of such hierarchies, while inner nodes contain linguistic generalizations based, e.g., on distribution patterns for parts of speech.

Viewed from the angle of the *parser*, whenever a lexical item is read from the textual input and it is identified as an entry in the lexicon, a lexical process, we refer to as *word actor*, is instantiated. Word actors combine lexicalized grammar knowledge and positional information from the text. They drive the parsing process by way of message passing using well-defined, linguistically motivated protocols (cf. Neuhaus and Hahn (1996), for a detailed description). Each of these protocols accounts for different functional relations lexical items may join, e.g., dependency relations (Neuhaus and Hahn, 1996), anaphoric relations (Hahn and Strube, 1996), etc. The parsing process is carried out incrementally, concurrently and, due to efficiency reasons, basically incomplete (for a motivation, cf. Neuhaus and Hahn, 1996). The target structure produced by the interactions of word actors at the syntax level are dependency trees labeled with functional relations such as SUBJ (subject), INDOBJ (indirect object) (cf. Fig. 2). The need for a *domain knowledge* base within a formally specified framework for natural language understanding is getting more and more accepted in the field of medical informatics (e.g., Campbell et al. 1994). It requires medical knowledge to be expressed in a sound and consistent manner. As a consequence, conceptual taxonomies, for instance, can be computed from formalized concept descriptions rather than being predefined. Though knowledge engineering in an ontologically coherent (sub)domain should make use of existing knowledge compilations, e.g., ICD-10, SNOMED, MESH, UMLS, as much as

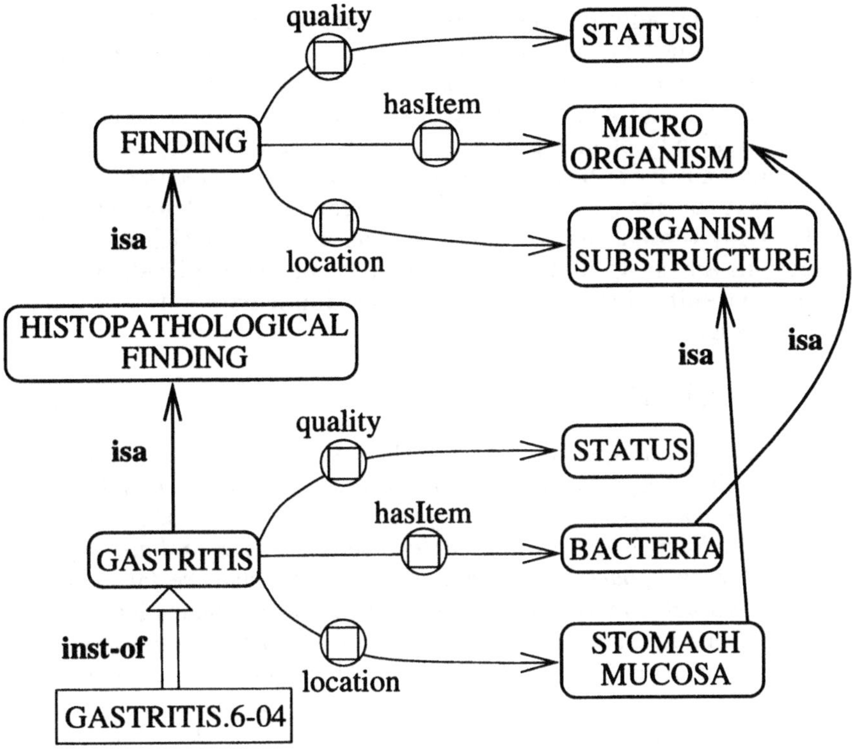

Figure 3: Fragment of the Conceptual Hierarchy of a Pathological Domain

possible, due to the lack of formal specification and an insufficient level of granularity only a limited amount of this knowledge can be reused even if it is judged transferable, in principle (Carenini and Moore, 1993).

In our approach, we chose a representation formalism based on terminological logics to incorporate semantic and conceptual constraints in the parsing process. The domain knowledge base is built from concept definitions that are organized in an acyclic directed graph by subsumption (subconcept) relations. (For a survey of major properties of terminological languages, cf. Woods and Schmolze, 1992).

In the example given in Fig. 3, the concept FINDING subsumes the concept GASTRITIS. Both have the same set of roles, but the subsumption relation is inferred from the range restrictions for the role LOCATION (the concept ORGANISMSUBSTRUCTURE is specialized by the concept STOMACHSUBSTRUCTURE) and for HASEVIDENCE (the concept MICROORGANISM is specialized by the concept BACTERIA). Linguistic and conceptual knowledge are linked in two ways. First, any valency specification contains a semantic compatibility check, whose evaluation is directly transferred to and executed in the domain knowledge base. Second, any semantic computation based on confirmed syntactic structures (e.g., the processing of intermediate structures for prepositional attachments, modal verbs, quantifiers or pronouns) is carried out in an autonomous context. Contexts are data structures which are fully embedded in the same terminological reasoning system we use for the representation and processing of domain knowledge. Since the coupling of the semantic and conceptual level of analysis is pretty tight, the overwhelming majority of interpretation processes can directly be executed in the domain knowledge base (more precisely, the *text knowledge* base constituting a text-specific instantiation of the original domain knowledge base; cf. Fig. 1). Only those semantic interpretation processes which require longer transactions are encapsulated at the semantic processing level. After termination, the results of these computations are immediately transferred to the text knowledge base.

4. A Sample Parse

In the following example, we focus on the interplay between syntactic analysis, semantic interpretation and conceptual reasoning. Consider the sentence: *"Der vorläufige Befund entspricht einer offensichtlich chronischen Gastritis mit Nachweis von Helicobacter-pylori." ("The preliminary findings correspond to an apparently chronic gastritis with evidence for helicobacter-pylori.")*. Syntactic analysis of the sentence produces a dependency structure as depicted in Fig. 2, where nodes correspond to word actors marked with an integer indicating their text position. Edges between nodes indicate an established dependency relation labeled by the corresponding relation tag. Note that due to a lacking lexical specification the text item *"vorläufige"* on position 2 has been skipped (cf. Neuhaus and Hahn, 1996 for a corresponding protocol description), while due to lacking conceptual specifications the text item *"offensichtlich"* on position 6 though still covered by the parse will, finally, receive no conceptual interpretation. Fig. 4 contains the conceptual representation from the text knowledge base as generated for the sample sentence. *Gastritis*.8-04 denotes an instance of the concept class CHRON-GASTRITIS; the first digit (8) relates this instance to the occurrence of the word "gastritis" at text position 8, while the second digit (04) is supplied by the name server of the knowledge base management system. (It guarantees the unique name condition). Compared to the dependency tree, the concept graph looks by far "slimmer" due to the normalizing operation of semantic rules.

Our aim in semantic interpretation is to categorize on the epistemological status of concepts in order to avoid an intractable number of interpretation rules. Therefore, all semantic interpretation rules are attached to categories which are inherited by concepts as with the conceptual counterpart CORRESPOND *("to correspond to")* of the surface lexeme *"entspricht"*. Furthermore, rules are attached to parts of speech that make exclusively part of a special semantic level (e.g., prepositions, predicative verbs). The semantic interpretation rule for CORRESPOND (cf. Fig. 6), for instance, establishes a co-reference relation between the fillers of the roles SUBJECT and INDIR-OBJECT at the conceptual level. Since (cf. Fig. 5) the filler of the role

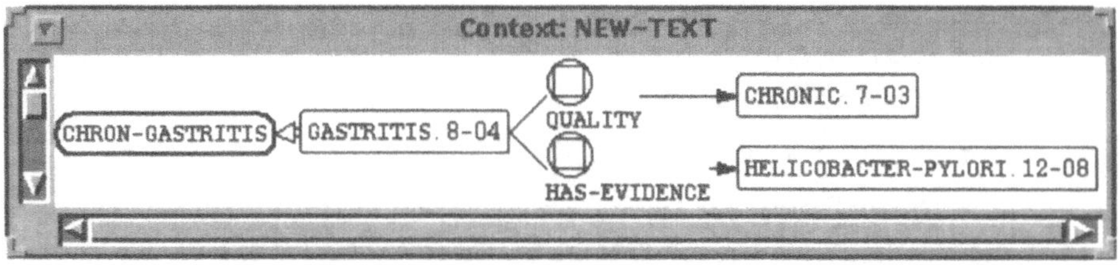

Figure 4: Sample Concept Graph

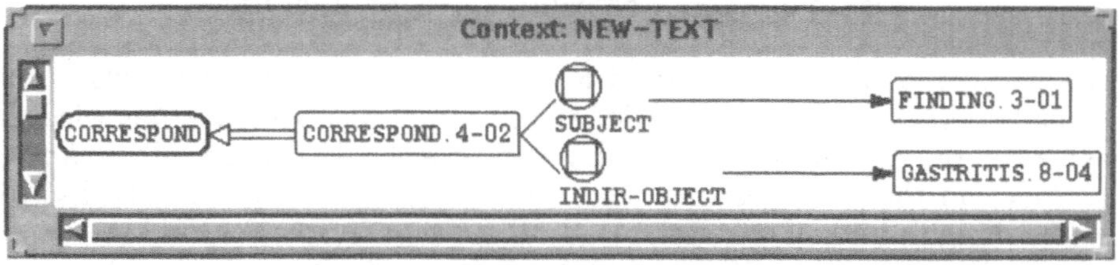

Figure 5: Sample Semantic Graph

INDIR-OBJECT, *Gastritis*.8-04, specializes (is subsumed by) the filler of the role SUBJECT, *Finding*.3-01 (according to the ISA relation stated in Fig. 3), both can be merged by a recursive attribute unification algorithm (the function **subsumes+**). Furthermore, the verb instance *Correspond*.4-02 can be removed, because its "meaning" is already encoded in the structures resulting from the merge. An analogous interpretation rule handles the conceptual attachment of the instance *Helicobacter-pylori*.12-08 to the instance *Gastritis*.8-04 via the role HASEVIDENCE. Finally, the classifier of the terminological knowledge representation system assigns *Gastritis*.8-04 as an instance of the concept CHRON-GASTRITIS *("chronic gastritis")*. As a final result of the text understanding process, various linguistic surface expressions are mapped to canonical representation structures from the underlying domain knowledge base. Continuously proceeding this way, sentence by sentence, the text knowledge base gets incrementally augmented by the processes of conceptual interpretation and knowledge integration.

5. Related Work

In the last few years, there has been a growing interest in developing text analysis system that yield semantic representations of medical narratives for further information access (Berrut, 1990; Volot et al., 1993; Zweigenbaum et al., 1995). The foundations in this area were already laid in the seventies by the LINGUISTIC STRING PROJECT (LSP). LSP provides a fully developed, broad coverage medical language processing system (Sager et al., 1987). Narrative texts are syntactically analyzed by taking semantic restrictions for possible constituent structures into consideration that are defined for the medical sublanguage. So-called *information formats* serve as templates for representing characteristic statement types for the sublan-

EXISTS v, s, o :
v : CORRESPOND $\sqcap$
v SUBJECT s $\sqcap$
v INDIR-OBJECT o $\Longrightarrow$
IF (subsumes+(s,o) OR subsumes+ (o,s) THEN
TELL s CO-REFERENT o

Figure 6: Sample Semantic Interpretation Rule

guage (Sager et al., 1994). In the LSP system, however, no strict separation is made between syntactic knowledge and semantic/conceptual knowledge. Semantic restrictions for establishing syntactical relations are all *explicitly* precoded in the grammar rules and must therefore be entirely anticipated in the design phase of the system. By the use of terminological knowledge representation systems, these restrictions can be inferred at run-time from the concept definitions.

(Baud et al., 1992) derive a semantic representation of medical narratives by exploiting the conceptual relations of proximate words in a sentence using a simple pattern matching method for syntactic analysis. Based on the clustering of words in sentences, conceptual graphs are used as knowledge respresentation formalism (Sowa, 1984) to determine the semantic relations. Although being very effective for the task the system is designed for, this approach is likely to run into problems when it has to deal with text phenomena and more complex sentence structures like coordination.

In the GALEN project (Rector et al., 1994) the task of knowledge acquisition from medical narratives has led to a division of the system into several logically independent components. While there are several source languages (English, German and French) to be parsed, the representational structures are considered language independent. A strict separation between the concept representation system (GRAIL) and the medical coding schemes is realized in GALEN. As in our approach, the KL-ONE-like knowledge representation language GRAIL prohibits syntactic structure building for semantically invalid statements on the basis of formally specified concept descriptions, though a fully developed natural language parsing module is currently not available within the GALEN system.

6. Conclusions

We sketched the major building blocks of a system for automatic knowledge acquisition from texts, which integrates heterogenous knowledge sources with different representation formalisms in its architecture. Emphasis was laid on the interplay between syntactic analysis, semantic interpretation and conceptual reasoning. Since we are dealing with "realistic" texts, ungrammatical and extragrammatical input must be processed. Hence, robustness turns out to be a primary issue for the design of realistic text understanding systems.

The knowledge acquisition prototype for German language medical texts we have implemented is fed with input from the medical documentation center of the Hospital of Freiburg University (Klar et al., 1991). We have concentrated so far on the domain of gastro-intestinal diseases. The lexicon currently contains about 1,500 entries, the medical knowledge base consists of approximately 550 concepts and 350 roles. The PARSETALK system is implemented in an actor dialect of Smalltalk, while the knowledge base is implemented in LOOM (MacGregor, 1994). The prototype so far has only been evaluated with respect to specialized functionalities, e.g., its potential for text coherence analysis (Strube and Hahn, 1996) or parsing efficiency (Neuhaus and Hahn, 1996), while a comprehensive and focused investigation of its information system capabilities is still under way.

Acknowledgments

We like to thank the members of our groups for fruitful discussions. We also gratefully acknowledge the provision of the LOOM system from USC/ISI.

References:

BAUD, R.; RASSONOUX, A. and SCHERRER, J. (1992): Natural language processing and semantical representation of medical texts. *Methods of Information in Medicine, 31(2), 117-125.*

BERRUT, C. (1990): Indexing medical reports: the RIME approach. *Information Processing and Management, 26(1), pp. 93-109.*

CAMPBELL, K.; DAS, A. and MUSEN M. (1994): A logical foundation for representation of clinical data. *Journal of the American Medical Informatics Association, 1(3), pp. 218-232.*

CARENINI, G. and MOORE, J. (1993): Using the UMLS Semantic Network as a basis for constructing a terminological knowledge base: a preliminary report. In: SCAMC'93 - Proc. 17th Annual Symposium on Computer Applications in Medical Care, pp. 725-729.

HAHN, U. (1990): Topic parsing: accounting for text macro structures in full-text analysis, *Information Processing & Management, 26(1),pp. 135-170*

HAHN, U.; KLENNER, M. and SCHNATTINGER, K. (1996): Learning from texts: a terminological metareasoning perspective. In: Wermter, S. and Riloff, E. and Scheler, G. (Eds.), Connectionist, Statistical and Symbolic Approaches to Learning for Natural Language Processing. Berlin: Springer, pp. 453-468. (LNAI, 1040)

HAHN, U. and SCHACHT, S. and BRÖKER, N. (1994): Concurrent, object-oriented dependency parsing: the PARSETALK model. *International Journal of Human-Computer Studies, 41(1-2), pp. 179-222*

HAHN, U. and STRUBE, M. (1996): PARSETALK about functional anaphora. In: G. McCalla (Ed.), Advances in Artificial Intelligence. Proc. 11th Biennial Conf. of the Canadian Society for Computational Studies of Intelligence (AI'96). Toronto, Ontario, Canada, May 21-24, 1996. Berlin: Springer, pp. 133-145. (LNAI, 1081)

KLAR, R.; ZAIß, A.; TIMMERMANN, A. and SCHRADER, U. (1991): The Information System of the Freiburger University Hospital. In: Adlassnig, K. et al. (Eds.): MIE - Proc. Medical Informatics Europe, Vienna, Austria, August 1991. Berlin: Springer, pp. 46-50.

MacGREGOR, R. (1994): A description classifier for the predicate calculus. In: AAAI'94 - Proc. 12th National Conf. on Artificial Intelligence. Seattle, Wash., July 31 - August 4, 1994. Menlo Park: AAAI Press/MIT Press, pp. 213-220.

NEUHAUS, P. and HAHN, U. (1996): Trading off completeness for efficiency: the PARSETALK performance grammar approach to real-world text parsing. In: FLAIRS'96 - Proc. 9th Florida Artificial Intelligence Research Symposium. Key West, Florida, May 20-22 pp. 60-65.

RECTOR, A. and NOWLAN, W. (1994): The GALEN project. *Computer Methods and Programs in Biomedicine, 45(1-2), pp. 75-78.*

SAGER, N.; FRIEDMAN, C. and LYMAN, M. (1987): Medical Language Processing. Computer Management of Narrative Text. Reading, MA: Addison-Wesley.

SAGER, N.; LYMAN, M.; BUCKNALL, C.; NHAN, N. and TICK, L. (1994): Natural language processing and the representation of clinical data. *Journal of the American Medical Informatics Association, 1(2), pp. 142-160.*

SOWA, J. (1984): Conceptual Structures: Information Processing in Mind and Machine. Reading, MA: Addison-Wesley.

STRUBE, M. and HAHN, U. (1996): Functional centering. In: ACL'96 - Proc. 34th Annual Meeting of the Association for Computational Linguistics. Santa Cruz, California, June 23-28, 1996. San Francisco, CA: Morgan Kaufmann, pp. 270-277.

VOLOT, F.; ZWEIGENBAUM, P.; BACHIMONT, B.; BEN SAID, M.; BOUAUD, J.; FIESCHI, M. and BOISVIEUX, J. (1993): Structuration and acquisition of medical knowledge. Using UMLS in the conceptual graph formalism. In: SCAMC'93 - Proc. 17th Annual Symposium on Computer Applications in Medical Care, pp. 710-714.

WOODS, W. and SCHMOLZE, J. (1992): The KL-ONE family. *Computers & Mathematics with Applications, 23(2-5), pp. 133-177*

ZWEIGENBAUM, P.; BACHIMONT, B.; BOUAUD, J.; CHARLET, J. and BOISVIEUX, J. (1995): A multi-lingual architecture for building a normalised conceptual representation from medical language. In: SCAMC'95 - Proc. 19th Annual Symposium on Computer Applications in Medical Care, pp. 357-361.

On the Development and Validation of Classification Schemes in Survival Data

Willi Sauerbrei

Institute of Medical Biometry and Informatics, University of Freiburg, Germany

Summary: The prognostic ability of classification schemes is often overestimated. Some of the reasons may be the tendency for too complex models which overfit the data and a lack of investigations on model stability. Using a breast cancer study, we will discuss aspects of model complexity, stability, shrinkage and overoptimism within the framework of Cox regression models and CART-trees. We controlled model complexity by the selection level of stepwise variable selection strategies and used crossvalidation for the further investigations.

1. Introduction

Classification schemes are important for the diagnosis and treatment of patients. In breast cancer, several schemes are published, but with the expection of the Nottingham Prognostic Index (NPI, Haybittle et al. 1982) none of them seems to be sufficiently validated in the literature. We used a study in patients with node negative breast cancer of the German Breast Cancer Study Group (GBSG) to validate some of the literature proposals and to develop new ones (Sauerbrei et al. 1996a). In contrast to the NPI, two classificaton schemes for node negative patients according to Glick et al. (1992) and Rubens (1992) did not discriminate the different groups as proposed in the original papers.

Concerning survival data, two common methods for the development of classification schemes are based either on Cox regression models or on classification and regression trees (CART). Also both approaches are conceptionally very different, both share the common problem of overoptimism concerning the predicitve ability, if the 'final' models are based on too many factors. Using crossvalidation, we will discuss the importance of aspects of model and tree complexity and stability for the development and validation of classification schemes.

2. Material and methods

In the Cox regression or proportional hazards model (Cox 1972), the risk function for a patient with values $(X_1, X_2, \ldots X_p)$ of the progostic factors is

$$\lambda_0(t)\ exp(\beta_1 X_1 + \beta_2 X_2 + \ \ldots \ + \beta_p X_p)$$

where $\lambda_0(t)$ is the baseline hazard function. The coefficients $\beta_1, \beta_2, \ldots, \beta_p$, which have to be estimated from the data, represent the influence of the different factors. If the number of factors is large, a more parsimonious model may be preferable, which is selected by using one of several variable selection strategies. We used backward elimination (BE) with different nominal selection levels to select our 'final' models, in which the estimated regression coefficients $\hat{\beta}_1, \ldots, \hat{\beta}_k$ represent the relative importance of the selected relevant factors, and $PI = \hat{\beta}_1 X_1 + \ldots + \hat{\beta}_k X_k$ is a prognostic index. Subgroups with different risk can then be defined by categorizing the values of the index using cutpoints.

To investigate complexity and stability of 'final' models, we used crossvalidation approaches. For Cox models, Verweij and van Houwelingen (1993) proposed a cross- validated log-likelihood l*, and the deviance D, which is the difference of - 2 loglikelihoods of a model to the model without covariates. For the approach with crossvalidation D* can be used for model selection. They further discuss a shrinkage factor for the prognostic index $PI = X\hat{\beta}$ based on cross-validation calibration. This shrinkage factor should be close to one if there is hardly any overfitting, whereas small values should indicate overfitting. Sauerbrei (1996b) extended this approach to parameterwise shrinkage factors (PWSF).

Based on a final model, a prognostic index $PI_i = X_i\hat{\beta}$ can be calculated for each patient i. Using one or more cutpoints for this index, patients can be categorized into groups with different risks. We investigated the ability of classification schemes based on the full model and the model selected with BE (selection level 0.157) - this level was chosen because it corresponds to the asymptotic significance level of the Akaike criterion in the linear model (Akaike (1969), Teräsvirta and Mellin (1986)) - to separate the patients in the GBSG study. For both models, we created classification schemes with four categories with increasing risk by using the 25 %, 50 %, and 75 % quantiles of the corresponding distribution function of PI^{orig} as cutpoints and another one with three categories by using the 33 % and 66 % quantiles. As a measure of the ability to separate groups with different prognosis, we used the parameter estimates $\hat{\beta}$ of the corresponding category indicator in comparison to the category with best prognosis as the reference group.

These estimates may be biased, because model selection and parameter estimation for PI and the estimation of the effect of the category indicator is based on the same data. This may lead to an overestimation of the prognostic ability of classification schemes. Therefore, we used 5-fold crossvalidation to estimate the regression parameters for a model based on the 80 % of the 'learning' data. For the remaining 20 % of the patients, we calculated the corresponding prognostic index. Repeating this procedure for all 5 parts gives for each patient i a prognostic index PI_i^{val} which is based on parameter estimates, which are independent of patient i. For the distribution of PI^{val}, we used the quantiles to define prognostic classification schemes in the same way as for PI^{orig} and estimated the prognostic ability by using the regression parameters of the category indicators based on the original data.

The CART approach (Breiman et al. 1984) splits the population repeatedly into two parts based on the variable which shows the largest influence on prognosis. Using a given cutpoint, an ordinal or continuous factor is split into two parts with 'low' and 'high' values and the influence of this binary factor on recurrence free survival time is tested by using a logrank test. Based on clinical judgement, we investigated only predefined cutpoints (e.g. age in 5 year intervals, size with cutpoints 10, 20, 30, 40) as possible splits for continous variables. For categorical variables, all possible binary categorizations were investigated. The minimal p-value over all variables and all possible binary categorizations were determined and the corresponding variable with its optimal binary categorization was used to split the (sub-) population. To overcome the problem of a different number of cutpoints and of a high probability of wrongly identifying a factor allowing many possible splits as influential we used the modification with corrected p-values proposed by Lausen et al. (1994). This procedure is repeated for the resulting two subpopulations until a stop criteria is reached. As stop criterion, we used a 5 % significance level and no further split was allowed for sample sizes less than 25.

The tree was developed using predefined cutpoints and corrected p-values for all variables with more than two categories. To investigate the importance of these decisions on the complexity of the tree, we further developped trees without p-value correction and allowed all possible cutpoints within the range of 10 % to 90 % quantiles of the distribution of the factor. Furthermore, the stability of these four situations (with and without p-value correction / predefined cutpoints) was investigated by using 5-fold crossvalidation.

All analyses were done using data of the German Breast Cancer Study Group (GSBG). Between 1984 and 1989 the GBSG entered 662 eligible patients into a prospective observational study for node negative breast cancer patients. 603 (91 %) of the 662 patients had complete data for seven 'traditional' prognostic factors. After a median follow-up of 5 years, 155 of the 603 patients had at least one event for recurrence free survival time (RFS). For the CART-approach, we used seven additional variables, which are sometimes considered in literature.

3. Results

3.1. Development of prognostic indices using Cox models

In table 1, we give the results of the full model and of three models selected with backward elimination using different selection levels. Besides the four binary variables age (decision for a cutpoint 40 because of dicussions in the literature, for more details on this aspect and the other cutpoints used see Sauerbrei et al. (1996a)), menopausal status, estrogen and progesteron receptor, there are two ordinal variables (tumor size and grade) and the categorical variable histologic type. Coding of the dummy variables $X_1 - X_{13}$ took into account the differences between categorical and ordinal variables. With the strongest level, a simple model including age, tumor size (≤ 2 cm vs >2 cm) and grade was selected (BE with selection level 0.01). With

		Model							
		full		BE(0.157)		BE(0.05)		BE(0.01)	
		$\hat{\beta}$	SE	$\hat{\beta}$	SE	$\hat{\beta}$	SE	$\hat{\beta}$	SE
Age									
≤ 40		0	-	0	-	0	-	0	-
> 40	X_1	-0.93	0.28	-0.94	0.27	-0.93	0.27	-0.56	0.22
Menopause									
pre		0	-	0	-	0	-		
post	X_2	0.44	0.22	0.49	0.22	0.51	0.22		
Tumorsize									
≤ 1 cm		0	-	0	-	0 (≤ 1 cm, ≤ 2 cm)	-	0	-
≤ 2 cm	X_3	-0.50	0.32	-0.48	0.32				
≤ 3 cm	X_4	0.70	0.20			0.61 (≤ 3 cm – ≤ 5 cm)	0.17	0.59 (≤ 3 cm – ≤ 5 cm)	0.17
≤ 4 cm	X_5	0.33	0.24	0.75 (≤ 4 cm, ≤ 5 cm)	0.19				
≤ 5 cm	X_6	-0.98	0.74						
> 5 cm	X_7	0.52	0.93						
Estrogen. Rec.									
< 20		0	-						
≥ 20	X_8	0.30	0.22						
Progest. Rec.									
< 20		0	-						
≥ 20	X_9	-0.16	0.21						
Grade									
I		0	-	0	-	0	-	0	-
II	X_{10}	0.64	0.25	0.58	0.25	0.59	0.24	0.59	0.24
III	X_{11}	0.98	0.29	0.89	0.27	0.88	0.27	0.85	0.27
Hist. type									
solid		0	-	0	-				
inv. duct / lob	X_{12}	-0.29	0.20	-0.30	0.20				

Table 1: Parameter estimates of the full model and three selected models using backward elimination with different selection levels

	full	BE(0.157)	BE(0.05)	BE(0.01)
# Parameter	13	8	5	4
D	54.4	49.4	42.8	36.7
D*	29.6	34.2	32.9	28.7
Shrinking factor				
global	0.72	0.81	0.86	0.87
parameterwise				
X_1	0.87	0.82	0.83	0.83
X_2	0.84	0.72	0.72	
X_3	0.54	0.54		
X_4	0.91	0.85	0.92	0.93
X_5	0.33			
X_6	-0.48			
X_7	-3.92			
X_8	-0.28			
X_9	-1.32			
X_{10}	0.75	0.76	0.76	0.76
X_{11}	0.81	0.81	0.82	0.83
X_{12}	0.47	0.46		
X_{13}	0.43	0.50		

Table 2: Comparison of the full and three selected Cox-models. Deviance without (D) and with (D*) crossvalidation and estimated global and parameterwise shrinkage factors.

BE (0.05) additionally menopausal status is included and the estimate for age is substantially changed because of the correlation between these two variables. With BE (0.157) additionally histological tumor type is included and a further dummy variable (X_3) of tumor size indicates the unusual effect that tumors of size 11 - 20 mm have a decreased risk in comparison to smaller ones.

3.2 Complexity and shrinkage for Cox models

For BE, we used the selection level to control model complexity. Whereas the deviance D increases for models with more variables (table 2), the deviance based on crossvalidation D* is maximal for BE (0.157), and the model with only 5 selected variables has a larger value of D* than the full model with 13 variables. For the standardized values of the variables shrinkage factors are given in table 2. All factors selected with BE (0.01) have an estimated PWSF between 0.76 and 0.93, and these factors hardly change in the models with more included variables. The three additional variables X_3, X_{12} and X_{13}, which are selected with BE (0.157), have shrinkage factors of about

Categories	Original data Full	Original data BE(0.157)	Crossvalidation Full	Crossvalidation BE(0.157)
a) 4 categories based on about 25 %, 50 % and 75 % of the distribution function of PI				
C1	0	0	0	0
C2	0.30	0.29	0.38	0.53
C3	0.80	0.92	0.56	0.67
C4	1.40	1.40	1.02	1.18
b) 3 categories based on about 33 % and 66 % of the distribution function of PI				
C1	0	0	0	0
C2	0.38	0.53	0.14	0.38
C3	1.27	1.35	0.85	1.08

Table 3: Based on prognostic indices of the full and a selected Cox-model estimates of the parameter $\hat{\beta}$ for category indicator using the category with the best prognosis (C1) as reference. Classification of patients based on original data and 5-fold crossvalidation

0.5, whichs indicates already a severe overestimation of the corresponding parameter estimates in table 1. Most of the PWSF for variables included in the full model only have a negative shrinkage factor. The global shrinkage factor of the full model is much smaller than the ones for the more parsimonious models.

3.3. Classification schemes based on Cox-models

The estimates based on PI of the original data show that the predicitve ability of the categorization schemes based on the full model and the BE model is about the same, with a slight advantage for the more parsimonious model (table 3). For the full model and four categories, 240 (40 %) patients were classified in another category for the scheme based on PI^{val} in contrast to PI^{orig}. 8 patients had even a difference of 2 categories. The differences between classifications into categories were less extrem for BE. No changers of two categories occured, and 189 (31 %) changed one category. Using only three categories, no change between the highest and lowest category occured, but 151 (25 %) changed one category in the schemes based on the full model, respectively 120 (20 %) by using BE.

These changes of the grouping has some important influence on the ability to discriminate prognostic groups as shown by the parameter estimates for the category indicator from the classification scheme given by PI^{val}. For both models and both schemes, the parameter estimates are substantially smaller than the ones based on the original data. Furthermore, it can be seen that the estimates from BE are higher than the ones from the full model.

3.4. Complexity and Stability of Trees

In table 4, the final trees are summarized for the analysis of the original data and the 5 CV runs. We give the split criteria of the first two levels and list the criteria entering the tree in further levels. For non-binary variables, the cutpoints are given in brackets. Using predefined cutpoints and p-value correction (table 4a), the results of CV confirm the dominating role of size and the cutpoint 20. As in the analysis of the original data, age (40) was selected as the most important factor for all 3 CV runs splitting by size at the first level. In one CV run a subpopulation of 115 patients with a grade 3 tumor could not be split any further. Somemore variations existed in the further levels.

Without the p-value correction, trees with more final nodes were selected. Size (20), age (40) and grade remained the most important variables, but CV exhibited already more variation, especially in the further levels. The analysis without predefined cutpoints and with p-value correction gives very small trees. In the original analysis and in 3 CV runs tree development was stopped after the first split into one subgroup, although the sample size was larger than 100 in all cases. Developing trees without predefined cutpoints and without p-value correction leads to complex trees with about 20 final nodes. The splitting criteria of the further levels demonstrate the obvious effect of overfitting as age was used in six subpopulations, always with another cutpoint. Generally, crossvalidation demonstrates that splitting criteria of the first two levels are relatively stable, but substantial differences exist in all of the runs in the further levels.

4. Discussion

Many classification schemes proposed in the literature are hardly validated and used in clinical practise. Concerning node negative breast cancer patients, Sauerbrei et al. (1996a) showed that the prognostic ability of several literature proposals was much worse than proposed in the original papers. Some of the reasons may be the tendency for too complex models which overfit the data, a lack of investigations on stability of selected models, and on the assessment of the overoptimism concerning the prognostic ability. For binary outcome variables, several proposals for a correction of this overoptimism based on computer intensive methods have been published (Efron (1983), Gong (1986)).

Developing a classification scheme based on the Cox model, we showed by using a crossvalidation likelihood approach (Verweij and van Houwelingen (1993)) that a more parsimonious model may be preferable to the full one. We used backward elimination and controlled the complexity of the model with the nominal selection level. We further estimated global and parameterwise shrinkage factors (PWSF) who gave a strong indication against the full model and who further indicated that even the model based on BE with selection level 0.157 may overfit the data. In the full model some of the PWSFs indicated that even the sign of the parameter estimate in table 1

a) Tree development with p-value correction and with predefined cutpoints

	final nodes	1. level	2. level a	2. level b	further level
orig.	7	S (20)	A (40)	G (1)	E (100), A (60), E (300)
1. CV	6	S (20)	A (40)	G (1)	A (55), E (300)
2. CV	7	S (20)	A (40)	G (2)	G (1), T (1), P (5)
3. CV	9	A (40)	R	S (20)	K (90), T (1), E (300), M, H
4. CV	4	G (2)	A (40)	–	M
5. CV	6	S (20)	A (40)	A (60)	G (1), E (100)

b) Tree development without p-value correction and with predefined cutpoints

	final nodes	1. level	2. level a	2. level b	further level
orig.	10	S(20)	A(40)	A(60)	G (1), E (100), A (40), G (2)
1. CV	10	A (40)	R	S (20)	T (1), G (2), G (1), A (55), A (60)
2. CV	9	S (20)	A (40)	G (2)	A (40), A (55), H, T (1), P (5)
3. CV	9	A (40)	R	S (20)	K (90), P (5), E (300), A (50), T (1)
4. CV	11	G (2)	A (40)	T (1)	A (50), A (60), S (20), E (10), P (300)
5. CV	7	S (20)	A (60)	A (60)	G (1), E (100), G (2)

c) Tree development with p-value correction and without predefined cutpoints

	final nodes	1. level	2. level a	2. level b	further level
orig.	6	S (19)	–	G (1)	E (187), G (2), M
1. CV	9	S (19)	H	T (0)	S (10), S (23), G (2), G (1), E (187)
2. CV	4	S (27)	S (19)	–	E (140)
3. CV	4	S (19)	–	T (0)	N (10)
4. CV	5	G (2)	G (1)	–	S(27), M
5. CV	8	S (19)	A (45)	A (62)	E (4), S (25), R

d) Tree development without p-value correction and without predefined cutpoints

	final nodes	1. level	2. level a	2. level b	further level
orig.	21	S (19)	E (6)	E (157)	S (109, S (24), S (27), A (40) A (45), A (59), A (62), A (64) P (9), P (131), P (350), E (3) E (19), N (7), N (9), G (2)
1. CV	26	S (19)	E (6)	E (158)	many
2. CV	19	S (27)	S (19)	G (1)	many
3. CV	16	S (19)	E (6)	E (157)	many
4. CV	17	S (27)	G (2)	G (1)	many
5. CV	18	S (19)	A (45)	A (62)	many

A – age, S – size, G – grading, M – menopause, E – estrogen, P – progesterone, T – type, R – resid. in ampu., K – Karnofsky-Index, H – hormone receptor, N – nodes invest.

Table 4: Influence of constraints on tree complexity and stability. Summary of final trees from original data and crossvalidation data sets may be wrong.

Using crossvalidation, our comparison of the diagnostic ability of classification schemes based on the full model and on a selected model gave further evidence for a preference of more parsimoneous models.

CART-trees are another popular method for the development of classifcation schemes. Although they are conceptually very different to regression models, they have the problem of overfitting the data and of overestimating the predictive ability in common. Serious problems concerning stability and overfitting arise with continuous variables if no restrictions are used to correct for multiple testing. Based on practical experiences, we prefer an approach with predefined cutpoints based on clinical reasons and use of p- values corrected for multiple testing. The influence of the stop criteria needs further investigations. Building of complex trees followed by pruning or amalgamation is even more questionable in survival data than in the analysis of binary outcomes. Investigations with crossvalidation showed that the splitting criteria at the first two levels are relatively stable for the prefered algorithm. Without any restriction, complex trees are derived, which are very instable and clinically uninterpretable.

A basic requirement for the clinical usefulness of a classification scheme is a sufficient validation in external studies. Using computer intensive methods as crossvalidation or bootstrapping, 'internal' validation is possible for the 'original' study and may lead to an improvement of proposed schemes. Investigations of stability, overestimation and overoptimism should become standard tools. This may lead to proposals of a more simple nature and to more realistic assessments concerning the prognostic ability (Sauerbrei (1996 b)).

References:

AKAIKE, H. (1969): Fitting autoregressive models for prediction, *Annals of the Institute of Statistical Mathematics, 21, 243-247.*

BREIMAN, L.; FRIEDMAN, J. H.; OHLSEN, R. A.; STONE, C. J. (1984): Classification and regression trees. Wadsworth. Belmont.

COX, D. R. (1972): Regression models and life tables (with discussion). *Journal of the Royal Statistical Society B 34: 187-220.*

EFRON, B. (1983): Estimating the Error Rate of a Prediction Rule: Improvement on Cross-Validation, *JASA, 78, 316-318.*

GLICK, J. H.; GELBER, R. D.; GOLDHIRSCH, A.; SENN, H. J. (1992): Meeting highlights: Adjuvant therapy for primary breast cancer. *JNCI 84: 1479-1485.*

GONG, G. (1986): Cross-Validation, the Jackknife, and the Bootstrap: Excess Error Estimation in Forward Logistic Regression, *JASA, 81, 108-113.*

HAYBITTLE, J. L.; BLAMEY, R. W.; ELSTON, C. W. et al. (1982): A prognostic index in primary breast cancer. *British Journal of Cancer 45: 361-366.*

LAUSEN, B.; SAUERBREI, W.; SCHUMACHER, M. (1994): Classification and regression trees (CART) used for the exploration of prognostic factors measured on different scales. In: Dierschedl R, Ostermann R (ed.), Computational statistics. Physica-Verlag. Berlin.

RUBENS, R. D. (1992): Management of early breast cancer. *British Medical Journal 304: 1361-1364.*

SAUERBREI, W.; HÜBNER, K.; SCHMOOR, C.; SCHUMACHER, M. (1996a): Validation of existing and development of new prognostic classification schemes in node negative breast cancer, Breast Cancer Research and Treatment, to appear.

SAUERBREI, W. (1996b): Resampling methods for the investigation of data driven variable selection - practical experience suggests more simplicty of a final regression model, submitted.

TERÄSVIRTA, T.; MELLIN, I. (1986): Model Selection Criteria and Model Selection Tests in Regression Models, *Scandinavian Journal of Statistics, 13, 159-171.*

VERWEIJ, P. J. M.; VAN HOUWELINGEN, H. C. (1993): Cross-validation in survival analysis, *Statistics in Medicine, 9, 487-503.*

Use of Crossvalidation to Assess Diagnostic Classification Schemes of Atopic Dermatitis

W. Sauerbrei[1], T. L. Diepgen[2]

[1]Institute of Medical Biometry and Informatics; University of Freiburg, Germany
[2]Department of Dermatology; University of Erlangen, Germany

Summary: With the goal of developing diagnostic classification schemes, we analysed 345 patients with atopic dermatitis (AD) and 618 controls with logistic regression models and CART trees. With 5-fold crossvalidation (CV), we investigated the stability of the selected final models and trees. Furthermore, we used CV to estimate the possible overoptimism concerning the estimates of sensitivity and specificity of the classification schemes, which may be present when the development of a classification scheme and the assessment of its diagnostic ability were based on the same data.

1. Introduction

For Atopic Dermatitis (AD) there exists no objective classification criteria for the disease, and the diagnosis is usually based on many variables like anamnestic, clinical, and laboratory findings proposed by different authors (Svensson et al. (1985), Rajka (1989)). These atopic features have not yet been investigated systematically. Available data of epidemiological and clinical studies are difficult to compare because of differences in definition of the disease, and the diagnostic value of "atopic features" still remains unclear. Although most typical cases of AD have a history or presence of a chronically relapsing course of eczema, especially in the antecubital and popliteal flexures (Rajka (1989)), which can be seen as a "golden standard" of AD, a flexural eczema can also be caused by other factors in non-atopic individuals, and more often subjects with an atopic diathesis can develop eczema without involvement of the flexures. In an unmatched case control study, we investigated 19 atopic binary features and two laboratory measurements of 345 AD patients and 618 controls aiming to develop diagnostic classification schemes, which can be used as a standardized tool for the diagnosis of AD. It will be useful for several tasks if the resulting classification schemes are reliable and reproducible by further investigators. For example, they can be applied to assessing the changes of the prevalence of a disease in a given population.

For the development of classification schemes, approaches based on logistic regression models or the derivation of classification and regression trees (CART) are commonly used. Although both approaches are conceptually different, both share the same problem of a possible overoptimism concerning the estimates of sensitivity and specificity, if the development of the classification scheme and the assessment of its diagnostic ability are based on the same data. For the assessment and correction of this overoptimism, computer intensive methods, such as bootstrap resampling and crossvalidation, have been proposed (Efron (1983), Gong (1986)). We developed and compared classification schemes based on logistic regression models and CART and used 5-fold crossvalidation to assess the stability of the classification schemes and the amount of overoptimism.

2. Material and methods

345 in- and out-patients with chronically relapsing course of flexural eczema (at least three relapses were required as the golden standard) were consecutively investigated in one clinic from 1988 to 1990 and 618 control subjects of the same age entered an unmatched case-control study. Each of the 19 binary atopic features were diagnosed independently by two experienced physicians. Based on their assessment and agreement, the variables were classified into 'subjective' or 'objective' criteria (table 1). Additionally two laboratory measures were investigated and classified as 'normal' or 'elevated'. A definition of all these atopic criteria is given elsewhere (Diepgen et al. (1996)). Because 'subjective' features may cause classification problems if a given scheme is used from other investigators and because the two laboratory measures may be difficult or even unethical to obtain in broad practical use, we additionally developed classification schemes under two constraints. In addition to sex and age, the following potential atopic features were considered as baseline variables for the three given situations:

(A)	no constraint:	objective, subjective and laboratory features
(B)	constraint objectivity:	objective and laboratory features
(C)	constraint simplicity:	only objective features

For the binary response variable (case / control status), we assumed a logistic regression model (LR)

$$\text{logit } P\ (Y = 1|X_1, \ldots, X_p) = \beta_0 + \beta_1 X_1 + \ \ldots \ + \beta_p\, X_p$$

where the logit of the probability for AD depended on the linear function of the atopic features $X_1, \ldots, X_p$ with $\beta_1, \ldots, \beta_p$ as the regression coefficients representing the influence of the features on AD. We used backward elimination (BE) to develop more parsimonious models (M_{1L}, M_{2L}, M_{3L}) with 1 % as the selection level in order to include only 'strong' criteria. For the final models the parameters were estimated using the same data, and the linear combination $\hat{\beta}'X$ (ignoring the indices for each model) defined the diagnostic scores S_{1L}, S_{2L}, S_{3L}.

Based on these scores, the conditional sampling probabilities for AD, given the features X and that a person was sampled for the study, were calculated for all subjects (Breslow and Day (1980)). Each subject was classified as Non-AD if the probability was less than or equal to a prespecified cutpoint and as AD if this probability was larger. This classification was used to estimate values for sensitivity and specificity for each score. Although the relative frequency of AD in our study was much higher (between 4 and 7 times as high) as it is in the general population, no correction was made for the intercept 0, because such a correction corresponds only to a different cutpoint. Several cutpoints were used and the result of the investigation of all possible cutpoints was presented as a receiver operating curve (ROC), which is a plot of (1-specificity) versus sensitivity (Hanley et al. 1982).

For the classification and regression tree approach (CART), we used a modification of the proposal in the standard book of Breiman et al. (1984). For all binary features a X^2-test of independence in a 2x2 table of the corresponding feature with the AD- status was calculated. With age being the only continous variable, the optimal cutpoint with the largest X^2-value was determined within the range of the 10 % to 90 % quantiles of the distribution. The corresponding p-value was corrected to account for multiple testing (Lausen et al. 1994). Based on the variable with the smallest corrected p-value, the population was split into two parts. In each part this process was repeated until one of the following stop criteria was reached:

- no significance at the 1 % level
- the size of a subpopulation is smaller than 20
- maximal split level is 6
- less than 5 AD or 5 Non-AD in the subpopulation

Subpopulations, which can not be further split are called final nodes. Every patient enters into one of the final nodes and based on the true AD-status, the probablility of AD was estimated by the percentage of AD-subjects. Using a cutpoint for this percentage, a diagnosis AD or Non-AD could be attached to the final nodes giving the classification scheme. The same cutpoints for the probability as for logistic regression were used. For a given cutpoint each subject could be classified as being correctly or wrongly diagnosed by the tree yielding the corresponding estimates of sensitivity and specificity.

To investigate the stability and to assess the amount of overoptimism concerning sensitivity and specificity of the resulting classification schemes, we used 5-fold crossvalidation (CV) in the following way. The data were randomly divided into 5 parts of equal size. Four parts were used as a 'training set' for building a final logistic regression model, respectively a CART-tree, and the remaining part was used to estimate sensitivity and specificity of the corresponding classification scheme. This process was repeated for all 5 parts, resulting in different classification schemes in each part and in the classification of each subject as AD or Non-AD based on a rule, which was developed independently of the subject itself. Based on the true AD-status, the cross-validated estimates of sensitivity and specificity were calculated and compared with the estimates from the usual classification schemes without cross-validation.

Var.		(A): no constrain $\hat{\beta}$	SE	(B): constrain objectivity $\hat{\beta}$	SE	(C): constrain simplicity $\hat{\beta}$	SE
Intercept		-7.73	0.69	-6.60	0.52	-5.72	0.43
sex	X_1						
age	X_2						
object.							
SEBOS	X_3	1.79	0.41	2.37	0.35	2.31	0.33
WOOL	X_4	1.78	0.39	1.89	0.33	1.73	0.31
PYTAL	X_5	2.71	0.65	2.93	0.56	2.79	0.55
PERL	X_6	1.15	0.39	1.22	0.32	1.38	0.30
KERAT	X_7	1.26	0.46	1.42	0.38	1.27	0.36
JUCK	X_8	2.51	0.45	2.39	0.38	2.43	0.35
DYSHI	X_9	2.74	0.44	2.57	0.38	2.52	0.37
DERMO	X_{10}	2.21	0.48	2.31	0.38	2.31	0.36
ASTH	X_{11}						
CHEIL	X_{12}						
FA	X_{13}						
EAR	X_{14}						
RHIN	X_{15}						
labor.							
IGE	X_{16}			1.18	0.39	nu	
PHAD	X_{17}	1.72	0.41	0.98	0.35	nu	
subj.							
HERTO	X_{18}	3.64	0.65	nu		nu	
DENNI	X_{19}	1.58	0.40	nu		nu	
PALMS	X_{20}	1.95	0.43	nu		nu	
LIGHT	X_{21}			nu		nu	
MILK	X_{22}			nu		nu	
FOOD	X_{23}			nu		nu	
-2logL		202.8		282.9		314.1	

Table 1: List of all features and selected final logistic regression models using backward elimination (0.01) for three situations (nu - not used)

3. Results

Concerning the logistic regression, Table 1 gives the final models with the parameter estimates for the three following situations: (A) without constraints and with the constraints (B) objectivity and (C) simplicity. From the 'objective' variables, X_3 to X_{10} were selected in all three situations, whereas X_{11} to X_{15} did not enter any model. The further variables which entered

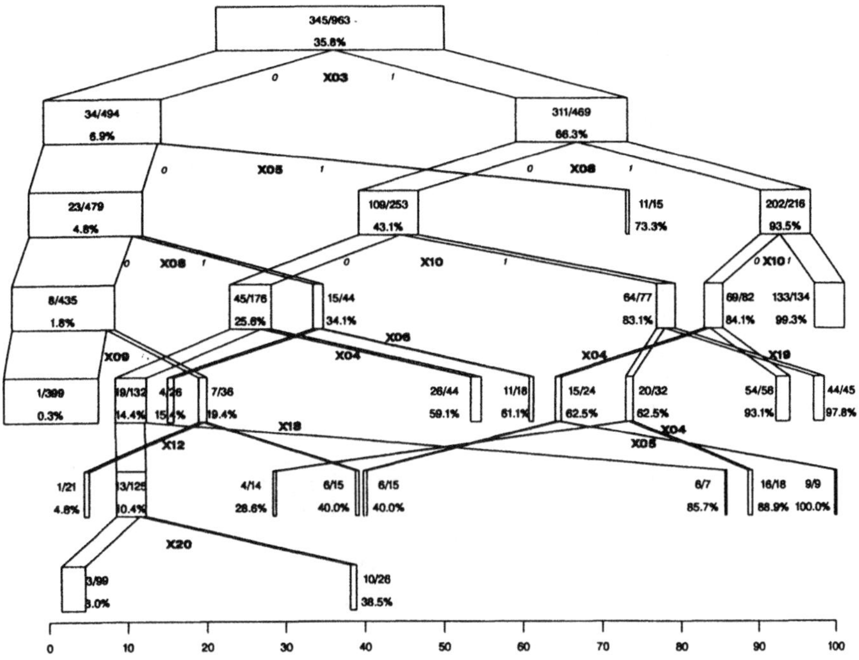

Figure 1: CART-tree for the situation (A) without constraints. Number AD and size of the (sub)populations are given in the boxes. Scale gives the percentage of AD. Nodes partly displaced because of clarity in graphical presentation. (0-feature not persent, 1-feature present)

into the final model in the less restrictive situations (A) and (B) were highly significant and led to a substantial improvement in the log-likelihood indicating a better fit of the model.

For situation (A) without constraints, the CART-tree is given in figure 1. In the first step, X_3 is the variable with the strongest association separating the population into two groups with 6.9 % AD if X_3 is not present respectively 66.3 % AD if X_3 is present ($X_2 = 370$, 1df). In the first subgroup, the prevalence of AD decreases further to 4.8 % if a subject did not have X_5, and the final tree shows that only 1 of the 399 subjects without the features X_3, X_5, X_8 and X_9 had AD. Furthermore, a small final node with 21 subjects and only one AD patient existed (X_9 present; X_3, X_5, X_8 and X_{12} not present). From the 469 subjects with the feature X_3, a subpopulation of 99 subjects (3 % AD) without the features X_8, X_{10}, X_4, X_{18} and X_{20} could be separated. In contrast to these final nodes with a very small prevalence of AD, the CART-tree separates 4 final nodes with an AD prevalence of more than 90 %. Altogether, 246 subjects of which 240 had AD (70 % of all 345 AD patients) entered into one of them, for example 133 of 134 subjects with the features X_3, X_8 and X_{10} had AD. There are 17 final nodes, from

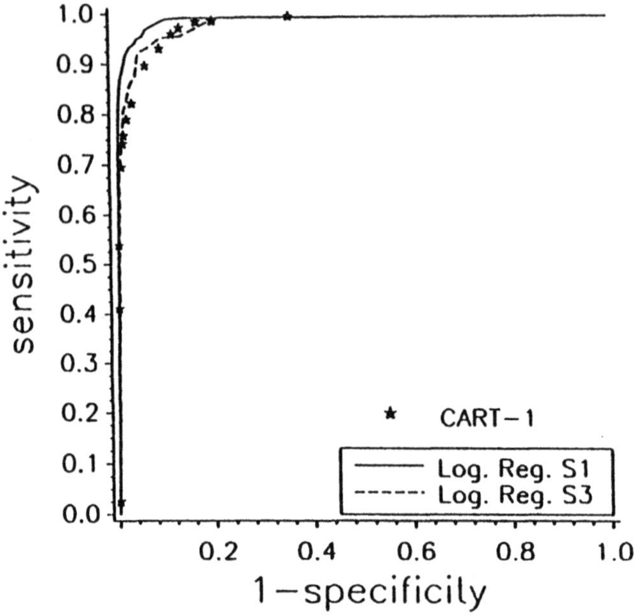

Figure 2: ROC curve for the classification schemes based on the logistic models (A,S_1) without constraints and (C,S_3) with the constraint simplicity and on the CART-tree for (A) without constraints

which the other ones have a prevalence of AD patients between 10 % and 90 % making a classification less obvious.

The tree developed under the constraint 'simplicity' was identical in the first steps as always 'objective' variables were selected. The 77 subjects at level 4 were divided by X_5 instead of X_{19} and in further levels by X_{14} and X_4. At level 5, X_9 was used to seperate the 132 subjects and later on X_7. In this paper, we will not further investigate the model with the constraint 'objectivity.' Further results for all three situations using logistic regression are given in Diepgen et al. (1996). In figure 2, the ROC-curves are given for the diagnostic scores S_{1L} and S_{3L} from the logistic regression models. Furthermore, the corresponding values for sensitivity and specificity for the CART-tree are shown. With the score S_{1L} a sensitivity of about 0.95 was reached with a corresponding specificity of about 0.95 when using a cutpoint of 0.36. The ROC-curve demonstates that the diagnostic ability of the simpler score S_{3L} is slightly inferior (e. g. sensitivity 0.928 and specificity 0.950 with a cutpoint of 0.36). The classification scheme based on the CART-tree has a similar diagnostic ability. As all three approaches lead to a high diagnostic ability for AD, aspects of practical usefullness may play an important role for the comparison. In this respect, CART has several advantages. It

Data	#	1.	sec	lev	third level				further levels
	nod	lev	a	b	aa	ab	ba	bb	
Orig.	17	X_3	X_5	X_8	X_8	N^1	X_{10}	X_{10}	X_4 X_5 X_6 X_9 X_{12} X_{18} X_{19} X_{20}
1. CV	17	X_3	X_{18}	X_4	X_8	N^1	X_{10}	X_{10}	X_2 X_5 X_8 X_9 X_{12} X_{17} X_{18} X_{20}
2. CV	15	X_3	X_5	X_8	X_8	N^1	X_{10}	X_{10}	X_4 X_9 X_{12} X_{16} X_{17} X_{20}
3. CV	15	X_3	X_5	X_8	X_8	N^1	X_{10}	X_{10}	X_2 X_4 X_6 X_9 X_{18} X_{19}
4. CV	17	X_3	X_5	X_8	X_8	N^1	X_{10}	X_{10}	X_2 X_4 X_6 X_9 X_{16} X_{18} X_{19}
5. CV	15	X_3	X_9	X_4	X_8	X_5	X_{18}	X_{10}	X_8 X_9 X_{12} X_{20}

[1]No further split; between 1.3 % and 2.3 % of the population in this final node; prevalence of AD between 67 % and 79 %

Table 2: Stability of selected CART-trees in the original data and in 5-fold CV sets (situation (A) without constraint)

is not necessary to calculate the scores and the sampling probabilities. Furthermore, the number of features is smaller and the hierarchical approach offers the possibility for principle thoughts about a better understanding of the disease. With 5-fold crossvalidation, we investigated the stability of the selected factors. Using the original data, our model included the twelve features $X_3 - X_{10}$ and $X_{17} - X_{20}$ in the situation without constraints. In the five CV runs, models with 10 or 11 variables were selected and no other variable was entered. X_6 and X_7 were excluded in two runs, X_7 and X_{17} in one run and X_6 respectively X_7 in the other two runs. This result demonstrates the stability of the selected models, although the inclusion of the factors X_6 and X may be questionable, despite their standardized estimates in the final model of the original data being 2.99 (X_6) respectively 2.75 (X_7). In CV, the sample size of the training part was reduced to 80 % of the original data which was associated with an increase of the standard errors for the parameter estimates. This is one important reason that in all CV runs the selected models included a smaller number of variables. In 4 CV runs for situation (C) with the constraint simplicity, the same model including the variables $X_3 - X_{10}$ was selected as for the original data, in one run only X_7 was excluded from this model with 8 variables.

A short summary of several CART-trees is much more complicated, because only subpopulations are investigated after the first split. For the CV training sets we give the results of the first three splitting levels in table 2, for further levels only variables which were used for further splitting of a node are listed.

The number of final nodes was always 15 or 17, and in the first level X_3 was always selected as the splitting criteria. That means that X_3 was the strongest variable in a univariate analysis in all CV-data sets. Note that in the final multivariate logistic regression model, 6 variables (X_4, X_8, X_8, X_{10}, X_{18}, X_{20}) had a larger standardized estimate than X_3. Already in the second level, some variation concerning the selected factors can be observed.

Cutpoint	LR				CART			
	Orig. data		CV		Orig. data		CV	
	sens.	spec.	sens.	spec.	sens.	spec.	sens.	spec.
(A) No constraint								
0.1	99.1	89.6	97.7	89.2	98.6	83.2	92.5	82.7
0.36	95.1	95.8	93.9	94.8	96.2	88.3	87.0	89.2
0.7	89.3	98.5	87.8	97.2	79.1	97.9	73.9	95.0
(C) Constraint simpl.								
0.1	99.4	79.1	96.8	80.72	98.0	81.9	95.1	77.5
0.36	92.8	95.0	91.9	94.7	95.1	88.2	90.1	87.2
0.7	82.3	98.1	82.6	97.1	76.5	98.2	75.1	95.0

Table 3: Comparison of estimated sensitivities and specificities based on the original data and on CV data sets for logistic regression models and CART

Surprisingly the splitting variables at the third level are identical in four CV-runs with the variables from the original data. Furthermore, it is obvious that the final node from the tree developed for the original data including the small group of patients (about 2 % of the population) with the feature X_5, but without the feature X_3, can not be seen as an artifact caused by the data. Obviously, more variation of selected splitting variables is present in the further levels of the tree where smaller subgroups were investigated. From the 12 variables selected with logistic regression, only X_7 was not used in any of the six trees as a splitting criterion. Furthermore, X_{17} was not used in 4 of the 6 trees, and X_6 and X_{19} were not used in 3 of them. The variable age (X_2) was selected in 3 of the crossvalidation runs in the further levels.

In table 3, we give estimates of sensitivity and specificity for several classification schemes using a small value (0.10), the prevalence of AD in the study (0.36) and a large value (0.70) as a cutpoint for the probability. The comparison of the estimates using all data with the corresponding CV ones can be used for an assessment of the overoptimism from the first approach. The CV estimates demonstrate that the proposed classification schemes have a high diagnostic ability to separate AD and Non-AD subjects as a sensitivity of more than 90 % is possible with a corresponding specificity of more than 90 %. Assessment of positive and negative predictive values from the schemes based on a logistic model and the original data is discussed in Diepgen et al. (1996).

For the original data the slight superiority of the classification scheme based on S_{1L} in comparison to the corresponding one from S_{3L} and the CART-approach was shown in figure 2. For the logistic regression the CV estimates show that the advantage of the estimates derived without constraints diminishes partly in comparison to the corresponding one using the constraint simplicity. With the exception of sensitivity and the cutpoint 0.1, the estimates of CV are nearly identical to the ones from the original data for the situation with the constraint simplicity, indicating that the amount of overoptimism is only small.

For the CART classification schemes the estimates for sensitivity from CV are substantially smaller for the situation (A), a decrease of the sensitivity is still present for the models from (C), but the difference is much smaller than for (A). Estimates of the specificity from the original data seem to be only slightly too high as the comparison to the values based on CV demonstrate.

4. Discussion

With logistic regression models and classification and regression trees we used two conceptually different approaches to derive classification schemes for atopic dermatitis. With both methods, we derived diagnostic schemes giving a probability for AD for each patient. Using a cutpoint for this probability, a diagnosis of AD or Non-AD could be given. Depending on the specific aim of a study, this cutpoint may be varied leading to a higher sensitivity with a lower specificity or vice versa.

The estimates for sensitivity and specificity are too optimistic if they are derived from data, which were already used for the development of the classification rule. Using 5-fold CV, we tried to get a more realistic assessment for the measures of the diagnostic ability. With our classification schemes, sensitivity and specificity can both reach about 90 %, with a slight advantage of the scheme based on the logistic model. This advantage in terms of diagnostic ability may be outweighted by the easier use of the CART-approach in many practical situations. Elimination of some variables because of several reasons (reproducibility in the assessment or costs involved for laboratory measurements) influenced the diagnostic ability of our proposals only slightly. In a validation study with the emphasis on external validation - 106 cases, 223 controls, approximately 15 % of the subjects examined by one of the investigators from the current study, the other by 8 investigators from other departments of dermatology - none of the 3 selected 'subjective' criteria ($X_{18} - X_{20}$) showed a strong influence on the AD status (Diepgen et al. 1996).

The investigations with CV showed the unusual stability of the selected models. One of the reasons may be our strategy to select only 'strong' factors (the selection level for backward elimination and for one of the stop criteria in the CART approach was 0.01) and for the CART-approach that we investigated besides age only 22 binary features. As we further corrected the p-value of the 'optimal' age cutpoint for this multiple testing problem, this variable hardly played a role in the development of the trees. Age was only selected in 3 of the CV-runs at split levels 5 (once) and 6 (twice). In a further analysis without p-value correction, age was used as a splitting criterion in the original data and in all 5 CV-runs. In two of these six runs age was used twice and with different cutpoints. In one CV-run, it was even selected at the third split level. A much more extreme example of the influence of the p-value correction on the final CART-tree is given in Lausen et al. (1994) for a brain tumor study. Without the correction, age was chosen each time with a different cutpoint as the splitting criterion in 7 subpopulations. Using the p-value correction age entered the tree only twice and the tree had only 11 final nodes in contrast to 17 final nodes for the tree without the correction.

The stop criteria plays a central role when using CART for the development of classification schemes. Often a complex tree is developed followed by a pruning algorithm. A CV approach is used to determine the final tree with a minimal misclassification rate. With a comparable approach, we got similar estimates of crossvalidated sensitivities and specificities to the ones given in table 3.

Based on one example we compared the ability to derive classification schemes for the conceptually different approaches of a logistic regression model and of a CART-tree. More investigations are needed to understand the advantages and disadvantages of both approaches and to investigate the importance of the complexity of a model respectively a tree for the question at hand. The general impression from further examples is that complex models or trees have the obvious advantage to fit the original data better and that they have a slightly better predictive ability, but, besides the severe disadvantages in the difficulties of a meaningful clinical interpretation, it seems that these 'statistical advantages' may be a result of overfitting the data and that they are no longer present in new data. An important impression concerning this aspect can be gained by using resampling methods as crossvalidation or bootstrapping for the investigation of stability and overoptimism.

References:

BREIMAN, L.; FRIEDMAN, J. H.; OLSHON, R. A. and STONE, C. J. (1984): Classification and regression trees. Montery, Wadsworth.

BRESLOW, N. E. and DAY, N. E. (1980): Statistical methods in Cancer Research 1: the Analysis of Case-Control Studies. Lyon, International Agency for Research on Cancer.

DIEPGEN, T. L.; SAUERBREI, W. and FARTASCH, M. (1996): Development and validation of diagnostic scores for atopic dermatitis incorporating criteria of data quality and practical usefulness, *J Clin Epidem, 49, 1031-1038*

EFRON, B. (1983): Estimating the error rates of a prediction rule: Improvement of cross-validation. *JASA 78, 316-331.*

GONG, G. (1986): Cross-validation, the Jackknife, and the bootstrap: Excess error estimation in forward logistic regression. *JASA 81, 108-113.*

HANLEY, J. A.; McNEIL, B. J. (1982): The meaning and use of the area under a receiver operating characteristic (ROC) curve. *Radiology, 143, 29-36.*

LAUSEN, B.; SAUERBREI, W. and SCHUMACHER, M. (1994): Classification and regression trees used for the exploration of prognostic factors measured on different scales, In: Dirschedl, P.; Ostermann, R. (eds.): Computational Statistics, Physika- Verlag, Heidelberg, 485-498.

RAJKA, G. (1989): Essential aspects of atopic dermatitis. Berlin, Springer.

SVENSSON, A.; EDMAN, B.; MÖLLER, H. (1985): A diagnostic tool for atopic dermatitis based on clinical criteria: *Acta Derm Venereol (Stockh). Suppl. 114, 33-40.*

Differences of Representing a Conventional Classification by the Compositional Concept Representation Languages BERNWARD and GRAIL

M. Schoop, D. Schoop

Medical Informatics Group, Department of Computer Science,
University of Manchester, Manchester M13 9PL, U.K.

Summary: In this paper two compositional concept representation languages, BERNWARD and GRAIL, will be compared by representing examples from the conventional AO-Classification of Fractures of Longbones. The different expressiveness of BERNWARD and GRAIL that results out of their different constructs will be investigated. Both BERNWARD and GRAIL contain restriction operators in order to control the creation of concept expressions. The question whether these restrictions are sufficient in order to reject all senseless concept expressions will be answered.

1. Introduction

The AO-Classification of Fractures of Longbones was developed in Switzerland and was published in 1990 by the working group for osteosynthesis (AO). It is a conventional mono-hierarchical classification that is used throughout Europe in the domains of surgery and orthopaedics to document fractures of longbones. The AO-Code is an alphanumeric code with 5 places combining topographical and morphological information (Müller et al. (1990)).

BERNWARD and GRAIL are compositional concept representation languages developed for the medical domain and based on semantic networks and conceptual graphs. BERNWARD was developed at the University of Hildesheim, Germany (Bernauer et al. (1995)), and GRAIL was developed by the Medical Informatics Group at the University of Manchester, U.K. (Rector et al. (1995)). Despite having the same bases, both languages have different features. These will be described using examples from the AO-Classification.

The motivation behind this work was to model a conventional classification with compositional concept representation languages in order to keep the standardised terminology but to use all of the advantages of compositional languages and thus be able to compute a multi-hierarchical classification (Schoop et al.(1995)).

2. Constructs of BERNWARD and GRAIL

A brief overview of the constructs of both BERNWARD and GRAIL is shown in Figure 1 below.

GRAIL	BERNWARD
Concept Operators	Concept Operators
(c which <r d ...>)	(c {(:r d) ... })
(c whichG <r d ...>)	
	(AND [c d])
	(OR [c d])
Role operators	
inverse role	
Statements	Statements
(c newSub a)	}
(c addSub a)	} (b $>_{es}$ a)
(c topicNecessarily r d)	}
(c name a)	def(a) = c
(r newAttribute s s^{-1} x)	
x ∈ {*oneOne*, *oneMany*, *manyOne*, *manyMany*}	
(r addSub s)	
(r specialisedBy s)	part-sensitive subsumption
	(a $>_{ep}$ b)

Figure 1: Constructs of GRAIL and BERNWARD

The basic entities of a compositional concept representation language are primitive concept expressions and roles. Concept operators are used to compose new concept expressions. GRAIL has the following notation: (*c which <r d>*) which means that the concept basis *c* which is itself a concept expression is related to concept expression *d* by role *r*. The pair *<r d>* is called the criterion of *c*. The concept expression *c* can be related to many other concept expressions by many roles, i.e. *c* can have a set of criteria. The syntax of BERNWARD differs from GRAIL but the semantics are the same. It is possible within BERNWARD to create coordinated concepts by combining concepts with the operators *AND* or *OR*. GRAIL does not have such constructs.

GRAIL implicitly contains a role operator. There is an inverse role for every existing role. BERNWARD does not have inverse roles.

Both languages contain statements. In GRAIL (*c newSub a*) introduces a new primitive concept expression a that will be subsumed by the concept expression c. The statement (c *addSub* a) means that the existing concept expressions c and a are set in an explicit subsumption relation. The concept expression c and all its subconcept expressions are necessarily related to the concept expression d or to one of its subconcept expressions by the role r by the statement (c *topicNecessarily* r d). Taken together, these three expressions correspond to thesingle expression ($b >_{es} a$) in BERNWARD which means that b explicitly subsumes a. Names can be given to concept expressions in both BERNWARD and GRAIL. The string a is the name for the concept expression c. GRAIL has the possibility of creating a role hierarchy by allowing roles to be subsumed by other roles. The statement (r *newAttribute* s s^{-1} x) creates the roles s and s^{-1} and sets s as a child of role r and s^{-1} as a child of the role r^{-1}. Furthermore, the cardinality x of both roles s and s^{-1} is specified, i.e. *oneOne, manyMany, oneMany, manyOne.* An existing role s is set into an explicit subsumption relation with role r by the statement (r *addSub* s). BERNWARD does not have such constructs. The statement (r *specialisedBy* s) means that the subsumption criterion for the role r takes also the role s into account. The philosophy of BERNWARD is that apart from the hierarchical subsumption relation there is the part-of relation which is also used for classification. Therefore, a part-sensitive subsumption is possible. This corresponds to (r *specialisedBy part-of*) for all roles r in GRAIL. In GRAIL pairs of roles can be selected by the construct *specialisedBy* but once they are selected there is no possibility of turning this specialisation off, whereas in BERNWARD there is always the choice between the normal subsumption and the part-sensitive subsumption. An example for the *specialisedBy* construct and the part-sensitive subsumption is that a fracture of the humerus diaphysis is also a fracture of the humerus. Here, the role *location* would be specialised by the role *part-of.* Beside having an explicit subsumption BERNWARD also gives the possibility of setting two concepts into an explicit part-of relation. Thus, we can build an explicit partonomy. An explicit partonomy cannot be built in GRAIL. This is no serious difference but because of the importance of the part-of-relation in medical concept systems it might be convenient to have both an explicit partonomy and an explicit taxonomy.

There are more operators and statements in both languages but they can all be created by using the constructs described so far.

Both BERNWARD and GRAIL have restriction statements and the different features of both languages will be discussed in detail using examples from the AO-Classification.

2.1 Conjunction of Criteria

Let us consider the following concept from the AO-classification: **complex segmental fracture of the femur diaphysis with one segmental and additional wedge fragment(s).** In both BERNWARD and GRAIL this concept can be modelled easily by expressing that this is a fracture which has the complexity complex, the form segmental, the location femur diaphysis, one fragment with the form segmental and one or more fragments with the form wedge. Formally, modelling this concept is done by conjunction of criteria, i.e. using the same role (*fragment*) twice.

GRAIL:
(*Fracture which*
<hasComplexity complex hasForm segmental hasLocation FemurDiaphysis
hasFragment (*Fragment which <hasForm segmental hasNumber 1>*)
hasFragment (*Fragment which <hasForm wedge hasNumber* $\geq$*1>*)*>*)

BERNWARD:
(*fracture*
{(*:complexity complex*) (*:form segmental*) (*:location FemurDiaphysis*)
(*:fragment* (*Fragment* {(*:form segmental*) (*:number 1*)})
(*:fragment* (*Fragment* {(*:form wedge*) (*:number* $\geq$*1*)})})

2.2 Conjunction of Concept Expressions

A fracture can have a direction, e.g. **downwards** or **inwards**. The two primitives *downwards* and *inwards* represent these two directions. A direction which is **downwards and inwards** shall be represented. This can be done directly in BERNWARD by using the *AND*-operator:

(*AND [downwards inwards]*).

One possiblity to model this concept expression in GRAIL is creating a primitive *downwardsAndInwards* and setting it under *downwards* and under *inwards* in the hierarchy by using explicit subsumption. This new concept expression is not in the same position as the coordinated concept expression would be as shown in Figure 2 below but since the coordinated concept expression will not exist in GRAIL the primitiveconcept expression behaves in a similar way.

A more correct effect can be produced in GRAIL by using *necessary* statements. The fact that it is possible to give both the role r and its inverse role r^{-1} the cardinality *one* is used here. The statements (*inwards topicNecessarily r inwards*) and (*inwards topicNecessarily* r^{-1} *inwards*) cause the role r tobehave w.r.t. *inwards* like an identity relation. Therefore, for this special case the GRAIL-expression (*downwards which r inwards*) has the same meaning as the BERNWARD-expression (*AND [downwards inwards]*). Fig-

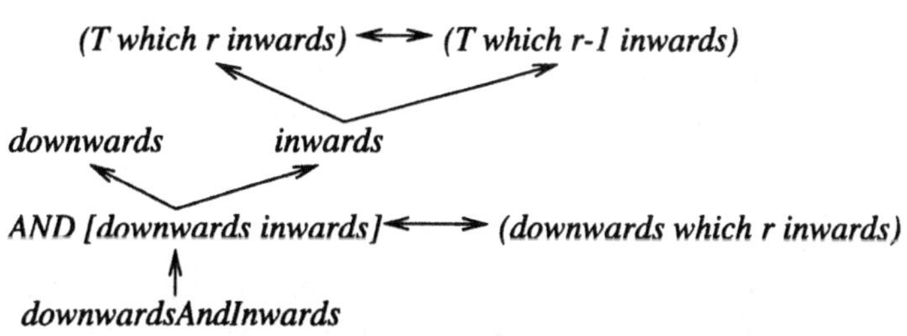

Figure 2: Concept Hierarchy

ure 2 shows the subsumption relation between these concept expressions. Because such a role *r* can be defined for all concept expressions, conjunction can be used between all concept expressions in GRAIL.

2.3 Conjunction of Values

The concept **extra-articular fracture of the distal humerus, metaphyseal simple, oblique downwards and inwards** can be modelled easily in BERNWARD by using the coordinated concept expression (*AND [downwards inwards]*) as the value for the role *:direction.* Thus, it can be expressed that the above concept represents a fracture with the articularity extra-articular, the location distal humerus, the complexity metaphyseal simple, the form oblique and the direction downwards and inwards.

For modelling this concept expression in GRAIL there are again two possibilities. The primitive *downwardsAndInwards* can be used as the value for the role *direction.* The other possibility is to move the value conjunction one step further so that is becomes a role conjunction. Both primitives *downwards* and *inwards* are explicit children of the concept expression *DirectionFeature.* The roles *hasDownUp* and *hasInOut* are defined as single valued and furthermore these roles are defined as identity roles by necessary statements and thus we obtain a concept expression which is equivalent to the one in BERNWARD.

GRAIL:
(*Fracture which*
<*hasArticularity extra-articular hasLocation DistalHumerus*
hasComplexity metaphyseal-simple hasForm oblique
hasDirection (*DirectionFeature which*
<*hasDownUp downwards hasInOut inwards*>)>)

2.4 Disjunction of Concept Expressions

Let us consider the AO-expression **simple or wedge**. BERNWARD will introduce a coordinated concept expression (*OR[simple wedge]*), whereas the only simple possibility of modelling this expression in GRAIL is creating a primitive *simpleOrWedge* and setting it as an explicit parent of both primitives *simple* and *wedge.* Compared to the coordinated concept expression, the primitive *simpleOrWedge* does not have the same position in the subsumption hierarchy because the coordinated concept expression would be a child of the primitive concept expression *simpleOrWedge* and would also subsume the two primitives *simple* and *wedge* but as mentioned before since GRAIL does not contain any coordinated concept expressions the primitive behaves in a similar way. The same can be done for arbitrary concept expressions by using *necessary* statements.

2.5 Disjunction of Values or Criteria

The AO-Classification introduces a **simple or wedge fracture of the ulna**. This represents a fracture with the complexity simple or wedge and the location ulna which is equivalent to a fracture with the complexity simple or the complexity wedge and the location ulna. In BERNWARD the coordinated concept expression (*OR[simple wedge]*) will be taken as the value for the role *complexity* and in GRAIL we will take the primitive *simpleOrWedge*as the value.

3. Missing Constructs

There are two useful features for representing the AO-Classification that are missing in both BERNWARD and GRAIL.

The AO-code 13-A represents an **extra-articular fracture of the distal humerus.** It should be expressed that **extra-articular** means not articular and that the two concepts are therefore disjoint. Obviously, a negation operator *NOT* is desirable here. Both BERNWARD and GRAIL do not contain such a construct.

Another construct missing in BERNWARD and GRAIL which would be useful for the AO-Classification is arbitrary cardinality restriction of roles. BERNWARD does not contain any cardinality and the cardinality in GRAIL is either *one* or *many.* An example for a concept expression using a more complex form of cardinality is a **complex spiral fracture with more than three intermediate fragments.**

4. All and Only Sensible Concept Expressions?

The main features for composing new concept expressions have been discussed so far. Compositional concept representation languages contain another main feature in order to restrict the composition of new concept expressions. Only sensible or relevant concept expressions should be allowed to be created. Therefore, both BERNWARD and GRAIL have the possibility to allow and forbid pairs of concept expressions to be connected by a role. These thoughts lead to the two questions whether it is possible to map every concept of an agreed set of concepts uniquely to a concept expression of the language and whether all possible concept expressions can be mapped to user's concepts w.r.t their world view.

4.1 All Sensible Concept Expressions?

As seen before the expressiveness of both languages is not powerful enough to express every concept in a direct and intuitive way. But this does not affect the question. We will be satisfied if a concept expression has a similar behaviour w.r.t. the subsumption relation in the language as the represented concept in the user's concept hierarchy. It can be shown that the intended behaviour of a concept expression in the subsumption hierarchy can be simulated in every case. If the formal subsumption relations of a concept expression do not mirror the behaviour of the mapped concept we can define the concept expression in a different way or introduce a primitive concept expression and use explicit subsumption. Therefore, the answer to the question whether it is possible to represent all agreed concepts in the languages is yes.

4.2 Only Sensible Concept Expressions?

In order to use the languages in applications it may be useful to restrict the set of all concept expressions which can be created in the language to only sensible concept expressions. Users may say that a concept expression is sensible if they can find a concept within their world view which can be mapped to the concept expression. Can this be done in any case by BERNWARD and GRAIL?

Both languages have constructs to allow and to forbid the connection of two concept expressions by a role. In GRAIL the statement (c *grammatically* r d_1) allows the concept expression c and each of its subconcept expressions to be linked to the the concept expression d_1 or a subconcept expression of d_1 by the role r or any subrole of r. Figure 3 shows these statements for the same c and the same r. It can be seen that these statements allow concept expressions that are within the related subtree. The statement (c *sensibly* r d_2) restricts the set of allowed concept expressions to more specific ones. The expression (c *mayNot* r d_3) states that neither the concept expression c nor any of its subconcept expressions can be linked to the concept expression d_3 or any of its subconcept expressions by r or a subrole of r. Therefore, this statement cuts out a subtree of the allowed tree.

BERNWARD has a similar set of statements which are tailored to its language operators. The statements ($RD(r)=\{c\}$) and ($RR(r)=\{d_1\}$) define the domain and the range of the role r and are therefore a generalization of the grammatical statement of GRAIL. The use of a role w.r.t. a concept expression can be restricted by the global role restriction ($GR(r,c)=\{d_2\}$). This means that the concept expression c can only be connected with d_2 or one of its subconcept expressions. It is not possible to "cut ou" a subtree of the concept expressions like in GRAIL.

Beside this restrictions BERNWARD offers a local role restriction which allows to restrict the values $\{a_1, \ldots, a_n\}$ of a role r for a concept expression a dependendent on the context c the expression is used in: ($LR(r,a,c)=\{a_1,\ldots, a_n\}$). Let us consider an example of the AO-Classification. A **wedge fracture of a diaphysis** can be specified further by expressing that the

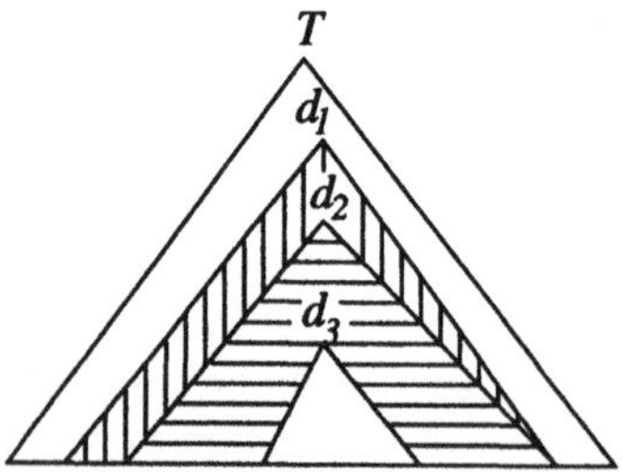

Figure 3: Sanctioning and Restriction

wedge can have the morphology **spiral, bending, fragmented, non-fragmented.** The **wedge** of a **wedge fracture of the radius diaphysis** can only be **fragmented** or **non-fragmented.** Thus, we want to forbid the values **spiral** and **bending** or allow only the values **fragmented** and **non-fragmented** for the, primitive **wedge** in the concept expression **wedge fracture of the radius diaphysis.** This is done by the following local role restriction: (*LR(:morph, wedge, wedge fracture of the radius diaphysis):= {fragmented, non-fragmented}*). It is not possible to restrict in such a context-sensitive way in GRAIL. Furthermore, the conjunction of concept expressions can be restricted by the coordination restriction ($CR(c)=\{a_1, ..., a_n\}$). This enables the user to allow certain conjunctions of concept expressions.

As seen before there is the restriction construct *LR* in BERNWARD which cannot be simulated in GRAIL and there is the restriction statement *mayNot* in GRAIL which cannot be simulated in general in BERNWARD. It follows directly that neither GRAIL nor BERNWARD can restrict the concept expressions to the only sensible ones in every case but we want to stress that the restriction mechanisms of both languages are powerful enough for most applications.

5. Conclusion

The different operators and statements in BERNWARD and GRAIL have been discussed. Examples from the AO-Classification were taken in order to show how conjunction of criteria, concept expressions and values and disjunction of primitives and values can be represented in these two languages.

It was demonstrated that a negation construct and arbitrary cardinality restriction of roles are useful for representing the AO-classification but that both BERNWARD and GRAIL do not have any of these constructs.

The constructs for restricting the composition of new concepts in BERNWARD and GRAIL were explained. It was demonstrated that there are senseless or irrelevant concept expressions that can be rejected by BERNWARD but not by GRAIL and vice versa.

It was shown that both systems differ slightly in their expressiveness and that they can both be taken for representing the AO-classification. Concepts can be modelled in different ways and thus the reality can be represented with different levels of detail. Furthermore, it was demonstrated that neither of the languages is superior to the other.

References:

BERNAUER, J.; FRANZ, M.; SCHOOP, M.; SCHOOP, D. and PRETSCHNER, D. P. (1995): The Compositional Approach For Representing Medical Concept Systems. *Proc. of the 8th World Congress on Medical Informatics, MedInfo 95, Vancouver, Canada, pp 70-74.*

MÜLLER, M. E.; NAZARIAN, S.; KOCH, P. and SCHATZKER, J. (1990): The Comprehensive Classification of Fractures of Long Bones. Springer Verlag.

RECTOR, A. L.; Bechhofer, S.; GOBLE, C. A.; HORROCKS, I.; NOWLAN, W. A. and SOLOMON, W. D. (1996): The GRAIL Concept Modelling Language for Medical Terminology. *Technical Report, The Galen Project.*

SCHOOP, M.; SCHOOP, D. and BERNAUER, J. (1995): A Compositional Classification Manager Exemplarily Shown for the AO/ASIF-Classification of Fractures of Longbones. *Proc. of the 8th World Congress on Medical Informatics, MedInfo 95, Vancouver, Canada, p. 1655.*

The Freiburg Center of Data Analysis and Model Building (FDM): An Interim Report about an Interdisciplinary Cooperation

Martin Schumacher

Institute of Medical Biometry and Informatics and
Center of Data Analysis and Model Building,
University of Freiburg, D - 79104 Freiburg, Germany

Summary: An outline of the concept and the aims of the Freiburg Center for Data Analysis and Model Building (FDM) is presented. The range of current projects is briefly reviewed; two projects featuring the development of stochastic models and application of statistical methods in biomedical research are discussed in more detail. Finally, a preliminary résumé about the interdisciplinary cooperation is given.

1. Introduction

Scientific data analysis and model building has become a substantial part in many areas of research. When modelling the behaviour of complex systems as environmental or biological systems, for example, and when analyzing data arising from the investigation of such systems large variability has usually taken into account. Therefore, stochastic and statistical methods are of fundamental importance in building adequate models where a sensible analysis and interpretation of data can be based on. This has been the starting point for the formation of an interdisciplinary Center for Data Analysis and Model Building at the University of Freiburg, Germany, that is described in this paper. After a brief review of general aims and organization of the center, an overview on current projects is given. Two projects, both from biomedical research due to the author's affinity, are presented in more detail. The specific responsibilities of the center are summarized and an preliminary résumé about the interdisciplinary cooperation between various disciplines is given.

2. Objectives and organization of the center

The center is intended as a "center of competence" for scientific data analysis and model building providing a platform for interdisciplinary cooperations between partners from various research institutions within and outside the university and from the industry. In project groups, solutions for problems arising in various fields of application are elaborated with methods from stochastics, statistics and physics.

Starting with a regular series of seminars on data analysis and model building in 1987, the center has been established in 1994 with support of the University of Freiburg and the State Ministry for Science and Research Baden-Württemberg. The core participating institutions (and scientific members) of the center are

- Institute of Mathematical Stochastics (E. Eberlein, R. Lerche, L. Rüschendorf, H. Witting)
- Faculty of Physics - Stochastic Dynamical Systems (J. Honerkamp)
- Institute of Medical Biometry and Informatics (M. Schumacher)
- Institute of Biophysics - Brain Research Unit (B. Fischer)

The center's location (Albertstraße 26-28) is very close to the core participating institutions. Currently, there are working 12 scientists, 2 graduates and 11 diploma students jointly within the center. Their work is partially supported by grants from the Deutsche Forschungsgemeinschaft and the industry. The center is equipped with a network of currently 14 SUN-Sparc stations connected to the Freiburg Data Network and the Internet. The scientific members of the center are organizing a series of seminars with approximately 6 invited speakers per term, other scientific meetings and a series of internal seminars. The results of the various research projects can be obtained as preprints before publication in peer-reviewed international journals.

3. Overview on current projects

There are four major areas of projects at present:

1. Characterization and modelling of time series,
2. Modelling of longitudinal data in clinical and epidemiological research,
3. Statistical analysis of stochastic processes,
4. Analysis and modelling of data in finance.

Topic 1 includes projects on "Acoustic quality control of motors" (Heitz, 1996), "Analysis of patch-clamp data of ion channels by means of Hidden Markov models" (Becker et al., 1994), "Analysis and modelling of the distribution of saccadic reaction times" (Fischer et al., 1995) and "Tremor classification and tremor time series" (Deuschl et al., 1995). Two projects of topic 2 are described in more detail below; other projects in this area are considering the development and validation of prognostic models in oncology (Altman et al. 1994) , the analysis of quality of life data in clinical research (Olschewski et al. 1994) or the treatment of incomplete data in clinical and epidemiological studies (Vach and Schumacher 1996). Topic 3 covers besides basic statistical research (e.g. Witting 1995) the development of statistical methods for various applications ranging from the stochastic analysis of algorithms (Cramer and Rüschendorf 1996) to methods of serial and spatial ordering with applications in archeology (Alt and Vach 1995). Under topic 4 projects on the investigation of the predictability of financial markets, modelling of option prices and development of deposit accounts are covered (Eberlein and Keller 1995; Beibel and Lerche 1996; Rüschendorf and Rachev 1994).

4. An example: Modelling occurence and impact of nosocomial infections

Nosocomial (or hospital aquired) infections are major complications arising in the patient's treatment in hospital. The prevalence of nosocomial infections in Germany is estimated as 3.5 %; it is highest (15.3 %) in intensive care units (Rüden et al. 1995). The problems to be addressed in this project, that has been carried through in cooperation with Drs. F. Daschner, I. Kappstein and A. Kropec, Dept. of Environmental Medicine and Hospital Epidemiology, Dr. K. Geiger, Dept. of Anesthesiology, and Dr. H. Just, Dept. of Cardiology of the University Hospital Freiburg, were as follows:

- Which factors determine the risk for the occurrence of nosocomial infections and how large is that risk?
- What is the impact of nosocomial infections with respect to mortality and length of hospital stay.

To get an answer to these questions, data have been collected prospectively in a cohort study comprising all patients (N=756) admitted to two intensive care units of the University Hospital Freiburg during two years. For modelling occurrence and impact of nosocomial infection a stochastic model has been developed representing the states and transitions relevant for the patients' course of disease. This model where only nosocomial pneumonia and nosocomial sepsis (or bloodstream infection) is considered is schematically displayed in Figure 1. Transition intensities and probabilities have been estimated using statisticals methods for counting processes (Andersen et al. 1993).

In addition to these data featuring relevant events in the patients'disease course, several other patient characteristics have been recorded for the investigation of potential risk factors. Among others, these comprise infection on admission, drainage of the thorax, acute or chronic renal failure, various diagnoses at admission, type of operation and various laboratory measurements at admission. When evaluating these factors and forming risk categories, several phenomena had to be taken into account:

- some factors affect the various types of noscocomial infections differently; so, for example, infection on admission appears to be protective for nosocomial pneumonia - probably because of the given antibiotics - whereas it increases the risk for nosocomial sepsis dramatically.
- Some factors are relevant for the occurence of nosocomial infections directly after admission, in the first week say, others for the second and third week. Thus time-dependent effects of potential risk factors have to be evaluated.
- The model building process leads to an overoptimistic judgement of the derived risk categories. Thus for example, the relative risk in the highest risk category of the pneumonia risk score (Kropec et al. 1996) with respect to the lowest is estimated as 7.3; cross-validation (Verweij and Houwelingen 1993) gives a more realistic value of only 5.4 that again needs confirmation by validation in an independent data set.

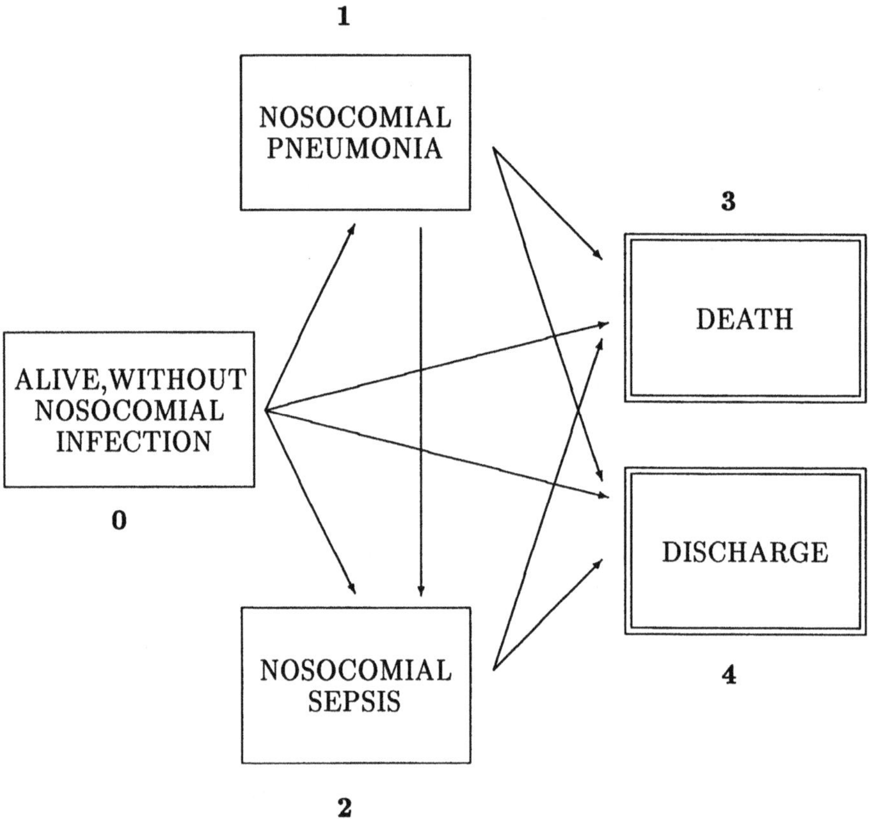

Figure 1: Potential states and transitions for occurence and impact of nosocomial infections

The impact of nosocomial infections with respect to prolongation of hospital stay is often investigated by simply comparing infected with non-infected patients. This "naive" approach that does not take into account that patients with a longer hospital stay and a probably more severe illness also have a higher chance of getting infected; this approach leads to an estimate of +14.4 extra days attributable to nosocomial pneumonia in our data. A more refined matching approach (Kappstein et al. 1992) comparing patients with similar characteristics and similar length of hospital stay until time of infection already leads to a reduced value of +8.2 extra days. Finally, a model-based approach (Schulgen and Schumacher 1995) using estimated transition intensities and probabilities derived within the stochastic model displayed in Figure 1 leads to a realistic estimate of +3.4 extra days attributable to nosocomial pneumonia.

5. Another project: Diagnosis of breast tumors by color Doppler sonography

In cooperation with Drs. H. Madjar and H. Prömpeler, Dept. of Gynecology of the University Hospital Freiburg, a study has been performed to examine the role of noninvasive sonographic measurements (color Doppler flow signals) for differentiation of benign and malignant breast tumors. Within approximately two years, N=458 women have been enrolled in the study and several sonographic measurements and patient characteristics have been recorded. Histological verification yielded 325 benign tumors and 133 malignant tumors. Logistic regression analysis yielded a simple classification rule using age, number of arteries in the tumor and number of arteries in the contralateral breast as predictor variables with a resulting sensitivity of 95.5 % and a specificity of 90.2 % (Sauerbrei et al. 1995); the percentage of correct classifications was 91.7 % . The additional consideration of several hemodynamic characteristics as maximum and average of peak systolic velocities per tumor and maximum and average resistance index per tumor did not yield any gain in information or in predictive accuracy since these characteristics showed a high correlation with each other and the number of arteries.

One part of the project was a comparison of this classification rule and a similar one obtained by the classification and regression tree (CART) methodology (Breiman et al. 1984) to classification rules based on neural networks (Schumacher et al. 1996). Starting with a logistic perceptron as the most simple neural network yielding identical results as the logistic regression analysis, the number of units in the hidden layer has been increased from J=1 up to J=40. The latter one corresponds to a nonlinear regression model with 200 parameters. Accordingly, the percentage of correct classifications increased to 99.6 % leading to a perfect classification of benign and malignant tumors. However, these values have been obtained as naive estimates from the same data where the classification rule has been derived from. When contrasting these values to those obtained by a 10-fold repetition of 10-fold cross-validation (Efron and Tibshirani 1993) as shown in table 1 it is clear that an uncritical application of neural networks by increasing the number of hidden units only results in a substantial overfitting of the data at hand; the results cannot be generalized. There is no gain in predictive accuracy at all as indicated by the percentages of correct classifications obtained by cross-validation. Thus for the "design" and the "training" of neural nets, similar considerations as in statistical model building are necessary. Validation in independent data sets or, at least, the use of cross-validation techniques is an essential prerequisite for an unambiguous judgement and an adequate interpretation of the results of this and similar studies.

6. Résumé

The specific objectives of the Freiburg Center for Data Analysis and Model Building can be summarized as follows:

Method	Percentage of correct classifactions	
	naive estimate	cross-validation estimate
Logistic regression	91,7 %	91,9 %
NN(1)	93,0 %	92,5 %
NN(2)	93,4 %	91,3 %
NN(3)	93,4 %	91,2 %
NN(4)	93,9 %	91,3 %
NN(6)	95,6 %	91,4 %
NN(8)	96,1 %	92,2 %
NN(10)	98,5 %	91,8 %
NN(15)	98,5 %	91,4 %
NN(20)	99,1 %	91,0 %
NN(40)	99,6 %	91,2 %

Table 1: Results of classification rules based on logistic regression and feed-forward neural networks with J hidden units (NN(J)) for the breast tumor data.

- Cooperation and support of partners from other disciplines during the planning phase of studies and experiments in order to ensure data in sufficient quantity, quality and completeness.
- Development and critical examination of stochastic models for the data at hand - guided by the specific subject-matter research problems.
- Estimation of model parameters by efficient statistical methods; if necessary, development of new, more adequate statistical methods.
- Interpretation of results and their critical appreciation jointly with partners from other disciplines.

Several advantages of such an interdisciplinary cooperation could have been observed so far. The direct exchange of ideas and methodology between the scientists working at the center leads to considerable synergistic effects for the treatment of problems from different areas of application. The graduates, Ph.D.- and diploma students have ample opportunity to work on interisting problems with high practical relevance and to participate in latest scientific findings of other projects. For several of the projects of the center adequate solutions have already been worked out and have been published in internationally recognized journals. Thus, the concept of such an interdisciplinary center has been proven successful and sound. Various further applications for potential cooperations by institutions within and outside the university that could not have been transfered into concrete projects so far because of limited resources led to the conclusion that there is still a considerable potential for further develoment.

References:

ALT, K. W. and VACH, W. (1995): Odontologic kinship analysis in skeletal remains: concepts, methods and relults. *Forensic Science International 74, 99–113.*

ALTMAN, D. G.; LAUSEN, B.; SAUERBREI, W. and SCHUMACHER, M. (1994): Danger of using "optimal" cutpoints in the evaluation of prognostic factors. *Journal of the National Cancer Institute 86, 829–835.*

ANDERSEN, P. K.; BORGAN, O; GILL, R. D. and KEIDING, N. (1993): Statistical models based on counting processes. New York: Springer.

BECKER, J. D.; HONERKAMP, J.; HIRSCH, J.; FRÖBE, U.; SCHLATTER, E. and GREGER, R. (1994): Analysing ion channels with hidden Markov models, *Pflügers Archiv 426, 328-332.*

BEIBEL, M. and LERCHE, H. R. (1996): A new look at warrant pricing and related optimal stopping problems. Statistica Sinica (in press).

BREIMAN, L.; FRIEDMAN, L.; OLSHEN, R. A. and STONE, C. J. (1984): Classification and regression trees. Wadsworth, Monterey.

CRAMER, M. and RÜSCHENDORF, L. (1996): Analysis of recursive algorithms by the contraction method. Proceedings Athens Conference Applied Prob. (in press).

DEUSCHL, G.; LAUK, M. and TIMMER, J. (1995): Tremor classification and tremor time series analysis. *Chaos, 5, 48–52.*

EBERLEIN, E. and KELLER, U. (1995): Hyperbolic distributions in finance. *Bernoulli 1, 281-299.*

EFRON, B. and TIBSHIRANI, R. J. (1993): An introduction to the bootstrap. Chapman and Hall, London, New York.

FISCHER, B.; GEZECK, S. and HUBER, W. (1993): The three-loop model: a neural network for the generation of saccadic reaction times. *Biol. Cybern. 72, 185-196.*

HEITZ, C. (1996): Classification of time series with optimized time-frequency representations. In: Bock, H. H., Polasek,W. (eds.): Data analysis and information systems. Stochastical and conceptual approaches. Studies in classification, data analysis and klowledge organization, volume 7, 41-51. Springer-Verlag, Heidelberg.

KAPPSTEIN, I.; SCHULGEN, G.; BEYER, U.; SCHUMACHER, M. and DASCHNER, F. D. (1992): Prolongation of hospital stay and extra costs due to ventilator-associated pneumonia in an intensive care unit. *European Journal of Clinical Microbiology & Infectious Diseases 11, 504–508.*

KROPEC, A.; SCHULGEN, G.; SCHUMACHER, M.; JUST, H.; GEIGER, K. and DASCHNER, F. (1996): Scoring system for nosocomial pneumonia in intensive cae units. Intensive Care Med. (in press)

OLSCHEWSKI, M.; SCHULGEN, G.; SCHUMACHER, M. and ALTMAN, D. G. (1994): Quality of life assessment in clinical research. *British Journal of Cancer,*

70, 1-5.

RÜDEN, H.; DASCHNER, F. and SCHUMACHER, M. (1995): Nosokomiale Infektionen in Deutschland – Erfassung und Prävention (NIDEP-Studie) Teil 1, Abschlußbericht und Anhänge, NOMOS Verlagsgesellschaft Baden-Baden, Bundesministerium für Gesundheit, Band 56.

RÜSCHENDORF, L. and RACHEV, S. T. (1994): Models for option pricing. Theory Probab. *Applications, 39, 150–199.*

SAUERBREI, W.; MADJAR, H. and PRÖMPELER, H. (1995): Development of diagnostic rules based on color Doppler flow signals for differentiation between benign and malignant breast tumors. Preprint. Freiburg.

SCHULGEN, G. and SCHUMACHER, M. (1995): Estimation of prolongation of hospital stay attributable to nosocomial infections. Preprint.

SCHUMACHER, M.; ROSSNER, R. and VACH, W. (1996): Neural networks and logistic regression: Part I. Computational Statistics and Data Analysis 21 (in press).

VACH, W. and SCHUMACHER, M. (1996): Behandlung von fehlenden Werten in den Einflußgrößen von Regressionsmodellen – Ein Überblick. *Informatik, Biometrie und Epidemiologie in Medizin und Biologie, 27 (1), 48–66.*

VERWEIJ, P. J. M. and VAN HOUWELINGEN, H. C. (1993): Cross-validation in survival analysis. *Statistics in Medicine, 12, 2305–14.*

WITTING, H.(1995): Rank tests for scale. Hajek's influence and recent developments. *Kybernetika 31, 269-291.*

Realization of a Medical Data Dictionary in a Relational Database Management System

J. Stausberg[1], M. Laghmouch[1], C. Moraga[2]

[1] Institut für Medizinische Informatik, Biometrie und Epidemiologie
Universitätsklinikum Essen, Hufelandstr. 55, D-45131 Essen

[2] Fachbereich Informatik, Universität Dortmund, D-44221 Dortmund

Summary: The importance of Medical Data Dictionaries in information and knowledge processing in medicine increased in the last years. To reach a practicable use in Health Care Information Systems, the availability of a Medical Data Dictionary in a Relational Database Management System (RDBMS) is necessary, providing a standardized access with the Structured Query Language. The objective of our project was the migration of a previously developed vocabulary to a RDBMS. Our Entity-Relationship-Model contains 4 entities: term, link, maintenance and library. Each element of the medical terminology has to be defined as a term. A link connects two terms or one existing link and a term through a link type. That link type itself is defined as a term. The Medical Data Dictionary was successfully implemented with ORACLE7. The database model is flexible enough to store existing taxonomic vocabularies and to support concept representation languages.

1. Introduction

Medical Data Dictionaries support the communication between and the integration of distributed information as well as knowledge-based systems on the terminology level (Stausberg and Ohmann (1996), Stausberg et al. (1995)). Data Dictionaries are a prerequisite for a couple of necessary developments in medical informatics, for example the sharing of patient data between different health care providers, the sharing of knowledge bases etc. In our view Medical Data Dictionaries have two main functions concerning aspects of integration. First, they provide a controlled vocabulary. Systems that use the concepts and denominations from that vocabulary could be integrated with less problems on the terminology level. Second, Medical Data Dictionaries could map between terminologies, for example between a clinical and an administrative terminology. Beside these aspects of integration a Medical Data Dictionary supports a predictive data entry into a computer-based patient record through the availability of sensible characteristics of clinical manifestations. In Hospital Information Systems a Medical Data Dictionary is a central service which enables a distributed information processing. Other central services could be a communication server, a computer-based patient record and a document management system. These central services are connected with departmental systems through a physical network.

Existing vocabularies are often stored in ASCII-files and imported in a computer system through a parser (Galen (1994), UMLS (1996)). To reach a practicable use in Health Care Information Systems, the availability of a Medical Data Dictionary in a Relational Database Management System (RDBMS) is necessary, providing a standardized access with the Structured Query Language. Thus, our objective was the migration of a previously developed vocabulary to ORACLE7. The database model should be as flexible as possible to support existing vocabulary structures and concept representation languages.

2. Methods

2.1 Conceptual Requirements

The database model has to provide the levels of commonality 1 to 4 which were introduced by Cimino (1994). Cimino mentioned five levels of commonality to achieve the ability to share data and knowledge between clinical workstations. The basic level is the definition of common data types. The next level is the agreement on general classes, for example complaints, physical findings etc. UMLS (1996) provides such general classes in its semantic network, called semantic types. The representation of a vocabulary is the third level, Cimino mentioned ASN.1. The fourth level is a common content for some domains. The last level is a common representation of medical knowledge. Common knowledge representation languages which stress the knowledge level, as KIF and Ontolingua (Gruber (1993)), are beyond our project.

The conceptual model is adopted from a previously introduced Data Dictionary (Stausberg et al. (1995)). The database model should support the following terminological structures:

- An identification of concepts.
- An assignation of denominations to concepts, that includes synonyms and homonyms.
- An intensional definition of concepts through characteristics, that includes attributes and possible values from a concept.
- A feature to build concept hierarchies as well as the use of non-hierarchical relationships between concepts.
- An ability to compose concepts from concepts.

2.2 CASE*Method

The entities of Medical Data Dictionaries are different to objects in the real world concering database modeling. A Medical Data Dictionary should provide the necessary terminological elements for a definition of real worl objects in information systems and knowledge bases. To reach the required flexibility on that abstract level we decided to use a special Entity-relationship Model, called CASE*Method, for data modeling. CASE*Method was developed by Barker (1990) and is part of the ORACLE CASE*Tools.

*Syntax of CASE*Method (cf. figure 1):* An entity is presented by a softbox with its name. The attributes are listed below the name of the entity. A mandatory attribute is pointed out by a star and an optional attribute is pointed out by a circle. A hash sign is used for the primary key. A relationship is represented by a line that joins two entity softboxes together. For a degree of many the relationship line joins a softbox at three points. For a degree of one the line joins at only one point. A mandatory end of a relationship has a solid line, an optional end has a broken line. A special element of CASE*Method is the exclusive arc that represents an either or relationship. Every relationship which is part of the exclusive arc meets that arc with a thick dot.

3. Database Model

Every unit in our Data Dictionary (cf. figure 1) has to be defined as a term. There is no difference between concepts, words, statements, relationships etc. A term is defined by a number and an optional denomination. Denominations can be words like pain, liver, intensity, values like low, the number 5, or the unit years.

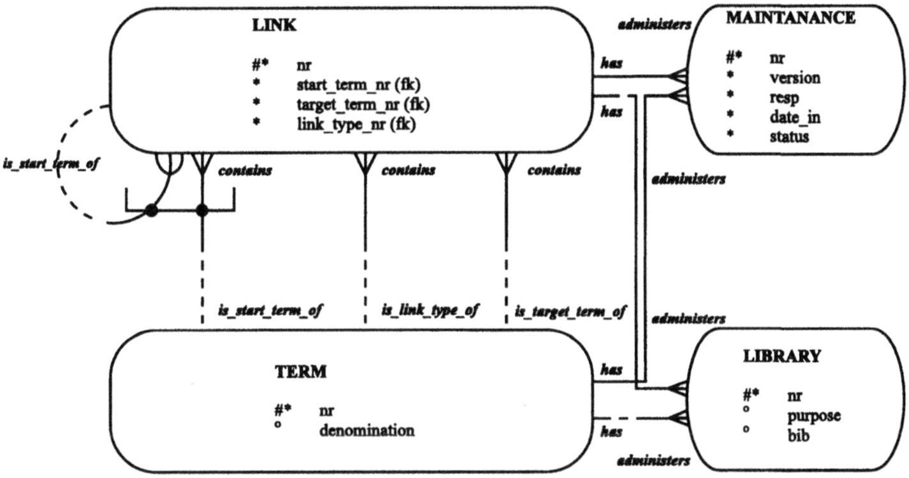

Figure 1: Database Model

The second main entity is a link. Two terms could be connected by another term and these three terms build up a link. A link is identified by a number but surely it has to include the connected terms. For a more distinct presentation the foreign keys (fk) are included in figure 1. The connected terms are called start_term and target_term because the connection is unidirectional. The term that is used to connect the two other terms is called link_type. For example a hierarchy could be defined by the link *pain is_a symptom*. An object - attribute - value triple could be represented with two links, for example *pain* as object, *intensity* as attribute and *low* as value. *Pain has_attribute intensity* and *intensity has_value low* are the defined links.

Note that *has_attribute* and *has_value* are terms itself. Part_of and other subsumptions could be realized as is_a, for example *finger is_part_of hand.* The entities term and link represent the terminological content of the Data Dictionary.

The entities maintenance and library hold additional information for a cooperative authoring, a version control and bibliographic purposes. The entity maintenance holds administrative information on terms and links, a version of terms and links, the responsible person, the date of input and a status. Permitted values for the attribute status are proposed, refused, validated, effective and inoperative. Library includes a purpose, for example a project in which a term or link is used, and bibliographic information. The attribute purpose enables different views on the Data Dictionary from different projects.

Each term and each link has to have at least one maintenance and may have one or more libraries. A term may be a start_term, a target_term or a link_type of a link. Note that both a term and a link could be a start_term of a link. To model the latter we used the exclusive arc of CASE*Method.

3.1 Graphical Representation

For the visualization of our approach we introduce a new graphical representation (cf. figure 2). A term is presented in a box with a gray upper and a white lower part, containing the optional denomination. A link is build with a line. That line meets the start_term with a circle, the link_type with a black box in the upper field and the target_term with an arrow. If we refer to a link as a start_term of another link, that link is surrounded by a gray box without a white lower part.

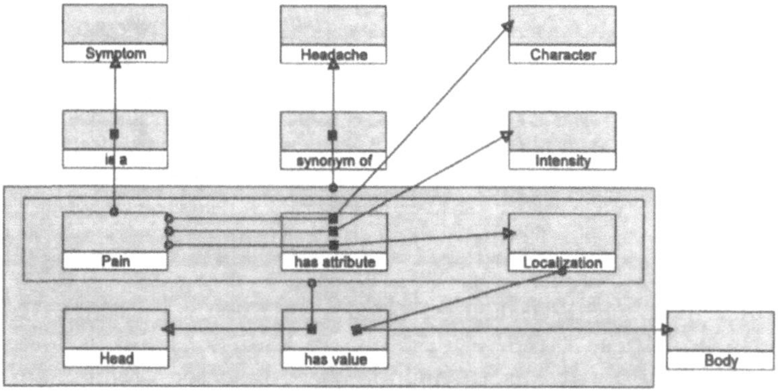

Figure 2: Graphical Representation

3.2 Compositional Structures

The main feature of our Data Dictionary is to define what is sensible to say, not to represent a concrete fact or statement. So the structure usually supports for example the definition of the body (and its parts) as an allowed localization of pain. We store that *pain has_attribute localization* and *localization has_value body* (cf. Figure 2). In the opposite we can define an instance of pain, pain with the localization head as a compositional structure. In that case we build localization of pain in the first link. Second, we use that concept as a start_term for a link and add the value head. In addition we can give this structure a new denomination like headache. With our database model we are able to represent such compositional structures.

3.3 Implementation

The Data Dictionary was implemented with ORACLE7 in a Client/Server-Environment. The server is running on an IBM RS/6000 Workstation. The Windows-Clients provide at the moment a simple user-interface which was implemented with SQL*FORMS. Our Data Dictionary includes about 16.000 terms, 26.000 links, one maintenance for every term and link, and a couple of libraries.

4. Support of Vocabulary Structures

A main requirement of our approach was the flexibility to store existing taxonomic vocabularies and to support concept representation languages. In the following we will present some examples to validate our database model. It was necessary to use our own consistent concept system in the following examples, because the presented taxonomic vocabularies provide different denominations for their conceptual structures. For example is_a is used for generic subsumption, and has_attribute and has_value are used to represent an object - attribute - value triple as well as characteristics of concepts.

4.1 Classification: OPS-301

A current problem in Germany is an adaptation of the ICPM, the *Operationenschlüssel nach §301 SGB V* (OPS-301) (DIMDI (1995)). It is a classification with a mono-hierarchy. As terms the classes of OPS-301 have to be defined, like appendectomy, as well as the hierarchical relationship is_a (cf. figure 3). We represent the hierarchy with links like *appendectomy is an operation at the appendix*. In addition we have to add the terms has_value

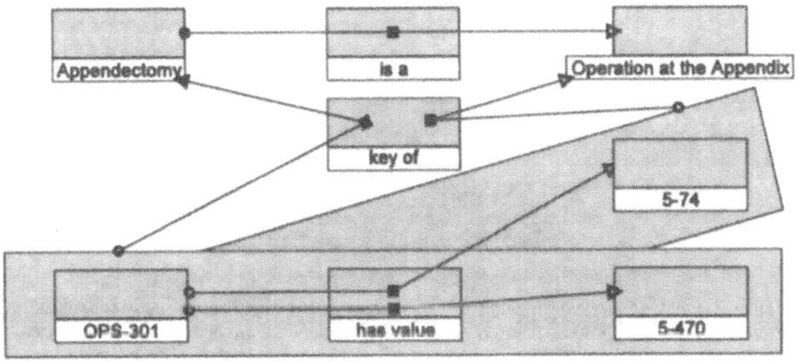

Figure 3: Example from OPS-301

and key_of for the coding. The value 5-470 of the OPS-301 is key of appendectomy for example.

4.2 Nomenclature: SNOMED

SNOMED provides a couple of axes that could be combined to build a disease, as morphology, localization and etiology (Côté (1984)). Parallel SNOMED includes an axis for diseases themselves.

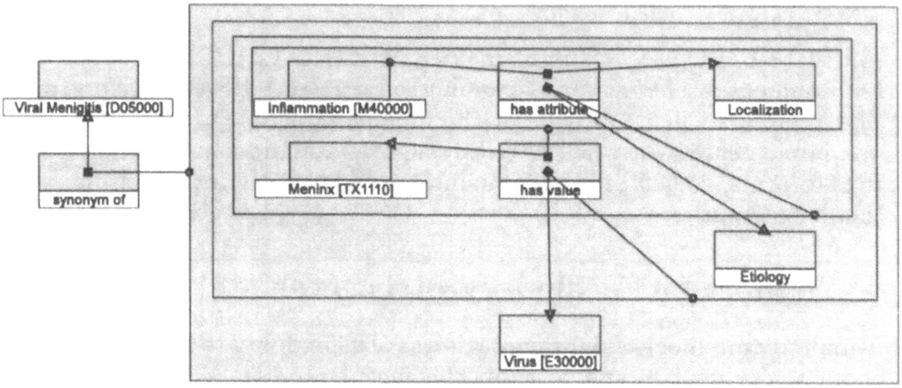

Figure 4: Example from SNOMED

In figure 4 terms from the axes morphology, localization and etiology are combined to compose the disease viral meningitis. The SNOMED-codes are included in brackets. But they have to be defined by other links, as in the previously example. The morphology *inflammation has_attribute localization.* That link is referenced to add the value meninx as concrete localization. So the concept meningitis is composed. *Meningitis has_attribute etiology* is referenced to add the value virus. With the last link viral meningitis is defined as the synonym of the composed structure.

4.3 Semantic Network: UMLS

The transformation of a semantic network into our database model could be carried out in a simple way, for example the conversion of the semantic network of UMLS (1996) (cf. figure 5).

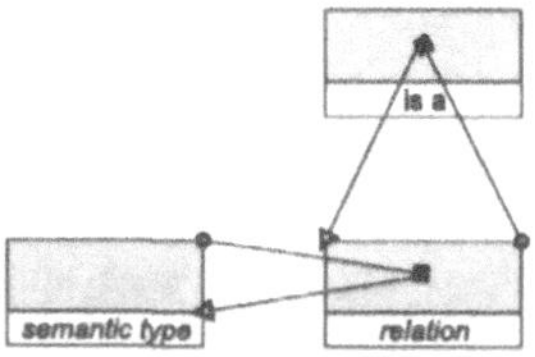

Figure 5: Structure of the Semantic Network of UMLS

In the semantic network of UMLS we have two hierarchies: the hierarchy of the semantic types and the hierarchy of the relations. We build these hierarchies with the term is_a. The relationships are equivalent to our links, so we reach the structure that a semantic type is a start_term, relation is a link_type and another semantic type is the target_term. An example for the hierarchy of semantic types is *virus is_an organism*. An example for the hierarchy of relations is *part_of is physically_related_to*. Typical non-hierarchical relationships are *cell component is_part_of cell* or *idea or concept is_result_of mental process*.

4.4 Concept Representation Language: GRAIL

The Galen Representation And Integration Language (GRAIL) is the concept representation language of the EU-Project Galen (1994). In the following a representation of central parts of GRAIL is presented and one concrete example is transformed into our approach. In figure 6 the core model of Galen is shown in the introduced graphical representation. That includes:

- A hierarchy of elementary entities.
- A composition of composite entities.
- A transformation of GRAIL's terminology (e. g. which, subsumes) in our representation.

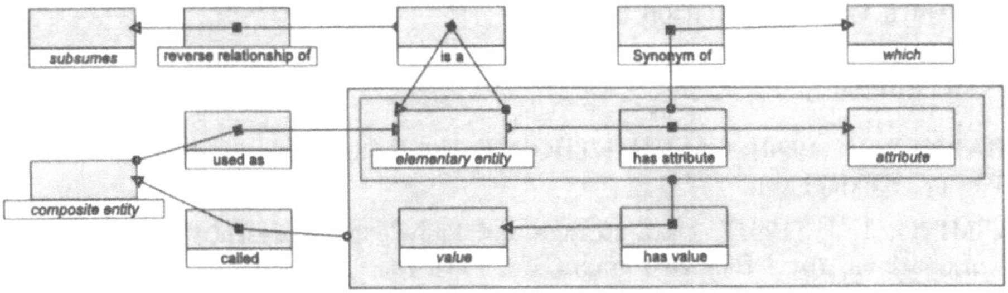

Figure 6: Structure of GRAIL

A typical statement of Galen is that there are spiral fractures of the femur. That statement is represented in GRAIL by *(Fractures which haveLocation-Femurs) which haveFractureForm-spiral.* In our database model we build the first composite entity with two links: *fracture has_attribute location* and *location of fracture has_value femur* (cf. figure 7).
The next criteria set of GRAIL (a criteria set is an attribute - value - tuple) is represented by two other links: *fracture of the femur has_attribute form* and *form of the fracture of the femur has_value spiral.*

5. Discussion

We developed a Data Dictionary in a Relational Database Management System that provides a high flexibility in the definition of medical terminology. The vocabulary is available through a standardized language, SQL. One might guess that the missing of semantic integrity rules in the database model is a limitaion of our approach. But on the one hand the integrity rules

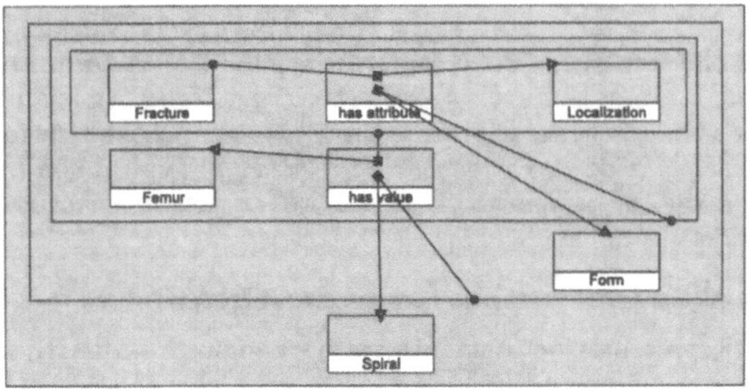

Figure 7: Example of Galen

have to be implemented by applications as well as the authors of this Data Dictionary have to care about the contents. On the other hand the existing vocabularies should provide integrity and consistency by themselves. To provide a central service in an routine environment our next step will be the development of standard interfaces to reach a "Plug and Pla"-functionality in Health Care Information Systems.

References:

BARKER, G. (1990): CASE*METHOD. Entity Relationship Modelling. Addison-Wesley, Wokingham.

CIMINO, J. J. (1994): Data storage and knowledge representation for clinical workstations. *Int J Biomed Comput, 33, 174-179.*

CÔTÉ, R. A. (1984): SNOMED - Systematisierte Nomenklatur der Medizin. Deutsche Ausgabe bearbeitet und adaptiert von F. Wingert. Springer, Berlin.

DIMDI (1995): Operationenschlssel nach S301 SGB V - Internationale Klassifikationen der Prozeduren in der Medizin. Version 1.1.

GALEN (1994): The Master Notation, version 2. Galen Deliverable 14.

GRUBER, T. R. (1993): A translation approach to portable ontology specifications. *Knowledge Acquisition 5, 199-220.*

STAUSBERG, J. and OHMANN, C. (1996): Integration of Knowledge-Based Systems: Methodical Foundations. *Proceedings of the Second World Automation Congress* (to be printed).

STAUSBERG, J.; WORMEK, A. and KRAUT, U. (1995): Terminological Reference of a Knowledge-based System: The Data Dictionary. In: Greenes, R.; Peterson, H.; Protti, D. (eds.): *Proceedings of the Eight World Congress on Medical Informatics.* IMIA, Edmonton, 157-161.

UMLS (1996): Knowledge Sources. 7th Experimental Edition. U. S. Department of Health and Human Services, National Library of Medicine, Bethesda.

Section 7

Applications in Economics and Social Sciences

Two-Mode Overlapping Clustering With Applications to Simultaneous Benefit Segmentation and Market Structuring

Daniel Baier[1], Wolfgang Gaul[1], Martin Schader[2]

[1]Institut für Entscheidungstheorie und Unternehmensforschung, Universität Karlsruhe (TH), Kollegium am Schloß, D-76128 Karlsruhe

[2]Lehrstuhl für Wirtschaftsinformatik III, Universität Mannheim, Schloß, D-68131 Mannheim

Summary: A new two-mode overlapping clustering procedure is presented. This procedure includes solution possibilities for two-mode (non-)overlapping additive clustering as well as (non-)overlapping clusterwise regression with conjoint experiments and can be used for simultaneous benefit segmentation and market structuring. Applications of various cases of the new procedure to conjoint data are used for comparisons.

1. Introduction

An important problem in competitive marketing environments is the definition of boundaries for product-markets, a typical two-mode clustering task resulting in possibly overlapping sets of products judged to be substitutes in prespecified usage situations and possibly overlapping sets of customers for whom such usages are relevant. (See, e.g., Day, Shocker, Srivastava (1979) for an early reference and Bauer (1989) for a more recent review.)

The traditional approach in this so-called benefit segmentation and market structuring context (see, e.g., Srivastava, Alpert, Shocker (1984)) consists in analyzing products-by-products similarity matrices via substitutability measures by one-mode additive clustering procedures, as, e.g., ADCLUS/MAPCLUS (Shepard, Arabie (1979) and Arabie, Carroll (1980)), its individual differences generalizations INDCLUS (Carroll, Arabie (1983)) and SINDCLUS (Chaturvedi, Carroll (1994)) as well as its fuzzy enlargements (Hruschka (1986)), or by two-mode additive clustering procedures, as, e.g., GENNCLUS (DeSarbo (1982)), PENCLUS (Both, Gaul (1987)) or the AE algorithm (Gaul, Schader (1996)).

However, major shortcomings may be observed when applying this traditional approach: First, when one-mode clustering is used, typically only sets of products are derived. The question which possibly overlapping sets of respondents might buy these possibly overlapping sets of products must be answered in an unrelated additional potentially suboptimal analysis step. Second, the collection and aggregation of customer data concerning substitutability is both time-consuming and expensive. More adequate would be an approach where sets of usage situations – in form of overlapping clusters of products – and classes of usage relevances – in form of overlapping clusters of respondents – could be derived from past buying behavior data/intentions or from preference judgments w.r.t. potentially competitive products using respondents-by-products two-mode data matrices. Additionally for computational reasons, most of the above procedures are not well suited for analyzing large data matrices (e.g., with 800 respondents and 30 products).

Thus, a new two-mode overlapping additive clustering procedure for analyzing buying behavior/intention or preference data has been developed, which allows to derive possibly overlapping sets of products and sets of respondents. Alternating least squares methodology is used for parameter estimation. (See, e.g., Chaturvedi, Carroll (1994), Chaturvedi, Carroll, Green (1994), Gaul, Schader (1996).) Applications which describe different possibilities for analyzing conjoint data show potential advantages of the new procedure.

2. A new two-mode overlapping clustering procedure

2.1 The model

The following notation is used

$i = 1, \ldots, I$	first mode objects (e.g., respondents),
$j = 1, \ldots, J$	second mode objects (e.g., brands, concepts, products),
$k = 1, \ldots, K$	first mode clusters,
$l = 1, \ldots, L$	second mode clusters,
$\mathbf{S} = (s_{ij})_{I \times J}$	matrix of observed responses ($s_{ij} \in \mathbf{R}\ \forall i, j$).

Model parameters are

$\mathbf{P} = (p_{ik})_{I \times K}$ matrix describing first mode cluster membership with

$$p_{ik} = \begin{cases} 1 & \text{if first mode object } i \text{ belongs to first mode cluster } k, \\ 0 & \text{otherwise,} \end{cases}$$

$\mathbf{Q} = (q_{jl})_{J \times L}$ matrix describing second mode cluster membership with

$$q_{jl} = \begin{cases} 1 & \text{if second mode object } j \text{ belongs to second mode cluster } l, \\ 0 & \text{otherwise,} \end{cases}$$

$\mathbf{W} = (w_{kl})_{K \times L}$ matrix of weights ($w_{kl} \in \mathbf{R}\ \forall k, l$) and

$\mathbf{C} = (c)_{I \times J}$ matrix, all of whose entries are equal to c ($c \in \mathbf{R}$),

which are determined in such a way to minimize the objective function

$$Z = \sum_{i=1}^{I} \sum_{j=1}^{J} (s_{ij} - \hat{s}_{ij})^2 \quad \text{with} \quad \hat{s}_{ij} = \sum_{k=1}^{K} \sum_{l=1}^{L} p_{ik} w_{kl} q_{jl} + c \quad \forall i, j, \tag{1}$$

or, equivalently, to maximize the variance accounted for

$$\mathrm{VAF} = 1 - \frac{\sum_{i=1}^{I}\sum_{j=1}^{J}(s_{ij} - \hat{s}_{ij})^2}{\sum_{i=1}^{I}\sum_{j=1}^{J}(s_{ij} - \bar{s})^2} \quad \text{with} \quad \bar{s} = \sum_{i=1}^{I}\sum_{j=1}^{J} s_{ij}/(IJ) \tag{2}$$

on the basis of the underlying model

$$\mathbf{S} = \mathbf{PWQ}' + \mathbf{C} + \text{error}. \tag{3}$$

2.2 Overlapping clustering

The different sets of model parameters **P**, **W**, **Q**, and **C** are alternatingly determined. We first discuss the iterative steps for obtaining improved estimates for selected model parameters when estimates for the remaining sets of model parameters are given. Finally, the complete procedure is presented.

a) Estimation of **P** *and* **W** *for given* **Q** *and* **C**:

Using

$$Z = \sum_{i=1}^{I}\sum_{j=1}^{J}\Big(\underbrace{s_{ij} - \sum_{\substack{k'=1 \\ k'\neq k}}^{K}\sum_{l=1}^{L} p_{ik'}w_{k'l}q_{jl} - c}_{=:\ s_{ijk}} - \sum_{l=1}^{L} p_{ik}w_{kl}q_{jl}\Big)^2, \tag{4}$$

(constant w.r.t. the parameters for first mode cluster k, i.e., $\{p_{1k}, \ldots, p_{Ik}, w_{k1}, \ldots, w_{kL}\}$)

estimates of **P** and **W** can be obtained by starting from initial values and alternatingly improving the parameter estimates for first mode cluster $k = 1, \ldots, K$ via

$$p_{ik} = \begin{cases} 1 & \text{if } \sum_{j=1}^{J}(s_{ijk} - \sum_{l=1}^{L} w_{kl}q_{jl})^2 < \sum_{j=1}^{J}(s_{ijk})^2 \\ 0 & \text{otherwise} \end{cases} \quad \forall i \tag{5}$$

and minimizing

$$\sum_{i=1}^{I}\sum_{j=1}^{J}(s_{ijk} - \sum_{l=1}^{L} p_{ik}q_{jl}w_{kl})^2 \quad \text{via OLS w.r.t. } \{w_{k1}, \ldots, w_{kL}\}. \tag{6}$$

b) Estimation of **Q** *and* **W** *for given* **P** *and* **C**:

Estimates for **Q** and **W** can be obtained analogous to **P** and **W** in *a)*.

c) Estimation of **C** *for given* **P**, **W**, *and* **Q**:

$$c = \sum_{i=1}^{I}\sum_{j=1}^{J}(s_{ij} - \sum_{k=1}^{K}\sum_{l=1}^{L} p_{ik}w_{kl}q_{jl})/(IJ). \tag{7}$$

Thus, our two-mode overlapping clustering procedure can be described as follows:

1. Determine initial estimates of **P**, **W**, **Q**, and **C**. Compute Z.
2. Repeat

   Improve the estimates of **P** and **W** using *a)*.

   Improve the estimates of **Q** and **W** using *b)*.

   Improve the estimate of **C** using *c)*.

   Until Z cannot be improved any more.

2.3 Non-overlapping clustering

In the non-overlapping case, the steps for improving **P**, **W**, **Q**, and **C** have to be replaced by steps where the clustering schemes are restricted to the non-overlapping conditions

$$\sum_{k=1}^{K} p_{ik} = 1 \ \forall i \quad \text{and} \quad \sum_{l=1}^{L} q_{jl} = 1 \ \forall j. \tag{8}$$

Moreover, without loss of generality, $\mathbf{C}=(c)_{I\times J}$ can be omitted, i.e., $c = 0$.

a) Estimation of **P** *for given* **W** *and* **Q**:

Set

$$p_{ik} = \begin{cases} 1 & \text{if } \sum_{j=1}^{J}(s_{ij}-\sum_{l=1}^{L} w_{kl}q_{jl})^2 = \min_{1\le\kappa\le K}\{\sum_{j=1}^{J}(s_{ij}-\sum_{l=1}^{L} w_{\kappa l}q_{jl})^2\} \\ 0 & \text{otherwise} \end{cases} \quad \forall i,k. \tag{9}$$

b) Estimation of **W** *for given* **P** *and* **Q**:

Set

$$w_{kl} = (\sum_{i=1}^{I}\sum_{j=1}^{J} p_{ik}s_{ij}q_{jl})/((\sum_{i=1}^{I} p_{ik})(\sum_{j=1}^{J} q_{jl})) \ \forall k,l. \tag{10}$$

c) Estimation of **Q** *for given* **P** *and* **W**:

Set

$$q_{jl} = \begin{cases} 1 & \text{if } \sum_{i=1}^{I}(s_{ij}-\sum_{k=1}^{K} p_{ik}w_{kl})^2 = \min_{1\le\lambda\le L}\{\sum_{i=1}^{I}(s_{ij}-\sum_{k=1}^{K} p_{ik}w_{k\lambda})^2\} \\ 0 & \text{otherwise} \end{cases} \quad \forall j,l. \tag{11}$$

Ties in *a)* and *c)* must be arbitrarily broken considering condition (8).

2.4 Constrained clustering

Important special cases can be handled by considering suitable constraints. Sometimes, a-priori information on cluster-membership is available. Such information can be incorporated by fixing corresponding p_{ik}- or q_{jl}-values or – if applicable – by omitting respective steps in the procedure.

A constrained case of special interest is the application of this procedure to one-stage benefit segmentation with conjoint experiments. (See Baier, Gaul (1995) for a recent review.) In this context, clusterwise (multivariate) regression approaches (see, e.g., Wedel, Steenkamp (1991)) are used for analyzing individual responses (e.g., buying intentions, preferences) w.r.t. identical sets of pre-determined attribute-level-combinations (stimuli). Input data consist of matrices **S** of observed individual responses w.r.t. the set of stimuli and **Q** which reproduces the dummy-coded stimuli description. Model parameters are the first mode cluster membership matrix **P** and the weights matrix **W** which in this special case can be interpreted as a matrix of segment-specific part-worths.

2.5 Model selection

If we assume that the error terms in formula (3) are independent and identical normally distributed according to N(0,σ^2), the likelihood function

$$L = \prod_{i=1}^{I} \prod_{j=1}^{J} \frac{1}{\sqrt{2\pi\sigma^2}} \exp\left(-\frac{1}{2\sigma^2}(s_{ij} - \hat{s}_{ij})^2\right) \tag{12}$$

or the concentrated log-likelihood function ($\hat{\sigma}^2 = \sum_{i=1}^{I}\sum_{j=1}^{J}(s_{ij}-\hat{s}_{ij})^2/(IJ)$)

$$\ln L = -\frac{IJ}{2}\ln\left(\frac{2\pi}{IJ}\sum_{i=1}^{I}\sum_{j=1}^{J}(s_{ij} - \hat{s}_{ij})^2\right) - \frac{IJ}{2} \tag{13}$$

can be used for model selection in the following way (Note that maximizing $\ln L$ is equivalent to minimizing Z.): W.r.t. the parameter estimates of **P**, **W**, **Q** and **C**, we calculate values of AIC (Akaike Information Criterion) (Akaike (1977)) or of CAIC (Consistent Akaike Information Criterion) (Bozdogan (1987)) according to

$$\text{AIC} = -2\ln L + 2\text{NEP} \quad \text{and} \quad \text{CAIC} = -2\ln L + (1 + \ln \text{NO})\text{NEP} \tag{14}$$

and choose the model with the smallest value of the underlying information criterion (see, e.g., Baier, Gaul (1996) for a similar usage of these criteria). NEP denotes the Number of Effective Parameters (for comparison reasons with the methodology used in Baier, Gaul (1996), only the elements of **W** and **C** are counted), NO is the Number of Observations.

			K=1	K=2	K=3	K=4	K=5	K=6	K=7	K=8
		VAF	0.1699	0.2929	0.3804	0.4349	0.4787	0.5185	0.5571	0.5826
constr.	n.	AIC	9408.99	9130.22	8909.42	8770.67	8655.28	8542.47	8421.17	8349.61
(L=25,		CAIC	9581.25	9468.12	9412.96	9439.86	9490.11	9542.94	9587.28	9681.35
Q=		VAF	0.2105	0.3243	0.4336	0.5181	0.5762	0.6186	0.6573	0.6946
unit	o.	AIC	9306.19	9037.10	8725.38	8444.18	8230.80	8064.71	7895.37	7709.14
matrix)		CAIC	9478.45	9375.01	9228.93	9113.36	9065.63	9065.17	9061.47	9040.89
		VAF	0.0000	0.2289	0.3234	0.3729	0.4293	0.4579	0.4958	0.5083
un-	n.	AIC	9742.71	9215.84	8957.83	8816.08	8640.89	8557.49	8434.91	8413.45
constr.		CAIC	9755.97	9248.97	9024.09	8928.72	8813.15	8802.64	8766.19	8844.11
(L=K)		VAF	0.1863	0.3361	0.3956	0.4818	0.5507	0.5907	0.6172	0.6425
	o.	AIC	9320.08	8908.99	8726.50	8425.06	8150.58	7981.44	7870.22	7760.04
		CAIC	9333.33	8942.11	8792.76	8537.69	8322.85	8226.58	8201.50	8190.71
		VAF	0.1659	0.2929	0.3808	0.4355	0.4784	0.5153	0.5525	0.5818
constr.	n.	AIC	9410.84	9114.22	8884.09	8736.49	8616.46	8508.05	8386.35	8289.53
(L=21,		CAIC	9556.60	9399.12	9308.13	9299.67	9318.78	9349.50	9366.94	9409.26
Q=		VAF	0.2301	0.3740	0.4608	0.5365	0.5762	0.6155	0.6440	0.6734
design	o.	AIC	9246.65	8864.48	8600.49	8332.37	8190.80	8033.30	7917.42	7782.73
matrix)		CAIC	9392.41	9149.39	9024.53	8895.55	8893.12	8874.75	8898.01	8902.45

Table 1: VAF-, AIC- and CAIC-values w.r.t. applications of different variants of the new procedure. Each result is based on 50 random starting solutions. (constr.=constrained, o.=overlapping, n.=non-overlapping)

3. Applications and comparisons

A major German producer of luxury bathroom armatures planned the introduction of a new product line of ecologically benefical water-taps. The authors investigated the potential customer-response and developed a typology of potential buyers on the basis of 25 concepts or stimuli described by attribute-level-combinations using an orthogonal design w.r.t. the five attributes "handling", "surface/material", "recycling", "savings", and "price", each with five levels. During autumn/winter 1994/95, data were collected from 82 respondents who were actually renovating their bathrooms or building an own one-family house. They were interviewed at sanitary exhibitions or at home with an average interviewing time of about 30 minutes. As main task each respondent had to indicate buying intentions for the 25 concepts on an eleven-point scale (1="definitely would not buy",...,6="might or might not buy",...,11="definitely would buy").

As can be seen from Table 1, all variants of the new procedure have been applied to these data: First, constrained clustering was performed where **Q** was set equal to a unit matrix with 25 columns which corresponds to one-mode additive clustering of the first mode objects. Second, unconstrained two-mode clustering was used where – for brevity – only runs with K=L=1,...,8 are reported. Third, **Q** was set to the standard dummy-coded conjoint analysis design matrix for the 25 concepts. In all cases, non-overlapping and overlapping clustering variants were considered. From the VAF-values we see that w.r.t. "fitting" terms the overlapping variants are superior to the respective non-overlapping ones as it should be. The second group of unconstrained two-mode clustering variants with much fewer second mode clusters and estimable weights compete well in this environment, a result which is supported by the respective CAIC-values. A more detailed discussion of the results is presented for the second group and the third group of variants. Table 2 shows findings for a variant from the third group which – as already mentioned – is quite similar to an application of non-overlapping clusterwise regression with conjoint experiments: We have L=21 overlapping second mode clusters with fixed membership matrix **Q**. Second mode cluster l=1 is a universal cluster, second mode cluster l=2 contains all concepts with "mixer tap" as handling alternative, second mode cluster l=3 comprises all "two-way tap" concepts, etc. By applying model selection w.r.t the CAIC-values, a solution with K=4 non-overlapping first mode clusters has been found (also supported by application of the elbow criterion w.r.t. the VAF-values). Note that the respondents' buying intentions for the presented concepts have been collected on an eleven-point scale and can be predicted by adding the weights from the respective first and second mode clusters to the constant c=5.704. From the weights we see that the first cluster of respondents shows an overall positive response (w_{11}=1.518) and could especially be satisfied by concepts in which "touchless", "mixer tap" or "sensor-op." occur as handling alternatives and in which moderate DM values are fixed as price alternatives. The label for this customer segment could be "technical interest". The second customer segment with a negative weight w.r.t. to the universal cluster l=1 (w_{21}=-2.174) prefers concepts in which "mixer tap" or "two-way tap" as well as "stainless steel" or "chromium plated" appear as attribute levels. We call this segment "conservative". In the same manner, the third customer segment could be labeled by "practical interest" and the fourth one by "price".

Finally, Table 3 shows results from unconstrained two-mode clustering. We get four overlapping clusters of respondents and four overlapping clusters of concepts with so far quite similar results as in Table 2: in the first cluster of concepts (l=1) "practical" handling alternatives are dominant, the second cluster comprises more "conservative" handling and material concepts whereas the third and the fourth cluster include overall less preferred concepts, e.g. "two-way tap" or high-priced "technically advanced" handling alternatives. Two clusters of respondents (k=3 and k=4) like especially the "practical" concepts (l=1) whereas cluster k=2 prefers "conservative" handling and material concepts (l=2).

	attribute	level	k=1 (28.0%)	k=2 (13.4%)	k=3 (34.2%)	k=4 (24.4%)
			weights			
l= 1 (100%)			1.518	-2.174	0.504	-1.256
l= 2 (20%)	handling	mixer tap	-0.147	4.236	0.450	-0.100
l= 3 (20%)		two-way tap	-5.207	3.527	-0.678	-1.220
l= 4 (20%)		foot-op.	-4.503	-0.746	-1.457	-0.580
l= 5 (20%)		sensor-op.	-1.590	0.199	-0.085	-0.320
		touchless	0.000	0.000	0.000	0.000
l= 6 (20%)	surface/	aluminium	-0.556	-0.968	0.223	-0.418
l= 7 (20%)	material	stainless steel	0.166	0.632	0.295	0.242
l= 8 (20%)		col./ann. lac	-0.278	0.269	-0.591	-1.118
l= 9 (20%)		chromium plated	-0.530	0.541	-0.334	-0.968
		plastic	0.000	0.000	0.000	0.000
l=10 (20%)	recycling	ecol. benefical cons.	-0.201	-0.342	-0.007	0.069
l=11 (20%)		return option	-0.001	-0.433	-0.035	-0.381
l=12 (20%)		part exchange	-0.132	-0.087	-0.249	0.099
l=13 (20%)		ecol.+return option	-0.201	-0.251	-0.057	-0.121
		ecol.+part exchange	0.000	0.000	0.000	0.000
l=14 (20%)	savings	timed switch off	0.093	-0.145	-0.087	-0.185
l=15 (20%)		flow meter	-0.046	-0.290	0.113	-0.265
l=16 (20%)		thermostat	0.154	0.091	-0.044	-0.015
l=17 (20%)		timed+flow meter	0.076	-0.127	0.006	0.095
		timed+thermostat	0.000	0.000	0.000	0.000
l=18 (20%)	price	350DM	1.093	0.006	0.904	1.702
l=19 (20%)		450DM	1.102	-0.048	0.726	1.302
l=20 (20%)		550DM	0.972	-0.030	0.561	1.162
l=21 (20%)		650DM	0.728	-0.066	0.447	0.872
		750DM	0.000	0.000	0.000	0.000
			technic. interest	conser-vative	practic. interest	price

Table 2: Results from the application of a constrained variant with K=4 non-overlapping first and L=21 overlapping second mode clusters (**Q**=design matrix) (c=5.704, VAF=.4355, AIC= 8736.49, CAIC=9299.67, op.=operated, col.=colored, ann.=annealed, ecol.=ecologically, cons.=constructed)

	l=1 (60%)	l=2 (48%)	l=3 (56%)	l=4 (44%)
	concept1,2,3,4,5, 16,17,18,19,20, 21,22,23,24,25	1,2,3,4,5, 6,7,8,9, 18,19,20,24	6,7,8,11,12, 13,14,15,16, 18,19,20,21,24	4,6,7,8,9, 10,11,13,14, 15,18
	weights			
k=1 (60.98%)	-1.049	0.192	-0.499	-0.810
k=2 (43.90%)	-1.126	2.596	-2.065	0.044
k=3 (50.00%)	1.929	-0.289	0.210	-0.081
k=4 (46.34%)	1.254	-1.474	0.112	-1.456
	mixer tap or sensor-op. or touchless	mixer tap or concepts with traditional materials	foot-/sensor-op./touchless or two-way tap	two-way tap or high-priced mixer tap/foot-/ sensor-op.

Table 3: Results from the application of unconstrained two-mode clustering with K=4 overlapping first and L=4 overlapping second mode clusters (c=6.002, VAF=.4818, AIC=8425.06, CAIC=8537.69)

4. Conclusions and outlook

In this paper we have shown that two-mode additive clustering procedures can be used for analyzing buying behavior/intention data for simultaneous benefit segmentation and market structuring. The new procedure is superior to other procedures in "fitting" terms. However, further comparisons with other overlapping clustering procedures are needed. (See, e.g., Opitz, Bausch (1986), Both, Gaul (1987), Gaul, Schader (1994,1996) and the references given there.)

References:

AKAIKE, H. (1977): On Entropy Maximization Principle. In P. R. Krishnaiah (ed.): *Applications of Statistics. North Holland, Amsterdam, 27–41.*

ARABIE, P. and CARROLL, J. D. (1980): MAPCLUS: A Mathematical Programming Approach to Fitting the ADCLUS Model. *Psychometrika, 45, 211-235.*

ARABIE, P.; CARROLL, J. D.; DESARBO, W. S. and WIND, J. (1981): Overlapping Clustering: A New Method for Product Positioning. *Journal of Marketing Research, 18, 310-317.*

BAIER, D. and GAUL, W. (1995): Classification and Representation Using Conjoint Data. In: W. Gaul, M. Schader (eds.): *From Data to Knowledge. Springer, Berlin, 298-307.*

BAIER, D. and GAUL, W. (1996): Analyzing Paired Comparisons Data Using Probabilistic Ideal Point and Vector Models. In: H.-H. Bock, W. Polasek (eds.): *Data Analysis and Information Systems. Springer, Berlin, 163-174.*

BAUER, H. H. (1989): Marktabgrenzung. Duncker & Humblot, Berlin.

BOTH, M. and GAUL, W. (1987): Ein Vergleich zweimodaler Clusteranalyseverfahren. *Methods of Operations Research, 57, 593-605.*

BOZDOGAN, H. (1987): Model Selection and Akaike's Information Criterion: The General Theory and Its Analytical Extensions. *Psychometrika, 52, 345-370.*

CARROLL, J. D. and ARABIE, P. (1983): An Individual Differences Generalization of the ADCLUS Model and the MAPCLUS Algorithm. *Psychometrika, 48, 157-169.*

CHATURVEDI, A and, CARROLL, J. D. (1994): An Alternating Combinatorial Optimization Approach to Fitting the INDCLUS and Generalized INDCLUS Models. *Journal of Classification, 11, 155-170.*

CHATURVEDI, A.; CARROLL, J. D. and GREEN, P. E. (1994): A Feature Based Approach to Market Segmentation via Overlapping k-Centroids Clustering. Discussion Paper, Wharton School, University of Pennsylvania.

DAY, G.S., SHOCKER, A.D., and SRIVASTAVA, R.K. (1979): Customer-Oriented Approaches to Identifying Product-Markets. *Journal of Marketing, 43, 8-19.*

DESARBO, W. S. (1982): GENNCLUS: New Models for General Nonhierarchical Clustering Analysis. *Psychometrika, 47, 449-475.*

GAUL, W. and SCHADER, M. (1994): Pyramidal Classification Based on Incomplete Dissimilarity Data. *Journal of Classification, 11, 171-193.*

GAUL, W and, SCHADER, M. (1996): A New Algorithm for Two-Mode Clustering. In: H.-H. Bock, W. Polasek (eds.): *Data Analysis and Information Systems. Springer, Berlin, to appear.*

HRUSCHKA, H. (1986): Market Definition and Segmentation Using Fuzzy Clustering Methods. *International Journal of Research in Marketing, 3, 117-134.*

OPITZ, O. and BAUSCH, T. (1986): Nichtdisjunkte Klassifikation mit qualitativen Daten. *Studien zur Klassifikation, 17, 211-220.*

SHEPARD, R. K. and ARABIE, P. (1979): Additive Clustering Representation of Similarities as Combinations of Discrete Overlapping Properties. *Psychological Review, 86, 87-123.*

SRIVASTAVA, R. K. and ALPERT, M. I., and SHOCKER, A.D. (1984): A Customer-Oriented Approach for Determining Market Structures. *Journal of Marketing, 48, 32-45.*

WEDEL, M. and STEENKAMP, J. B. (1991): A Clusterwise Regression Method for Simultaneous Fuzzy Market Structuring and Benefit Segmentation, *Journal of Marketing Research, 28, 385-396.*

An Application of Two-Mode Classification to Analyze the Statistical Software Market

U. Bankhofer, A. Hilbert

Institut für Statistik und Mathematische Wirtschaftstheorie,
Universität Augsburg, D-86135 Augsburg

Summary: This paper presents the results of a market research concerning statistical software packages for Microsoft Windows. On the one hand, the implemented features as well as the numerical accuracy and the computing speed of the software packages are analyzed. On the other hand, a segmentation of the market using a two-mode classification method is performed.

1. Introduction

This investigation has the object to analyze the existing statistical software packages in the German market concerning their implemented features and numerical accuracy. Thereby, only those software packages are considered which run under the operating system MS-Windows[1]. Programs which can only be started in a DOS-box do not satisfy this requirement and are not analyzed. To ensure the comparability of the statistical software packages, only programs with a common and wide spectrum of statistical procedures are considered. According to other studies on this topic (cp. e.g. Bausch and Bankhofer (1992)), a basic equipment of required statistical procedures was defined. By this, software packages for special statistical methods, multifunction packages and expert systems will not be analyzed in this study.

2. Database of the Analysis

The statistical software packages shall be characterized by several variables. Thereby, the following nine subsets of variables can be formed: *general system characteristics, documentation and user support, import/export of data and system interfaces, user interface, data management, numerical accuracy and computing speed, variety of statistical procedures, graphics, possibilities of presentation.*

To ensure the comparability and an easy aggregation of the individual variables, the scaling of all variables is standardized into the interval $[0;10]$. This scale can be interpreted as a percental performance index of the single features ($10 = 100$ percent). A comprehensive representation of all considered variables and the used scaling method is given in Hilbert et al. (1995).

Based on several publications in computer magazines like Koch and Haag (1995), Sarnow (1994) and publications in the Academic Software Coopera-

[1]In this paper MS-Windows is simply called Windows.

tion (1994), a first secondary statistical analysis delivered a list of statistical software packages which were available on the German market and designed for Windows. Considering the defined basic equipment of statistical procedures, the producers or distributors of the corresponding software were requested to put their software at disposal. Finally, the packages represented in line 1 of table 1 could be analyzed in this study. The inquiry of

	Almo 32	Sigma-stat 1.0	S-Plus 3.2	SPSS 6.1	Stat-gra-phics 1.1	Sta-tis-tica 4.5	Sy-stat 5.03	Uni-stat 3.0	Win-stat 3.0
General system characteristics	8.36	4.73	7.94	8.03	7.36	7.68	5.65	3.96	3.55
Documentation and user support	5.81	6.38	4.93	9.06	7.00	8.35	7.70	7.63	5.29
Import/export of data and system interfaces	5.33	8.10	2.90	8.20	6.03	9.55	8.88	8.43	7.75
User interface	6.05	7.88	3.05	8.80	8.25	8.88	6.53	7.28	5.95
Data management	7.87	7.44	7.06	9.89	9.05	8.96	8.62	8.83	7.93
Numerical accuracy and computing speed	7.12	7.67	8.18	8.02	7.65	8.14	7.63	7.39	6.31
Variety of statistical procedures	7.19	4.54	7.44	9.08	6.49	9.22	8.07	8.34	5.83
Graphics	4.32	6.52	9.67	7.76	7.36	9.48	9.12	9.88	4.78
Possibilities of presentation	6.20	7.80	7.00	8.20	6.30	9.00	8.00	8.40	6.30

Table 1: Aggregated data matrix

the data was made with the help of secondary statistical as well as primary statistical methods (cp. Hilbert et al. (1995)). The compendiums of the tested products and several publications in magazines served as secondary sources. The primary statistical inquiry consists of the observation and the experiment. The experiments to the numerical accuracy and the computing speed were realized in February 1995.

The resulting data matrix has the dimension 369x9. Due to the small sample size and a simplification of the data analysis, an aggregation to the nine subsets of variables was made, whereas the loss of information was accepted. The resulting data matrix with nine variables and nine objects is described in table 1. The used weights of aggregation were chosen with regard to the relative importance of the variables (cp. Hilbert et al. (1995)). Since all values are measured as percentage performance, the problem of different scales does not exist.

	Minimum	Mean	Maximum	Range
General system characteristics	3.55	6.36	8.36	4.81
Documentation and user support	4.93	6.91	9.06	4.13
Import/export of data and system interfaces	2.90	7.24	9.55	6.65
User interface	3.05	6.96	8.88	5.83
Data management	7.06	8.41	9.89	2.83
Numerical accuracy and computing speed	6.31	7.57	8.18	1.87
Variety of statistical procedures	4.54	7.36	9.22	4.68
Graphics	4.32	7.65	9.88	5.56
Possibilities of presentation	6.20	7.47	9.00	2.80

Table 2: Structure of the nine aggregated variables

3. Market Survey

3.1 Structure Analysis

To get an impression of the structure of the aggregated variables and to detect outliers, several basic statistical evaluations are made. The results in table 2 show that five variables have an arithmetic mean greater than seven. However, only one variable has a mean greater than eight. This indicates a generally good equipment of the packages. The maximum values show that packages exist which nearly fulfill the demanded performances.

Furthermore, the range of the variables is mostly high. Above all, the tested products differ concerning the variables *general system characteristics, documentation and user support, import/export and system interfaces, user interface, variety of methods* and *graphics*. Altogether, there are great differences among the nine packages. A product inhomogeneity might be assumed.

The star plots in figure 1 confirm the heterogeneity of the software packages. The length of the star point represents the relative size of the corresponding variable value. The programs STATISTICA and SPSS possess the comparatively best equipment and numerical accuracy. Above all, WINSTAT has in comparision to the other packages a small equipment and a low performance.

Figure 2 shows the star plots of the software packages concerning the numerical accuracy and computing speed. Once more, the length of the star point represents the relative size of the corresponding variable value. The programs have their strong and weak points in different topics. For example, STATISTICA delivers comparatively bad results on the experiments concerning the test procedures, while the results concerning the other topics are very good in comparision to the other packages.

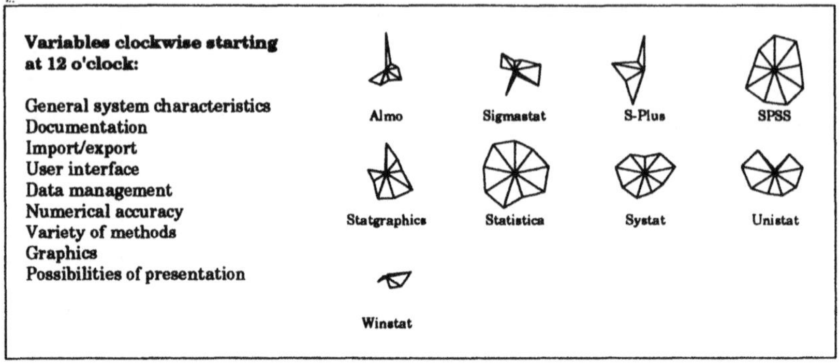

Figure 1: Star plots of the software packages (aggregated variables)

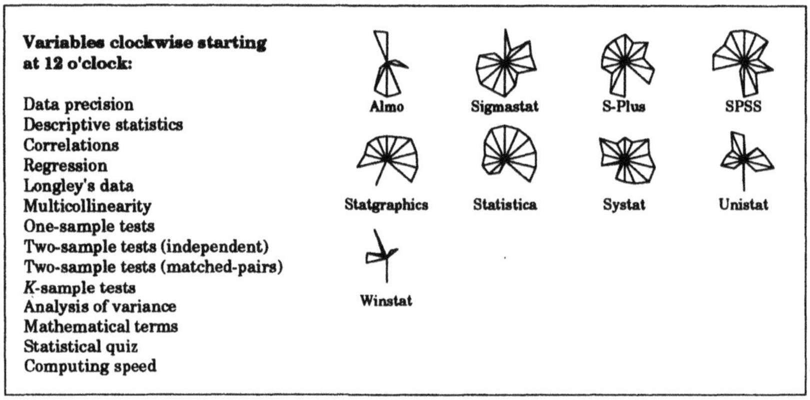

Figure 2: Star plots of the software packages (numerical accuracy and computing speed)

3.2 Two-Mode Market Segmentation

The descriptive analysis shows differences among the packages. Based on the aggregated data matrix, a two-mode classification tries to find out similarities among the objects and variables simultaneously. The resulting clusters contain objects and variables at the same time. Therefore the interpretation of the clusters is easier in comparision to the one-mode classification results. In this study the missing-value-algorithm of Espejo and Gaul (1986), implemented in $\mathcal{MARK}^2\mathcal{MAN}$ (Gaul and Baier (1993)), is used. Other two-mode clustering algorithms are described e.g. in McCormick et al. (1972), Both and Gaul (1985) and Eckes and Orlik (1991).

The use of the missing-value-algorithm leads to the complete linkage dendrogram represented in figure 3.

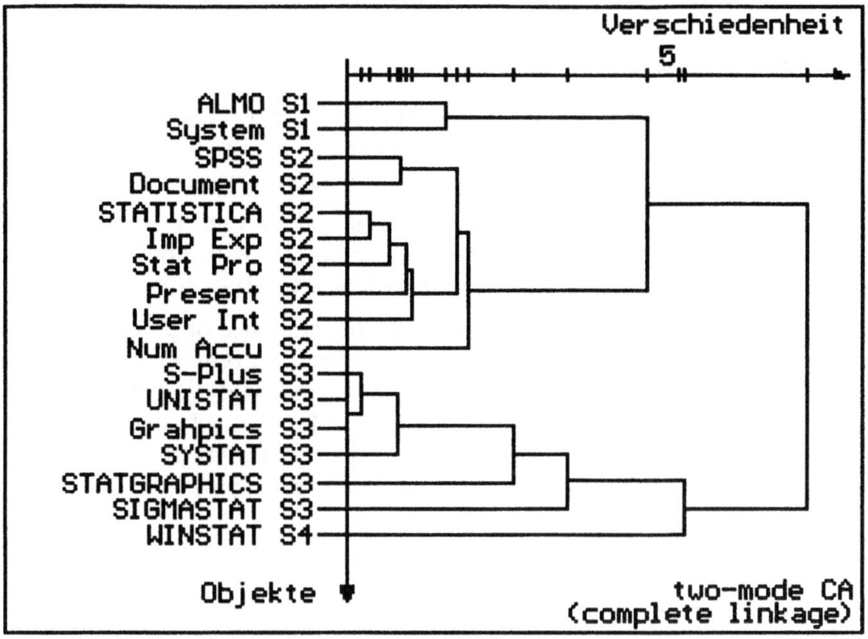

Figure 3: Two-mode clustering solution

The solution votes for two or four classes. Due to the few objects, the two-class solution makes more sense but the two programs WINSTAT and ALMO are outliers which should be regarded as proper classes. Therefore, the four-class solution is described in the following:

- **Class 1**: ALMO, *general system characteristics*
- **Class 2**: SPSS, *documentation and user support*, STATISTICA, *import/export and system interfaces, variety of statistical procedures, possibilities of presentation, user interface, numerical accuracy and computing speed*
- **Class 3**: SIGMASTAT, S-PLUS, UNISTAT, *graphics*, SYSTAT, STATGRAPHICS
- **Class 4**: WINSTAT

The results of the two-mode cluster analysis show that e.g. the programs STATISTICA and SPSS are similiar. SPSS has a very good *documentation and user support*, while STATISTICA is characterized by a high scoring concerning the variable *import/export and system interface*. Furthermore, these products possess comparatively high scoring values concerning the variables *variety of statistical procedures, possibilities of presentation, user interface* and *numerical accuracy*. The other classes can be interpreted analogously.

The results of the cluster analysis confirm the already noticed heterogeneity of the market. There are programs with a extensive *variety of statistical methods* and *graphics*, good *numerical accuracy and computing speed* which are interesting for professional users. For those users which do not need many methods but the Windows handling, the products of the first and fourth class represent an alternative.

4. Conclusions

In respect of quality and performance, this investigation considered only a small portion of the statistical software packages available on the German market. Finally, some results of the study are summarized:

The tests concerning the *numerical accuracy and computing speed* produce good results. Except Winstat, all programs compute with double precision. Furthermore, messages in respect of a problem are rarely given.

Based on the results of the market survey, the existence of a heterogeneous market is obvious. Each producer realizes an own program philosophy in his software. Based upon the small sample, the market analysis presents only a part of the reality. Nevertheless, the variety of the products is recognizable. Each user is able to get a product for his own requirements, whereas the packages for professional users with an extensive variety of methods prevail.

References:

AKADEMIC SOFTWARE COOPERATION (1994): Software-Führer 94/95: Lehre und Forschung. Ingenieurwissenschaften, Springer, Berlin.

BAUSCH, T. and BANKHOFER, U. (1992): Statistical Software Packages for PCs - a Market Survey. *Statistical Papers, 33, 283–306.*

BOTH, M. and GAUL, W. (1985): PENCLUS: Penalty Clustering for Marketing Applications, Discussion Paper No. 82, Institut für Entscheidungstheorie und Unternehmensforschung, Universität Karlsruhe.

ECKES, T. and ORLIK, P. (1991): An Agglomerative Method for Two-Mode Hierarchical Clustering. In: H. Bock and P. Ihm (eds.): *Classification, Data Analysis and Knowledge Organisation*, North-Holland, Amsterdam, 3–8.

ESPEJO, E. and GAUL, W. (1986): Two-Mode Hierarchical Clustering as an Instrument for Marketing Research. In: W. Gaul and M. Schader (eds.): *Classification as a Tool of Research*, North-Holland, Amsterdam, 121–128.

GAUL, W. and BAIER, D. (1993): Marktforschung und Marketing Management, Oldenbourg, München.

HILBERT, A.; BANKHOFER, U.; ELSHOLZ, B. and NIEDERÖCKER, U. (1995): Statistiksoftware unter Windows: Eine Marktanalyse. Arbeitspapiere zur Mathematischen Wirtschaftsforschung, Heft 124, Universität Augsburg.

KOCH, A. and HAAG, U. (1995): The Statistical Software Guide 94/95. *Computational Statistics & Data Analysis, 19, 237–245.*

MCCORMICK, W. T.; SCHWEITZER, P. J. and WHITE, T.W . (1972): Problem Decomposition and Data Reorganization by a Clustering Technique. *Operations Research 20, 993–1009.*

SARNOW, K. (1994): Abstraktionsinstrumente, Statistiksysteme: von Shareware bis High-End. *c't 3/1994, 174–188.*

Correspondence Analysis of Square Tables, with an Application to Social Mobility

Michael Greenacre[1], Jörg Blasius[2]

[1] Departament d'Economia,
Universitat Pompeu Fabra, E-08008 Barcelona, Spain
[2] Zentralarchiv für Empirische Sozialforschung,
Universität zu Köln, D-50931 Köln, Germany

Summary: The application of correspondence analysis to square tables is often unsuccessful because of the strong role played by the diagonal entries of the matrix. We illustrate how a simple modification of the centering of the matrix allows a decomposition into symmetric and skew-symmetric matrices, with a corresponding decomposition of the total inertia into two parts. The symmetric and skew-symmetric matrix components can then be analyzed using the same masses and metrics, yielding two graphical displays each containing one set of points. The methodology is illustrated using an example from social mobility, showing how the percentage explanation of inertia as well as the substantive interpretation is enhanced.

1. Introduction

Correspondence analysis (CA) is a technique for visualizing tables of frequencies as well as nonnegative data on a commensurate set of ratio scale variables - see Benzécri *et al.* (1973), Greenacre (1984, 1993), Lebart *et al.* (1984, 1995). Here we consider the special case of a square table of frequencies where the rows and columns refer to the same set of objects at two different states, for example at two different time points. Examples of such tables are frequently found, for example social mobility tables, confusion matrices in psychology, brand switching tables in marketing, cross-citations between journals, transition matrices between behavioural states and migration tables. Our approach can be considered the equivalent in a correspondence analysis conteext of the bilinear modelling approach considered recently by van der Heijden and Mooijaart (1995).

We start by introducing the example of social mobility used to illustrate our approach (Section 2). Then in Section 3 we describe the application of ordinary CA to the mobility table and show the deficiencies of this analysis. In Section 4 we describe how a table can be recentred to allow a decomposition into symmetric and skew-symmetric parts which can be jointly analyzed using CA with the same masses and distances. The paper ends with a discussion (Section 5).

2. "Frankfurter Nationalversammlung" Mobility Data

In the 1980s Best collected a data set of the members of the "Frankfurter Nationalversammlung", considered to be the first democratic government in Germany, which was in session from May 1848 to May 1849. The candidates were notabilities only, and 80 percent of the male population were allowed to take part in the election (for more details, see Langewiesche (1981)).

One of Best's (1990) main points of interest was the analysis of mobility between the jobs of the parliamentarians when entering the labour market and their main jobs in May 1848 when the government started its session. According to functional differences, Best classified the different jobs of the parliamentarians into ten main categories. According to Parsons' (1971) system theory these ten groups belong to four clusters: within these clusters there should be a high intragenerational mobility whereas between them the mobility should be low.

According to Best (1990) and Parsons (1971) the first cluster should include persons working in "education" (schools, universities) as well as persons working for the church and self-employees, especially artists and journalists. In the middle of the last century the members of the clergy as well as journalists and artists had a major influence in educating the population; they were all part of the "educational system". Parsons called this cluster "pattern maintenance". The second cluster is labelled "integration" and it includes persons working in the field of "social control", e.g. lawyers and the members of administration and justice. The third cluster, "goal attainment", combined farmers and military. Note that all farmers were landlords having enough time and money for engaging in politics, and usually a farmer's first son inherited the farm and the second followed a military career (Best (1990)). A fourth cluster called "adaption" includes all members of the "economic subsystem", especially businessmen. Because of their very high occupational stability we have chosen to exclude this group in our present study.

In this paper we consider one of the tables from Best's study, which crosstabulates "job when entering the labour market" against "main occupation in May 1848". The data matrix is given in Table 1.

The analysis of mobility data such as these has a long history in social science research. In a famous study Blau and Duncan (1967) described occupational mobility in the United States, using mostly multidimensional scaling techniques. The initial data were transformed into dissimilitaries which can be interpreted as distances, leading to the visualization of the set of occupations as points in a low-dimensional space. CA can itself be considered a special case of multidimensional scaling, since it leads to a map of the occupations based on interpoint distances.

Abbreviations refer to original German names: rows are labelled "e" for "erst" and columns "h" for "haupt" for the first letter; the second letter refers to occupations "justiz" (justice), "verwaltung" (administration), "bildung" (education), "militär" (military), "kirche" (church), "gutsbesitzer" (farmer), "advokat" (lawyer) and "freiberuflich" (self-employed).

Job when entering		*Main occupation in May 1848*							
labour market		hj	hv	hb	hm	hk	hg	ha	hf
Justice	ej	117	83	19	0	1	17	76	6
Administration	ev	11	37	6	0	0	7	7	6
Education	eb	4	3	67	1	4	3	5	13
Military	em	0	5	1	16	0	8	1	0
Church	ek	0	1	11	0	26	0	1	0
Farmer	eg	0	4	0	0	0	19	1	0
Lawyer	ea	8	2	1	0	0	0	22	1
Self-employed	ef	0	7	20	0	4	1	0	37

Table 1: Mobility table from "Frankfurter Nationalversammlung"

3. Application of Simple Correspondence Analysis

We first summarize the theory of simple CA in a form which will be used in the next section. Let $\mathbf{P}$ be the correspondence matrix corresponding to the square frequency table $\mathbf{N}$, i.e. $\mathbf{P}$ consists of the elements of $\mathbf{N}$ divided by the grand total of $\mathbf{N}$. Let $\mathbf{D}_r$ and $\mathbf{D}_c$ be the diagonal matrices of the row and column sum vectors $\mathbf{r}$ and $\mathbf{c}$, respectively. Then the ordinary CA solution is obtained from the generalized singular value decomposition (GSVD) of the centered matrix $\mathbf{P} - \mathbf{r}\mathbf{c}^{\mathsf{T}}$ in the metrics $\mathbf{D}_r^{-1}$ and $\mathbf{D}_c^{-1}$, respectively:

$$\mathbf{P} - \mathbf{r}\mathbf{c}^{\mathsf{T}} = \mathbf{U}\mathbf{D}_\alpha\mathbf{V}^{\mathsf{T}} \qquad \text{where } \mathbf{U}^{\mathsf{T}}\mathbf{D}_r^{-1}\mathbf{U} = \mathbf{V}^{\mathsf{T}}\mathbf{D}_c^{-1}\mathbf{V} = \mathbf{I} \tag{1}$$

(this differs from the usual SVD, which is a GSVD with identity metrics – see Greenacre (1984, Appendix A)). The total inertia of the data table is the weighted sum of squares of the matrix $\mathbf{D}_r^{-1/2}(\mathbf{P} - \mathbf{r}\mathbf{c}^{\mathsf{T}})\mathbf{D}_c^{-1/2}$, which is equal to the sum of squared singular values $\sum_k \alpha_k^2$. Thus each cell of the table contributes to the total inertia. The "principal coordinates" of the row and column points are given by $\mathbf{F} = \mathbf{D}_r^{-1}\mathbf{U}\mathbf{D}_\alpha$ and $\mathbf{G} = \mathbf{D}_c^{-1}\mathbf{V}\mathbf{D}_\alpha$, respectively. To plot the rows and columns in two dimensions, the rank 2 solution given by the first two columns of $\mathbf{F}$ and $\mathbf{G}$ is used and the respective dimensions account for an amount α_1^2 and α_2^2 of the inertia, usually expressed as a percentage. The resulting plot is called the *symmetric map*.

The symmetric map of the first two dimensions of the CA solution to Table 1 is shown in Figure 1. Inspection of the contributions to total inertia in each cell (not shown here) shows that the highest contributions are from the cells along the diagonal of the table, and that they alone account for 81.2%

of the total inertia of the table. Since in CA the total inertia is decomposed under the condition that the highest proportion of inertia is explained by the first axis of the vector subspace, the second highest amount of inertia by the second axis and so on, it can be assumed that the first axes would show mostly the effects of occupational stability.

In Figure 1 most of the row-column pairs are very close together, with only one pair (military) separating. Since the CA of a symmetric matrix would give coincident pairs of rows and columns, we might conclude that the data are close to symmetric.

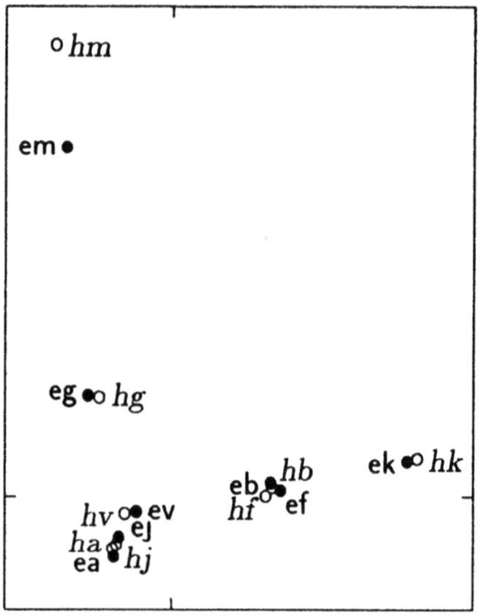

Figure 1: Correspondence analysis of Table 1

The total inertia of the data is equal to 2.057, of which the first axis explains 0.644 (31.3%). The main difference is between the categories of "pattern maintenance", especially "clergy", versus the categories of "integration" and "goal attainment". The second axis explains an additional 0.531 (25.8%) of the total inertia and can be described mainly by the categories of "military". In general, the remaining categories of both row and column variables load either on the same sides of the same axes or they are both uncorrelated with one of the two axes shown in Figure 1. It can thus be concluded empirically that occupational stability and the equality of flux to and from occupational categories is the most important for describing the mobility from initial to main occupations.

Notice that the three expected clusters "integration", "pattern maintenance", and "goal attainment" are well separated in the two-dimensional map. This means, if somebody changed his occupation he moved to another occupation in his particular cluster, for example from "justice" either to "administration" or to "lawyer", or from "self employment" to "education". Somewhat outside of the cluster "pattern maintenance" is "clergy" – changes to and from this occupation are relatively rare. The same holds for "military" which is also at an outlying position compared to the other member of its cluster, "farmer".

Apart from this general view of occupational stability visualized in Figure 1, we are also interested in differential fluxes within the mobility table, for example tendencies for changes of career away from certain occupations towards others. In technical terms this means that we are interested also in the deviations from symmetry in the data table, which leads us to the next section.

4. Symmetric and Skew-symmetric Components

The decomposition of a square asymmetric matrix into the sum of a symmetric matrix and skew-symmetric matrix is described by Constantine and Gower (1978) and Gower (1980). If the original table is denoted by $\mathbf{A}$, then we can decompose $\mathbf{A}$ into two components, a symmetric table and a skew-symmetric table:

$$\mathbf{A} = \mathbf{Q} + \mathbf{R} \tag{2}$$

where $\mathbf{Q} = \frac{1}{2}(\mathbf{A} + \mathbf{A}^{\mathsf{T}})$ and $\mathbf{R} = \frac{1}{2}(\mathbf{A} - \mathbf{A}^{\mathsf{T}})$. The sum-of-squares is similarly partitioned:

$$\mathrm{tr}(\mathbf{A}\mathbf{A}^{\mathsf{T}}) = \mathrm{tr}(\mathbf{Q}\mathbf{Q}^{\mathsf{T}}) + \mathrm{tr}(\mathbf{R}\mathbf{R}^{\mathsf{T}}) \tag{3}$$

since the sum-of-cross-products vanishes: $\mathrm{tr}(\mathbf{Q}\mathbf{R}^{\mathsf{T}}) = 0$.

The difference between the simple case above and the CA case here is the centering of the matrices and the presence of the row and column weighting factors: in CA we want to decompose the sum-of-squares (total inertia) $\mathrm{tr}(\mathbf{D}_r^{-1}(\mathbf{P} - \mathbf{r}\mathbf{c}^{\mathsf{T}})\mathbf{D}_c^{-1}(\mathbf{P} - \mathbf{r}\mathbf{c}^{\mathsf{T}})^{\mathsf{T}})$. Greenacre (1996) shows how CA can be adapted to accommodate the matrix decomposition (2). The essential idea is to change the centering and weighting of the correspondence matrix slightly. Instead of the usual centered matrix $\mathbf{P} - \mathbf{r}\mathbf{c}^{\mathsf{T}}$, which is decomposed in the metrics $\mathbf{D}_r^{-1}$ and $\mathbf{D}_c^{-1}$, we analyze $\mathbf{P} - \mathbf{w}\mathbf{w}^{\mathsf{T}}$ in the metrics $\mathbf{D}_w^{-1}$ and $\mathbf{D}_w^{-1}$, where $\mathbf{w} = \frac{1}{2}(\mathbf{r} + \mathbf{c})$. Thus the centered matrix can be written:

$$\mathbf{P} - \mathbf{w}\mathbf{w}^{\mathsf{T}} = \mathbf{S} - \mathbf{w}\mathbf{w}^{\mathsf{T}} + \mathbf{T} \tag{4}$$

where $\mathbf{S} = \frac{1}{2}(\mathbf{P}+\mathbf{P}^{\mathsf{T}})$ and $\mathbf{T} = \frac{1}{2}(\mathbf{P}-\mathbf{P}^{\mathsf{T}})$. The corresponding decomposition of inertia is

$$\begin{aligned} &\mathrm{tr}(\mathbf{D}_w^{-1}(\mathbf{P} - \mathbf{w}\mathbf{w}^{\mathsf{T}})\mathbf{D}_w^{-1}(\mathbf{P} - \mathbf{w}\mathbf{w}^{\mathsf{T}})^{\mathsf{T}}) \\ &= \mathrm{tr}(\mathbf{D}_w^{-1}(\mathbf{S} - \mathbf{w}\mathbf{w}^{\mathsf{T}})\mathbf{D}_w^{-1}(\mathbf{S} - \mathbf{w}\mathbf{w}^{\mathsf{T}})^{\mathsf{T}}) + \mathrm{tr}(\mathbf{D}_w^{-1}\mathbf{T}\mathbf{D}_w^{-1}\mathbf{T}^{\mathsf{T}}) \end{aligned} \tag{5}$$

Because $\mathbf{w}\mathbf{w}^{\mathsf{T}}$ is not the closest rank 1 matrix to $\mathbf{P}$ we expect the inertia on the left hand side of (5) to be higher than the usual inertia in the ordinary CA – how much higher will depend on how different $\mathbf{r}$ and $\mathbf{c}$ are.

Displaying the inertia terms on the right hand side of (5) can now be achieved by performing a GSVD, again in the metrics $\mathbf{D}_w^{-1}$ and $\mathbf{D}_w^{-1}$, on the centred symmetric table $\mathbf{S} - \mathbf{w}\mathbf{w}^{\mathsf{T}}$ and the already centred skew-symmetric $\mathbf{T}$. The advantage of "homogenizing" the metrics throughout equation (5) is now clear – the analysis of $\mathbf{S} - \mathbf{w}\mathbf{w}^{\mathsf{T}}$ is symmetric in the rows and columns and yields one set of points, while the analysis of $\mathbf{T}$, the skew-symmetric component, also yields just one set of points (see Constantine and Gower 1978). Greenacre (1996) shows how both the symmetric and skew-symmetric analyses may be performed by one simple CA of the block matrix

$$\begin{bmatrix} \mathbf{N} & \mathbf{N}^{\mathsf{T}} \\ \mathbf{N}^{\mathsf{T}} & \mathbf{N} \end{bmatrix}$$

where $\mathbf{N}$ is the original table of frequencies.

Labels refer to the occupation abbreviations given in Table 1.

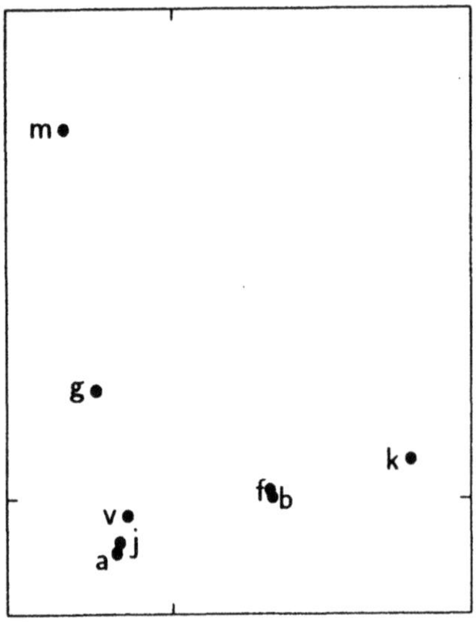

Figure 2: Correspondence analysis of symmetric component $\mathbf{S}$

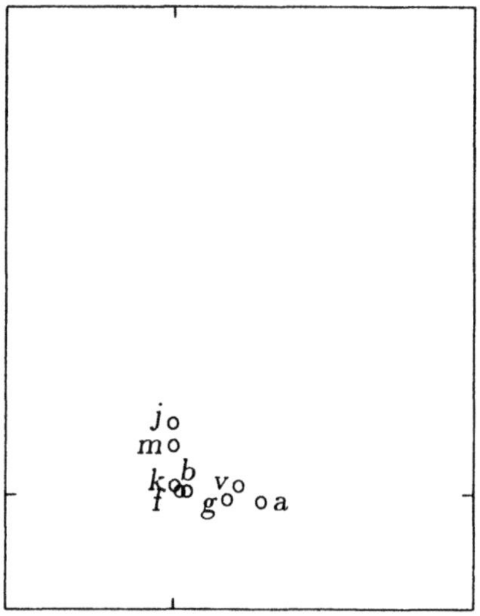

Figure 3: Correspondence analysis of skew-symmetric component $\mathbf{T}$

The displays of the two matrix components with respect to the first two dimensions are shown in Figures 2 and 3, which are drawn to the same scale. The components of inertia in (5) have values $2.222 = 1.902 + 0.320$. The percentages of inertia are expressed with respect to the total inertia 2.222. Thus the first two axes of the symmetric analysis explain 0.702 (28.8%) and 0.528 (22.7%), respectively, of the total inertia, while the first two axes of the skew-symmetric analysis explain 0.139 (6.4%) each (in the latter case the two axes explain 87% of the skew-symmetric inertia component 0.320). In total, therefore, Figures 2 and 3 explain 64.3% of the total inertia, compared to the percentage of 57.1% explained by the simple CA in Figure 1.

Apart from the symmetry property that there is no distinction between initial and main occupations, Figure 2 is practically identical to Figure 1.

Figure 3 shows the more interesting pattern of fluxes between occupational categories which we could not see previously. Notice first that only 12.6% of the total inertia is explained in this map, as indicated by the closeness of the categories to the origin, but the interpretation is very clear. The categories close to the centroid, "clergy", "self employment" and "education" have influx and outflux that nearly balance. Larger unidirectional changes take place between the occupations which are more distant from the origin. The direction of change can be deduced from the relative orientations of the points – in this case the change is interpreted in a clockwise direction so that the categories lying more or less along the vertical axis in Figure 3 are losing people to the categories on the horizontal axis.[1]

Upwards in the vertical direction we find occupations that had an outflux greater than their influx, i.e. they lost people in the time between first and main occupations. Occupations in which the number of persons increased appear in the horizontal direction to the right. One can see that the occupations "administration" and "lawyer" received a relatively high number of persons from "justice" and that many farmers started their career as officers. Therefore, work in "justice" seems to be used for collecting first experience in the field of "integration" before the members of the "Frankfurter Nationalversammlung" started to work as lawyers or in "administration". Landlords started their careers very often as a member of the military, then after getting the fields from their parents they went back to their roots.

5. Discussion

The use of a crosstable such as "job at time point one" and "job at time point two" as input data for correspondence analysis leads essentially to a visualization of occupational "stability". Because the likelihood to stay in a job is higher than to leave (in this case there were seven alternatives), the differences between expected and empirical values are relatively high, their part of the total variation in the data is high. Decomposing this variation must reflect the high values in the main diagonal of the input data on the first dimensions.

However, if one wishes to inspect changes between the occupations one has to consider higher dimensions. Using the method mentioned above allows us to seperate the dimensions into symmetric and skew-symmetric parts. Whereas the symmetric part confirms mainly the solution of the correspondence analysis of the original crosstable, the skew-symmetric part shows the changes in the jobs. Both parts together help to understand occupational mobility in more detail.

[1]There is a rotational indeterminacy in each pair of dimensions of the skew-symmetric analysis since the explained inertia is identical for each dimension. The change between occupational groups A and B is proportional to the area of the triangle formed by the origin and the positions A and B, and any orientation of the axes gives the same interpretation – see Constantine and Gower (1978) or Gower (1980) for more details.

References:

BENZÉCRI, J.-P. (1973): L'Analyse des Données. Tome 1: La Taxinomie. Tome 2: L'Analyse des Correspondances. Dunod, Paris.

BEST, H. (1990): Die Männer von Bildung und Besitz. Econ,Düsseldorf.

BLASIUS, J. (1994): Correspondence analysis in social science research. In: M. J. Greenacre and J. Blasius (eds.): *Correspondence Analysis in the Social Sciences.* Academic Press, London, 23–52.

BLAU, P. and DUNCAN, O. D. (1967): The American Occupational Structure. Wiley, New York.

CONSTANTINE, A. G. and GOWER, J. C. (1978): Graphical representation of asymmetry. *Applied Statistics, 27, 297–304.*

GOWER, J. C. (1980): Problems in interpreting asymmetrical chemical relationships. In: F.A. Bisby, J.G. Vaughan and C.A. Wright (eds.): *Chemosystematics: Principles and Practice.* Academic Press, London, 399–409.

GREENACRE, M. J. (1984): Theory and Applications of Correspondence Analysis. Academic Press, London.

GREENACRE, M. J. (1993): Correspondence Analysis in Practice. Academic Press, London.

GREENACRE, M. J. (1996): An adaptation of correspondence analysis for square asymmetric matrices. Working Paper, Department of Economics and Business, Pompeu Fabra University, Barcelona, submitted for publication.

LANGEWIESCHE, D. (1981): Die deutsche Revolution von 1848/49 und die vorrevolutionäre Gesellschaft: Forschungsstand und Forschungsperspektiven. *Archiv für Sozialgeschichte, 21, 458–498.*

LEBART, L.; MORINEAU, A. and WARWICK, K. (1984): Multivariate Descriptive Statistical Analysis. Wiley, New York.

LEBART, L., MORINEAU, A., and PIRON, M. (1995): Statistique Exploratoire Multidimensionelle. Dunod, Paris.

PARSONS, T. (1971): The System of Modern Societies. Englewood Cliffs, New Jersey.

VAN DER HEIJDEN, P. and MOOIJAART, A. (1995): Some new log-bilinear models for the analysis of asymmetry in a square contingency table. *Sociological Methods and Research, 24, 7–29.*

Identifying Benchmarking-Partners Using Two-Mode Classification

R. Lasch

Institut für Statistik und Mathematische Wirtschaftstheorie,
Universität Augsburg, D-86135 Augsburg

Summary: Benchmarking is the search for and implementation of best practices. The adoption or adaption of the best practices allows an organization to raise the performance of its products, services and processes to leadership levels. Market surveys show that the search for appropriate benchmarking-partners is often limited to the same industry. The search for partners outside the industry guaranties on the one hand that the most innovative practices and processes are found but this results on the other hand in the confrontation with big and bad structured data. Many organizations are overcharged in handling this data. Two-mode classification can help to structure this data and find benchmarking-partners outside the industry.

1. Benchmarking Basics

Many definitions have been proposed for benchmarking, but one developed at APQC[1] represents a consensus among some 100 companies:

Benchmarking is a systematic and continuous measurement process; a process of continuously measuring and comparing an organization's business processes against business process leaders anywhere in the world to gain information which will help the organization take action to improve its performance.

Benchmarks may be descriptive, as in the description of a best industry practice. They may be converted to a performance measurement that shows the effect of incorporating or adopting the practice. Developing new products, introducing new technologies and more effective market strategies are all essential components of a successful strategy. Benchmarking can identify competitors' strategies, strength and weakness, determine the key factors of success and utilize this information to surpass the competition.

Benchmarking performance measurements are useful means to identify organizations whose performance levels are significantly better and who may have best practices. The real benefit of benchmarking, however, comes from understanding the practices that permit the performance and the reasoned transfer to the organization. There are three types of benchmarking and each is defined by the target or object of the benchmarking activity. Table 1 represents a summary of the three types of benchmarking.

[1]American Productivity & Quality Center, Houston, Texas

Type	Definition	Advantages	Disadvantages
Internal	Similar activities in different locations, departments, operating units, countries, etc.	• Data easy to collect • Good results for diversified excellent companies	• Limited focus • Internal bias
Competitive	Direct competitors selling to same customer base.	• Information relevant to business results • Comparable practices • Check own market position	• Data-collection difficulties • Sector oriented newline view
Functional	Organizations recognized as having best in class products, services, processes.	• Highest potential for discovering innovative practices • Readily transferable practices/technology • Stimulating results	• Time-consuming • Difficulties to transfer practices into different environments

Table 1: Types of Benchmarking

Benchmarking follows a structural three-phases approach involving ten steps which are independent of the benchmarking type.

Planning

1. *Identify what is to be benchmarked.* Identify the largest opportunity to improve performance in the organization. This requires identifying and prioritizing the key work processes and flowcharting them for analysis and comparison of practices.
2. *Form a benchmarking-team.* It is necessary to form a team of 5 to 6 specialists concerning the benchmarking object because the most benchmarking efforts are team activities.
3. *Identify benchmarking-partners.* The competitor or the industry leaders are certainly prime candidates to benchmark, although benchmarking can also be conducted against leadership companies or organizations regardless of industry.
4. *Determine data collection method.* During this stage of process, specific information-collection methods are selected.

Analysis

5. *Determine current performance gap.* Comparing company functions against those of firms that perform the function best will identify any performance gap.
6. *Understand reasons for the gap.* Understanding how the companies you study achieve their results is usually more important and valuable than obtaining some precisely quantified metrics. The data should focus on the processes and practices rather than just the results.
7. *Project future performance levels.* Decide how much the performance gap will narrow or widen in the near future and what repercussions this has for the organization.

Action

8. *Establish goals and develop action plans.* Establishing operational goals for improvement involves careful planning to incorporate processes and practices.

9. *Implement actions and monitor progress.* Implementation of findings should involve periodic measurements and assessment of attempts to reach stated goals.
10. *Recalibrate benchmarks.* To incorporate changes in the competitive environment there must be a built-in provision for recalibration the benchmarks. Determine where the organization is in its quality pursuits and the implications for benchmarking.

Benchmarking is a continuous process, not an one-off exercise. It is necessary to repeat the three-phases approach until a leadership position is attained.

Companies realizing benchmarking successfully say that step 3 "Identify benchmarking-partners" is very important because the time spent for identifying an appropriate benchmarking-partner is about 30% of the whole benchmarking-process. Useful information sources for partners are companies receiving recognition or awards for some outstanding efforts, citations from others such as financial analysts or company watchers, lists of companies that are best in class as judged by a periodical's research, internal know-how and company/literature data bases. Companies breaking off the benchmarking-process argue that the main problem is to find an appropriate partner because they are overcharged in handling this amount of bad structured data (Lasch et al., 1995). Two-mode classification can help to structure this data.

2. Two-Mode Clustering Methods

The objective of clustering is to reveal structures within and between sets of objects by grouping those objects being similar with respect to certain properties into subsets called clusters. One-mode classification is employed to reveal an object-to-object similarity structure. Two-mode data describe the relationship between objects of two sets of different modes. A two-mode classification showing simultaneously the structure of two sets of entities and of their interrelationships may help in finding an adequate interpretation for the clusters.

After a short review of existing two-mode clustering procedures, three algorithms, which are used to identify benchmarking-partners in section 3, are presented more detailed. Two-mode approaches for jointly clustering n objects and m attributes in a data matrix can be classified into three categories.

The first category is composed by methods performing a reorganization of the data matrix by reordering rows and columns. The resulting clusters are interpretable directly on the input data. An example is the Bond Energy Algorithm by McCormick et al. (1972) which permutes the rows and columns of a data matrix in such a way that numerically larger matrix elements are pushed together.

The second category contains methods which are generalizations of the ADCLUS model proposed by Shepard/Arabie (1979). The basic assumption of this methods is that the similarity of two objects is considered to be describable as an additive function of weights which are associated with salient properties which both of the objects have in common. An example is the GENNCLUS algorithm developed by DeSarbo (1982) which is presented in section 2.1.

Methods belonging to the third category are fitting additive or ultramatric tree structures to two-mode data. An algorithm constructing additive trees is given by De Soete et al. (1984). Models estimating an ultrametric tree are distinguished in direct methods, using only the information contained within the input data matrix and indirect methods which transform the data matrix into a matrix satisfying the two-mode ultrametric inequality. A direct algorithm is given by Eckes/Orlik (1991) in section 2.2 and an indirect algorithm by Espejo/Gaul (1986) presented in section 2.3.

2.1 GENNCLUS Model

A generalization of the ADCLUS model to the case of nonsymmetric or two-mode proximity data is the GENNCLUS methodology (DeSarbo, 1982). The GENNCLUS model allows for a given number k of clusters a simultaneous disjoint or nondisjoint classification of two modes. Given a rectangular two-mode $(n \times m)$-matrix $S = (s_{ij})$ of similarities the problem is to find a best fitting $(n \times m)$-matrix $\hat{S} = (\hat{s}_{ij})$ of the form

$$\hat{S} = PWQ^t + C,$$

where

$P = (p_{il})$ is a $(n \times k)$-matrix designating membership of the n row objects in k clusters,
$W = (w_{ll'})$ is a $(k \times k)$-matrix of weights,
$Q = (q_{jl'})$ is a $(m \times k)$-matrix designating membership of the m column attributes in k clusters,
$C = (c)$ is a $(n \times m)$-matrix acting as an additive constant.

Here, given S, the GENNCLUS algorithm attempts to estimate P, W, Q and C. The estimation is done using an alternating least-squares algorithm which minimizes the function

$$z = \sum_{i=1}^{n}\sum_{j=1}^{m}(s_{ij} - \sum_{l=1}^{k}\sum_{l'=1}^{k} p_{il}w_{ll'}q_{l'j} - c)^2.$$

2.2 Centroid Effect Method

In contrast to the GENNCLUS model the centroid effect method (Eckes/ Orlik, 1991) constructs a two-mode hierarchical clustering system with clusters having minimal internal heterogeneity. Given a similarity matrix $S = (s_{ij})$ the mean squared deviation is used as the internal heterogeneity measure of a two-mode cluster r which can be expressed as

$$MSD_r = Var_r + (\bar{s}_r - \max_{i,j} s_{ij})^2.$$

The heterogeneity measure of a cluster r is equal to the variance Var_r of the similarities in cluster r and the squared difference between the mean $\bar{s}_r$ of the similarities in cluster r and the maximum entry $\max_{i,j} s_{ij}$ in the input matrix S. This squared difference is called the "centroid effect" of cluster k. At each step of the agglomerative algorithm those two clusters are merged which result in a minimal increase in the internal heterogeneity measure.

2.3 Missing Value Approach

The missing value approach (Espejo/Gaul, 1986) belongs to the indirect construction methods and generates tree structures which are ultrametric to two-mode data. Starting from a two-mode $(n \times m)$-matrix the so called $(n+m \times n+m)$-grand matrix is constructed, containing (dis-)similarities between objects/objects, objects/attributes and attributes/attributes. Using the grand matrix one-mode clustering algorithms are applied. The missing value approach starts with a given two-mode dissimilarity matrix and treats the unknown object/object- and attribute/attribute-distances in the grand matrix as missing values.

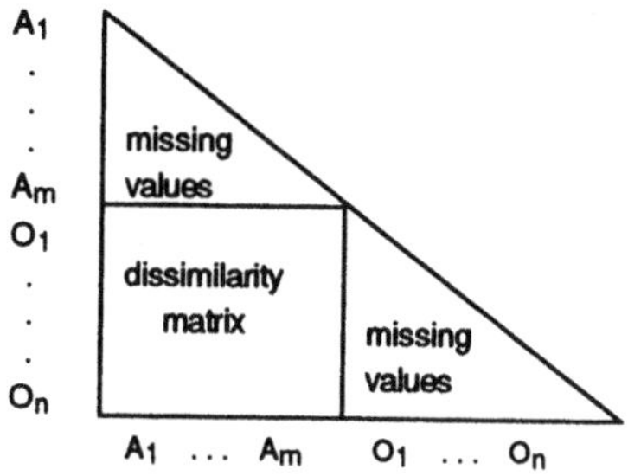

Fig. 2.1: Grand matrix

3. Empirical Results

Four companies (C1, C2, C4, C9 in Table 3.1) of different industries, all having problems with their stores, are looking for appropriate benchmarking-partners. To identify approriate benchmarking-partners it is necessary to differentiate the stores in a structural part and a process part. Only if the relevant structural factors have similar values a direct comparison of the process data is possible. The following structural factors are examined:

Storage area (SA), number of palettes (NP), storemen full-time (SF), storemen part-time (SP), number of articles (NA), used articles per day (UAD), customer orders per day (COD), average number of positions per order (NPO), average number of picks per position (NPP), number of palette-picks per day (NPD), number of carton-picks per day (NCD).

Table 3.1 shows the standardized values of the structural factors for the four companies with problems and five other companies of different industries. Figure 3.1 shows the dendrogram resulting from the missing value complete linkage algorithm. Four clusters containing companies and structural factors

	SA	SF	SP	NA	UAD	NPO	NPP	NP	NPD	NCD	COD
C1	1.34	0.80	-0.90	-1.12	-0.75	-0.66	1.32	1.03	1.05	-0.34	-0.24
C2	-0.03	0.23	1.97	0.33	1.26	1.15	-0.88	0.51	-0.56	1.75	1.82
C3	-0.81	-0.60	-0.44	0.97	1.85	1.99	-0.80	-0.47	-0.23	-0.07	-0.89
C4	0.37	-0.91	-0.74	1.13	-0.55	-0.70	0.77	-0.10	-0.52	-0.99	0.53
C5	1.64	1.58	-0.44	-0.70	-0.55	-0.65	1.88	1.99	2.42	-0.22	-1.26
C6	-1.46	-1.00	0.62	-1.42	-0.97	-0.69	-0.88	-1.52	-0.75	1.16	0.48
C7	-0.60	-0.91	1.37	-1.07	-0.84	-0.69	-0.61	-0.89	-0.15	0.96	0.88
C8	-0.97	-0.70	-0.54	1.05	-0.44	-0.71	-0.06	-0.60	-0.71	-1.21	0.08
C9	0.53	1.51	-0.90	0.81	0.98	0.94	-0.75	0.04	-0.54	-1.03	-1.40

Table 3.1: Data matrix (standardized values)

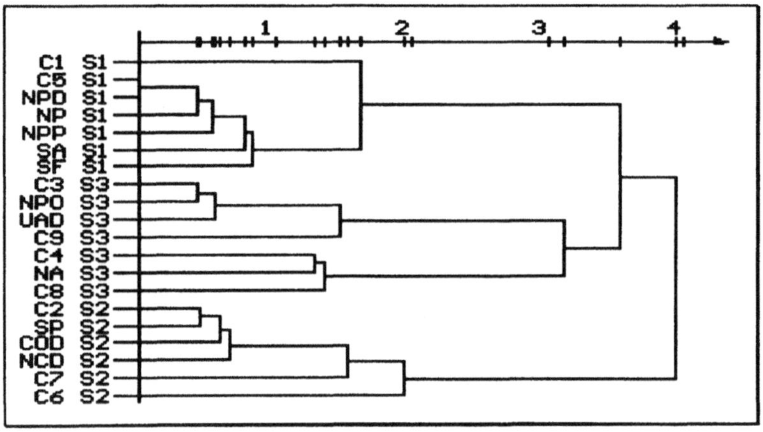

Fig. 3.1: Missing value complete linkage dendrogram

can be separated. Using the centroid effect method the same four clusters can be detected. Also the non-hierarchical GENNCLUS algorithm – given four non-overlapping classes – generates the same clusters so that one can speak of a robust classification result.

The 4-classes-result of figure 3.1 shows the similarities in the structural factors between the companies. Companies in the same class are possible benchmarking-partners because their structural factors are comparable. A comparison of the correspondent processes of innerclass companies identifies benchmarking-objects which can be improved. Possible benchmarking-objects concerning the process part are also given in the following interpretation of the four classes.

Class 1: C1, C5, SA, SF, NP, NPD, NPP.
Company 1 and company 5 have similarities in the storage area, storemen full-time, number of palettes and palette-picks per day and number of picks per position but show significant differences in commission-times so that the commission-process should be compared.

Class 2: C3, C9, NPO, UAD.
Company 3 and company 9 have similarities in the number of positions per order and used articles per day. But company 3 employs only 25% of the worker of company 9 because of the implementation of many computer-aided processes.

Class 3: C4, C8, NA.
Company 4 and company 8 have nearly the same number of articles but differ considerable in the cycle-time between recept of goods and storage. The reason is the usage of bar codes by company 8 which reduces the cycle-time.

Class 4: C2, C6, C7, SP, COD, NCD.
Company 2, company 6 and company 7 have similarities in the number of part-time storemen, customer orders per day and number of carton-picks per day. Company 2 has a lot of faulty orders in contrast to company 6 or 7. The reason is the better organized change of shift of the part-time storemen in company 6 and 7.

References:

CAMP, R. (1989): Benchmarking: The Search for Industry Best Practices That Lead to Superior Performance. ASQC Quality Press, Milwaukee, Wisconsin.

DESARBO, W. (1982): New Models for General Nonhierarchical Clustering Analysis. *Psychometrika, 47, 449-475.*

DE SOETE, G.; DESARBO, W.; FURNAS, G. and CARROLL, J. (1984): The Estimation of Ultrametric and Path Length Trees from Rectangular Proximity Data. *Psychometrika, 49, 289-310.*

ECKES, T. and ORLIK, P. (1991): An Agglomerative Method for Two-Mode Hierarchical Clustering. In: H. Bock and P. Ihm (eds.): *Classification, Data Analysis and Knowledge Organization.* North-Holland, Amsterdam, 3-8.

ESPEJO, E. and GAUL, W. (1986): Two-Mode Hierarchical Clustering as an Instrument for Marketing Research. In: W. Gaul and M. Schader (eds.): *Classification as a Tool of Research.* North-Holland, Amsterdam, 121-128.

LASCH, R.; EIBER, M. and HIRSCHLE, A. (1995): Benchmarking in der Logistik. *Arbeitspapiere zur mathematischen Wirtschaftsforschung, Heft 125.*

McCORMICK, W.; SCHWEITZER, P. and WHITE, T. (1972): Problem Decomposition and Data Reorganization by a Clustering Technique. *Operations Research, 20, 993-1009.*

SHEPARD, R. and ARABIE, P. (1979): Additive Clustering: Representation of Similarities as Combinations of Discrete Overlapping Properties. *Psychological Review, 86, 87-123.*

Supporting the Search for Final Scenarios by the Fuzzy-C-Means Algorithm

M. Missler-Behr

Institut für Statistik und Mathematische Wirtschaftstheorie,
Universität Augsburg, D-86135 Augsburg, Germany

Summary: In scenario analysis we are looking for some consistent, but quite different pictures of the future concerning a given problem. Not just the pictures themselves but also the way from today to a forward-looking situation shall be explained. Agglomerative methods in classification are already used to find the right number of final scenarios and to determine the final scenarios themselves. Here we examine, whether these questions can better be answered with the help of fuzzy-c-means cluster algorithm or whether fuzzy classification results can support interpretation and decision making in that context.

1. Introduction to Scenario Analysis

Scenario analysis is more and more used in strategic planning as a systematic method to develop possible pictures of the future called scenarios. Not only the pictures themselves but the logical process of the scenarios from today to a future timepoint are the main aspects in scenario analysis. An advantage of this method to well-known forecasting methods is the possibility to use quantitative and qualitative variables to enlighten the future.

Scenario analysis is based on the idea of multiple futures. Future can not be predicted any more in an unique way. We have to think about several future possibilities. Also we have to reflect upon the complex networking structures companies are placed in nowadays, as well as the diverse relationships a company has with its internal and external environment (Gausemeier et al. (1995)). Not to predict the future but to think of possible, different futures in advance is the main task of scenario analysis.

2. Scenario Process

2.1 General approach

The scenario process can be subdivided into five basic steps (Gausemeier et al. (1995), von Reibnitz (1991)). Fig. 1 shows the general course of the process and states the most important actions at each step. In this paper we deal with the second part of step 4.

The Scenario Creation and Selection in step 4 can be divided into two parts:

1. **Preselection out of the total scenario space**
   Out of all theoretically possible scenarios the most consistent and stable scenarios are selected. Consistent means that there are no contra-

Step 1	**Scenario Preparation** ↓	definition, distinguishing and analysis of the investigated problem
Step 2	**Scenario Environment Analysis** ↓	description of the complex problem and determination of the critical key items
Step 3	**Scenario Forecasting** ↓	analysing the critical key items and describing their possible future developments (features)
Step 4	**Scenario Creation and Selection** ↓	creating all possible feature combinations of the complete key items (scenario space), looking for few, consistent, different and representative scenarios
Step 5	**Scenario Transfer**	integration of the selected scenarios in strategic planning

Fig. 1: The Scenario Process

dictions between different scenario items. Stable means that a slight change in a scenario item does not change a goodness of fit criterion used to evaluate scenarios. Methods of preselection are described by Missler-Behr (1993).

2. **Final selection out of the preselected scenarios**
Out of the preselected consistent scenarios just a few (2-5), the most different and representative scenarios are selected for further analysis in strategic planning. Here classification methods can be used (Missler-Behr (1993), Opitz, Missler-Behr (1993)). But with fuzzy clustering more detailed results can be achieved.

2.2 Formal Aspects

A scenario can be described in a formal way as a binary vector of the features of all critical key items. Do we have m key items with n_i $(i = 1, \dots, m)$ features each, we can describe a scenario as a $(1 \times n)$ vector of 0/1 components, whereby $n = \sum_{i=1}^{m} n_i$. The value 1 in the i-th vector component means that the i-th feature has occured, 0 means that it has not occured. Only one feature of each key item can occur. Therefore the scenario vector contains m-times the value 1 and $(n-m)$-times the value 0. Tab. 1 shows an example. All considered scenarios can be summerized in a binary $(\mu \times n)$-matrix X, where μ is the number of considered scenarios.

The total scenario space can be imagined as a n-dimensional unit cube.

	$m = 8$, $n = 21$, vector dimension: (1×21)							
n_i	2	3	2	2	2	3	4	3
scenario	01	100	10	01	01	010	0010	001

Tab. 1: Formal, binary description of an examplary scenario

Each edge of this cube symbolizes a n-dimensional binary scenario vector. Because the scenarios are regularly arranged in the cube we can not find any structure in the scenario space at all. The preselection methods have the task to cut holes in the cube. Some parts of the cube will be further analysed and other parts will be neglected. Now we are able to apply successfully classification methods.

3. Applying Classification Methods

Two questions have to be answered by classification methods in scenario analysis:

1. What is the equivalent number of classes, which means the equivalent number of scenarios to represent the total scenario space?
2. What are the representative scenarios?

3.1 Agglomerative Methods

Up to now mainly agglomerative classification methods are used in scenario analysis: starting heuristics and maximal c-cliques (Missler-Behr (1993), Opitz, Missler-Behr (1993)). In the following the preselected scenarios are summerized in a scenario set $\mathcal{S}$.

In starting heuristics we are looking for a classification $\mathcal{K} = \{C_1, \ldots, C_s\}$ of $\mathcal{S}$ in maximal different, non-empty classes. For a given number s of classes the algorithm determines maximal different class centers starting from a given scenario as center one. All remaining scenarios of $\mathcal{S}$ are assigned to their nearest center.

To measure the distance between two scenarios we use a distance index d with the following properties:

$$d : \mathcal{S} \times \mathcal{S} \to \mathbb{R}_+ \quad \text{with} \quad \begin{aligned} d_{ii} &= 0 \qquad \forall i \in \mathcal{S} \\ d_{ij} &= d_{ji} \qquad \forall i, j \in \mathcal{S} \end{aligned} \tag{1}$$

The first given center i_1 will be the most consistent scenario out of $\mathcal{S}$. To determine all other centers $i_2, \ldots, i_s$, we have to solve the following optimization problem:

$$\max_{i \in \mathcal{S}} \min_{r=1,\ldots,\sigma} d_{ii_r} \qquad \text{for} \quad \sigma = 1, 2, \ldots, s-1 \tag{2}$$

i_σ is the center of class C_σ of $\mathcal{K}$. All other scenarios of $\mathcal{S}$ will be assigned to the nearest center which can be described in the following way:

$$C_\sigma = \{i \in \mathcal{S} \ : \ d_{ii_\sigma} = \min_{r=1,\dots,s} d_{ii_r}\} \qquad \text{for} \quad \sigma = 1,\dots,s \tag{3}$$

The centers i_σ $(\sigma = 1,\dots,s)$ are the representatives of their classes. We get no hints for further representatives by the starting heuristics. Because of construction the representatives of the classes can be regarded as maximal different.

To determine the proper number of classes for a given application we can compare maximal and minimal distances between two classes. If the separation of the classes seems not sharp enough any more by those criterions, no further center should be determined. With the help of maximal c-cliques the results of starting heuristics can be reviewed. Essentially new knowledge can not be gained. Missler-Behr (1993) describes the use of maximal c-cliques in detail.

3.2 Fuzzy-C-Means (FCM) Algorithm

Classical classification techniques suffer from the lack of possibility to assign each object to more than one unique class. Fuzzy clustering allows a much more realistic assignment of partial membership to all classes.

In fuzzy clustering each class C_σ $(\sigma = 1,\dots,s)$ can be characterized by a vector $p_\sigma = (p_{1\sigma},\dots,p_{\mu\sigma})$ $(\mu = |\mathcal{S}|)$ which describes the grades of membership $p_{i\sigma}$. $p_{i\sigma}$ measures how strongly scenario i adopts the typical characteristics of class σ (Deimer (1986)). C_σ is called a fuzzy cluster if the following conditions hold:

$$p_{i\sigma} \ \geq \ 0 \tag{4}$$

$$\sum_{\sigma=1}^{s} p_{i\sigma} \ = \ 1 \qquad \text{for} \quad i = 1,\dots,\mu \tag{5}$$

The membership vectors p_σ can be summerized in a $(\mu \times s)$-matrix P of $\mathcal{S}$ with $P = (p_1^T,\dots,p_s^T)$. P is called the fuzzy s-partition of $\mathcal{S}$.

A scenario can now belong to each class with different grades of membership. Typical representatives of C_σ have a grade of membership $0.7 \leq p_{i\sigma} \leq 1$. They are called fuzzy centers (Bock (1979)).

Starting from a given, mostly random, fuzzy s-partition the FCM-algorithm determines an optimal fuzzy s-partition P^* by minimizing a variance criterion (Bezdek (1981), Zimmermann (1991)):

$$b(P) = \sum_{\sigma=1}^{s}\sum_{i=1}^{\mu} p_{i\sigma}^r \sum_{k=1}^{n}(x_{ik} - z_{\sigma k})^2 \rightarrow \min_P \tag{6}$$

such that

$$z_{\sigma k} = \sum_{i=1}^{\mu} p_{i\sigma}^r x_{ik} \Big/ \sum_{i=1}^{\mu} p_{i\sigma}^r \,. \tag{7}$$

r is an exponential weight for the grades of membership with $r \in (1, \infty)$. For large values of r only $p_{i\sigma}$'s ≈ 1 will have any influence in (6) whereas the influence of small $p_{i\sigma}$'s will nearly vanish. For $r \approx 1$ all values of $p_{i\sigma}$ are equally valued. $z_{\sigma k}$ is the mean of the k-th feature r-weighted by the grades of membership of class σ.

The FCM-algorithm works iteratively. A detailed description can be found in Bezdek (1981) or Bock (1979). It has been used already in a lot of different application areas.

3.3 Fuzzy Clustering and Scenario Analysis

When applying fuzzy clustering in scenario analysis to find good representatives for the scenario space we are only interested in the fuzzy centers of the different classes. We want to use fuzzy clustering now for two aims: to review the results of starting heuristics and to find new, objective representatives of the classes formed with starting heuristics.

To do so, we have to take into account the follwing aspects by using the FCM-algorithm:

- All calculations are based on the results of starting heuristics.
- The classes resulted from starting heuristics are reproduced with the help of the grades of membership. We determine the starting fuzzy s-partition P in the following way: assume that starting heuristics assign scenario i to class j. Set p_{ij} quite high and $p_{i\sigma}$ quite low $((\sigma = 1, \ldots, s)$ and $(\sigma \neq j))$. An example to put the grades of membership for scenario i with $j = 2$ in a three clusters solution is (0.25/0.50/0.25).
- The exponent r has to be chosen in such a way, that the centers of classes from starting heuristics are also fuzzy centers after running the FCM-algorithm. r depends on P and vice versa. In chapter 5 recommendations how to choose r and P are given.
- Only fuzzy centers are used for interpretation.

4. Example

The application shall be described with the help of an example: A regional bank in Germany asks about its possible future developments up to the year 2000. A detailed description of the problem and of the used database is given by Missler-Behr (1993) and von Reibnitz (1991). 17 critical key items with 37 features in total were determined. Therefore a scenario can be written as a binary vector of the dimension (1×37). The total scenario space contains 442,368 scenarios. After applying preselection methods 89 consistent scenarios were chosen for the final selection (Missler-Behr (1993)). The 89 scenarios are numbered consecutively.

Based on the most consistent scenario numbered 9 from the preselection procedure starting heuristics result in scenarios with numbers 9, 88, 31, and 46 as cluster centers for a one to four clusters solution. Cluster 3 and cluster

$p_{i\sigma} \in \{0.65, 0.35\}$ $r = 6$			
cluster 1		cluster 2	
scenario no.	grade of membership	scenario no.	grade of membership
53	0.920	89	0.920
9	0.918	88	0.910
50	0.881		
60	0.876		
17	0.875		

Tab. 2: Maximal grades of membership for the fuzzy two clusters solution

4 split of the huge cluster 1. A solution with three clusters seems to be optimal in our example by using maximal and minimal distances between the classes as a goodness of fit criterion. A detailed description of this analysis is given by Missler-Behr (1993).

To initialize the FCM-algorithm for a solution with two classes we choose 0.65 as grade of membership for cluster elements of the starting heuristics and 0.35 for non-cluster elements. The exponent r is fixed at 6 ($r = 6$). Tab. 2 shows the maximal grades of membership for the two classes after running the FCM-algorithm.[1]

After defuzzyfying 87 scenarios are assigned to cluster 1 and 2 scenarios to cluster 2. This result equals to the two clusters solution of the starting heuristics. In cluster 1 we can clearly see that beside scenario 9 other good representatives exist. Candidates for those are scenarios with number 53, 50, 60, and 17. In cluster 2 both elements with number 89 and 88 seem to be equally valued to represent the cluster.

To initialize the FCM-algorithm for the three clusters solution we choose 0.5 as grade of membership for cluster elements of the starting heuristics and 0.25 for non-cluster elements. The exponent r is fixed at 8 ($r = 8$). Tab. 3 shows the maximal grades of membership for the three classes after running the FCM-algorithm.

In the three clusters solution 73 scenarios are assigned to cluster 1, 2 scenarios to cluster 2, and 14 scenarios to cluster 3. Cluster 3 arises totally out of cluster 1. Scenarios numbered 53, 9, and 50 extract clearly from cluster 1 as representatives. They have not changed compared to the two clusters solution. There is no change in the results for cluster 2. Cluster 3 is represented by scenarios with number 31, 74, 32, and 30. Here we can already recognize a larger difference between the highest and the second highest grade of membership.

To initialize the FCM-algorithm for the four clusters solution we choose 0.4

[1]All of the following calculations were done by running the software package MSTAT (Bausch, Opitz (1993)).

$p_{i\sigma} \in \{0.5, 0.25\}$ $r = 8$					
cluster 1		cluster 2		cluster 3	
s. no.	grade of mem.	s. no.	grade of mem.	s. no.	grade of mem.
53	0.850	89	0.920	31	0.872
9	0.810	88	0.913	74	0.777
50	0.810			32	0.747
75	0.741			30	0.731
7	0.739				

s. no.: scenario number
grade of mem.: grade of membership

Tab. 3: Maximal grades of membership for the fuzzy three clusters solution

$p_{i\sigma} \in \{0.4, 0.2\}$ $r = 10$							
cluster 1		cluster 2		cluster 3		cluster 4	
s. no.	g.o.m	s. no.	g.o.m	s. no.	g.o.m	s. no.	g.o.m
53	0.796	88	0.889	31	0.893	46	0.952
9	0.782	89	0.882	74	0.678	3	0.639
50	0.733			32	0.641		
7	0.652			30	0.601		
60	0.649						

s. no.: scenario number
g.o.m.: grade of membership

Tab. 4: Maximal grades of membership for the fuzzy four clusters solution

as grade of membership for cluster elements of the starting heuristics and 0.2 for non-cluster elements. The exponent r is fixed at 10 ($r = 10$). Tab. 4 shows the maximal grades of membership for the four classes after running the FCM-algorithm.

In the four clusters solution 73 scenarios are assigned to cluster 1, 2 scenarios to cluster 2, 11 scenarios to cluster 3 and 3 scenarios to cluster 4. No element changed in cluster 1 or 2 compared to the three clusters solution. Cluster 4 arises totally out of cluster 3 in contrast to the solution of starting heuristics. Scenarios numbered 53, 9, and 50 confirmed there role as representatives of cluster 1, also did scenario 88 and 89 for cluster 2. In cluster 3 and 4 we can see, that only one scenario is left as fuzzy center. It is scenario 31 and 46. The grades of membership show, that these two classes can not be seperated clearly any more. Therefore the results of fuzzy clustering also indicate a three clusters solution as do the results of starting heuristics.

5. Recommendations for the Parameters

To apply the FCM-algorithm we have to determine two types of parameters: the grades of membership $p_{i\sigma}$ and the exponent r. About the behaviour of these parameters we can make a general statement: the smaller the given maximal grade of membership for a cluster element of starting heuristics, the greater the exponent r has to be chosen or vice versa. On the basis of empirical research we can put the above statement in concrete forms: For a cluster element of the starting heuristics the grade of membership should be greater or equal to $2/(s+1)$ and for non-cluster elements less or equal to $1/(s+1)$. For the exponent r the relation $r \geq 2s+2$ should hold.

6. Conclusions

Finally we can state that fuzzy clustering facilitates a much more objective selection of the representatives of the scenario space. On the one hand the results of the starting heuristics are reviewed, on the other hand possible further representatives are found. In this sense the quality of the results is considerably improved compared to the results of crisp classification methods. By the recommendations for the parameters concrete support is given for a successful application of the FCM-algorithm in scenario analysis.

References:

BAUSCH, T. and OPITZ, O. (1993): PC-gestützte Datenanalyse mit Fallstudien aus der Marktforschung. Vahlen, München.

BEZDEK, J. C. (1981): Pattern Recognition with Fuzzy Objective Function Algorithms. Plenum Press, New York, London.

BOCK, H. H. (1979): Clusteranalyse mit unscharfen Partitionen. In: H. H. Bock (ed.): *Klassifikation und Erkenntnis III: Numerische Klassifikation.* Studien zur Klassifikation, Band 6, 137–163.

DEIMER, R. (1986): Unscharfe Clusteranalysemethoden. Schulz-Kirchner, Idstein.

GAUSEMEIER, J.; FINK, A. and SCHLAKE, O. (1995): Szenario-Management: Planen und Führen mit Szenarien. Hanser, München, Wien.

MISSLER-BEHR, M. (1993): Methoden der Szenarioanalyse. DUV, Wiesbaden.

OPITZ, O. and MISSLER-BEHR, M. (1993): Klassifikationsverfahren in der Szenarioanalyse. In: J. Niedereichholz and W. Schuhmann (eds.): *Wirtschaftsinformatik — Beiträge zur modernen Unternehmensführung.* Campus, Frankfurt, New York.

REIBNITZ, U. VON (1991): Szenario-Technik: Instrumente für die unternehmerische und persönliche Erfolgsplanung. Gabler, Wiesbaden.

ZIMMERMANN, H.-J. (1991): Fuzzy Set Theory and its Applications. Kluwer, Boston, Dordrecht, London.

Two-Mode Classification in Advertising Research

M. Schwaiger

Institut für Statistik und Mathematische Wirtschaftstheorie,
Universität Augsburg, D-86135 Augsburg, Germany

Summary: In the past years two-mode clustering methods were applied to marketing subjects in order to represent relevant marketing information. The clustering methods just mentioned can be roughly divided into methods using grand matrices and other methods, such as the centroid-effect-method or the GENNCLUS-algorithm. This paper focusses another way of constructing a grand matrix from recall and recognition data. Using the grand matrix the well-known one-dimensional clustering methods can be applied to gain more detailed information about cognitive communication effects in advertising research.

1. Some Questions in Advertising Impact Research

Scrolling through newer survey articles on advertising research (e.g. Erichson and Maretzki (1993), Nieschlag et al. (1994)) one finds out that in general multivariate techniques are not used to analyze advertising impacts, except for the well known multidimensional scaling technique (Dichtl and Schobert (1979)). Awareness impacts, cognitive and emotional impacts as well as attitude impacts are mainly being controlled by calculating single ratios based on recall and recognition data.

Espejo and Gaul (1986) have shown that two-mode clustering algorithms may serve as a an appropriate tool for analyzing data resulting from unaided recall tests.
The goal of the present article is to elaborate this idea and to point out other advertising research sections, where two-mode clustering can be applied to achieve more qualified answers to interesting questions. More qualified in our sense means that, providing the information, we are

- blanketing the entire market instead of performing partial analysis
- considering similarities between objects and items (instead of analyzing specific attributes of a single object) and
- revealing connexions that would remain hidden using univariate techniques.

In a first step, some questions related to advertising impact research are presented. Then, existing two-mode classification algorithms are briefly described, before a new approach in constructing a grand matrix is shown. Finally, some recommendations are given, revealing which algorithm to apply on which advertising impact test.

Concerning awareness impacts of ads researchers may know, which print-ads/spots show high awareness rates in which segments (e.g. customer typologies, buyer/non-buyer etc.) compared to concurrent ads. Available test methods are folder-, spot- and day-after-recall tests.

Instead of just calculating the percentage of people recalling ad j one could imagine creating a database $\mathbf{X}_{(n \times m)} = (x_{ij})$ containing the percentage of persons from group i, who recall ad/spot j ($i = 1, \ldots, n; \quad j = 1, \ldots, m$).

After performing a two-mode cluster analysis on this database, one may look for clusters of segments containing only one special ad or for clusters of ads containing only one special customer segment. Any way the marketer may achieve hints for the relative awareness impact of his own ad/spot looking at the clusters resulting.

Concerning cognitive impacts of ads, special emphasis is laid on the correct assignment of masked ads. Questions like "Are masked ads assigned to the correct brand?, What impact on brand-recognition results from slogans?" can be answered by performing two-mode cluster analysis on confusion data resulting from impact tests. The database $\mathbf{X}_{(n \times m)} = (x_{ij})$ contains the number of persons who join ad i to brand j.

As far as emotional and attitude impacts are concerned, emotional-ratio-scales or semantic differentials provide a database $\mathbf{X}_{(n \times m)} = (x_{ij})$ showing the averaged rating-scores of ad i concerning item j, $i = 1, \ldots, n$; $j = 1, \ldots, m$).

Questions that can be answered by analyzing the two-mode clustering process are

- Which dimensions of perceptions are associated with brands/products/ companies?
- Do communicated messages/pictures fit the brand image?
- Which proposed ads/spots may be choosen best (pretest)?

just to name a few.

2. Two-Mode Classification Algorithms

The main objective of two-mode classification is a simultaneous clustering of row and column elements of a two- mode data matrix $\mathbf{X}$ as described above. In general the resulting clusters contain objects **and** elements at the same time, so the interpretation of specific clusters is to be done a lot easier than using one- mode classification.

The roots of two-mode classification go back to the bond-energy-algorithm (BEA) of McCormick et al. (1972), who arrange rows and columns of an association matrix in a way that a matrix of dense blocks containing large numeric values will result. Due to the non-hierarchical procedure, the clustering process remains invisible. Moreover, detailed instructions for grouping are missing and proposed permutation algorithms guarantee only sub-optimum solutions. Hence, applying the BEA should be avoided.

Another group of algorithms is formed by GENNCLUS (DeSarbo 1982) and PENCLUS (Both and Gaul (1985)). These non-hierarchical procedures are based on ADCLUS (Shepard and Arabie (1979)), thus representing interobject proximities as combinations of discrete and possibly overlapping properties. The difference between the two methods consists in solving the optimization problem: whereas GENNCLUS uses an ALS algorithm, PENCLUS uses penalty functions. It has to be considered, that the number of clusters must be specified in advance, the clustering process is invisible and cardinally scaled data are compulsory. On the other hand, specific constraints allow for overlapping clusters, which is a great advantage.

A hierarchical approach is presented by Eckes and Orlik (1991). The agglomerativ Centroid-Effect-Method uses an internal heterogeneity measure similar to the variance. Objects and attributes are merged according to the minimum increase of the described heterogeneity measure. The problem resulting from this method is, that objects or attributes can only be merged with an existing two-mode cluster. As a result, no clusters containing objects **or** attributes only may exist. Thus, a contradiction to similarities may arise.

As a last group, algorithms using grand matrices should be discussed. Starting from a two-mode matrix (e.g. association or confusion data) the so called grand matrix, containing (dis-)similarities between objects/objects, attributes/attributes and objects/attributes is constructed. Using the grand matrix one-mode clustering algorithms are applied.

Whereas Espejo and Gaul (1986) treat object/object- and attribute/attribute-distances as missing values, DeSoete et al. (1984) calculate, starting from a data matrix $\mathbf{X}_{(n \times m)}$, a matrix $\mathbf{T}_{(n \times m)}$ satisfying the bimodal ultrametric inequality. $\mathbf{T}$ is then transformed into a grand matrix $\mathbf{Z}_{(n+m)\times(n+m)}$ satisfying the unimodal ultrametric inequality. Though less complex, the missing value algorithms of Espejo and Gaul (1986) have competed favourably with the 1MLS (DeSoete (1984)) and the 2MLS algorithm of DeSoete et al. (1984).

Nevertheless, clusters containing objects only or attributes only aren't allowed here as well. Hence, even **identic** objects or attributes are not merged at a minimum fusion level, although the priciple of cluster analysis (forming homogenous clusters) would require to do so. Thus, in the following chapter, a procedure is proposed, which allows for strong orientation on similiarities between the elements of the two modes.

3. A New Approach Constructing a Grand Matrix

Starting from a confusion or association matrix $\mathbf{X}_{(n \times m)}$ with n objects (rows) and m attributes (columns), we first calculate distances between object i and object i' by

$$d_{ii'} = \frac{1}{m} \sum_{j=1}^{m} |x_{ij} - x_{i'j}| \,, \quad i, i' = 1, \ldots, n, \tag{1}$$

and distances between attribute j and attribute j' by

$$d_{jj'} = \frac{1}{n} \sum_{i=1}^{n} |x_{ij} - x_{ij'}| , \quad j, j' = 1, \ldots, m. \tag{2}$$

Moreover, distances from object i to attribute j are calculated by

$$d_{ij} = \max_{i,j} x_{ij} - x_{ij} , \quad i = 1, \ldots, n, \; j = 1, \ldots, m. \tag{3}$$

This procedure requires identically scaled attributes and similar ranges in attribute values. The latter is obtainable by standardizing the matrix $\mathbf{X}$.

To ensure identical influence of all distances on the clustering process, distances calculated by (1)– (3) are divided by the respective block maxima.

$$d^*_{ii'} = \frac{d_{ii'}}{\max_{i,i'} d_{ii'}}, \quad d^*_{jj'} = \frac{d_{jj'}}{\max_{j,j'} d_{jj'}}, \quad d^*_{ij} = \frac{d_{ij}}{\max_{i,j} d_{ij}} , \tag{4}$$
$$i, i' = 1, \ldots, n, \qquad j, j' = 1, \ldots, m.$$

The so constructed grand matrix is shown in figure 1 and serves now as symmetric distance matrix which is treated with one-mode clustering algorithms.

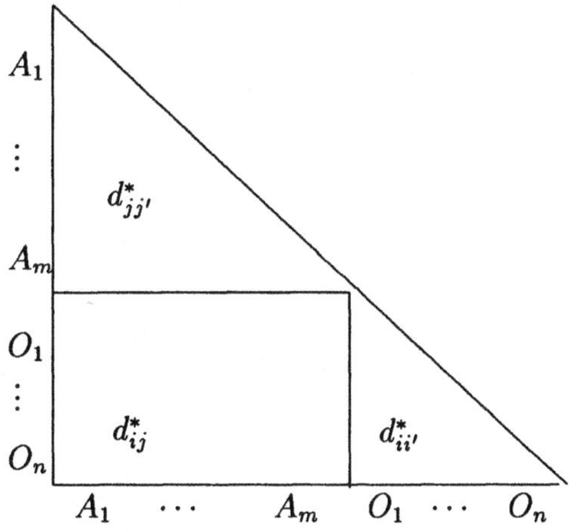

Figure 1: Grand Matrix

4. Empirical Results and Recommendations

Table 1 shows confusion data published in Gaul and Baier (1993, p. 340f.). 65 persons were presented print ads of eight cigarette brands (two ads for each brand, brand name masked) and asked to name the corresponding brand. Rows in table 1 denote ads, columns denote brands. Thus, the first value '60' means, 60 persons assigned the first camel print ad 'CAMEL(1)' to the brand 'CAMEL'.

	Camel	Dunhill	Ernte	HB	Kim	Krone	Lord	Marlboro	Philip Morris	R6	Stuyvesant	West
Camel(1)	60	0	0	1	0	1	0	1	1	0	0	0
Camel(2)	58	0	0	0	0	1	1	3	0	2	0	0
HB(1)	12	0	6	20	2	2	1	5	2	3	4	7
HB(2)	4	1	16	19	1	2	2	4	2	9	2	3
Krone(1)	0	14	5	2	10	6	6	1	11	6	2	0
Krone(2)	0	5	7	8	2	21	6	0	6	4	2	0
Lord(1)	0	0	4	8	4	7	13	1	2	6	18	0
Lord(2)	1	2	1	1	23	4	5	0	7	14	2	0
Marlboro(1)	4	1	0	0	1	0	1	54	1	0	0	1
Marlboro(2)	3	0	1	0	0	0	0	42	0	1	1	16
R6(1)	1	2	4	4	1	15	8	3	1	5	14	2
R6(2)	2	1	1	1	28	0	4	0	3	13	9	0
Stuyvesant(1)	0	5	1	0	1	0	3	7	12	4	20	11
Stuyvesant(2)	1	1	6	1	0	9	8	1	11	5	16	2
West(1)	35	0	5	1	0	2	1	12	1	1	0	6
West(2)	0	2	2	2	1	1	1	5	3	1	8	35

Table 1: Confusion Data

Figures 2 and 3 show the dendrograms resulting from the missing value average linkage algorithm (Espejo and Gaul (1986)) and from the complete linkage algorithm based on a grand matrix as described in section 3.

We have chosen these linkage types, because of their goodness-of-fit measures (see table 2) Variance Accounted For (VAF, see Arabie et al. (1987)) and Cophenetic Correlation Coefficient (CCC, see Sokal and Rolph (1962)).

Clustering Method		VAF	CCC
MV-Algorithm	Single Linkage	–	0,846
	Average Linkage	0,801	0,896
	Complete Linkage	0,305	0,867
New Approach	Complete Linkage	0,993	0,834
	Average Linkage	0,992	0,787
	Single Linkage	0,693	0,751

Table 2: Goodness-of-fit Measures

We see the MVAL-Algorithm in lead concerning CCC and the new approach showing a higher VAF. Comparing the results one finds out, that three separated clusters can be detected in both dendrograms: both camel-ads and WEST(1) form the 'CAMEL-cluster', MARLBORO ads and brand form another one, and finally WEST(2) and WEST is a separate group.

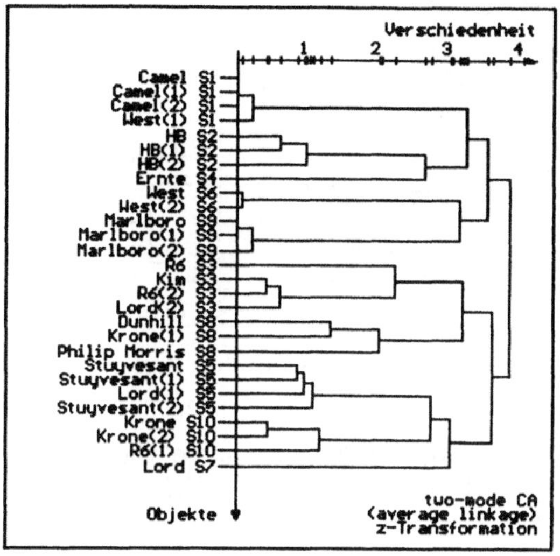

Figure 2: Missing Value Average Linkage - Dendrogram

Analyzing the differences between the two figures it comes clearly, that in figure 3 two clusters remain containing brands only (ERNTE – R6) and ads only (HB(1) – LORD(2)), whereas in figure 2 – except for the singletons LORD and ERNTE – there are no clusters containing brands only or ads only. So the MVAL joins e.g. HB-ads with the HB-brand comparatively early, although correct assignments of the HB-ads are below 30%. The same fact can be observed regarding STUYVESANT-ads.

From the marketing point of view a strong orientation on similarities is appropriate when analyzing confusion data. It seems that those ads and those brands who were grouped into the named two clusters are in lack of an independent emotional or cognitive position and could be substituted by other cluster brands or ads respectively. Unless this state is desired, brand managers of the affected brands/ads should draw the conclusion, that repositioning the brand or redesigning the ad would be helpful.

Considering the properties of the discussed two-mode algorithms some recommendations reaching beyond the comparison of goodness-of-fit measures can be given. The example in chapter 4 has shown that for analyzing **cognitive impacts** a strong orientation on similarities can be adequate. So the preferable algorithm here may be the one proposed in chapter 3.

To control **awareness impacts** the centroid-effect-methods serves well, because we have cardinally scaled data (percentages) and we are controlling impacts in relation to concurrent ads, so that single clusters containing ads or customer segments only are not useful. Concerning **attitude impacts** each cluster of ads should contain at least one attribute and vice versa. Thus,

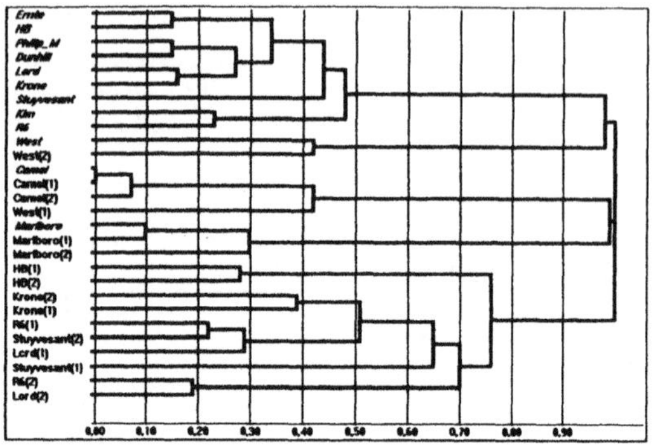

Figure 3: New Approach - Complete-Linkage- Dendrogram

due to the variety of linkage methods available, the missing value algorithms or – if overlapping clusters are desired – GENNCLUS and PENCLUS are a good choice.

References:

ARABIE, P.; CARROLL, J. D. and DeSARBO W. (1987): Three-Way Scaling and Clustering, Newbury Park BOTH, M. and GAUL, W. (1985): PENCLUS: Penalty Clustering for Marketing Applications, Discussion Paper No. 82, Institut für Entscheidungstheorie und Unternehmensforschung, Universität Karlsruhe

DeSARBO, W. S. (1982): New Models for General Nonhierarchical Clustering Analysis. *Psychometrica 47, 449–475.*

DeSOETE, G.; DeSARBO, W. S.; FURNAS, G. W. and CARROLL, J. D. (1984): The Estimation of Ultrametric and Path Length Trees from Rectangular Proximity Data, *Psychometrika 49, 289–310*

DeSOETE, G. (1984): Ultrametric Tree Representations of Incomplete Dissimilarity Data, *Journal of Classification 1, 235–242*

DICHTL, E. and SCHOBERT, R. (1979): Mehrdimensionale Skalierung, Methodische Grundlagen und betriebswirtschaftliche Anwendungen, München 1979

ECKES, T. and ORLIK, P. (1991): An Agglomerative Method for Two-Mode Hierarchical Clustering. In: H. Bock and P. Ihm (eds.): *Classification, Data Analysis and Knowledge Organisation*, North-Holland, Amsterdam, 3–8.

ERICHSON, B. and MARETZKI, J. (1993): Werbeerfolgskontrolle, in: Berndt, R.; Hermanns, A. (Hrsg.): *Handbuch Marketing-Kommunikation, Strategien – Instrumente – Perspektiven*, Wiesbaden, 521–560

ESPEJO, E. and GAUL, W. (1986): Two-Mode Hierarchical Clustering as an Instrument for Marketing Research. In: W. Gaul and M. Schader (eds.): *Classification as a Tool of Research*, North- Holland, Amsterdam, 121–128.

GAUL, W. and BAIER, D. (1993): Marktforschung und Marketing Management, München

McCORMICK, W. T.; SCHWEITZER, P. J. and WHITE, T. W. (1972): Problem Decomposition and Data Reorganization by a Clustering Technique, *Operations Research 20, 993–1009*

NIESCHLAG, R.; DICHTL, E. and HÖRSCHGEN, H. (1994): Marketing, 17., neu bearbeitete Auflage, Berlin

SHEPARD, R. N. and ARABIE, P. (1979): Additive Clustering: Representation of Similarities as Combinations of Discrete Overlapping Properties, *Psychological Review 86, 87–123*

SOKAL, R. R. and ROHLF, F. J. (1962): The Comparison of Dendrograms by Objective Methods, *Taxon, Vol. 11, 1962, S. 33-40*

Neural Networks as Instruments for Automotive Market Segmentation

K.-P. Wiedmann[1], H.-H. Jung[2]

[1] Lehrstuhl für Allgemeine Betriebswirtschaftslehre und Marketing II, Universität Hannover, D-30167 Hannover
[2] ISUMA Consulting GmbH, Prof. Dr. Wiedmann & Partner, D-68131 Mannheim

Summary: This contribution demonstrates the efficiency of neural networks using the example of automotive market segmentation and, more specifically, the case of the *a priori* classification of new car buyers. At its centre is a methodological comparison between a conventional and a neural discriminance analysis within the meaning of the probability concept (Bayesian approach).
This methodological comparison is carried out against the background of a brief sketch of the current challenges facing the Automotive industry, designed to emphasize the necessity of implementing innovative concepts of market segmentation (mass customizing) with substantially higher demands as regards the taking into account of complex, non-linear and time-sensitive contexts.
The results of a broadly based study – results which can only be reproduced in extracts here – demonstrate that compared with conventional methods, neural-based networks are clearly better able to represent the hybrid behaviour of new car buyers, and they can thus support innovative segmentation concepts in a much better way.

1. Current challenges faced by the Automotive Industry

A strategic analysis of the situation in the automotive industry shows that exclusively **market-oriented** marketing strategies and programmes (e.g. product or competition orientation) are only in part sufficient when meeting the current challenges. In stagnating markets or markets with slow growth[1], in order to be able to market increasingly homogenous products in individual market segments by meeting consumer demand and/or being event-oriented[2], it is necessary to record and analyse the reasons for buying and the intended use for present as well as potential customers on a more differentiated basis than was hitherto the case. This is true not least because the explosive nature of **social conflicts** (e.g. "ecology" versus "auto-mobility"), and their regulation in a political and legal context influences the purchasing behaviour / use pattern (e.g. willingness of certain segments of car buyers

[1]Both the German Automotive Association (Verband der Automobilindustrie, VDA) and the Association des Constructeurs Européens d'Automobiles expect the German and European automotive markets to grow slowly (Knust 1994).

[2]Numerous studies of the automotive market clearly show the **changing role of the car** from a status symbol to a medium for leading a particular lifestyle and for realising a particular event orientation (Spiegel 1993).

to pay more for cars meeting environmental standards; Jung et al. 1995).

The basic pattern of this situation, which has been briefly outlined here, has to be faced not just by automotive companies; it can – at least at first glance – be termed as the "new confusion":

- The thoughts and actions of decision makers in numerous companies are determined by **insular solutions**. Thus the analysis of competition constellations and the planning of strategies remains closely linked to **purely monocausal, linear and time-invariant management models**. As a result, the demands of market and social environments are not sufficiently taken into account. Especially in the latter area, the resulting challenges necessitate the deepening and broadening of marketing in the automotive sector as it is understood today, leading to a *society-oriented approach* (Wiedmann 1989).
- Even in companies where modern and holistic approaches are being adopted by the management there is a lack of a corresponding quantitative basis for decisions, not least as a result of the *limitations of currently used, obsolete methods*. A number of conventional statistical methods reduce the prevailing complexity, non-linearity and time sensitivity, thus simplifying the representation of reality in an unwarranted manner.

The current reasons for the new misery, which is described by using the image of a new confusion, are thus found at the heart of the very "think tanks" of companies where, in view of the already existing competitive challenges – i.e. to be better, cheaper, faster and more responsible (outpacing competition, Wiedmann 1992) – the growth nuclei for new solutions should be found. In view of these challenges it becomes all the more important to avoid past mistakes and to use those opportunities resulting from the process integration of innovative instruments (neural networks) and from new management approaches (mass customizing).

2. Process-oriented integration of innovative instruments and of new management approaches in automotive industry segmentatio

2.1 Foundations of neural modelling

On the one hand, neural networks are representations of biological brain cell complexes and for that reason – albeit to a somewhat restricted degree – they can be understood as **explanatory models of individual decision making behaviours and/or of market interactive processes** (Zimmermann 1994, Jung et al. 1995). On the other hand, neural networks allow for a particularly efficient and effective way of formulating **universal approximators** (Hornik 1991), and in the early 1990's this literally led to a breakthrough for neural calculation models in a number of practical applications. For some years now the efficiency of neural explanation and calculation models has been successfully used to solve economic problems in operational practice (e.g. analysis, forecasts, simulation, optimization).

Yet a metanalysis of **published feasibility studies of neural networks** shows (Wiedmann and Jung 1996) that, despite the existence of efficient neurosimulators and the development of neurospecific modelling designs in science as well as in an operational context, some fundamental difficulties (e.g. Gigo effect, trial-and-error approach) can still be observed.

Only the **choice of an appropriate philosophy of neural modelling** (Refenes 1995) makes it possible to fully exhaust the performance potential of the innovative basis for evaluation. This is demonstrated in the example of the **dilemma of optimum parametrisation in a neural network** (Hecht-Nielsen 1989). At present, bf three different approaches solving this dilemma (Occam's razor problem) can be distinguished:

- analytical techniques
- constructive techniques
- pruning techniques

Analytical techniques attempt to determine the size of a network using heuristic, statistical and/or algebraic methods (e.g. conflict estimation, information criteria, sample size). The **deficiencies** of analytical techniques are, amongst others, that they only permit rough estimates, for example for the purposes of dimensioning a network. The main problem, however, is the **nature** of this approach as it merely quantifies the start of the iterative process of neural modelling. The techniques in the second category – **constructive techniques** (e.g. cascade correlation, neural decision tree) – start from a minimal initial configuration, gradually enlarging the network's hidden layer as part of the modelling process. Nevertheless, constructive techniques show **deficiencies** both in respect of their **generalisation and stability characteristics.** In addition, as well as the analytical techniques they can be classified as "black box techniques" as they **do not permit a sufficient economic interpretation of neural models.**

Pruning techniques in particular have shown to be a suitable basis for the development of neural classification models as regards the demands of marketing. They work in the reverse way compared to constructive techniques, by starting with a large network at the beginning of the iterative modelling process which is then gradually thinned out in order to achieve the optimum configuration (e.g. regulation and elimination procedures). Certain neuro- simulators provide suitable **neuralbased evaluation tests** for this purpose, which are specifically geared to neural modelling (Zimmermann 1994). Evaluation studies show that such pruning techniques significantly improve the generalisation ability of a neural model (Hergert et al. 1992). **The systematic application of pruning techniques as well as of the relevant neuralbased evaluation tests** (Fig. 1) constitutes an adequate neuro- philosophy, particularly as systematic optimization at all levels of neural modelling enables the **performance potential** of the neural paradigm **to be fully exhausted** (avoiding Occam's razor problem). In addition, the gradual thinning out of non-relevant modelling determinants, using input pruning as integral part of optimization, permits **meaningful economic interpretations** of neural representational mechanisms.

Fig. 1: Optimization of neural models by pruning techniques and neuralbased tests

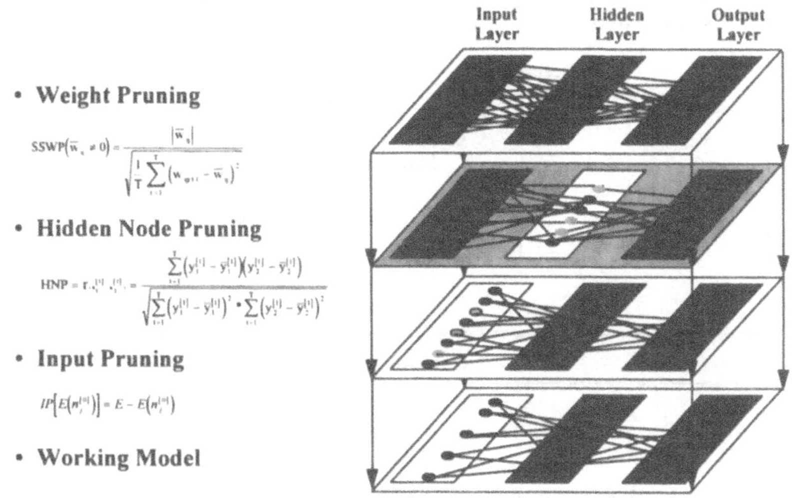

2.2 Mass customizing as innovative model of market segmentation

The fundamental concept behind mass customizing (Pine 1993) – initially discussed in connection with business-to-business marketing – is the simultaneous identification and realization of standardization and individualization potentials in a company's entire modular marketing program Compared to other market segmentation approaches, the mass customizing model offers significant advantages:

- In many cases **segment-oriented marketing** (target group concept) coupled with orienting the entire range of services towards target groups' demands leads to an insufficient depth of segmentation and is thus of insufficient use to current and potential customers (underspending effect).
- Individually tailored marketing, on the other hand, bears the potential risk of segmentation which is too far-reaching. This results not least in the problem of exponentially increasing service costs and the risk of overoptimizing the offer from the customer's point of view (overspending effect).

Mass customizing, however, is able to overcome the limitations of current models of market segmentation. Unlike other segmentation approaches it does not mean **focussing on just one dimension of the strategic triangle, i.e. the customer**. The narrow outlook characteristic of traditional approaches considers neither the **dimension of company orientation** nor that of the **competition orientation** of market segmentation. The ques-

Fig. 2: Models of Market Segmentation

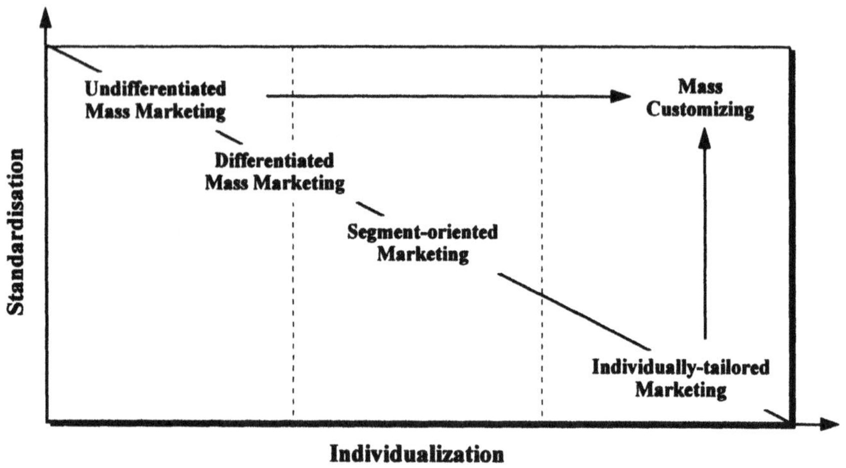

tion of the ratio between standardized and individualized orientation and implementation of marketing activities can only be answered in the light of a company's potential and the relevant competitive conditions.

Linked with mass customizing are higher demands with regard to the instrumental level of market segmentation. Unlike conventional methods neural networks are suitable for solving the **complexity of a segmentation task resulting from the multidimensional nature of mass customizing within the meaning of the strategic triangle.**

The study from which the case study below was extracted aims to demonstrate, using select issues of Automotive market segmentation (*a priori* classification of new car buyers) how enlarging the potential is linked with a process-oriented linking of mass customizing with neural discriminance models, as compared with conventional segmentation approaches (multi-variate statistical target group concept).

3. Case study: multi-variate statistical target group concept versus neuralbased mass customizing on the basis of the study "New Car Buyer Survey (NCBS)"

The study was carried out in conjunction with an automotive producer as part of a research project; it is based on the NBCS market research study set up as joint study of the global players in the automotive market. This study has been implemented in the present form in almost every volume markets throughout the world since 1982. In NCBS, the essential structural data for the offer and demand side of an automotive market are represented.

Fig. 3: Correspondence-Analysis: Category Quantifications

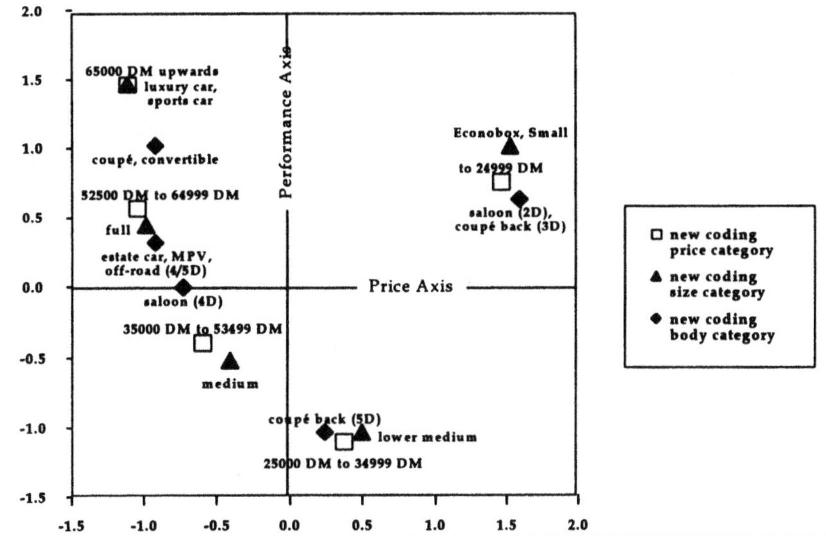

This contribution cannot present all findings of the comprehensive research study (Wiedmann and Jung 1996). However, some of the core tasks and their findings are presented here in the form of an outline summary.

As a precondition for carrying out a methodological comparison between conventional and neural discriminance models on the basis of the **New Car Buyer Survey NCBS 1994**, the existing concept for recording the vehicle types available on the market (correspondence analysis, Greenacre 1984) had to be developed further in a methodological way. Fig. 3 shows examples of five typical product concepts and the dimensions by which they are defined (price, size and body category). **Target customer profiles** for typical product concepts containing vehicles from the main volume and niche markets were deduced using a discriminance analysis.

For this purpose, corresponding to a confirmation-based new car buyer model **35 variables from NCBS** were applied as a first step. Using a (rotating) factor analysis these raw variables were then condensed to about **12 factors** reflecting **central mobility patterns** (e.g. work-related mobility), **family lifecycle phases** e.g. Full Nest I/II) and **significant automotive values** (e.g. ecological, security and innovative orientation).

In the following we will merely reproduce the results of the methodological comparison based on the factor model. In order to realize a conventional analysis (step-by-step methods using Wilks Lambda, Backhaus et al. 1994) and a neural discriminance analysis (multilayer perceptron using back-propagation learning and pruning techniques optimization, Richard and Lippmann 1991, Hastie and Tibshirani 1994, Zimmermann 1994) methods were selected in each case designed to solve the task at hand within the meaning of the Bayesian probability approach.

Method-oriented benchmarks (e.g. hit rate) were used to evaluate conven- tional and neural market segmentation instruments, in order to determine the quality of discriminance models. Given the different demands of models of market segmentation the evaluation framework was, however, further expanded to include **benchmarks within the meaning of confirmatory modelling** (e.g. α-mistakes of statistical decision theory).

The use of method-oriented benchmarks shows that the neural discriminance

	Conventional Discriminance Model	Neural Discriminance Model
hit rate modelling	59 %	69 %
hit rate generalization	57 %	63 %
α-mistakes (12 factors)	2 factors	0 factors

Table 1: Overview of Results of Study (factor model)

model **correctly allocates a higher number of current and potential new car buyers to the relevant and *a priori* defined categories**. The higher rate of correct allocation is particularly due to the ability of neural networks to **represent the discriminatory potential of explanatory determinants** – provided to both methods for the purposes of forming corresponding discriminance functions – **in an undistorted and consistent manner**.

By **including non-linearity** in the neural discriminance model the aggregate feature variables are appropriately represented as regards their ability to explain and separate, whereas conventional discriminance analysis leads to a substantial number of incorrect evaluations. The example of the factor 'family lifecycle phase Full Nest I/II' outlines the extent of incorrect evaluations using conventional discriminance analysis (cDMA) (Fig. 4). Here the deficits of conventional modelling are particularly noticeable as the linear approximation particularly in category 4 (e.g. Audi Avant) shows the most significant deviation. In reality, however, belonging to a family lifecycle phase where children and teenagers live under the same roof as the car buyer proves to be one of the central reasons for purchasing an estate car (category 4).

This example shows clearly that the insufficient representation of nonlinearity using conventional methods means that the management of automotive companies **set the wrong priorities when differentiating strategies and marketing program**. In this context, **mass customizing** with its emphasis on identifying **individualization and standardization potentials**, sets high standards as to the exact representation of the determi-

Fig. 4: Non-linearity: example of factor family lifecycle phase (FLC) Full Nest I/II

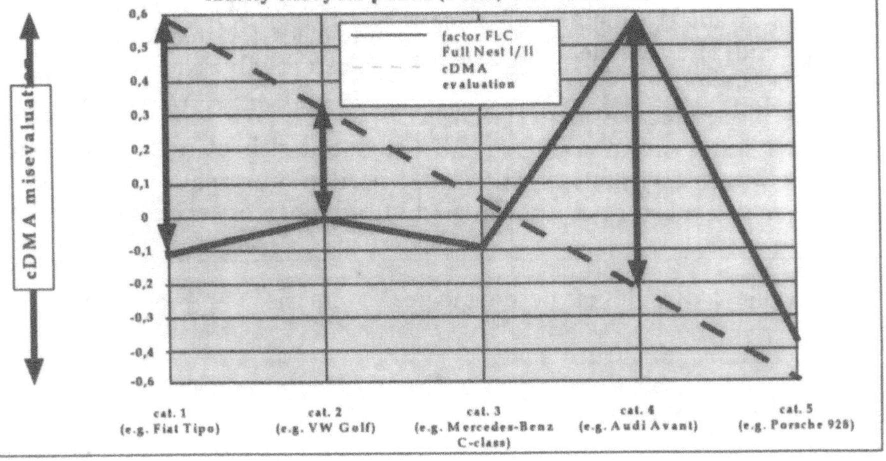

Fig. 5: Corporate Research & Intelligence Management-Approach

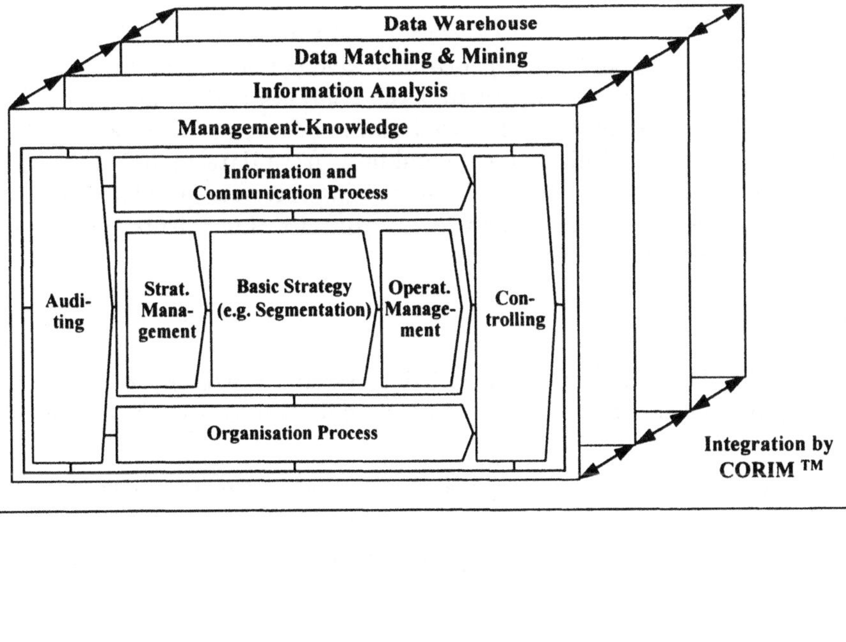

Results of mehodological comparison	Marketing Consequences
Neural discriminance analysis achieves higher hit rate compared to cDMA.	Marketing strategies and programs can be more precisely targeted than before
Unlike cDMA, neural discriminance analysis is able to include all non-linear and multicolinear determinants in the models.	The risk of not including essential approaches for marketing activities is substantially reduced.
Neural discriminance analysis correctly assesses all non-linear and multicolinear determinants whereas cDMA makes substantial assessment errors.	Reduced risk of setting the wrong priorities in orientation and design of the range of services on offer.

Table 2: Results of methodological comparison and marketing

nants of buyers' behavior.

4. CORIM as integrative concept of neuralbased automotive market segmentation

The selected results on their own show clearly that **neural network clearly form a more suitable methodological basis for putting orientation and design concepts for companies on a sufficiently strong footing.** It will, however, be possible to successfully apply the higher performance of neural market segmentation models to a significantly larger extent once they are integrated into a comprehensive management concept.

Corporate Research and Intelligence Management (CORIM^{TM}) constitutes an appropriate approach, integrating the contents and technological levels of information systems (from files to data models), creating cross-functional conventional or neural methodical combines (from data models to decision-oriented information models) and overcoming interface problems of all relevant processes on a management level (from decision-oriented information models to management knowledge).

References:

BACKHAUS, K.; ERICHSON, B.; PLINKE, W. and WEIBER, R. (1994): Multivariate Analysemethoden – Eine anwendungsorientierte Einführung, 7th Ed., Berlin.

GREENACRE, M. (1984): Theory and Applications of Correspondence Analysis, London.

HASTIE, T. T. and TIBSHIRANI, R. J. (1994): Nonparametric Regression and

Classification – Part II: Nonparametric Classification, in: Cherkassky, V. / Friedman, J.H. / Wechsler, H. (Eds.): From Statistics to Neural Networks, Berlin, pp. 70-82.

HECHT-NIELSEN, R. (1989): Neurocomputing, Reading.

HERGERT, F. B.; FINNOF, W. and ZIMMERMANN, H. G. (1992): Evaluation of Pruning Techniques, Technischer Report der Zentralen Forschungs- und Entwicklungsabteilung (ZFE) der Siemens AG, München (Technical Report of the Central Research and Development Department of Siemens AG, Munich).

HORNIK, K. (1991): Approximation Capabilities of Multilayer Feedforward Networks, *Neural Networks No. 4, pp. 251-257.*

JUNG, H.-H.; WIEDMANN, K.-P. and RAFFÉE, H. (1995): Eignung neuronaler Netze als Berechnungsansatz der Marketingforschung, Arbeitspapier Nr. 107 des Instituts für Marketing der Universität Mannheim (Working Paper No. 107 of the Institute for Marketing, Mannheim University).

KNUST, C. (1994): Die Autobauer in der langen Kehre, *Frankfurter Allgemeine Zeitung, 23rd July, p. 11.* PINE, J. B. (1993): Mass Customization – The New Frontier in Business Competition, Boston.

REFENES, A.-P. (1995): Methods for Optimal Network Design, in: Refenes, A.-P. (Ed.): Neural Networks in the Capital Markets, Chichester u.a., pp. 33-54.

RICHARD, M. D. and LIPPMANN, R. P. (1991): Neural Network Classifiers Estimate Bayesian a posteriori Probabilities, *Neural Computation No. 4, pp. 461-483.*

SPIEGEL (1993): Auto, Verkehr und Umwelt, Hamburg.

WIEDMANN, K.-P. (1989): Gesellschaft und Marketing: Zur Neuorientierung der Marketingkonzeption im Zeichen des gesellschaftlichen Wandels, in: Specht, G., Silberer, G. Engelhardt, W. H. (Eds.): Marketing-Schnittstellen - Herausforderungen für das Management, Stuttgart, pp. 227-246.

WIEDMANN, K.-P. (1992): PR. Wandel – Wege – Wagnisse, Arbeitspapier Nr. 96 des Instituts für Marketing der Universität Mannheim (Working Paper No. 96 of the Institute for Marketing, Mannheim University).

WIEDMANN, K.-P. and JUNG, H.-H. (1996, in preparation): Neuronale Netze als Instrument der Automobilmarktsegmentierung.

ZIMMERMANN, H. G. (1994): Neuronale Netze als Entscheidungskalkül – Grundlagen und ihre ökonometrische Realisierung, in: Rehkugler, H. / Zimmer- mann, H. G. (Eds.): Neuronale Netze in der Ökonomie – Grundlagen und finanzwirtschaftliche Anwendungen, München, pp. 3-87.

Section 8

Applications in Archaeology, Biology, Linguistics and Dialectometry

Seriation in Archaeology: Modelling, Methods and Prior Information

R. R. Laxton

Department of Mathematics,
University of Nottingham, Nottingham, NG 7 2RD, England

Summary: The standard model of seriation for chronologically ordering archaeological provenances from their profiles of contents will be introduced together with a method which attempts to carry this out. Various difficulties and the nature of archaeological data, which usually precludes further sampling, suggest that it is necessary to incorporate into the methodology both estimates of the original site profiles (as opposed to the sample profiles) and prior information about the provenances, their contents and their chronological order. How this is carried out, to a limited extent, in the traditional method used here will be explained. Ideally, several seriated orders should be produced and submitted to the archaeologists as possible candidates for the chronological order of the set of provenances, together with some measure of their relative merits. A Bayesian approach to seriation seems to be indicated by these remarks. Some comments will be made on this method and on the difficulties involved in implementing it. The examples used have been selected to illustrate these difficulties to the exclusion of other considerations.

1. The problem

Seriation in archaeology is concerned with a group of related *provenances*, such as graves in a cemetery, which contain different *types*, or forms, of an artefact, such as the different types of a brooch present in the graves. The problem in its simplest form is to place the provenances in a *relative chronological order* using only the variation in the types of the artefact they contain (see, for example, the early papers of Robinson (1951), Phillips et al. (1951) and Kendall (1969)). As an example, consider the variation of five different (colour) types of ceramics found in ten layers of trash at Awatovi (Burgh (1959)) as shown in Table 1. In this case the layers have a stratigraphic order, the one shown with the earliest, m, at the bottom. So, presumably, this is their relative chronological order also. But can we obtain this *relative* chronological ordering of the layers using only their associated *profiles* of counts, or percentages, of the various types of pottery they contain and without knowing the stratigraphic order? Can we recapture this order from the profiles if the relative chronological order of the corresponding provenances had been unknown to begin with? This is the problem of seriation. In order to be able to attempt this we have to make the basic assumption that

the only reason why the different types change is that they evolve over time

– and not because of other, non-temporal, factors. For example, the different types of brooches in the graves are the result of changes in fashion as time passes and to which everyone in that society is subject, a phenomena we witness in every day of our lives and are not, for example, because some graves are of males with their specific types of brooches and the rest are of females with their own types (see, for example, Hodson (1977)). The above assumption does give rise to difficulties, both theoretically and practically, but we will leave them and refer the interested reader to the literature. (An excellent introduction to the nature of the practical problems can be found in the articles in Madsen (1988).)

The data for seriation is commonly in one of two forms; as an *abundance matrix* where the rows are the profiles of the original counts or, more usually, of percentages, as in Table1, or as an *incidence matrix* where the rows are profiles of incidences and where a 1 means a particular type is present in the provenance and a 0 means it is absent. We will only consider abundance data here.

2. The model

In order to solve the archaeologist's problem, we must have a model for the way artefacts evolve over time.

Consider the Awatovi data of Table 1 with the profiles of percentages of five types of ceramics in the stratigraphic order of the corresponding layers. We see that the percentages of type 1 pottery decreases steadily from earlier to later, type 2 pottery rises and then falls steadily, type 3 pottery rises after being absent in the earlier layers, type 4 pottery also rises then falls and, finally, type 5 pottery rises a little in the very latest layers Let us assume that the pottery types are evolving with time and that the different proportions of the various types of pottery in the layers are good approximations to the ceramic profiles at the archaeological site at the times of breakage, ie. to the (original) *site* profiles. Furthermore, assume that the stratigraphic order of the layers is their relative chronological order and that there has been little mixing of the contents between layers since they were laid down (see Burgh (1959) on this). Then we make the following empirical observation (Robinson (1951) and Kendall (1969; 1971)).

Single-Peak Model. With the profiles (rows) in the relative chronological order of their corresponding provenances, the percentages of all the types (columns) never strictly decrease and then strictly increase again.

Other examples of artefact types satisfying the single-peak model can be found in the literature (see, for example, Clarke (1978) for a discussion from the archaeological point of view). We view this model as an ideal representation of artefact change in a society; it seems to be reasonably well approximated in practice by abundance data, at least when the artefact is not rare or only sparcely represented in the profiles.

Notice that if a set of profiles satisfies the single-peak model in a particular

Proflile	Type 1	Type 2	Type 3	Type 4	Type 5
d	4	6	86	0	4
e	6	14	76	0	4
f	8	19	70	1	2
g	18	49	30	3	0
h	23	54	20	3	0
i	32	49	5	14	0
j	39	43	0	18	0
k	49	30	0	21	0
l	62	20	0	18	0
m	88	3	0	9	0

Table 1: The ten profiles (rows) of percentages of five types of painted pottery (columns) from ten stratigraphically ordered layers in the Western Mound at Awatovi (Burgh 1959). The profiles are in the order of the stratigraphy with layer m lowest.

order, then it is still single-peaked if some of the profiles are *removed* or if some of the profiles adjacent in the order are *averaged* (Laxton and Restorick (1989)). For after such changes the percentages will still not go strictly down and then strictly up again. These properties are important in applications.

The single-peak model is *symmetric* in that if the single-peak model holds with the profiles in one order, then it also holds in the reverse order. Usually other archaeological evidence has to be introduced to determine the chronological *orientation* of the order as a whole. For convenience, we identify these two orders; by two different seriated orders of a set of profiles we mean that one is not the reverse order of the other.

3. A method of seriating

3.1 Profiles that satisfy the single-peak model

It is, in fact, relatively easy to seriate such profiles usng the similarity measure of *common content*, Com. For example, consider the last three profiles d, e, f and the first profile m in Table 1;

$$\begin{aligned} d &= (4, 6, 86, 0, 4) \\ e &= (6, 14, 76, 0, 4) \\ f &= (8, 19, 70, 1, 2) \\ m &= (88, 3, 0, 9, 0). \end{aligned}$$

Their common contents are

$$Com(d,e) = 4 + 6 + 76 + 0 + 4 = 90$$
$$Com(d,f) = 4 + 6 + 70 + 0 = 2 = 82$$
$$Com(e,f) = 6 + 14 + 70 + 0 + 2 = 92$$
$$Com(f,m) = 8 + 3 + 0 + 1 + 0 = 12.$$

If, then, an artefact is evolving over time with various types satisfying the single-peak model, we can assume that provenances which were formed close together in real time will have more in common than those which were formed at more widely separated times, and vice versa. So from these values the seriation might be started as follows.

First put e and f next each other to form the group $[e, f]$as these have most in common. Secondly add d to this group next to e since it has more in common with e than with f. The result is the seriated group $[d, e, f]$.

Proceeding in this way with all ten of the profiles, an intermediate stage of seriation (partial seriation)

$$[d, e, f],\ [g, h,],\ [i, j, k, l,],\ m$$

is obtained and then, finally, the total seriation $[d, e, f, g, h, i, j, k, l, m]$, ie. the stratigraphic order of the corresponding provenances.

Other measures of *similarity* S may be used to seriate but, whichever is employed, if the model is the single-peak one, then the following rule of Kendall's should hold.

If P, Q, R are three profiles and they satisfy the single-peak model in this order, then S(P,Q) and S(R,Q) are not greater than S(P,R). That is, the profile in the middle is at least as S- similar to the two outside ones as the outside ones are to each other.

Kendall proved that this was satisfied by the common-content similarity. This simple method can be used to seriate sets of profiles which satisfy the single-peak model (see Gelfand (1971)). There can be problems with its use in practice when there are ties among the similarities of pairs of profiles, eg, if the types are rare and the data is sparce. However, this is not the problem that we are concerned with here. The serious problem we are concerned with is when the profiles are *sample* profiles, which they normally are; for them the single-peak model is not necessarily satisfied or even applicable!

3.2 Profiles that do not satisfy the single-peak model in any order

In the original Awatovi data displayed in Table 1 each profile is the average of profiles in many layers at the same level across many 'room-columns' within the mound. We have assumed that they accurately record breakage within the site at the times of breakage and so are good approximations to the site profiles. Table 2 contains the profiles of percentages of the same five ceramic types in the ten layers in *one* of these rooms. It is clear that the profiles are not the same level-by-level as the overall averages given in Table 1. Indeed, the profiles do not satisfy the single-peak model in the stratigraphic or any other order. We regard these as sample profiles of the ceramic change at the Awatovi site and so these percentages need not be exactly, or even close to, the percentages of the various types being broken at the site at the time the rubbish was being abandoned in the mound. The

Proflile	Type 1	Type 2	Type 3	Type 4	Type 5
d	2	0	87	0	11
e	2	4	82	0	12
f	17	60	**19**	4	**0**
g	22	60	**14**	**3**	1
h	50	**30**	**19**	**1**	0
i	**9**	59	27	5	0
j	36	**41**	**0**	23	0
k	**29**	**44**	**0**	27	0
l	35	47	**0**	**18**	0
m	44	30	2	24	0

Table 2: Profiles of percentage of the five types of painted pottery in one room-column within the Western Mound at Awatovi.

percentages in bold show where the single-peak model is not satisfied with the profiles in this stratigraphic order. Several infringements are minor but, for example, the fit is very poor for Type 1 in profile i and Type 2 in h.

The simple method described above in 3.1 has difficulties in seriating such data because it puts together profiles that satisfy the single-peak model in some order and at some point it will come across (sample) profiles that do not. A modification we have employed in the past to seriate in these circumstances has been described in reasonable detail elsewhere (Laxton and Restorick (1989)) and so we will only illustrate it here to bring out the points we are concentrating on in this article. The basic idea is for two or more profiles whose common contents are high to be *averaged* in the expectation that this average approximates the corresponding site profile of the longer period better that the individual profiles do their site profiles of their respective shorter periods. To illustrate this, consider the following intermediate stage in the seriation of the above set of sample profiles using this method. The four seriated groups

$$h,\ [e,d],\ [g,f,i],\ [l,j,k,m]$$

have been formed and now some or all of these are to be merged into a larger seriated group. The average of Group 1 is, of course, h itself. The average of Group 4 is shown at the bottom of the table. It turns out that the two averages of the four with most in common are h and Av[4], with a value of 67.5. To determine where profile h should be placed relative to the four profiles in Group 4, the averages of the so-called 'half-groups' are calculated.

Proflile	Type 1	Type 2	Type 3	Type 4	Type 5
Group 1					
h	50	30	19	1	0
Group 4					
l	35	47	0	18	0
j	36	41	0	23	0
k	29	44	0	27	0
m	44	30	2	24	0
Av[4]	36	40.5	0.5	23	0

Table 3: Intermediate stage of the seriation of the profiles in Table 2

This has been done in Table 4.

The first ordering of these averages satisfies the single-peak model whereas the second does not (altough it only just fails). Consequently the first is selected for the larger seriated group $[l, j, k, m, h]$.

This ordering of the five individual profiles in this group does not satisfy the single peak model although the corresponding ordering with the 'half-group' averages and h does. If the above seriation is continued to the end to a total seriation of all ten profiles, then the result is as shown in Table 6. Again, the places in the columns where the single-peak model is not satisfied are put in bold type. It is clear that the most serious infringment is with Type 3 in profile h.

We remark that a *reordering* of already seriated profiles is always possible with this method if an alternative order of them is a better approximation to the single-peak model (again we refer to Laxton and Restorick (1989)).

4. Prior evidence

The fact that this final, total seriation is not a particulary good result is, in our view, because we are dealing with sample profiles and not site ones. Usually it is impossible to resample archaeological provenances to attempt to improve the fit of the data to the model. Consequently, any other information the archaeologists have on the provenances will be valuable. Now the fact that the program described above goes through stages where various small ordered groups are formed first and then, possibly, enlarged step-by-step, enables us to attempt to do this. For the program can be *started*

Proflile	Type 1	Type 2	Type 3	Type 4	Type 5
Av[l,j]	35.5	44	0	20.5	0
Av[k,m]	36.5	37	1	25	0
h	50	30	19	1	0
h	50	30	19	1	0
Av[l,j]	**35.5**	44	0	20.5	0
Av[k,m]	36.5	37	1	25.5	0

Table 4: Testing two possible orders of h relative to the averages of the two 'half-groups' of Group 4, Av[l,j] and Av[k,m]

Proflile	Type 1	Type 2	Type 3	Type 4	Type 5
Group 5					
l	35	47	0	18	0
j	36	**41**	0	23	0
k	**29**	44	0	27	0
m	44	30	2	24	0
h	50	30	19	1	0

Table 5: The seriation of the merged Groups 1 and 4

with various sets of small ordered groups, with the orders *chosen by the archaeologists.* The formal seriation would then continue from this beginning. These initial partial orders chosen would, presumably, reflect the archaeologists' *prior evidence* concerning the chronological orders of some of the corresponding provenances. This evidence would have to be independent of the evidence of the seriation of the profiles. For example, it could be based on stratigraphy among some of the provenances, or on the Carbon-14 dating of a few of them, etc. Provided this evidence is independent and clearly stated, together with some discussion of the confidence in it, we regard this as a perfectly legitimate approach. The subsequent merging and reorderings within the ordered groups based on the single-peak model will then *interact* with the archaeologists' prior evidence.

For example, although the complete stratigraphy of these ten layers may not be known, the archaeologists may have offered the following evidence.

There is some evidence to suggest that provenances h and i are more or less

Proflile	Type 1	Type 2	Type 3	Type 4	Type 5
d	2	0	87	0	11
e	2	4	82	0	12
i	9	59	27	5	**0**
f	17	60	19	**4**	**0**
g	22	60	**14**	**3**	1
l	35	47	**0**	18	0
j	36	**41**	**0**	23	0
k	**29**	44	**0**	27	0
m	44	30	**2**	24	0
h	50	30	19	1	0

Table 6: Total seriation of the profiles in Table 2, the profiles from a room-column in the mound at Awatovi.

contemporary and chronologically positioned somewhere in the middle of the others. However, there is also some evidence that they have been disturbed at some time.

This evidence is about the relative chronology of some of the provenances and about the site activity (disturbance) and its effect on the relation of the sample profiles found on excavation to the original site profiles formed in the society. It is independent of the above seriation. This evidence can be incorporated into a seriation by *starting* with the following ordered groups;

$$[h, i],\ d,\ e,\ f,\ g,\ j,\ k,\ l,\ m.$$

So to start the seriation there is one ordered group of two profiles and the rest are single profiles with no information about them. Partial seriation and reordering within the subgroups proceeds as before and the final result is the following. The order is different from the one obtained previously in Table 6, where no prior information was used. Notice the difference in the position of profile h. Before it was next to profile m whilst now it is much closer to profile i as the archaeologists had suggested in their evidence. However, it is not next to i as the archaeologists had suggested. Obviously there has been an interaction with both the prior evidence and the single-peak model having an effect on the final ordering. (h has been reordered away from i at some point in the seriation process.) Provided the archaeologists' evidence is discussed and supported there is, in our view, no *a priori* objection to its use. The archaeologists may well choose the order in Table 7 in preference to that in Table 6 as giving a better relative chronological order of the original ceramic site activity. If some other archaeologists do not agree with the prior evidence used here, then the order without prior evidence can be chosen or,

Proflile	Type 1	Type 2	Type 3	Type 4	Type 5
d	2	0	87	0	11
e	2	4	82	0	12
i	9	59	27	5	**0**
f	17	60	19	**4**	**0**
g	22	60	**14**	**3**	1
h	50	**30**	19	**1**	0
l	35	47	**0**	18	0
k	**29**	44	**0**	27	0
j	36	41	**0**	**23**	0
m	44	30	2	24	0

Table 7: Total ordering of the profiles of Table 2 using prior information.

if they have other evidence, this too can be incorporated into the seriation process in a similar fashion. The method and evidence used is clear to all; it uses the model but allows other, *very local*, knowledge about the site and the site profiles to enter into the final assessment of the chronology.

What are the weaknesses in the above approach? Certainly the method of averaging to approximate the site profiles is crude at best. The method of incorporating prior evidence, though not its use, may be inadequate too. To our mind the real weakness in the above is that *degrees of confidence* in the prior evidence cannot be incorporated into the procedure and that its interaction with the subsequent formal seriation using the model cannot be *assessed* adequately in a rational, quantitative manner.

5. Bayesian methods

What other means of estimating the original site profiles are there and how otherwise can we assess the archaeologists evidence together with the evidence of the data and the model? Given that we feel that prior evidence has to be used at times in archaeological analyses, is there a reasonable way of incorporating it with the data and model in an analysis, so that the result is rational and *avoids the accusation of circularity*? One way forward to answer both these questions would appear to be to employ *Bayesian Theory.* This uses personal probabilities about the prior evidence and conditional probabilities concerning the data given a model to arrive at posterior probabilities on the parameters given the data. These are connected by the Bayes' Equation,

$$p(\theta|\mathbf{x}) \propto p(\mathbf{x}|\theta).p(\theta),$$

where;
$p(\theta)$ is the *prior* probability and represents 'how much our belief in the values attached to the parameters, θ, of the model *before* the data, $\mathbf{x}$, was observed',

$p(\mathbf{x}|\theta)$ is *the likelihood* function and is interpreted as to 'how probable is the data set, given specific values of the unknown parameters'

and

$p(\theta|\mathbf{x})$ is the *posterior* probability and is 'how much belief we attach to the possible values of the parameters *after* observing the data.'

It is only relatively recently that reasonable computational methods have become available for estimating the parameters involved in this equation (Gibbs sampler / Monte Carlo methodology) and the theory is only just being introduced into archaeology, by Buck, Cavanagh and Litton (1996) (but see also Cowgill (1993)). Its use in seriation has hardly begun; all we offer here is an illustration of Buck and Litton (1990) for estimating the site profiles from the samples using the single-peak model within the Bayesian methodology. Consider the abundance data given in Table 8; these are

Proflile/Site	Type 1	Type 2	Type 3	Type 4	Type 5	Type 6	Type 7
$1/p_1$	20	3	4	42	18	0	13
$2/p_2$	85	3	12	0	0	0	0
$3/p_3$	26	40	8	0	0	26	0
$4/p_4$	20	1	4	13	58	0	4
$5/p_5$	67	10	23	0	0	0	0
$6/p_6$	26	29	8	3	0	33	1

Table 8: Profiles of percentages of seven types of microliths found at sites in South-East England (see Laxton (1989)).

sample profiles for seven different types of microliths found in six sites in South-East England. The single-peak model is satisfied in the order

$$2, 5, 3, 6, 1, 4,$$

and only this order, and the six profiles can be readily seriated into this order using the simple method outlined first. Because of this we might accept this, or the reverse order, as the chronological order of the original sites. But these sites were not sealed immediately on abandonment, so the proportions of flints actually found and used to construct these sample profiles may not be accurate estimates of the proportions left at the sites when abandoned, ie. the original site profiles. Let, for example,

$$\mathbf{n_1} = (n_{11}\ , n_{12}\ , n_{13}\ , n_{14}\ , n_{15}\ , n_{16}\ , n_{17})$$

be the actual sample found on excavation of Site 1 and

$$\mathbf{p_1} = (p_{11}\ , p_{12}\ , p_{13}\ , p_{14}\ , p_{15}\ , p_{16}\ , p_{17})$$

be the corresponding sample profile of proportions (ie, the first row of the Table 8). This may not be the same as the *unknown* site profile

$$\theta_1 = (\theta_{11}\ , \theta_{12}\ , \theta_{13}\ , \theta_{14}\ , \theta_{15}\ , \theta_{16}\ , \theta_{17})$$

ie. proportions at Site 1 at the time of abandonment several thousands of years ago. Traditionally we seriate the six profiles $\mathbf{p_1}$, $\mathbf{p_2}$, $\mathbf{p_3}$, $\mathbf{p_4}$, $\mathbf{p_5}$, $\mathbf{p_6}$ of the sample data, whereas what we should do is seriate the six site profiles, θ_1, θ_2, θ_3, θ_4, θ_5, θ_6 . It is they that we assert satisfy the single-peak model and not necessarily the sample profiles! But, of course, we do not know the values of the θ_i However, if we can *simulate* them in some manner, given that we do at least know the sample data, then we could seriate them instead if they satisfy the single-peak model in some order. However, in this case the answer may no longer be the sample-profile order, p_2, p_5, p_3, p_6, p_1, p_4, the only in which these six satisfy the model.

Furthermore, if we are able to simulate as suggested, then different sets of six simulated site profiles will be formed satisfying the single-peak model in, possibly, different orders. Hence we will arrive at a whole collection of orders satisfying the single-peak model each of which we envisage could correspond to the original profiles at the site. Consequently, we will not obtain one answer for the order for the six profiles but, rather, a whole distribution of them with associated *probabilities* calculated from the frequences at which they occured in the simulations.

How are these simulations carried out? We use Bayes' rule

$$p(\theta_\mathrm{i}|\mathbf{n}_\mathrm{i}) \propto p(\mathbf{n}_\mathrm{i}|\theta_\mathrm{i})p(\theta_i)$$

for each profile i = 1, ... ,6. A beta distribution is used for the likelihood function;

$$p(\mathbf{n}_\mathrm{i}|\theta_\mathrm{i}) = \prod(\theta_{ij})^{\mathbf{n}}_{\mathbf{ij}}$$

where $\sum_{j=1}^{7}\theta_{ij} = 1$ and $\sum_{j=1}^{7} n_{ij} = n_i$, the total number of all types sampled at site i. For the prior probabilities we choose the Dirichlet distribution (conjugate prior) with parameter vector α_i

$$p(\theta_\mathrm{i}) = \prod \theta_{ij}^{\alpha_{ij}^{-1}}$$

Hence the distribution functions for the seven posterior distributions are

$$p(\theta_\mathrm{i}|\mathbf{x}_\mathrm{i}) \propto \prod \theta_{ij}^{n_{ij}+\alpha_{ij}^{-1}}$$

again a Dirichlet distribution. If, as here in the Buck-Litton example, *vague* or *non-informative* priors are used for the $p(\theta_\mathrm{i})$, then $\alpha_i = 0.5$ is selected for all $i = 1, \ldots, 6$ (see, for example, Gelman et al. (1995), Ch 2). The results are Dirichlet distributions for the posteriors.

Now the Gibbs-sampler methodology is used to simulate distributions for each $\theta_i, i = 1, \ldots, 6$ (Gelman et al. (1995), Ch 11). Then random selections are made from each of these in turn. The selected site profiles $\theta_1, \theta_2, \theta_3, \theta_4, \theta_5, \theta_6$, are then tested to see if they satisfy the single- peak model in some order and, if they do, this seriated order is recorded. If they do not satisfy the model in some order, the selection is discarded. In this particular example the selections were repeated until 1000 orders satisfying the single-peak model were obtained. These orders are listed in Table10 together with their frequency and posterior probabilities. So the unique order in which the orig-

Ordering	Frepuency	Posterior Probability
2, 5, 3, 6, 1, 4	679	0.68 (0.77)
2, 5, 3, 6, 4, 1	128	0.13 (0.04)
2, 5, 6, 3, 1, 4	86	0.09 (0.10)
4, 1, 6, 3, 2, 5	46	0.05 (0.06)
2, 5, 6, 3, 4, 1	28	0.03 (0.01)
3, 6, 5, 2, 1, 4	16	0.02 (0.02)

Table 9: Seriations of the simulated site profiles which satisfy the single-peak model, together with their frequencies of occurrence and posterior probabilities (the six most frequent only are used here).

inal sample profiles satisfied the single-peak model is the most likely one for the simulated site profiles with a probability of 0.68 but it is not the only one order to satisfy the model. There are a number of other orders which represent possible seriations of the site profiles given this self-same data. The next most probable has a value of 0.13 and, as we remarked earlier, the sample profiles do not satisfy the single-peak model in this or any of the other six orders listed in the table.

At present the major drawback to implementing this technique is sampling from the various Dirichlet distributions in such a way that the resulting profiles satisfy the single-peak model; an efficient way of doing this is needed. Also, we have not yet incorporated into the example *non-vague* priors concerning the parameters, though this is not a problem. Finally, we have to include prior information on the chronological order of the corresponding provenances. However, this is not a problem. For example, let us say that the archaeologists judge that they have some evidence, not based on the above seriations, that site 1 was occupied and abandoned at some time between the occupation and abandonment of sites 6 and 4 in the above example. Furthermore, they think that the probability of this is 0.8. Then after a selection which satisfies the model and has θ_1 between θ_6 and θ_4 a random choice with probability 0.8 is made to retain it while for those in which θ_1 is not between θ_6 and θ_4 a random choice with probability 0.2 is made to keep it. The result will be another set of posterior probabilities; in this case those given in brackets in the table. As to whether all prior evidence can be incorporated in this way is an open question. What is also open is how to *elicit* from the archaeologists their prior knowledge in such a way as to convert this into prior probabilities; it is very problem specific!

Finally we point out that we have confined ourselves to the above simple examples to underline the need for a Bayesian approach and how this method works in principle. If, for example, the archaeological data consists of much sparcer profiles, then new difficulties emerge are present (see the forthcoming article by Halekoh and Vach).

References:

BUCK, C. E.; CAVANAGH, W. G. and LITTON, C. D. (1996): The Bayesian Approach to Interpreting Archaeological Data. Wiley, London.

BUCK, C. E. and LITTON, C. D. (1990): A computational Bayes approach to some common archaeological problems. In: K. Lockyear and S. Rahtz (eds.): Computer Applications and Quantitative Methods in Archaeology, BAR International Series 565. *Tempus Reparatum, Oxford, 93-99.*

BURGH, R. F. (1959): Ceramic profiles in the Western mound at Awatovi, Northeastern Arizona. *American Antiquity, 25, 184-202.*

CLARKE, D. L. (1978): Analytical Archaeology. 2nd Ed. Methuen, London.

COWGILL, G. L. (1993): Distinguished Lecture in Archaeology: Beyond Critizing New Archaeology. *American Anthropology, 95(3), 551-573.*

GELFAND, A. E. (1971): Rapid seriation methods with archaeological applications. In: Hodson, F. R.; Kendall, D. G. and Tautu, P. (eds.): Mathematics in the Archaeological and Historical Sciences. Edinburgh University Press, Edinburgh, 186-201.

GELMAN, A.; CARLIN, J. B.; STERN, H. S. and RUBIN, D. B. (1995): Bayesian Data Analysis. Chapman and Hall, London.

HALEKOH, U. and VACH, W. (1996): A Bayesian Approach to Seriation Problems in Archaeology. Technical report, Center for Data Analysis and Model Building, University of Freiburg, to appear.

HODSON, F. R. (1977): Quantifying Hallstatt: some initial results. *American Antiquity, 42, 394-412.*

KENDALL, D. G. (1969): Some problems in statistical archaeology. *World Archaeology, 1, 68-76.*

KENDALL, D. G. (1971): Abundance matrices and seriation in archaeology. *Z. Wahrscheinlichkeitstheorie verw. Geb., 17, 104-112.*

LAXTON, R. R. (1989): Methods of Chronological Ordering. In: Voorrips, A and Ottoway, B. S. (eds.): New Tools in Archaeology, Proc. Conference of I. U.

P. P., Cracow, Poland.

LAXTON, R. R. and RESTORICK, J. (1989): Seriation by similarity and consistency. In: Rahtz, S. and Richards, J. (eds.): Computer Applications and Quantitative Methods in Archaeology, BAR International Series 548, Oxford, England, 215-226.

MADSEN, T. (1988): Multivariate Archaeology, Numerical Approaches to Scandinavian Archaeology, Jutland Archaeological Society Publications, XXI. Aarhus University Press, Aarhus.

PHILLIPS, P.; FORD, J. A. and GRIFFIN, J. B. (1951): Archaeological Survey in Lower Mississippi Alluvial Valley. Peabody Museum Papers, 25.

ROBINSON, W. S. (1951): A method for chronologically ordering archaeological deposits. *American Antiquity, 16, 293-301.*

Acknowledgements

The author would like to thank Dr. C. D. Litton and J. Restorick for their help in preparing this paper.

Probabilistic Aspects of Sequence Repeats and Sequencing by Hybridization

Gesine Reinert

Joint work with Richard Arratia, Daniela Martin and Michael S. Waterman

Department of Mathematics,
University of Southern California, Los Angeles CA 90089-1113, USA

Summary: Sequence repeats are studied in two different contexts: Repeats between two (or more) DNA sequences are employed for sequence comparison; repeats within a single DNA sequence determine whether a sequence is uniquely recoverable from "sequencing by hybridization".

Sequencing by hybridization is a fast and inexpensive tool to determine a DNA sequence from the unordered list of all l-tuples contained in this sequence; typical numbers for l are $l = 8, 10, 12$. For theoretical purposes, we assume that the multiset of all l-tuples is known. This set determines the DNA sequence uniquely if none of the so-called Ukkonen transformations are possible. These transformations require repeats of $(l-1)$-tuples in the sequence, and that these repeats occur in certain spatial patterns.

Using the Chen-Stein method for Poisson process approximation, we approximate the probability that a given DNA sequence is uniquely recoverable from its list of l-tuples; we bound the error and give some numerical examples.

1. Sequencing by Hybridization

One of the primary goals of the Human Genome Project is to increase the rate of DNA sequencing and to reduce its costs. Around 1988, several groups of biologists proposed, independently of each other, a novel technique called "Sequencing by hybridization" (SBH) (see Pevzner and Lipshutz (1994) for references). DNA sequences are built out of four nucleotides, abbreviated A, C, G, and T. These nucleotides bind or hybridize with each other: A and T hybridize, and C and G hybridize. For example, the sequence TGTGT GAGTG hybridizes with ACACA CTCAC. In a sequencing chip, all 4^l possible oligonucleotides ("probes") of length l are attached to the surface of a substrate, each fragment at a distinct location. Today, chips with $l = 8$ are in use, chips with $l = 10$ are under construction, and chips with $l = 12$ may be possible in the near future. To use an SBH chip, the single-stranded target DNA is amplified, labelled by a fluorescent, and exposed to the sequencing chip. The probes on the chip will hybridize to a copy of the single-stranded target DNA if the substring complementary to the probe exists in the target. These probes are then detected with a spectroscopic detector. For example, if $l = 4$, the sequence TGTGT GAGTG will hybridize to the probes ACAC, ACTC, CACA, CACT, CTCA, and TCAC.

As chips can be washed out and used again and due to automatization, this method is not only fast but also inexpensive. Originally developed to sequence DNA, as for today its main use has been to compare different DNA strings in order to detect mutations. In particular, SBH chips are employed in the analysis of HIV blood samples, to decide whether a virus in the sample is a known form of HIV or a new mutation.

There are still technical difficulties in producing an error-free chip; moreover the SBH picture may be diffficult to read. However, even if these sources of errors may be eliminated, a main drawback of the SBH procedure is that more than one sequence may produce the same SBH data. For example, if $l = 4$, the sequence ACACT CACAC will hybridize to the same probes as the sequence ACACA CTCAC.

To control this error resulting from non-unique recoverability, we are interested in an estimate for the probability that a sequence is uniquely recoverable. This probability will depend on the probe length l, on the length m of the target sequence, and on the frequencies of the different nucleotides, A, C, G, and T, in the sequence. Furthermore we need to bound the error made in estimating the probability of unique recoverability in order to make assertions about the reliability of the chip.

2. Unique Recoverability

As a simplification, we assume that we not only know the set of all l-tuples in the sequence but also their multiplicity (but not the order in which they occur). This multiset is called the l-spectrum of the sequence. In the sequel, unique recoverability is understood to mean unique recoverability of a sequence from its l-spectrum.

Pevzner (1989) characterizes unique recoverability from the l-spectrum using the de Bruijn-graph (see van Lint and Wilson (1992)) whose vertices are the $(l-1)$-tuples in the sequence. Two vertices v and w are joined by a directed edge from v to w if the l-spectrum contains an l-tuple for which the first $(l-1)$ nucleotides coincide with v and the last $(l-1)$ nucleotides coincide with w. Pevzner (1989) showed that a sequence is uniquely recoverable from its l-spectrum if and only if there is a unique (Eulerian) path connecting all the vertices. Ukkonen (1992) conjectured and Pevzner (1995) proved that there are exactly three structures that spoil unique recoverability:

1. Rotation. The sequence starts and ends with the same $(l-1)$-tuple. In this case, the de Bruijn-graph is a cycle, and any vertex could be chosen as starting point.

2. Transposition with a three-way repeat. If an $(l-1)$-tuple occurs three times in the sequence, then the de Bruijn-graph has two loops at this vertex, and the order in which these loops are passed is not fixed.

3. Transposition with two interleaved pairs of repeats. There are two "interleaved" pairs of $(l-1)$-tuple repeats, i.e. in the de Bruijn-graph there are two vertices x and y connected by a path of the form $\ldots c \ldots y \ldots x \ldots y \ldots$, where we described a path connecting all the vertices by listing the vertices in the order they are used in the path. This implies that there are two ways of going from x to y in the graph.

Example. The sequence ACACA CTCAC has as 4-spectrum the multiset $\{ACAC, ACAC, CACA, CACT, ACTC, CTCA, TCAC\}$. The competing sequence ACACT CACAC has the same 4-spectrum. The de Bruijn-graph for the sequence ACACA CTCAC has as vertices ACA, CAC, ACT, CTC, and TCA. There are two directed edges from ACA to CAC, and one directed edge each from CAC to ACA, from CAC to ACT, from ACT to CTC, from CTC to TCA, and from TCA to CAC. The competing sequence ACACT CACAC has the same de Bruijn-graph. For the sequence ACACA CTCAC, a path connecting all vertices is ACA, CAC, ACA, (CAC, ACT, CTC, TCA,) CAC. The alternate path ACA, (CAC, ACT, CTC, TCA,) CAC, ACA, CAC, also connecting all the vertices, corresponds to the sequence, ACACT CACAC, with the same 4-spectrum.

Thus unique recoverability can be described in terms of possibly overlapping repeats of $(l-1)$-tuples within a single sequence.

Heuristics for unique recoverability. We model DNA as a random sequence $\mathbf{A} = A_1 A_2 \ldots A_m$, where $A_1, A_2, \ldots$ are independent and identically distributed over the alphabet $\{A, C, G, T\}$. For a sequence to be uniquely recoverable, the event to have an $(l-1)$-tuple repeat should be a rare event. This implies that we consider the occurrence of $(l-1)$-tuples under a Poisson regime. (Note that we are interested in the configuration in which the repeats occur; hence we need a Poisson process approximation for the process of repeats rather than a Poisson approximation for the number of repeats.) If repeats are rare, then three-way repeats are negligible, and so is the probability that a sequence starts and ends with the same $(l-1)$-tuple. After bounding these probabilities, we thus restrict our attention to interleaved pairs of repeats. Under the Poisson regime, if there are k pairs of repeats, then the occurrences of these repeats are discrete uniform. Additional randomization makes the position of the repeats continuously uniform, so that all orderings of these pairs will be approximately equally likely. This allows to apply a combinatorial argument using ! Catalan numbers to obtain that the number of interleaved pairs of repeats, if k repeats are present, is approximately $2^k/(k+1)!$. If λ is the expected number of repeats of l-tuples in a single sequence, we hence get

$$\mathbf{P}(\mathbf{A} \text{ is uniquely recoverable from its } l\text{-spectrum}) \approx e^{-\lambda} \sum_{k \geq 0} \frac{(2\lambda)^k}{k!(k+1)!}. \quad (1)$$

Using the Chen-Stein method for Poisson approximation, we will obtain explicit bounds for the error terms in this approximation and thus make the result applicable for SBH chips. The method is closely related to the treatment of sequence matching in Arratia, Gordon, and Waterman (1990) and Waterman (1995). As it is of independent interest, here is an overview.

3. Sequence Matching

Sequence matching is the problem of finding long matches between two independent sequences. A typical situation is to compare a newly sequenced gene to sequences in a database (using BLAST, e.g.), in order to find statistically significant relationships to other genes. Statistically significant relationships can be used as a hint for biologically relevant relationships, although there is no exact correspondence between the two. For simplicity, we concentrate here on pure matches; for alignments with some mismatches and insertions/deletions see Vingron and Waterman (1994), Waterman(1995).

Erdös and Rényi (1970) give a heuristic that can be applied to sequence matching as follows. Let p be the probability that two random letters match. (Thus $p = 1/4$ if the four letters A, C, G, and T are equally likely.) At any two fixed starting positions, a perfect match of length k (or more) between two independent random sequences $\mathbf{A} = A_1 \dots A_m$ and $\mathbf{B} = B_1 \dots B_m$ has probability p^k. There are roughly $\binom{m}{2}$ starting points for matches of length (at least) k, so the expected number of matches of length k is approximately $\frac{m^2}{2}p^k$. As the longest match should be essentially unique, for the length R_m of the longest match, we should have $1 \approx \frac{m^2}{2}p^{R_m}$, i.e. $R_m \approx 2log_{1/p}(m) + c$, where c is a constant.

To make this intuition precise, Arratia, Gordon, and Waterman (1990) used a Poisson process approximation. The main tool, which we will also employ for the treatment of repeats within a single sequence, is the Chen-Stein method. It originates from Chen (1975); a friendly description is Arratia, Goldstein, and Gordon (1990), and an extensive treatment is Barbour, Holst, and Janson (1992).

The Chen-Stein method. For any two random processes $\mathbf{X}, \mathbf{Y}$ both with values in the same space T, the total variation distance is defined by

$$d_{TV}(\mathbf{X}, \mathbf{Y}) = \sup |\mathbf{P}(\mathbf{X} \in B) - \mathbf{P}(\mathbf{Y} \in B)|,$$

where the supremum is taken over all (measurable) subsets $B \subset T$.

Theorem 1 *Let* $\mathbf{X} \equiv (X_\alpha)_{\alpha \in I}$ *is a process of indicator random variables with* $\mathbf{E}X_\alpha = \mathbf{P}(X_\alpha = 1) = 1 - \mathbf{P}(X_\alpha = 0)$. *Let* $\mathbf{Y} \equiv (Y_\alpha)_{\alpha \in I}$ *be a process with independent Poisson distributed coordinates* Y_α, *with* $\mathbf{E}Y_\alpha = \mathbf{E}X_\alpha$. *Suppose for each* $\alpha \in I$ *there is a "neighborhood of dependence"* $B_\alpha \subset I$ *such that* X_α *is independent of the sigma-algebra generated by all* $X_\beta, \beta \in I \setminus B_\alpha$. *Let*

$$b_1 = \sum_{\alpha \in I} \sum_{\beta \in B_\alpha} \mathbf{E}X_\alpha \mathbf{E}X_\beta, \qquad b_2 = \sum_{\alpha \in I} \sum_{\beta \in B_\alpha, \beta \neq \alpha} \mathbf{E}(X_\alpha X_\beta).$$

Then

$$d_{TV}(\mathbf{X}, \mathbf{Y}) \leq b_1 + b_2.$$

Let t be the test value for the longest pure match between the random sequences $\mathbf{A}$ and $\mathbf{B}$. We are interested in the process of matches: Choose as index set

$$I = \{(i,j) : 0 \le i \le j \le m-t\}, \tag{2}$$

and define $Y_{i,j} = \mathbf{1}(A_{i+1}A_{i+2}\dots A_{i+t} = B_{i+1}B_{j+2}\dots B_{i+t})$. Thus $Y_{i,j} = 1$ is and only if there is a pure match on length t between $\mathbf{A}$ and $\mathbf{B}$ starting at $i+1$ in $\mathbf{A}$ and at $j+1$ in $\mathbf{B}$.

There is a general phenomenon here: matches occur in clumps (see Aldous (1989)). If there is a match at (i,j), then the probability of a match at $(i+1, j+1)$ is p, which is very high compared to the probability p^t of a match at any two given positions. For a Poisson approximation, we will hence only consider the number of clumps of repeats. The distribution of the clump size is independent of the number of clumps (and geometric, see Waterman (1995)). We "declump" by considering only leftmost matches:

$$X_{i,j} = \begin{cases} Y_{i,j} & \text{if } i=0 \text{ or } j=0 \\ (1-\mathbf{1}(A_i = B_j))Y_{i,j} & \text{otherwise.} \end{cases}$$

Then the expected number λ of clumps is $\lambda = (2m-2t+1)p^t + (m-t)^2(1-p)p^t$. For $\alpha = (i,j) \in I$, define as neighborhood of dependence

$$B_\alpha = \{(i',j') \in I : |i-i'| \le t \text{ or } |j-j'| \le t \text{ or } |i-j'| \le t \text{ or } |j-i'| \le t\}. \tag{3}$$

Then $|B_\alpha| \le 2(2t+1)m$, and B_α satisfies the condition of Theorem 1, so that to establish a Poisson process approximation for the process $\mathbf{X} = (X_\alpha)_{\alpha\in I}$ we only have to bound b_1 and b_2. Bounding b_1 is immediate:

$$b_1 = \sum_{\alpha\in I}\sum_{\beta\in B_\alpha} \mathbf{E}X_\alpha \mathbf{E}X_\beta < \sum_{\alpha\in I} 2(2t+1)mp^t = 2\lambda(2t+1)mp^t.$$

For b_2, the situation is more complicated. Pick an $\alpha = (i,j) \in I$ and a $\beta = (i',j') \in B_\alpha$. We classify two different situations according to their "degree of overlap"; see Waterman (1995), p. 284 for the corresponding pictures.

1. The "crabgrass" case. If $|i-i'| \le t$ and $|j-j'| > t$, or if $|i-i'| > t$ and $|j-j'| \le t$, then we are in the so- called "crabgrass" case. Suppose $|i-i'| \le t$ and $|j-j'| > t$ (the other cases can be treated similarly). For $X_\alpha X_\beta$ to be 1, each index in $\{i, i+1, \dots, i+t-1\} \cap \{i', i'+1, \dots, i'+t-1\}$ has to satisfy two matching conditions, the remaining indices have to satisfy one matching condition each. Let $\theta = |\{i, i+1, \dots, i+t-1\} \cap \{i', i'+1, \dots, i'+t-1\}|$, and let $p_3 = P($ three random letters match$)$. Ignoring the declumping factor and using Hölder's inequality (see Waterman (1995), p. 285), we obtain the inequality $\mathbf{E}(X_\alpha X_\beta) \le p_3^\gamma p^{2t-2\theta} \le p^{3t(\frac{1}{2}+3\delta)}$, where $\delta \in (0, \frac{1}{6}]$ is a constant that arises from Hölder's inequality. As the number of indices in the crabgrass case can be bounded by $2(2t+1)m$,

$$\sum_{\alpha,\beta\in \text{"crabgrass"}} \mathbf{E}(X_\alpha X_\beta) \leq 2(2t+1)mp^{3t(\frac{1}{2}+3\delta)}. \tag{4}$$

2. The "accordion" case. If both $|i-i'| \leq t$ and $|j-j'| \leq t$, then we are in the so-called "accordion" case. In this case, both $\{i, i+1, \ldots, i+t-1\} \cap \{i', i'+1, \ldots, i'+t-1\}$ and $\{j, j+1, \ldots, j+t-1\} \cap \{j', j'+1, \ldots, j'+t-1\}$ are non-empty. Now we cannot so easily count the number of matchings required for the different indices for $X_\alpha X_\beta$ to be 1; the upper bound for the matches required for one letter is $2t+1$. Using Hölder's inequality again, we obtain $\mathbf{E}(X_\alpha X_\beta) \leq p^{t(\frac{1}{2}+2\gamma)+\frac{1}{2}+\gamma}$, where $\gamma \in (0, \frac{1}{2} - \frac{1}{2t+1}]$ is a constant. Note that there are at most $(2t+1)^2$ indices in the accordion case; hence

$$\sum_{\alpha,\beta\in \text{"accordion"}} \mathbf{E}(X_\alpha X_\beta) \leq (2t+1)^2 p^{t(\frac{1}{2}+2\gamma)+\frac{1}{2}+\gamma}. \tag{5}$$

Combining (4) and (5), we get

$$b_2 < 2(2t+1)mp^{3t(\frac{1}{2}+3\delta)} + (2t+1)^2 p^{t(\frac{1}{2}+2\gamma)+\frac{1}{2}+\gamma}.$$

If $t = 2log_{\frac{1}{p}} m + c$, as suggested in the heuristic, then we obtain

$$b_1 + b_2 = O((logn)/n) + O((logn)^2/(n^{2\gamma})) + O((logn)/(n^{6\delta})). \tag{6}$$

This estimate on the total variation distance between the process of leftmost matchings and a corresponding Poisson process can be given in explicit form involving the above constants, see Waterman (1995), p. 288.

For sequence repeats, we will follow the same strategy.

4. Poisson Approximation for Sequence Repeats

Poisson approximation for sequence repeats have been obtained by Zubkov and Mikhailov (1974), and by Karlin and Ost (1987), but neither of them give bounds on the error in the approximation. In order to make our approximation of the probability of unique recoverability applicable (l being small), we need bounds. To this purpose, we apply the Chen-Stein method as above. Let again $\mathbf{A} = A_1 \ldots A_m$ be a random sequence consisting of independent identically distributed letters, and let p be the probability that two random letters match. We write t for $l-1$, as we are interested in $(l-1)$-repeats. As index set we choose I as in (2), and again we have to declump: Define $X_{i,i} = 0$ for all i, and

$$X_{i,j} = \begin{cases} \mathbf{1}(A_1 \cdots A_t = A_{j+1} \cdots A_{j+t}) & \text{if } i = 0 \\ (1 - \mathbf{1}(A_i = A_j))\mathbf{1}(A_{i+1} \cdots A_{i+t} = A_{j+1} \cdots A_{j+t}) & \text{otherwise.} \end{cases}$$

Thus $X_{i,j} = 1$ if and only if there is a leftmost repeat starting after i and j. A careful analysis (see Arratia, Martin, Reinert and Waterman (1996)) yields that the process $\mathbf{X} = (X_\alpha)_{\alpha\in I}$ is sufficient to decide whether a sequence is uniquely recoverable from its l=spectrum (although $\mathbf{X}$ contains strictly less information than the process $(R_\alpha)_{\alpha\in I})$).

For a Poisson process approximation, we first identify the expected number λ of leftmost repeats. If $\alpha = (i,j)$ does not have self-overlap, that is, if $j - i > t$, then

$$\mathbf{E}(X_\alpha) = \begin{cases} p^t & \text{if } i = 0 \\ (1-p)p^t & \text{otherwise.} \end{cases}$$

Hence the expected number λ^* of repeats without self-overlap is

$$\lambda^* = \binom{m-2t}{2}(1-p)p^t + (m-2t)p^t.$$

If α does have self-overlap, we are in a situation similar to the crabgrass case above: In order to have a leftmost repeat at α, for indices in the overlapping set, two matches are required, and for indices in the non-overlapping set, one match is required. Let $d = j - i$, then $E(X_\alpha)$ depends on the decomposition of $t + d$ into a quotient q of d and a remainder r (such that $t + d = qd + r$): if p_q is the probability that q random letters match, then

$$\mathbf{E}(X_\alpha) = \begin{cases} p_{q+1}^r p_q^{d-r} & \text{if } i = 0 \\ (p_q - p_{q+1})^r p_q^{d-r} & \text{otherwise.} \end{cases}$$

If λ^* is bounded away from 0 and infinity, which corresponds to having $t = 2log_{1/p}(m) + c$ for some constant c, then it can be seen that

$$\lambda \approx \frac{m^2}{2}(1-p)p^t.$$

The next ingredient needed to apply Theorem 1 is the neighborhood of dependence. For $\alpha = (i,j)$, we define the neighborhood B_α as in (3). Again, B_α satisfies the condition of Theorem 1. For a Poisson process approximation, it remains to bound b_1 and b_2 from Theorem 1. In comparison to sequence matching there are additional complications. We have seen that even for calculating the expected number of repeats we are already in the crabgrass case, due to repeats allowing self-overlap. This makes the bounds on b_1 more complicated; Hölder's inequality gets employed as well as, for the rare cases with overlap of all indices, the crude bound

$$\mathbf{E}(X_\alpha) \leq \xi_*^t,$$

where $\xi_* \equiv \max_{a\in\{A,C,G,T\}} P(A_1 = a)$. The situation gets worse for b_2. Not only do we have to deal with accordion structures, but it is also possible that "cycles" occur: If $i = 1, j = 2 = i', j' = 4$, and $t = 4$, for example, for $E(X_{i,j}X_{i',j'})$ to be 1 we need in particular that $A_2 = A_3 = A_4$ for $X_{i,j}$ to be 1, and that $A_2 = A_4$ for $X_{i',j'}$ to be 1. Thus the number of matches required between k letters is not always $k-1$, as in the accordion case, but may also be k (for cycles). To still obtain applicable bounds, a detailed analysis of the structure of these cycles is needed, and it results in a lengthy expression. The final result on $b_1 + b_2$ is in Arratia, Martin, Reinert and Waterman (1996). The calculations simplify in the uniform case (all letters are equally likely) (see also Dyer, Frieze and Suen (1994)), because a lot of additional independence results from the uniformity. Under the regime that λ is bounded away from 0 and infinity, here is a short version of the general result in Arratia, Martin, Reinert and Waterman (1996).

Theorem 2 *Let* $\mathbf{Y} \equiv (Y_\alpha)_{\alpha \in I}$ *be a process with independent Poisson distributed coordinates* Y_α, *with* $\mathbf{E}Y_\alpha = \mathbf{E}X_\alpha, \alpha \in I$. *Then*

$$d_{TV}(\mathbf{X}, \mathbf{Y}) \le b(m,t) \sim \begin{cases} 16\lambda^2 \frac{t}{m} & \text{in the uniform case} \\ m\xi_*^t & \text{in the nonuniform case.} \end{cases}$$

In Arratia, Martin, Reinert and Waterman (1996), a more general result is derived for general alphabets, and explicit bounds are obtained. These bounds can be used to approximate the probability of unique recoverability.

5. Numerical Results for Unique Recoverability

After making the heuristic for unique recoverability precise and bounding the other errors in the approximation, we get a computable estimate $f(\lambda)$, given by (1), for the probability that a random sequence is uniquely recoverable from its l-spectrum, and we get computable bounds on the error associated with this approximation. Recall that λ is the expected number of leftmost repeats, m is the sequence length, and l is the length of the probes on the SBH chip. We denote by R the overall bound on the error in the approximation and by R_2 the contribution of the Poisson process approximation to the error. The first table summarizes the approximation for some values of m and l in the uniform case, so that all letters A, C, G, and T, are equally likely to be chosen. For l we are interested in $l = 8, 10$, or 12, as these chips either exist or may come into existence in the near future. In the first block, we aim for m such that we can give a 95% guarantee of unique recoverability. Due to the discreteness of the problem, 95% cannot be reached exactly; we chose the first value of m such that $f(\lambda) - R \ge .95$. For the second block, the focus lies in the actual limit probability $f(\lambda)$ to be roughly .95. The last block gives the results for $m = 180, l = 8$ and $m = 2450, l = 12$ that are obtained in Pevzner et al. (1991) using simulations.

In the next table, we consider a rather nonuniform case. The distribution of A, C, G, and T in the complete chloroplast genome of the liverwort Marchantia Polymorpha (see Arratia, Gordon and Waterman (1990)) is as follows: A

m	l	$f(\lambda) - R$	λ	$f(\lambda)$	R	R_2
85	8	.9525	.1422	.9967	.0442	.0346
469	10	.9503	.3037	.9861	.0357	.0320
2288	12	.9501	.4638	.9693	.0191	.0179
169	8	.6458	.6068	.9499	.3040	.2808
659	10	.8559	.6059	.9500	.0941	.0883
2615	12	.9218	.6066	.9499	.0281	.0266
180	8	.5707	.6916	.9368	.3660	.3406
2450	12	.9372	.5322	.9605	.0232	.0219

Table 1: The uniform case

m	l	$f(\lambda) - R$	λ	$f(\lambda)$	R	R_2
17	8	.9556	.0123	.99997	.0442	.0360
69	10	.9507	.0267	.9998	.0491	.0401
378	12	.9501	.0746	.99909	.0489	.0393
97	8	-1.523	.6043	.9503	2.474	2.367
321	10	.0093	.6069	.9499	.9405	.8828
1081	12	.6239	.6068	.9499	.3260	.2964

Table 2: A nonuniform case

has probability .3544, C has probability .1430, G has probability .1451, and T has probability .3575. Again we report guaranteed and estimated 95% probabilities. There are no simulated values in the literature to compare our results with.

More values (and more quantiles) can be found in Arratia, Martin, Reinert and Waterman (1996).

6. Discussion

In the uniform case, the computed value for $l = 12$ is relatively close to simulated values; moreover, our values are mathematically rigorous. The situation is less satisfying in a strongly nonuniform case, but, as these are the only results available so far, we still consider them of value. The decomposition of the error terms shows that the main contribution comes from the Poisson process approximation; we do not see how to considerably improve our bounds using the Chen-Stein method.

Furthermore, in our analysis we make some idealizing assumptions Firstly, we suppose to know not only the SBH data but the whole underlying multiset of l-tuples. However, from the structure of the de Bruijn-graph it is plausible that multiplicities are reconstructable. Moreover, we have not yet addressed data with errors. As even the underlying SBH chip may not be error-free, and both false-positives and false-negatives may occur, this is a very complex problem. Instead of using and i.i.d. model, a Markov model might be better to model DNA sequences, but the computations would become even more difficult and hence we expect the bounds to become much worse and thus be without practical relevance.

For future research, a natural extension of our result would be to consider "partial"recovery of sequences, such as the length of the longest segment that is uniquely recoverable. Moreover, SBH chips described above have a high redundancy. In practice, often different chip designs are used, employing cross-grouping of the nucleotides (see Pevzner and Lipshutz (1994)); the probabilistic analysis of these chips remains to be done.

References:

ALDOUS, D. (1989): Probability Approximations via the Poisson Clumping Heuristic. Springer, New York.

ARRATIA, R.; GOLDSTEIN, L. and GORDON, L. (1990): Poisson approximation and the Chen-Stein method. *Statistical Science, 5, 403–434.*

ARRATIA, R.; GORDON, L. and WATERMAN, M. S. (1990): The Erdös-Rényi law in distribution, for coin tossing and sequence matching, *Ann. Statist., 18, 539–570.*

ARRATIA, R.; MARTIN, D.; REINERT, G. and WATERMAN, M. S. (1996): Poisson process approximation for sequence repeats, and sequencing by hybridization. *J. Comp. Biology, 3, 425–463.*

BARBOUR, A. D.; HOLST, L. and JANSON, S. (1992): Poisson Approximation. Clarendon, Oxford.

CHEN, L. H. Y. (1975): Poisson approximation for dependent trials. *Ann. Probab., 3, 534-545.*

DYER, M.; FRIEZE, A. and SUEN, S. (1994): The probability of unique solutions of sequencing by hybridization. *J. Comp. Biology, 1, 105–110.*

ERDÖS, P. and RÉNYI, A. (1970): On a new law of large numbers. *J. Anal. Math., 22, 103- 111.*

KARLIN, S. and OST, F. (1987): Counts of long aligned word matches along random letter sequences. *Adv. Appl. Prob., 19, 293–351.*

VAN LINT, J. H. and WILSON, R. M. (1992): A Course in Combinatorics. Cambridge University Press.

PEVZNER, P. et al. (1991): Improved chips for sequencing by hybridization. *J. Biomolecular Structure & Dynamics, 9, 399–410.*

PEVZNER, P. (1995): DNA physical mapping and alternating Eulerian cycles in colored graphs. *Algorithmica, 13, 77–105.*

PEVZNER, P. and LIPSHUTZ, R. J. (1994): Towards DNA sequencing chips. *19th Symposium on Mathematical Foundations fo Computer Science, Kosice, Slocakia, Lecture Notes in Computer Science, 841, 143–158.*

UKKONEN, E. (1992): Approximate string-matching with q-grams and maximal matches. *Theoretical Computer Science, 92, 191–211.*

VINGRON, M. and WATERMAN, M. S. (1994): Sequence comparison significance and Poisson approximation. *Statistical Science, 9, 367–381.*

WATERMAN, M. S. (1995): Introduction to Computational Biology. Chapman & Hall, Cambridge.

ZUBKOV, A. M. and MIKHAILOV, V. G. (1974): Limit distributions of random variables associated with long duplications in a sequence of independent trials. *Theory Prob. Appl., 19, 172–179.*

On the Equivalence of Two Tree Mapping Measures

O. Eulenstein[1] and M. Vingron[2]

[1] Dept. of Computer Science, Research Group of Prof. Lengauer, University of Bonn, D-53117 Bonn, Germany.

[2] Theoretical Bioinformatics Devision, German National Cancer Research Center (DKFZ), D-69120 Heidelberg, Germany

Summary: *Tree-mapping* is a method of deducing the evolutionary history of a set of genes, given two seemingly contradictory phylogenetic trees for the genes and for the species in which the genes were found. The method is based on the postulation of gene duplication events which then allow to resolve the contradiction between the phylogenetic trees. To quantify the difficulty of this reconciliation two biologically meaningful measures have been introduced in the literature. Mirkin, Muchnik, and Smith conjecture that the two measures are in fact equivalent.
We introduce the problem of comparing trees under the assumption of duplication events. Finally a short overview of the proof of the Mirkin-Muchnik-Smith conjecture will be given.

1. Introduction

It is today generally accepted that any two forms of life on earth have evolved from a common ancestor (Smith (1958), Li and Graur (1991)). One aim of evolutionary biology is the reconstruction of the evolutionary history of current species. Based on the assumption of common ancestors this history can be depicted as a tree, generally called a phylogenetic tree. Its nodes correspond to hypothetical ancestral species and its edges are lines of descent.

Together with a species, its various characters evolved according to the same evolutionary tree. Thus every character, e.g. the shape of the skull, reflects the evolutionary history of the species. Studying the history of a character is the main source of information for the reconstruction of evolutionary relationships between species. However, it is of prime importance to study characters which are in fact comparable. A fly and a bird both have wings and yet the bird is not more closely related to insects than to other vertebrates. The wings of birds and flies are incomparable characters because they do not share a common evolutionary origin. Deducing history from this selection of characters will lead to results contradicting our current knowledge about evolution.

With the rise of molecular biology the DNA sequences of genes have become available. They can be treated as characters from which to deduce phylogenetic trees (see e.g. Fitch and Margoliash (1967), Nei (1987), Felsenstein (1988)). However, deciding which genes actually are comparable may be problematic. There exist large families of genes having several distinct copies in many species. In order to derive a tree that correctly reflects the evolution of species and not of the genes of this particular family, one needs precise knowledge as to which copies of the gene are the comparable ones. Since this is generally only possible after careful study of the entire family, the tree derived from a selection of genes from a gene family and the tree describing the evolution of species will frequently disagree. We will call a tree derived from a set of genes the *gene tree* and the tree according to which we assume that species have evolved the *species tree.* We assume that the gene tree contains exactly one gene from each of the species in the species tree.

Gene families evolve through duplication of genes. Thus, possible discrepancies between a gene tree and a species tree can be explained by postulating duplication events that gave rise to different copies of a gene. For alternative explanation see Nei (1987). *Tree-mapping* compares a gene tree and species tree on postulating duplications. The method was first applied by Czelusniak et al. (1979) and Nelson et al. (1981). More recently it has been carefully developed by Page (1994). Consider the example in Fig. 1 (see, e.g. Graur and Li (1991)): The family of hemoglobin genes in vertebrates contains, among others, two types of genes: α-hemoglobin and β-hemoglobin. Both types evolved from an ancestral hemoglobin that existed prior to the vertebrates. This ancestral gene then was duplicated and the two new copies gave rise to vertebrate α- and β-hemoglobins, respectively. A researcher studying the α-hemoglobins from human, chimpanzee, and horse will find that human and chimpanzee have a common ancestor which in turn has a common ancestor with the horse. If the researcher studied β-hemoglobins from the same set of species he would find the same result. Were this family not as well-studied as it is today, the researcher might, however, have chosen a β gene from chimp and α genes from human and horse as the set to base his analysis on. Consequently he would have found that human and horse group together versus the chimp which appears to be of older evolutionary origin. While this is correct for the particular selection of genes, it is an uncommon view as to the development of species.

2. Mapping function & duplication nodes

Tree-mapping makes use of gene tree and species tree to postulate one or more duplication events in the history of the gene. These duplications are chosen such that the gene tree then becomes interpretable in terms of the evolution of species as given in the species tree and these duplication events. Duplication events are postulated based on a function M, called the *mapping function.* M maps a gene (node in the gene tree), be it contemporary or ancestral, to the most recent species that must have contained that gene. This species will be represented by the node of the species tree (node in the species tree) which is followed by all species of which we know that they contained the gene in question or its descendent genes. If gene b is a child of

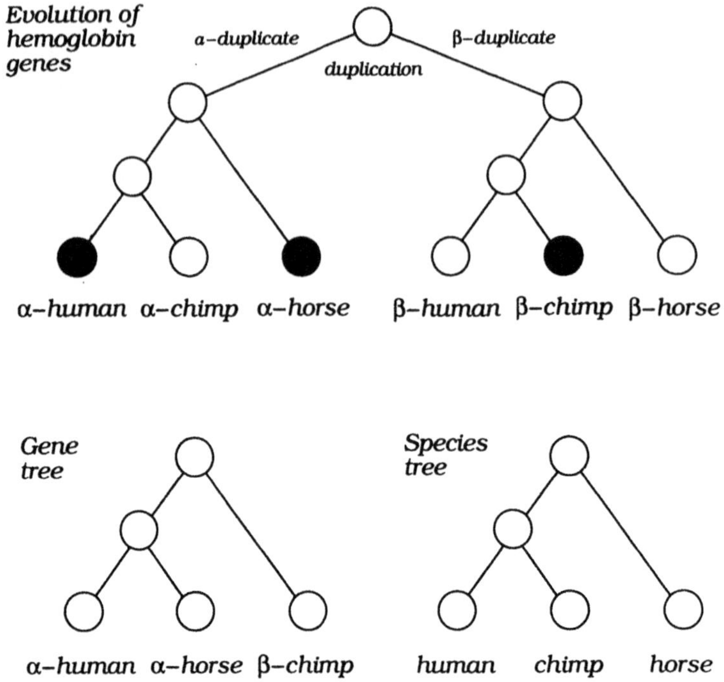

Figure 1: A duplicated gene and its history

gene a in the gene tree and $M(a)$ and $M(b)$ coincide, a duplication must have occurred and b descended from a duplicate of a. We therefore call a a *duplication node.* Such a situation is depicted in Fig. 2, taking up the hemoglobin example again. Suppose a researcher has derived the gene tree clustering together two hemoglobin genes from human and horse versus a third one from chimpanzee. Of course the ancestral gene a to all three hemoglobin genes must have been present in the ancestor x to human, chimpanzee, and horse. From the comparison of gene tree and species tree the researcher can also deduce that gene b, the gene ancestral to human and horse genes, must have been present already in that same species x. Formally, M maps both a and b to species x, thus implying the existence of two hemoglobin genes that are copies of a in x (as depicted in Fig. 1). From one of the copies (b) the α-hemoglobin has evolved while the other one gives rise to the chimp β-hemoglobin.

When dealing with many genes and species the number of duplications needed in order to explain a gene tree from a given species tree may be large. In fact, the number of duplications is used as an asymmetric distance measure between two trees by Czelusniak et al. (1979). It has subsequently been recognized, though, that the postulation of a duplication also implies the existence of a certain set of genes which are not observed in the given data set (see, e.g., Page (1994)). In our example those are the β-hemoglobin genes from human and horse and the α-hemoglobin gene from chimp. Were the example more involved the unobserved genes might not constitute sin-

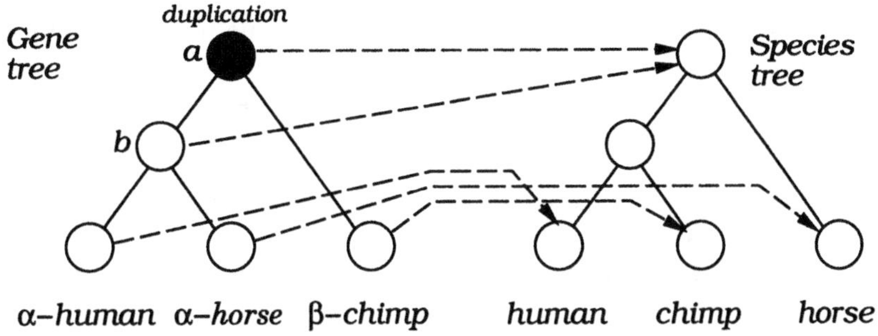

Figure 2: Mapping M between a gene tree and a species tree.

gle lineages but entire subtrees. We adopt the convention to count such a subtree only once, independent of its size. It is thus only the development of the unobserved ancestral gene that is accounted for, not necessarily each unobserved contemporary gene. Of course, the number of unobserved genes will grow with the number of duplications. Mirkin et al. (1996) study two (asymmetric) dissimilarity measures for a gene and a species tree. Both account for the number of duplications as well as for the number of unobserved subtrees. Section 3 introduces the two measures and gives an example.

Although formally different the measures agree on many examples. This lead Mirkin et al. (1996) to conjecture that they are in fact identical. Eulenstein and Vingron (1995) present a proof of their conjecture. Section 4 gives an overview of this (rather lengthy) proof.

3. The two measures

Given a gene tree G and a species tree S, Mirkin et al. (1996) discuss two inconsistency measures L and C to quantify the difference of G and S. Both measures have a biologically different interpretation. One summand of either measure is the number of postulated duplications. L and C differ in how they account for deduced but unobserved gene lineages. Section 3.1 and 3.2 give an informal description of the two measures. In section 3.3 we present an example for these measures.

3.1 The measure L

For the L-measure we count the number of postulated duplications and add for each inner node a of G a value $l(a)$. $l(a)$ counts possibly unobserved genes in the lineages from a to both of its children. In the following the two children of a are called a_c and $a_{\bar{c}}$, respectively. a is contained in species $M(a)$ and therefore develops in the subtree of $M(a)$. For example the descendent species $M(a_c)$ of $M(a)$ contains a_c. Consider the path in the species tree $p(M(a), M(a_c))$ from $M(a)$ to $M(a_c)$ leaving out $M(a)$ and $M(a_c)$. A species on $p(M(a), M(a_c))$ possibly contains an unobserved gene that is descending from a and an ancestor of a_c. These possibly unobserved genes are counted by $l(a)$. There is possibly another unobserved gene if a was duplicate which is counted by 1 as defined below. The following three cases need to be distinguished:

(i) $M(a) = M(a_c) = M(a_{\bar{c}})$.
$l(a) := 0$.
a is called a *double copy node* because both inferred copies of the gene a are observed in the gene tree.

(ii) For one of the children of a, say a_c, $M(a) = M(a_c)$ and $M(a_{\bar{c}}) \neq M(a)$.
$l(a)$ is defined as 1 plus the number of nodes between $M(a)$ and $M(a_{\bar{c}})$ (extending $M(a)$, $M(a_{\bar{c}})$ themselves).
a is called a *single copy node* because exactly one of its inferred copies is observed in the gene tree.

(iii) a is not a duplication node, i.e. $M(a) \neq M(a_c)$ and $M(a) \neq M(a_{\bar{c}})$.
$l(a)$ is defined as the number of nodes on $p(M(a), M(a_c))$ and $p(M(a), M(a_{\bar{c}}))$.

3.2 The measure C

The C-measure, like the L-measure, accounts for the number of postulated duplications. Additionally, for each duplication node a a value $c(a)$ is added. We call the duplicated genes of a its *copies*. For calculating $c(a)$ consider the subtree of S rooted at $M(a)$. We denote this subtree by $T_S(M(a))$. Since a is a duplication node, its copies must be contained in $M(a)$. We use "+" to mark genes descending from one of the copies and "−" for genes descending from the other copy. Due to the one-to-one correspondence between leaves of G and S one can label the leaves of $T_S(M(a))$ with "+" or "−" according to which of the copies its given gene has descended from. This is achieved by assigning "+" to, say, a_c and letting it percolate down to the leaves under a_c. Analogously one proceeds for "−" and $a_{\bar{c}}$. The resulting labeling of leaves in $T_G(M(a))$ can be carried over to yield the correct labeling of leaves of $T_S(M(a))$. Thus, a subset of the leaves of $T_S(M(a))$ will be lapelled either "+" or "−". $c(a)$ counts the maximal subtrees of $T_S(M(a))$ containing either no "+"-leaf or "−"-leaf.

3.3 An example

We will present an example to show that the equivalence of the two measures is not trivial.

We denote inner nodes of the species tree with capitals and inner nodes of the gene tree in lower case. Nodes of the species tree are annotated with the mapping M: Next to the node there is a list of nodes of the gene tree which all map to this node in the species tree. Nodes a, b, c, and d in G are duplication nodes. We use the same convention as in Mirkin et al. (1996) of designating one duplicate of a gene with "+" and the other with "−".

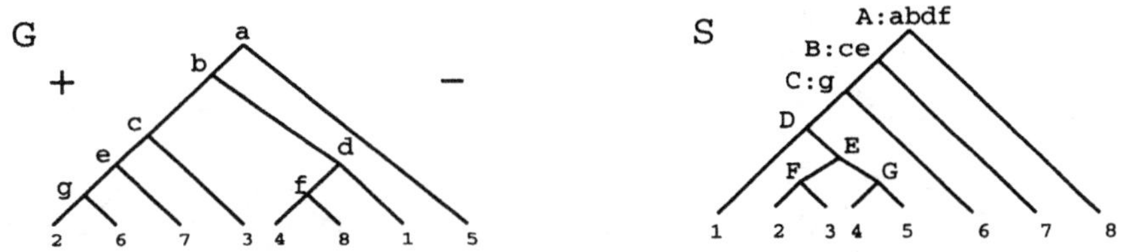

Figure 3: Gene and species tree

3.3.1 Calculating L(G, S): In Table 1 we calculate the measure $l(a)$ for each inner node of G. Consider the first line, corresponding to node a. The children of a are $a_c = 5$ and $a_{\bar{c}} = b$. These map to A and 5, respectively (Fig. 3). Since $M(a) = M(a_c) \neq M(a_{\bar{c}})$, case (ii) of the definition in Section *3.1* applies. This is denoted in the second column of the table. 1 will be added to the path-measure p. The path length $p(a_c)$ counts the nodes within the path $A - B - C - D - E - G - 5$, not counting the ends A and 5. Thus according to case (ii) $l(a) = 5 + 1 = 6$. Table 1 shows the results for the remaining nodes. Summing over the last column yields 24.

3.3.2 Calculating C(G, S): Figure 4 shows four copies of S. Consider the first one of these, associated with the duplication node a. The gene a is duplicated, giving rise to a "+"-copy and a "−"-copy. The gene tree tells us that species 5 inherits a successor of the "−"-copy while all others have a successor of the "+"-copy (cf. the "+/−"-pattern at the leaves of the tree). Now, nodes A, B, C, D, E, and G have both "+" and "−" copies underneath them. The subtrees having leaves with either "+" or "−" are the one connecting 2, 3 and F, the one consisting only of species 4, and so on. In the figure these subtrees have been marked by leading to a full square. There are 7 subtrees, thus, $c(a) = 7$ (see Table 1, last column). The sum of these for all duplication nodes is 24.

	$L - measure$				$C - measure$
Vi_G	$\in V_1$	$p(a_c)$	$p(a_{\bar{c}})$	$l(a)$	$c(a)$
a	1	5	0	6	7
b	1	0	0	1	6
c	1	4	0	5	6
d	1	3	0	4	5
e	0	0	0	0	0
f	0	5	0	5	0
g	0	3	0	3	0
$\sum_{a \in Vi_G}$				24	24

Table 1: Calculating $L(G, S)$ and $C(G, S)$. V_1 represents the set of single duplications.

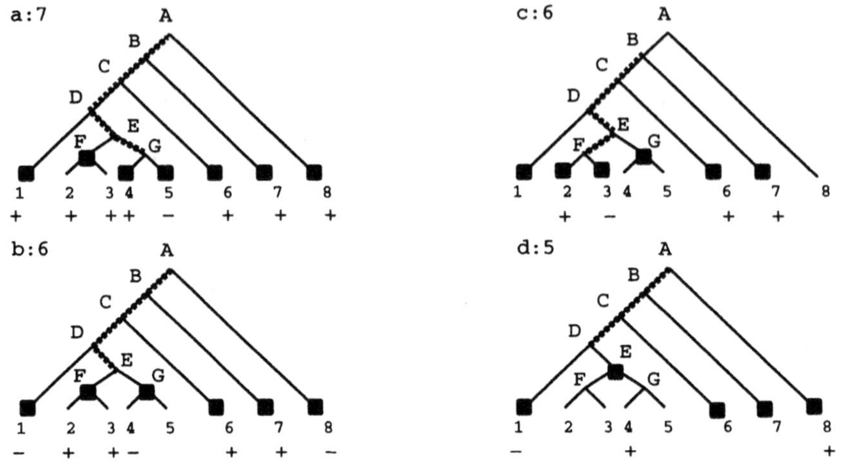

Figure 4: Calculating $C(G,S)$

4. Equivalence of L and C measures

We now give a short outline of the proof of equivalence of the two measures. The first idea in proving the equivalence would be to check whether $l(a)$ and $c(a)$ agree for all nodes a in the gene tree. Table 1 shows that this is not the case. In particular, a node which is not a duplication node (e.g. node f) will have a positive contribution in the measure L and contribute nothing in C.

In our approach (Eulenstein and Vingron (1995)) we first reformulate both measures such that for a given node in G specific edges in S are counted. Both c and l thus imply mappings from the nodes of G to the edges of S. Instead of keeping the node in G fixed while counting edges in S we keep an edge in S fixed and ask how many nodes in G are mapped onto this edge. The thrust of the proof is to show that under both l and c these numbers agree. This is done using certain structural properties of the node sets in G which are mapped onto a specific edge by either of the measures. For example, a certain important subset of nodes in the gene tree which count an edge under the L–measure form an antichain in G. On the other hand, a certain important subset of nodes in the gene tree counting an edge under the C–measure has a succeeding node in the L–measure counting the same edge. Based on such observations the cardinality of these node sets under L and under C can be shown. Thus after summing over all edges in S this yields the proof of the equivalence of the two measures.

References:

CZELUSNIAK, J.; GOODMAN, M.; MATUSTA, G. MOORE, G. and ROMERO–HERRERA, A. (1979): Fitting the gene lineage into its species lineage: A parsimony strategy illustrated by cladograms constructed from globin sequences. *Syst. Zool., 28, 132–168.*

EULENSTEIN, O. and VINGRON, M. (1995): On the equivalence of two tree mapping measures. "Arbeitspapiere der GMD", No. 936, Germany.

FELSENSTEIN, J. (1988): Phylogenies from molecular sequences: Inference and reliability. *Annu. Rev. Genet., 22, 521–565.*

FITCH, W. and MARGOLIASH, E. (1967): Construction of Phylogenetic Trees. *Science, 155, 279–284.*

GORDON, A. (1987): A Review of Hierarchical Classification. *Journal of Royal Statistical Society, 150, 119–137.*

LI, W. and GRAUR, D. (1991): Fundamentals of Molecular Evolution. Sinauer Associates, Inc., Massachusetts.

GUIGÓ, R.; MUCHNIK, J. and SMITH, T. F. (1996): Reconstruction of Ancient Molecular Phylogeny. *Mol. Phylog. Evol.* (to appear).

MIRKIN, B.; MUCHNIK, I. and SMITH, T. F. (1996): A Biologically Consistent Model for Comparing Molecular Phylogenies. *Journal of Computational Biology* (to appear).

NEI, M. (1987): Molecular Evolution Genetics. Columbia University Press, New York.

NELSON, G. and PLATNIK, N. L. (1981): Cladistics and vicariance. Columbia University Press.

PAGE, R. D. (1994): Maps between trees and cladistic analysis of historical associations among genes, organisms, and areas. *Systematic Biology, 43, 58–77.*

SMITH, J. M. (1958): The Theory of Evolution. Penguin Books Ltd, England.

Deriving Grammars from Large Corpora

Sven Naumann, Kathrin Gieseking

Department of Computational Linguistics,
University of Trier, D-54286 Trier, Germany

Summary: We propose a method for generating grammars for natural language that applies an inductive learning algorithm to large corpora. This learning procedure incrementally parses the corpora using a linguistically motivated basic grammar. For structures not describable by this grammar, rule hypotheses are generated, evaluated, and finally integrated into the grammar. Both statistical information and linguistic constraints are employed during the generation and evaluation process.

1. Introduction

Numerous research goals in linguistics, psychology, artificial intelligence, and related disciplines depend on the availability of large, syntactically analysed text corpora. A prerequisite for parsing large corpora is the existence of a grammar that covers most phenomena found in written natural language. There are two principle approaches to developing such grammars:

1. The first approach, which has dominated linguistic research for more than thirty years, aims at formulating grammars which reflect the linguistic competence of the ideal speaker/hearer. In their most recent forms, *competence grammars* tend to be highly modular systems consisting of a small number of general principles and constraints as well as language-specific information. In spite of their high degree of abstractness these grammars are characterized by their appeal to linguistic intuition and by their theoretical embeddedness. However, these manually established rule systems often have an unsufficient coverage of the data found in real texts.
2. The second approach has become popular with the re-emergence of empiricist conceptions in the late eighties. It proposes the use of statistical methods for constructing grammars from huge amounts of data, typically n-gram techniques. While these methods tend to produce *performance grammars* that have a good coverage of the corpus data, the rules generated by them often seem to contradict linguistic intuition.

As an alternative to either of these two approaches we present a procedure that extends the grammar and thereby its coverage of the corpus data by applying an inductive learning algorithm to a corpus which uses both statistical information and linguistic contraints. It avoids the shortcomings of version space-based approaches to grammar learning (cf. VanLehn/Ball(1987) and Giardano(1993)) which suffer from high time and space complexity. As in Berwick(1985), Wolff(1991), and Osborne(1994), learning is initiated by parsing failure, and linguistic background knowledge is used to keep the number of hypotheses for grammar extension small.

2. A method for learning grammar rules

The area of machine learning offers a wide variety of approaches and methods. In order to be considered suitable for our purposes a learning method should fulfil the following criteria:

- It should allow the aquisition of lexical and syntactic knowledge.
- It should generate linguistically plausible hypotheses.
- Its complexity should not exceed an acceptable degree.
- It should require no or only little external control.

A learning method that fulfils these conditions is sketched in Fig. 1.

First, a small basic corpus C_B is completely analysed by hand in order

	START			END
Lexicon	L,	L_1,	...,	L_n
Grammar	G,	G_1,	...,	G_n
Corpora	C_B		C_L	

Figure 1: A sketch of the learning method

to formulate a start-up grammar G. It is also ensured that the lexicon L contains all word forms appearing in C_B. As a result, the knowledge base is complete regarding this basic corpus, i.e. G covers all syntactic structures found in C_B and L contains an entry for every lexical item found in C_B. C_B is also used to compute transition probabilities between lexical categories, later to be used for generating hypotheses about unknown word forms.

With respect to a much bigger learning corpus C_L, both G and L are incomplete. As the system tries to analyse the sentences of C_L one by one, unknown constructions and word forms are encountered. The learning algorithm step by step adapts the grammar and the lexicon to these new constructions and word forms, leading to the final grammar G_n and the lexicon L_n, which, ideally, completely describe C_L.

The characteristics of the learning method used can be summed up as follows:

(i) The learning method is inductive.
(ii) Learning takes place on the basis of positive examples only (since all sentences found in the corpus will be considered grammatical).
(iii) The grammar is expanded incrementally.
(iv) Learning is constructive: The grammar is expanded by forming new rules which cover unfamiliar constructions that the parser encountered when trying to parse the sentence in question.

(v) Learning is supervised in the sense that whenever the learning process fails to come up with at least one hypothesis or is unable to select one out of several competing hypotheses, intervention by a linguist is called for.
(vi) The knowledge base is extended monotonously.
(vii) Learning takes place parser-driven.

3. The learning algorithm

Fig. 2 illustrates this incremental process of building up the grammar in more detail.

```
WHILE (not all sentences are parsed) DO
    (1) Select the next appropriate sentence α.
    (2) Parse α.
    (3) IF (α is parsed successfully) THEN
            GOTO (1)
        ELSE
            (3-1) Generate lexical and syntactic hypotheses for α.
            (3-2) IF (these hypotheses allow the parsing of α) THEN
                      Select best hypothesis and modify lexicon and
                      syntax accordingly.
                  ELSE
                      Defer the analysis of α.
```

Figure 2: The learning algorithm

In the following sections we will elaborate upon what is meant by an *appropriate* sentence in Fig. 2 (1), how hypotheses are generated (Fig. 2 (3-1)), and how they are evaluated in order to choose the best one of them (Fig. 2 (3-2)).

4. Selecting an *appropriate* sentence

The most difficult problem of inductive learning methods that rely on only positive data as input is the danger of producing irrecoverable overgeneralizations (cf. Gold (1967) and Angluin (1980)). One way to avoid such overgeneralizations is to sort the data in a way that keeps the inductive generalizations necessary in each learning step as small as possible.

To facilitate these small learning steps, it would be desirable to have a criterion at hand which allows to determine the *distance* between the current grammar and a sentence by evaluating the available morpho-syntactic clues. However, since such a measure is currently not available to us, we approximate it by taking sentence length (counted in words) as a sorting criterium, assuming that on average longer sentences are syntactically more complex than shorter ones, and sort C_L by increasing sentence length.

5. Parser-driven learning

Like in most other systems for constructive grammar induction (cf. Berwick (1985), Wolff (1991), Naumann and Schrepp (1992), and Osborne (1994)), the learning process is triggered by the the parser, or, to be precise, by the *errors* which turn up when the system tries to parse a sentence not covered by the current grammar.

The parser we use is an Earley-based chart parser which is extended in two ways:

- While sentence processing normally proceeds from left to right, the parser can switch to a bidirectional analysis if necessary. When the parser fails to parse a sentence, partial information about its structure is gathered in the chart. Hypotheses about how to extend and to complete this information in order to get a complete structural analysis of the sentence are used as *islands* for the continuation of the analysing process.
- The parser processes ID/LP-grammars, which form a specific variant of context-free grammars. ID/LP-grammars contain two kinds of rules:
  - I*mmediate* d*ominance* rules[1]determine which categories form a constituent without specifying their order.
  - L*inear* p*recedence* rules are word order or constituent order rules. They regulate the linear order of the categories within a *local tree*.

6. Generating hypotheses

Two kinds of hypotheses are formed during the parsing process: unknown words in the sentences lead to hypotheses concerning the word class they belong to, while syntactic structures not covered by the current grammar cause the generation of hypotheses for grammar rules.

6.1 Lexical hypotheses

For each unknown word form a number of hypotheses about its part-of-speech are generated. Currently all open word classes are considered valid hypotheses. Bigram information taken from the completely analysed corpus C_B provides the best n hypotheses for the part-of-speech. While developing the system, we restricted n to 3. These n hypotheses are tested by parsing the sentence using the alternative tags. For those tags that lead to a parsable structure, provisional lexical entries are generated, consisting of the word form w, its morpho-syntactic features f, and a confidence value c mirroring the strength of the bigram data. Depending on its positive or negative effect on the ability of the system to analyse new sentences, the confidence value c of a provisional lexical entry is raised or lowered. If c is above a predetermined threshold value, the provisional entry is accepted and turned

[1]To distinguish ID-rules from context-free rules, the symbols on the right side of an ID-rule are separated by commas.

into a regular lexical entry. Otherwise it is deleted.

6.2 Syntactic hypotheses

Whenever the parser fails to parse a sentence α, its mechanism for extending the current grammar G by generating new rules is called. The hypotheses which finally lead to the formulation of new rules correspond to the following three types:

- *Hypotheses for structure expansion*
  A structure for α results from lexical material that is inserted into the structure of a sentence α' which is covered by G.
- *Hypotheses for structure reduction*
  A structure for α is gained by deleting constituents of the structure of a sentence α' which is covered by G.
- *Hypotheses for LP-constraint relaxation*
  In case α cannot be parsed because some of the LP-rules of G block a complete analysis, the LP-constraints responsible for this parsing failure are relaxed as far as necessary.

The problems that arise in generating syntactic hypotheses are twofold. First, the number of generated hypotheses tends to be very large, so that a way must be found to limit this number. Second, from this reduced number of valid hypotheses the best one must be selected and integrated into the grammar.

6.3 Restricting the number of syntactic hypotheses

The solution of the first problem, the restriction of the number of hypotheses, is achieved by observing three conditions:

- *General well-formedness conditions for ID-rules*
  The X-bar convention is used as a well-formedness condition for ID-rules. This step drastically reduces the number of rules proposed by the learning algorithm.
- *Maximally specific LP-rules*
  The LP-rules employed by the current grammar are as specific as possible (i.e., as restrictive as possible).
- *Heuristic principles for operations of structure building*
  When considering a hypothesis for structure expansion, the possibilities of inserting lexical material are constrained by two requirements:
  - The inserted material has only *local* effects on the structure into which it is inserted (*locality constraint*).
  - The lexical material is interpreted in a way which is compatible to the information that is already contained about it in the chart (*principle of information preservation*).

6.4 Evaluation of syntactic hypotheses

For the evaluation of the generated hypotheses, the different grammars that result from integrating the different rule hypotheses into the current grammar are compared on the corpus level and grammar-internally.

On the corpus level, it is surveyed which of the alternative grammars covers more sentences of the corpus than the others by parsing the corpus successively with different grammars. The grammar with the best result (i.e. the grammar which covers most sentences of the corpus) is kept, the others are discarded. If this parsing behaviour does not lead to a clear decision because two or more grammars are empirically equivalent, the entropy of each grammar is computed and serves as the decision criterium. The entropy of a grammar G is computed by the following formula (cf. Cook et al. (1976)):

$$E(G) = \sum_{i=1}^{n}(\sum_{j=1}^{m_i} E(x_{ij})),$$

with $E(x) = (|x|+1)log(|x|+1) - \sum_{i=1}^{s} k_i \ log \ k_i$. x_{ij} denotes the jth symbol and m_i the last symbol on the right side of production i.

6.5 Examples

The following two examples illustrate the rule generation and evaluation procedure:

(1) In the first example, the current grammar contains the two rules applying to noun phrases listed in Fig. 3a. These rules are able to analyse the tag sequence **<det adj n>**. If a word tagged **adv** is found in a new sentence between **<det>** and **<adj n>** (see Fig. 3b), the five rule hypotheses shown in Fig. 3c are able to integrate the adverb into the noun phrase. The first of these hypotheses will be eliminated by the X-bar convention, because no non-terminal symbol is available that contains either **det** or **adv** as its head. To decide which of the other hypotheses should be accepted, the pars-

(a) Current grammar

⋮
NP → det, N1
N1 → adj, n
⋮

(b) Structure

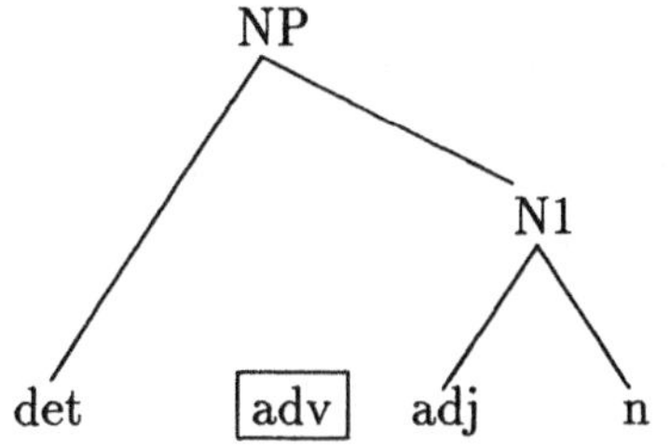

(c) Rule hypotheses

? → det, adv
NP → det, adv, N1
N1 → adv, adj, n
AP → adv, adj
AP → adv, A1 / A1 → adj

Figure 3: Extending the NP-structure

(a) Current grammar

⋮

VP → v, NP

⋮

(b) Structure

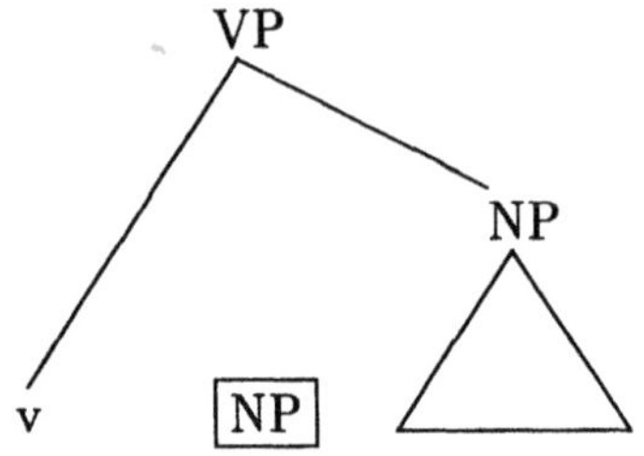

(c) Rule hypotheses

VP → VP, NP

VP → v, NP, NP

Figure 4: Extending the VP-structure

ing behaviour of these hypotheses in connection with the current grammar is evaluated. For that, a set of sentences S containing one or more <adv adj> sequences is collected from C_L by a search routine. Then, four competing grammars G_{h1}, G_{h2}, G_{h3}, G_{h4} are formed by extending the current grammar with one of the hypotheses each. The sentences in S are parsed with each grammar one after another. The grammar that covers the highest number of sentences from S will serve as the basis for the next incremental learning step. For our learning corpus, G_{h4} is selected. However, other corpora might lead to different selections.

(2) The second example shows a situation in which the existing grammar contains, among others, the verb phrase rule in Fig. 4a. If a new sentence contains the structure given in Fig. 4b, the two rule hypotheses in Fig. 4c are valid. This is a starting point quite different from the one in the previous example, since the available data either do not allow a decision based on parsing behaviour, or would wrongly lead to the acceptance of the more general, recursive rule. Here, heuristic principles guarantee that the first hypothesis is rejected in favour of the second, non-recursive, more specific rule (the second rule in Fig. 4c).

7. Further work

The described system is still in its beginning stages. Though first results look promising, a number of changes and additions are being planned: With regard to the lexicon, the simple word form lexicon currently used will be replaced by a lexicon containing lemmas as entries and inheritance hierarchies for efficient storage. For word forms not found in the lexicon, better hypotheses will be generated by the use of a morphological word parser. Both the basic grammar and the quality of the generated rule hypotheses will be enhanced by unification procedures. In addition, unsatisfying restrictions on formulating LP-rules will be removed by adhering to methods proposed by the theory of word order domains.

References:

ANGLUIN, D. (1980): Inductive inference of formal languages from positive data. *Information and Control, 45, 117-35.*

BERWICK, R. C. (1985): The Acquisition of Syntactic Knowledge. MIT Press, Cambridge, Massachusetts.

COOK, C. M.; ROSENFELD, A. and ARONSON, A. R. (1976): Grammatical inference by hill-climbing. *Informational Sciences, 10, 59-80.*

GIORDANO, J.-Y. (1993): Version space for learning context-free grammars. *Grammatical Inference: Theory, Applications ans Alternatives, 17-25.*

GOLD, E. M. (1967): Language identification in the limit. *Information and Control, 10, 447-474.*

NAUMANN, S. and SCHREPP, J. (1992): An empirical approach to syntax learning. *Proceedings of KONVENS 1992, 209-217.*

OSBORNE, M. (1994): Learning unification-based natural language grammars. PhD thesis, University of York.

VANLEHN, K. and BALL, W. (1987): A version space approach to learning context-free grammars. *Machine Learning, 2, 39-74.*

WOLFF, J. G. (1991): Towards a Theory of Cognition and Computing. Ellis Horwood, New York.

Investigation of the Language in Germany and Austria Using Statistical Methods

P. Filzmoser

Department of Statistics and Probability Theory,
University of Technology Vienna, A-1040 Vienna, Austria

Summary: On the basis of texts from the 18th century the German language is investigated. We emphasize not the use of words but the grammatical distribution of the words distinguished by codes. The relative frequencies of codes in the texts may be compared statistically, and a similarity measure can be deduced. These values are analyzed with a fuzzy-cluster method and the result is shown in maps. The specially adapted graphical presentation enables a good insight into the data and hence allows interesting interpretations of the results.

1. Introduction

Basis of this investigation are text cuttings from books and periodica, each with about 500 words. The texts had to be published between 1782 and 1792, in the period of softer censoring. Moreover, the birth-place of each author had to be known and to be attached to a certain district. About 500 texts could be found, they were distributed as to German and Austrian districts (countries) and as to sorts of texts. Words and punctuation marks were divided into 110 groups from a grammatical point of view. These groups were distinguished by means of three-figure codes (Adel et al. (1994)).

Word frequencies, use of words, prefixes and suffixes, sequences of words, transition of word classes, lengths of words, lengths of main clauses and clauses were investigated for the regions Germany and Austria (Adel et al. (1994)). With the grammatical distribution of the words the frequencies of the codes may be analysed. A graphical presentation with bar charts makes it possible to get a first impression of the frequency distribution of the codes in Germany and Austria or in the districts. Since the texts are not equally distributed over the districts, the bar width may be chosen to be proportional to the number of texts in each district. Code frequencies in different texts, districts or regions may be statistically compared using the binomial distribution model and approximation with the normal distribution. We want to point out in this paper a new method, namely the comparison of code frequencies in 2 texts (or groups of texts) by means of the p-value and the further investigation with fuzzy-cluster analysis.

2. Comparison of frequencies

Let m be the number of codes under consideration. The i-th code is denoted by c_i ($i = 1,\ldots,m$). We consider the frequencies of the same code c_i in 2 texts (or text groups) and denote the relative frequencies by p_i and

q_i, respectively. The total numbers of the codes considered in the respective text we denote by n_1 and n_2. Suppose that n_1 and n_2 are large and p_i and q_i not too extreme (neither close to 0 nor to 1) then we may assume an approximately normal distribution of $\hat{p}_i$ and $\hat{q}_i$ with mean values p_i and q_i and variances $p_i(1-p_i)/n_1$ and $q_i(1-q_i)/n_2$, respectively.
If we consider the quantity

$$T = \frac{\hat{p}_i - \hat{q}_i}{s} \tag{1}$$

with the estimated standard error s of $\hat{p}_i - \hat{q}_i$, namely $s = \sqrt{\frac{p_i(1-p_i)}{n_1} + \frac{q_i(1-q_i)}{n_2}}$, as approximately t-distributed with n_1+n_2-2 degrees of freedom, the p-value of the statistics T is $p = P(\mid T \mid \geq \mid t \mid)$, with t as the obtained (computed) value of T.

These p-values in formal tests are compared with a significance level α ($p < \alpha$ means *significance*). On the other hand p-values can be used as inverse (complementary) distance measures between $\hat{p}_i$ and $\hat{q}_i$.

3. Cluster analysis

We denote the number of districts in Germany and Austria by N. Let $f_i(l_j)$ be the relative frequency of code c_i in district l_j (summary of all texts in this district), $j = 1, \ldots, N$. We can compare two districts each by calculating the p-values defined above. The comparison of the relative frequencies of code c_i in the districts l_j and l_k ($j, k = 1, \ldots, N$) gives p-values $p_i(l_j, l_k)$. This means, for each code c_i we obtain a square matrix of p-values, which may be used as similarity measure of the districts. The dissimilarities required for cluster analysis are simply $1 - p_i(l_j, l_k)$.

For clustering we use a fuzzy-cluster method (Kaufman and Rousseeuw (1990)). Fuzzy-clustering grants the possibility that an object which can not definitely be classified may be distributed among some clusters. A function is minimized which includes a similarity measure of the objects and so-called membership coefficients. The membership coefficients are a measure for membership of the objects to the clusters, where the sum for each object over all clusters has to be 1.

4. Example

We consider Code *101* (*original adverbs*). Calculation of the p-values for the districts gives the similarities which are used for fuzzy-clustering. With a system of 4 clusters we get the most convincing separation. The membership coefficients, which give here the membership of the districts to the clusters, are shown in shades of gray. The darker the pattern of a district in the map the higher is the certainty that it belongs to the respective cluster. In cluster 1 of Figure 1 there is a large related area, this means that these districts coincide with reference to a distinct quality, in this case to frequencies of codes. Clusters 2-4 have no interpretation at first view.

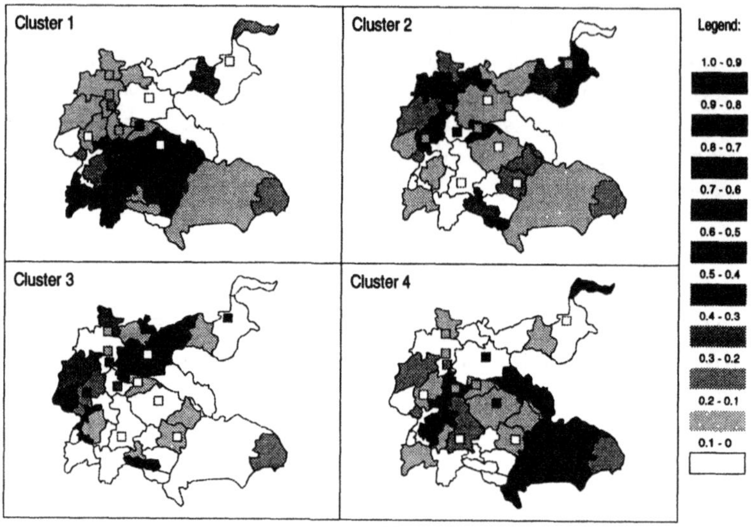

Figure 1: Fuzzy-clustering for Code *101* (*original adverbs*)

5. Conclusions

The investigation of all codes shows, like in the previous example, a lot of interesting results. In this application fuzzy-clustering has a big advantage to "hard" clustering. While hard cluster methods produce a fixed separation of districts, fuzzy-clustering makes a "soft" crossing from one cluster to another, expressed by the darkness of the shades of gray. Maximum information is preserved if each cluster is shown in an own map.

This method can also be applied to other statistical procedures like principal component analysis or factor analysis. From the correlation matrix of code frequencies in districts we may determine new components or factors. Instead of visualization of the estimated factor scores in planes spanned by the new co-ordinate system, the distances of the scores in the plane can be used as dissimilarity measure for fuzzy-clustering. A graphical presentation of the membership coefficients in a map gives a better view.

References:

ADEL, K.; DUTTER, R.; FILZMOSER, H. and FILZMOSER, P. (1994): Tiefenstrukturen der Sprache: Untersuchung regionaler Unterschiede mit statistischen Methoden. WUV-Universitätsverlag, Vienna.

ROUSSEEUW, P. J.; TRAUWAERT, E. and KAUFMAN, L. (1995): Fuzzy Clustering with High Contrast. *Journal of Computational and Applied Mathematics, 64, 81-90.*

Current Trends in Dialectometry: The Handling of Synonym Feature Realizations

Guillaume Schiltz

Institut für Romanistik, Universität Salzburg, A-5020 Salzburg, Austria

Summary: Dialectometry has to deal to a certain extent with synonym feature realizations on the nominal scale. Generating a similarity matrix out of these data outlines some fundamental problems in handling synonyms. Apart from a general discussion of possible approaches, this paper describes all similarity indexes actually used in dialectometry, which include an appropriate handling.

1. Introduction

The main purpose of dialectometric analysis lies in building up a similarity relation between geographically fixed dialects, where each dialect is defined by its specific realizations of a set of linguistic features.

The term *dialect* is extrordinarly problematic and gives rise to a multitude of diverging definitions. For our purpose we outline *dialect* as beeing the part of the linguistic communication system of a small, mostly rural, speaker community (cf. Steger (1983), Schiltz (1996)). *Dialect* and *local dialect* should therefore be considered as synonyms and lead to a simple geographic localization, namely the locality of a fixed speaker community. Collecting dialect data is based on visiting these communities and investigating their liguistic communication system by gathering responses to a fixed set of dialect features. These features may be of phonetic, morphological, syntactic or lexical nature, e.g. the actual expression for "the young castrated male cattle" (steer, ox, bullock etc.). Usually features are published by means of dialect maps and collected in dialect atlases.

All categorial data items are gathered in a data matrix (dialects $\times$ features, cf. table 1) and may now be computed into numerical similarity values by using a certain similarity index. The resulting values are stored in a similarity matrix (dialects $\times$ dialects), which is used for all further dialectometric analysis, such as clustering and similarity visualization.

But dialect features and thereby dialect maps are characterized by a large amount of null responses (missing data) and multiple responses (synonym feature realizations). Whereas missing data is the result of unknown expressions for a particular feature, multiple responses are normally based on different expressions uttered by different dialect speakers within the same community. In calculating similarity values missing data are not very estimated, but they may be incorporated in some pliant way (Bock (1974), Schader and Gaul (1992)). The handling of synonyms on the contrary has hardly been discussed in taxometric literature and there seems to be no adequate solution to this problem. In the following we discuss some basic operational possibilities and concentrate on specific dialectometric solutions.

2. General Discussion

E. Schneider (1984) first outlined the given problem and offered a list of general approaches, which are repeated here with further extensions and some dialectometric considerations.

(a) Features containing synonym realizations should be excluded from similarity measurement. If synonyms are scarce and if the corresponding features turn out to be negligible this method seems to be the easiest way to deal with the problem. But dialect corpora, especially if resulting from German or English investigations, include synonyms to a great extent. Up to 7 % of the data may be covered by synonyms, which are spread over almost every feature. So eliminating these features would often result into a meaningless subcorpus.

(b) All synonyms to a given expression are handeld as a single (uniform) feature realization. The origin of synonyms is highly associated with the investigation method. Their occurence depends on the number of interviewed speakers within a community (normally 1 to 4) and on the personal cooperation of these speakers (cf. König (1975)). This approach would thereby reflect alterning interview situations and should be excluded from dialectometric analysis.

(c) Among synonyms a single realization is chosen and all remaining synonyms are neglected. This choice may be purely random or performed by frequency considerations. The first method is applicable for data with scarcely documented synonyms as in Romance dialect corpora; H. Goebl used this procedure for all his Romance dialectometric investigations. But in the case of German or English corpora, where synonyms are extremely frequent, the random selection of a single synonym may heavily distort the dialectometric results. Alternatively, the choice can be made depending on the overall frequency of the realizations in question. So only the most or the least frequent realization may be chosen. By choosing the most frequent realization, the whole procedure gains some error-tolerance, as less frequent often single existing realizations are due to erroneous responses.

(d) The identity weight fixed for non-synonym realizations (normally 1) is split up between synonyms. This is based on the assumption, that synonyms are less "valuable" than a single mononym and reflect to some extent the beginning of sociolinguistic desintegration of the local dialect (Goebl (1996)). Two different approaches may then be adopted to value synonyms. First, if homogeneous additional criterions (e.g. social or historic statements) are added to synonyms, a differentiated valuation may be deduced from this information. Second, the valuation may be calculated, depending on the number of participating synonyms in a single realization. The more synonyms are found for a certain realization, the less must be their respective

weight. Whereas the first procedure is hardly practicable, because of inhomogenous criterions, the second approach was performed in dialectometry by Schneider (1984) and Goebl (1996).

(e) For a given feature the corresponding realization in a specific dialect is considered as a set in the mathematical sense. Mononyms then give rise to single-element sets and synonyms result in a multi-element set. The identity between dialects is performed by set-operations such as intersection and discrepancy (symmetric difference). All identity weights are kept equal and a feature identity between two dialects is noticed, if the intersection of the considered realization sets is different from the empty set. Alternatively, dissimilarity is defined by the presence of at least one realization, that is not included in both sets simultaneously. From a communication point of view, this approach may be founded on the fact, that dialects, bearing synonym feature realizations, function as a mediator between those dialects with a corresponding single realizations. Practice confirms this assumption by showing, that these dialects are often located adjacent to each other. Thereby, the enlarged communication range of mediator dialects is naturally integrated in the similarity measurement.

3. Similarity Indexes Handling Synonyms

In dialectometric analysis three of the above discussed approaches were actually implemented and will be introduced below. All measure examples are based on the data matrix illustrated in table 1, where the feature realizations **l** and **m** resp. **i** and **j** are synonyms.

D	n	n	n	n	n	n
C	l	**l,m**	l	m	X	m
B	i	**i,j**	j	X	i	k
A	e	e	e	f	g	h
	1	2	3	4	5	6

dialects

features

feature-vector B

dialect-vector 6

X = missing data

Table 1: data matrix

3.1 Alternative Relative Identity Index (RIWa)

This similarity index is based on the commonly used RIW (cf. Goebl (1984)). It was introduced by Schiltz (1996a) and handles synonyms by maintaining the realization with the highest frequency within the given feature-vector and by neglecting all remaining synonyms (cf. 2. approach (c)). The synonyms of table 1 are therfore reduced to **m** for feature C of dialect 2 and to **i** for feature B of dialect 2.

$$\mathsf{RIWa}_{pq} = 100 \cdot \frac{\sum_{i=1}^{\tilde{n}} (\mathsf{COI}_{pq})_i}{\tilde{n}} \tag{1}$$

with RIWa_{pq} alternative RIW between the dialects p and q.

$(\mathsf{COI}_{pq})_i$ co-identity: 1 if the realizations for feature i within the dialect-vectors p and q are identical, otherwise 0. Missing values are excluded from the comparision.

$\tilde{n}$ number of features taken into account for the comparision.

Based on the data of table 1, similarity values with RIWa are e.g.

$\mathsf{RIWa}_{1,2} = 100 \cdot \frac{1+1+0+1}{4} = 75\%$

$\mathsf{RIWa}_{2,3} = 100 \cdot \frac{1+0+0+1}{4} = 50\%$

$\mathsf{RIWa}_{2,5} = 100 \cdot \frac{0+1+1}{3} = 66.67\%$

3.2 Standardized Relative Identity Index (RIWs)

The next similarity index, also introduced by Schiltz (1996a), handles synonyms in set considerations (cf. 2. approach (e)). Thereby the dialect-vector 2 of table 1 consists of the following feature realizations: {e}, {i,j}, {l,m}, {n}. The formula of RIWs is identical to RIWa but COI is now defined as:

$(\mathsf{COI}_{pq})_i$ co-identity: 1 if $\mathcal{T}_p(i) \cap \mathcal{T}_q(i) \neq \{\emptyset\}$, otherwise 0. Missing values are excluded from the comparision.

and $\mathcal{T}_p(i)$ set of realizations for feature i inside the dialect-vector p.

The similarity values calculated from table 1 are now e.g.

$\mathsf{RIWs}_{1,2} = 100 \cdot \frac{1+1+1+1}{4} = 100\%$

$\mathsf{RIWs}_{2,3} = 100 \cdot \frac{1+1+1+1}{4} = 100\%$

$\mathsf{RIWs}_{2,5} = 100 \cdot \frac{0+1+1}{3} = 66.67\%$

3.3 Variable Valuing Ochiai-Index (VBOM)

In order to compute the similarity matrix for the English CLAE data (Computer Developed Linguistic Atlas of England), H. Goebl (1996) introduced this similarity index. It requires a binarization of the original data matrix and handles synonyms by giving them a less weighty identity value than mononyms (cf. 2. approach (d)). The binarization can be easily performed as shown in table 2, where synonyms (highlighted by circles) are inserted the same way as mononyms. Within the matrix + marks a feature realization existing in the given dialect, – marks it as non-realized. The sign X stands for missing data.

features	realizations	1	2	3	4	5	6
D	n	+	+	+	+	+	+
C	m	–	⊕	–	+	X	+
	l	+	⊕	+	–	X	–
B	k	–	–	–	X	–	+
	j	–	⊕	+	X	–	–
	i	+	⊕	–	X	+	–
A	h	–	–	–	–	–	+
	g	–	–	–	–	+	–
	f	–	–	–	+	–	–
	e	+	+	+	–	–	–

dialects

Table 2: binarized data matrix (cf. tab. 1)

Using the Four-Field-Table of table 3, the Variable Valuing Ochiai-Index between the dialects p and q is defined by

		q: +	q: –
p	+	a	b
	–	c	d

Table 3

$$\mathsf{VBOM}_{pq} = \frac{100 \cdot a}{\sqrt{(a+b)(a+c)}} \tag{2}$$

and results in the following similarity values:

$$\mathsf{VBOM}_{1,2} = \frac{100 \cdot 4}{\sqrt{(4+0)(4+2)}} = 81.65\%$$

$$\mathsf{VBOM}_{2,3} = \frac{100 \cdot 4}{\sqrt{(4+2)(4+0)}} = 81.65\%$$

$$\mathsf{VBOM}_{2,5} = \frac{100 \cdot 2}{\sqrt{(2+2)(2+1)}} = 57.73\%$$

4. Conclusion

All three indexes supply highly variated similarity values, as could be shown in the above examples. They generate different similarity matrices and, thus, affect the later geolinguistic analysis.

By comparing clusteranalysis results based on the the VBOM and on an RIW with random selection, H. Goebl (1996) found notable diverging clusters at the lower agglomerative level. Similar if not as grave differences can be observed in comparing cluster results calculated upon RIWa and RIWs (cf. maps 1-2). These maps are based on the DASB corpus (Schiltz 1996a) and cover dialects of Southwest Germany.

An appropriate choice of the similarity index is influenced on one hand by the underlying corpus and its investigation method, on the other hand by geolinguistic aspects. Thus a dialectometric analysis may focus on sociolinguistic facts and handle synonyms as markers of language change, indexes such as VBOM are suitable here. If however the main focus lies on communication models, then indexes comparable to RIWs can be considered. Nevertheless it should always be reasonable to perform different similarity measurements and compare their results.

References:

BOCK, H.-H. (1974): Automatische Klassifikation. Vandenhoeck & Ruprecht, Göttingen.

GOEBL, H. (1984): Dialektometrische Studien. Anhand italoromanischer, rätoromanischer und galloromanischer Sprachmaterialien aus AIS und ALF. Niemeyer, Tübingen.

GOEBL, H. (1996): Some Dendrographic Classifications of the Data of CLAE I and CLAE II. In: W. Viereck, H. Ramisch (eds.): *CLAE II Computer Developed Linguistic Atlas of England.* Niemeyer, Tübingen.

KÖNIG, W. (1975): Überlegungen zur Beschreibung von Aufnahmesituation und Informant bei sprachgeographischen Erhebungen. *Deutsche Sprache, Heft 4, 346-364.*

SCHADER, M. and GAUL W. (1992): The MVL (Missing Values Linkage) Approach for Hierarchical Classification when Data are Incomplete. In: M. Schader (ed.): *Analyzing and Modeling Data and Knowledge. Proc. 15th Ann. Conf. of the GfKl.* Springer, Berlin, Heidelberg, 107-115.

SCHILTZ, G. (1996): German Dialectometry. In: H.-H. Bock and W. Polasek (eds.): *Data Analysis and Information Systems. Proc. 19th Ann. Conf. of the GfKl.* Springer, Berlin, Heidelberg, 526-539.

SCHILTZ, G. (1996a): Konzepte eines dialektometrischen Informationssystems. Der dialektometrische Atlas von Südwest-Baden (DASB). N.G. Elwert, Marburg.

SCHNEIDER, E. (1984): Methodologische Probleme der Dialektometrie. In: H. Goebl (ed.): *Dialectology.* Studienverlag Dr. N. Brockmeyer, Bochum, 314-333.

STEGER, H. (1983): Raumgliederung der Mundarten. W. Kohlhammer, Stuttgart.

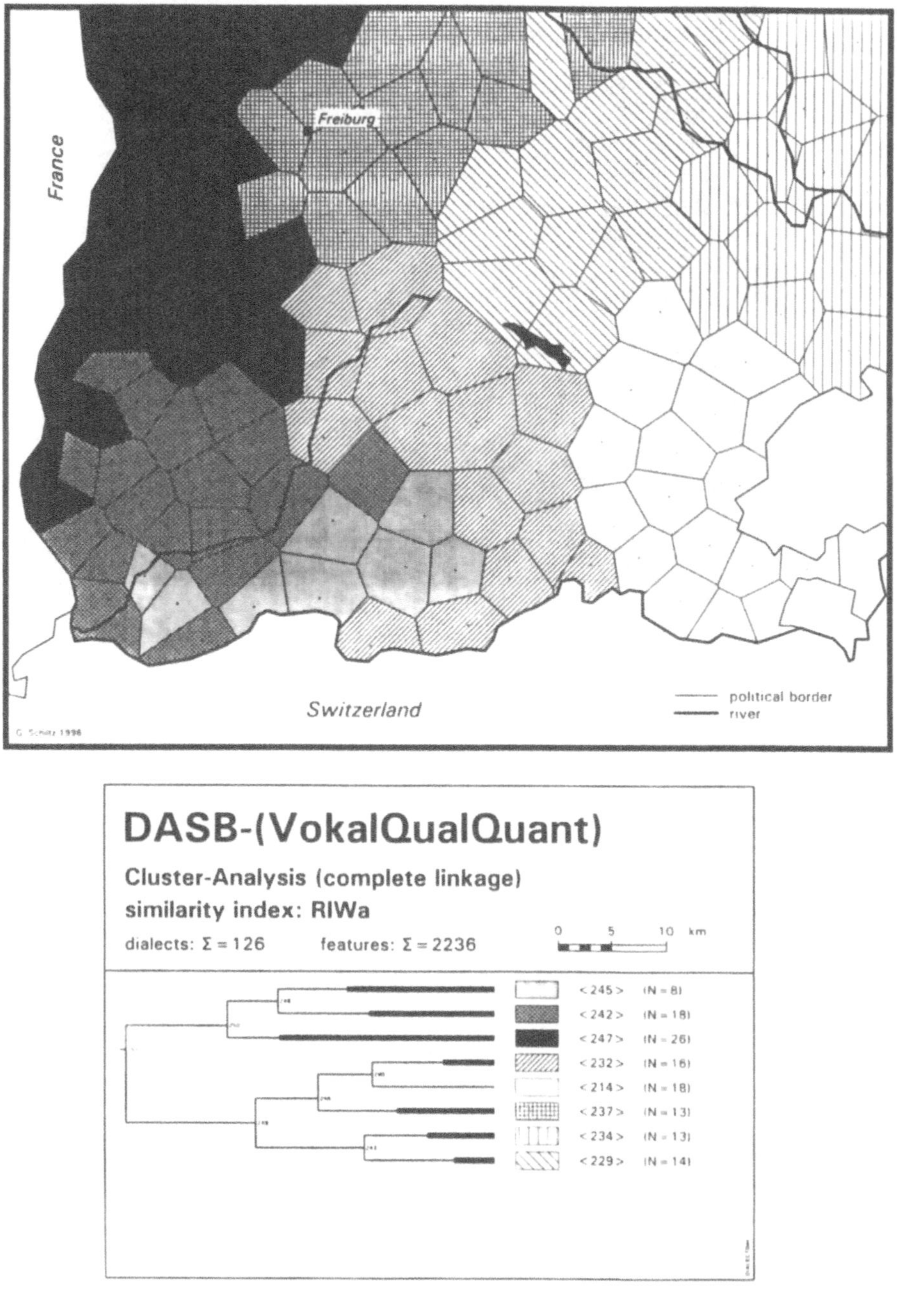

Map 1: Clusteranalysis performed upon the similarity index RIWa

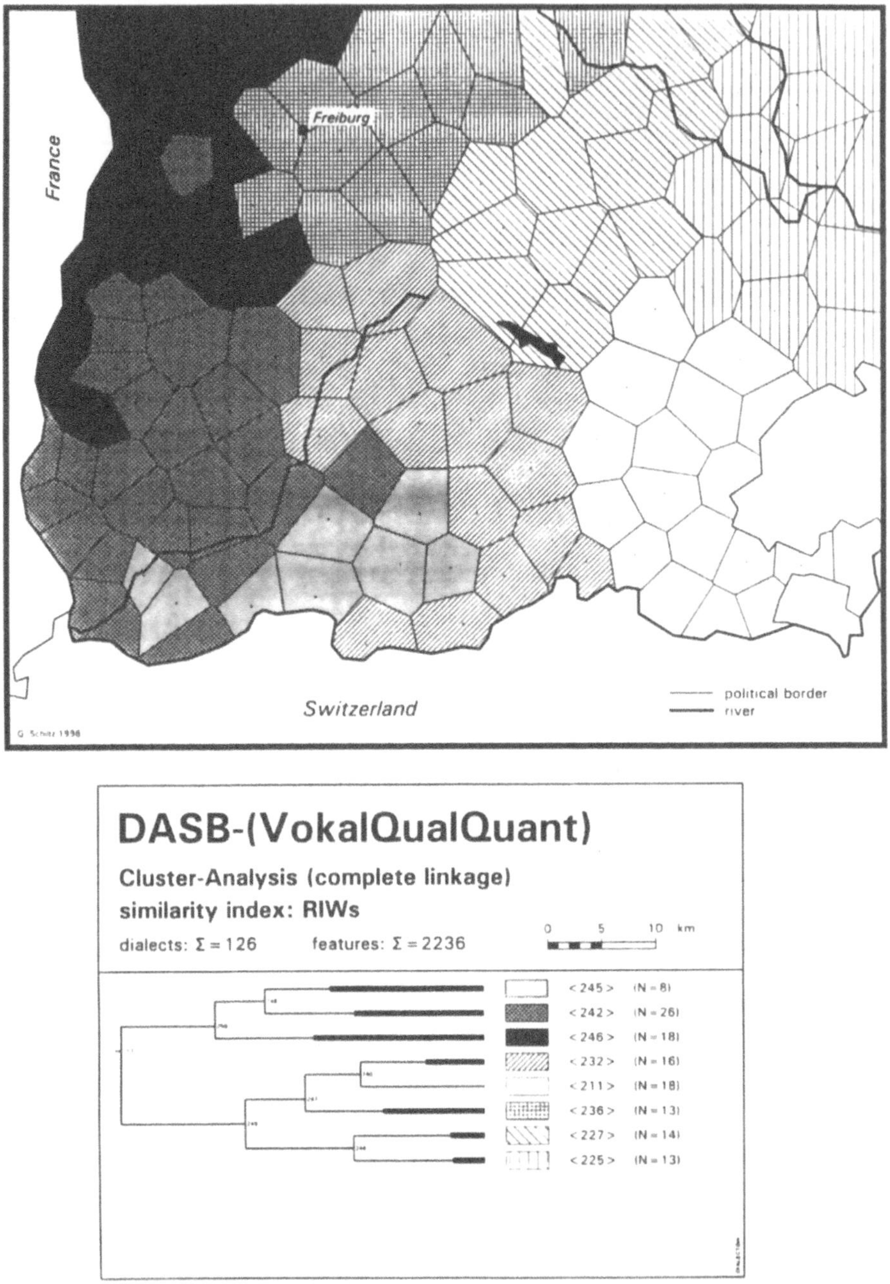

Map 2: Clusteranalysis performed upon the similarity index RIWs

Interactive Graphical Analysis of Regional Dialects

Adalbert Wilhelm, Meike Sander

Institut für Mathematik,
Universität Augsburg, D-86135 Augsburg, Germany

Summary: One aim of linguistics is to classify different locations into dialectical regions according to the terminology used. Since language data are usually nominal, common classification procedures can not be applied directly. Instead of transforming the data into a numerical form to make them tractable for common classification procedures we apply interactive graphical methods to the original raw data. Using data from the Bavarian-Swabian language atlas we show the usefulness of interactive graphical methods as they are provided in the software packages REGARD and MANET.

1. Exploratory data analysis and linguistic data

Traditionally, dialect surveys are directed towards the publication of maps in hard copy known as 'atlases'. A conventional folio atlas includes 500 - 1000 maps with descriptive volumes of word geography, verb morphology and pronunciation. There have been some attempts to publish an electronic atlas that allows the users to interact live and on-line with a database. One attempt was made by Kirk and Kretzschmar (1992) who provided two prototypes of an appropriate retrieval and display software. Dialectologists have also used GIS (Geographic Information System) software as a tool in linguistic map making.

Analysis of dialect surveys, however, consists not only in mapping but also in determining regions with similar dialects and classifying objects into main dialect groups. In dialectometry it is common to transform the nominal data into numerical similarities that can be used for cluster analysis procedures, see Schiltz (1995). In contrast, interactive statistical graphics can deal with the data in its original nominal form.

The questions posed in dialectology are usually exploratory in nature and hence interactive graphical methods as an exploratory data analysis tool are very suitable.

Interest in applying interactive statistical graphics to linguistic data has not only grown because confirmatory analysis tools can hardly be applied to these data, but also because interactive graphics offer a new view of the data. They allow the exploration of structure and patterns of data sets in many different ways from many different angles at the click of a mouse. Linking of graphs and maps opens a way to multidimensional analysis and also permits the inclusion of geographical structures into the investigation.

Dialectologists are experienced in the use of maps and plots as tools for data analysis, see König (1994) for example. On one hand, this experience may stimulate the use of interactive graphics, on the other hand, it will reduce the danger of misinterpreting graphical displays.

2. Dialect survey in Bavarian-Swabia

The data stem from the Bavarian-Swabian language atlas, one of six projects currently running in Bavaria that aim to document regional dialects, so to create a kind of dialect museum and to show regional differences in the dialects. Therefore, in the years 1984 until 1989 three questioners gathered data in 272 villages in the region of Bavarian-Swabia. They spent about a week in each village asking between two and seven informants. These persons were selected according to certain conditions: they all had to be born in the village, both parents should come from the same village and they should have spent almost all their life in the village. The investigators used a questionnaire with 2267 items covering the use and meaning of about 800 words, differences in grammatic forms as well as different pronunciations.

The data used in the present investigation are taken from the topics:

- human body
- physical and emotional expressions
- human society
- clothing.

3. Interactive statistical graphics

Interactive statistical graphics are one of the most powerful tools for exploratory data analysis. The main principles of interactive graphics are selection and linking. Selection in its widest sense means that interesting points can be marked, clearly identified and interrogated as well as separately analysed. Selection comes to its full glory when it is combined with linking that means every action performed on the data in one display is immediately transmitted to all other displays showing the same data set. The principle of linking was first realised in scatterplot-brushing, as introduced in Becker, Cleveland and Wilks (1987). In today's software linking is not restricted to scatterplots but is extended to all displays.

The present investigation was carried out with the software tools REGARD and MANET. The former, developed for geographical data, offers a multi-layer concept so that a line-layer with geographical information like rivers and boundaries can be combined with point layers for all variables, see Unwin (1994). MANET is research software for the concise treatment of missing values in exploratory data analysis software. In addition, it serves as a platform for new ideas on interactive statistical graphics, see Unwin et al. (1996).

4. Univariate view of the data

We demonstrate the basic principles of interactive statistical graphics by presenting some univariate views of the data. A clear clustering arises with the word 'Kirchweih', see Figure 1. There are mainly three different forms

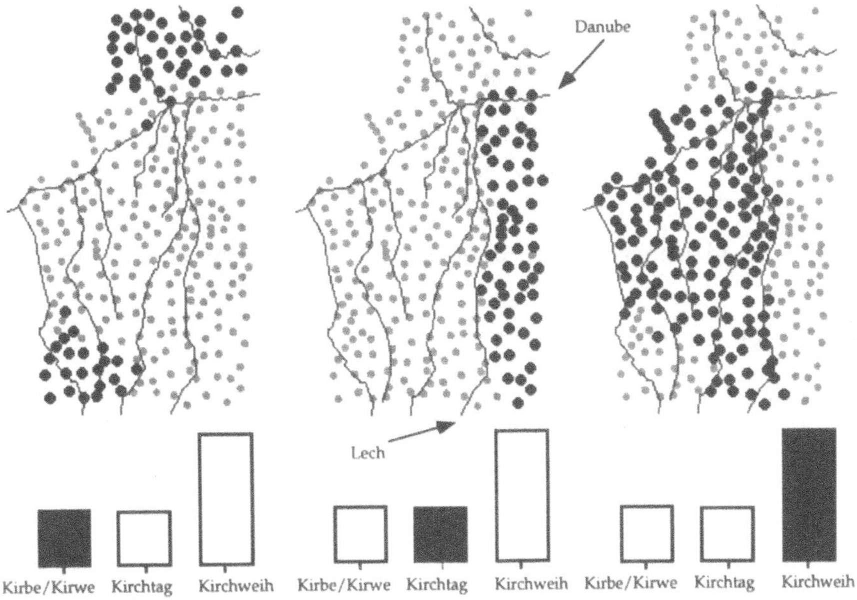

Figure 1: Geographic pattern of the word 'Kirchweih'.

of this word, 'Kirbe/Kirwe' is used in the south-west and in the north-east, 'Kirchtag' is used in the region east of the Lech river and south of the Danube, in all other villages 'Kirchweih' is common.
In contrast to maps from a traditional dialectic atlas, Figure 1 displays, besides the geographic structure, the proportions at which the different forms occur. In about 50 percent of the investigated locations the form 'Kirchweih' is used, whereas the other two forms occur at a rate of about 25 percent each.

The cluster corresponding to 'Kirchtag', see the map in the middle of Figure 1, indicates that the Lech river constitutes a sharp boundary between the Swabian and Bavarian dialect. There is only a single point far east from the Lech not belonging to this cluster. Using the interactive information view shows that this was Mammendorf, a village where this variable was not recorded, see Figure 2.

Not all variables carry such a clear structure. The main reason for this is that in many villages the informants used more than one form of some words. To deal with this multiple responses we introduced new artificial categories for these variables. Hence the word 'Kruste' denoting the crust on a sore splits into 18 categories resulting in an inhomogeneous spatial pattern, see

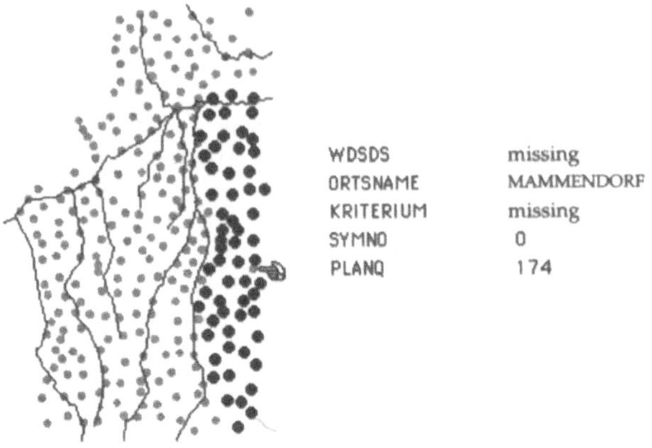

Figure 2: Identification and interrogation of an exceptional observation.

the map on the left in Figure 3.

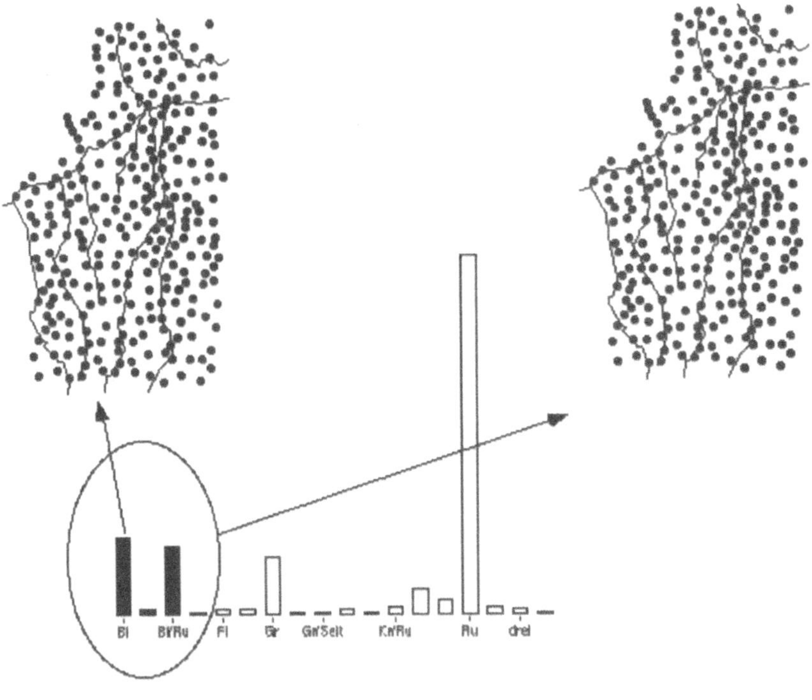

Figure 3: Detecting common regions by interactively combining categories.

However, as can be seen from the map on the right in Figure 3, interactive combination of similar categories – all categories including the form 'Bletzen' – results again in a common region east from the Lech and south from the

Danube.

So far we have focused on ideas that could – with more or less effort – also have been realized using static maps. In the next section we switch towards multivariate aspects exemplifying the benefits of interactive graphics.

5. Linked bar charts and mosaic plots

Mapping each variable separately is sufficient for presenting a dialect museum. For analysing the structure and for classification purposes, a multivariate approach is needed. Obviously, one cannot expect to end up with any cluster at all by combining all 800 word variables, some bearing a very fine microstructure. We will mainly focus on up to four variables.

The use of the intersection mode is demonstrated with the variables 'niesen', 'Schluckauf', 'schau' and 'schubsen'. The word 'niesen' only occurs in two forms, 'niesen' and 'pflitzgen', abbreviated with 'n' and 'p', see Figure 4. In

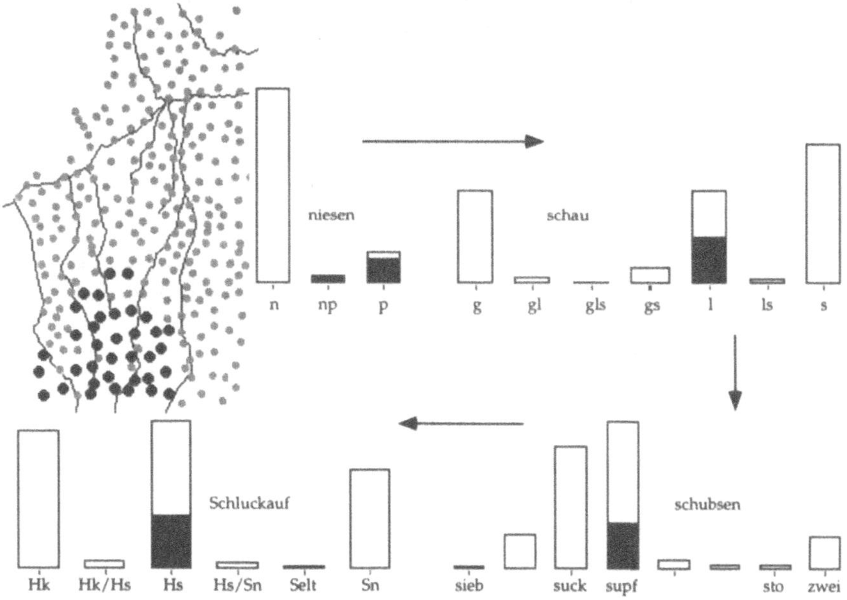

Figure 4: Finding multivariate clusters using intersection mode.

some villages both forms are used, a separate category was introduced for these cases (above in the corresponding bar chart the category in the middle denoted 'np'). Selecting all locations where at least the form 'p' is used and then using the intersection mode to select those villages that also use 'l' (=luag) for 'schau', 'Supf' (=schupfen) for 'schubsen' and 'Hs' (= Hescher) for 'Schluckauf', we end up with a cluster in the south-west of our region of interest, see Figure 4.

The main disadvantage of this approach is the restriction to one selection sequence at the time. Therefore, one has to have some preposition in mind to have good odds for finding a cluster. Cross-classifying some variables immediately yields all combinations and their frequencies. Visualisation of a cross-classification is best done with mosaic plots, as introduced in Hartigan and Kleiner (1981). They represent each cell of a contingency table directly by a rectangle whose area is proportional to the cell frequency. The resulting display is a mixture of rectangles of various sizes and, commonly, a large number of empty cells which are marked by lines with a zero through them. The pattern of rectangles shown by the mosaic plot is useful for making visual comparisons across portions of a contingency table, and highlighting large and small counts. Mosaic plots are built-up in a recursive way. The categories of the first variable divide up the horizontal axis into columns whose widths are proportional to their numbers. The categories of the second variable divide up the vertical axis similarly but conditional on the first variable. The third variable further divides up the horizontal axis within each class of the first variable and so on. We cross-classified the three variables 'Kirchweih', 'schau', and 'niesen', the corresponding mosaic plot with 160 cells is shown in Figure 5, the main categories of the first two variables are named. The highlighting in this figure is caused by the selection of all

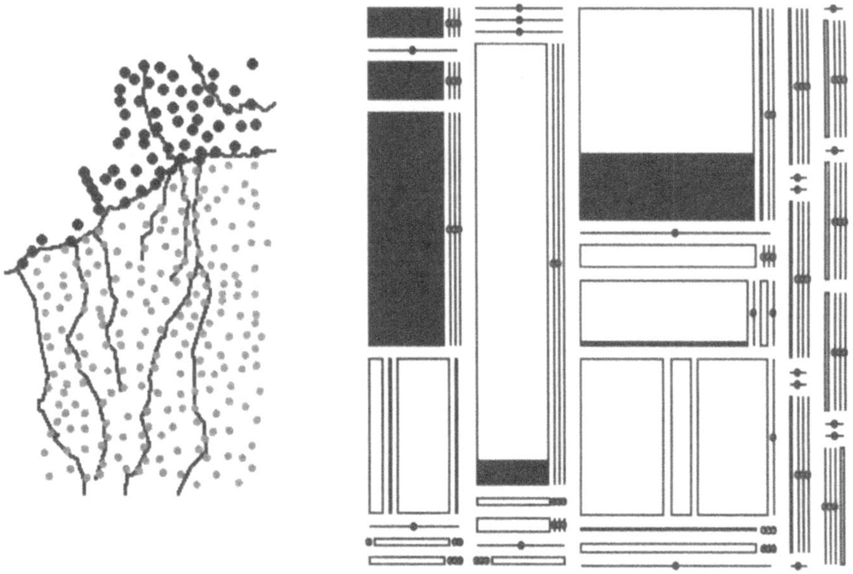

Figure 5: Mosaic plot for variables 'Kirchweih','schau', and 'niesen'; all villages north to the Danube highlighted.

villages north to the Danube. Most of the cells in the mosaic plot are empty, only 31 are not. The many empty cells at the right margin are expected because they represent combinations of missing values and some special and seldom-used words, but all the other empty cells indicate an inverse clustering in the sense that the corresponding combinations of words do not appear.

For example, the form 'Kirchtag' never occurs in combination with 'guck' or 'pflitzgen'. When embedded in an interactive environment mosaic plots make it easy to find connected regions in the map by simply grouping neighbouring rectangles. In Figure 6 adiacent rectangles were grouped such that

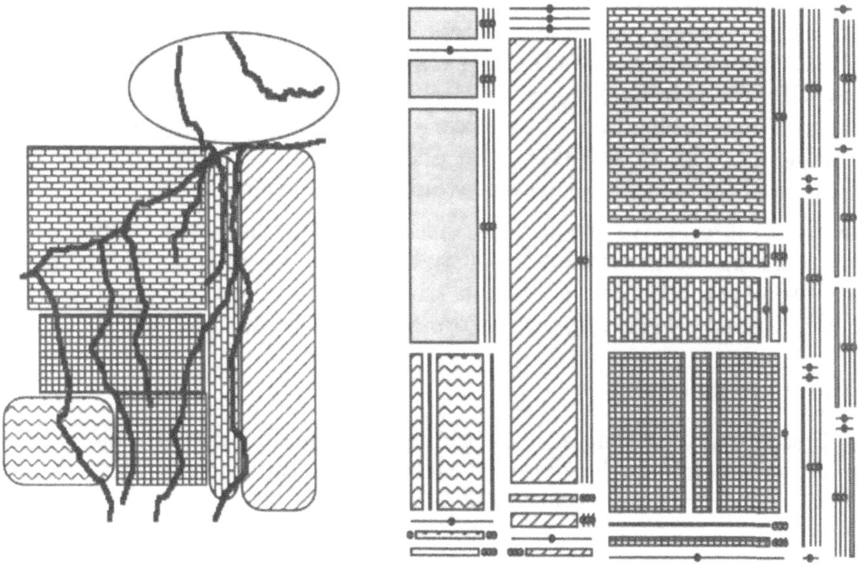

Figure 6: Finding multivariate clusters using mosaic plots.

the corresponding datapoints coincide in at least two variables. The same shading is used for rectangles in the same group. The result is a division of the whole region in six big clusters.

6. Concluding remarks

In this article we have given a brief insight into the possibilities offered by interactive graphics for the analysis of dialectical survey data. Although, it is impossible to depict the fascination of interacting live and on-line with graphics, since for presenting this paper only static views are available, we think that it is possible to imagine the rich opportunities offered by these exploratory analysis tools. Beside the classification of objects presented in this paper interactive statistical graphics allow also the classification of variables. The latter aspect is the main point of view in a follow-up paper by Sander and Wilhelm (1996). The results of classifying objects by interactive graphics are fuzzy clusters in the sense that not all points have to be classified at the end. Looking at the data we would claim that the region under investigation could be divided into six groups, see Figure 6.

A cardinal problem of dialectical survey data is the occurrence of multiple responses for one variable at one location. In our investigation we introduced a new artificial category for these cases. The flexible manner, in which interactive graphics act, allows toggling between inclusion and exclusion of these multi-response objects to avoid the dangers associated with the generation of artificial categories.

Future work should focus on two topics: on the software side the geographic facilities like the layer concept of REGARD should be combined with the rich and increasing choice of different interactive plots offered in MANET; on the statistical side the obtained clusters should be compared and evaluated with the results of more classical tools like cluster analysis and factor analysis. An application of the latter two methods to linguistic data can be found in Adel et al. (1995). In addition, a link to the database of the informants' socio-economic status would be desirable.

The use of interactive graphics as part of an electronic atlas is surely useful as a teaching instrument and will make it easy for unexperienced users to find interesting features in regional dialects. More sophisticated users will be able to make maximum use of the full potential of the data.

Acknowledgement

We thank Werner König and his group for providing us with the data and the dialectological background. We also thank the referee for the valuable comments made on an earlier draft of the paper.

References:

ADEL, K.; DUTTER, R.; FILZMOSER, H. and FILZMOSER, P. (1994): Tiefenstrukturen der Sprache – Untersuchung regionaler Unterschiede mit statistischen Methoden. WUV-Universitätsverlag, Wien.

BECKER, R.; CLEVELAND, W. S. and WILKS, A. R. (1987): Dynamic Graphics for Data Analysis. *Statistical Science, 2, 355-395.*

HARTIGAN, J. A. and KLEINER, B. (1981): In: W.F. Eddy (ed.): *Computer Science and Statistics: Proceedings of the 13th Conference on the Interface.* Wiley, New York, 268-273.

KIRK, J. M. and KRETZSCHMAR, W. A. (1992): Interactive linguistic mapping of dialect features. *Literary and Linguistic Computing, 7, 168-175.*

KÖNIG, W. (1994): dtv-Atlas zur deutschen Sprache, 10. Aufl. dtv, München.

SANDER, M. and WILHELM, A. (1996): Interactive statistical analysis of dialect features. Submitted to *The Statistician.*

SCHILTZ, G. (1995): German dialectometry. In: H.H. Bock and W. Polasek (eds): *Data Analysis and Information Systems*, Springer, Berlin, 526-539.

UNWIN, A. R. (1994): REGARDing geographic data. In: P. Dirschedl and R. Ostermann (eds.): *Computational Statistics. Physica, Heidelberg*, 315-326.

UNWIN, A. R.; HAWKINS, G.; HOFMANN, H. and SIEGL, B. (1996): Interactive graphics for data sets with missing values MANET." *Journal of Computational and Graphical Statistics, 5, 113-122.*

Subject Index
(including List of Authors)